# Neutron Stars 1

ASTROPHYSICS AND SPACE SCIENCE LIBRARY

VOLUME 326

P. Haensel, A.Y. Potekhin, and D.G. Yakovlev

# Neutron Stars 1

## Equation of State and Structure

Springer

P. Haensel
Nicholaus Copernicus
Astronomical Center
Bartycka 18
Warszawa 00-716
POLAND
haensel@camk.edu.pl

A.Y. Potekhin
Ioffe Physico-Technical
Institute
Polytekhnicheskaya 26
St. Petersburg 194021
RUSSIA
palex@astro.ioffe.ru

D.G. Yakovlev
Ioffe Physico-Technical
Institute
Polytekhnicheskaya 26
St. Petersburg 194021
RUSSIA
yak@astro.ioffe.ru

Cover illustration: The background is an image of the Crab Nebula -- the remnant of a star that exploded as a supernova in 1054. The Crab Pulsar is at the center of a shining whirl of the central coccoon. This image is a courtesy of NASA, Hubble Space Telescope/Chandra galleries (http://www.skyimagelab.com/m1chacrabne.html).

Library of Congress Control Number: 2006923828

ISBN-10: 0-387-33543-9 e-ISBN-10: 0-387-47301-7
ISBN-13: 978-0-387-33543-8 e-ISBN-13: 978-0-387-47301-7

Printed on acid-free paper.



9 8 7 6 5 4 3 2 1

springer.com

# TABLE OF CONTENTS

# INTRODUCTION

Neutron stars play a unique role in physics and astrophysics. On the one hand, they contain matter under extreme physical conditions, and their theories are based on risky and far extrapolations of what we consider reliable physical theories of the structure of matter tested in laboratory. On the other hand, their observations offer the unique opportunity to test these theories. Moreover, neutron stars are important *dramatic personae* on the stage of modern astrophysics; they participate in many astronomical phenomena.

Neutron stars contain the matter of density ranging from a few $\mathrm{g\ cm^{-3}}$ at their surface, where the pressure is small, to more than $10^{15}$ $\mathrm{g\ cm^{-3}}$ at the center, where the pressure exceeds $10^{36}$ $\mathrm{dyn\ cm^{-2}}$. To calculate neutron star structure, one needs the dependence of the pressure on density, the so called *equation of state* (EOS), in this huge density range, taking due account of temperature, more than $10^9$ K in young neutron stars, and magnetic fields, sometimes above $10^{15}$ G.

The present book is mainly devoted to the theory of the EOS of neutron star matter and its consequences for neutron star structure. As one moves from the neutron star surface to the center, the methods to calculate the EOS change. Atomic structure and plasma theories are used for the surface stellar layers. Deeper layers of the neutron star crust require nuclear theory combined with plasma physics, both in very exotic density-temperature regimes. Finally, the neutron star core necessitates many-body theory of dense strongly interacting systems, together with the physics of strong interactions of elementary particles.

Several not too old books treating the topic of the EOS of neutron stars are available in the literature. Many aspects of the EOS problem are considered by Shapiro & Teukolsky (1983) in the monograph *Black Holes, White Dwarfs, and Neutron Stars*. However, this excellent book reflects the state of the art by the beginning of the 1980s. Some aspects of the EOS problem, stellar stability and equilibrium are described in *Stellar Physics* by Bisnovatyi-Kogan (2001, 2002). Selected EOS models, based on the relativistic theories of hadronic matter, are discussed in detail by Glendenning (2000) in the monograph *Compact Stars: Nuclear Physics, Particle Physics, and General Relativity* and in the book by Weber (1999), *Pulsars as Astrophysical Laboratories for Nuclear and Particle Physics*. However, neither of these two monographs gives the detailed description of the variety of possible EOSs in stellar cores and crusts.

The present monograph is complementary to those cited above. We describe all neutron star layers paying special attention to the theoretical basis of calculations, physical properties (especially the EOS) and observational manifestations. First we give an overview of neutron star theory and observations. Then we analyze current EOS models in the neutron star crusts and cores. We show that when the density increases the models become more numerous and different but less reliable. The most interesting density range is a few times the standard nuclear density and higher, where the composition of the matter becomes largely unknown. According to the different models, this dense matter may contain nucleons, hyperons, pion or kaon condensates, deconfined quarks or the mixture of these components. We analyze all these possibilities. Then we study the variety of models of neutron stars composed of the matter with the different EOSs. Finally, we compare theoretical neutron star models with observations and formulate current constraints on the EOS which result from this comparison.

By the time of this writing (2006) the EOS of the neutron star core remains not strongly constrained and the problem is thus open. This state of the art makes the problem especially exciting and intriguing. The progress in the neutron star theory and observations is tremendous. Forthcoming observations of neutron stars, combined with new theoretical achievements, should be crucial to strongly constrain theoretical models and solve thus the main mystery of neutron star physics — the composition and equation of state in neutron star cores. We expect that the monograph will be a useful guide-book in achieving this goal.

This book benefited from our teaching experiences at the University of Warsaw and University of Toruń in Poland, at Université de Paris and Observatoire de Paris in France (*PH*), and at the Saint-Petersburg State Polytechnical University in Russia (*DGY, AYP*), as well as at several international schools for young astrophysicists and physicists.

In our presentation we tried to get to the forefront of theoretical calculations. Nevertheless, we supplement the analysis of modern techniques and recent results with the description of the historical development of the ideas. On many occasions, re-establishing the true sequence of events turned out to be a fascinating experience. We hope that that the reader will find these fragments of the text interesting and entertaining.

We have made an effort to present the results of calculations in the form of formulae, which are easier to use than tables. Sometimes we give a critical analysis of theories and our suggestions how to improve them. Both the degree of criticism and the details of suggestions are of course subjective. The prophecy is a dangerous activity, but we took risks.

We have tried to make this book complete, but of course our selection of topics reflects our limited research interests and competence. The list of references is long and reflects the huge work done in the past. We apologize to those authors whose work was not cited because of the natural limitation of space. The book was mainly completed in November 2005 and reflects the state of the neutron star physics by that time. Some minor addenda were made in August 2006.

Many discoveries related to neutron stars, for instance, the discovery of the first pulsar or the first binary pulsar, were serendipitous. However, Pasteur once said that in research "Chance favors only the prepared mind" (in French: "La chance ne sourit qu'aux esprits bien préparés," Pasteur 1854). We hope that this book will serve the cause of preparing the readers for future discoveries. The book can be useful to those theoreticians and observers who are working in the field of neutron stars and related fields of physics and astrophysics. We expect that it can also be useful to students and to many scientists who are just interested in the fascinating world of neutron stars.

We are grateful to our colleagues who, over the years of book writing, helped us through their collaboration, discussions, and expertise: M.K. Abubekerov, N.E. Alexandrovich, D.A. Baiko, D.P. Barsukov, M. Bejger, V.S. Beskin, G.S. Bisnovatyi-Kogan, S. Bonazzola, A.M. Bykov, B. Carter, G. Chabrier, N. Chamel, A.M. Cherepashchuk, A.I. Chugunov, H.E. DeWitt, M.E. Gusakov, D. Gondek-Rosińska, E. Gourgoulhon, V.S. Imshennik, I.L. Iosilevskiy, J. Kałużny, A.D. Kaminker, R.P. Kirshner, A.M. Krassilchtchikov, D. Lai, J.M. Lattimer, K.P. Levenfish, B. Paczyński, G.G. Pavlov, C.J. Pethick, Yu.A. Shibanov, P.S. Shternin, R.A. Sunyaev, T. Takatsuka, R. Tamagaki, L. Titarchuk, A.I. Tsygan, V.A. Urpin, D.A. Varshalovich, J. Ventura, J.L. Zdunik, J. Ziółkowski. Our special thanks are to Olga Burstein, Małgosia Haensel, and Marina Potekhina for their careful attention, permanent support and encouragement.

# Chapter 1

# OVERVIEW

## 1.1. Neutron stars

*Neutron stars* are compact stars which contain matter of supranuclear density in their interiors (presumably with a large fraction of *neutrons*). They have typical masses $M \sim 1.4\,M_\odot$ and radii $R \sim 10$ km. Thus, their masses are close to the solar mass $M_\odot = 1.989 \times 10^{33}$ g, but their radii are $\sim 10^5$ times smaller than the solar radius $R_\odot = 6.96 \times 10^5$ km. Accordingly, neutron stars possess an enormous gravitational energy $E_{\rm grav}$ and surface gravity $g$,

$$\begin{aligned} E_{\rm grav} &\sim GM^2/R \sim 5 \times 10^{53}\ {\rm erg} \sim 0.2\,Mc^2, \\ g &\sim GM/R^2 \sim 2 \times 10^{14}\ {\rm cm\ s^{-2}}, \end{aligned} \tag{1.1}$$

where $G$ is the gravitational constant and $c$ is the speed of light. Clearly, neutron stars are very dense. Their mean mass density is

$$\bar{\rho} \simeq 3M/(4\pi R^3) \simeq 7 \times 10^{14}\ {\rm g\ cm^{-3}} \sim (2-3)\,\rho_0, \tag{1.2}$$

where $\rho_0 = 2.8 \times 10^{14}$ g cm$^{-3}$ is the so called *normal nuclear density*, the mass density of nucleon matter in heavy atomic nuclei. The central density of neutron stars is even larger, reaching $(10-20)\,\rho_0$. By all means, neutron stars are *the most compact stars* known in the Universe.

This book consists of the overview (this Chapter) and main Chapters 2–9. In the overview we outline the neutron star history (§ 1.2), the current theoretical and observational status of the neutron star physics (§§ 1.3 and 1.4), the relation of the neutron star physics to other branches of physics and astrophysics (§ 1.5), and the contents of the main chapters (§ 1.6). We hope that the overview can serve as an introduction to the neutron star physics and can be read independently of the main chapters.

## 1.2. History: from prediction to discovery

**Landau – anticipation.** One commonly relates neutron stars with *neutrons*, their main ingredients. J. Chadwick announced his discovery of a neutron in a paper published in *Nature* on February 27, 1932 (Chadwick 1932). But a year before Lev Landau had written a paper devoted to dense stars. Landau was 23 at that time; he was a graduate student of the Leningrad Physico-Technical Institute (now the Ioffe Physico-Technical Institute, St.-Petersburg). The paper was completed in February 1931 in Zürich (as stated in the original publication, Landau 1932). The first part was really brilliant. Landau calculated the maximum mass of white dwarfs, somewhat later than Chandrasekhar (1931) and independently of him. The second part was more questionable. Landau speculated on a possible existence of stars more compact than white dwarfs, containing matter of nuclear density. At that time it was a problem to construct atomic nuclei without neutrons, from protons and electrons, because the Heisenberg principle forbids to localize the electrons within the nuclei. The only solution which Landau could propose was to violate the quantum mechanics. Thus, he suggested that "all stars heavier than 1.5 $M_\odot$ certainly possess regions in which the laws of quantum mechanics (and therefore of quantum statistics) are violated." That questionable suggestion might have prevented readers from reading the article to the end, whereas the last part of the paper was very important because Landau made a conclusion that in such stars "*the density of matter becomes so great that atomic nuclei come in close contact, forming one gigantic nucleus.*" The last quotation is a concise description of dense matter in neutron star interiors superficially anticipated by Landau before the discovery of the neutron.

From Zürich Landau went to Copenhagen, and stayed there from February 25 till March 19, 1931, before returning to Leningrad. It is highly probable that he discussed his paper with Bohr and Rosenfeld during his stay. The paper was submitted to *Physikalische Zeitschrift der Sowjetunion* on January 7, 1932, and published in the February issue of this journal (Landau 1932). The coincidence of the publication date with the announcement of the neutron discovery seems accidental but produced some contradictory comments (see, e.g., Rosenfeld 1974).

**Baade and Zwicky – prediction.** The actual theoretical prediction of neutron stars was made by W. Baade (Mt. Wilson Observatory) and F. Zwicky (Caltech) who analyzed observations of supernova explosions and proposed an explanation of an enormous energy release in these explosions. Their results were presented at the meeting of the American Physical Society at Stanford (December 15–16, 1933) and published in the 15 January issue of the *Physical Review* (Baade & Zwicky, 1934a). They wrote, in particular: "*With all reserve we advance the view that supernovae represent the transitions from ordinary*

*Figure 1.1.* A brief summary of the neutron star prediction presented to non-expert readers in the form of a cartoon in the *Los Angeles Times* on January 19th, 1934 (reproduced by Kirshner 1977). The sentence in the left box: "Cosmic rays are caused by exploding stars which burn with a fire equal to 100 million suns and then shrivel from 1/2 million mile diameters to little spheres 14 miles thick, says Prof. Fritz Zwicky, Swiss Physicist." Used with the permission of The Associated Press Copyright ©1934. All rights reserved.

*stars to* ***neutron stars****, which in their final stages consist of extremely closely packed neutrons.*" It is impossible to formulate the idea more precisely (Fig. 1.1). In the next publication Baade & Zwicky (1934b) explain further that a neutron star "*. . . may possess a very small radius and an extremely high density. As neutrons can be packed much more closely than ordinary nuclei and electrons, the 'gravitational packing' energy in a* ***cold*** *neutron star may become very large, and, under certain circumstances, may far exceed the ordinary nuclear packing fractions.*" This is again an exact statement. To avoid an illusion that the authors understood the problem exactly in the same way as we do it now, we cite their next paper (Baade & Zwicky, 1934c), where they speculate on the transition of an ordinary star into a neutron star: "If neutrons are produced on the surface of an ordinary star they will "rain" down towards the center if we assume that the light pressure on neutrons is practically zero."

**Theoretical progress before the Second World War.** We will be brief here; more historical remarks can be found in the main part of the book (particularly, in § 5.1). The neutron star history is also described in a comprehensive review by Baym (1982).

The most crucial ingredient of the neutron star theory is the *equation of state* (EOS) of dense matter in neutron star interiors. The term "EOS" has different meanings. It often means the dependence of the pressure $P$ on mass density $\rho$ (or associated energy density $\mathcal{E} = \rho c^2$) and on temperature $T$ of matter. Because neutron stars are mainly composed of strongly degenerate fermions (neutrons, protons, electrons, and others), the temperature dependence is mostly negligible and the EOS can be calculated at $T = 0$. The dependence $P(\rho)$ is needed to build neutron star models (§§ 1.3.3, 6.4). Sometimes the term "EOS" means not only $P(\rho)$ but also the composition of the matter and an underlying microphysical model.

When an ordinary star transforms to a neutron star, the stellar matter undergoes strong compression accompanied by beta captures of atomic nuclei with the creation of neutrons. The first attempt to construct the EOS of nuclear matter in equilibrium with respect to beta-capture and beta-decay processes was most probably made by Sterne (1933). He considered the matter composed of electrons, protons, several species of atomic nuclei, and neutrons, particularly at $T \to 0$. His analysis was naturally oversimplified, according to the today level, but he predicted the neutronization of matter with increasing $\rho$.

A splash of interest in neutron stars in 1937–1938 was caused by the problem of stellar energy. The energy source was unknown then. Gamow (1937) and Landau (1937) independently suggested that any (normal) star could contain a neutron star in its core. This would have initiated a slow "accretion" of stellar matter within the normal star onto its neutron star core, so that the stellar energy could have been supplied by the gravitational energy release in the course of that accretion. However, very soon Bethe & Critchfield (1938) showed that the energy of normal stars is provided by thermonuclear reactions, and the idea was almost forgotten.

Let us mention the paper by Zwicky (1938) who estimated the maximum binding energy of a neutron star of mass $M$ (and obtained $\sim 0.42\, Mc^2$). Thanks to Zwicky, we understand that the term "neutron star mass" is ambiguous. One should distinguish between the so called *baryon mass* $M_{\rm b}$ (the sum of baryon masses in a neutron star interior) and *gravitational mass* $M$ obtained from $M_{\rm b}$ by subtracting the gravitational binding energy (§ 6.2). Zwicky (1938) also noticed that on the surface of the star "... the acceleration of gravity [should be] very high, and light coming from this surface should be subject to enormous *gravitational red shifts* [of wavelengths]." He developed the idea of redshifts in a subsequent lengthy paper (Zwicky, 1939).

The next most important step was done by R.C. Tolman from Caltech and by J.R. Oppenheimer & G.M. Volkoff from the University of California, Berkeley. Their papers (Tolman, 1939; Oppenheimer & Volkoff, 1939) were received in the *Physical Review* the same day, January 3, 1939, and appeared in the same February 15 issue. Both papers contained the derivation of the equation of hydrostatic equilibrium for a spherically symmetric star in the framework of General Relativity. This is *the basic equation for building neutron star models.* It is called the *Tolman-Oppenheimer-Volkoff* (or, "for short", the *Oppenheimer-Volkoff*) equation (see § 6.4 for more details). The gravitational neutron star energy is a sizable fraction of its rest-mass energy, $Mc^2$; neutron stars are so compact that space-time is essentially curved around and within them, and the effects of General Relativity are most important.

Tolman and Oppenheimer & Volkoff derived their equation independently but they certainly communicated and discussed their results. Tolman (1939) obtained eight exact solutions of the new equation. They do not correspond to any realistic EOS of the neutron star matter, although they enable one (Oppenheimer & Volkoff, 1939) to understand the existence of a maximum mass of neutron stars. Oppenheimer & Volkoff (1939) used their equation for solving the most important problem. They numerically calculated neutron star models for the simplest EOS of stellar matter composed of a noninteracting strongly degenerate relativistic gas of neutrons. They showed that *stable static neutron stars have the maximum (gravitational) mass*, $M_{\rm max} \approx 0.71\,M_\odot$, which is often called *the Oppenheimer-Volkoff* mass limit. They worried that their mass limit is lower than the Chandrasekhar mass limit of white dwarfs, $1.44\,M_\odot$: this would hamper the formation of neutron stars from ordinary stars. They understood the simplicity of their model of noninteracting neutrons and discussed a possible repulsive component of neutron-neutron interaction which may stiffen the EOS and increase the maximum mass. However, after some discussion, they pessimistically concluded: "It seems likely that our limit of $\sim 0.7\,M_\odot$ is near the truth." Fortunately, their conclusion has turned out to be wrong, although their mass limit is extremely important. Combined with very accurately measured masses $(1.25-1.44)\,M_\odot$ of some neutron stars (§ 9.1) this limit gives a *direct astrophysical evidence* of strong repulsive interaction in dense matter at supranuclear density.

It is worth mentioning that von Neumann and Chandrasekhar obtained the same general relativistic equation of hydrostatic equilibrium several years earlier, in 1934, to study highly collapsed stars, but their result was not published (Baym, 1982). This was done at the Trinity College (Cambridge, England) after Chandrasekhar had constructed models of white dwarfs.

Thus, before the Second World War some elementary EOSs of dense matter were obtained, and the main equation of the neutron star structure was derived. More realistic EOSs could not be proposed at that time, because the properties of strong interactions and nuclear matter were poorly known. The theoretical basis was supplemented by the idea that neutron stars should be born in supernova explosions, relating neutron stars to observational astronomy. With the outbreak of the War, the interest to neutron stars naturally faded and was renewed much later.

**Theoretical progress: After the Second World War to the discovery of neutron stars.** Until the beginning of the 1960s neutron stars had been treated as the work of imagination of some (weird) theoreticians, and the theory had been developing slowly. The situation started changing later, with the hope to discover neutron stars in observations. Let us outline four main lines of theoretical studies (additional explanations can be found in § 1.3).

*(1) EOS of dense matter.* The main field of neutron star theory prior to the discovery was concerned with constructing model EOSs of dense stellar matter. The progress in this field is described in § 5.1. Here we stress the important contribution of Wheeler and his collaborators in the 1950s (reviewed by Harrison *et al.* 1958). They constructed a model of a neutron star crust and calculated the EOS of neutron star cores composed of free neutrons, protons, and electrons in beta equilibrium (§ 5.1). It was Cameron (1959) who emphasized the utmost importance of nuclear forces for the neutron star structure. He showed that the inclusion of nuclear forces can considerably stiffen the EOS. This can increase the maximum mass of neutron stars from the Oppenheimer-Volkoff limit of $\sim 0.7\,M_\odot$ to about $2\,M_\odot$, making the formation of neutron stars in supernova explosions quite realistic. Zeldovich (1961) used a model of baryon interaction through a massive vector field and constructed a very stiff EOS with the sound speed lower than the speed of light $c$, and tending to $c$ in the very high density limit.

Eventually, it was understood that neutron star cores may contain not only neutrons, protons and electrons, but also other particles, such as muons, mesons, and hyperons. First arguments in favor of hyperons were put forward by Cameron (1959) and Salpeter (1960); some EOSs of hyperonic matter and associated neutron star models were calculated by several authors, particularly, by Ambartsumyan & Saakyan (1960) and Tsuruta & Cameron (1966b). Ivanenko & Kurdgelaidze (1965, 1969) considered hypothetical quark cores of neutron stars.

*(2) Superfluidity of neutron star matter.* Another important step was the theoretical prediction of superfluidity of neutron star interiors. This activity was initiated by the theory of electron superconductivity in metals developed by Bardeen, Cooper, & Schrieffer (1957) (BCS theory). The electron superconductivity is explained by Cooper pairing of electrons under a weak attraction induced by the electron-phonon interaction. A superconducting state appears with decreasing temperature as a result of a second-order phase transition;

the typical critical temperatures are $T_c \sim$ (1–10) K. Microscopically, the phenomenon consists in the appearance of an energy gap $\Delta$ in the electron energy spectrum near the Fermi level. One year after the publication of the BCS theory Bohr *et al.* (1958) suggested that the phenomenon that had much in common with superconductivity could appear in atomic nuclei. Cooper pairing of nucleons could occur due to the attractive part of their interaction.

Migdal (1959) was one of the first who applied the BCS theory to atomic nuclei. He remarked also that neutron superfluidity could occur in neutron star interiors; he predicted the superfluid gap $\Delta \sim 1$ MeV and the associated critical temperature $T_c \sim 10^{10}$ K. Five years later Ginzburg & Kirzhnits (1964) in a brief article estimated the gap produced by the singlet-state pairing of neutrons at the densities $\rho = 10^{13}$–$10^{15}$ g cm$^{-3}$ and obtained $\Delta \sim (5 - 20)$ MeV. A very serious step was made by Wolf (1966). He showed that the singlet-state neutron pairing operates at subnuclear densities in the inner neutron star crust, but disappears in the core, since the singlet-state neutron-neutron interaction becomes repulsive at supranuclear densities. The number density of protons in the core is smaller than that of neutrons. Accordingly, the singlet-state proton-proton interaction is mostly attractive there, and it can lead to proton pairing. The possibility of neutron pairing in the core due to the attractive part of the triplet-state neutron-neutron interaction was understood later (see § 1.3.4). The results obtained and well established before the discovery of pulsars were reviewed by Ginzburg (1969). Further progress is outlined in § 1.3.4.

It is thought that superfluidity is important for explaining pulsar glitches (§ 1.4.4). It affects also the heat capacity and neutrino emission of neutron stars (§ 1.3.5), and hence their thermal evolution (§ 1.3.7). The effect of neutron superfluidity on the neutrino emission (in the modified Urca process) was first studied by Wolf (1966).

*(3) Neutrino emission from neutron stars.* Another line of theoretical studies was inspired by the expectations at the beginning of the 1960s to discover neutron stars by detecting the thermal radiation from their surfaces (some attempts are described below). Born hot in supernova explosions (§ 1.4.2), neutron stars cool down by the thermal emission of photons from stellar surfaces and the neutrino emission from stellar interiors. This makes the neutrino processes vitally important. These processes are outlined in § 1.3.5.

Some neutrino processes were proposed in the 1960s. Let us mention the seminal paper by Chiu & Salpeter (1964) who suggested the modified Urca process [explained in § 1.3.5, Eq. (1.9)] and estimated its neutrino emissivity. It is the leading neutrino process in the cores of not too massive neutron stars. First detailed calculations of the neutrino emissivity were performed by several authors, particularly by Bahcall & Wolf (1965a) (who also gave detailed discussion of earlier work).

Bahcall & Wolf (1965a) also studied a model of dense matter which contains free pions. It was a progenitor of models of pion-condensed matter introduced into the neutron star physics in the beginning of the 1970s (§§ 1.3.2, 7.3). Bahcall & Wolf considered the neutrino process which consists of two reactions: neutron beta decay followed by beta capture in the presence of pions. The neutrino emissivity appeared to be much higher than in the modified Urca process, giving the first example of the enhanced neutrino emission. Similar enhancement in a more realistic model of pion-condensed matter was analyzed much later by Maxwell *et al.* (1977). More references to earlier papers on the neutrino emission from neutron stars can be found in Yakovlev *et al.* (2001).

*(4) Neutron star thermal evolution.* These studies were also motivated by the attempts to discover neutron stars. The first estimates of the thermal emission from cooling neutron stars were most probably done by Stabler (1960). Four years later Chiu (1964) repeated the estimates and theoretically proved the possibility to discover neutron stars from their thermal emission. First, simplified calculations of the neutron star cooling were done by Morton (1964), Chiu & Salpeter (1964), and Bahcall & Wolf (1965a,b). The latter authors emphasized the strong dependence of the cooling rate on neutrino emission processes and pointed out that this dependence can be used to explore the EOS of dense matter by comparing theoretical cooling models with observations of thermal radiation from neutron stars. The foundation of the strict cooling theory was made by Tsuruta & Cameron (1966a). In particular, these authors formulated the main elements of the cooling theory such as the neutrino and photon cooling stages, the relation between the internal and surface temperatures. A more detailed description of the history of neutron star cooling is given, for instance, by Yakovlev *et al.* (1999). The current state of the art is mentioned in § 1.3.7.

**A search and the discovery.** The first serious attempts to discover neutron stars were made when the era of practical X-ray astronomy began in the 1960s. It was expected to detect the thermal radiation from surfaces of cooling (isolated) neutron stars. A star with the surface temperature of $\sim 10^6$ K would mainly emit soft X-rays which cannot be detected by ground-based facilities. In the beginning of the 1960s the first X-ray detectors were launched on rockets and balloons.

The first cosmic X-ray source of non-solar origin, Sco X-1 (in the Scorpius constellation), was discovered in rocket experiments by Giacconi *et al.* (1962). Soon after the same group discovered several X-ray sources.[1] The discovery initiated a great interest in neutron stars but the first attempts failed to prove the relation between neutron stars and newly discovered compact X-ray sources. In particular, Bowyer *et al.* (1964) measured the size of the X-ray source in

[1] In 2002 R. Giacconi was awarded the Nobel Prize for outstanding contribution into the X-ray astronomy.

the Crab Nebula from observations during a lunar occultation on July 7, 1964. Their result, $\sim 10^{13}$ km, indicated that the source was much larger than a neutron star should be. Ironically, the Crab Nebula turned out to be an exception: *there was a neutron star there*, the famous Crab pulsar, but it was hidden within a compact plerion pulsar nebula. The Crab Nebula is actually the archetype of a *plerion* – a supernova remnant with an active pulsar at its center, powering the nebula expansion and radiation. Interestingly, just at the same time Kardashev (1964) considered a collapse of a magnetized rotating star into a compact object (collapsar) with the appearance of a surrounding envelope (nebula). He emphasized that the collapsar can gain rapid rotation during its birth, and its spin energy can be transferred to the surrounding nebula by the magnetic field. He assumed that this mechanism can power the Crab Nebula.

Many scientists proposed different methods of discovering neutron stars (as described, e.g., by Zeldovich & Novikov 1971, Shapiro & Teukolsky 1983, and Lyne & Graham-Smith 1998). For instance, Zeldovich & Guseynov (1966) suggested to observe some selected binaries with optical primary components and invisible secondary components, assuming that neutron stars represent the invisible components. Pacini (1967), in a paper published in *Nature*, showed that a rapidly rotating neutron star with a strong dipole magnetic field could efficiently transform its rotational energy into electromagnetic radiation and, subsequently, accelerate particles to high energies. He suggested that the rotational energy loss rate is the same as produced by a magnetic dipole rotating in vacuum (very close to the modern point of view, see § 1.4.4). He, like Kardashev, anticipated that such a star could power a surrounding nebula, particularly, the Crab Nebula.

It was important that Sandage *et al.* (1966) had identified Sco X-1, the first detected and the brightest X-ray source, as an optical object of 13th magnitude. Analyzing those observations, Shklovsky (1967) concluded that the source "... corresponds to a neutron star in a state of accretion ... " and "... the natural and very efficient supply of gas for such a accretion is a stream of gas, which flows from a secondary component of a close binary system toward the primary component which is a neutron star." Now we know that Sco X-1 is, indeed, an X-ray binary containing an accreting neutron star but at that time the Shklovsky's arguments were mostly ignored by the astrophysical community. By 1968, about twenty compact X-ray sources had been observed but their association with neutron stars seemed to be not convincing. It happened so that neutron stars were discovered unexpectedly and by absolutely different methods.

In 1965 Antony Hewish (Cavendish Laboratory, Cambridge, England) started to construct a new radio telescope. Its wavelength was 3.7 m; it was an array of 2048 dipole antennae that covered an area of about 18,000 square meters. The telescope differed from the other ones by a good temporal resolution:

it was designed to study scintillations of radio sources while their radiation passes through inhomogeneities of solar wind in the interplanetary space (similar twinklings of stars are seen by naked eye). The telescope was inexpensive, about £10,000, because it was mainly built by volunteers. It was completed in July 1967.

On August 6, 1967, Jocelyn Bell – a graduate student supervised by Hewish since 1965 – discovered a weak variable radio source (Hewish, 1975). It was observed at night time whereas scintillations of ordinary radio sources are stronger at day time, when a telescope is directed closer to the Sun. By the end of September the source had been observed several times; it was suggested to be a flare star. By November 28 the observations had indicated that the source had emitted strictly periodic pulses. The accurate measurement of the period started on December 11. The period turned out to be wonderfully stable, 1.3373012 s. This produced a suspicion that the signals were of artificial origin, created, for instance, by space satellites or even by an extraterrestrial civilization. The signals were even called "LGM" ("little green men"), and it was decided to postpone the publication of the discovery till the situation would clarify. It took several weeks to understand that the rapidly pulsating source, *the pulsar*, was well outside the solar system. Now it is called PSR B1919+21.[2] Pilkington, Scott, and Collins – the colleagues of Hewish and Bell – studied the drift of the pulsar radio frequency during a pulse. By the beginning of February, 1968, three other pulsars had already been discovered (PSR B1133+16, PSR B0834+06, and PSR B0950+08). It was suggested that the sources represented either oscillating white dwarfs or neutron stars.

The discovery was announced in the 24 February 1968 issue of *Nature* (Hewish *et al.*, 1968) and produced a sensation. By the end of 1968, in ten remaining months, over 100 pulsar articles had been published (Will, 1994). The hypotheses on white dwarfs and oscillating neutron stars were ruled out. In a strong competition, the idea of Gold (1968) won, that *pulsars are rotating magnetized neutron stars*.[3] The important piece of evidence was given by the discovery of the Crab pulsar in 1968. Its pulsation (spin) period was measured in November 1968 (Comella *et al.*, 1969) and appeared to be very short, $P = 33$ ms. White dwarfs could not sustain such a rapid rotation: they would be destroyed by centrifugal forces.

[2]The right ascension of 19 hours 19 minutes by Greenwich time, with the 21° declination; 'B' means that the pulsar position is given according to the B1950 coordinate system; 'J' instead of 'B' would mean the use of the J2000 coordinate system.

[3]Submitting his paper, Gold was unaware of the earlier paper by Pacini (1967), where the pulsar model had been suggested (see above). The collaboration between Gold and Pacini "... was, of course, soon established ... These two men should clearly share the credit for establishing the linkage between pulsars and neutron stars" (Lyne & Graham-Smith, 1998).

Thus, pulsars are spinning neutron stars with their magnetic moments inclined to spin axes. Their radio emission is generated outside a star, in the *magnetosphere* (§ 1.4.4); it is beamed along the magnetic axis. The beamed radiation rotates with the star, and a pulsar is detected if its beam crosses the Earth. Hence, any pulsar operates like a *lighthouse*, and we detect strictly regular sequences of radio pulses. The emitted electromagnetic radiation carries away the rotational energy and momentum, and produces a slow regular spindown of the pulsar (an increase of the pulse period). The spindown effect is really observed. It rules out the model of stellar oscillations (because oscillation periods are expected to be stable) and confirms thus the pulsar model.

Curiously, in 1965 Hewish and his student Okoye discovered a scintillating radio source in the Crab nebula (Hewish & Okoye, 1965). Its nature was not clear and they suggested that it "might be the remains of the original star which had exploded and which still showed activity in the form of flare-type radio emission... This source later turned out to be none other than the famous Crab pulsar" (Hewish, 1975). They looked into archives and discovered that the source had been observed earlier, in April 1962.

In 1974 Hewish was awarded the Nobel Prize for the discovery of pulsars, but "In some circles, controversy still lingers over the decision of the Swedish Academy not to include Ms Bell in the award" (Will, 1994).

## 1.3. Internal structure and processes – Theoretical outlook

After the discovery of pulsars, observations and theory of neutron stars have been developing at such a rapid pace that we are unable to review the wealth of outstanding works due to the natural limitation of space. Thus, we briefly outline the basic contemporary (2006) theoretical ideas (§ 1.3), and their relations to observations (§ 1.4).

### 1.3.1 Structure

According to current theories, a neutron star can be subdivided into the *atmosphere* and four main internal regions: the *outer crust*, the *inner crust*, the *outer core*, and the *inner core* as shown in Fig. 1.2. These regions can further be subdivided into subregions, discussed in the main chapters of this book (for instance, see Fig. 2.1 in Chapter 2).

**The atmosphere** is a thin plasma layer, where the spectrum of thermal electromagnetic neutron star radiation is formed. The spectrum, beaming and polarization of emerging radiation can be determined theoretically by solving the radiation transfer problem in atmospheric layers. This radiation contains valuable information on the parameters of the surface layer (on the effective surface temperature, surface gravity, chemical composition, strength and geometry of the surface magnetic field) and on the masses and radii of neutron stars.

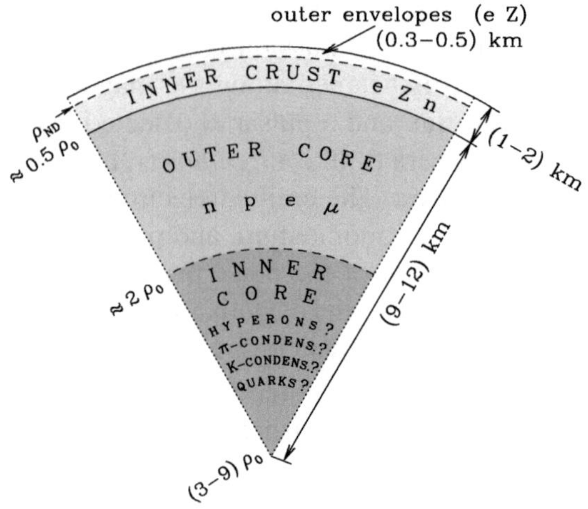


*Figure 1.2.* Schematic structure of a neutron star. Stellar parameters strongly depend on the EOS in a neutron star core.

Some attempts to extract this information from observations are described in Chapter 9. The atmosphere thickness varies from some ten centimeters in a hot neutron star (with the effective surface temperature $T_s \sim 3 \times 10^6$ K) to a few millimeters in a cold one ($T_s \sim 3 \times 10^5$ K). Very cold or ultramagnetized neutron stars may have a solid or liquid surface. Neutron star atmospheres have been studied theoretically by many authors (see, e.g., Zavlin & Pavlov 2002 and references therein). Current atmosphere models, especially for neutron stars with surface temperatures $T_s \lesssim 10^6$ K and strong magnetic fields $B \gtrsim 10^{11}$ G, are far from being complete. The most serious problems consist in calculating the EOS, ionization equilibrium, and spectral opacity of the atmospheric plasma (Chapters 2 and 4).

If the radiation flux is too strong, the radiative force exceeds the gravitational one and makes the atmosphere unstable with respect to a plasma outflow. In a hot nonmagnetized atmosphere, where the radiative force is produced by Thomson scattering, this happens whenever the stellar luminosity $L$ exceeds the *Eddington limit*

$$L_{\rm Edd} = 4\pi c G M m_p / \sigma_{\rm T} \approx 1.3 \times 10^{38}\,(M/M_\odot) \quad {\rm erg\ s^{-1}}, \tag{1.3}$$

where $\sigma_{\rm T}$ is the Thomson scattering cross section and $m_p$ the proton mass.

**The outer crust (the outer envelope)** extends from the atmosphere bottom to the layer of the density $\rho = \rho_{\rm ND} \approx 4 \times 10^{11}$ g cm$^{-3}$. Its thickness is some hundred meters (Chapter 6). Its matter consists of ions $Z$ and electrons $e$ (Chapters 2 and 3). A very thin surface layer (up to few meters in a hot star) contains a non-degenerate electron gas. In deeper layers the electrons constitute a strongly degenerate, almost ideal gas, which becomes ultrarelativistic at $\rho \gg 10^6$ g cm$^{-3}$. The pressure is mainly provided by electrons. For $\rho \gtrsim 10^4$ g cm$^{-3}$, atoms are fully ionized by the electron pressure. In the outer atmosphere layers the ions may constitute a Boltzmann gas, but in deeper layers they form a strongly coupled Coulomb system (liquid or solid). A larger fraction of the envelope is usually solidified; hence, the envelope is often called *the crust*. The electron Fermi energy grows with increasing $\rho$. This induces beta captures in atomic nuclei and enriches the nuclei with neutrons. At the base of the outer crust the neutrons start to drip out from the nuclei producing a free neutron gas.

**The inner crust (the inner envelope)** may be about one kilometer thick. The density $\rho$ in the inner crust varies from $\rho_{\rm ND}$ at the upper boundary to $\sim 0.5\rho_0$ at the base. Here, $\rho_0$ is the saturation nuclear matter density introduced in § 1.1. The matter of the inner crust consists of electrons, free neutrons $n$, and neutron-rich atomic nuclei (Chapter 3). The fraction of free neutrons increases with growing $\rho$. The neutronization at $\rho \approx \rho_{\rm ND}$ greatly softens the EOS, but at the crust bottom the repulsive short-range component of the neutron-neutron interaction comes into play and introduces a considerable stiffness. In the bottom layers of the crust, in the density range from $\approx \frac{1}{3}\rho_0$ to $\approx \frac{1}{2}\rho_0$, the nuclei may become essentially nonspherical and form a "mantle", but this result is model dependent. The nuclei disappear at the crust-core interface. Free neutrons in the inner crust and nucleons confined in the atomic nuclei can be in superfluid state.

**The outer core** occupies the density range $0.5\rho_0 \lesssim \rho \lesssim 2\rho_0$ and is several kilometers thick (Chapters 5 and 6). Its matter consists of neutrons with several per cent admixture of protons $p$, electrons, and possibly muons $\mu$ (the so called $npe\mu$ composition). The state of this matter is determined by the conditions of electric neutrality and beta equilibrium, supplemented by a microscopic model of many-body nucleon interaction. The beta equilibrium implies the equilibrium with respect to the beta (muon) decay of neutrons and inverse processes. All $npe\mu$-plasma components are strongly degenerate. The electrons and muons form almost ideal Fermi gases. The neutrons and protons, which interact via nuclear forces, constitute a strongly interacting Fermi liquid and can be in superfluid state.

**The inner core**, where $\rho \gtrsim 2\,\rho_0$, occupies the central regions of massive neutron stars (and does not occur in low-mass stars whose outer core extends to the very center). Its radius can reach several kilometers, and its central density can be as high as $(10-15)\rho_0$ (Chapter 6). Its composition and the EOS are very model dependent (Chapter 5). Several hypotheses have been put forward, predicting the appearance of new fermions and/or boson condensates. The main four hypotheses are (Chapters 5 and 7):

(*1*) *Hyperonization* of matter – the appearance of hyperons, first of all $\Sigma^-$ and $\Lambda$ hyperons.

(2) *Pion condensation* – the appearance of a boson condensate of pion-like excitations with a strong renormalization and mixing of nucleon states.

(3) *Kaon condensation* – the Bose-Einstein condensation of kaon-like excitations which, like real kaons, possess strangeness.

(4) A phase transition to the *quark matter* composed of deconfined light $u$ and $d$ quarks and strange $s$ quarks, and a small admixture of electrons, or even no electrons at all.

Nucleon and nucleon/hyperon matter, called respectively *nuclear* and *hypernuclear* matter, have been studied experimentally in ordinary nuclei and hypernuclei. Pion and kaon condensations have not been discovered in laboratory so far. Some very tentative signatures of quark deconfinement have been recently detected in the relativistic heavy-ion collisions. The models (2) – (4) are often called *exotic* models of dense matter. A new phase may appear via a first-order or a second-order phase transition. Its appearance is accompanied by the softening of the EOS. One cannot exclude the existence of mixed phases of dense matter.

Let us mention a special hypothetical class of compact stars, which are called *strange stars* (Chapter 8). They could exist only if the absolute ground state of hadronic matter is a self-bound quark matter. Strange stars entirely (or nearly entirely) consist of strange quark matter. In some models, this matter extends to the very surface; such stars are called *bare strange stars*. In other models, strange stars have a normal crust extending from the surface not deeper than to the neutron-drip density $\rho_{\rm ND}$.

### 1.3.2 The main mystery: The equation of state in neutron star cores

The pressure of the matter in neutron star interiors is mainly produced by highly degenerate fermions and can be calculated assuming $T = 0$. It is the *bulk* property of the matter, provided by entire Fermi seas of fermions. The EOS in a neutron star crust has been calculated with an accuracy, sufficient to construct neutron star models, although some theoretical problems are unsolved (Chapters 2–4). The theory is based on reliable experimental data on atomic nuclei, nucleon scattering, and on the well elaborated theory of strongly coupled Coulomb systems.

By contrast, the EOS at $\rho \gtrsim \rho_0$ cannot be reproduced in laboratory, and it cannot be calculated exactly because of the lack of the precise relativistic many-body theory of strongly interacting particles. Instead of the exact theory, there are many theoretical models (Chapter 5). The reliability of these models decreases with growing $\rho$. Thus, by the time of this writing (2006) the EOS in neutron star cores is largely unknown. *The fundamental problem of the EOS of superdense matter constitutes the main mystery of neutron stars.*

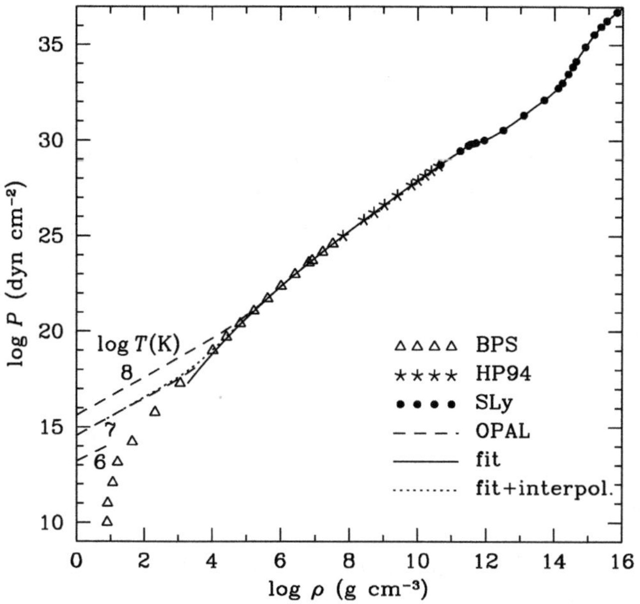


*Figure 1.3.* EOS of neutron star matter: BPS (triangles), Haensel & Pichon (1994) (HP94, stars), SLy (dots), OPAL at $T = 10^6$, $10^7$, and $10^8$ K (dashed lines), the fit (C.2) (the solid line) and the fit modified at low $\rho$ (the dotted line). From Haensel & Potekhin (2004).

The model EOSs can be subdivided into the *soft*, *moderate* and *stiff* ones with respect to the *compressibility* of the matter. Basing on different EOSs, one obtains (Chapter 6) different stellar models, particularly, different maximum masses, from $M_{\rm max} \sim 1.4\,M_\odot$ for the softest EOSs to $M_{\rm max} \sim 2.5\,M_\odot$ for the stiffest ones. The EOSs can also be subdivided with respect to the *composition* of the matter (as already outlined in § 1.3.1). Very stiff EOSs can possibly be attributed only to nucleon matter.

Figure 1.3 gives an example of the EOS in a neutron star, from the core to the surface. We show a moderately stiff EOS (SLy, dots, §§ 3.6 and 5.12) of dense nucleon matter derived by Douchin & Haensel (2001) based on a Skyrme-type energy density functional. It is equally valid in the neutron star core and the crust, and is plotted for $\rho > 5 \times 10^{10}$ g cm$^{-3}$. At lower densities in the crust, $10^8$ g cm$^{-3} \lesssim \rho < 5 \times 10^{10}$ g cm$^{-3}$, we plot the EOS of Haensel & Pichon (1994) (HP94, stars, § 3.2), based on experimental nuclear data. At still lower $\rho$ we replace it by the EOS of Baym *et al.* (1971b) (BPS, triangles, § 3.2) for the ground state of the matter at zero temperature. However, at $\rho \lesssim 10^5$ g cm$^{-3}$ the actual EOS becomes temperature dependent as shown by the dashed lines. These lines present the EOS of iron matter for $T = 10^6$, $10^7$ and $10^8$ K provided by the Opacity Library (OPAL; Rogers *et al.* 1996; see § 2.4).[4] The solid and dotted lines in Fig. 1.3 are analytic interpolations explained in Appendix C.

### 1.3.3 Neutron star models

Once the EOS is given, one can compute stellar models (Chapter 6). This should be done in the framework of General Relativity. For non-rotating stars, the computation is performed by solving the Tolman-Oppenheimer-Volkoff equation of hydrostatic equilibrium (§ 1.2). In this way one gets a family of neutron star models parameterized by the central density $\rho_c$, so that the gravitational mass $M = M(\rho_c)$ and the circumferential radius $R = R(\rho_c)$ (with $2\pi R$ being a proper length of the neutron star equator; § 6.1).

The models *stable* with respect to small perturbations correspond to such a part of the $M(\rho_c)$ curve, where $M$ increases with growing $\rho_c$. From $M(\rho_c)$ and $R(\rho_c)$ one gets a *mass-radius ($M - R$) relation*, $M = M(R)$ (§ 6.6). The stability condition leads to the existence of *minimum* and *maximum* masses of neutron stars, $M_{\rm min} \sim 0.1\, M_\odot$ and $M_{\rm max} \sim (1.4 - 2.5)\, M_\odot$ (§ 6.5). Neutron star models, the $M - R$ relation, and $M_{\rm max}$ depend on the EOS of dense matter. This opens an attractive possibility to study the EOS by comparing theory with observations (Chapter 9). The models of strange stars (Chapter 8) strongly differ from the models of neutron stars at $M \lesssim M_\odot$. In contrast to neutron stars, strange stars can have masses $M \ll M_\odot$ and radii $R \ll 10$ km.

For a given neutron star, the effects of General Relativity can be characterized by the *compactness parameter*

$$x_{\rm GR} = r_{\rm g}/R, \quad r_{\rm g} = 2GM/c^2 \approx 2.95\, M/M_\odot \;\; {\rm km}, \tag{1.4}$$

where $r_{\rm g}$ is the Schwarzschild radius. The surface gravity $g$ and the gravitational redshift $z_{\rm surf}$ (§ 6.6.3) for frequency of of photons emitted from the stellar surface can be expressed as

$$g = GMR^{-2}/\sqrt{1 - r_{\rm g}/R}, \quad \omega_\infty = \omega_0/(1 + z_{\rm surf}) = \omega_0 \sqrt{1 - r_{\rm g}/R}. \tag{1.5}$$

Here, $\omega_0$ is the frequency of photons emitted from the stellar surface (in a local reference frame), while $\omega_\infty$ is the frequency of these photons as detected by a distant observer.

Instead of $R$ one often introduces the apparent (radiation) radius which the distant observer would measure if the telescope could resolve the star: $R_\infty = R/\sqrt{1 - r_{\rm g}/R}$.

Let us mention the so called *canonical neutron star model* used by many authors. This artificial model does not imply any specific EOS, but just assumes

[4] The OPAL table for iron has been kindly provided by F. J. Rogers.

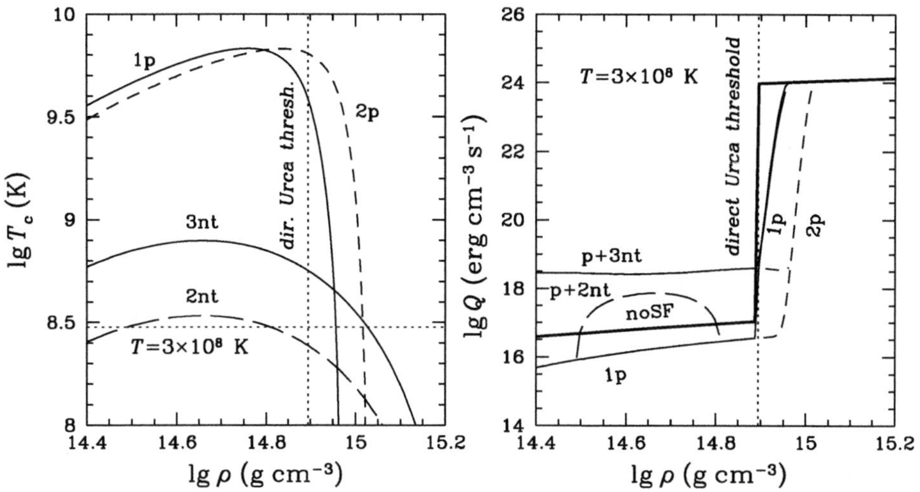


*Figure 1.4. Left:* Superfluid transition temperatures versus density for protons (models 1p and 2p) and neutrons (models 2nt and 3nt for triplet-state pairing) in a neutron star core. *Right:* Density profiles of the neutrino emissivity $Q$ at $T = 3 \times 10^8$ K in nonsuperfluid matter (noSF) and in the presence of superfluidity. Vertical dotted line shows the threshold for the direct Urca process (see the text for details).

the typical values of $M$ and $R$:

$$M = 1.4\, M_\odot, \quad R = 10 \text{ km}, \quad R_\infty = 13 \text{ km}, \quad g = 2.43{\times}10^{14} \text{ cm s}^{-2}. \quad (1.6)$$

The models of rotating neutron stars are more complicated (§ 6.12). The theory predicts that rotation noticeably affects the stellar structure at spin periods $P \lesssim 3$ ms. The fastest rotators may have spin periods $P \sim 0.7$ ms; their shape may deviate from axial symmetry, allowing them to emit gravitational radiation.

### 1.3.4 Superfluidity

Various baryons in neutron star matter can be in *superfluid* state produced by Cooper pairing of baryons due to an attractive component of baryon-baryon interaction (as already mentioned in § 1.2). Superfluidity of a particular baryon species is switched on when the temperature $T$ falls below some critical temperature $T_c$. Superfluidity is a Fermi-surface phenomenon; it has almost no effect on the EOS, neutron stars masses and radii.

The theory predicts superfluidity of free neutrons and of nucleons in atomic nuclei in the inner neutron star crust. Neutrons, protons and other baryons in the stellar core can also be superfluid. Superfluidity of charged particles (for instance, protons) means superconductivity.

As outlined in § 1.2, free neutrons in the stellar crust undergo *singlet-state* ($^1S_0$) *pairing* which disappears in the core (Wolf 1966). However, neutrons in the core can be superfluid due to a weaker *triplet-state* ($^3P_2$) *pairing*. The idea of such pairing in neutron star cores, first estimates and simplified calculations of the superfluid gap have been published by a number of authors (particularly, by Ruderman 1967, Maekawa & Tamagaki 1968, Tamagaki 1969, Hoffberg *et al.* 1970), mostly in conference proceedings. Usually these results are solely attributed to the first journal publication (Hoffberg *et al.*, 1970). The foundation of the strict relevant theory was laid by Tamagaki (1970). Proton superfluidity in the core is thought to be mainly produced by singlet-state proton pairing. One has also invented superfluidity of hyperons (Balberg & Barnea, 1998) and quarks (Bailin & Love, 1984). Pion and kaon condensates affect superfluidity of nucleons (Takatsuka & Tamagaki, 1995, 1997a,b).

Critical temperatures $T_c$ of various particle species have been calculated by many authors as reviewed by Lombardo & Schulze (2001) (more references can be found in Yakovlev *et al.* 1999). The results are extremely sensitive to strong interaction models and many-body theories employed. In all the cases mentioned above microscopic calculations give density-dependent critical temperatures $T_c \lesssim 10^{10}$ K and lower. As a rule, superfluidities weaken and disappear at essentially supranuclear densities, where the attractive part of strong interaction becomes inefficient. For example, in the left panel of Fig. 1.4 we present $T_c(\rho)$ in a neutron star core composed of $npe$ matter with a moderately stiff EOS of Prakash *et al.* (1988) (after Yakovlev *et al.* 2002). We plot four purely phenomenological models: models 1p and 2p for single-state proton pairing and models 2nt and 3nt for triplet-state neutron pairing. The curves in the right panel are explained in § 1.3.5.

In addition, Alford *et al.* (1998) proposed a new type of quark superfluidity associated with *color superconductivity* (§ 8.8.3). For a typical Fermi energy of quarks $\sim 500$ MeV, one may expect $T_c \sim 50$ MeV $\sim 5 \times 10^{11}$ K.

Superfluidity affects the heat capacity and neutrino emission of neutron stars. It induces also a number of macroscopic quantum phenomena. For instance, consider the core of a rotating neutron star composed of neutrons, protons and electrons. A rotation of neutron superfluid is realized in the form of quantized (Feynman-Onsager) vortices parallel to the spin axis (Ginzburg & Kirzhnits 1964, Baym *et al.* 1969). The total amount of vortices in the star is estimated as $\sim 2 \times 10^{16}/P$, where $P$ is the stellar spin period in seconds. The vortex motion of neutron superfluid, averaged over small macroscopic fluid elements, reproduces a solid-body rotation. The vortices occur also in the inner crust, where free neutrons are superfluid. A pulsar spindown induces the outward drift of vortices and their disappearance at the boundary of the superfluid region.

Below the critical temperature, superconductivity of protons (and other charged baryons) in the neutron-star core is described by the Ginzburg-Landau theory. The proton coherence length (2–6 fm) is typically much smaller than the London screening length (100–300 fm) of electric currents in superconducting medium. This probably means type II superconductivity. If an initially normal hot core contained a quasi-uniform magnetic field $\boldsymbol{B}$, superconductivity splits the field into fluxoids (Abrikosov vortices), which are thin quantized magnetic flux tubes parallel to the initial field. The total number of fluxoids is $\sim 10^{31}\,(B/10^{12}$ G).

Neutron vortices may pin to atomic nuclei or lattice defects in the crust and to fluxoids in the core. The pinning may be accompanied by vortex creep. These and related phenomena are invoked, for instance, to explain observations of pulsar glitches and to study the evolution of internal magnetic fields (§ 1.4.4).

### 1.3.5 Neutrino processes

Neutrinos are generated in numerous reactions in neutron star interiors as reviewed, for instance, by Pethick (1992) and Yakovlev *et al.* (2001). Only neutrinos emitted during supernova explosions can be detected by neutrino observatories. In several minutes after the explosion the neutrino flux is already too small to be detected. However, neutrinos carry away energy and provide an efficient cooling of warm neutron stars, with an internal temperature $T \gtrsim 10^7$ K. The most powerful neutrino emission is produced in the stellar core. Typical neutrino energies are $\gtrsim k_{\rm B}T$, much higher than possible neutrino rest energies. These neutrinos can be treated as massless particles. The emissivity of a neutrino process in a nonsuperfluid core can be written as

$$Q(T,\rho) = Q_0(\rho)\,T_9^k \;\; \mathrm{erg\;cm^{-3}\;s^{-1}}, \tag{1.7}$$

where $Q_0(\rho)$ is a slowly varying function of $\rho$, $k = 6$ or 8 (see below), and $T_9 = T/(10^9$ K).

The most powerful neutrino emission is produced by the *direct Urca* processes. In the simplest model of the $npe\mu$ dense matter, there are two such processes. Each process is a sequence of neutron decay and inverse reaction associated with either electron or muon ($\ell = e$ or $\mu$):

$$n \to p + \ell + \bar{\nu}_\ell, \quad p + \ell \to n + \nu_\ell. \tag{1.8}$$

These processes are forbidden in the outer cores of neutron stars due to insufficiently high number densities of protons and leptons $\ell$ (that violates momentum conservation for reacting particles). However, the direct Urca processes operate in the inner cores of massive stars with those model EOSs which give rather large amount of protons and leptons. This circumstance was noted by Boguta (1981) but his paper has remained unnoticed. The nucleon direct Urca processes have been taken into account only after the paper of Lattimer *et al.* (1991), who introduced them into neutron star astrophysics and calculated the associated neutrino emissivity. If both, electron and muon, processes are allowed, they have the same emissivity given by Eq. (1.7) with $k = 6$ and $Q_0 \sim 10^{26} - 3 \times 10^{27}$. If open, the process produces an extremely strong neutrino cooling. In hyperonic matter, similar and also very intense processes with hyperons can be allowed (Prakash *et al.*, 1992).

For example, in the right panel of Fig. 1.4 we show the density dependence of the neutrino emissivity in a neutron star core (composed of $npe$ matter) at $T = 3 \times 10^8$ K. The vertical dotted line indicates the direct Urca threshold for the given EOS. The thick solid line marked 'noSF' refers to nonsuperfluid matter. The huge emissivity after the threshold is determined by the direct Urca process.

The neutrino emission from exotic phases of matter in the inner cores of massive neutron stars can be produced by direct-Urca-like processes with $k = 6$ in Eq. (1.7) (see Pethick 1992 for details). Such processes are also efficient, but somewhat weaker than the nucleon direct Urca processes. In a pion or kaon condensed matter, the process is a sequence of two reactions: $\widetilde{B}_1 \to \widetilde{B}_2 \ell \bar{\nu}_\ell$ and $\widetilde{B}_2 \ell \to \widetilde{B}_1 \nu_\ell$, where $\widetilde{B}_1$ and $\widetilde{B}_2$ stand for quasibaryons (baryon states mixed by the condensate fields). The leading process in the quark matter is $d \to u e \bar{\nu}_e$ followed by $ue \to d\nu_e$. The calculations using various models give the emissivity (1.7) with $Q_0 \sim 10^{23-26}$ for a pion-condensed matter, and $Q_0 \sim 10^{23-24}$ for a kaon-condensed or quark matter (in all cases $k = 6$).

In addition to these efficient neutrino processes, there are weaker ones. Their advantage is that they operate everywhere in the stellar core, particularly in its outer part. They are dominant neutrino processes in low-mass neutron stars, where the inner core is absent. The most important processes of such a type in the $npe\mu$ matter are the *modified Urca processes*,

$$n + N \to p + N + \ell + \bar{\nu}_\ell, \quad p + N + \ell \to n + N + \nu_\ell \tag{1.9}$$

(where $N$ stands for a nucleon, neutron or proton), and the nucleon-nucleon bremsstrahlung processes

$$N + N \to N + N + \nu + \bar{\nu}. \tag{1.10}$$

The latter processes produce all neutrino flavors. Modified Urca processes differ from their direct Urca progenitors by an additional nucleon-spectator $N$ required for momentum conservation. There are three bremsstrahlung processes ($nn$, $np$, and $pp$) in the $npe\mu$ matter. The emissivities of the processes (1.9) and (1.10) can be written in the form (1.7) with $k = 8$. One gets $Q_0 \sim 10^{20} - 3 \times 10^{21}$ for modified Urca processes and $Q_0 \sim 10^{19-20}$ for bremsstrahlung processes. Similar processes involving various particles may occur in hyperon or quark matter.

The neutrino emissivity (thick solid line 'noSF') in Fig. 1.4 before the direct Urca threshold is mainly produced by the modified Urca processes in the $npe$ matter. It is more than 7 orders of magnitude weaker than after the threshold.

The neutrino reactions are drastically affected by baryon superfluidity in neutron star interiors as reviewed by Yakovlev *et al.* (1999, 2001). Because the baryons are strongly degenerate, they can actively participate in a reaction only if their energies lie within the thermal widths of their Fermi levels. When the temperature $T$ drops much below the critical temperature $T_c$ of given baryons, the energy gap in the baryon energy spectrum makes these baryons inactive, greatly (as a rule, exponentially) suppressing all reactions involving these baryons. For instance, a strong superfluidity of protons in the $npe$ matter switches off all Urca processes, but does not affect the $nn$ bremsstrahlung. Both model proton superfluidities in Fig. 1.4, 1p and 2p, are really strong before the direct Urca threshold. If present, they lower the neutrino emission before the threshold to the same solid line denoted by '1p'. They reduce also the direct Urca process (after the threshold). Superfluidity 2p extends deeper behind the threshold and reduces the direct Urca stronger (the short-dashed line).

Suppressing traditional processes, superfluidity initiates a new specific neutrino process associated with *Cooper pairing of baryons* (Flowers *et al.*, 1976). In a quasiparticle language, it can be described as annihilation of quasibaryons into a neutrino pair (all neutrino flavors), for instance, $\widetilde{n} + \widetilde{n} \to \nu + \bar{\nu}$, for superfluidity of neutrons (where $\widetilde{n}$ is a quasineutron). In a cooling star the process switches on at $T = T_c$, produces the maximum emissivity at $T \sim 0.8\,T_c$, and becomes exponentially suppressed at $T \ll T_c$. For realistic density profiles $T_c(\rho)$, the integral neutrino luminosity of the star due to this process can be one to two orders of magnitude higher than the luminosity provided by the modified Urca processes in a nonsuperfluid star. This Cooper pairing neutrino process operates in neutron star cores and inner crusts.

For the conditions, presented in Fig. 1.4, both proton superfluidities are too strong to produce a noticeable Cooper-pairing emission. If, in addition to strong proton superfluidity (1p or 2p), we introduce weaker neutron superfluidity (2nt or 3nt) we will raise the neutrino emissivity (to the long-dashed 'p+2nt' line or the solid 'p+3nt' line) due to Cooper pairing of neutrons. Before the direct Urca threshold this raise is the same for proton superfluids 1p and 2p.

## 1.3.6 Thermodynamics and kinetics

Thermodynamics and kinetics of neutron star matter are very rich in physics due to huge ranges of densities and temperatures available in neutron stars.

The most important thermodynamic property, the EOS, has already been discussed in § 1.3.2. One often needs additional thermodynamic quantities, such as particle chemical potentials or heat capacity. The heat capacity is reviewed, for instance, by Yakovlev *et al.* (1999, 2001). The main heat reservoir is provided by strongly degenerate fermions in the stellar core. The thermal energy of a nonsuperfluid star with the internal temperature $T$ is estimated as $\sim 10^{48}\, T_9^2$ erg.

The kinetic theory of neutron star crusts has been reviewed by Yakovlev & Kaminker (1994) and Ventura & Potekhin (2001). The theory deals with transport coefficients, such as the thermal and electrical conductivities ($\kappa$ and $\sigma$) as well as the shear and bulk viscosities ($\eta$ and $\zeta$). In the crust, the conductivities $\kappa$ and $\sigma$ and the viscosity $\eta$ are mostly provided by electrons (although the thermal conductivity of the outermost nondegenerate surface layers is predominantly radiative). The major electron scattering mechanism is Coulomb scattering by ions (or, equivalently, scattering by phonons, which quantify vibrations of ions in solidified matter). At rather low temperatures (e.g., $T \lesssim 3\times10^7$ K in the inner crust) the electron-ion scattering is strongly suppressed. In this case the electron scattering by charged impurities (by a small amount of ions whose charges are different from charges of most abundant ions) and/or electron-electron scattering can be more important, particularly due to efficient Landau damping of transverse plasmons in the electron-electron interaction (Shternin & Yakovlev, 2006). For illustration, in the left panel of Fig. 1.5 we plot the thermal conductivity throughout the neutron star crust and core at $T = 10^7$, $10^8$, and $10^9$ K. The conductivity in the crust is taken from Gnedin *et al.* (2001) (for the electron-ion scattering) and from Shternin & Yakovlev (2006) (for the electron-electron scattering which becomes important for $T = 10^7$ K and $\rho \gtrsim 10^{11}$ g cm$^{-3}$).

The foundation of the kinetic theory of neutron star cores was laid by Baym *et al.* (1969). An important contribution was made by Flowers & Itoh (1976, 1981). Some results have been briefly reviewed by Yakovlev (1991, 1993). In a nonsuperfluid $npe\mu$ matter (as well as in a hyperonic one), $\kappa$, $\sigma$, and $\eta$ are expected to be not too sensitive to the composition and EOS of neutron star cores. Their density dependence should not be too strong; the leading transport carriers and their scattering mechanisms can be different. Very crude estimates are

$$\sigma \sim \frac{3\times10^{29}}{T_8^2}\ \mathrm{s}^{-1}, \quad \kappa \sim \frac{3\times10^{23}}{T_8}\ \frac{\mathrm{erg}}{\mathrm{cm\ s\ K}}, \quad \eta \sim \frac{10^{19}}{T_8^2}\ \frac{\mathrm{g}}{\mathrm{cm\ s}}, \tag{1.11}$$

where $T_8$ is the temperature in units of $10^8$ K. The density dependence of the thermal conductivity of neutrons in the stellar core composed of nonsuperfluid $npe$ matter is shown in the left panel of Fig. 1.5 (from Baiko *et al.* 2001a; the same version as the long-dashed curve in their Fig. 3).

Generally, neutron stars are excellent conductors with very high electron and thermal conductivities. As a result, neutron star interiors are usually almost isothermal. However, $\kappa$ and $\sigma$ become much lower in the very surface layer, no more than 10 meters under the surface. This layer produces the thermal insulation of the stellar interior, making the interior warmer than the surface

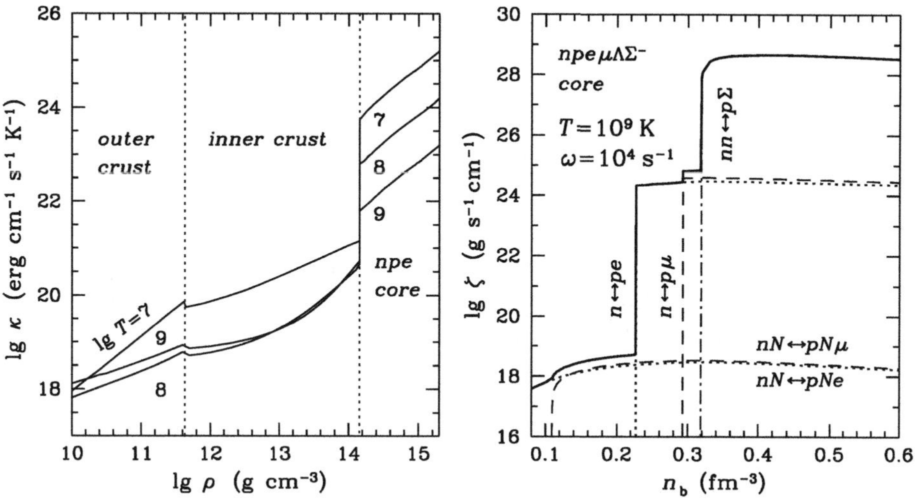


*Figure 1.5.* *Left:* Density dependence of the thermal conductivity in a neutron star crust and a nonsuperfluid $npe$ core at $T = 10^7$, $10^8$, and $10^9$ K. *Right:* Partial bulk viscosities associated with various reactions (indicated near the curves) and the total bulk viscosity (the thick solid line) versus number density $n_\mathrm{b}$ at $T = 10^9$ K in a nonsuperfluid $npe\mu\Lambda\Sigma^-$ core which undergoes subthermal pulsations with frequency $\omega = 10^4\ \mathrm{s}^{-1}$ (see the text for details).

(for the surface temperatures $T_\mathrm{s} \gtrsim 10^3$ K). The main temperature gradient is located there. There is a strong (several orders of magnitude) jump of $\kappa$, $\sigma$, and $\eta$, while passing from the inner crust to the core (left panel of Fig. 1.5). The jump is produced by the change of transport mechanism (from a very efficient electron transport due to Coulomb scattering in the crust to a slower neutron transport due to scattering between strongly degenerate nucleons in the core). As a result, the thermal relaxation in the inner crust lasts longer than in the core.

The bulk viscosity $\zeta$ is drastically different from other transport coefficients. It is associated with particle transformations ("chemical reactions") under compressions and rarefactions of the matter. The bulk viscosity of the core can vary by many orders of magnitude, depending on the composition of dense matter. The description of different regimes and bibliography can be found, for instance, in Haensel *et al.* (2002a,b). In an oscillating star $\zeta$ depends on an oscillation frequency $\omega$ and on oscillation amplitudes $\delta\mu$ of chemical potentials of particles. *Suprathermal* oscillations with $\delta\mu \gtrsim k_\mathrm{B}T$ dramatically enhance the bulk viscosity in comparison to the case of *subthermal* oscillations with $\delta\mu \lesssim k_\mathrm{B}T$.

In the right panel of Fig. 1.5 we show the density dependence of $\zeta$ in the stellar core at $T = 10^9$ K for subthermal pulsations with $\omega = 10^4\ \mathrm{s}^{-1}$ (from Haensel *et al.* 2002a). The density $\rho$ is parameterized by the baryon number density $n_b$ ($\rho = \rho_0$ corresponds to $n_b \approx 0.16$ fm$^{-3}$). The adopted EOS is taken from Glendenning (1985); it assumes the presence of neutrons, protons, and electrons in the outermost stellar core, and the appearance of muons, $\Sigma^-$ and $\Lambda$ hyperons at higher densities. Thin lines of various types show partial bulk viscosities produced by different particle transformations. The thick solid line is the total $\zeta$.

In the outer stellar core composed of $npe\mu$ matter, $\zeta$ is determined by the nonequilibrium modified Urca processes (1.9) involving electrons ($nN \rightleftharpoons pNe$, the lower dotted line) and muons ($nN \rightleftharpoons pN\mu$, the lower short-dashed line). In deeper layers, where the direct Urca processes (1.8) are open ($n \rightleftharpoons pe$ and $n \rightleftharpoons p\mu$, the higher dotted and short-dashed lines), they become the dominant ones, enhancing $\zeta$ by 5–7 orders of magnitude. In hyperon or quark matter $\zeta$ can be determined by hyperon or quark reactions, which change strangeness but do not involve leptons. In this case $\zeta$ is additionally enhanced by 3–5 orders of magnitude. In Fig. 1.5 such an enhancement is produced by the $nn \rightleftharpoons p\Sigma^-$ process (the dot-dashed line).

The transport coefficients are affected by magnetic fields. A magnetic field makes transport properties anisotropic. For instance, the electron electric or thermal transport in a magnetized plasma is characterized by three transport coefficients, which determine the conductivities along and across the field, and the Hall conductivity (see, e.g., Yakovlev & Kaminker 1994 and Ventura & Potekhin 2001). The anisotropy is strong if the magnetization parameter is large, $\omega_c \tau \gtrsim 1$, where $\omega_c$ is the electron gyrofrequency and $\tau$ is the effective electron relaxation time. For instance, the electron thermal conductivity of the layer with $\rho \lesssim 10^6$ g cm$^{-3}$ is strongly affected by the fields $B \gtrsim 10^{10}$ G. These fields greatly reduce the transverse conductivity (because electrons rapidly rotate about magnetic field lines). In higher fields, the effects associated with the quantization of transverse electron motion (Landau levels) become important (see Chapter 4 for details). They induce de Haas-van Alphen oscillations of the transport coefficients (if electrons populate a few or many Landau levels) or greatly modify these coefficients (if the ground Landau level is populated alone). Strong magnetic fields can affect the transport properties of neutron star cores (as reviewed, e.g., by Yakovlev 1991, 1993).

The kinetic properties are also affected by superfluidity of dense matter, especially in neutron star cores. We have already described the effects of superfluidity on neutrino emission (§ 1.3.5). Superfluidity reduces particle collision rates and affects the transport directly, for instance, by modifying the diffusive thermal conductivity and bulk viscosity (see, e.g., Baiko *et al.* 2001a and Haensel *et al.* 2002a). In addition, the transport can be affected by macroscopic superfluid quantum phenomena. For instance, a convective counterflow of normal particles, induced by a flow of superfluid liquid in the presence of a temperature gradient, carries the heat. This effect was mentioned, e.g., by

Flowers & Itoh (1976), but has not been studied in the neutron star physics. Superfluidity affects momentum transfer (viscosity) and hydrodynamics of neutron star interiors. Dynamics of mixtures of normal and superfluid liquids is a complicated problem. In particular, it involves specific mutual friction and entrainment forces. The evolution of perturbations in internal neutron star layers can be strongly affected by nonequilibrium beta processes and related phenomena (Reisenegger, 1995).

### 1.3.7 Cooling of neutron stars

Neutron stars are born hot in supernova explosions (§ 1.4.2). In about one minute after the birth the star becomes transparent to neutrinos. Let us outline the cooling of an isolated star during the subsequent neutrino-transparent stage. The subject has been reviewed, for instance, by Yakovlev & Pethick (2004), and Page *et al.* (2004, 2005), who give the bibliography of earlier work. The cooling is realized via two channels – by neutrino emission from the entire stellar body and by heat diffusion from the internal layers to the surface resulting in the thermal emission of photons. The stellar interior stays nonisothermal for the first 10–100 years. The neutrino emission dominates the thermal photon emission for $\sim 10^5$ years, while the photon emission dominates later.

The cooling of middle-aged neutron stars (of age $t \sim 10^4 - 10^5$ years) is mainly regulated by neutrino emission, heat capacity, superfluidity of their interiors, and by heat-insulating properties of the outermost layer (affected by the surface magnetic fields and by the presence of light elements on the surface). The cooling of older stars may also be regulated by some reheating mechanisms, for instance, by Ohmic dissipation of internal magnetic fields.

For example, in Fig. 1.6 (from Yakovlev *et al.* 2004) we present theoretical cooling curves of neutron stars of several masses for three models of neutron and proton superfluidity in the stellar core (composed of neutrons, protons and electrons). Any cooling curve gives the effective surface temperature $T_s^\infty$, as detected by a distant observer, versus stellar age $t$. The superfluid models are the same as those displayed in Fig. 1.4. The dotted regions are filled by cooling curves of neutron stars of masses from $\sim M_\odot$ to the maximum mass ($1.977\,M_\odot$, for the given model EOS in the stellar core). Theoretical curves are compared with observations of nine isolated middle-aged neutron stars (see Yakovlev *et al.* 2004 for details).

In the left and middle panels we adopt, respectively, models 1p and 2p for strong proton superfluidity in cooling stars. Low-mass stars cool very slowly via neutrino bremsstrahlung in neutron-neutron collisions, whereas other neutrino processes are suppressed by proton superfluidity (§ 1.3.5). High-mass stars show fast cooling via the direct Urca process in their inner cores (nonsuppressed by superfluidity, as shown in Fig. 1.4). Both superfluid models (1p or 2p) can explain the observations.

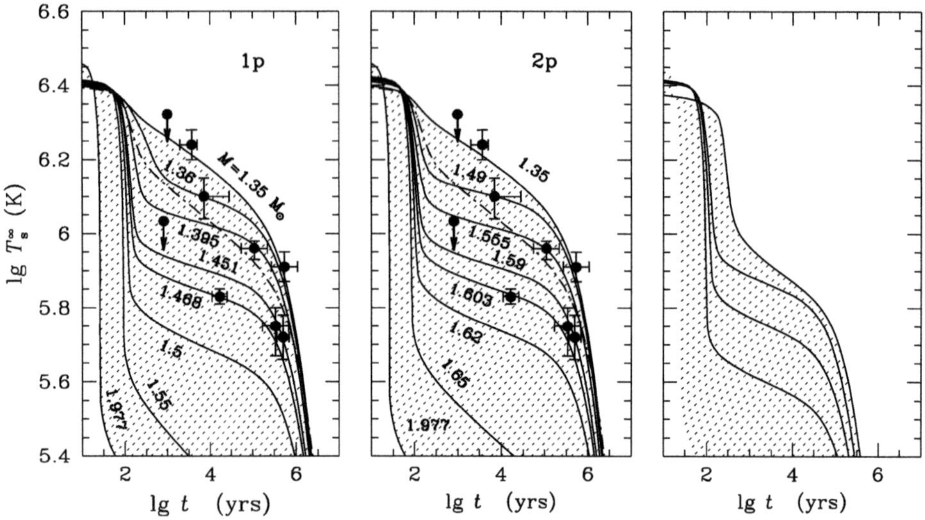


*Figure 1.6.* Theoretical cooling curves of neutron stars of several masses $M$ (numbers near curves) for three superfluid models in the stellar cores (*left*: 1p; *middle*: 2p; *right*: 1p+3nt). Dot-and-dashed line: cooling of a nonsuperfluid 1.35 $M_\odot$ star. Any point in a shaded region can be explained by a cooling neutron star with a certain mass $M$ and given superfluid model. The theory is compared with observations of 9 isolated neutron stars. From Yakovlev *et al.* (2004).

In the right panel we adopt model 1p for proton superfluidity and model 3nt for mild neutron superfluidity in the stellar core. Strong neutrino emission due to Cooper pairing of neutrons accelerates cooling of low-mass stars and hampers the interpretation of neutron stars hottest for their age. This superfluidity model (1p+3nt) is inconsistent with the observations.

### 1.3.8 Magnetic fields

Many neutron stars possess very strong surface magnetic fields. As we have seen in § 1.2, the very discovery of neutron stars was possible because neutron stars operate as pulsars owing to their strong magnetic fields. Comparison of observations with theoretical models of pulsar radiation shows that the surface magnetic fields in the majority of pulsars is $B \sim 10^{11}$–$10^{13}$ G (Manchester *et al.*, 2005), but may reach up to $B \lesssim 10^{14}$ G in radio pulsars and up to $B \lesssim 10^{15}$ G in magnetars (see § 1.4, particularly, §§ 1.4.4 and 1.4.5). Naturally, their internal magnetic fields may be higher than the surface ones. Magnetic fields strongly affect observational manifestations of neutron stars.

In some cases an indirect evidence of strong magnetic fields, based on theoretical models, is confirmed by *direct measurements*, based on X-ray observations of the electron cyclotron lines in neutron star spectra. The first cyclotron line was discovered by Trümper *et al.* (1978) in the spectrum of the X-ray binary Hercules X-1. For several neutron stars, two or three cyclotron harmonics have been registered; Santangelo *et al.* (1999) observed even four harmonics from the X-ray transient X0115+63, confirming the magnetic field $B > 10^{12}$ G.

Most complicated problems of the origin and evolution of these fields have been studied by many authors (as reviewed, e.g., by Reisenegger 2003 and Reisenegger *et al.* 2005). The fields may be inherited from presupernova stars and amplified during the gravitational collapse owing to magnetic flux conservation. This argument even led to the *prediction* of the fields $B \sim 10^{12}$ G in neutron stars a few years before the discovery of pulsars (Ginzburg, 1964).[5] However, there is no detailed physical picture of such a flux-conserving collapse. Alternatively, magnetic fields may be generated during supernova explosions or in young neutron stars.

The evolution of neutron star magnetic fields can be regulated by many factors, first of all by Ohmic decay. The magnetic evolution is coupled to the thermal evolution (because the electric conductivity and associated Ohmic decay rate depend on temperature). The Ohmic decay time of a not too strong field ($B \lesssim 10^{12}$ G) in a nonsuperconducting stellar core is larger than the Universe age (Baym *et al.*, 1969) because of a very high electric conductivity in the core, Eq. (1.11). The decay of stronger field in the core may be accelerated by the enhancement of the electric resistivity across the field lines (Haensel *et al.*, 1990b). In a superconducting core, the field splits into magnetic fluxoids whose evolution is determined by many factors, particularly, by interaction of the fluxoids with rotational vortices (§§ 1.3.4 and 1.3.6). The Ohmic decay of the crustal magnetic field (see, e.g., Cumming *et al.* 2004 and references therein) can be much faster than in the core due to lower electric conductivity in the crust and shorter Ohmic diffusion time-scales. The crustal and core magnetic fields may be closely related or almost independent. The magnetic field evolution can be affected by the Hall effect. The magnetic field of an accreting neutron star can be greatly affected by accretion (e.g., Cumming *et al.*, 2001).

What is the maximum filed strength in a neutron star? The estimates based on the virial theorem show that a field $B \gtrsim 10^{18} - 10^{19}$ G cannot be sustained in a considerable fraction of the star because the magnetic energy ($\sim R^3B^2/6$) would exceed the gravitational binding energy ($\sim 3\,GM^2/5R$), and the field would induce a dynamical instability of a hydrostatic configuration (Chandrasekhar & Fermi, 1953; Shapiro & Teukolsky, 1983; Lai & Shapiro, 1991). These estimates agree with detailed numerical simulations (Bocquet *et al.*, 1995;

[5] It is interesting to note that the same order of magnitude for the neutron-star field strength could have been expected on the basis of the approximate proportionality of stellar angular momenta and magnetic moments. Similar arguments led Blackett (1947) to the prediction of magnetic white dwarfs.

Cardall *et al.*, 2001) which indicate that the internal magnetic field of a neutron star cannot exceed a few times $10^{18}$ G.

Strong magnetic fields induce a large variety of phenomena in neutron star atmospheres and magnetospheres, reviewed, e.g., by Harding & Lai (2006).

### 1.3.9 Seismology

Like other stars, neutron stars are good resonators, where oscillation modes can be excited. These modes can be numerous, including, particularly, fundamental (f), pressure (p), gravity (g), Rossby (r) modes (see, e.g., McDermott *et al.* 1988a; Stergioulas 2003). The period of fundamental radial oscillation is $\sim$(0.5–1) ms. Observations of oscillations have potential to explore the internal structure of neutron stars (see, e.g., Kokkotas *et al.* 2001 and references therein). Some modes can be unstable and grow with time till they saturate due to nonlinear effects. In particular, these can be $r$-modes which are generically unstable in rotating stars composed of perfect fluid (see, e.g., Andersson & Kokkotas 2001, Lee & Yoshida 2003 and references therein). They are extensively studied as possible sources of gravitational radiation from rapidly rotating neutron stars.

## 1.4. Observations versus theory

Neutron stars are observed in all bands of electromagnetic spectrum in our Galaxy and in nearby satellite galaxies (such as the Large Magellanic Cloud and Small Magellanic Cloud). The neutron star astronomy is extending into the Local group of galaxies and even further. For instance, X-ray bursters (§ 1.4.6) have been observed in the galaxy M31 (Pietsch & Haberl, 2005), and an ecpipcing X-ray binary has been discovered in the galaxy M101 outside the Local group (Liu *et al.*, 2006). We outline main observational manifestations of neutron stars and mention also some theoretical work on interpretation of these manifestations. The same star can manifest itself in different ways. Our description will inevitably be schematic and bibliography incomplete. We summarize numerous observational manifestations of neutron stars in Table 1.1. We hope that the table will simplify reading of this section. The table heading is just the title of the conference which took place from September 30 till October 11, 1996, in Italy on Lipari, one of the Aeolian Islands in the Tyrrhenian Sea off the north coast of Sicily.

### 1.4.1 Telescopes

Observational technique is fascinating and rapidly developing. The best way to learn on details is to use the Internet; every telescope has its own web site.

*Table 1.1.* The many faces of neutron stars

| Objects | Manifestations |
|---|---|
| Neutron stars (NSs): | emitters of radio, infrared, optical, ultraviolet, X-ray, gamma-ray radiation; high-energy particles; neutrinos; gravitational waves |
| Neutron stars: | associated or not directly associated with supernova remnants (SNRs) |
| Neutron stars: | isolated stars or binary members |
| Pulsars (PSRs): | radio, classical X-ray, anomalous X-ray (AXPs), gamma-ray, millisecond, glitching, noising, powering pulsar wind nebulae (PWNe) and bow shock nebulae (BSNe) |
| Pulsars: | powered by rotation, accretion, magnetic field (magnetars) |
| Magnetars: | AXPs and soft gamma-ray repeaters (SGRs) |
| Isolated neutron stars: | pulsars, compact central objects in SNRs (CCOs), dim objects, old objects accreting interstellar medium, magnetars |
| NSs in binaries: | with nondegenerate stars, white dwarfs, NSs |
| NSs in binaries: | transient and persistent sources |
| NSs in binaries: | low-mass and high-mass X-ray binaries (LMXBs and HMXBs) |
| NSs in binaries: | soft X-ray transients (SXTs), hard X-ray transients, X-ray bursters, (classical) X-ray pulsars, sources of quasiperiodic oscillations (QPOs) |

**Electromagnetic emission.** Neutron stars are multiwavelength emitters. They are observed in radio, infrared, optical, ultraviolet, X-ray and gamma-ray spectral bands.

Radio observations are conducted with ground based telescopes or their arrays throughout the world. The examples are the *Arecibo* (Fig. 1.7) and *Parkes* telescopes. Near infrared and optical observations can be performed with large ground based telescopes (such as the *Keck* telescope, the *Very Large Telescope (VLT)*, the *Subaru* telescope) because neutron stars are weak in these bands. Ultraviolet (and optical) observations are performed with the *Hubble Space Telescope (HST)*.

Extreme ultraviolet, X-ray and gamma-ray observations of neutron stars are conducted with space observatories. The best X-ray observatories suited for such observations by 2006 are the *Chandra* (Fig. 1.8) and *X-ray Multi Mirror (XMM-Newton)* observatories, and also the *Rossi X-ray Timing Explorer (RXTE)*. The examples of gamma-ray observatories operating by 2006 are the *High Energy Transient Explorer (HETE-2)* and the *International Gamma-Ray Astrophysics Laboratory (INTEGRAL)*.

*Figure 1.7.* The *Arecibo* radio telescope; 305 meters in diameter. The platform suspended above the reflector weighs 500 tons (photo courtesy of the NAIC – Arecibo Observatory, a facility of the NSF).

**Neutrino observatories.** The neutrino luminosity of neutron stars in several minutes after their birth becomes too week to be detected directly. However, one expects to detect a splash of neutrino emission for a few ten seconds preceding a visual supernova explosion at a neutron star birth (§ 1.4.2). The examples of currently operating neutrino observatories are the *Sudbury Neutrino Observatory (SNO)* and the *Kamioka Liquid-scintillator Anti-Neutrino Detector (KamLAND)*.

**Gravitational-wave observatories.** Rapidly rotating neutron stars may lose axial symmetry (e.g., due to $r$-mode instability, see § 1.3.9) and emit gravitational waves. Compact binaries containing neutron stars are even much more efficient sources of gravitational radiation (especially at the final inspiral stage; see § 9.1). Gravitational radiation of double neutron star binaries and neutron star – white dwarf binaries has already been observed indirectly, by detecting relativistic decay of pulsar orbits. It is hoped that direct detection of gravitational

*Figure 1.8.* Artistic view of the *Chandra* X-ray satellite in orbit (illustration courtesy of NGST).

waves will be made soon after 2006 with the new generation of gravitational observatories, such as the ground-based *Laser Interferometer Gravitational-Wave Observatory (LIGO)*, *European Gravitational Observatory* (*VIRGO* detector), and the *Large Interferometer Space Antenna (LISA)*.

## 1.4.2 Birth in supernova explosions. Supernova connection

Neutron stars are final products of stellar evolution. It is widely accepted that they are born in supernova explosions after their presupernova progenitors (giant or supergiant stars) exhaust nuclear fuel in their cores. The cores undergo *gravitational collapse* into neutron stars (or black holes), while outer presupernova layers are blown away by an expanding shock wave, producing supernova remnants. The whole event is usually referred to as a *core-collapse* (type II) supernova explosion (see, e.g., Imshennik & Nadyozhin 1988; Arnett 1996, and references therein). The neutron star – supernova connection was suggested by Baade and Zwicky in 1933 as described in § 1.2.

The explosion, which occurs in the presupernova core, triggers a shock wave propagating outward (after bouncing off the dense core). It takes several hours for the shock to travel through extended presupernova outer layers. At this stage the presupernova, observed from outside, looks just as usual, as if nothing happened in its interior. After the shock reaches the surface, it produces a splash of radiation in all bands of electromagnetic spectrum to be observed as a supernova event. In addition, the core collapse itself should be accompanied by a powerful outburst of neutrino emission and, possibly, of gravitational radiation. These events could be detectable by neutrino and gravitational observatories prior to the electromagnetic outburst.

Supernova explosions are accompanied by an enormous energy release, a few times $10^{53}$ erg in total (of the order of the gravitational energy of a neutron star, Eq. (1.1)). It is expected that the energy is mostly released in the form of neutrinos. About 1% of the total energy transforms into the kinetic energy of the explosion ejecta, and only a minor part ($\sim 10^{49}$ erg) into electromagnetic radiation; a smaller part can be emitted in the form of gravitational waves.

Theoretical simulations of gravitational collapse are extremely complicated because they should generally involve three-dimensional hydrodynamics with neutrino transport and convection. Many attempts to simulate the collapse (in inevitably restricted formulations) failed to reproduce a neutron star birth accompanied by the formation of a powerful outgoing shock wave (see, e.g., Janka 2004 and references therein). The collapse and supernova explosion can be strongly affected by the combined effect of stellar rotation and magnetic field (see, e.g., Moiseenko *et al.* 2003, Akiyama *et al.* 2003, and references therein).

A gravitational collapse of a degenerate stellar core occurs on time scales of 0.1 s. If the shock wave produced by the core bounce is successful in ejecting the outer layers, it should result in the appearance of a *protoneutron star* with the internal temperature $T \sim 10^{11}$ K (see, e.g., Pons *et al.* 2001 and references therein). This protoneutron star is very special. It is hot, opaque to neutrinos, and larger than an ordinary neutron star. It lives for about one minute and transforms then into an ordinary neutron star which is transparent for neutrinos.

Current estimates of supernova explosion rate in the Galaxy are uncertain and give one event per 60–1000 years (e.g., Arzoumanian *et al.* 2002). Electromagnetic radiation from some of these explosions cannot be observed from the Earth, being hidden by gaseous and dust-grain clouds in the Galactic plane. The total number of neutron stars in the Galaxy is estimated as $10^8 - 10^9$. Only a very limited fraction of these stars can be observed.

Supernova explosions in the Galaxy were observed by naked eye centuries ago (for recent reviews of such *historical supernovae* see Green & Stephenson 2002, 2003). The most prominent observation is dated back to 1054. It was the birth of the Crab Nebula and the Crab pulsar at its heart. In modern times, it was sinologist Édouard Biot (the son of famous physicist J.-B. Biot) who first paid attention to a "guest star" reported in Chinese chronicles for AD 1054 (Biot, 1846). Other Chinese, Japanese, Arabic, and European historical records were discovered in the 20th century. The extraordinary bright star appeared probably in April,[6] remained visible in daylight till August of 1054 and gradually faded away to 1056. Lundmark (1921) listed this star as a "suspected nova," Hubble (1928) noticed that its position was at the center of the Crab Nebula, and Mayall (1939) identified it as a supernova.

Nowadays astronomers detect several tens of type II supernovae per year from distant galaxies (too far away to study collapsars).

Any association of a neutron star with a supernova remnant has a great importance: the age and distance to the star are those of the remnant. Many neutron stars acquire large proper velocities (kicks, § 1.4.7) during their births and leave quickly their parental supernova remnants. Supernova remnants themselves dissolve in $\sim 10^5$ years after explosion. Thus, the majority of neutron stars are not related to observable supernova remnants.

A newly born neutron star remains hidden behind an expanding supernova envelope for several years. This prevents direct observation of very young neutron stars.

The most famous supernova detected in the present epoch is the supernova 1987A. It was discovered in the nearby Large Magellanic Cloud on February 23, 1987, really close ($\sim$ 50 kpc) to us (see, e.g., Imshennik & Nadyozhin 1988; Arnett 1996, and references therein). This is the first (and still the only one) supernova from which the neutrino outburst was observed. All the attempts to find a collapsar (a neutron star or a black hole) in this supernova remnant have failed. Nearby supernova explosions are rare. However, we should be ready to witness a new event.

Let us remark that neutron stars can also be formed via a collapse of accreting white dwarfs in binary systems, after the white dwarf mass exceeds the Chandrasekhar limit. This *accretion induced collapse* occurs only under specific conditions when electron captures effectively decrease the Chandrasekhar mass limit (Nomoto 1987; Nomoto & Kondo 1991; for a review see Canal 1994). The number of neutron stars formed in this way is expected to be small (Fryer *et al.*, 1999), but it may be the only viable scenario of their formation in some binaries (see, e.g., Nomoto & Kondo 1991; van Paradijs *et al.* 1997, and references therein). Let us remind that the majority of accreting white dwarfs, whose masses become close to the Chandrasekhar limit, are disrupted by a thermonuclear explosion, producing supernova Ia events (see Nomoto *et al.* 1994 for a review).

[6] There were debates about the date of its appearance and about credibility of different records. The most detailed Chinese report indicates July 4, but other historical sources point to earlier dates. For the list of the historical observations, references, and discussion, see Collins *et al.* (1999) and Polcaro & Martocchia (2006).

### 1.4.3 Evolutionary scenarios: Three driving forces

Many neutron stars are thought to be born as isolated stars from isolated presupernovae (in neutron star community "isolated" is used for "solitary"). However, many other neutron stars are born in binary systems (also see § 1.4.6). Their evolution can be drastically different from the evolution of isolated stars (see, e.g., Lipunov 1992 and references therein). Let us assume that initially a binary contains two ordinary stars. Depending on their masses and orbital parameters, both companions undergo their nuclear evolution which can be strongly affected by mass exchange (accretion from one companion to the other, especially in compact binaries). Eventually, one of the companions may explode as a supernova producing a neutron star (or a black hole). The explosion may or may not destroy the binary. If survived, the binary continues its evolution and the second component may explode as a supernova (although there are other possibilities). If a compact double neutron star binary forms and survives, its evolution is accompanied by intense gravitational radiation (§ 9.1.2); the companions fall onto each other and finally merge producing a powerful outburst of gravitational, neutrino, and electromagnetic radiation.

In addition to accretion, the evolution of neutron stars can be strongly affected by their rotation and magnetic fields. Thus, the evolution and observational manifestations of neutron stars are regulated by the *three main factors*:

Rotation, Accretion, and Magnetic Field.

We will illustrate the effects of these factors in the next sections (also see Table 1.1). The effects are often superimposed. For instance, the magnetic field and rotation affect the regime of accretion on a neutron star and, hence, its X-ray emission (§ 1.4.6). On the other hand, an intense accretion can spin up the neutron star and affect its magnetic field. All in all, there are vastly different scenarios of neutron star evolution. Statistical studies of the evolution can be performed with population synthesis codes (see, e.g., Lipunov *et al.* 1996).

### 1.4.4 Pulsars

*Pulsars* are rotating neutron stars which produce pulsed emission (modulated by their rotation). It is thought that the emission is generated in a neutron star atmosphere and/or magnetosphere. Naturally, the emission should be anisotropic to be detected as pulsed. The anisotropy is caused by a neutron star magnetic field.

Pulsars are subdivided into *radio pulsars*, *X-ray pulsars* and *gamma-ray pulsars*, depending on the spectral range where pulsations are observed. Some neutron stars, for instance, the Geminga and the Crab pulsars, show pulsations in all spectral bands.

Pulsars can also be subdivided with respect to mechanisms which power their activity. One can distinguish *rotation powered*, *accretion powered*, and *magnetically powered* pulsars. We outline them in the next subsections. Pulsars can be single *isolated* stars or enter *binary systems*.

**Radio pulsars.** *Radio pulsars* (often called *pulsars*, for short) are spinning magnetized neutron stars which emit radio waves and operate as *light houses* (§ 1.2). The radio emission is generated in pulsar magnetospheres (§ 1.3) and is peaked with respect to magnetic poles. It can be observed only if a pulsar beam crosses the Earth.

The majority of known neutron stars are observed as radio pulsars. By 2006 $\sim$ 1700 radio pulsars have been observed. The information on radio pulsars can be found in catalogues, for instance, in the Australia Telescope National Facility (ATNF) catalogue [http://www.atnf.csiro.au/research/pulsar/psrcat/].

Many radio pulsars demonstrate very stable rotation and serve as superprecise clocks. Radio telescopes allow observers to perform highly accurate pulsar timing and measure the pulsar spin period $P$, its time derivatives, $\dot{P}$, and sometimes $\ddot{P}$ (see, e.g, Lyne & Graham-Smith 1998; Lorimer 2001; Livingstone *et al.* 2006, and references therein). Isolated radio pulsars show regular spindown ($\dot{P} > 0$) and spend their rotational energy to emit electromagnetic radiation and accelerate particles in their magnetospheres; they belong to the class of rotation-powered pulsars (see below). Profiles, spectra, polarization and other properties of radio emission give information on dispersion measure, useful for estimating distances to pulsars, and on physical properties of pulsar magnetospheres. Timing of pulsars in binary systems is useful for measuring masses of binary components and for studying the orbital parameters (§ 9.1).

Some radio pulsars show very rapid rotation (see § 9.4). Pulsars with spin periods $P \lesssim 30$ ms form a special class of *millisecond pulsars*. PSR B1937+21 with $P = 1.558$ ms, the first discovered millisecond pulsar (Backer *et al.*, 1982), remained the fastest rotator till 2005, where still faster PSR J1748−2446ad with $P = 1.396$ ms was discovered (Hessels *et al.*, 2006). Many millisecond pulsars are thought to be old ($t \gtrsim 10^9$ years) neutron stars with small magnetic fields ($B \lesssim 10^8 - 10^9$ G) recycled by accretion in binary systems.

Some radio pulsars, including the Crab pulsar, rarely show *giant pulses* of radio emission. The intensity of giant pulses exceeds the intensity of ordinary pulses by a factor of a few hundreds or even thousands. Giant pulses can be very short (as short as 2 ns, once observed from the Crab pulsar, Hankins *et al.* 2003). They can have extremely high brightness temperature (for instance, exceeding $5 \times 10^{39}$ K for PSR B1937+21, Soglasnov *et al.* 2004).

**Pulsar glitches.** *Glitches* are observed as sudden jumps $\Delta\Omega$ of pulsar spin frequencies $\Omega = 2\pi/P$ followed by a slow partial relaxation to a pre-glitch regular spindown ($\dot{P} > 0$). Relaxation time scales range over the period of days, months or years. The relative frequency jumps $\Delta\Omega/\Omega$ vary from $\sim 10^{-10}$ to

$\sim 5 \times 10^{-6}$. Glitches are mainly observed from relatively young radio pulsars. The first glitches were detected from the Crab and Vela pulsars (Boynton *et al.* 1969, Radhakrishnan & Manchester 1969, Reichley & Downs 1969). By 2006 one has observed $\sim 100$ glitches from $\sim 30$ pulsars. The strongest glitch with $\Delta\Omega/\Omega \approx 1.6 \times 10^{-5}$ was detected by Hobbs *et al.* (2002) from PSR J1806–2125. Some glitches have been discovered in X-rays with the *Rossi X-ray Timing Explorer*. In particular, such glitches have been observed from a millisecond pulsar (PSR J0537–6910, Marshall *et al.* 2004), an accreting pulsar (KS 1947+300, Galloway *et al.* 2004), and an anomalous X-ray pulsar (1RXS J170849.0–400910, Dall'Osso *et al.* 2003).

The most popular interpretation of glitches associates them with a sudden unpinning of vortices of neutron superfluid in the neutron star crust. The idea was proposed by Anderson & Itoh (1975) and Alpar (1977). The interpretation requires the moment of inertia of crustal neutron superfluid to be $\sim 1\%$ of the total moment of inertia of the star (see § 9.7). The theory of pulsar glitches (see, e.g., Larson & Link 2002 and references therein) has not been very elaborated yet; comparison of observation and theory has potential to explore internal structure of neutron stars.

**Timing noise.** In addition to glitches, pulsar timing reveals slow irregular or quasiregular variations of pulses (over time scales of months, years, and longer) called *pulsar timing noise*. They look like random walks in pulsar rotation rate (with relative variations of spin period $\lesssim 10^{-10} - 10^{-8}$), spindown rate, or pulse phase. Timing noise was discovered by Boynton *et al.* (1972) in optical timing observations of the Crab pulsar. Theoretical hypotheses on the nature of timing noise are numerous (see, e.g., Cordes & Greenstein 1981).

**Free precession.** Timing of a few isolated radio pulsars indicates that they may undergo free precession (see, e.g., Link 2003 and references therein). The most convincing evidence is provided by the observations of PSR B1642–03 (Cordes 1993, Shabanova *et al.* 2001) and PSR B1828–11 (Stairs *et al.*, 2000). In particular, the observations of PSR B1828–11 over 13 years show regular variations of the pulsar emission with periods of $\approx 500$ and $\approx 250$ days. These variations are thought to be produced by a free precession of pulsar spin axis. The precession period can be $\approx$500 days (whereas the 250-day periodicity can be interpreted as the second harmonic). If this interpretation is true, it gives the direct evidence for the existence of a rigid neutron star crust (without which free precession would be impossible). Recently, a much longer free-precession period of $\approx 7$ years has been inferred from X-ray observations of the isolated neutron star RX J0720.4–3125 by Haberl *et al.* (2006).

**Rotation-powered pulsars.** The pulsar rotational energy is $E_{\rm rot} = I\Omega^2/2$, where $I$ is the moment of inertia. This energy can be very high. For the Crab pulsar ($P = 33.08$ ms) at a typical value $I = 10^{45}$ g cm$^2$ one gets $E_{\rm rot} \sim 2 \times 10^{49}$ erg. Because $\dot{P}$ and hence $\dot{\Omega}$ can be measured, the rotational energy release can be determined as $\dot{E}_{\rm rot} = I\Omega\dot{\Omega}$. For the Crab pulsar we have $\dot{P} = 4.22 \times 10^{-13}$ and $\dot{E}_{\rm rot} \sim 5 \times 10^{38}$ erg s$^{-1}$.

The values of $P$ and $\dot{P}$ inferred from pulsar timing are widely used to estimate pulsar magnetic fields and ages. The magnetic field is estimated assuming that $\dot{E}_{\rm rot}$ is equal to the magnetic dipole energy loss rate (as produced by the emission of electromagnetic waves by a rotating magnetic dipole in vacuum). In this model,

$$I\dot{\Omega} = \mathcal{N}, \qquad \mathcal{N} = -\frac{2\Omega^3 B_{\rm eff}^2 R^6}{3c^3}, \tag{1.12}$$

where $\mathcal{N}$ is the torque acting on the pulsar due to the magnetic dipole radiation, $R$ is the stellar radius, $B_{\rm eff} = B_{\rm eq} \sin\alpha$, $B_{\rm eq}$ is the magnetic field at the magnetic equator, and $\alpha$ is the angle between the spin and magnetic axes. From Eq. (1.12) one immediately has

$$B_{\rm eff} = \left(\frac{3Ic^3}{8\pi^2 R^6}\, P\dot{P}\right)^{1/2} \approx 3.2 \times 10^{19}\, \frac{I_{45}^{1/2}}{R_6^3}\, \sqrt{P\dot{P}}\ \ {\rm G}, \tag{1.13}$$

where $I_{45}$ is $I$ in units of $10^{45}$ g cm$^2$, and $R_6 = R/10^6$ cm. In the last expression $P$ is meant to be in seconds. One usually assumes $I_{45} = 1$, $R_6 = 1$, uses the measured values of $P$ and $\dot{P}$, and treats the calculated value of $B_{\rm eff}$ as the *characteristic pulsar magnetic field.* For instance, one gets $B_{\rm eff} \approx 3.8 \times 10^{12}$ G for the Crab pulsar.

The basic equation for estimating the pulsar age comes from Eq. (1.12). More generally, one can write the spindown torque as $\mathcal{N} = -A\Omega^n$, where $n$ is the *braking index.* Using this torque and assuming a constant $n$, from Eq. (1.12) one immediately gets: $P^{n-1}\dot{P} = {\rm const}$. Integrating from the pulsar birth $t = 0$ to a current age $t$, one obtains

$$t = \frac{P(t)}{(n-1)\,\dot{P}(t)} \left\{1 - \left(\frac{P(0)}{P(t)}\right)^{n-1}\right\}. \tag{1.14}$$

If a newly born pulsar rotated much faster than now ($P(0) \ll P(t)$), then $t = P/[(n-1)\dot{P}]$. For the magnetic dipole model (with the braking index $n = 3$), the last expression reduces to the very well known formula

$$t = P/(2\dot{P}). \tag{1.15}$$

The age estimated in this way is called the *characteristic pulsar age.* For instance, the characteristic age of the Crab pulsar is $t = 1240$ years, in a qualitative agreement with its true age (see §§ 1.4.2 and 9.5).

From Eq. (1.12) with $\mathcal{N} = -A\Omega^n$ one gets $n = \Omega\ddot{\Omega}/\dot{\Omega}^2$. If $P$, $\dot{P}$, and $\ddot{P}$ are measured from pulsar timing, this equation allows one to determine $n$. The braking indices, measured for several pulsars, are smaller than $n = 3$ (Livingstone *et al.*, 2006). For instance, Lyne *et al.* (1988) reported $n = 2.5$ for the Crab pulsar. Taking this value and the true age, from Eq. (1.14) we obtain the initial spin period $P(0) \approx 18.7$ ms.

**Pulsar magnetospheres.** It is clear that characteristic magnetic fields and characteristic pulsar ages may differ from real fields and ages. The simplicity of the models used in the preceding paragraph is illusive. In fact, the pulsar energy losses occur due to complicated processes in pulsar *magnetospheres*, which occupy space above atmospheres (see, e.g., Beskin *et al.* 1993, Beskin 1999, Melrose 2000, and references therein). Magnetized spinning neutron stars operate as unipolar inductors and generate magnetospheric electric fields. In those places of the stellar surface, where the electric field has a favorable direction and sufficient strength, it can pull electrons (or perhaps even ions) from the surface into the magnetosphere filling it by a rarefied plasma.

The magnetosphere is an important place, where magnetospheric electromagnetic emission is formed and plasma particles are accelerated. It is convenient to introduce a *light cylinder* of radius $cP/(2\pi)$ around the pulsar spin axis. At this cylinder, the velocity of a hypothetical reference frame corotating with the star would reach speed of light. The space within this cylinder is called the *corotation* zone. In this zone the outer stellar magnetic and electric fields are nearly stationary in the reference frame corotating with the star. Outside the light cylinder the electromagnetic fields become essentially dynamic.

The corotation zone is further subdivided into the regions of *closed* (inside this zone) and *open* magnetic field lines. It is believed that the space of closed field lines is filled by the "Goldreich-Julian" plasma distributed in such a way that the electric field created by charged particles in the pulsar-corotating reference frame compensates the projection of the electric field, induced by the pulsar magnetic field, onto magnetic field lines. The space of closed magnetic field lines is inactive for particle acceleration and radiation generation.[7] Therefore, the main magnetospheric activity is limited by the region of open magnetic field lines in the corotation zone. Charged particles are moving along magnetic lines and accelerated by the electric field component parallel to $\boldsymbol{B}$.

---

[7]For estimates, one usually assumes the charge density $\simeq -\boldsymbol{\Omega} \cdot \boldsymbol{B}/(2\pi c)$, derived by Goldreich & Julian (1969) for a steady-state, axially symmetric configuration with a dipole magnetic field aligned with the pulsar spin axis. A more general formula for the charge density around a rotating body with the dipole field (not necessarily aligned with the spin axis) had been derived by Hones & Bergeson (1965) four years earlier. One should note, however, that the Goldreich-Julian magnetosphere model suffers from theoretical difficulties, and the corresponding plasma density distribution is likely unstable. Thus, the real charge density can differ from the Goldreich-Julian model (for review, see Michel 2004 and references therein).

The electric field can be strongly affected, among other things, by general relativistic dragging of inertial frame around the spinning star (Muslimov & Tsygan, 1990, 1992). The accelerated electrons emit quanta due to curvature of magnetic field lines. These quanta may produce electron-positron pairs; newly born electrons undergo further acceleration and curvature radiation, leading to new pair production and inducing thus pair cascading. It is also important to take into account inverse Compton scattering. The flow of electrons pulled away from the star is likely superimposed with the flow of positrons moving toward the stellar surface. In different places of the magnetosphere the charged particles emit radiation in different spectral bands, with different beaming and polarization. As a result, a powerful (nonthermal) magnetospheric emission is created in the widest spectral range, from radio waves to hard gamma-rays, accompanied by a flow of highly energetic particles. The latter flow produces a *pulsar wind* which feeds up *pulsar wind nebulae*. Many theoretical aspects of this schematic model of rotation-powered pulsars are still not elaborated.

Depending on neutron star parameters (particularly, on magnetic field strength and geometry, relative orientation of magnetic and spin axes) and on a theoretical description of magnetospheric processes, one can obtain different models of active magnetospheric zones. In some models, which are called *polar cap* models, these zones are placed in the vicinity of magnetic poles near the pulsar surface. In other models, the active magnetospheric zones are located much higher above the surface, in the outer magnetosphere. They are the so called *outer gap* models. In both types of models the inverse positron current can be sufficiently strong to heat (magnetic) polar cap regions and create hot spots on the stellar surface.

As a result, the total pulsar spindown loss rate $\dot{E}_{\rm spin}$ is thought to be approximately the same as produced by the magnetic dipole radiation in vacuum, but the entire physics is different. In particular, only a small fraction ($\sim 10^{-6}$) of $\dot{E}_{\rm spin}$ is radiated away at radio frequencies while the rest is emitted in high-energy bands and in the form of pulsar wind, which may create a pulsar wind nebula. For instance, the energy generated by the Crab pulsar is sufficient to feed up the entire plerion Crab Nebula. Synchrotron radiation detected from the Nebula gives a strong evidence of particle acceleration in the pulsar magnetosphere. An X-ray image of the Nebula obtained with the *Chandra* observatory is shown in Fig. 1.9.

Some pulsars have very high velocities (§ 1.4.7). The interaction of their winds with the interstellar medium can create *bow-shock nebulae* in which the pulsar wind is confined by ram pressure. A spectacular example is provided by the Guitar nebula (see, e.g., Chatterjee & Cordes 2004).

**$P - \dot{P}$ diagram. The death line.** The evolution of pulsars is often studied by plotting them on the $P - \dot{P}$ diagram (see, e.g., Lyne & Graham-Smith 1998 and

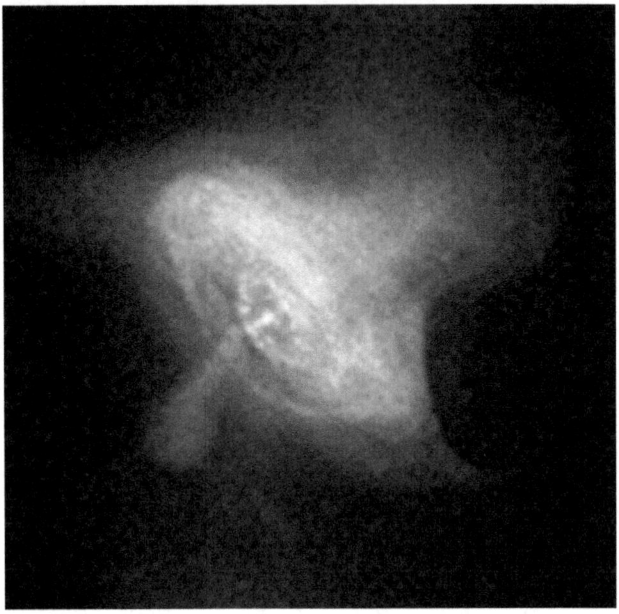

*Figure 1.9.* *Chandra* X-ray image of the Crab Nebula with rings of high-energy particles supplied by the pulsar (in the center) at a distance of more than one light year from it. Jet like structures perpendicular to the rings are also produced by high-energy particles moving away from the pulsar. Photo credit: NASA/CXC/SAO.

Lorimer 2001, and references therein). These studies do not show any strong evidence of the magnetic field evolution. The majority of radio pulsars have $B_{\rm eff} \sim 10^{12}$ G, although a few of them have much higher $B_{\rm eff}$. For instance, PSR J1847–0130 has $B_{\rm eff} \approx 0.94 \times 10^{14}$ G, as high as in magnetars (see below). Old and slowly rotating isolated neutron stars induce much weaker electric fields (in their comoving reference frames) and cannot produce powerful outflow of charged particles from their surfaces. In this way they cannot support their magnetospheric activity and cease to operate as radio pulsars. These neutron stars (dead pulsars) have large $P$ and small $\dot{P}$ and could have been placed in the corresponding corner of the $P - \dot{P}$ diagram, separated from the main part of the diagram by the so called *pulsar death line*.

### 1.4.5 Isolated neutron stars

The majority of isolated neutron stars are observed only as radio pulsars. However, more than 100 of them have been detected by 2006 in other spectral bands.[8] Let us describe them in more detail.

**Cooling isolated neutron stars.** If an isolated neutron star possesses a strong magnetic field and rotates rapidly, it behaves as a rotation-powered pulsar (§ 1.4.4), emitting a non-thermal multiwavelength magnetospheric radiation *and* thermal-like radiation from hot polar caps (spots) on the surface. The examples of such multiwavelength sources are the Vela and Geminga pulsars. In addition, young ($t \lesssim 10^3$ years) or middle-aged ($t \lesssim 10^5 - 10^6$ years) neutron stars can be sufficiently hot inside (§ 1.3.7). The thermal flux emerging from their interiors is radiated away through their atmospheres. This thermal radiation is thus powered by the internal thermal energy of the star. The expected surface temperatures $T_s$ of middle-aged stars range from $\sim 2 \times 10^5$ to $\sim 10^6$ K. The thermal radiation of such stars is mainly emitted in soft X-rays and extreme ultraviolet; its Rayleigh-Jeans tail may be detected in optics and far ultraviolet.

The attempts to discover thermal X-ray radiation from isolated neutron stars were made even in the pre-pulsar era but failed (§ 1.2). The next attack was undertaken with the launch of the *Einstein* X-ray orbital observatory (1978–1981). The important contribution was made by the *Röntgen Satellite (ROSAT,* 1990–1998). The current observational status is reviewed, for instance, by Pavlov *et al.* (2002), Pavlov & Zavlin (2003), and Kaspi *et al.* (2004); also see § 9.3.1.

The magnetospheric and/or polar-cap radiation of young stars is expected to be strong enough to outshine their thermal surface radiation. The radiation from polar caps of old pulsars ($t \gtrsim 10^5 - 10^6$ years) can also be stronger than the thermal radiation from their entire surfaces. Nevertheless, the thermal radiation from the entire surface has been detected or constrained from a dozen of isolated middle-aged neutron stars. In particular, the X-ray observations of the Vela pulsar (the characteristic age $t = 1.1 \times 10^4$ years) can be described using a hydrogen atmosphere model and $T_s \sim 7 \times 10^5$ K (Fig. 1.10).

Apart from cooling neutron stars, which are observed as radio pulsars (like the Vela pulsar), there are radio quiet cooling isolated neutron stars of different types.

In particular, there is an interesting class of several radio-quiet neutron stars dubbed *compact central objects (CCOs)* in supernova remnants. Their distances and ages are thought to be those of host supernova remnants. The best investi-

[8] These sources are listed in the online catalog http://www.ioffe.ru/astro1/psr-catalog/Catalog.php.

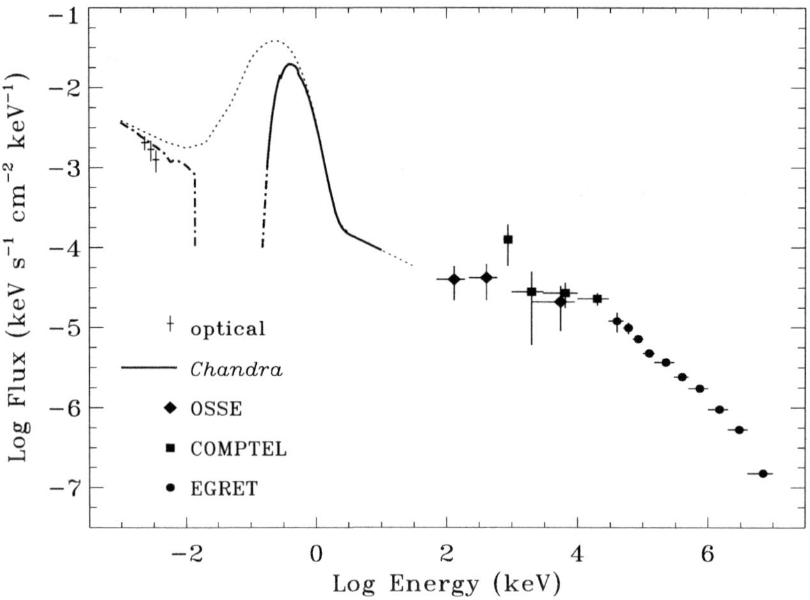


*Figure 1.10.* Multiwavelength spectrum of the Vela pulsar (from Pavlov *et al.* 2002 with the kind permission of the authors). The solid line shows the fit to the *Chandra* X-ray observations with the model spectrum of the thermal (neutron-star hydrogen atmosphere model) plus nonthermal (power-law) radiation components (including the effect of interstellar absorption). The dot-and-dashed line is the extrapolation of the fit to the ultraviolet and optical bands. The dotted line is the same extrapolated fit but excluding the effect of interstellar absorption. The various symbols show the spectrum detected with other observatories in the optical and gamma-ray bands.

gated object is J1210–5226 (=1E 1207.4–5209) in the center of the supernova remnant G296.5+10.0 ($t \approx (3-20) \times 10^3$ years). It is the first isolated neutron star found to exhibit pronounced spectral features (X-ray absorption spectral lines) in its radiation spectrum (Sanwal *et al.*, 2002) although the interpretation of these features seems ambiguous. Its radiation contains the thermal component which can be interpreted with the aid of a hydrogen atmosphere model as the thermal radiation from the surface of the neutron star with the effective temperature $\sim (1.4 - 1.9) \times 10^6$ K. Some other objects (e.g., J2323+5848 in Cassiopeia A, $t \sim 320$ years) show thermal-like radiation which cannot be emitted from the entire stellar surface. The radius of the emission region, inferred from observations, is 0.5–1 km, much smaller than the expected neutron star radius. This radiation may be produced by a spot on the neutron star surface, but in this case the absence of pulsations of the observed radiation requires explanation.

There is also a class of *dim isolated neutron stars* not associated directly with supernova remnants. Their emission is characterized by black-body X-ray spectra with the effective temperatures $\sim (0.5-1) \times 10^6$ K. They are probably nearby isolated neutron stars. The most famous is RX J1856.5–3754 discovered by Walter *et al.* (1996). Parallax measurements give the distance $140 \pm 40$ pc which makes this object one of the closest observed neutron star (Kaplan *et al.*, 2002)[9] the age $\sim 5 \times 10^5$ years is estimated from kinematics of proper motion. Observations show no spectral lines and no pulsations. The effective surface temperature is estimated to be $T_{\rm s} \approx 4.3 \times 10^5$ K (Ho *et al.*, 2006). Another example – RX J0720.4–3125 – is a dim object which shows periodic variations with a long period $P = 8.4$ s. Its characteristic age is $t \sim 1.3 \times 10^5$ years, and $T_{\rm s} \sim 5 \times 10^5$ K. This object shows a phase-dependent absorption feature (Haberl *et al.*, 2004) and precession (Haberl *et al.*, 2006).

**Soft gamma repeaters + anomalous X-ray pulsars = magnetars.** Soft gamma repeaters and anomalous X-ray pulsars are two other types of isolated neutron stars. They seem to form a larger class of *magnetars* (see, e.g., Thompson 2002, Kaspi 2004, and references therein).

*Soft gamma repeaters (SGRs)* are sources of repeating soft gamma-ray and X-ray bursts. Typical bursts last for $\sim 0.1$ s and have energies $\sim 10^{41}$ erg. Their bursting activity is highly irregular. Years of quiet states are interlaced with weeks of hundreds of bursts. By 2006 four soft gamma repeaters and two candidates have been discovered. The first discovered object, SGR 0525–66, is in the Large Magellanic Cloud, whereas other ones are in the Galactic plane. The most remarkable events were three gigantic gamma-ray bursts, much stronger than typical bursts. The first one was detected from SGR 0525–66 on March 5, 1979 (Mazets *et al.*, 1979a), the second one was detected from SGR 1900+14 on August 27, 1998 (Hurley *et al.*, 1999) and the third from SGR 1806–20 on December 27, 2004 (e.g., Hurley *et al.* 2005; Mazets *et al.* 2005). The energy of the third burst was especially huge and exceeded $10^{46}$ erg.

Periodic pulsations with large periods, from 5 to 8 s, have been detected in X-rays from the three sources. Two of them show pulsations in quiescent states which have enabled one to measure $\dot{P}$. In particular, one has got $P = 5.2$ s and $\dot{P} = 6.1 \times 10^{-11}$ for SGR 1900+14. Then Eq. (1.15) gives the characteristic age $t \sim 1.3 \times 10^3$ years. Using Eq. (1.13) (with all the reservations about its validity!), we immediately obtain an enormous characteristic magnetic field $B_{\rm eff} \sim 5.7 \times 10^{14}$ G. There are other arguments that soft gamma repeaters are young, slowly rotating and rapidly spinning down neutron stars with *superstrong magnetic fields* $B \sim 10^{14}$–$10^{15}$ G.

---

[9]By the time of this writing, the most recent parallax measurements give the distance $\sim 160$ pc (D.L. Kaplan *et al.*, in preparation).

*Anomalous X-ray pulsars (AXPs)* are sources of pulsed X-ray emission. The pulsation periods range from 6 to 12 s, and the X-ray luminosities range from $\sim 10^{33}$ to $\sim 10^{35}$ erg s$^{-1}$. These pulsars differ from the classical X-ray pulsars in X-ray binaries (§ 1.4.6) by the absence of any evidence that they enter binary systems. By 2005 five AXPs were discovered, together with several candidates. Some of them have been detected in optical. In most of the cases pulsar timing has been performed and the values of $\dot{P}$ have been measured. The estimated characteristic ages are slightly higher than for soft gamma repeaters, but the characteristic magnetic fields are of the same order of magnitude. For instance, for 1E 1048.1–5937 one has $P = 6.4$ s, $\dot{P} = 3.3 \times 10^{-11}$, $t \sim 3.1 \times 10^3$ years, and $B_{\rm eff} \sim 4.7 \times 10^{14}$ G.

Therefore, AXPs have much in common with soft gamma repeaters. A solid piece of evidence that these sources are related was provided by the discovery of bursting activity of AXPs (in particular, two bursts, separated by 16 days, from 1E 1048.1–5937, Gavriil *et al.* 2002; and over 80 bursts detected in June 2002 from 1E 2259+586, Kaspi *et al.* 2003). It is currently assumed that soft gamma repeaters and AXPs belong to the same class of neutron stars, which are called *magnetically powered pulsars* or *magnetars* – neutron stars with superstrong magnetic fields. The magnetar hypothesis was put forward, on theoretical grounds, by Duncan & Thompson (1992) and Paczyński (1992). Soft gamma repeaters are thought to be younger and transform into a AXPs in the course of their evolution. The sources of both types can be powered by huge magnetic fields located in neutron star interiors. Bursts are thought to be associated with episodic releases of stresses caused by the evolution of magnetic fields in neutron star crusts. The superstrong magnetic field is estimated to decay in $\sim 10^4$ years hampering the activity of these sources when they become older. Further observations are required to confirm these ideas.

### 1.4.6 Neutron stars in binary systems – X-ray binaries

Neutron stars have been observed in binaries with other neutron stars, white dwarfs, and nondegenerate stars. These systems are useful for measuring neutron star masses, testing theories of stellar evolution, and solving many other problems (see, e.g., § 9.1). We are still waiting for a discovery of a neutron star in binary with a black hole. Observed binaries can be divided into wide systems (without mass exchange) and more compact systems (with mass transfer, which often results in accretion onto a neutron star). If the mass transfer is absent, a neutron star behaves usually as an isolated object (§ 1.4.5). A mass transfer in a compact binary makes this binary an X-ray source. Such systems are called *X-ray binaries*, and they are outlined below. Many of them are observed not only in X-rays but also in other spectral bands. Some X-ray binaries contain black holes rather than neutron stars.

There is a rich phenomenology of X-ray binaries containing neutron stars (see, e.g., Lipunov 1992 and Lewin *et al.* 1997; also see Table 1.1). Their X-ray emission is generated either at (near) neutron star surfaces and/or in accretion disks. Generally, these binaries are divided into *high-mass X-ray binaries (HMXBs)*, $M_2 \gtrsim (2-3)\, M_\odot$, and *low-mass X-ray binaries (LMXBs)*, $M_2 \lesssim M_\odot$, with respect to companion masses $M_2$. By 2006, about 100 low-mass X-ray binaries and about 40 high-mass X-ray binaries containing neutron stars have been discovered. Neutron star companions in high-mass binaries are usually massive O-B stars, while in low-mass binaries they are dwarf stars (particularly, red dwarfs). Massive O-B stars produce strong stellar wind; its accretion on a neutron star may be nearly spherical. Life times of massive main-sequence stars, and hence life times of high-mass X-ray binaries are sufficiently short. A strong accretion from dwarf companions in compact low-mass X-ray binaries occurs if a dwarf star fills its Roche lobe and the plasma outflows through the first Lagrange point. This accretion regime is favorable for the formation of an accretion disk. X-ray emission produced by a population of (unresolved) high-mass X-ray binaries in distant galaxies serves as the indicator of the star formation rate in these galaxies (see, e.g., Grimm *et al.* 2003).

X-ray binaries can be sources of *regular* (periodic) and *irregular* emission; they can also be subdivided into *persistent* and *transient* sources. The latter sources are called *X-ray transients*. X-ray binaries can be observed as *X-ray pulsars*, *X-ray bursters*, sources of *quasiperiodic X-ray oscillations* (QPOs), etc. X-ray pulsars are powered by accretion. These accretion-powered pulsars should not be confused with rotation-powered pulsars or AXPs, §§ 1.4.4 and 1.4.5. The complicated phenomenology of X-ray binaries reflects the complex nature of these sources which is far from being clear. The same source can manifest itself in different ways. For instance, Vela X-1 is a classical (accretion powered) X-ray pulsar, a persistent source of regular X-ray pulsations. A 0538–66 is also an X-ray pulsar whose activity is transient. XTE J2123–058 (§ 9.1.1 c) is an X-ray transient which demonstrates X-ray bursts and quasiperiodic oscillations.

**X-ray transients.** They are X-ray sources which go from active (or 'on') to quiescent (or 'off') states and back on timescales of some hours and longer. As a rule, quiescent states last longer than active ones. The first transient, Cen X-2, was discovered by Harries *et al.* (1967) in April 1967 with rocket-born X-ray detectors.

X-ray transients do not form a uniform class of objects. For instance, the X-ray pulsar A 0538–66 with the spin period $P = 69$ ms demonstrates transient X-ray activity with the well determined period $P_b = 16.66$ days, which is the orbital period in a highly eccentric binary. Strong accretion, which powers the pulsar, occurs only near periastron passages and turns the system into active states for short periods of time. However, the majority of transients show irregular sequence of active states (weeks–months or even years) interspersed with longer periods of quiescence (months–years or even decades). Active states can be switched on by many mechanisms, particularly, by instabilities in accretion disks, irregular outflow of matter from a donor star, or by the changes of the accretion regime near a neutron star surface.

Some X-ray transients, for instance 4U 0115+63, have hard spectra in active states, with spectral fluxes extended to some tens keV. They are called *hard X-ray transients (HXTs)*, and they are usually identified with high-mass X-ray binaries (with Be companions). Other transients have softer spectra (extended to $\lesssim 1-2$ keV in active states). Accordingly, they are called *soft X-ray transients*. They are compact low-mass X-ray binaries. Some soft X-ray transients in quiescence (for instance, Aql X-1) show thermal-like radiation component which can be fitted by neutron-star atmosphere models. As suggested by Brown *et al.* (1998), this radiation emerges from the interiors of warm neutron stars (with the surface temperatures $\sim 10^6$ K) being produced by deep crustal heating in the inner crust. The heating mechanism proposed by Haensel & Zdunik (1990a) consists in pycnonuclear burning of accreted matter sinking in the crust under the weight of newly accreted material. There is a close correspondence (see, e.g., Yakovlev & Pethick 2004 and references therein) between the theory of thermal states of transiently accreting neutron stars and the theory of neutron star cooling (§ 1.3.7). Some X-ray transients have short active states (hours–days).

**X-ray pulsars in X-ray binaries.** They are accretion-powered rotating and strongly magnetized neutron stars in compact binaries. The first source recognized as an X-ray pulsar was Cen X-3 (in observations of Schreier *et al.* 1972 with the *Uhuru* X-ray satellite). There are about 35 X-ray pulsars in our Galaxy known by 2006. They mainly enter high-mass X-ray binaries (HMXBs) with intense accretion. An example is Vela X-1, a binary with the orbital period $P_b \approx 9$ days. It consists of a neutron star and a companion, GP Vel, a B0.5 Ib supergiant whose mass is (23–28) $M_\odot$ (§ 9.1.1 b). The pulsar spin period is $P = 283$ s. GP Vel nearly fills its Roche lobe and produces a powerful stellar wind. The star is bulky and creates eclipses of the X-ray source. Many X-ray pulsars are very slow rotators ($P \gtrsim 100$ s) but not all (e.g., $P = 69$ ms for A 0538–66).

It is thought that the accretion is channeled by the pulsar magnetic field into a thin *accretion column* near the neutron star surface. Strong X-ray emission from these columns corotating with neutron stars creates regular X-ray pulsations. The accretion energy release rate can be estimated as

$$\dot{E}_{\rm acc} \approx GM\dot{M}/R \approx 8.4 \times 10^{35} \dot{M}_{-10}\,(M/M_\odot)/R_6 \quad {\rm erg\,s^{-1}}, \tag{1.16}$$

where $\dot{M}_{-10}$ is the mass accretion rate $\dot{M}$ in units of $10^{-10} M_\odot$ yr$^{-1}$. X-ray luminosities of X-ray pulsars range from $\sim 10^{35}$ to $\sim 10^{39}$ erg s$^{-1}$. As a rule, an intense accretion spins up a neutron star, decreasing the pulsar spin period $P$ (in contrast to rotation-powered pulsars). However, rotation of some pulsars (e.g., Vela X-1) is nearly spun up to the equilibrium limit; their spin periods seem to undergo variations around these steady-state values.

Spectra of some X-ray pulsars show prominent electron cyclotron lines which serve to directly measure pulsar magnetic field $B$. The cyclotron lines are observed at photon energies $\sim \hbar\omega_c \sim (30-50)$ keV (where $\omega_c$ is the electron cyclotron frequency) and give $B \sim (3-5) \times 10^{12}$ G. These lines were predicted by Gnedin & Sunyaev (1974) and discovered by Trümper *et al.* (1978) in the spectrum of Her X-1.

**X-ray bursters.** On September 28, 1975 Grindlay *et al.* (1976) discovered two X-ray bursts from the X-ray source 4U 1820–30 in the globular cluster NGC 6624 with the *Astronomical Netherlands Satellite (ANS)*. Since then *X-ray bursters* have been observed many times (see, e.g., Strohmayer & Bildsten 2004). The total number of detected X-ray bursters is about 50. An X-ray burst lasts usually from a few to a few tens of seconds. Bursts repeat quasiperiodically with the recurrence time of several hours. X-ray bursts are thought to occur on the surfaces of neutron stars in compact low-mass X-ray binaries (LMXBs). A low-mass companion fills its Roche lobe and ejects matter, which is accreted by a neutron star. In some cases, X-ray eclipses have been detected. X-ray bursters concentrate within the Galactic bulge; many (but not all) are observed in globular clusters. The bursts are detected in soft X-rays, their spectra are much softer than the spectra of X-ray pulsars. Soft X-ray transients in active states are usually bursting sources.

All X-ray bursts are subdivided into two nonequal parts: type I and type II bursts. Type I bursts are widespread phenomena. X-ray luminosity $L_X$ in burst maxima often reaches the Eddington limit, $L_{\rm Edd} \sim 10^{38}$ erg s$^{-1}$, Eq. (1.3). An X-ray energy emitted during one burst constitutes typically $\sim 10^{-2}$ of the energy emitted during a recurrence (quasi)period. Type I bursts are explained by explosive thermonuclear burning of accreted matter on the surfaces of neutron stars with low magnetic fields ($B \lesssim 10^8 - 10^9$ G). The nuclear energy release ($\sim$ 5 MeV per one accreted nucleon) is just $\sim 10^{-2}$ of the accretion energy ($\sim$ 200 MeV per nucleon) responsible for the persistent X-ray emission. The instability of accreted matter with respect to nuclear burning was predicted by Hansen & Van Horn (1975); it was related to X-ray bursters by Woosley & Taam (1976). The state of the theory by 2004 is described, for instance, by Strohmayer & Bildsten (2004) and Woosley *et al.* (2004).

Among all X-ray bursts, one can clearly distinguish the so called *superbursts*. They were discovered in the system 4U 1735–444 by Cornelisse *et al.* (2000).

They are rare events, but very strong. They last 2–12 hours and their recurrence times are of several years. The total energy release in a superburst can be as high as $\sim 10^{42}$ erg, several orders of magnitude higher than in an ordinary X-ray burst (see Strohmayer & Bildsten 2004 for review). They are usually explained by unstable carbon burning in deep layers of the outer crust of an accreting neutron star (although this explanation meets some difficulties; see, e.g., Page & Cumming 2005). Such a burning has been studied theoretically (Woosley & Taam, 1976; Taam & Picklum, 1978; Brown & Bildsten, 1998) before the discovery of superbursts.

Type II X-ray bursts are demonstrated by two sources – neutron stars in transiently accreting LMXBs. These bursts are very frequent (with variable burst intervals which can be as short as tens of seconds); the burst energy correlates with burst intervals. It is most likely that these bursts are associated with the nonstationary accretion onto a neutron star (Lamb & Lamb, 1977), and the burst energy is supplied by accretion. The first source, MXB 1730–335, is the famous *rapid burster* discovered in March 1976 during observations with the *Small Astronomical Satellite* (*SAS 3*) observatory (Lewin *et al.*, 1976). It shows also type I bursts. The second source, GRO J1744–28, was discovered in December 1995 with the Burst and Transient Source Experiment (BATSE) aboard the *Compton Gamma Ray Observatory* (Fishman *et al.* 1995, Kouveliotou *et al.* 1996). In addition to type II X-ray bursts it demonstrates periodic X-ray pulsations (revealing neutron star spin period $P = 0.467$ s). It is called the *bursting pulsar.*

It was a long-standing problem to detect periodic X-ray pulsations associated with neutron star rotation in X-ray bursters of type I, but it was solved. Wijnands & van der Klis (1998) discovered regular $P = 2.5$ ms pulsations from an outburst of the X-ray transient SAX J1808.4–3658 in the observations with the *Rossi X-ray Timing Explorer*. That was the first discovered accreting millisecond pulsar and the first observational evidence that millisecond pulsars are associated with LMXBs. Now we know other examples (e.g., XTE J1814–338, Strohmayer *et al.* 2003). Another example – neutron-star spin pulsations were observed during a superburst of 4U 1636–53 (Strohmayer & Markwardt, 2002). These observations indicate the presence of nonuniform regions (hot spots) on the surfaces of X-ray bursters. Comparing theoretical models of these spots with observations one will be able to obtain useful constraints on neutron star masses and radii (e.g., Strohmayer 2004, Bhattacharyya *et al.* 2005).

Very powerful X-ray bursts have super-Eddington luminosities, so that the radiative pressure in the neutron star atmosphere exceeds the gravity. Such a burst initiates a huge expansion of the neutron star atmosphere by the radiative pressure (up to a few hundred kilometers) followed by a contraction to to the initial state. These bursts serve as nearly standard candles (with $L_X$ slightly higher than $L_{\rm Edd}$) useful to estimate distances to the bursters.

**Sources of quasiperiodic X-ray oscillations.** Some X-ray binaries are the sources of *quasiperiodic X-ray oscillations (QPOs).* These oscillations are not exactly periodic – not pulsar clocks; also see § 9.3.2. They were discovered by van der Klis *et al.* (1985) in observations of the LMXB GX 5-1 (4U 1758–25) with the *European X-ray Observatory Satellite (EXOSAT)*, operated from 1983 to 1986. Great progress in observations of quasiperiodic oscillations has been made with the *Ginga* (Japanese for *galaxy*) satellite (1987–1991) and with the *Rossi X-ray Timing Explorer* (launched in 1996).

Quasiperiodic X-ray oscillations have been observed from X-ray binaries containing neutron stars, black holes and white dwarfs. We focus on the binaries with neutron stars (see, e.g., van der Klis 2000). By 2006 about 20 such objects were discovered and identified as compact LMXBs; some of them are X-ray bursters. In 16 systems, the neutron star spin period has been determined. Drifts of quasiperiodic oscillation frequencies have been analyzed, as well as the tracks of the sources on the so called color-color diagram (hardness ratio in harder X-ray channels versus hardness ratio in softer channels). One distinguishes "atoll" and "Z"sources which have corresponding tracks. It is especially important to analyze the power spectrum of X-ray flux fluctuations which extends from tens of Hz to about 1 kHz (to 1.330 kHz for 4U 0614+09) with a cutoff afterwards. A power spectrum may contain several (up to three) pronounced peaks, which may drift from one observation to another.

Theoretical interpretation of a zoo of observational properties of quasiperiodic oscillations is not simple. It is likely that oscillations occur in the accretion disks around neutron stars with low magnetic fields ($B \lesssim 10^8 - 10^9$ G). Oscillation frequencies can be associated with the Keplerian frequency of the innermost stable orbit of matter elements in a disk, or with some resonant frequency in the disk itself, or with combination of these frequencies and neutron-star spin frequency. After understanding the real nature of quasiperiodic X-ray oscillations, their observation will be very useful to put stringent constraints on neutron star masses and radii.

## 1.4.7 Neutron stars in the Galaxy

X-ray binaries in the Galaxy concentrate to the galactic bulge and the galactic plane. By contrast, the distribution of radio pulsars (see, e.g., Arzoumanian *et al.* 2002, and references therein) is drastically different from the distribution of X-ray binaries and normal stars. Some radio pulsars are observed at high galactic latitudes, and many of them demonstrate strong proper motion, with the velocities $v \gtrsim 500$ km s$^{-1}$. Thus, neutron stars populate much wider space and move with much higher velocities than other stars. Observational constraints on radio pulsar (velocity and spatial) distribution in the Galaxy are still rather uncertain. Nevertheless, there are strong indications of the two-component velocity distribution with characteristic velocities of $\sim$ 100 km s$^{-1}$ and $\sim$

500 km $s^{-1}$, respectively. Both components contain a comparable number of sources, and $\sim$ 10% of radio pulsars have velocities $\gtrsim$ 1000 km $s^{-1}$. The escape velocity from the Galactic potential is estimated to range from 450 km $s^{-1}$ to 650 km $s^{-1}$ (Leonard & Tremaine, 1990). Thus, a sizable fraction of all radio pulsars are sufficiently fast to escape from the Galaxy. This means that the Galaxy possesses an extended halo of radio pulsars which are evaporated into the intergalactic space. The fastest is PSR B2224+65 in the Guitar Nebula whose projected (perpendicular to line of sight) velocity $\sim$ 1600 km $s^{-1}$ is nearly parallel to the Galactic plane (Cordes *et al.*, 1993). Another pulsar, PSR J1740+1000, may be moving much (about twice) faster but its velocity measurement (McLaughlin *et al.*, 2002) is still ambiguous.

It is believed that high pulsar velocities are gained at neutron star birth due to *pulsar kicks*. The origin of huge kick velocities is a subject of debates (see, e.g., Lai *et al.* 2001, Arzoumanian *et al.* 2002 and references therein).

## 1.5. Neutron stars as "superstars" in physics and astrophysics

**The Pines theorem.** The theorem was formulated by David Pines in a talk given at the conference on "Neutron Stars: Theory and Observation" (The NATO Advanced Study Institute, Crete, Greece, September 3–14, 1990). The formulation is fairly simple:

**Neutron Stars are Superstars.**

**Proof.** After reading §§ 1.1–1.4 the proof is trivial. Indeed, neutron stars are *super*dense objects; *super*fast rotators; *super*fluid and *super*conducting inside; *super*accelerators of high-energy particles; sources of *super*strong magnetic fields; *super*precise timers; *super*glitching objects; *super*rich in the range of physics involved. Neutron stars are related to many branches of contemporary physics and astrophysics, particularly to nuclear physics; particle physics; condensed matter physics; plasma physics; general theory of relativity; hydrodynamics; quantum electrodynamics in superstrong magnetic fields; quantum chromodynamics; radio-, optical-, X-ray and gamma-ray astronomy; neutrino astronomy; gravitational-wave astronomy; physics of stellar structure and evolution, etc.

Let us stress that neutron stars contain the matter of essentially supranuclear density in their interiors (§ 1.3.1). This state of matter cannot be reproduced in laboratory because the nuclear matter is known to be highly incompressible under laboratory conditions. The compression in neutron stars is produced by enormous gravitational forces. This enables one to treat neutron stars as *unique natural laboratories of superdense matter* under the most extreme conditions.

One can test theoretical models of dense matter by comparing observations of neutron stars with theoretical predictions.

Observational manifestations of neutron stars are really numerous (§ 1.4). The need to observe these objects has triggered the development of foremost telescopes and detectors, from the best modern radio telescopes to laser interferometers for detecting gravitational waves. One cannot imagine the modern observational astrophysics without the astrophysics of neutron stars.

In other words, neutron stars are fascinating objects to observe and to study theoretically.

## 1.6. About the rest of this book

We hope that, after reading this chapter, the reader has become familiar with the main ideas of neutron star physics and can decide if it is worthwhile to read further. Naturally, the subject is too wide to be discussed in one book. Moreover, the field is rapidly developing; many problems have not been solved yet; many outstanding discoveries could be expected soon after this writing.

In the next chapters we will focus on the internal structure of neutron stars. We will discuss thermodynamic properties of the matter in all neutron star layers, from the surface to the center (Chapters 2–5, 7 and 8). Our main concern will be to consider the structure, composition and equation of state of dense matter, particularly, the basic problem – the equation of state in inner neutron star cores (§ 1.3.2). We will also discuss the models of neutron star structure, masses and radii of neutron stars (Chapters 6 and 8), and observational tests for these models (Chapter 9). We will try to be pedagogical and describe not only the results of sophisticated theories but explain these theories and underlying ideas. Although the book is written by theoreticians, we will summarize the necessary data coming from the nuclear physics experiments and from some neutron star observations.

In the present book we do *not* discuss in detail kinetic properties of neutron star matter, neutrino emission mechanisms, thermal evolution of neutron stars, evolution of their magnetic fields, and associated observational problems. We hope to address these issues in a separate book. The present book also does not touch many other problems, for instance, neutron star birth in supernova explosions, the structure and evolution of protoneutron stars, the evolution of neutron stars in binary systems, the physics of neutron star magnetospheres. Some references to these subjects can be found in this and subsequent chapters of the present book.

Chapter 2

# EQUILIBRIUM PLASMA PROPERTIES. OUTER ENVELOPES

After an overall introduction into the neutron star physics, we begin a systematic description of neutron star layers (Fig. 1.2). Chapters 2–4 are devoted to neutron star envelopes. These envelopes are non-uniform. The structure of an envelope for a star with the effective surface temperature $\sim 10^6$ K is schematically shown in Fig. 2.1. We indicate characteristic densities and geometrical depths of interfaces between adjoining layers. One can distinguish the atmosphere, the ocean of ion liquid, followed by solidified layers denoted as the outer and inner crusts. At the bottom of the envelope there may be a mantle containing a liquid crystal of nonspherical nuclei (§ 3.7.2). These layers are already outlined in § 1.3.1 and will be studied in more detail below. The properties of the atmosphere, ocean, and the upper edge of the solidified layer strongly depend on temperature. For the assumed surface temperature $\sim 10^6$ K the internal temperature of a cooling (isolated) star would be $\sim 10^8$ K; the main temperature gradient would occur in the ocean, while deeper layers would be nearly isothermal. When the star cools, the atmosphere and the ocean become thinner. In a sufficiently cold star they may be absent and the solidified matter will extend to the very surface.

Let us remark that neutron star envelopes (which we define as layers containing atomic nuclei or non-uniform nuclear structures, in contrast to the uniform matter in the liquid stellar core) are often called collectively crusts. Strictly speaking, this is inaccurate because the envelopes may contain such layers as the ocean and the mantle, which are not solid. We will mainly use the term "envelope" in this chapter but we will often use the widespread term "crust" (meaning the entire envelope) in subsequent chapters.

The neutron star envelopes are important in many ways. For instance, they provide thermal insulation of the stellar interior and control thus the cooling of

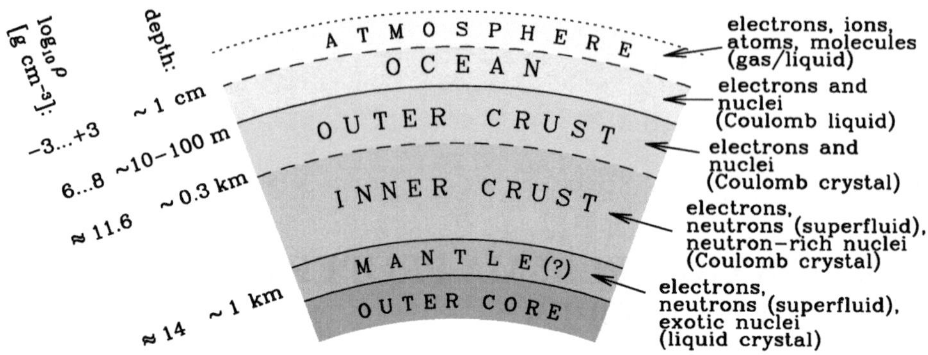


*Figure 2.1.* Schematic structure of an envelope of a neutron star with the internal temperature $\sim 10^8$ K (see text for more details).

neutron stars (§ 1.3.7). To study the cooling, one needs to know the equation of state (EOS) and transport properties of the envelopes. The evolution of neutron star magnetic fields is often associated with Ohmic decay of the fields confined in the envelopes. The bursting activity of X-ray bursters (§ 1.4.6) is explained by nuclear explosions of accreted matter in the outer envelopes. The glitches of radio pulsars (§ 1.4.4) are commonly associated with depinning of superfluid neutron vortices in the inner crust. Finally, the EOS and opacities of neutron star atmospheres control the spectra of radiation emitted from the surface (e.g., Romani 1987; Pavlov *et al.* 1995; also see § 1.3.1).

While studying the envelopes, one should take into account Coulomb interaction of charged particles (electrons and ions) and strong interaction of nucleons (neutrons and protons in atomic nuclei and free neutron gas between the nuclei in the inner envelope). The importance of the nuclear interaction increases with the growth of the density. In this chapter we focus on the plasma physics effects associated with Coulomb interactions, with the emphasis on the outer envelopes (i.e., for densities lower than the neutron drip density, where Coulomb interactions are most important). Note, however, that many results of the present chapter are valid for the inner envelope. The EOS in the inner envelope (after the neutron drip) will be mainly studied in Chapter 3. Because physical conditions in outer envelopes are similar to those in white dwarfs, the results of the present chapter are equally useful for the physics of white dwarfs.

Here we restrict ourselves to the *nonmagnetic* envelopes. The equilibrium thermodynamics of charged particles is not affected by the magnetic field[1]

[1]This statement, proven independently in thesis works by Niels Bohr (Copenhagen, 1911) and by J.H. van Leeuwen (Leiden, 1919; published by van Leeuwen 1921) is known as the Bohr-van Leeuwen theorem. It is often formulated as follows: "At any finite temperature, and in all finite applied magnetic or thermal

unless the field is *quantizing* (see § 4.1.1). In particular, the results of the present chapter are applicable to the envelopes of those neutron stars whose magnetic field does not exceed $\sim 10^{10} - 10^{11}$ G [more accurate conditions will be given by Eqs. (4.29), (4.30)]. A stronger field can significantly affect thermodynamic functions in the outer envelope. These effects will be considered in Chapter 4.

In § 2.1 we introduce the plasma parameters, and in § 2.2 describe the basic thermodynamic quantities. Since the conditions of full ionization are satisfied in the largest part of the envelope, we are primarily concerned with the thermodynamics of fully ionized electron-ion plasmas, considered in detail in § 2.3 and § 2.4. In many applications to neutron stars, one can assume that the electrons constitute an ideal, uniform Fermi gas, and the ions form an *one-component plasma* (OCP) or a *multi-component plasma* on the uniform ("rigid") charge-neutralizing electron background. This case is considered in § 2.3. In § 2.4 we describe the effects of the Coulomb interactions associated with a compressible electron background under the assumption that the internal structure of bound species is unimportant. Finally, in § 2.5 we outline the various approaches to include bound species in the EOS of the stellar atmosphere, and consider partially ionized atomic hydrogen as an example.

## 2.1. Plasma parameters

### 2.1.1 Overall parameters

Let $n_e$ be the electron number density and $n_j$ the number density of ion species $j$=1,2,... (with mass and charge numbers $A_j$ and $Z_j$, respectively). The subscript '$j$' enumerates the ions of different chemical elements or the same element in different ionization states. In the latter case $Z_j$ means the charge number of the atomic nucleus minus the number of bound electrons. The total number density of atomic nuclei (ions) in the plasma is $n_{\mathcal{N}} = \sum_j n_j$. The electric neutrality of the matter implies $n_e = \langle Z \rangle n_{\mathcal{N}}$. Here and in what follows the brackets $\langle \ldots \rangle$ denote averaging over ion species $j$ with statistical weights proportional to $n_j$:

$$\langle f \rangle = \frac{1}{n_{\mathcal{N}}} \sum_j n_j f_j. \tag{2.1}$$

In the OCP, the averaging is unnecessary, so that we shall omit the subscript '$j$' and the brackets whenever possible.

fields, the net magnetization of a collection of pointlike charged particles in thermal equilibrium vanishes identically." In other words, the free energy does not depend on the magnetic field. This follows from the fact that a particle momentum $\boldsymbol{p}$ and the vector potential $\boldsymbol{A}$ of the magnetic field are independent variables in a classical Hamiltonian function of an ensemble of particles. Thus, while integrating over $\boldsymbol{p}$ one can eliminate $\boldsymbol{A}$ from the classical partition function by replacing $\boldsymbol{p} \to \boldsymbol{p} - (q/c)\boldsymbol{A}$ ($q$ being the particle charge).

An essential parameter is the number density of baryons, $n_\mathrm{b}$. In the outer envelope, one has $n_\mathrm{b} \approx \langle A \rangle \, n_\mathcal{N}$, while in the inner envelope $n_\mathrm{b} \approx A' \, n_\mathcal{N}$, Here $A' \equiv \langle A \rangle + A''$, $\langle A \rangle$ is the average number of nucleons bound in one nucleus, and $A''$ is the number of free neutrons per one atomic nucleus. We have $A'' n_\mathcal{N} = n_n \, (1 - w)$, where $n_n$ is the local number density of free neutrons (in space between atomic nuclei) and $w$ is the fraction of volume occupied by the atomic nuclei. Clearly, $A' = \langle A \rangle$ in the outer envelope, but $A' > \langle A \rangle$ in the inner envelope.

The mass density of the matter of relativistic stellar objects is generally determined as $\rho = \mathcal{E}/c^2$, where $\mathcal{E}$ is the total energy density. However, for evaluating $\rho$ in the neutron star envelopes it is often sufficient to set $\rho \approx m_\mathrm{u} n_\mathrm{b}$, where $m_\mathrm{u} = 1.6605 \times 10^{-24}$ g is the atomic mass unit.

The characteristic temperatures, which separate the different domains of plasma parameters, are shown in Fig. 2.2 for typical densities in the outer envelope (for the inner envelope, see Fig. 3.17).

At low densities and/or high temperatures the matter contains also non-negligible amount of positrons and photons which contribute to thermodynamic properties. We mainly neglect their contributions because they are unimportant for the problems considered in this book. The only exception is § 2.5, where we include the contribution of photons.

### 2.1.2 Electrons

The state of a free electron gas is determined by $n_e$ and $T$. Instead of $n_e$ it is convenient to introduce the relativity parameter (Salpeter, 1961)

$$x_\mathrm{r} \equiv \frac{p_\mathrm{F}}{m_e c} \approx 1.00884 \left( \frac{\rho_6 \langle Z \rangle}{A'} \right)^{1/3}, \tag{2.2}$$

where

$$p_\mathrm{F} = \hbar k_\mathrm{F} = \hbar \, (3\pi^2 n_e)^{1/3}, \tag{2.3}$$

is the electron Fermi momentum and $\rho_6 \equiv \rho/10^6$ g cm$^{-3}$. The electron Fermi momentum is a measure of electron number density; it limits electron momenta in the Fermi distribution at $T = 0$. Accordingly, the Fermi energy

$$\epsilon_\mathrm{F} = c^2 \sqrt{(m_e c)^2 + p_\mathrm{F}^2} \tag{2.4}$$

is a convenient energy parameter which has the meaning of the chemical potential in a strongly degenerate electron gas. Here and hereafter, we include the electron rest energy, $m_e c^2$, in the electron energy $\epsilon$ (unless the opposite is stated explicitly). The electron Fermi temperature is

$$T_\mathrm{F} = T_\mathrm{r} \, (\gamma_\mathrm{r} - 1), \tag{2.5}$$

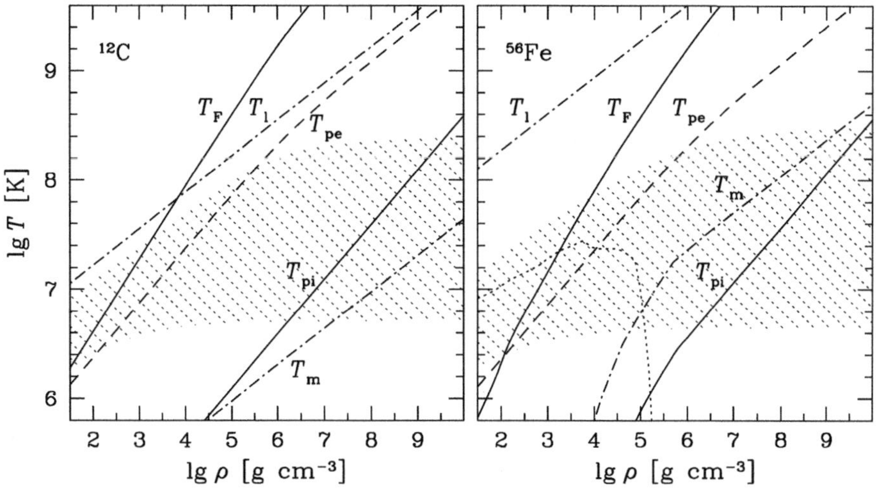


*Figure 2.2.* Density-temperature diagram for the outer envelope composed of carbon (*left*) or iron (*right*). We show the electron Fermi temperature ($T_\mathrm{F}$), the electron and ion plasma temperatures ($T_\mathrm{pe}$ and $T_\mathrm{pi}$), the temperature of the gradual gas-liquid transition ($T_\mathrm{l}$), and the temperature of the sharp liquid-solid phase transition ($T_\mathrm{m}$). Shaded are the regions of typical temperatures in the outer envelopes of middle-aged cooling neutron stars (which are $\sim 10^4 - 10^6$ years old). The lower left domain on the right panel, separated by the dotted line, is characterized by strong electron response or bound-state formation (specifically, it corresponds to the effective charge number $Z_\mathrm{eff} < 20$ determined as described on p. 91).

where

$$T_\mathrm{r} = m_e c^2/k_\mathrm{B} \approx 5.930 \times 10^9 \text{ K} \tag{2.6}$$

is the relativistic temperature unit, $\gamma_\mathrm{r} = \sqrt{1 + x_\mathrm{r}{}^2}$, and $k_\mathrm{B}$ is the Boltzmann constant. The electron gas is non-relativistic at $T \ll T_\mathrm{r}$ and $x_\mathrm{r} \ll 1$, and it is ultrarelativistic at $x_\mathrm{r} \gg 1$ or $T \gg T_\mathrm{r}$. It is nondegenerate at $T \gg T_\mathrm{F}$ and strongly degenerate at $T \ll T_\mathrm{F}$. For brevity, the density or temperature are called *relativistic* if the respective parameter $x_\mathrm{r}$ or

$$t_\mathrm{r} = T/T_\mathrm{r} \tag{2.7}$$

is large.

The Fermi temperature for carbon and iron plasmas at densities $10 \text{ g cm}^{-3} \lesssim \rho \lesssim 10^{10} \text{ g cm}^{-3}$ in the outer envelope is shown in Fig. 2.2. The shaded bands overlap typical temperature profiles of middle-aged cooling neutron stars. At $T \gtrsim 10^7$ K, the degeneracy temperature $T_\mathrm{F}(\rho)$ is nearly the same for both carbon and iron. However, at $T \lesssim 10^7$ K and $\rho \lesssim 10^4 \text{ g cm}^{-3}$ the curve for iron (on the right panel) bends downward, because the formation of bound species lowers the effective charge of iron ions.

It is also convenient to introduce the electron Fermi velocity $v_F = \partial\epsilon_F/\partial p_F$ and the corresponding dimensionless parameter

$$\beta_r = v_F/c = x_r/\gamma_r. \tag{2.8}$$

Sometimes it is useful to introduce the density parameter

$$r_s \equiv \frac{a_e}{a_0} = \left(\frac{9\pi}{4}\right)^{1/3} \frac{\alpha_f}{x_r} \approx \frac{0.01400}{x_r}, \tag{2.9}$$

where

$$a_e = [3/(4\pi n_e)]^{1/3} \tag{2.10}$$

is the electron-sphere radius, $a_0 = \hbar^2/(m_e e^2)$ is the Bohr radius, and $\alpha_f = e^2/(\hbar c) = 1/137.036$ is the fine-structure constant. One can write $r_s \approx 1.172\, n_{24}^{-1/3}$, where $n_{24} \equiv n_e/10^{24}$ cm$^{-3}$. It is also convenient to write $r_s = (\rho_{0s}/\rho)^{1/3}$, where $\rho_{0s} = 2.696$ g cm$^{-3}$ for the fully ionized hydrogen plasma and $\rho_{0s} = 2.675(A'/\langle Z\rangle)$ g cm$^{-3}$ for the plasma of heavier elements.

The strength of the electron Coulomb interaction in a plasma of nondegenerate electrons can be characterized by the electron Coulomb coupling parameter

$$\Gamma_e = \frac{e^2}{a_e k_B T} \approx \frac{22.75}{T_6} \left(\rho_6 \frac{\langle Z\rangle}{A'}\right)^{1/3}, \tag{2.11}$$

where $T_6 \equiv T/10^6$ K. We shall use this parameter $\Gamma_e$ for any plasma conditions though $\Gamma_e$ has no transparent physical meaning for degenerate electrons.

In the regions of partially degenerate electrons, one often uses the degeneracy parameter

$$\theta = T/T_F. \tag{2.12}$$

In the non-relativistic and ultrarelativistic cases ($x_r \ll 1$ and $x_r \gg 1$), we have $\theta = 0.543\, r_s/\Gamma_e$ and $\theta \approx (263\,\Gamma_e)^{-1}$, respectively.

Another important parameter is the *screening* (*Thomas-Fermi*) wave number[2] $k_{TF}$. It is the inverse length $r_e$ of electron screening in a plasma. In the linear plasma response approximation, a test charge $q$ embedded in the plasma (at $\boldsymbol{r} = 0$) creates a perturbation $\tilde{n}_e(r) = (\partial n_e/\partial\mu_e)\tilde{\mu}(r)$ of the electron number density, where $\tilde{\mu}(r) = -e\phi(r)$ is the perturbation of the electron chemical potential $\mu_e$ (§ 2.3.1), and $\phi(r)$ is the excess electrostatic potential determined by the Poisson equation $\nabla^2\phi(r) = -4\pi[q\delta^3(\boldsymbol{r}) - e\tilde{n}(r)]$. Then we come to the equation $(\nabla^2 + k_{TF}^2)\,\phi(r) = -4\pi q\,\delta^3(\boldsymbol{r})$, with the solution $\phi(r) = (q/r)\exp(-r/r_e)$, where

$$r_e^{-1} = k_{TF} = \left(4\pi e^2\, \partial n_e/\partial\mu_e\right)^{1/2}. \tag{2.13}$$

[2]Sometimes the term "Thomas-Fermi wave number" is applied only to $k_{TF}$ in the limit of strong degeneracy, Eq. (2.15).

For a nondegenerate electron gas ($T \gg T_\mathrm{F}$) we have $\partial n_e/\partial\mu_e = n_e/(k_\mathrm{B}T)$, and $r_e$ equals the Debye-Hückel electron screening length (which neglects the contribution of plasma ions, that will be considered below):

$$r_e \approx a_e/\sqrt{3\Gamma_e}. \tag{2.14}$$

For a strongly degenerate electron gas ($T \ll T_\mathrm{F}$),

$$k_\mathrm{TF} = 2\sqrt{\alpha_\mathrm{f}/(\pi\beta_\mathrm{r})}\, k_\mathrm{F} = \left(0.185/\sqrt{\beta_\mathrm{r}}\right)\, a_e^{-1}. \tag{2.15}$$

For further use, it is convenient to introduce the electron plasma temperature

$$T_\mathrm{pe} = \hbar\omega_\mathrm{pe}/k_\mathrm{B} \approx 3.300\times 10^8\, x_\mathrm{r}\, \sqrt{\beta_\mathrm{r}}\ \mathrm{K}, \tag{2.16}$$

where

$$\omega_\mathrm{pe} = \left(4\pi e^2 n_e/m_e^*\right)^{1/2} \tag{2.17}$$

is the electron plasma frequency and $m_e^* \equiv \epsilon_\mathrm{F}/c^2 = m_e\gamma_\mathrm{r}$ is the effective dynamical mass of an electron at the Fermi surface. In Fig. 2.2, where $T_\mathrm{pe}$ is displayed as a function of density in the outer envelope, the corresponding curves for carbon (left panel) and iron (right panel) plasmas at $\rho \gtrsim 10^3$ g cm$^{-3}$ nearly coincide, because the ratio $Z/A' \approx \frac{1}{2}$, that determines $n_e(\rho)$ in Eq. (2.17), is almost the same.

The polarization properties of the electron gas are described by the longitudinal dielectric function $\varepsilon(k,\omega)$, where $k$ is the wave number and $\omega$ is the frequency. Since the electrons respond almost instantly to ion motion, we shall mainly be interested in the static dielectric function $\varepsilon(k,0) = \varepsilon(k)$. For the fully degenerate free electron gas, Jancovici (1962) obtained in the random-phase approximation:

$$\begin{aligned}\varepsilon(k) &= 1 + \frac{k_\mathrm{TF}^2}{k^2}\left\{\frac{2}{3} - \frac{2}{3}\frac{y^2 x_\mathrm{r}}{\gamma_\mathrm{r}}\ln(x_\mathrm{r}+\gamma_\mathrm{r})\right.\\ &\quad + \frac{{x_\mathrm{r}}^2 + 1 - 3{x_\mathrm{r}}^2y^2}{6y{x_\mathrm{r}}^2}\ln\left|\frac{1+y}{1-y}\right|\\ &\quad \left. + \frac{2y^2{x_\mathrm{r}}^2 - 1}{6y{x_\mathrm{r}}^2}\frac{\sqrt{1+{x_\mathrm{r}}^2y^2}}{\gamma_\mathrm{r}}\ln\left|\frac{y\gamma_\mathrm{r} + \sqrt{1+{x_\mathrm{r}}^2y^2}}{y\gamma_\mathrm{r} - \sqrt{1+{x_\mathrm{r}}^2y^2}}\right|\right\},\end{aligned} \tag{2.18}$$

where $y = k/(2k_\mathrm{F})$. In the non-relativistic case ($x_\mathrm{r} \ll 1$), Eq. (2.18) reduces to the Lindhard (1954) dielectric function,

$$\varepsilon(k) = 1 + \frac{k_\mathrm{TF}^2}{2k^2}\left(1 + \frac{1-y^2}{2y}\ln\left|\frac{1+y}{1-y}\right|\right). \tag{2.19}$$

For any $x_{\rm r}$ and $y \ll 1$, Eq. (2.18) yields

$$\varepsilon(k) = 1 + \frac{k_{\rm TF}^2}{k^2}. \qquad (2.20)$$

Equation (2.20) (unlike Eq. (2.18)) is actually valid at any $T$, provided that $\omega$ and $k$ are sufficiently small and $k_{\rm TF}$ is calculated from the general Eq. (2.13) (see, e.g., Chabrier, 1990).

The random-phase approximation does not take into account short range correlations in electron motion. These correlations modify the dielectric function. The modification can be assessed by taking into account the polarization effects in the electron-electron interaction in the density response function. For the non-relativistic electron plasma this correction was described, for instance, by Chabrier (1990):

$$\varepsilon(k) = 1 + \frac{\varepsilon_{\rm L}(k) - 1}{1 - [\varepsilon_{\rm L}(k) - 1]G(k)}, \qquad (2.21)$$

where $\varepsilon_{\rm L}(k)$ is the Lindhard dielectric function (2.19) and $G(k) = G(k, \omega \to 0)$ is the so-called *local field correction* which can be evaluated numerically.

At high densities, $x_{\rm r} \gtrsim 1$, the electrons become relativistic. In this regime, $\epsilon_{\rm F}$ is so high that the electron gas is almost ideal and the electron correlations are negligible. Thus, at $x_{\rm r} \gtrsim 1$ one can use the zero-temperature dielectric function (2.18) in thermodynamic calculations.

At low temperatures, when the ions crystallize (§§ 2.1.3 and 2.3.4), the electron energy spectrum acquires gaps (forbidden energy bands) associated with crystalline structure. The electron band structure is of utmost importance in the solid-state physics (e.g., see, Kittel 1963, 1986) but is relatively tiny and insignificant in high-density stellar matter. However, this structure can be important for some kinetic and neutrino processes in neutron star envelopes (e.g., Raikh & Yakovlev 1982; Kaminker *et al.* 1999). The effects of electron band structure on thermodynamics of dense stellar matter are almost not explored and will not be discussed below.

### 2.1.3 Ions

At any temperature and density throughout the envelope, the ions are non-relativistic. Their state is determined by Coulomb and nuclear interactions. The strength of the Coulomb interaction of ion species $j$ is characterized by the Coulomb coupling parameter,

$$\Gamma_j = \Gamma_e Z_j^{5/3} = (Z_j e)^2 / (a_{\rm i}^{(j)} k_{\rm B} T), \qquad (2.22)$$

where

$$a_{\rm i}^{(j)} = a_e Z_j^{1/3} \qquad (2.23)$$

is the *ion sphere* radius. In an OCP, we will omit the index '$j$'. In a multicomponent ion plasma, it is useful to introduce the mean ion-coupling parameter $\Gamma = \Gamma_e \langle Z^{5/3} \rangle$.

At sufficiently high temperatures, the ions form a classical Boltzmann gas. With decreasing $T$, the gas gradually, without a phase transition, becomes a strongly coupled Coulomb liquid, and then (with a phase transition) a Coulomb crystal. The gas and liquid constitute the neutron star ocean, while the crystal is formed under the ocean (Fig. 2.1).

The gaseous regime of an OCP occurs for $\Gamma \ll 1$, or $T \gg T_{\rm l}$, where

$$T_{\rm l} = \frac{Z^2 e^2}{a_{\rm i} k_{\rm B}} \approx 2.275 \times 10^7 \, Z^2 \left( \frac{\rho_6}{A'} \right)^{1/3} \ {\rm K}. \tag{2.24}$$

For a plasma of light elements one has $T_{\rm l} < T_{\rm F}$, whereas for heavy elements $T_{\rm l} \gtrsim T_{\rm F}$ (Fig. 2.2).

The ion-sphere radius $a_{\rm i}^{(j)}$, defined by Eq. (2.23), is the main ingredient of the *ion-sphere model* for strongly coupled Coulomb systems ($T \lesssim T_{\rm l}$) with nearly uniform electron background. In this model (see, e.g., Shapiro & Teukolsky 1983), the matter is considered as an ensemble of ion spheres filled by the uniform electron background. It is supposed that any ion sphere contains an ion in its center. The radius $a_{\rm i}^{(j)}$ is chosen in such a way for the electron charge within the sphere to compensate the ion charge. Then the spheres are electrically neutral and can be treated as non-interacting elements of the ensemble. We will see that the ion-sphere model is very successful in explaining many properties of strongly coupled Coulomb systems.

In the gaseous regime, the characteristic length of the Coulomb screening by ions is the Debye length

$$r_{\rm D} = q_{\rm D}^{-1} = \left[ \frac{4\pi}{k_{\rm B} T} \sum_j n_j (Z_j e)^2 \right]^{-1/2}, \tag{2.25}$$

where $q_{\rm D}$ is the Debye wavenumber. In the OCP, $r_{\rm D} = a_{\rm i}/\sqrt{3\Gamma}$.

In the gas of ions ($T \gg T_{\rm l}$), the cumulative electron-ion screening wave number is

$$q'_{\rm D} = \left( q_{\rm D}^2 + k_{\rm TF}^2 \right)^{1/2}, \tag{2.26}$$

where $k_{\rm TF}$ is given by Eq. (2.13). If the electrons are non-degenerate and non-relativistic ($T \gg T_{\rm F}$ and $x_{\rm r} \ll 1$), then $q'_{\rm D}$ equals $q_{\rm D}$ from Eq. (2.25) with the summation extended over all particle types (ions *and* electrons).

The scale-length that characterizes quantum-mechanical effects on thermodynamics of liquid and gaseous phases and controls ionization equilibria is the thermal (de Broglie) wavelength

$$\lambda_j = \left( \frac{2\pi\hbar^2}{m_j k_\mathrm{B} T} \right)^{1/2}, \tag{2.27}$$

where $m_j$ is the particle mass. In the following, we shall also use the *electron* thermal length $\lambda_e$, given by Eq. (2.27) with $m_j$ replaced by $m_e$.

As was noted, for instance, by Abrikosov (1960), the ions crystallize at sufficiently high densities. The crystallization of a Coulomb plasma occurs at the temperature

$$T_\mathrm{m} = \frac{Z^2 e^2}{a_\mathrm{i} k_\mathrm{B} \Gamma_\mathrm{m}} \approx 1.3 \times 10^5 \, Z^2 \left( \frac{\rho_6}{A'} \right)^{1/3} \frac{175}{\Gamma_\mathrm{m}} \ \mathrm{K}, \tag{2.28}$$

where $\Gamma_\mathrm{m}$ is the melting value of $\Gamma$. For a classical OCP, $\Gamma_\mathrm{m} \approx 175$ (§ 2.3.4).

Let us stress the principal difference between the ions and the electrons at low temperatures. Because the electrons are nearly ideal, they become degenerate. In contrast, the ions become strongly coupled by Coulomb forces.

At low $T$ the quantum effects on ion motion become important. They are especially pronounced at $T \ll T_\mathrm{pi}$, where

$$T_\mathrm{pi} \equiv \frac{\hbar \omega_\mathrm{pi}}{k_\mathrm{B}} \approx 7.832 \times 10^6 \left( \frac{\rho_6}{A'} \left\langle \frac{Z^2}{A} \right\rangle \right)^{1/2} \ \mathrm{K} \tag{2.29}$$

is the plasma temperature of ions determined by the ion plasma frequency

$$\omega_\mathrm{pi} = \left( 4\pi e^2 \, n_{\scriptscriptstyle N} \, \langle Z^2 / m_\mathrm{i} \rangle \right)^{1/2}. \tag{2.30}$$

The corresponding dimensionless parameter is

$$t_\mathrm{p} = T/T_\mathrm{pi} \approx 2.92 \, \langle Z \rangle^{1/2} \langle Z^2/A \rangle^{-1/2} / (\Gamma_\mathrm{e} \sqrt{x_\mathrm{r}}). \tag{2.31}$$

In the OCP, $t_\mathrm{p} = \sqrt{R_S/3}/\Gamma$, where

$$R_S = a_\mathrm{i} m_\mathrm{i} (Ze)^2/\hbar^2 = r_s (m_\mathrm{i}/m_e) Z^{7/3} \tag{2.32}$$

is the ion density parameter, analogous to the electron density parameter $r_s$ defined by Eq. (2.9). If $T_\mathrm{pi} \gg T_\mathrm{m}$, huge zero-point ion vibrations suppress crystallization and reduce the melting temperature in comparison with that given by Eq. (2.29). Under these conditions, the actual melting temperature decreases with growing $\rho$ and drops to zero at a certain $\rho = \rho_\mathrm{m}$ (see § 2.3.4).

The density profiles of the melting and plasma temperatures are plotted in Fig. 2.2 for the outer envelope composed of carbon and iron. For the envelopes composed of the ground-state matter, $T_\mathrm{m}(\rho)$ and $T_\mathrm{pi}(\rho)$ will be presented in Fig. 3.17 of Chapter 3. On the left panel of Fig. 2.2 (for carbon), the functions $T_\mathrm{m}(\rho)$

and $T_{\rm pi}(\rho)$ are straight lines (in the logarithmic scale), because carbon is fully ionized in the displayed domain, implying $Z = 6$ and $A = A' = 12$ in Eqs. (2.28) and (2.29). On the right panel (for iron), these lines bend downward at the lowest displayed densities and temperatures, because of the effective charge lowering.

In a neutron star envelope just beneath the ocean (Fig. 2.1) the crystal can be classical while in deeper layers it is usually in the quantum regime ($T \ll T_{\rm pi}$).

## 2.2. Free energy and thermodynamic quantities

While studying the plasma thermodynamics, it is convenient to choose the volume $V$ and the temperature $T$ as independent variables and the Helmholtz free energy $F(V, T, \{N_j\})$ as the thermodynamic potential. Here, $\{N_j\}$ is the set of numbers of various ions $j$. The free energy is obtained from microscopic physics through the well-known relation (e.g., Landau & Lifshitz 1993)

$$F(\{N_j\}, V, T) = -k_{\rm B}T \ln \mathcal{Z}(\{N_j\}, V, T), \tag{2.33}$$

where $\mathcal{Z} = \mathrm{Tr}\, \exp(-\hat{H}/k_{\rm B}T)$ is the canonical partition function of the system, with the total Hamiltonian operator $\hat{H}$. The equilibrium state can be found by minimizing $F$ at given $V$ and $T$ with respect to the numbers $\{N_j\}$, subject to stoichiometric constraints. For instance, if a reaction $A \rightleftarrows B + C$ (e.g., ionization or dissociation) is allowed in the system, the corresponding stoichiometric relation reads

$$\partial F/\partial N_A = \partial F/\partial N_B + \partial F/\partial N_C. \tag{2.34}$$

The main difficulty is that $\mathcal{Z}(\{N_j\}, V, T)$, as a rule, is not exactly tractable for a real many-body system. Thus one should find such approximations of $\mathcal{Z}$ (or, equivalently, of $F$) which, on the one hand, would be calculable and, on the other hand, would preserve the principal physical features of the real system.

One often employs the factorization of the many-body partition function into *ideal* (corresponding to non-interacting particles) and *excess* (or configurational) parts. If the Hamiltonian can be separated into the kinetic part $\hat{K}$ and the potential, momentum-independent part $\mathcal{U}$, then a *translational* factor can be extracted from $\mathcal{Z}$. According to Eq. (2.33), this factorization leads to the splitting of the free energy into additive parts:

$$F(V, T, \{N_j\}) = F_{\rm id} + F_{\rm ex} = F_{\rm trans} + F_{\rm int} + F_{\rm ex}. \tag{2.35}$$

The internal structure of a composite particle is generally affected by its environment. Hence, the separation (2.35) is approximate. In different layers of the star, it may be applied to different plasma constituents (atoms, molecules, ions, and electrons in the atmosphere; nuclei, free baryons, and leptons in the inner regions). In some layers, one can neglect internal quantum degrees of freedom of composite particles (then $F_{\rm int}$ is dropped) or non-ideality effects (then only $F_{\rm trans}$ is retained).

The pressure $P$, the internal energy $U$, and the entropy $S$ can be obtained using the thermodynamic relations (e.g., Landau & Lifshitz 1993)

$$P = -\left(\frac{\partial F}{\partial V}\right)_{T,\{N_j\}}, \quad U = \left(\frac{\partial (F/T)}{\partial (1/T)}\right)_{V,\{N_j\}}, \quad S = -\left(\frac{\partial F}{\partial T}\right)_{V,\{N_j\}}. \tag{2.36}$$

Whenever the excess part of the free energy vanishes at high $T$ (which is usually the case), it can be obtained from the internal energy by integrating the second equation,

$$F_{\rm ex}(\Gamma, r_s) = \int_0^\Gamma \frac{U_{\rm ex}(\Gamma', r_s)}{\Gamma'}\,\mathrm{d}\Gamma'. \tag{2.37}$$

Second-order thermodynamic quantities are derived by differentiating first-order ones. These are, for example, the logarithmic derivatives of the pressure,

$$\chi_T = (\partial \ln P/\partial \ln T)_V, \quad \chi_\rho = -(\partial \ln P/\partial \ln V)_T, \tag{2.38}$$

and the heat capacities at constant volume or at constant pressure,

$$C_V = (\partial S/\partial \ln T)_V, \quad C_P = (\partial S/\partial \ln T)_P. \tag{2.39}$$

An important astrophysical quantity is the adiabatic gradient

$$\nabla_{\rm ad} = (\partial \log T/\partial \log P)_S, \tag{2.40}$$

that determines the *Schwarzschild condition* for convective stability (Schwarzschild, 1958); a layer can be convectively unstable if

$$\mathrm{d}\ln T/\,\mathrm{d}\ln P > \nabla_{\rm ad}. \tag{2.41}$$

Note that the partial derivatives in Eqs. (2.38)–(2.40) imply a free adjustment of the equilibrium particle numbers $N_j$ to changing conditions. In contrast, the derivatives in Eq. (2.36) assume that these numbers are kept fixed.

From Eqs. (2.38)–(2.40), using the Maxwell relations (e.g., Landau & Lifshitz 1993), we get

$$C_P = C_V + \frac{PV}{T}\frac{\chi_T^2}{\chi_\rho}, \tag{2.42}$$

$$\nabla_{\rm ad} = \frac{\chi_T}{\chi_T^2 + \chi_\rho C_V T/(PV)}. \tag{2.43}$$

If the free energy is known exactly, different ways to determine second-order quantities [e.g., Eq. (2.40) and Eq. (2.43) for $\nabla_{\rm ad}$] are equivalent. In numerical models, however, this is not always true. If a numerical EOS does not satisfy the Maxwell relations in some region of the phase diagram, it is called *thermodynamically inconsistent* in this region. A check for the thermodynamic consistency is an important probe for the accuracy of an EOS (e.g., Saumon *et al.* 1995).

Plasma physics studies are facilitated by the fact that ions are much heavier than electrons. This justifies the *adiabatic (Born-Oppenheimer) approximation*: a configuration of electrons is assumed to be instantaneously adjusted to a given ion configuration, so that the heavy particles (ions, atoms, and molecules) interact via effective (electron-screened) potentials. This approximation will be adopted throughout this chapter.

## 2.2.1 Fully ionized dense plasma. The basic decomposition

Neutron star envelopes are mostly fully ionized. This regime can be defined by the requirement that the spacing $a_{\rm i}$ between ions is small compared to the Thomas-Fermi radius of the atomic core, $r_a \sim a_0/Z^{1/3}$, which leads to the condition $\rho \gg \rho_{\rm eip} = (m_{\rm u}/a_0^3)AZ \approx 11\,AZ$ g cm$^{-3}$ (e.g., Pethick & Ravenhall 1995). At $\rho \gg \rho_{\rm eip}$, a model of an *electron-ion plasma* (eip) of bare pointlike nuclei with the charge $Z = Z_{\rm nuc}$ on an electron-liquid background can provide a good approximation for thermodynamic functions. Generally, the background is compressible, and Coulomb correlation of ions and electrons should be taken into account. The excess free energy $F_{\rm ex}$ in Eq. (2.35) can be represented as a sum of the ionic Coulomb $F_{\rm ii}$, the electronic exchange-correlation $F_{\rm xc}$, and the ion-electron $F_{\rm ie}$ parts. Thus

$$F = F_{\rm id}^{(\rm i)} + F_{\rm id}^{(e)} + F_{\rm xc} + F_{\rm ii} + F_{\rm ie}. \tag{2.44}$$

To avoid misunderstandings let us stress that the decomposition of $F$ is not unique. In $F_{\rm ii}$ we include the Coulomb interaction between ions as well as the interaction between ions and electrons and between electrons in the approximation of rigid electron background (this uniform negative background compensates the part of $F_{\rm ii}$ that goes to infinity in the thermodynamic limit $N_{\rm i} \to \infty$). Accordingly, $F_{\rm ie}$ includes the interaction between the ions and *polarized* electron background, associated with *deviations* from the rigid background. The term $F_{\rm xc}$ includes the exchange and correlation corrections in a purely electron subsystem.

The free-energy decomposition induces decompositions of other thermodynamic quantities. For instance, the pressure can be presented as

$$P = P_{\rm id}^{(\rm i)} + P_{\rm id}^{(e)} + P_{\rm xc} + P_{\rm ii} + P_{\rm ie}. \tag{2.45}$$

To guide the reader through a subsequent analysis of the different terms, in Table 2.1 we give order-of-magnitude estimates of the various pressure components in a strongly degenerate electron gas ($T \ll T_{\rm F}$) and fully ionized ($\rho \gg \rho_{\rm eip}$), strongly coupled ($T \lesssim T_{\rm l}$) ion system.

*Table 2.1.* Order-of-magnitude of the pressure components in Eq. (2.45) for the matter containing strongly degenerate electrons and fully ionized, strongly coupled ions (after Yakovlev & Shalybkov 1989). The parameter $\alpha_B = \alpha_f/\beta_r$ is small at $\rho \gg 1$ g cm$^{-3}$.

| $P_{\rm part}$ | $\lvert P_{\rm part}\rvert / P_{\rm id}^{(e)}$ | Comment |
|---|---|---|
| $P_{\rm id}^{(e)}$ | 1 | Ideal degenerate electron gas, leading term |
| $P_{\rm xc}$ | $\lesssim \alpha_B$ | Exchange-correlation corrections in electron gas |
| $P_{\rm id}^{(i)}$ | $\sim T/(ZT_F) \sim \alpha_B Z^{2/3}/\Gamma$ | Ideal ion gas contribution |
| $P_{\rm ii}$ | $\sim \alpha_B Z^{2/3}$ | Coulomb corrections in the rigid electron background |
| $P_{\rm ie}$ | $\sim \alpha_B^2 Z^{4/3}$ | Coulomb corrections owing to electron polarization |

The leading contribution comes from the pressure $P_{\rm id}^{(e)}$ of the ideal degenerate electron gas. Other partial pressures are smaller; their ratios to $P_{\rm id}^{(e)}$ contain the parameter $\alpha_B = \alpha_f/\beta_r = e^2/(\hbar v_F)$. It is essentially the fine-structure constant in the ultrarelativistic electron gas, but it grows with decreasing density in the non-relativistic electron gas. The parameter $\alpha_B$ can be regarded as a small parameter of the theory. Its smallness determines the smallness of partial pressures in comparison with $P_{\rm id}^{(e)}$. Note that contributions of the different terms in second-oder thermodynamic quantities can be different than in the pressure. For instance, we will see (§ 2.4.6) that the heat capacity can be determined by ions (by the ii term). In any case the exchange-correlation corrections (xc) and polarization corrections (ei) under the formulated conditions remain to be small. Therefore, we shall drop these corrections in § 2.3. However, because the parameter $\alpha_B$ ceases to be small in a low-density matter (surface layers of neutron stars; partial or removed electron degeneracy; incomplete ionization), all decomposition terms may become important. For this reason we will study the exchange-correlation and polarization corrections in §§ 2.4 and 2.5.

## 2.3. Fully ionized plasma without electron correlations

If the density is so high that the spacing between electrons is considerably smaller than the $K$-shell radius of an atom, one has a fully ionized plasma of ions immersed in the "rigid" electron background. Then the electron-ion correlations can be neglected. This regime operates at $r_s \lesssim 1/(2Z)$, or $\rho \gtrsim \rho_{\rm rigid} \approx 22\, Z^2 A$ g cm$^{-3}$. The same condition can also be obtained from the requirement for a typical attractive electron-ion Coulomb potential per electron $\sim Ze^2/a_i$ to be much smaller than the electron Fermi energy $\sim \hbar^2/(m_e a_e^2)$. In this section, we assume that the latter condition is fulfilled and consider the plasma of ions immersed in the ideal electron gas.

## 2.3.1 Ideal electron gas. Fermi-Dirac integrals

### 2.3.1 a Ideal electron gas

The electrons in thermodynamic equilibrium are described by the Fermi-Dirac distribution function

$$f^{(0)}(\epsilon - \mu_e, T) \equiv \frac{1}{\exp[(\epsilon - \mu_e)/k_\mathrm{B}T] + 1}, \tag{2.46}$$

where

$$\epsilon = \sqrt{m_e^2 c^4 + p^2 c^2} \tag{2.47}$$

is the electron energy and $p$ is the momentum. It is also convenient to introduce the quantity

$$\chi = (\mu_e - m_e c^2)/k_\mathrm{B}T. \tag{2.48}$$

The free energy of the electron gas can be found from the thermodynamic relation

$$F_\mathrm{id}^{(e)} = \mu_e N_e - P_\mathrm{id}^{(e)}\, V, \tag{2.49}$$

where $P_\mathrm{id}^{(e)}$ is the ideal-gas pressure. The pressure and the number density are, in turn, functions of $\mu_e$ and $T$:

$$\begin{aligned} P_\mathrm{id}^{(e)} &= 2k_\mathrm{B}T \int \ln\left[1 + \exp\left(\frac{\mu_e - \epsilon}{k_\mathrm{B}T}\right)\right] \frac{\mathrm{d}^3p}{(2\pi\hbar)^3} \\ &= \frac{8}{3\sqrt{\pi}} \frac{k_\mathrm{B}T}{\lambda_e^3} \left[I_{3/2}(\chi, t_\mathrm{r}) + \frac{t_\mathrm{r}}{2} I_{5/2}(\chi, t_\mathrm{r})\right], \end{aligned} \tag{2.50}$$

$$n_e = 2 \int f^{(0)}(\epsilon - \mu_e, T) \frac{\mathrm{d}^3p}{(2\pi\hbar)^3} = \frac{4}{\sqrt{\pi}\,\lambda_e^3} \left[I_{1/2}(\chi, t_\mathrm{r}) + t_\mathrm{r} I_{3/2}(\chi, t_\mathrm{r})\right], \tag{2.51}$$

where $\lambda_e$ is the electron thermal wavelength (2.27), and

$$I_\nu(\chi, \tau) \equiv \int_0^\infty \frac{x^\nu\, (1 + \tau x/2)^{1/2}}{\exp(x - \chi) + 1}\, \mathrm{d}x \tag{2.52}$$

is a Fermi-Dirac integral. The internal energy is given by

$$\frac{U_\mathrm{id}^{(e)}}{V} = \frac{4}{\sqrt{\pi}} \frac{k_\mathrm{B}T}{\lambda_e^3} \left[I_{3/2}(\chi, t_\mathrm{r}) + t_\mathrm{r}\, I_{5/2}(\chi, t_\mathrm{r})\right]. \tag{2.53}$$

### 2.3.1 b Fermi-Dirac integrals

In the limit $t_\mathrm{r} \to 0$, the Fermi-Dirac integrals (2.52) turn into the standard non-relativistic Fermi integrals $I_\nu(\chi)$, which can be calculated using highly accurate Padé approximations presented by Antia (1993) for $\nu = -\frac{1}{2}, \frac{1}{2}, \frac{3}{2}$, and $\frac{5}{2}$.

*Table 2.2.* Parameters of Eqs. (2.54) and (2.55). The powers of 10 are given in square brackets

| $i$ | 1 | 2 | 3 | 4 | 5 |
|---|---|---|---|---|---|
| $c_i^{(0)}$ | 0.370 450 57 | 0.412 584 37 | 0.097 779 82 | 5.373 4153 [−3] | 3.8746281 [−5] |
| $c_i^{(1)}$ | 0.396 031 09 | 0.694 687 95 | 0.223 227 60 | 1.526 2934 [−2] | 1.308 1939 [−4] |
| $c_i^{(2)}$ | 0.769 346 19 | 1.789 1437 | 0.707 549 74 | 5.675 5672 [−2] | 5.557 1480 [−4] |
| $\chi_i^{(0)}$ | 0.431 398 81 | 1.759 7537 | 4.104 4654 | 7.746 7038 | 13.457 678 |
| $\chi_i^{(1)}$ | 0.817 631 76 | 2.4723339 | 5.116 0061 | 9.044 1465 | 15.049 882 |
| $\chi_i^{(2)}$ | 1.255 8461 | 3.2070406 | 6.123 9082 | 10.316 126 | 16.597 079 |
| $x_i$ | 0.072 653 51 | 0.2694608 | 0.533 122 | 0.786 8801 | 0.956 9313 |
| $\xi_i$ | 0.263 560 32 | 1.413 4031 | 3.596 4258 | 7.085 8100 | 12.640 801 |
| $h_i$ | 0.038 187 35 | 0.125 6732 | 0.198 6308 | 0.197 6334 | 0.106 5420 |
| $v_i$ | 0.295 058 69 | 0.320 648 56 | 0.073 915 57 | 3.608 7389 [−3] | 2.336 9894 [−5] |

The accuracy of non-relativistic formulae decreases rapidly as $T$ goes above $10^7$ K. Fortunately, there exist fully relativistic piecewise fitting formulae for each of the integrals $I_\nu(\chi, t_\mathrm{r})$ with $\nu = \frac{1}{2}, \frac{3}{2}$, and $\frac{5}{2}$ (Blinnikov *et al.*, 1996; Chabrier & Potekhin, 1998):

$$I_{k+1/2}(\chi \le 0.6, t_\mathrm{r}) = \sum_{i=1}^{5} c_i^{(k)} \frac{\left(1 + \chi_i^{(k)} t_\mathrm{r}/2\right)^{1/2}}{\exp(-\chi_i^{(k)}) + \exp(-\chi)}, \tag{2.54}$$

$$\begin{aligned} I_{k+1/2}(0.6 < \chi < 14, t_\mathrm{r}) = \sum_{i=1}^{5} \Bigg[ & h_i x_i^k \frac{\chi^{k+3/2} \left(1 + \chi x_i t_\mathrm{r}/2\right)^{1/2}}{1 + \exp(\chi x_i - \chi)} \\ & + v_i (\xi_i + \chi)^{k+1/2} \left(1 + (\xi_i + \chi) t_\mathrm{r}/2\right)^{1/2} \Bigg], \end{aligned} \tag{2.55}$$

$$I_{k+1/2}(\chi \ge 14, t_\mathrm{r}) = J_k(\chi, t_\mathrm{r}) + \frac{\pi^2}{6} \chi^k \frac{k + 1/2 + (k+1)\chi t_\mathrm{r}/2}{R}, \tag{2.56}$$

where $R \equiv \chi^{1/2}(1 + \chi t_\mathrm{r}/2)^{1/2}$,

$$J_0(\chi, t_\mathrm{r}) = (\chi + t_\mathrm{r}^{-1}) R/2 - (2t_\mathrm{r})^{-3/2} \ln(1 + t_\mathrm{r}\chi + \sqrt{2t_\mathrm{r}}\, R), \tag{2.57}$$

$$J_1(\chi, t_\mathrm{r}) = [2\,R^3/3 - J_0(\chi, t_\mathrm{r})]/t_\mathrm{r}, \tag{2.58}$$

$$J_2(\chi, t_\mathrm{r}) = [2\,\chi\,R^3 - 5J_1(\chi, t_\mathrm{r})]/(4t_\mathrm{r}). \tag{2.59}$$

If $\mu_e/(m_e c^2) - 1 \equiv \chi t_\mathrm{r} \ll 1$, then it is advisable to use the non-relativistic limits, $J_k \to \chi^{k+3/2}/(k+3/2)$. The constants $c_i^{(k)}$, $\chi_i^{(k)}$, $x_i$, $\xi_i$, $h_i$, and $v_i$ are listed in Table 2.2. Relative error of Eqs. (2.54)–(2.56) does not exceed 0.2% at $T < 10^{11}$ K (any $\rho$), being typically a few parts in $10^4$.

*Table 2.3.* Parameters of Eq. (2.61)

| $i$ | 1 | 2 | 3 | 4 | 5 |
|---|---|---|---|---|---|
| $a_{\rm i}^{(1)}$ | 44.593 646 | 11.288 764 | 39.519 346 | −5.751 746 4 | 0.265 942 91 |
| $a_{\rm i}^{(2)}$ | 34.873 722 | −26.922 515 | 26.612 832 | −20.452 930 | 11.808 945 |

### 2.3.1 c The electron chemical potential

The chemical potential can be found numerically from Eq. (2.51), using Eqs. (2.54)–(2.56). In the nonrelativistic regime, it can also be found from the equation

$$\chi = X_{1/2}\left(\frac{2}{3}\,\theta^{-3/2}\right), \tag{2.60}$$

where $\theta$ is the degeneracy parameter and $X_\nu(I)$ is the inverse function to the Fermi integral, also fitted by Antia (1993) for $\nu = -\frac{1}{2}, \frac{1}{2}, \frac{3}{2}$, and $\frac{5}{2}$. The lowest-order (but still accurate within 0.003%) fit to the function in Eq. (2.60) reads

$$X_{1/2}(z) = \begin{cases} \ln(z\,R^{(1)}(z)) & \text{for} \quad z < 4 \\ z^{2/3}\,R^{(2)}(z^{-2/3}) & \text{otherwise,} \end{cases} \tag{2.61}$$

where

$$R^{(n)}(z) \equiv \frac{a_1^{(n)} + a_2^{(n)} z + z^2}{a_3^{(n)} + a_4^{(n)} z + a_5^{(n)} z^2},$$

and the coefficients $a_{\rm i}^{(n)}$ are given in Table 2.3.

For arbitrary relativism, the following analytic fit can be used (Chabrier & Potekhin, 1998):

$$\chi = \chi_{\rm nonrel} - \frac{3}{2}\ln\left[1 + \frac{t_{\rm r}\,(1 + q_1\sqrt{t_{\rm r}} + q_2 q_3 t_{\rm r})}{[1 + t_{\rm r}/(2\theta)](1 + q_2 t_{\rm r})}\right]. \tag{2.62}$$

Here, $\chi_{\rm nonrel}$ is given by the non-relativistic formula (2.60), and the coefficients $q_i$ are functions of $\theta$:

$$q_1 = \frac{3}{2}(\mathrm{e}^{\theta} - 1)^{-1}, \qquad q_2 = 12 + 8\theta^{-3/2}, \tag{2.63}$$

$$q_3 = \frac{2}{\pi^{1/3}} - \frac{\mathrm{e}^{-\theta} + 1.612\,\mathrm{e}^{\theta}}{6.192\,\theta^{0.0944}\,\mathrm{e}^{-\theta} + 5.535\,\theta^{0.698}\,\mathrm{e}^{\theta}}. \tag{2.64}$$

The relative error $\delta\chi/\chi$ turns to infinity at $\chi = 0$. However, since thermodynamic quantities are expressed through $\chi$ by virtue of thermal averaging [similar to that given by Eq. (2.52)], a natural measure of the relative error of the fit is $\delta\chi/\max(|\chi|, 1)$. Thus determined, the error lies within 0.4% (reached in the range $0.7 < \theta < 1.5$, $T > T_{\rm r}$), and it is smaller than 0.2% for $T < T_{\rm r}$. Another measure of the inaccuracy is the relative difference between the densities $n_e$ calculated with the exact and fitted values of $\mu_e$ in Eq. (2.51). This difference lies within 0.4% at large $T > T_{\rm r}$ and within 0.1% at $T \leq T_{\rm r}$.

### 2.3.1 d The limits of nondegenerate and strongly degenerate electrons

In the limiting case of nonrelativistic classical Boltzmann electron gas ($T \gg T_{\rm F}$) we have $P_{\rm id}^{(e)} = n_e k_{\rm B} T$ and $\chi = \ln(n_e \lambda_e^3/2)$ (the factor 2 takes account of the spin degeneracy).

In the opposite limiting case of very strong degeneracy, $T \ll T_{\rm F}$, the Fermi-Dirac integrals are given by Eqs. (2.54)–(2.59). It is important to note that in this case first-order thermodynamic functions depend on $T$ very weakly. Therefore, the derivatives of $F_{\rm id}^{(e)}$ over $T$ cannot be evaluated numerically at $\theta \gg 1$ with high accuracy but should be calculated from direct analytic expressions. For example, it is disadvantageous to evaluate the internal energy numerically by taken derivative in Eq. (2.36); the explicit expression (2.53) should be used instead. It is useful to employ the Sommerfeld expansion in powers of $T^2$ in order to evaluate the thermodynamic quantities in the regime $\theta \gg 1$ (e.g., Yakovlev & Shalybkov 1989). To the first order, the free energy density is

$$\frac{F_{\rm id}^{(e)}}{V} = \frac{P_{\rm r}}{8\pi^2}\left[x_{\rm r}\,(1+2{x_{\rm r}}^2)\,\gamma_{\rm r} - \ln(x_{\rm r}+\gamma_{\rm r}) - \frac{4\pi^2}{3}\,t_{\rm r}^2\,x_{\rm r}\,\gamma_{\rm r}\right], \qquad (2.65)$$

where

$$P_{\rm r} \equiv m_e c^2/{\lambda\!\!\!^{-}_{\rm C}}^3 = 1.4218 \times 10^{25}\ {\rm dyn\ cm}^{-2} \qquad (2.66)$$

is the relativistic unit of the electron pressure, and $\lambda\!\!\!^{-}_{\rm C} \equiv \hbar/(m_e c) = 386.16$ fm is the electron Compton wavelength. The derivatives of Eq. (2.65) calculated using Eqs. (2.36) and (2.39) yield the pressure and specific heat at $T \ll T_{\rm F}$:

$$\begin{aligned} P_{\rm id}^{(e)} &= \frac{P_{\rm r}}{8\pi^2}\left[x_{\rm r}\left(\frac{2}{3}{x_{\rm r}}^2 - 1\right)\gamma_{\rm r} + \ln(x_{\rm r}+\gamma_{\rm r})\right. \\ &\quad \left. + \frac{4\pi^2}{9}\,t_{\rm r}^2\,x_{\rm r}\,(\gamma_{\rm r}+\gamma_{\rm r}^{-1})\right], \qquad (2.67) \\ C_V^{(e)} &= \pi^2 k_{\rm B}\,N_e\,\frac{\gamma_{\rm r} t_{\rm r}}{x_{\rm r}^2} = \frac{k_{\rm B}^2 T m_e^* p_{\rm F} V}{3\hbar^3}. \qquad (2.68) \end{aligned}$$

Equations (2.65) and (2.67) without thermal corrections (for $T = 0$) were obtained by Frenkel (1928), but his paper remained unnoticed by the astrophysical community. Later these equations were derived independently by Stoner (1930, 1932) and used by Chandrasekhar for constructing models of white dwarfs (e.g., Chandrasekhar 1939).

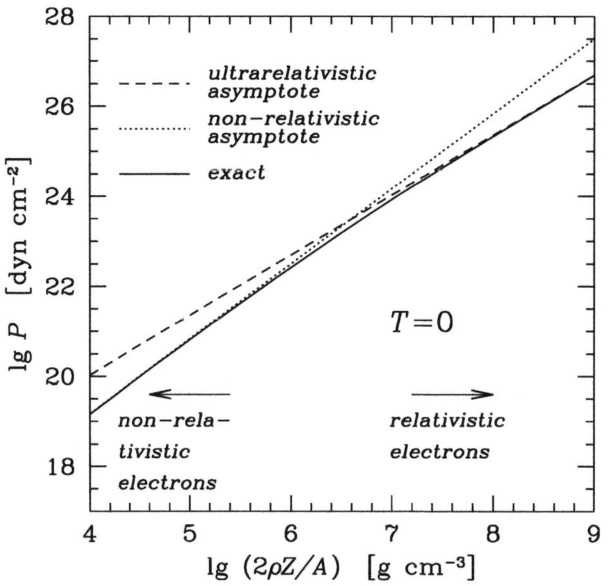


*Figure 2.3.* The electron pressure $P = P_{\rm id}^{(e)}$ versus density for an ideal strongly degenerate electron gas (the solid line) together with non-relativistic and ultrarelativistic asymptotes (the dotted and dashed lines).

### 2.3.1 e The electron pressure in outer neutron star envelopes and white dwarf cores

In a strongly degenerate electron gas the electron pressure $P_{\rm id}^{(e)}$ can be accurately approximated by Eq. (2.67) at $T = 0$. In this case $P_{\rm id}^{(e)}$ is determined only by the electron number density $n_e$. It makes the major contribution to the pressure in degenerate layers of outer envelopes of neutron stars and cores of white dwarfs (see page 66, Table 2.1). Therefore, the EOS and hydrostatic structure of these layers are determined by degenerate electrons. This is especially important for white dwarfs whose overall hydrostatic equilibrium is mainly supported by the electron pressure (e.g, Shapiro & Teukolsky 1983). This "universal" EOS is plotted in Fig. 2.3 as a function of $2\rho Z/A \propto n_e$ (let us remind that $A \approx 2Z$ for astrophysically important atomic nuclei heavier than H).

In the limit of non-relativistic and ultrarelativistic degenerate electrons $P_{\rm id}^{(e)}$ takes the *polytropic form*

$$P = K\rho^{\gamma_{\rm ad}} = K\rho^{1+1/n}, \tag{2.69}$$

where $K$ is a constant, $\gamma_{\rm ad}$ is the adiabatic index and $n$ is the polytropic index (see, e.g., Shapiro & Teukolsky 1983 for a description of polytropic stars).

More precisely,

$$P_{\rm id}^{(e)} \approx P_{\rm r} {x_{\rm r}}^{3\gamma_{\rm ad}}/(9\pi^2\gamma_{\rm ad}), \tag{2.70}$$

where one has $\gamma_{\rm ad} = 5/3$ ($n = 3/2$) for nonrelativistic electrons ($x_{\rm r} \ll 1$), and $\gamma_{\rm ad} = 4/3$ ($n = 3$) for ultrarelativistic electrons $x_{\rm r} \gg 1$ (recall that $x_{\rm r} \propto \rho^{1/3}$). In the latter case $P_{\rm id}^{(e)} \to (\hbar c/4)\,(3\pi^2 n_{\rm e}^4)^{1/3}$ becomes independent of the electron mass. In the mildly relativistic gas ($x_{\rm r} \sim 1$, $\rho \sim 10^6$ g cm$^{-3}$) we have $P_{\rm id}^{(e)} \sim 10^{23}$ dyn cm$^{-2}$ (seven orders of magnitude higher than the pressure in the center of the Sun).

From Fig. 2.3 we see that when the density grows up and the non-relativistic regime ($x_{\rm r} \ll 1$) is replaced by the ultrarelativistic one ($x_{\rm r} \gg 1$), the EOS becomes softer. This is important (§ 6.9) for the hydrostatic and thermal structure of outer neutron star envelopes. This softening is even more important for white dwarfs (§ 6.5.3) because it determines the Chandrasekhar limit of their mass (e.g, Shapiro & Teukolsky 1983).

## 2.3.2 Coulomb liquid of ions

### 2.3.2 a Theory

The free energy of an *ideal* gas of $N = N_{\mathcal{N}}$ non-relativistic classical ions is

$$F_{\rm id}^{({\rm i})} = N k_{\rm B} T \left[\ln(n_{\mathcal{N}} \lambda_{\rm i}^3/g_{\rm i}) - 1\right], \tag{2.71}$$

where $g_{\rm i}$ is the spin degeneracy (obviously, the rest-mass energy is not included here). The internal energy and pressure contributions from the ideal gas of ions are simply

$$U_{\rm id}^{({\rm i})} = \frac{3}{2} N k_{\rm B} T, \quad P_{\rm id}^{({\rm i})} = n_{\mathcal{N}} k_{\rm B} T. \tag{2.72}$$

A study of the *excess* terms is more complicated. We mark these terms (which arise from the Coulomb interactions assuming a rigid electron background, § 2.2) by the subscript "ii." In a classical OCP the quantity $F_{\rm ii}/N k_{\rm B} T$ is a function of a single argument $\Gamma$. This function determines all other excess quantities which can also be expressed via some universal functions of $\Gamma$ (e.g., Yakovlev & Shalybkov 1989).

In the weak coupling regime ($\Gamma \ll 1$), the theory of Debye & Hückel (1923) is applicable. The Debye-Hückel formula for the free energy of a mixture of non-relativistic weakly coupled ions with charges $Z_j e$ and number densities $n_j$ reads (e.g., Landau & Lifshitz 1993, § 78)

$$\frac{F_{\rm ex}(\Gamma \to 0)}{V} = -\frac{2e^3}{3} \left(\frac{\pi}{k_{\rm B} T}\right)^{1/2} \left(\sum_j n_j Z_j^2\right)^{3/2} . \tag{2.73}$$

For the OCP, it gives

$$U_{\rm ii,DH} = -(\sqrt{3}/2)\, N k_{\rm B} T\, \Gamma^{3/2}, \quad F_{\rm ii,DH} = -(1/\sqrt{3})\, N k_{\rm B} T\, \Gamma^{3/2}. \qquad (2.74)$$

Abe (1959) used the perturbation theory to obtain corrections of the order $\Gamma^3$ to Eq. (2.74). The next correction of the order $\Gamma^{9/2}$ was calculated by Cohen & Murphy (1969). The low-$\Gamma$ expansion of the internal energy reads

$$\begin{aligned} \frac{U_{\rm ii}}{k_{\rm B}TN} &= -\frac{\sqrt{3}}{2}\,\Gamma^{3/2} - 3\Gamma^3 \left[\frac{3}{8}\ln(3\Gamma) + \frac{C_E}{2} - \frac{1}{3}\right] \\ &\quad -\Gamma^{9/2}(1.6875\,\sqrt{3}\,\ln\Gamma - 0.23511) + \ldots, \end{aligned} \qquad (2.75)$$

where $C_E = 0.57721\ldots$ is the Euler's constant. Accordingly, Eq. (2.37) yields

$$\begin{aligned} \frac{F_{\rm ii}}{k_{\rm B}TN} &= -\frac{\Gamma^{3/2}}{\sqrt{3}} - \Gamma^3 \left(\frac{3}{8}\ln\Gamma + 0.24225\right) \\ &\quad -\Gamma^{9/2}(0.64952\,\ln\Gamma - 0.19658) + \ldots \end{aligned} \qquad (2.76)$$

Liquid and solid phases of the OCP of ions at large $\Gamma$ were extensively studied by various numerical methods (for reviews, see Baus & Hansen 1980; Ichimaru *et al.* 1987). The most efficient of them proved to be the Monte Carlo method (Brush *et al.*, 1966) and the family of techniques based on the so-called hypernetted chain (HNC) equations (e.g., Hansen & McDonald 1976).

Monte Carlo calculations start from some initial configuration of $N$ ions and allow subsequent ion displacements. The potential energy $\mathcal{U}(\boldsymbol{r}_1, \ldots, \boldsymbol{r}_N)$ of the Coulomb interaction of ions with other ions and with the uniform electron background is calculated at every step and contributes (with a proper weight) to the excess internal energy

$$U_{\rm ii} = \frac{1}{V^N \mathcal{Z}_{\rm conf}} \int \mathcal{U} \exp\left(-\frac{\mathcal{U}}{k_{\rm B}T}\right) \prod_{q=1}^{N} {\rm d}\boldsymbol{r}_q, \qquad (2.77)$$

where

$$\mathcal{Z}_{\rm conf} = \frac{1}{V^N} \int \exp\left(-\frac{\mathcal{U}}{k_{\rm B}T}\right) \prod_{q=1}^{N} {\rm d}\boldsymbol{r}_q \qquad (2.78)$$

is the configurational part of the partition function.

The HNC method is based on integral equations for the radial pair distribution function

$$g(|\boldsymbol{r}_1 - \boldsymbol{r}_2|) = \frac{N(N-1)}{N^2 \mathcal{Z}_{\rm conf} V^{N-2}} \int \exp\left(-\frac{\mathcal{U}}{k_{\rm B}T}\right) \prod_{q=3}^{N} {\rm d}\boldsymbol{r}_q, \qquad (2.79)$$

derived in various approximations. The classical HNC approximation consists in neglecting the contribution of the so-called bridge diagrams (which involve an infinite series of multiple integrals) in the $N$-body general diagrammatic resummations (Hansen & McDonald, 1976). The pair correlation function $g(r)$ is of much shorter range than the Coulomb potential $V(r)$ (Hansen & McDonald, 1976). The HNC theory is particularly suitable for such long-range systems. Its resulting inaccuracies in the free energy, the internal energy, and the pressure are at most of the order of 1% (see, e.g., DeWitt *et al.* 1996; Chabrier 1990).

The advantage of the HNC technique is that it requires less computer resources and directly provides the free energy along with the internal energy. Its disadvantage is the intrinsic inaccuracy owing to the approximations in the HNC equations.

The ion correlations in momentum space are described by the dynamical ion structure factor

$$S(\boldsymbol{q},\omega) = \frac{1}{2\pi N}\int_{-\infty}^{+\infty} \mathrm{d}t \int \mathrm{d}\boldsymbol{x}\,\mathrm{d}\boldsymbol{x}'\, e^{i\boldsymbol{q}\cdot(\boldsymbol{x}-\boldsymbol{x}')-i\omega t} \left\langle \tilde{n}(\boldsymbol{x},t)\,\tilde{n}(\boldsymbol{x}',0)\right\rangle_{\mathrm{th}}, \tag{2.80}$$

where

$$\tilde{n}(\boldsymbol{x},t) = \sum_{\boldsymbol{r}_q} \delta(\boldsymbol{x}-\boldsymbol{r}_q) - n_N$$

is an instant charge density in units of $Ze$, and $\langle\ldots\rangle_{\mathrm{th}}$ denotes the average over the Gibbs ensemble of ions (which will be called the *thermal* or *thermodynamic average*). It is often useful to introduce the *static structure factor*

$$S(\boldsymbol{q}) = \int_{-\infty}^{+\infty} S(\boldsymbol{q},\omega)\,\mathrm{d}\omega. \tag{2.81}$$

In this chapter we assume that the plasma is isotropic (in particular, there is no macroscopic magnetic field). Then the static structure factor can be written in the form

$$S(q) = 1 + \frac{3\Gamma}{a_{\mathrm{i}}^3}\int_0^{\infty} \frac{\sin(qr)}{qr}\,[g(r)-1]\,r^2\,\mathrm{d}r. \tag{2.82}$$

In the limit $q \to 0$, one has $S(q) \to (qa_{\mathrm{i}})^2/(3\Gamma)$ (e.g., Hansen 1973). The Debye-Hückel screening corresponds to $S(q) = q^2/(q^2 + 3\Gamma a_{\mathrm{i}}^{-2})$. Generally, $S(q)$ is calculated directly from Eq. (2.80), for instance, by the Monte Carlo method (Hansen, 1973; Itoh *et al.*, 1983) or by the HNC technique. A very useful analytic fit to OCP $S(q)$ was derived by Young *et al.* (1991).

The contribution of the ion-ion interactions into the internal energy can be obtained from $g(r)$ or $S(q)$ as

$$U_{\mathrm{ii}} = \frac{N^2}{2V}\int \frac{(Ze)^2}{r}[g(r)-1]\,\mathrm{d}^3r = Nk_{\mathrm{B}}T\,\frac{\Gamma a_{\mathrm{i}}}{\pi}\int_0^{\infty}[S(q)-1]\,\mathrm{d}q. \tag{2.83}$$

Accurate Monte Carlo and HNC results for the internal and free energies of the OCP of ions at $1 \leq \Gamma \leq 160$ were presented by DeWitt *et al.* (1996) (see references therein for earlier results).

The *quantum corrections* make the thermodynamic functions of the OCP dependent not only on $\Gamma$ but also on $r_s$. The lowest-order quantum correction to the Helmholtz free energy is given by the Wigner formula (e.g., Landau & Lifshitz 1993):

$$F_{\mathrm{q}} = \frac{\hbar^2}{24(k_{\mathrm{B}}T)^2} \sum_{j=1}^{N} \frac{1}{m_{\mathrm{i}}} \left\langle \left( \frac{\partial \mathcal{U}}{\partial \boldsymbol{r}_j} \right)^2 \right\rangle_{\mathrm{th}} . \tag{2.84}$$

For the OCP of ions in the rigid electron background this averaging gives (Hansen, 1973)

$$F_{\mathrm{q}} = \frac{\pi \hbar^2 N n_{\mathcal{N}} (Ze)^2}{6\, m_{\mathrm{i}} k_{\mathrm{B}} T} = N \frac{k_{\mathrm{B}} T_{\mathrm{pi}}^2}{24\, T} . \tag{2.85}$$

This is the so called *quantum diffraction* or *Wigner-Kirkwood* term in the free energy of the OCP. It is the same for the liquid and solid phases.

Quantum effects can be studied numerically by the path-integral Monte Carlo technique (see, e.g., Jones & Ceperley 1996; also see Militzer & Pollock 2005 and reference therein) which requires extensive computer resources.

### 2.3.2 b Analytic formulae and discussion

The internal energy of the Coulomb liquid ($\Gamma \leq \Gamma_{\mathrm{m}}$) calculated by DeWitt *et al.* (1996) can be fitted by

$$\frac{U_{\mathrm{ii}}}{k_{\mathrm{B}} T N_{\mathcal{N}}} = \Gamma^{3/2} \left( \frac{A_1}{\sqrt{A_2 + \Gamma}} + \frac{A_3}{1 + \Gamma} \right) + \frac{B_1 \Gamma^2}{B_2 + \Gamma} + \frac{B_3 \Gamma^2}{B_4 + \Gamma^2}, \tag{2.86}$$

where $A_3 = -\sqrt{3}/2 - A_1/\sqrt{A_2}$. The best-fit parameters are $A_1 = -0.9070$, $A_2 = 0.62954$, $B_1 = 0.00456$, $B_2 = 211.6$, $B_3 = -0.0001$, and $B_4 = 0.00462$. The difference between the values given by Eq. (2.86) and by DeWitt *et al.* (1996) is smaller than 5 parts in $10^5$. In addition, Eq. (2.86) reproduces (within 0.03%) numerical values provided by Eq. (2.75) for $\Gamma < 0.1$. Figure 2.4 compares the fit (2.86) with the Monte Carlo results (DeWitt & Slattery, 1999), HNC calculations (Chabrier & Potekhin, 1998) and the Debye-Hückel limit (2.75) including the Abe and Cohen-Murphy corrections.

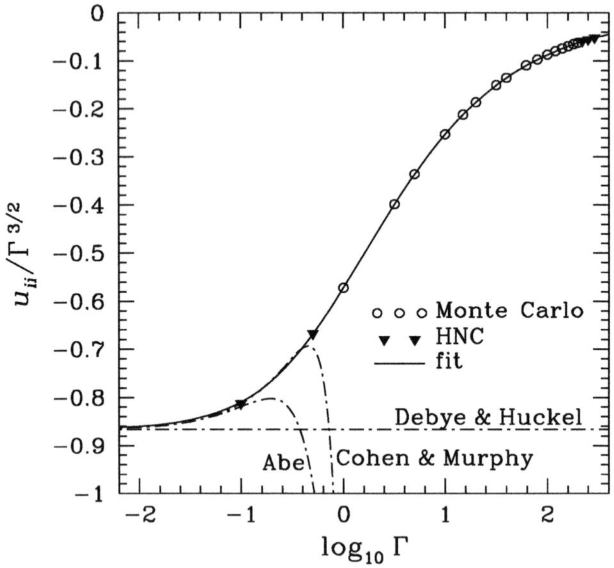


*Figure 2.4.* Comparison of the fit (2.86) with the Debye-Hückel, Abe, and Cohen-Murphy asymptotes at small $\Gamma$ and with Monte Carlo and HNC calculations.

The integration of Eq. (2.86) over $\ln\Gamma$ according to Eq. (2.37) yields the excess free energy of the Coulomb liquid:

$$\begin{aligned} \frac{F_{\mathrm{ii,liq}}}{N_{\mathcal{N}}k_{\mathrm{B}}T} &= A_1\left[\sqrt{\Gamma(A_2+\Gamma)} - A_2\ln\left(\sqrt{\Gamma/A_2}+\sqrt{1+\Gamma/A_2}\right)\right] \\ &+ 2A_3\left[\sqrt{\Gamma} - \arctan\left(\sqrt{\Gamma}\right)\right] \\ &+ B_1\left[\Gamma - B_2\ln\left(1+\frac{\Gamma}{B_2}\right)\right] + \frac{B_3}{2}\ln\left(1+\frac{\Gamma^2}{B_4}\right). \end{aligned} \tag{2.87}$$

This formula approximates $F_{\mathrm{ii}}$ within 0.02% for any $\Gamma$ in the gaseous and liquid phases.

The corresponding contributions to the pressure and the heat capacity are

$$P_{\mathrm{ii}} = \frac{1}{3}\frac{U_{\mathrm{ii}}}{V}, \tag{2.88}$$

$$\begin{aligned} C_{V,\mathrm{ii}} &= N_{\mathcal{N}}k_{\mathrm{B}}\left[\frac{\Gamma^{3/2}}{2}\left(-\frac{A_1A_2}{(A_2+\Gamma)^{3/2}} + \frac{A_3\,(\Gamma-1)}{(1+\Gamma)^2}\right)\right. \\ &\left. -\frac{B_1B_2\Gamma^2}{(B_2+\Gamma)^2} + \frac{B_3(\Gamma^4-B_4\Gamma^2)}{(B_4+\Gamma^2)^2}\right]. \end{aligned} \tag{2.89}$$

The Coulomb pressure $P_{\rm ii}$ is exactly one third of the Coulomb energy density, because $U_{\rm ii}/V$ depends on the single parameter $\Gamma \propto V^{1/3}/T$ (for a classical Coulomb plasma, § 2.3.2 a). The Coulomb pressure and energy are negative, because Coulomb interactions introduce an additional binding of ions.

In the gaseous regime the Coulomb energy is small, $|U_{\rm ii}| \ll N k_{\rm B} T$, so that the Coulomb binding is lower than the thermal energy of ions, and the ions are nearly free. In the strongly coupled liquid ($\Gamma \gtrsim 1$) one has $|U_{\rm ii}| \gtrsim N k_{\rm B} T$ and the Coulomb energy exceeds the thermal one. In this regime the ions are bound in Coulomb potential wells (whose depth is $\sim Z^2 e^2/a_{\rm i}$) and slightly oscillate there due to thermal motion. In a Coulomb liquid there is no exact order and the wells slowly migrate within the liquid. In the limit of $\Gamma \gg 1$ from Eq. (2.86) one obtains the linear $\Gamma$ dependence, $U_{\rm ii}/N k_{\rm B} T \approx A_1 \Gamma$. In this limit $U_{\rm ii} \approx N A_1 Z^2 e^2/a_{\rm i}$ and $P_{\rm ii} \approx A_1 n_{\mathcal{N}} Z^2 e^2/(3a_{\rm i})$ are nearly independent of temperature. The numerical factor $A_1$ is very close to the value $A_1^{\rm IS} = -0.9$ predicted by the ion-sphere model (§ 2.1.3). In that model the Coulomb energy of one ion sphere is equal to $-0.9 Z^2 e^2/a_{\rm i}$, which includes the electron-electron interaction, $0.6 Z^2 e^2/a_{\rm i}$, and the electron-ion interaction, $-1.5 Z^2 e^2/a_{\rm i}$ (e.g., Shapiro & Teukolsky 1983).

## 2.3.3 Coulomb crystal

### 2.3.3 a Basic concepts

Below the melting temperature, at $T < T_{\rm m}$, an infinite ion motion is replaced by oscillations near equilibrium positions, which means that a crystal is formed. Let us recall the basic concepts of the physics of crystals (e.g., Kittel 1986). A three-dimensional crystal lattice can be mapped onto itself through a translation by a vector $\boldsymbol{T} = l_1 \boldsymbol{a}_1 + l_2 \boldsymbol{a}_2 + l_3 \boldsymbol{a}_3$, where $\boldsymbol{a}_j$ are called *primitive translation vectors* and $l_j$ are arbitrary integers. The lattice can be figured out as composed of *primitive cells* that fill all space by repetition of translations $\boldsymbol{T}$. A primitive cell centered at an equilibrium ion position is a *Wigner-Seitz cell*. It is often convenient to consider a *reciprocal* lattice, defined by the primitive vectors $\boldsymbol{b}_l$ such that $\boldsymbol{b}_l \cdot \boldsymbol{a}_j = 2\pi\delta_{lj}$. The reciprocal lattice is mapped on itself by a *reciprocal lattice vector* $\boldsymbol{G} = l_1 \boldsymbol{b}_1 + l_2 \boldsymbol{b}_2 + l_3 \boldsymbol{b}_3$, where $l_j$ are again some integers. A Wigner-Seitz cell for the reciprocal lattice is known as a *Brillouin zone*. Its volume equals $(2\pi)^3/n_{\mathcal{N}}$.

The ground state of the OCP of ions corresponds to the body-centered cubic (bcc) lattice. In a real crystal, this happens at high densities (at $r_s \ll 1$), whereas other types of lattices (face-centered cubic, fcc; hexagonal close-packed, hcp) may form the ground state at $r_s \gtrsim 1$ (e.g., Kohanoff & Hansen 1996). In the following, we focus on the cubic lattices (bcc and fcc). Note that the simple cubic Coulomb lattice is unstable (as any simple cubic lattice of particles interacting via central forces; Born 1940).

In this section we describe thermodynamic functions of Coulomb crystals. Another important aspect is the elasticity of crystallized matter. Although elastic properties of the neutron star crust are fully determined by Coulomb interactions, they will be considered in § 3.7 of Chapter 3.

While studying strongly coupled Coulomb systems, it is often very useful to approximate the actual Wigner-Seitz cell by a sphere of the same volume. In simple lattices, the radius of the Wigner-Seitz cell is equal to the ion-sphere radius $a_{\rm i}$ [see Eq. (2.23)]. The same approximation in reciprocal lattice results in a spherical Brillouin zone with an equivalent radius

$$q_{\rm BZ} = (6\pi n_{\mathcal N})^{1/3}. \tag{2.90}$$

### 2.3.3 b Harmonic approximation

Let us outline the standard method to study ion vibrations (e.g., Kittel 1963, 1986). Let the ions oscillate near equilibrium positions $\boldsymbol{R}_j$. An ion position can be written as $\boldsymbol{r}_j = \boldsymbol{R}_j + \boldsymbol{u}_j$, where $\boldsymbol{u}_j$ is a small ion displacement. Since the ions are considered as pointlike, their *local* number density can be written as

$$n(\boldsymbol{x}, t) = \sum_j \delta(\boldsymbol{x} - \boldsymbol{R}_j - \boldsymbol{u}_j(t)). \tag{2.91}$$

Expanding the Coulomb energy of an ion (say, $j = 1$) in Taylor series with respect to $\boldsymbol{u}_j$ and keeping the terms up to $\boldsymbol{u}_j^2$, we get the Coulomb energy in the harmonic approximation. In particular, the potential energy increase due to the ion displacements is

$$\Delta E_1 = \frac{(Ze)^2}{2} \sum_{j=2}^{N_{\mathcal N}} \frac{3(\boldsymbol{u}_{1j} \cdot \boldsymbol{R}_j)^2 - \boldsymbol{u}_{1j}^2 R_j^2}{R_j^5} - \frac{2\pi}{3} n_{\mathcal N} (Ze)^2 \boldsymbol{u}_1^2, \tag{2.92}$$

where $\boldsymbol{u}_{1j} = \boldsymbol{u}_1 - \boldsymbol{u}_j$. The sum comes from the ion-ion interactions, while the last term comes from the ion interaction with the electron background. Taking ion displacements in the form

$$\boldsymbol{u}_j(t) = A \boldsymbol{e}_{\boldsymbol{k}s} \exp(\mathrm{i}\boldsymbol{k}\boldsymbol{R}_j - \mathrm{i}\omega_{\boldsymbol{k}s} t) \tag{2.93}$$

and using the Newtonian equation of ion motion, one obtains the dynamical equations (e.g., Coldwell-Horsfall & Maradudin 1960),

$$\omega_{\boldsymbol{k}s}^2 e_{\boldsymbol{k}s\alpha} = D_{\alpha\beta}(\boldsymbol{k}) e_{\boldsymbol{k}s\beta}, \tag{2.94}$$

where $\alpha$ and $\beta$ are Cartesian indices. These equations determine the eigenfrequencies $\omega_{\boldsymbol{k}s}$ and polarization vectors $\boldsymbol{e}_{\boldsymbol{k}s}$. Here, $s$ enumerates branches of crystal oscillation modes,

$$D_{\alpha\beta}(\boldsymbol{k}) = \frac{\omega_{\rm pi}^2}{3}\,\delta_{\alpha\beta} - \frac{\omega_{\rm pi}^2}{\pi} \sum_{\boldsymbol{l}\neq 0} {\rm e}^{{\rm i}\boldsymbol{k}\boldsymbol{R}_j}\, \frac{3l_\alpha l_\beta - l^2\delta_{\alpha\beta}}{l^5} \tag{2.95}$$

is the dynamical matrix, and $\boldsymbol{l}$ is a vector with integer components (either all odd or all even for the bcc lattice). The convergence of sums over lattice sites can be improved using the Ewald summation technique (Ewald, 1921). In the OCP model of a simple Coulomb crystal (OCP of ions on the rigid electron background), there are three oscillation branches, two transverse acoustic ones ($s = 1, 2$, with a linear dispersion relation $\omega(\boldsymbol{k}) \propto k$ at small wave numbers $k$) and one longitudinal optical branch ($s = 3$, whose frequency goes to $\omega_{\rm pi}$ at $k \to 0$). It is easy to see from Eqs. (2.94) and (2.95) that the eigenfrequencies obey the Kohn's sum rule

$$\sum_s \omega_{\boldsymbol{k}s}^2 = \omega_{\rm pi}^2. \tag{2.96}$$

Taking into account a finite compressibility of the ambient electron gas, one finds that this rule is not exact at small $k$. When $k$ decreases below $k_{\rm TF}$, the frequency of the vibration branch $s = 3$ tends to zero with a limiting dispersion law $\omega \to \omega_{\rm pi}\, k/k_{\rm TF}$ (Pollock & Hansen, 1973). Thus the branch $s = 3$ is not the true optical branch. It becomes acoustic at large wavelengths owing to a finite electron screening.

### 2.3.3c Quantization

In order to quantize ion motion, which is important at $T \lesssim T_{\rm pi}$, one has to consider lattice dynamics in terms of elementary quantum excitations – phonons (e.g., Landau & Lifshitz 1993; Kittel 1986). Accordingly, the local ion density given by Eq. (2.91) should be replaced by the density operator $\hat{\rho}$, which acts in space of quantum states characterized by phonon numbers; analogously, $\boldsymbol{u}_j$ must be replaced by the operator $\hat{\boldsymbol{u}}_j$ of ion displacement. Here we restrict ourselves to the most important case in which the effects of quantum statistics of ions (exchange corrections) can be neglected. In this case one should take into account the ion spin degeneracy but, otherwise, one can treat the ions as spinless particles [3].

The harmonic approximation proves to be very useful to describe quantum crystals. In this approximation (e.g., Landau & Lifshitz 1993) the free energy of the ion lattice is

$$F_{\rm i} = U_0 + 3N_{\mathcal{N}} k_{\rm B} T \left\langle \ln(1 - {\rm e}^{-z_{\boldsymbol{k}s}}) \right\rangle_{\rm ph}, \tag{2.97}$$

[3] Such particles are sometimes called "boltzmannons".

where

$$z_{\boldsymbol{k}s} \equiv \hbar\omega_{\boldsymbol{k}s}/k_{\rm B}T, \tag{2.98}$$

and $U_0$ is the zero-point lattice energy. The averaging $\langle\ldots\rangle_{\rm ph}$ is performed over phonon wave vectors $\boldsymbol{k}$ in the first Brillouin zone and over phonon polarizations $s$:

$$\langle f_s(\boldsymbol{k})\rangle_{\rm ph} = \frac{1}{3V_{\rm BZ}}\sum_s \int_{V_{\rm BZ}} {\rm d}\boldsymbol{k}\, f_s(\boldsymbol{k}), \tag{2.99}$$

$V_{\rm BZ} = \frac{4}{3}\pi q_{\rm BZ}^3$ being the volume of the Brillouin zone. The function $\langle\ldots\rangle_{\rm ph}$ depends on the only one parameter $t_{\rm p} = T/T_{\rm pi}$ and determines the temperature dependence of all thermodynamic functions. The zero-point energy consists of two parts,

$$U_0 = U_{\rm M} + U_{\rm q}, \tag{2.100}$$

where

$$U_{\rm M} = -N_{\mathcal N} C_{\rm M}(Ze)^2/a_{\rm i}, \qquad U_{\rm q} = \frac{3}{2} N_{\mathcal N}\hbar\omega_{\rm pi}u_1. \tag{2.101}$$

Here, $U_{\rm M}$ is the classical static-lattice energy, $C_{\rm M}$ is the *Madelung constant* (Table 2.4), $U_{\rm q}$ accounts for the zero-point quantum vibrations, and $u_1$ belongs to the family of average phonon frequency moments

$$u_p \equiv \langle\omega_{\boldsymbol{k}s}^p\rangle_{\rm ph}/\omega_{\rm pi}^p. \tag{2.102}$$

From Eq. (2.97), one obtains the internal energy

$$U_{\rm i} = U_0 + 3N_{\mathcal N}k_{\rm B}T\left\langle z_{\boldsymbol{k}s}({\rm e}^{z_{\boldsymbol{k}s}}-1)^{-1}\right\rangle_{\rm ph} \tag{2.103}$$

and the heat capacity

$$C_{V{\rm i}} = 3k_{\rm B}N_{\mathcal N}\left\langle z_{\boldsymbol{k}s}^2\,{\rm e}^{z_{\boldsymbol{k}s}}({\rm e}^{z_{\boldsymbol{k}s}}-1)^{-2}\right\rangle_{\rm ph}. \tag{2.104}$$

As a rule, the main contribution into the internal energy comes from the classical static-lattice energy. The Madelung constant $C_{\rm M}$, which determines this energy, is remarkably close to 0.9 for all lattice types. [4] Let us recall that the value 0.9 is predicted by the ion-sphere model (§ 2.3.2 b). Therefore, the ion-sphere expression $U_{\rm M} = -0.9\,N_{\mathcal N}\,Z^2e^2/a_{\rm i}$ is sufficiently accurate in the ion liquid and solid, as long as the ions are strongly coupled.

The low-order frequency moments of bcc, fcc, and hcp crystals are very similar (see Table 2.4). According to Eq. (2.96), $u_2 = \frac{1}{3}$ for any OCP lattice. Many thermodynamic and kinetic properties of Coulomb crystals are remarkably independent of the crystal type and are similar to those in a strongly coupled Coulomb liquid. For instance, the transport properties of degenerate electrons which scatter off ions are almost the same in the liquid and crystal phases (Baiko *et al.*, 1998). The difference between the liquid and crystal and between different crystal types is really tiny (also see §§ 2.3.4 and 2.4.6).

[4]For the bcc Coulomb lattice, this constant was first calculated by Fuchs (1935).

*Table 2.4.* Parameters of Coulomb crystals: Madelung constant $C_{\rm M}$, frequency moments $u_n$, and $\bar{l} = \langle \ln(\omega_{\boldsymbol{k}s}/\omega_{\rm pi}) \rangle_{\rm ph}$ (Baiko *et al.*, 2001b).

| Lattice | $C_{\rm M}$ | $u_{-2}$ | $u_{-1}$ | $u_1$ | $u_3$ | $\bar{l}$ |
|---|---|---|---|---|---|---|
| bcc | 0.895 929 255 682 | 12.972 | 2.798 55 | 0.511 3875 | 0.250 31 | −0.831 298 |
| fcc | 0.895 873 615 195 | 12.143 | 2.719 82 | 0.513 1940 | 0.249 84 | −0.817 908 |
| hcp | 0.895 838 120 459 | 12.015 | 2.702 6 | 0.513 33 | 0.249 84 | −0.815 97 |

### 2.3.3 d Analytic formulae for thermodynamic functions

**Classical Coulomb crystal.** Asymptotically, at $\Gamma \to \infty$, one has $U_{\rm i} \approx U_{\rm M}$, where $U_{\rm M}$ is given by Eq. (2.101). A thermal correction

$$U_{\rm th} \equiv U_{\rm i} - U_{\rm M} = 3N_{\mathcal{N}} k_{\rm B}T + U_{\rm anharm} \qquad (2.105)$$

to the internal energy of the OCP at $\Gamma \gg 1$ is very small. The term $3N_{\mathcal{N}} k_{\rm B}T$ comes from the harmonic lattice vibrations and $U_{\rm anharm}$ represents anharmonic corrections discussed in § 2.3.3 e.

In the classical regime ($T \gtrsim T_{\rm pi}$), one has $\hbar\omega_{\boldsymbol{k}s} \ll k_{\rm B}T$, and Eq. (2.97) reduces to

$$F_{\rm i} = -N_{\mathcal{N}} C_{\rm M} \frac{(Ze)^2}{a_{\rm i}} + 3N_{\mathcal{N}} k_{\rm B}T \left[ \ln\left(\frac{\hbar\omega_{\rm pi}}{k_{\rm B}T}\right) + \bar{l} \right], \qquad (2.106)$$

with the term

$$\bar{l} \equiv \langle \ln(\omega_{\boldsymbol{k}s}/\omega_{\rm pi}) \rangle_{\rm ph} \qquad (2.107)$$

presented in Table 2.4. Subtracting the ideal-gas free energy (2.71) from $F_{\rm i}$ given by Eq. (2.106), one obtains the excess free energy of ions in the harmonic approximation,

$$F_{\rm ii,harm} = U_{\rm M} + N_{\mathcal{N}} k_{\rm B}T \left( \frac{3}{2}\ln\Gamma + 1 + 3\bar{l} + \frac{1}{2}\ln\frac{6}{\pi} \right). \qquad (2.108)$$

**Strong quantum limit.** Now consider the regime of $T \ll T_{\rm pi}$. In this case, only a small central part of the Brillouin zone, where $z_{\boldsymbol{k}s} \lesssim 1$, contributes actually to the averaging in Eq. (2.97). In addition, the contribution from the optical (longitudinal) phonons ($s = 3$) is much smaller than that from the acoustic ones ($s = 1, 2$). Carr (1961) evaluated $C_{V{\rm i}}$ and obtained $C_{V{\rm i}} = \zeta N_{\mathcal{N}} k_{\rm B} (T/T_{\rm pi})^3$, where $\zeta$ is a numerical constant (its accurate value is given on p. 83). This yields

$$F_{\mathrm{i}} = U_0 - \frac{\zeta N_{\mathcal{N}} k_{\mathrm{B}} T}{12} \left( \frac{T}{T_{\mathrm{pi}}} \right)^3 . \tag{2.109}$$

Because of the rapid decrease of the $T$-dependent terms at low $T$, the ion heat capacity becomes smaller than the electron one in the strong quantum limit, $t_{\mathrm{p}} \to 0$ (§ 2.4.6).

**Harmonic Coulomb crystal at arbitrary temperature.** At intermediate temperatures $T \sim T_{\mathrm{pi}}$ the asymptotic expressions (2.106) and (2.109) become inaccurate and thermodynamic quantities should be evaluated numerically. For estimates, however, one can use a simple model (Chabrier *et al.*, 1992) that treats two acoustic modes as degenerate Debye modes with $\omega_{\boldsymbol{k}s} = \alpha_t \omega_{\mathrm{pi}} k / q_{\mathrm{BZ}}$ and the longitudinal mode as an Einstein mode with $\omega_{\boldsymbol{k}s} = \alpha_l \omega_{\mathrm{pi}}$. We shall call it D$^2$E ("2 Debye + 1 Einstein modes") model. In this model

$$3 \left\langle \ln(1 - \mathrm{e}^{-z_{\boldsymbol{k}s}}) \right\rangle_{\mathrm{ph}} = -\frac{2}{3} D_3(\alpha_t \eta) + 2 \ln(1 - \mathrm{e}^{-\alpha_t \eta}) + \ln(1 - \mathrm{e}^{-\alpha_l \eta}), \tag{2.110}$$

where

$$\eta = T_{\mathrm{pi}}/T = t_{\mathrm{p}}^{-1} \tag{2.111}$$

and

$$D_n(x) = \frac{n}{x^n} \int_0^x \frac{t^n}{\mathrm{e}^t - 1} \, \mathrm{d}t \tag{2.112}$$

is a Debye function. We can propose the interpolation formulae for $D_1$ and $D_3$ which are accurate within 7 parts in $10^4$ and reproduce the small-$x$ [up to $o(x^2)$] and large-$x$ asymptotes,

$$\begin{aligned} D_1(x) &\approx \frac{1 + R_1(x)/x}{1 + \frac{1}{4}x + \frac{5}{144}x^2 + 0.00139\,x^4 + (6/\pi^2)\,R_1(x)}, \\ R_1(x) &= (440x + 9\mathrm{e}^x)\,(0.1x)^6; \end{aligned} \tag{2.113}$$

$$D_3(x) \approx \frac{1 + R_3(x)/x^3}{1 + \frac{3}{8}x + \frac{29}{320}x^2 + 0.01656\,x^{3.2} + 5.9 \times 10^{-5}\,x^6 + (5/\pi^4)\,R_3(x)}, \tag{2.114}$$

$$R_3(x) = 4.7 \left[ \mathrm{e}^x - \sum_{n=0}^{4} \frac{x^n}{n!} \right] (0.1x)^4 .$$

Chabrier (1993) checked that the D$^2$E model is reasonably accurate for evaluating the thermodynamic functions. Let us choose the parameters $\alpha_t$ and $\alpha_l$ in such a way as to reproduce exactly $u_2 = \frac{1}{3}$ and the value of $\bar{l}$ given in Table 2.4. We find $\alpha_t = 0.426548$ and $\alpha_l = 0.88412$ for the bcc lattice, $\alpha_t = 0.436671$

and $\alpha_l = 0.87817$ for fcc, and $\alpha_t = 0.436671$ and $\alpha_l = 0.87817$ for hcp. For example, Monte Carlo results of Pollock & Hansen (1973) for the normalized specific heat $C_V/(3N_{\mathcal{N}}k_{\rm B})$ of the bcc lattice are reproduced with the maximum error of 0.015 at $t_{\rm p} \approx 0.14$.

More accurate analytic formulae for thermodynamic functions of a Coulomb crystal in the harmonic approximation were constructed by Baiko *et al.* (2001b). The authors calculated several lowest phonon frequency moments and the Madelung constant (presented in Table 2.4), and also the free energy, internal energy, and heat capacity. The latter functions were fitted by analytic formulae. For instance, the free energy is given by Eq. (2.97), where

$$3\left\langle \ln(1-\mathrm{e}^{-z_{\boldsymbol{k}s}})\right\rangle_{\rm ph} = \sum_{n=1}^{3} \ln\left(1-e^{-\alpha_n\eta}\right) - \frac{\sum_{n=0}^{8} a_n\,\eta^n}{\sum_{n=0}^{7} b_n\,\eta^n + c_6\,a_6\,\eta^9 + c_8\,a_8\eta^{11}}, \tag{2.115}$$

with the parameters $\alpha_n$, $a_n$, and $b_n$ presented in Table 2.5. The accuracy of Eq. (2.115) is a few parts in $10^6$. Other thermodynamic functions can be obtained from Eq. (2.115) by differentiation. For instance, $U_{\rm i}$ and $C_{V{\rm i}}$ are written explicitly by Baiko *et al.* (2001b); any differentiation increases the relative error by a factor of $\approx 3$. At $T \ll T_{\rm pi}$, Eq. (2.115) is consistent with Eq. (2.109) if $\zeta = 12/c_8$.

### 2.3.3 e Anharmonic corrections

**Classical Coulomb crystal.** Anharmonic corrections for classical Coulomb lattices were numerically evaluated in many papers (e.g., Pollock & Hansen 1973; Albers & Gubernatis 1986; Nagara *et al.* 1987; Farouki & Hamaguchi 1993). They can be expanded over negative powers of $\Gamma$. For instance, the internal energy can be written as

$$\frac{U_{\rm anharm,cl}}{N_{\mathcal{N}}k_{\rm B}T} = \sum_{p\geq 1}\frac{a_p}{\Gamma^p}. \tag{2.116}$$

Assuming that the anharmonic part of the free energy $F_{\rm ii,anharm}$ vanishes in the strong-coupling limit $\Gamma \to \infty$ (Pollock & Hansen, 1973), from Eqs. (2.37) and (2.116) one obtains

$$\frac{F_{\rm ii,anharm,cl}}{N_{\mathcal{N}}k_{\rm B}T} = -\sum_{p\geq 1}\frac{a_p}{p\,\Gamma^p}. \tag{2.117}$$

Farouki & Hamaguchi (1993) performed numerical simulations of the classical Coulomb solid for $\Gamma$ up to 3000 and approximated the anharmonic corrections in various $\Gamma$ ranges by keeping 2–4 terms in Eq. (2.116). The most reliable

*Table 2.5.* Parameters of the analytic approximation (2.115) of the thermal free energies of bcc and fcc Coulomb lattices. Powers of 10 are given in square brackets.

| | bcc lattice | | fcc lattice | |
|---|---|---|---|---|
| $\alpha_1$ | 0.932446 | | 0.916707 | |
| $\alpha_2$ | 0.334547 | | 0.365284 | |
| $\alpha_3$ | 0.265764 | | 0.257591 | |
| $c_6$ | 4.7570140 [−3] | | 4.92387 [−3] | |
| $c_8$ | 4.7770935 [−3] | | 4.37506 [−3] | |
| $n$ | $a_n$ | $b_n$ | $a_n$ | $b_n$ |
| 0 | 1 | 261.66 | 1 | 303.20 |
| 1 | 0.1839 | 0 | 0 | 0 |
| 2 | 0.593586 | 7.07997 | 0.532535 | 7.7255 |
| 3 | 5.4814 [−3] | 0 | 0 | 0 |
| 4 | 5.01813 [−4] | 0.0409484 | 3.76545 [−4] | 0.0439597 |
| 5 | 0 | 3.97355 [−4] | 0 | 1.14295 [−4] |
| 6 | 3.9247 [−7] | 5.11148 [−5] | 2.63013 [−7] | 5.63434 [−5] |
| 7 | 0 | 2.19749 [−6] | 0 | 1.36488 [−6] |
| 8 | 5.8356 [−11] | – | 6.6318 [−11] | – |

results, which agree with the value of $A_1$ given by the perturbation theory (Nagara *et al.*, 1987), are provided by a three-term fit in the range $170 < \Gamma < 400$, with $a_1 = 10.9$, $a_2 = 247$, and $a_3 = 1.765 \times 10^5$.

Finally, Eqs. (2.108) and (2.117) give the total free energy of the classical OCP of ions,

$$F_{\mathrm{ii,solid}} = F_{\mathrm{ii,harm}} + F_{\mathrm{ii,anharm}}. \tag{2.118}$$

**Zero-temperature limit.** In the limit $T = 0$, the expansion of the Coulomb potential in powers of ion displacements from their lattice points leads to the expansion of the lattice energy in powers of $R_S^{-1/2}$, where $R_S$ is given by Eq. (2.32) and is treated as a large parameter. The coefficients of this expansion were calculated, e.g., by Carr (1961) and Albers & Gubernatis (1986). Their result can be written as $F_{\mathrm{i}} = U_{\mathrm{i}} = U_{\mathrm{i,harm}} + U_{\mathrm{i,anharm}}$. The harmonic part $U_{\mathrm{i,harm}} = U_0$ is given by Eqs. (2.100) and (2.101). It contains the familiar classical static-lattice and quantum zero-point vibration energies,

$$\begin{aligned} U_{\rm i,harm} &= -C_{\rm M}\, N\, \frac{Z^2 e^2}{a_{\rm i}} + \frac{3}{2}\, u_1 N\, \hbar\omega_{\rm pi} \\ &= -\frac{m_{\rm i}(Ze)^4}{2\hbar^2}\, N \left( \frac{2C_M}{R_S} - \frac{3\sqrt{3}u_1}{R_S^{3/2}} \right). \end{aligned} \tag{2.119}$$

The anharmonic zero-temperature part is

$$\begin{aligned} U_{\rm i,anharm} &= -\hbar\omega_{\rm pi}\, N \left( 0.0391 + 0.03262\, \frac{\hbar\omega_{\rm pi}}{Z^2 e^2/a_{\rm i}} + \ldots \right) \\ &= -N\, \frac{m_{\rm i}(Ze)^4}{2\hbar^2} \left( \frac{0.703}{R_S^2} + \frac{1.017}{R_S^{5/2}} + \ldots \right). \end{aligned} \tag{2.120}$$

Anharmonic corrections in quantum crystals at finite temperatures ($0 < T \lesssim T_{\rm pi}$) are currently unknown.

### 2.3.3f The structure factor of a Coulomb crystal

In the general case of a quantum Coulomb crystal, Eq. (2.80) for the dynamical structure factor should be generalized:

$$S(\boldsymbol{q},\omega) = \frac{1}{2\pi N_{\mathcal{N}}} \int_{-\infty}^{+\infty} {\rm d}t \int {\rm d}\boldsymbol{x}\; {\rm d}\boldsymbol{x}'\, {\rm e}^{{\rm i}\boldsymbol{q}\cdot(\boldsymbol{x}-\boldsymbol{x}')-{\rm i}\omega t} \left\langle \hat{\rho}^\dagger(\boldsymbol{x},t)\, \hat{\rho}(\boldsymbol{x}',0) \right\rangle_{\rm th}, \tag{2.121}$$

where $\hat{\rho}(\boldsymbol{x},t)$ is the operator of charge density in units of $Ze$. The thermodynamic average $\langle \ldots \rangle_{\rm th}$ must be distinguished from the average over the phonon spectrum defined by Eq. (2.99). Using Eq. (2.91) and the quantization prescription (p. 79), we can integrate over $\boldsymbol{x}$ and $\boldsymbol{x}'$ in Eq. (2.121) and obtain

$$\begin{aligned} S(\boldsymbol{q},\omega) &= \frac{1}{2\pi N_{\mathcal{N}}} \int_{-\infty}^{+\infty} {\rm d}t\, {\rm e}^{-{\rm i}\omega t} \\ &\times \left\langle \sum_{i,j} {\rm e}^{{\rm i}\boldsymbol{q}\cdot(\boldsymbol{R}_i-\boldsymbol{R}_j)} \left[ {\rm e}^{{\rm i}\boldsymbol{q}\cdot\hat{\boldsymbol{u}}_i(t)} - 1 \right] \left[ {\rm e}^{-{\rm i}\boldsymbol{q}\cdot\hat{\boldsymbol{u}}_j(0)} - 1 \right] \right\rangle_{\rm th}. \end{aligned} \tag{2.122}$$

Expanding the ion displacement operator $\hat{\boldsymbol{u}}_i$ in the phonon creation-annihilation operators and using the Weyl operator identity ${\rm e}^A {\rm e}^B = {\rm e}^{A+B} {\rm e}^{[A,B]/2}$ (e.g., Kittel 1986), we can decompose the structure factor into two parts,

$$S(\boldsymbol{q},\omega) = S'(\boldsymbol{q})\, \delta(\omega) + S''(\boldsymbol{q},\omega), \tag{2.123}$$

where $S'(\boldsymbol{q})$ is the partial static structure factor associated with ion ordering (the Bragg-diffraction or elastic part), while the "inelastic" part $S''(\boldsymbol{q},\omega)$ is associated with ion fluctuations around equilibrium lattice sites. In the following we shall use the harmonic-lattice approximation which gives (e.g., Kittel 1963)

$$S'(\boldsymbol{q}) = (2\pi)^3 n_{\mathcal{N}} \mathrm{e}^{-2W(\boldsymbol{q})} \sum_{\boldsymbol{G}\neq 0} \delta(\boldsymbol{q}-\boldsymbol{G}), \tag{2.124}$$

where

$$\exp(-2W(\boldsymbol{q})) \equiv \langle \exp(i\boldsymbol{q}\cdot\hat{\boldsymbol{u}}_j)\rangle^2_{\mathrm{th}} \tag{2.125}$$

is the *Debye-Waller factor*. In Eq. (2.124), $\sum_{\boldsymbol{G}\neq 0}$ denotes a summation over all reciprocal lattice vectors $\boldsymbol{G}$ except $\boldsymbol{G}=0$.

In isotropic (e.g., cubic) crystals, one has (cf. Kittel 1986)

$$W(\boldsymbol{q}) = r_T^2 q^2/6, \tag{2.126}$$

where $r_T^2 = \langle \hat{\boldsymbol{u}}^2\rangle_{\mathrm{th}}$ is the mean-squared ion displacement. In a harmonic crystal,

$$r_T^2 = \frac{3\hbar}{2m_{\mathrm{i}}} \left\langle \frac{1}{\omega_{\boldsymbol{k}s}\tanh(z_{\boldsymbol{k}s}/2)} \right\rangle_{\mathrm{ph}}, \tag{2.127}$$

where $z_{\boldsymbol{k}s}$ is defined by Eq. (2.98). In a classical crystal, where $T \gtrsim T_{\mathrm{pi}}$, characteristic values of $z_{\boldsymbol{k}s}$ are small, and Eq. (2.127) simplifies to

$$r_T^2(t_{\mathrm{p}} \gtrsim 1) \approx \frac{3k_{\mathrm{B}}Tu_{-2}}{4\pi n_{\mathcal{N}} Z^2 e^2} = \frac{u_{-2}\, a_{\mathrm{i}}^2}{\Gamma}. \tag{2.128}$$

Formally, in this case one can write $r_T^2 = 3u_{-2}r_{\mathrm{D}}^2$, where $r_{\mathrm{D}}$ is the Debye ion radius, Eq. (2.25). In the opposite limit

$$r_T^2(t_{\mathrm{p}} \ll 1) \approx \frac{3\hbar u_{-1}}{2m_{\mathrm{i}}\omega_{\mathrm{pi}}} = \frac{u_{-1}\, a_{\mathrm{i}}^2}{\Gamma t_{\mathrm{p}}}. \tag{2.129}$$

For any $\eta$ the quantity $r_T^2$ can be estimated from the D$^2$E model (§ 2.3.3 d),

$$r_T^2 \approx \frac{a_{\mathrm{i}}^2\eta}{\Gamma}\left[\frac{u_{-1}}{2} + \frac{\eta}{3\alpha_l(\mathrm{e}^{\alpha_l\eta}-1)} + \frac{2}{\alpha_t^2}D_1(\alpha_t\eta)\right]. \tag{2.130}$$

The inelastic part of the structure factor can be presented as (Baiko *et al.*, 1998)

$$S''(\boldsymbol{q},\omega) = \frac{1}{2\pi N_{\mathcal{N}}}\int_{-\infty}^{+\infty} \mathrm{d}t\, \mathrm{e}^{-\mathrm{i}\omega t - 2W(\boldsymbol{q})} \sum_{ij} \mathrm{e}^{\mathrm{i}\boldsymbol{q}\cdot(\boldsymbol{R}_i-\boldsymbol{R}_j)} \sum_{n=1}^{\infty}\frac{1}{n!}$$
$$\times \left\{ \frac{\hbar}{2m_{\mathrm{i}}N_{\mathcal{N}}} \sum_{\boldsymbol{k}s} \frac{(\boldsymbol{q}\cdot\boldsymbol{e}_{\boldsymbol{k}s})^2}{\omega_{\boldsymbol{k}s}} \left[\alpha_{ij\boldsymbol{k}s}(\bar{n}_{\boldsymbol{k}s}+1) + \alpha^*_{ij\boldsymbol{k}s}\bar{n}_{\boldsymbol{k}s}\right] \right\}^n, \tag{2.131}$$

where $\alpha_{ij\boldsymbol{k}s} \equiv \exp[m_{\rm i}\boldsymbol{k}\cdot(\boldsymbol{R}_i - \boldsymbol{R}_j) - m_{\rm i}\omega_{\boldsymbol{k}s}t]$, and

$$\bar{n}_{\boldsymbol{k}s} = \frac{1}{\mathrm{e}^{z_{\boldsymbol{k}s}} - 1} \tag{2.132}$$

is the mean number of phonons ($\boldsymbol{k}$, $s$).

In § 2.3.2 a, we have introduced the static structure factor of a Coulomb plasma, Eq. (2.81). For a crystal, the static structure factor becomes

$$S(\boldsymbol{q}) = \frac{1}{N_{\mathcal{N}}}\sum_{i,j}\mathrm{e}^{\mathrm{i}\boldsymbol{q}\cdot(\boldsymbol{R}_i - \boldsymbol{R}_j)}\left\langle \mathrm{e}^{\mathrm{i}\boldsymbol{q}\cdot\hat{\boldsymbol{u}}_i}\mathrm{e}^{-\mathrm{i}\boldsymbol{q}\cdot\hat{\boldsymbol{u}}_j}\right\rangle_{\rm th}. \tag{2.133}$$

It can also be decomposed into the elastic (Bragg, or static-lattice) and inelastic parts, $S(\boldsymbol{q}) = S'(\boldsymbol{q}) + S''(\boldsymbol{q})$, where

$$S''(\boldsymbol{q}) = \mathrm{e}^{-2W(q)}\sum_{\boldsymbol{R}_j}\mathrm{e}^{\mathrm{i}\boldsymbol{q}\cdot\boldsymbol{R}_j}\left[\mathrm{e}^{v(\boldsymbol{q},\boldsymbol{R}_j)q^2} - 1\right], \tag{2.134}$$

and

$$v(\boldsymbol{q},\boldsymbol{R}) = \frac{3\hbar}{2m_{\rm i}}\left\langle \frac{(\boldsymbol{q}\cdot\boldsymbol{e}_{\boldsymbol{k}s})^2}{q^2}\,\frac{\cos(\boldsymbol{k}\cdot\boldsymbol{R})}{\omega_{\boldsymbol{k}s}\tanh(z_{\boldsymbol{k}s}/2)}\right\rangle_{\rm ph}. \tag{2.135}$$

In general, the function $S''(\boldsymbol{q})$ has sharp peaks (e.g., Baiko *et al.* 2000), whose calculation is difficult. However, in integrals over $\boldsymbol{q}$ at $q \gg q_{\rm BZ}$ the peaks can often be smeared out. In such cases, a method introduced by Raikh & Yakovlev (1982) proves to be useful. Let us note that in these integrals $\boldsymbol{q}$ runs over many Brillouin zones, each having its own principal polarization vectors $\boldsymbol{e}_{\boldsymbol{k}s}$. Thus the polarization vectors change rather rapidly on scales $q \sim q_{\rm BZ}$, and for $q \gtrsim q_{\rm BZ}$ one can approximately replace

$$\sum_s(\boldsymbol{q}\cdot\boldsymbol{e}_{\boldsymbol{k}s})^2 f(\omega_{\boldsymbol{k}s}) \to q^2\langle f(\omega_{\boldsymbol{k}s})\rangle_{\rm ph}, \tag{2.136}$$

provided that $f(\omega)$ is a sufficiently smooth function. Then from Eq. (2.131) after averaging over orientations of $\boldsymbol{q}$ one gets (Baiko *et al.*, 1998)

$$S''(q,\omega) = \exp\left[-2W(q) - \frac{\hbar\omega}{2k_{\rm B}T}\right]\int_{-\infty}^{+\infty}\frac{\mathrm{d}t}{2\pi}\,\mathrm{e}^{-i\omega t}K(q,T,t), \tag{2.137}$$

$$K(q,T,t) = \exp\left[\frac{\hbar q^2}{2m_{\rm i}}\left\langle\frac{\cos\omega_{\boldsymbol{k}s}t}{\omega_{\boldsymbol{k}s}\sinh\left(z_{\boldsymbol{k}s}/2\right)}\right\rangle_{\rm ph}\right] - 1. \tag{2.138}$$

Equation (2.137) takes into account ion correlations associated with absorption and emission of any number of phonons. In the solid state theory, one often linearizes quantum-mechanical amplitudes of interactions with respect to $\hat{\boldsymbol{u}}$, neglecting thus multi-phonon processes. The structure factor in this *one-phonon approximation* simplifies to (e.g., Flowers & Itoh 1976)

$$S''_{1\mathrm{ph}}(q,\omega) = \mathrm{e}^{-2W(q)}\,\frac{\hbar q^2}{2m_\mathrm{i}}\left\langle \frac{\delta(\omega-\omega_{\boldsymbol{k}s})+\delta(\omega+\omega_{\boldsymbol{k}s})}{\omega_{\boldsymbol{k}s}|\exp(\hbar\omega/k_\mathrm{B}T)-1|}\right\rangle_\mathrm{ph}, \tag{2.139}$$

which is equivalent to replacing $\{\exp[\ldots]-1\}\to[\ldots]$ in Eq. (2.138). This approximation is justified only at $T \ll T_\mathrm{pi}$. In dense astrophysical plasmas, *multi-phonon processes* appear to be important near the melting temperature (Baiko *et al.*, 1998).

The approximate inelastic part of the static structure factor, corresponding to Eq. (2.137), reads (Baiko *et al.*, 1998)

$$S''_0(q) = 1 - \mathrm{e}^{-2W(q)}. \tag{2.140}$$

This approximate expression is quite accurate for $q \gtrsim 3/a_\mathrm{i}$ (Baiko *et al.*, 2000).

### 2.3.4 Melting

The solid-liquid phase transition occurs at $\Gamma = \Gamma_\mathrm{m}$, where the free energies $F(\Gamma)$ of liquid and solid phases intersect. It is difficult to find $\Gamma_\mathrm{m}$ with high precision, because the intersection of the free energy curves is strongly affected by the thermal corrections which are small at $\Gamma \sim \Gamma_\mathrm{m}$ (the free energy curves are nearly parallel). For instance, a 0.1% error in $F_\mathrm{ii}$ shifts the intersection by $\Delta\Gamma \approx 15$ (Pollock & Hansen, 1973).

According to Fig. 2.2, the melting in the envelopes of middle-aged neutron stars takes place at $\rho \gtrsim 10^6$ g cm$^{-3}$. Therefore, the model of the rigid electron background presents a good starting approximation for calculating $\Gamma_\mathrm{m}$ in most cases. The first realistic estimate, $\Gamma_\mathrm{m} = 170 \pm 10$, was obtained by Van Horn (1969). He used the *Lindemann criterion* according to which a crystal melts when the amplitude $r_T$ of root-mean-square ion vibrations in a lattice site (defined on p. 86) becomes a sizable fraction of inter-ion distance. Specifically, Van Horn equated the ratio $r_T/a_\mathrm{i}$ to the experimental value $r_T/a_\mathrm{i} = 0.070 \pm 0.004$ for alkali metals. This value of $\Gamma_\mathrm{m}$ was considerably larger than numerical estimates available at that time, but it proved to be remarkably close to the value obtained twenty years later from sophisticated theoretical studies. Those studies included numerical evaluation of the internal energy of the classical liquid and solid OCP and used quantum expansions of thermodynamic functions of OCP, as well as Monte Carlo simulations.

By equating $F_\mathrm{ii,liq}$ given by our fit (2.87) to $F_\mathrm{ii,solid}$ given by Eqs. (2.108) and (2.118), we obtain $\Gamma_\mathrm{m} = 175$. This value, however, may substantially change if we take into account an electron polarization (see § 2.4.5 c below).

The phase transition is first-order, but very weak. Strictly speaking, if we fix the number density of ions and decrease the temperature, we obtain discontinuous changes of the ion energy $U_\mathrm{i}$ and the ion pressure $P_\mathrm{i}$ at the melting point. One can easily show that both quantities undergo sudden drops. The drop of the ion energy density $U_\mathrm{i}/V$, associated with the crystallization to the bcc lattice at $\Gamma_\mathrm{m} = 175$, is $\Delta U_\mathrm{i}/V \approx 0.76\, n_{\scriptscriptstyle N} k_\mathrm{B} T_\mathrm{m}$, with $\Delta U_\mathrm{i}/U_\mathrm{i} \approx 0.5\%$, and the drop of the ion pressure $\Delta P_\mathrm{i}/P_\mathrm{i}$ is approximately three times smaller. Because the ion pressure is only a small part of the total pressure $P$ (while the leading part $P_\mathrm{e}$ is provided by the electrons), the fractional drop of the total pressure $\Delta P/P$ is much smaller than the fractional drop of the ion pressure $\Delta P_\mathrm{i}/P_\mathrm{i}$. For instance, for the $^{56}$Fe plasma at $\rho = 10^8$ g cm$^{-3}$ or for the $^{12}$C plasma at $\rho = 10^5$ g cm$^{-3}$ (the cases discussed further in § 2.4.6 and presented in the left and right panels of Fig. 2.7 on p. 103) one gets $\Delta P/P \approx 2 \times 10^{-4}$. However, even a small $\Delta P/P$ means that the thermodynamic equilibrium between the two phases (which requires $P$ =constant) should be accompanied by a density discontinuity $\Delta\rho/\rho = (\Delta P/P)/\chi_\rho \sim \Delta P/P$, an unambiguous manifestation of the first-order phase transition (recall that for a strongly degenerate electron gas the pressure exponent $\chi_\rho$ varies from 5/3 at $x_\mathrm{r} \ll 1$ to 4/3 at $x_\mathrm{r} \gg 1$). The density jump is so small that it can be neglected in most applications.

Nevertheless, these discontinuities are important for astrophysical modeling. For example, they determine the latent heat released at freezing. It equals $Q_\mathrm{latent} = T(S_\mathrm{m,solid} - S_\mathrm{m,liq}) = U_\mathrm{m,liq} - U_\mathrm{m,solid} \approx \Delta U_\mathrm{i}$, where $S_\mathrm{m,liq}$ and $S_\mathrm{m,solid}$ are the entropy values at $T = T_\mathrm{m}$ in the liquid and solid phases, respectively, and $U_\mathrm{m,liq}$ and $U_\mathrm{m,solid}$ are the corresponding values of the internal energy. Thus, $Q_\mathrm{latent}$ is only $\approx 0.5\%$ of the ion contribution $U_\mathrm{i}$, which is only a tiny fraction of the total internal energy $U$ (mainly determined by degenerate electrons at the conditions mentioned above). Nevertheless, the absolute value of $Q_\mathrm{latent}$ ($\approx 0.76 k_\mathrm{B} T$ per ion) is a substantial fraction of thermal thermal energy of the ions. It is sufficient, for instance, to delay the cooling of old white dwarfs with crystallizing cores (Chabrier, 1999; Hansen, 2004, and references therein), that is important for the white dwarf chronology. Note also that dense matter may form a supercooled liquid and freeze at $T < T_\mathrm{m}$ as will be discussed in § 2.4.6.

At high densities zero-point quantum vibrations of ions may become comparable to the lattice constant and hamper lattice ordering. This effect was studied by many authors (e.g., Mochkovich & Hansen 1979; Ceperley & Alder 1980). In particular, Nagara *et al.* (1987) used an expansion technique, Chabrier (1993) invented a simple analytic model, and Jones & Ceperley (1996) applied a path-integral Monte Carlo simulation. These studies confirm that at high densities the melting temperature $T_\mathrm{m}(\rho)$ becomes essentially lower than the classical melting temperature given by Eq. (2.28), and at some critical density $\rho_\mathrm{m}$ zero-point vibrations destroy the lattice even at $T = 0$ ($T_\mathrm{m}(\rho_\mathrm{m}) = 0$). Hence at $\rho > \rho_\mathrm{m}$ the liquid does not solidify at all, and one obtains the *quantum liquid* domain at $T \lesssim T_\mathrm{pi}$ and $\rho \gtrsim \rho_\mathrm{m}$. In that domain spin statistics of ions is important. Numerical values of $\rho_{\rm m}$ correspond to $R_S \approx 140-160$ for bosons and $R_S \approx 90-110$ for fermions (Ceperley & Alder, 1980; Chabrier, 1993; Jones & Ceperley, 1996).

The applicability of the latter results to neutron star envelopes is restricted. For astrophysically abundant carbon, oxygen, or nitrogen ions, $\rho_{\rm m}$ is not reached in the outer crust. On the other hand, such ions cannot survive in the inner crust; they would immediately disappear there owing to beta captures and pycnonuclear reactions (see Shapiro & Teukolsky 1983 and Chapter 3). Therefore, the appearance of quantum liquid is actually important only for hydrogen and helium in the outer neutron star envelope, where zero-point motion drastically suppresses the crystallization of these ions.

### 2.3.5 The pressure of degenerate electrons with electrostatic corrections

For many applications it is sufficient to use a fairly simple EOS of the outer envelope of a neutron star. One can take the pressure $P_{\rm id}^{(e)}$ of the ideal fully degenerate electron gas from Eq. (2.67) and supplement it with the Coulomb correction calculated within the ion-sphere model, $P_{\rm ii} \approx -0.3\, n_{\mathcal{N}}\, Z^2 e^2/a_{\rm i}$. (§§ 2.1.3 and 2.3.3 c). This gives

$$P = \frac{P_{\rm r}}{8\pi^2}\left[x_{\rm r}\left(\frac{2}{3}\,{x_{\rm r}}^2 - 1\right)\gamma_{\rm r} + \ln(x_{\rm r} + \gamma_{\rm r})\right] - 0.3\, n_{\mathcal{N}}\,\frac{Z^2 e^2}{a_{\rm i}}. \qquad (2.141)$$

This EOS can be used in the outer envelope not too close to the surface. It is also widely used in white dwarf cores. It is temperature independent and applies to any composition of the matter. Its derivation implies that the electron contribution $P_{\rm id}^{(e)}$ should be much larger than the Coulomb correction $P_{\rm ii}$. In the ultrarelativistic electron gas ($x_{\rm r} \gg 1$) both pressures have the same (polytropic) density dependence, $P_{\rm id}^{(e)} \propto P_{\rm ii} \propto {x_{\rm r}}^4$. Therefore, their ratio is density independent being determined only by the ion charge number $Z$,

$$\frac{P_{\rm ii}}{P_{\rm id}^{(e)}} = -\frac{6}{5}\left(\frac{4}{9\pi}\right)^{1/3}\frac{Z^{2/3}e^2}{\hbar c} \approx -0.0046\, Z^{2/3}, \qquad (2.142)$$

which gives $P_{\rm ii}/P_{\rm id}^{(e)} \approx -0.04$ for the matter composed of iron.

In the nonrelativistic electron gas one has $P_{\rm id}^{(e)} \propto {x_{\rm r}}^5$ and $P_{\rm ii} \propto {x_{\rm r}}^4$. Accordingly, the ratio

$$\frac{P_{\rm ii}}{P_{\rm id}^{(e)}} = -\frac{3}{2}\left(\frac{4}{9\pi}\right)^{1/3}\frac{Z^{2/3}e^2}{\hbar c \beta_{\rm r}} \approx -0.0057\,\frac{Z^{2/3}}{x_{\rm r}} \qquad (2.143)$$

depends on $\rho$. Both estimates agree with those presented in § 2.2.1 (Table 2.1). With decreasing $\rho$ the Coulomb pressure becomes more important. Formally, the pressure $P$ given by Eq. (2.141) vanishes at $\rho = \rho_* \approx 0.18\,ZA$ g cm$^{-3}$. However, Eq. (2.141) is invalid at such low $\rho$ because at least the approximation of the rigid electron background (implied in the ion-sphere model) breaks down. The applicability of Eq. (2.141) is limited by the conditions $T \ll T_{\rm F}$, $\rho \gg \rho_{\rm rigid}$ (see p. 66) and $T \lesssim T_{\rm l}$. The latter inequality ensures the regime of strong ion coupling. Its violation at high temperatures, where the Coulomb corrections are nevertheless small, does not introduce serious errors into the pressure that is mainly determined by electrons. Therefore, Eq. (2.141) can be used disregarding this inequality in many applications.

## 2.4. Intermediate-density regime. Electron response

The most complicated region of plasma parameters refers to moderate temperatures and intermediate densities between the high densities, where the plasma is fully ionized, and low densities, where the plasma is nondegenerate and nearly ideal (if it is warm). At intermediate densities one often employs the mean-ion approximation, in which a real plasma is modeled as the electron-ion plasma containing a single ion species with an effective charge $Z_{\rm eff}$. The effective charge can be chosen in different ways. One often uses a simple average, $Z_{\rm eff} = \langle Z \rangle$.

### 2.4.1 Interpolation

The simplest way to obtain a reasonable EOS in the intermediate density regime is an interpolation between the low and high densities. For hydrogen and helium plasmas this was done by Fontaine *et al.* (1977), who interpolated pressure logarithms along isotherms. A modification of this technique, that makes use of the Opacity Library (OPAL) EOS (see, e.g., Rogers *et al.* 1996) for iron plasma at intermediate densities and high temperatures, was presented by Potekhin *et al.* (1997a). The interpolation covers the gap between maximum densities $\rho_{\rm OPAL}$, presented in the OPAL tables, and the minimum densities $\rho_{\rm rigid}$, at which the approximation of the fully ionized plasma with the rigid electron background holds. The OPAL tables are approximately restricted by $\rho_{\rm OPAL} \approx 10\,T_6^3$ g cm$^{-3}$. The density $\rho_{\rm rigid}$, defined in §2.3, is $\rho_{\rm rigid} \approx 8 \times 10^5$ g cm$^{-3}$ for the iron plasma. The resulting EOS, constructed by Potekhin *et al.* (1997a), satisfies the normality condition $(\partial P/\partial T)_V > 0$ and preserves some signatures of the Fe shell structure exhibited by the OPAL EOS. Once an interpolated EOS is constructed, the values of $Z_{\rm eff}$ in the mean-ion approximation can be evaluated by adjusting the effective number density of free electrons to the pressure $P_{\rm id}^{(e)} + P_{\rm id}^{(\rm i)}$, calculated, e.g., from Eqs. (2.50) and (2.72), to reproduce the pressure $P$ given by the accurate EOS at the same $\rho$ and $T$. Then $Z_{\rm eff}$ is set equal to the mean ion charge, required to produce the electron density.

The EOS at low, intermediate and high densities is shown in Fig. 1.3 (in Chapter 1, p. 15). It combines the EOSs developed by different authors for different density ranges as described in § 1.3.2. In particular, the dashed lines show the OPAL EOS for $T = 10^6, 10^7$, and $10^8$ K. At densities $\rho \gtrsim 10^5$ g cm$^{-3}$ the EOS is almost temperature independent. However, at lower densities the thermal effects become significant and noticeably increase the pressure. The dotted line is the analytic interpolation expression for $T = 10^7$ K described in Appendix C.

More elaborate calculations of thermodynamic functions of the electron-ion plasmas with allowance for the $ee$ and $ie$ correlations start from the first principles, but then make use of some simplifying assumptions and numerical schemes. Some models are briefly reviewed below.

## 2.4.2 Thomas-Fermi-like theories

In the intermediate density region, the statistical (Thomas-Fermi) model is often used, which is particularly advantageous for heavy elements. Its simplest version was first constructed by Frenkel (1928), but his paper remained unnoticed by astrophysicists. Later is was proposed in the famous paper by Feynman *et al.* (1949). This version assumes that the electrons in a spherical Wigner-Seitz cell behave like a partially degenerate non-relativistic gas in a spherically symmetric partially screened attractive potential $\phi(r)$ of the central ion. Therefore, the electron local number density $n_e(r)$ is related to $\phi(r)$ by Eq. (2.51) with $\chi$ replaced by $\chi_r = \chi + e\phi(r)/k_{\rm B}T$:

$$n_e = (4/\sqrt{\pi})\, \lambda_{\rm e}^{-3} I_{1/2}(\chi_r), \tag{2.144}$$

where $I_\nu(\chi_r) \equiv I_\nu(\chi_r, 0)$ is the non-relativistic Fermi integral. For $\phi(r)$, the Poisson equation reads

$$\nabla^2 \phi(r) = -4\pi \left[ Ze\delta(\boldsymbol{r}) - en_e(r) \right]. \tag{2.145}$$

The obvious boundary condition and normalization are

$$\phi'(a_{\rm i}) = 0, \qquad \int_0^{a_{\rm i}} n_e(r)\, 4\pi r^2\, {\rm d}r = Z. \tag{2.146}$$

The internal energy is the sum of contributions of the ideal ions [Eq. (2.72)], the electron gas [see Eq. (2.53)], and the electron-ion electrostatic interaction:

$$\begin{aligned} U &= U_{\rm id}^{(\rm i)} + N_{\mathcal N} k_{\rm B} T \int_0^{a_{\rm i}} \frac{4}{\sqrt{\pi}\, \lambda_{\rm e}^3} I_{3/2}(\chi_r)\, 4\pi r^2\, {\rm d}r \\ &\quad - \frac{N_{\mathcal N}}{2} \int_0^{a_{\rm i}} \left( \frac{Ze^2}{r} + e\phi(r) \right) n_e(r)\, 4\pi r^2\, {\rm d}r. \end{aligned} \tag{2.147}$$

The pressure is given by

$$P = P_{\rm id}^{(\rm i)} + \frac{8}{3\sqrt{\pi}\,\lambda_e^3} I_{3/2}(\chi_{a_{\rm i}}). \tag{2.148}$$

The important modifications of the Thomas-Fermi theory consist in including the electron exchange (Thomas-Fermi-Dirac model, Cowan & Ashkin 1957), the electron correlations (Kirzhnits, 1958), and the gradient correction (Weizsäcker, 1935) which takes into account that the electron wave functions are locally not pure plane waves. A comparison of different versions of the Thomas-Fermi EOS is given, e.g., by Lai *et al.* (1991). Note also Thomas-Fermi-like calculations of thermodynamical properties of crystalline matter based on real (nonspherical) Wigner-Seitz cells (see, e.g., Hubbard & Macfarlane 1985 and references therein).

The Thomas-Fermi theory can describe a number of thermodynamic effects, such as pressure ionization of cold plasmas. Early versions of the theory missed some quantum-mechanical effects, for instance, the shell structure of bound species, and showed clear discrepancy with experiment (e.g., Kirzhnits 1967). In order to eliminate such a discrepancy, one often incorporates new experimental data by introducing phenomenological modifications into the theory. For example, the "mean-ion model" constructed by Basko (1985) takes into account some thermodynamic characteristics of real substances by virtue of a number of empirical parameters. This ensures a satisfactory agreement with high-pressure experiments.

Zink (1968) incorporated the shell effects into the Thomas-Fermi theory, using a solution of the Schrödinger equation for the inner electrons. This "hybrid" model was extended by Lai *et al.* (1991), who found it to be not essentially better than the Thomas-Fermi-Dirac-Weizsäcker model for calculating thermodynamic quantities.

A reasonable compromise between simplicity and accuracy in the description of high-pressure matter, especially composed of heavy elements, is provided by the "refined Thomas-Fermi" (RTF) model (Kirzhnits & Shpatakovskaya, 1995). It supplements the classical Thomas-Fermi method with several semiclassical, exchange, and shell corrections whose parameters are chosen to satisfy experimental data.

Since the Thomas-Fermi-like models statistically treat the electrons in a Wigner-Seitz cell, they are inaccurate in the case where the number of these electrons is not large.

## 2.4.3 Electron exchange and correlation effects

At extremely low $T$ and $\rho$, corresponding to large $\Gamma_e$ and $r_s$ (§ 2.1.2), the OCP of electrons behaves similarly to that of ions, and the results of § 2.3.2 can be applied. However, the exchange effects at finite $r_s$ imposed by the Pauli principle are often as important as the Coulomb correlation effects associated with non-zero $\Gamma_e$. Therefore, the excess part of the electron free energy $F_{\rm xc}$ in Eq. (2.44) is often called the exchange-correlation part.

The exchange-correlation effects were studied by many authors. Pokrant (1977) evaluated the Slater sum using a variational method. Perrot & Dharma-wardana (1984) used the local-density approximation of the density functional theory (DFT), with the effective screened potential calculated in the random-phase approximation (that yields the internal energy at small $\Gamma_e$ and arbitrary $\theta$). Nagara *et al.* (1987) developed an expansion scheme, more powerful than the Wigner-Kirkwood expansion, and showed that the melting value $\Gamma_{\rm m}$ increases with growing $r_s$ and turns to infinity at $r_s \lesssim 230$. Hence the electron liquid does not freeze at any $T$ if $n_e \gtrsim 1.3 \times 10^{17}$ cm$^{-3}$. Tanaka *et al.* (1985) calculated the interaction energy of the electron fluid at intermediate $\Gamma_e$ and $\theta$ using the approximation suggested by Singwi *et al.* (1968).

Tanaka *et al.* (1985) presented a fitting formula that reproduces their numerical results as well as the results of other authors in various limits (Pokrant, 1977; Ceperley & Alder, 1980; Dandrea *et al.*, 1986) with deviations less than 1%. In particular, their formula incorporates the parameterization of the exchange (Hartree-Fock) energy by Perrot & Dharma-wardana (1984). It reads $F_{\rm HF}(\theta) = (e^2/a_0)\, a(\theta)/r_s$, where

$$a \equiv \left(\frac{9}{4\pi^2}\right)^{1/3} \frac{0.75 + 3.04363\theta^2 - 0.09227\theta^3 + 1.7035\theta^4}{1 + 8.31051\theta^2 + 5.1105\theta^4} \tanh\frac{1}{\theta}. \quad (2.149)$$

The result reads: [5]

$$\frac{F_{\rm xc}^{\rm (nonrel)}}{k_{\rm B} T N_e} = g\Gamma + \frac{2A}{f}\sqrt{\Gamma} + \left(\frac{B}{f} - \frac{dA}{f^2}\right)\ln|f\Gamma + d\sqrt{\Gamma} + 1|$$
$$- \frac{2(dB + CA)}{fD}\left[\arctan\left(\frac{2f\sqrt{\Gamma} + d}{D}\right) - \arctan\left(\frac{d}{D}\right)\right], \quad (2.150)$$

where $A = b - gd$, $B = a - g$, $C = 2 - d^2/f$, $D = \sqrt{4f - d^2}$, $a = a(\theta)$ is given by Eq. (2.149), and $b$, $d$, $f$, and $g$ are the following functions of $\theta$

[5] Let us remark that at a finite degeneracy $\theta$, the interaction energy considered by Tanaka *et al.* (1985) differs from the thermodynamic internal energy arising from the electron non-ideality.

(Ichimaru *et al.*, 1987):

$$
\begin{aligned}
b &= \sqrt{\theta}\,\tanh\left(\frac{1}{\sqrt{\theta}}\right)\frac{0.341308+12.0708\,\theta^2+1.148889\,\theta^4}{1+10.495346\,\theta^2+1.326623\,\theta^4},\\
d &= \sqrt{\theta}\,\tanh\left(\frac{1}{\sqrt{\theta}}\right)\frac{0.614925+16.996055\,\theta^2+1.489056\,\theta^4}{1+10.10935\,\theta^2+1.22184\,\theta^4},\\
f &= \theta\,\tanh\left(\frac{1}{\theta}\right)\frac{0.539409+2.522206\,\theta^2+0.178484\,\theta^4}{1+2.555501\,\theta^2+0.146319\,\theta^4},\\
g &= 0.872496+0.025248\,\exp(-1/\theta).
\end{aligned}
$$

The results of Ichimaru *et al.* (1987) are non-relativistic. They seem to be sufficient for most applications, because the electron gas is almost ideal at densities and temperatures, where the electrons are relativistic.

The relativistic results for the exchange energy at $\theta < 1$ were derived by Shaviv & Kovetz (1972). Spurious misprints, which persisted in that and many subsequent publications, were eliminated by Stolzmann & Blöcker (1996), whose result can be written as

$$
\frac{F_{\rm X}^{(\rm rel)}}{N_e k_{\rm B} T} = \frac{3\alpha_{\rm f}}{4\pi}\,\frac{1}{{x_{\rm r}}^3 t_{\rm r}}\left[f_0(x_{\rm r}) + f_2(x_{\rm r},t_{\rm r})\,t_{\rm r}^2 + f_4(x_{\rm r})\,t_{\rm r}^4\right], \tag{2.151}
$$

where

$$
\begin{aligned}
f_0(x) &= \frac{3}{2}\,\frac{L^2(x)}{1+x^2} - 3xL(x) + \frac{3}{2}x^2 + \frac{x^4}{2},\\
f_2(x,\tau) &= \frac{\pi^2}{3}\left[2\ln\left(\frac{2x^2}{\tau}\right) + x^2 - \frac{3L(x)}{x} - 0.7046\right],\\
f_4(x) &= \frac{\pi^4}{18}\left(1 - \frac{1.1}{x^2} - \frac{3.7}{x^4} - \frac{6.3}{x^5}\,L(x)\right),\\
L(x) &= \sqrt{1+x^2}\,\ln\left(x+\sqrt{1+x^2}\right).
\end{aligned}
$$

Using Eq. (2.151), one can obtain the exchange contribution to the heat capacity of the relativistic degenerate electron gas:

$$
\frac{C_{V,x}^{(\rm rel)}}{N_e k_{\rm B}} = -\frac{3\alpha_{\rm f}}{4\pi}\,\frac{t_{\rm r}}{{x_{\rm r}}^3}\left[2f_2(x_{\rm r}) - \frac{2\pi^2}{3}\,t_{\rm r} + 12\,f_4(x_{\rm r})\,t_{\rm r}^3\right]. \tag{2.152}
$$

The correlation corrections in the relativistic gas are negligible, so that one can set $F_{\rm xc} \approx F_{\rm x}$ in this case. In general, the exchange and correlation corrections are of little importance in neutron star matter as long as the electrons are relativistic.

### 2.4.4 Electron polarization in ion liquid

In §§ 2.3.2–2.3.3 we considered the Coulomb plasma of ions with the rigid electron background. This approximation is justified in many applications, especially in dense layers of neutron star envelopes (§ 2.2.1, Table 2.1). However, in some cases the polarization ion-electron ($ie$) interaction associated with deviations from the rigid electron background can be noticeable (for example, in a neutron star atmosphere, where the density is relatively low, § 2.5.2). In § 2.4.2 we have already considered the electron polarization effect in the Thomas-Fermi approximation (valid for high-$Z$ ions). Here we present a more accurate treatment of the problem.

#### 2.4.4 a Theory

In a plasma of ions and electrons, the ion interactions are partly screened by the electron polarization. As long as the polarization interaction is weak compared to the kinetic energy of the electrons (for instance, at $Ze^2/a_e \ll k_B T_F$, for degenerate electrons), this interaction can be treated within the linear response theory. Under these conditions, the exact Hamiltonian of the electron-ion plasma can be separated into a Hamiltonian for a fluid of ions screened by electrons and a Hamiltonian for a *rigid* electron background, the so-called "jelly" Hamiltonian (Galam & Hansen, 1976).

At $r_s \ll 1$ and $T \ll T_F$, the polarization corrections due to the electron screening can be obtained in the zero-temperature random-phase approximation [Jancovici 1962; cf. § 2.1.2]. If $r_s \lesssim 1$ and $T \lesssim T_F$, the nonlinear effects due to the local field correction (2.21) and a finite electron degeneracy should be taken into account. The electrons can be treated by the DFT in various approximations or by the Monte Carlo technique (e.g., Pierleoni *et al.* 1994 and references therein). The Monte Carlo simulations are the most rigorous ones, but require excessive computer resources.

The screening contribution can also be evaluated by the method of effective potentials (see, e.g., Ichimaru *et al.* 1987). In this method, the bare Coulomb potential in the expression for the electrostatic energy $\mathcal{U}$ is replaced by the effective potential screened by electrons. Its Fourier transform is

$$V^{\text{eff}}(k) = \frac{4\pi(Ze)^2}{k^2\varepsilon(k)}, \tag{2.153}$$

where $\varepsilon(k)$ is the static screening function of the electron gas [Eq. (2.18)] or fluid [Eq. (2.21)].

A numerical scheme based on the effective potential technique and the HNC equations was developed and applied to the hydrogen plasma by Chabrier (1990) who considered an arbitrary electron degeneracy and a broad range of $\Gamma$, from the strong-coupling limit $\Gamma \gg 1$ to the Debye-Hückel limit $\Gamma \ll 1$. Chabrier & Potekhin (1998) extended these calculations to ion charges up to $Z = 26$.

### 2.4.4 b Analytic formulae

The numerical HNC results were fitted by the following expression (Chabrier & Potekhin, 1998; Potekhin & Chabrier, 2000)

$$f_{\rm ie} \equiv \frac{F_{\rm ie}}{N_{\mathcal N} k_{\rm B}T} = -\Gamma_e \frac{c_{\rm DH}\sqrt{\Gamma_e} + c_{\rm TF} a \Gamma_e^{\nu} g_1(r_s) g_3(x_{\rm r})}{1 + \left[b\sqrt{\Gamma_e} + a g_2(r_s)\Gamma_e^{\nu}/r_s\right]\gamma_{\rm r}^{-1}}. \tag{2.154}$$

The parameter

$$c_{\rm DH} = \frac{Z}{\sqrt{3}}\left[(1+Z)^{3/2} - 1 - Z^{3/2}\right] \tag{2.155}$$

ensures the transition of the excess free energy to the Debye-Hückel limit at small $\Gamma$. Indeed,

$$\begin{aligned} F_{\rm ex}(\Gamma \to 0) &= -N_{\mathcal N} k_{\rm B}T\,\Gamma^{3/2}/\sqrt{3} - N_{\rm e} k_{\rm B}T\,\Gamma_e^{3/2}/\sqrt{3} - N_{\mathcal N} k_{\rm B}T\, c_{\rm DH}\,\Gamma_e^{3/2} \\ &= -N_{\mathcal N} k_{\rm B}T\, Z\left[(1+Z)\,\Gamma_e\right]^{3/2}/\sqrt{3}, \end{aligned} \tag{2.156}$$

in agreement with Eq. (2.73).

The parameter

$$c_{\rm TF} = c_\infty Z^{7/3}\left(1 - Z^{-1/3} + 0.2\, Z^{-1/2}\right) \tag{2.157}$$

determines the screening contribution in the limit of large $\Gamma_e$ and small $r_s$. The factor $c_\infty = (18/175)(12/\pi)^{2/3} = 0.2513$ is consistent with the zero-temperature Thomas-Fermi approximation (Salpeter, 1961),

$$F_{\rm ie}^{\rm TF}/N_e = U_{\rm ie}^{\rm TF}/N_e = -2c_\infty \gamma_{\rm r}\, Z^{7/3}\,{\rm Ry}, \tag{2.158}$$

which becomes exact at small $r_s$ and very large $Z$. Here, we introduce the Rydberg energy Ry$=\frac{1}{2}m_e e^4/\hbar^2 = 13.6$ eV. The parameters

$$\begin{aligned} a &= 1.11\, Z^{0.475}, \\ b &= 0.2 + 0.078\,(\ln Z)^2, \\ \nu &= 1.16 + 0.08 \ln Z \end{aligned}$$

provide some starting approximation to $F_{\rm ie}$ (with a maximum error up to 30% at large $Z$ and $r_s$), while the functions

$$g_1(r_s) = 1 + \frac{0.78}{21 + \Gamma_e (Z/r_s)^3}\left(\frac{\Gamma_e}{Z}\right)^{1/2}, \tag{2.159}$$

$$g_2(r_s) = 1 + \frac{Z-1}{9}\left(1 + \frac{1}{0.001\, Z^2 + 2\Gamma_e}\right)\frac{r_s}{4 + r_s^{-2}} \tag{2.160}$$

improve the fit at relatively large $r_s$ and reduce the maximum relative fit error of $F_{\rm ie}$ to within 4.3%.

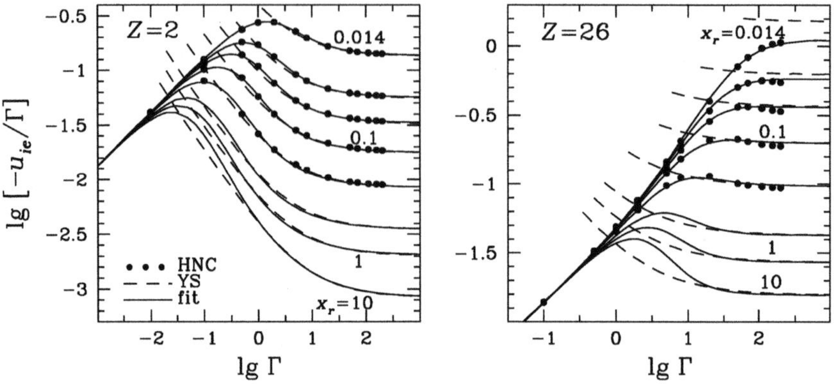


*Figure 2.5.* Screening part of the free energy for an electron-ion plasma with $Z = 2$ (*left*) and $Z = 26$ (*right*) versus $\Gamma$. Numerical finite-temperature non-relativistic HNC calculations by Chabrier & Potekhin (1998) (dots) and zero-temperature relativistic perturbation results by Yakovlev & Shalybkov (1989) (valid at small $r_s$ and large $\Gamma$; dashed lines) are compared with the fit (2.154). Numbers at the curves indicate the values of $x_\mathrm{r} = 0.014/r_s$ (from top to bottom, $x_\mathrm{r} = 0.014$, 0.034, 0.058, 0.1, 0.22, 0.55, 1, 10).

Note that the asymptotic behavior of Eq. (2.154) at $\Gamma \to \infty$ should change from $f_\mathrm{ie} \propto \Gamma r_s$ to $f_\mathrm{ie} \propto \Gamma r_s \gamma_\mathrm{r}$ if the electrons are relativistic. This is ensured by the factor $\gamma_\mathrm{r}$ inserted in the denominator of Eq. (2.154). The factor $g_3$ corrects the fit at relatively low $Z$ in the relativistic domain,

$$g_3(x_\mathrm{r}) = \frac{1 + {x_\mathrm{r}}^2/5}{1 + 0.18\, Z^{-1/4} x_\mathrm{r} + 0.37\, Z^{-1/2} {x_\mathrm{r}}^2 + {x_\mathrm{r}}^2/5}. \tag{2.161}$$

The calculated and fitted values of $F_\mathrm{ie}$ are shown in Fig. 2.5. For comparison, we present also the results of Yakovlev & Shalybkov (1989), obtained from the equation

$$f_\mathrm{ie} = -\frac{\Gamma a_\mathrm{i}}{\pi} \int_0^\infty S(q)\, [\varepsilon(q, x_\mathrm{r}) - 1]\, \mathrm{d}q. \tag{2.162}$$

This equation was derived by Galam & Hansen (1976) using the formalism of electron dielectric function and treating polarization corrections as small perturbations.

Equations (2.87) and (2.154) ensure the relative accuracy of the fit to $F_\mathrm{ii} + F_\mathrm{ie}$ to be $\lesssim 2\%$ for $\Gamma < 170$ and $0.0625 \leq r_s \leq 2$ in the non-relativistic calculations (Chabrier & Potekhin, 1998) and they agree with the relativistic zero-temperature results of Yakovlev & Shalybkov (1989) at $r_s \ll 1$ and $T \ll T_\mathrm{F}$.

### 2.4.5 Electron polarization in ion solid

Crystallization of ions affects the electron screening functions. A calculation of thermodynamic functions for a Coulomb solid with the allowance for the electron polarization is a complex problem. For classical ions, the simplest screening model consists in replacing the Coulomb potential by the Yukawa potential. Molecular dynamics simulations of Yukawa systems have been performed by Hamaguchi *et al.* (1997). They found that, with decreasing the Yukawa screening radius, the melting point shifts to higher $\Gamma$. Moreover, when the Yukawa screening radius becomes smaller than the Debye radius (2.25), there appears a second phase transition (at still higher $\Gamma$) from bcc to fcc lattice; at still smaller screening radius there is a direct transition from liquid to fcc crystal. Note that Hamaguchi *et al.* (1997) simulated a classical Yukawa system with a fixed screening parameter (independent of density and temperature). However, this is not exactly true for an electron-ion plasma; the effective screening does depend on plasma parameters.

A rigorous treatment of the ionic Coulomb crystal with allowance for the electron screening would consist in calculating the dynamical matrix and solving a corresponding dispersion relation for the phonon spectrum. The first-order perturbation approximation for the dynamical matrix of a classical Coulomb solid with the polarization corrections, based on an *effective* inter-ion potential, was developed by Pollock & Hansen (1973). The phonon spectrum in a quantum crystal could be calculated from a quantum-mechanical treatment of the electron-phonon interactions. This has not been done yet, except in the harmonic approximation (Baiko, 2002). However, the harmonic approximation *in principle* cannot provide, for example, a correct classical (i.e., $T \gg T_{\rm pi}$) limit of the $ie$ contribution to the heat capacity (as discussed below in §§ 2.4.5 c and 2.4.6). Here we present a simpler, semiclassical perturbation approach for evaluating the polarization corrections, proposed by Potekhin & Chabrier (2000).

#### 2.4.5 a Perturbation treatment

An ion crystal with the rigid electron background is a natural starting approximation. Let us note that the effective inter-ion potential in the adiabatic perturbation approximation (Pollock & Hansen, 1973) is just the electrostatic potential, the same in the liquid and solid phases. The difference of this potential from the bare Coulomb potential can be considered as perturbation. Then we can apply the perturbation theory of Galam & Hansen (1976), based on the *exact* expression for the free energy. It involves an integration over a coupling parameter which determines the perturbation "strength." In this way we arrive at Eq. (2.162) with $S(q) = (4\pi)^{-1} \int S(\boldsymbol{q})\, \mathrm{d}\Omega_{\boldsymbol{q}}$, where $\mathrm{d}\Omega_{\boldsymbol{q}}$ is a solid angle

*Table 2.6.* Parameters of Eq. (2.167).

| lattice type | $a_1$ | $a_2$ | $a_3$ | $a_4$ |
|---|---|---|---|---|
| bcc | 1.1866 | 0.684 | 17.9 | 41.5 |
| fcc | 1.1857 | 0.663 | 17.1 | 40.0 |

element in the direction of $\boldsymbol{q}$; the structure factors $S(\boldsymbol{q})$ and $S(q)$ depend on $\Gamma$ and $t_\mathrm{p}$.

Using the decomposition of $S(q)$ into the elastic and inelastic parts (§ 2.3.3 f), we can rewrite Eq. (2.162) as

$$f_\mathrm{ie} = f'_\mathrm{ie} + f''_\mathrm{ie}, \tag{2.163}$$

$$f'_\mathrm{ie} = -\frac{3\Gamma}{2}\sum_{\boldsymbol{G}\neq 0} \frac{\varepsilon(G, x_\mathrm{r}) - 1}{(Ga_\mathrm{i})^2}\,\mathrm{e}^{-2W(G)}, \tag{2.164}$$

$$f''_\mathrm{ie} = -\frac{\Gamma a_\mathrm{i}}{\pi}\int_0^\infty S''(q)\,[\varepsilon(q, x_\mathrm{r}) - 1]\,\mathrm{d}q. \tag{2.165}$$

For the inelastic part of the structure factor, an exact expression (2.134) is impractical because of a slow convergence of the sum. Nevertheless, a simple estimate may be obtained using Eq. (2.140). As argued in § 2.3.3 f, this approximation is good for integrals over $\boldsymbol{q}$ at $q > q_\mathrm{BZ}$. In Eq. (2.165), however, it is important to recover the correct limiting behavior of $S''(q)$ at $q \to 0$, because $(\varepsilon - 1) \propto q^{-2}$ becomes large in this limit. For this reason, Potekhin & Chabrier (2000) used a piecewise approximation: $S''(q) = S''_0(q)$ at $q > q_1$, and $S''(q) = S''_1(q)$ at $q \le q_1$ where $S''_1(q) = \left[\partial^2 S''/\partial q^2\right]_{q\to 0}(q^2/2)$ and the free parameter $q_1$ was set $q_1 = 0.94\,q_\mathrm{BZ}$. The exact result for classical Coulomb plasmas (Vieillefosse & Hansen, 1975) at $qa_\mathrm{i} \ll 1$ yields $S''_1(q) = (qa_\mathrm{i})^2/(3\Gamma)$. Generally, $S''_1(q)$ can be found from Eq. (2.134):

$$S''_1(q) = \frac{(qa_\mathrm{i})^2}{6\Gamma}\,\frac{t_\mathrm{p}^{-1}}{\tanh(t_\mathrm{p}^{-1}/2)}. \tag{2.166}$$

Using Eqs. (2.140) and (2.163)–(2.166), Potekhin & Chabrier (2000) calculated the polarization correction $f_\mathrm{ie}$ in a wide range of plasma parameters $\Gamma$, $x_\mathrm{r}$, and $Z$.

At $t_\mathrm{p} \lesssim 1$ the quantization of ionic vibrations becomes important, which modifies the structure factor. In the formalism of Potekhin & Chabrier (2000) this effect was taken into account by the factor $t_\mathrm{p}^{-1}/\tanh(t_\mathrm{p}^{-1}/2)$ in (2.166). As

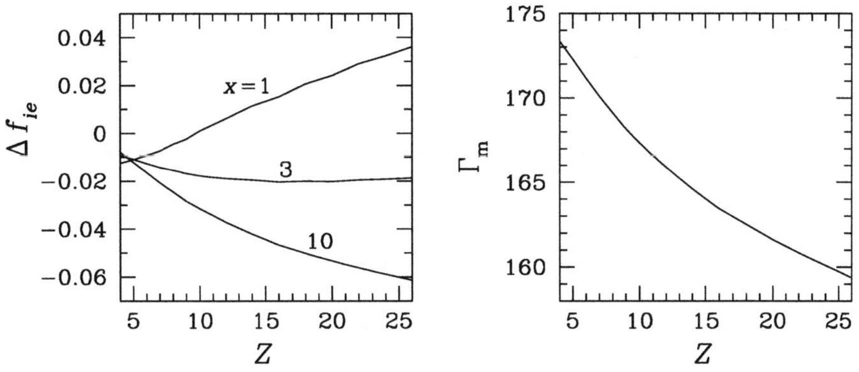


*Figure 2.6. Left*: The difference between polarization corrections to the free energy in the solid and liquid phases at $\Gamma = 175$ and $x_\mathrm{r} = 1, 3, 10$. *Right*: The Coulomb coupling parameter at the melting point with allowance for polarization corrections at $x_\mathrm{r} \gg 1$.

a consequence, the polarization contribution $C_{V,ie}$ to the specific heat goes to zero at $T \to 0$. However, at $t_\mathrm{p} \to 0$ it was found to become larger than the ion and electron contributions, thus invalidating the perturbation approximation. This indicates that the employed approximation for the inelastic part of the structure factor is inapplicable in the strong quantum regime.

### 2.4.5 b Analytic formula

In the *classical* domain, the numerical results can be fitted by the expression

$$f_\mathrm{ie} = -f_\infty(x_\mathrm{r})\,\Gamma\,[1 + A(x_\mathrm{r})/\Gamma^s]\,, \tag{2.167}$$

where

$$f_\infty(x) = a_\mathrm{TF} Z^{2/3} b_1 \sqrt{1 + b_2/x^2},$$

$$A(x) = \frac{b_3 + a_3 x^2}{1 + b_4 x^2}, \qquad s = \left[1 + 0.01\,(\ln Z)^{3/2} + 0.097\,Z^{-2}\right]^{-1},$$

$$b_1 = 1 - a_1\,Z^{-0.267} + 0.27\,Z^{-1}, \quad b_2 = 1 + \frac{2.25}{Z^{1/3}}\,\frac{1 + a_2\,Z^5 + 0.222\,Z^6}{1 + 0.222\,Z^6},$$

$$b_3 = a_4/(1 + \ln Z), \qquad b_4 = 0.395\ln Z + 0.347 Z^{-3/2}.$$

The parameter $a_\mathrm{TF}$ [as well as $c_\mathrm{TF}$ in Eq. (2.154)] is chosen in such a way to reproduce the Thomas-Fermi limit, Eq. (2.158), at $Z \to \infty$:

$$a_\mathrm{TF} = \left(\frac{9\pi}{4}\right)^{1/3} \frac{\alpha_\mathrm{f}}{Z^{7/3}}\, c_\infty = \frac{54}{175}\left(\frac{12}{\pi}\right)^{1/3} \alpha_\mathrm{f} \approx 0.00352.$$

The coefficients $a_1$–$a_4$ are given in Table 2.6; they are slightly different for bcc and fcc crystals. The fit error does not exceed 3%.

### 2.4.5 c Discussion

The results presented above indicate that, although the polarization in a Coulomb crystal is typically weak, it does not vanish even at very large $\Gamma$ and $x_{\mathrm{r}}$. It introduces an additional binding into the system. In a strongly coupled Coulomb liquid (§ 2.4.4), it is roughly proportional to $(k_{\mathrm{TF}}a)^2$, which tends to a finite limit at relativistic densities. More accurately, the order of magnitude of the screening correction $F_{\mathrm{ie}}/F_{\mathrm{i}}$ for a classical Coulomb plasma at arbitrarily high densities is given by the Thomas-Fermi result (Salpeter, 1961), $F_{\mathrm{ie}}/F_{\mathrm{i}} \approx 0.004\, Z^{2/3}$. It is reproduced by Eq. (2.167) at $\Gamma \to \infty$ and extremely large $Z$. The quantitative difference of the perturbation result at finite $Z$ from the Thomas-Fermi limit is quite perceptible, $\sim Z^{-0.3}$.

At relativistic densities the contribution of the electron polarization to the pressure, $P_{\mathrm{ei}}$ (see Eq. (2.45)), has the same density dependence ($P_{\mathrm{ei}} \propto \rho^{4/3}$) as the leading electron pressure $P_{\mathrm{id}}^{(e)}$ and the electrostatic correction $P_{\mathrm{ii}}$ discussed in § 2.3.5. Thus, the relative contribution of the electrostatic correction does not vanish at ultrarelativistic densities. However, this correction is small by itself ($|P_{\mathrm{ii}}|/P_{\mathrm{id}}^{(e)} \sim \alpha_{\mathrm{f}} Z^{2/3}$). Accordingly, the polarization correction is much smaller ($|P_{\mathrm{ei}}|/P_{\mathrm{id}}^{(e)} \sim \alpha_{\mathrm{f}}^2 Z^{4/3}$, see Table 2.1) and can be neglected in the EOS at high densities. For instance for the iron matter at relativistic densities we have $|P_{\mathrm{ii}}|/P_{\mathrm{id}}^{(e)} \approx 4\%$, and $|P_{\mathrm{ei}}|/P_{\mathrm{id}}^{(e)} \approx 0.07\%$.

In spite of their smallness, the polarization corrections can affect the melting temperature of the Coulomb crystal, as illustrated in Fig. 2.6. Although the difference $\Delta f_{\mathrm{ie}}$ between the $f_{\mathrm{ie}}$ values in the solid and liquid phases at the OCP melting point is small as compared to the total excess free energy, it noticeably affects $\Gamma_{\mathrm{m}}$ (Potekhin & Chabrier, 2000). Because $\Delta f_{\mathrm{ie}}$ remains finite at any $x_{\mathrm{r}}$, the classical OCP value $\Gamma_{\mathrm{m}} = 175$ is never recovered, even at very large densities. In addition, the polarization corrections can change the type of the most stable Coulomb lattice in dense matter at zero temperature, from bcc to fcc (Hamaguchi *et al.*, 1997; Baiko, 2002).

Finally, note that the approximation presented in this section grasps some effects beyond the harmonic approximation for the crystal of ions interacting through an effective potential. For example, the $ie$ contribution to the specific heat in the harmonic approximation (Baiko, 2002) vanishes at $T \gg T_{\mathrm{pi}}$, because $C_V = 3Nk_{\mathrm{B}}$ for any harmonic crystal, whether or not the effective ion-ion interaction potential includes the polarization correction. On the contrary, Eq. (2.162), used above, gives a finite polarization correction to $C_V$ at $T \gg T_{\mathrm{pi}}$.

However, one should bear in mind that the above approach is not exact because the ion structure factor is taken in the harmonic approximation. Moreover, the derivation of Eq. (2.162) (based on the semiclassical first-order perturbation theory) does not apply at $T \ll T_{\mathrm{pi}}$.

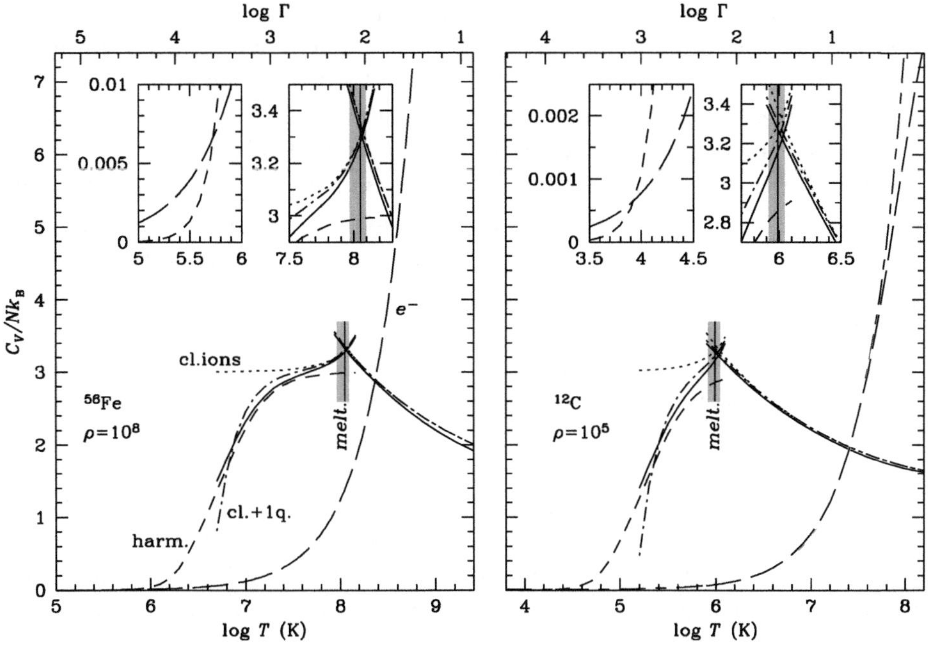


*Figure 2.7.* Heat capacity per ion versus $T$ (*bottom axis*) and $\Gamma$ (*top axis*) for iron plasma at $\rho = 10^8$ g cm$^{-3}$ (*left*) and carbon plasma at $\rho = 10^5$ g cm$^{-3}$ (*right*). The dotted line is for classical ions; the dot-dash line additionally includes the first quantum correction; the short-dash line is the harmonic-lattice approximation in the solid phase; the solid line is the same plus anharmonic and electron polarization corrections in the solid phase and the sum of classical, quantum, and electron polarization contributions in the liquid. The long-dashed line is the ideal electron gas contribution, and the short-dash–long-dash line on the right panel is the approximation of strongly degenerate electrons. The vertical line is the melting point, and the grey band shows the range $\Gamma = 150$–$200$. The insets present the competition between the electron and ion contributions at low $T$ (*left inset* on each panel) and the behavior near the melting point (*right inset* on each panel).

In reality, the picture discussed above is further complicated owing to the band structure in the electron energy spectrum in a crystal (§ 2.1.2). Thus the problem deserves further study.

## 2.4.6 The heat capacity

To illustrate the temperature dependence of thermodynamic quantities, let us outline the main features of the heat capacity of the Coulomb plasma. In Fig. 2.7 we show the temperature dependence of the electron and ion heat capacities (calculated per one ion at constant volume). The left panel shows various contributions to $C_V$ for the iron plasma at $\rho = 10^8$ g cm$^{-3}$. In this case, the plasma parameters are $x_r = 3.626$, $T_F = 1.627 \times 10^{10}$ K, that is, the electrons are relativistic and strongly degenerate ($T \ll T_F$ in the shown range of $T$). The right panel is for the carbon plasma at $\rho = 10^5$ g cm$^{-3}$, where the electrons are nonrelativistic ($x_r = 0.372$) and become mildly degenerate as $T$ approaches $T_F = 3.963 \times 10^8$ K. The ion plasma temperature is $T_{pi} = 3.636 \times 10^7$ K and $1.238 \times 10^6$ K for the iron and carbon matter, respectively.

The electron contribution (long-dashed lines) is given by Eq. (2.39), where the entropy $S = (-\partial F_{id}^{(e)}/\partial T)_{V,N_e}$ is calculated using Eq. (2.49) and the fitting expressions (2.54)–(2.59). At $T \ll T_F$ this contribution is accurately given by Eq. (2.68), but at higher $T$ it becomes invalid (as shown by the short-dash–long-dash line on the right panel).

The ion contribution is calculated in various approximations for the ion liquid (to the right of the melting point) and the bcc ion crystal (to the left of the melting point).

The dotted line displays the heat capacity for a classical Coulomb plasma of ions. It is calculated using Eq. (2.89) in the ion liquid, and from Eqs. (2.105) and (2.116) in the crystal. When the temperature decreases to the melting temperature $T_m$, the dimensionless heat capacity per ion, $c_i \equiv C_{V,i}/N_{\mathcal{N}} k_B$ increases from $c_i = \frac{3}{2}$ in the gaseous phase to $c_i \approx 3$ in the strongly coupled liquid (where the ions mainly oscillate in their slowly migrating potential wells, producing the heat capacity of oscillatory type). This increase is not abrupt, but gradual, due to the nonideal term $C_{V,ii}$, as first mentioned by Mestel & Ruderman (1967) and Van Horn (1968). At $\Gamma = 175$ in the liquid the classical heat capacity of ions (including the Coulomb term but without electron polarization and quantum contributions) is $c_i^{cl} = 3.39$. Just after the crystallization, at the same $\Gamma$, it drops by 4% to $c_i^{cl} = 3.27$ (as shown in the right insets in more details). However, this drop is so small that it can be ignored in the majority of applications. The drop would entirely disappear if the bcc lattice crystallized at $\Gamma \approx 160$. Note in passing that for an fcc crystal at $\Gamma = 175$ we would have $c_i^{cl} = 3.33$ (instead of 3.27) and the drop would disappear at $\Gamma \approx 170$.

With decreasing $T$ below $T_m$ the heat capacity of the classical crystal decreases to the value $c_i = 3$ predicted by the harmonic-lattice theory. The 9% difference between the real value $c_i = 3.27$ for the bcc crystal with a non-polarized electron background at $T = T_m$ and the harmonic-lattice value $c_i = 3$ is the measure of the anharmonicity of the classical bcc Coulomb crystal at the melting point. For lower $T$ the anharmonicity becomes less important. At $T \ll T_m$ the classical theory breaks down and highly overestimates the heat capacity (cannot be used in applications).

The dot-dashed line in Fig. 2.7 presents the same classical heat capacity of ions with the added quantum diffraction correction. The correction $\delta c_q = -\frac{1}{12}(T_p/T)^2$ is calculated using Eq. (2.85); it is the same for the liquid and solid. We see that the quantum effects only slightly lower the heat capacity in the liquid, but become large at $T < T_m$. For the matter composed of lighter elements the quantum correction in the liquid is more important (compare the difference between the dotted and dot-dashed lines on the left and right panels of Fig. 2.7).

The short-dashed line is the ion heat capacity of the crystal calculated in the harmonic-lattice approximation from Eq. (2.110). It is expected to be a valid approximation at low temperatures, but underestimates the heat capacity by $\sim$10% at the melting point.

Finally, the solid line in Fig. 2.7 includes the classical and quantum ion contributions plus electron polarization corrections. In the liquid, it is calculated using Eqs. (2.89), (2.85), and (2.154). In the solid, it is the sum of the harmonic approximation, classical anharmonic correction, and the polarization correction obtained using Eqs. (2.110), (2.116), and (2.167), respectively. As discussed in §§ 2.3.3 e and 2.4.5 c, the latter two corrections are not known at $T \lesssim T_{\rm pi}$, which does not allow us to continue the solid line to very low temperatures. Thus, a correct theory of a Coulomb crystal which combines the quantum and classical regimes still has to be elaborated.

The electron polarization effects can shift the melting temperature (§ 2.4.5 c). Therefore, the phase transition may occur anywhere within the grey-shadowed regions in Fig. 2.7 ($150 \leq \Gamma \leq 200$). For the bcc crystal and polarized electron background, we have $\Gamma_{\rm m} = 166$ and 170 for the conditions in the left and right panel of Fig. 2.7, respectively. These melting points are shown by the vertical lines within the grey bands. In our two examples, the total (electron and ion, including all nonideal terms) heat capacity per ion $C_V/N_{\!\scriptscriptstyle N} k_{\rm B}$ drops from 4.70 to 4.62 on the left panel and from 3.28 to 3.09 on the right panel of Fig. 2.7.

At high temperatures the electron heat capacity dominates over the ion one (Fig. 2.7) because of the large amount of electrons per one ion in the iron and carbon matter. However, with decreasing $T$ the electron heat capacity decreases proportionally to $T$, while the ion heat capacity is a slow function of $T$ as long as the quantum effects are unimportant. Accordingly, the ion heat capacity dominates in a large temperature interval. However, when the temperature becomes much lower than $T_{\rm pi}$, the quantum effects strongly reduce the ion heat capacity, and the electron heat capacity dominates again, as seen in the left insets of both panels of Fig. 2.7.

While studying the temperature dependence of the heat capacity (Fig. 2.7) one should not forget about the latent heat released at the crystallization point $T = T_{\rm m}$ (§ 2.3.4). Strictly speaking, the latent heat adds a delta-function contribution to $C_V(T)$, in the form of an infinitely narrow and high peak at $T = T_{\rm m}$ (not shown in the figure). One should bear in mind, that real freezing of stellar matter can be more complicated than an idealized freezing at the thermodynamic phase transition. Monte Carlo simulations of freezing of classical OCP (e.g., Dewitt *et al.* 1993) indicate that, as a rule, the ions do not freeze at $T = T_{\rm m}$ but form a supercooled fluid and freeze at lower $T$ (depending on initial conditions and other parameters). This happens because the phase transition is really tiny. If the same takes place in stellar matter, various fluid elements solidity at different $T$ below $T_{\rm m}$, and the average heat capacity $C_V(T)$ can contain a bump, associated with latent heat releases.

### 2.4.7 Multi-ionic mixtures

A multi-ionic mixture is a straightforward generalization of the single-ion model. In this case the effective Hamiltonian reads:

$$H^{\rm eff} = \hat{K} + \frac{1}{2V} \sum_{\boldsymbol{k} \neq 0} \frac{4\pi e^2}{k^2} \left[ \frac{\rho_{Z\boldsymbol{k}} \rho^*_{Z\boldsymbol{k}}}{\varepsilon(k)} - N_{\mathcal{N}} \langle Z^2 \rangle \right], \tag{2.168}$$

where $\hat{K}$ is the kinetic energy operator and $\rho_{Z\boldsymbol{k}} = \sum_j Z_j \rho_{j\boldsymbol{k}}$ is a Fourier component for a fluctuation of the ion charge number.

For a classical binary ionic liquid with a rigid electron background (where $\varepsilon(k) \equiv 1$), the excess free energy, as well as related thermodynamic quantities, can be expressed with high accuracy by the so-called "linear mixing rule" in terms of the free energies of pure phases,

$$f_{\rm ex}(Z_1, Z_2, \Gamma_e, x_1) \approx x_1\, f_{\rm ex}(\Gamma_1, x_1 = 1) + x_2\, f_{\rm ex}(\Gamma_2, x_1 = 0), \tag{2.169}$$

where $x_1 = N_1/(N_1+N_2) = 1-x_2$. A very high accuracy of the linear mixing rule (2.169) was first demonstrated by Hansen *et al.* (1977) and confirmed later by several authors, using very accurate Monte Carlo calculations (e.g., DeWitt *et al.* 1996). Notice, that the linear mixing for the excess free energy breaks down in the gaseous regime. However, the excess free energy in that regime is small by itself, while the ideal gas contribution, dominant in the gaseous regime, satisfies the linear mixing rule. Therefore, Coulomb ion mixtures seem to respect the linear mixing with very good precision.

The validity of the linear mixing rule for the excess free energy in the case of an ionic mixture immersed in a *compressible* finite-temperature electron background, described by the Hamiltonian (2.168), was examined by Hansen *et al.* (1977) in the first-order thermodynamic perturbation approximation and later by Chabrier & Ashcroft (1990), who solved the HNC equations with the effective Coulomb screened potentials for arbitrary ion charge ratios ranging from a symmetric case ($Z_2/Z_1 \sim 1$) to a highly asymmetric case ($Z_2/Z_1 \gg 1$). These authors found that the linear mixing rule remains very accurate in the classical binary Coulomb liquids with the electron response, except possibly for highly asymmetric mixtures in the region of weak electron degeneracy (where the departure from the linearity can reach a few percent).

## 2.5. Low densities. Partial ionization

A partially ionized plasma is present in the outer parts of neutron star atmospheres, at relatively low $\rho$ and $T$. The larger the ion charge number $Z$, the higher the values of $\rho$ and $T$ up to which the plasma is partially ionized. For instance, the partial ionization is unimportant in the left panel of Fig. 2.2, which shows the $\rho$–$T$ diagram for a carbon plasma, but it is important in the lower left corner of the diagram for an iron plasma in the right panel.

### 2.5.1 Ionization equilibria in the "chemical picture"

A theoretical description of partially ionized plasmas can be based either on the physical picture or on the chemical picture of the plasma. In the chemical picture, bound objects (atoms, molecules, ions) are treated as elementary members of the thermodynamic ensemble, along with free electrons and nuclei. In the physical picture, nuclei and electrons (free and bound) are the only constituents of the ensemble. For instance, Rogers & DeWitt (1973) proposed an approach based on the activity (=fugacity) expansion (e.g., Landau & Lifshitz 1993, §74) in the framework of the physical picture of Coulomb plasmas. It was employed in the Opacity Library (OPAL) project and resulted in the EOS presented by Rogers *et al.* (1996).

Both pictures can be thermodynamically self-consistent, but the chemical picture has a limited microscopic consistency. For instance, it does not provide a proper treatment of such cluster configurations as "an atom + a close alien ion" (a pseudo-molecular ion). With increasing density, the ionization of electrons bound to particular nuclei proceeds through a progressive delocalization involving cluster ("hopping") states (Perrot & Dharma-wardana, 1994). Since an inclusion of these clusters as new members of the thermodynamic ensemble would overcomplicate the free energy model, one usually restricts oneself to the basic chemical species which dominate at low densities.

On the other hand, the physical picture is commonly based on diagrammatic expansions which converge only at low densities. In particular, the OPAL tables are bound from the low-$T$–high-$\rho$ side by the line $\rho = \rho_{\rm OPAL} \approx 10\, T_6^3$ g cm$^{-3}$. In addition, the physical picture, unlike the chemical one, does not automatically provide occupation numbers for the bound species, which are needed to obtain radiative opacities. A special procedure is required to evaluate the actual occupation numbers (Rogers, 1986), which is not free of ambiguities.

Thus one has to resort to additional assumptions in the framework of either picture, in order to progress to the region of plasma non-ideality. Relative merits of the two approaches were discussed by Däppen (1992). The chemical picture, combined with the free energy minimization method, represents a reasonable compromise between the rigorous treatment and the practical application.

In the chemical picture at low densities four basic approximations are usually employed (Fontaine *et al.*, 1977). The *first* (quasiclassical) approximation replaces a summation over all quantum states in the canonical partition function $\mathcal{Z}$ by an integral over phase space. It is justified if the thermal wavelength $\lambda_j$ is much smaller than the average distance between particles. The *second* approximation consists in the separation of the kinetic (translational) part (§ 2.2.1, Eq. (2.44)). The *third* approximation reduces the intractability of the volume integral by making certain assumptions about the interaction energy $\mathcal{U}(\boldsymbol{r}_1, \ldots, \boldsymbol{r}_N)$ among those particles which are chosen as elementary members of the statistical ensemble in the chemical picture (in the stellar atmosphere, these are electrons, protons, and bound species, such as atoms, molecules, and ions). The *fourth* approximation consists in taking the term representing the internal energies of the bound species outside the volume integral, which results in the separation of $F_{\rm int}$ in Eq. (2.33). This is possible, for example, if one assumes that the internal energy of a particle does not depend on its position relative to other particles.

In reality, any of the above approximations can be inadequate at sufficiently high density *or* for a sufficiently high bound level. The main difficulty in the treatment of partial ionization in the framework of the chemical picture is the uncertainty of the criterion for the partial or full ionization. In particular, considering a partially ionized atmosphere, one must specify, which non-ideality effects are strong enough to destroy the atoms. The inherent necessity to allow for such a "pressure destruction" of bound species is already signaled by the well known divergence of the internal partition function of atoms in case these effects are neglected.

Clearly, a bound state cannot be populated, if the corresponding "size" of the electron "orbit" (the electron cloud) is too large. In other words, an electron becomes unbound either in the absence of a free space or in the presence of strong perturbations. The problem is how to include the unbinding into the chemical picture in a self-consistent manner. The simplest way would be to truncate the internal partition function at an appropriate "critical" quantum number $\kappa$, – for instance, when a quantum-mechanical size of an atom becomes comparable with the mean inter-particle distance. However, an abrupt cutoff produces unrealistic discontinuities in the free energy. Continuous truncation procedures imply the introduction of *occupation probabilities* $w_\kappa$, which suppress the contribution of higher states and ensure the convergence. For example, let us consider nondegenerate electrons in equilibrium with classical ions of a $Z$th chemical element with $j$ unbound electrons and mass $m_j$, which can be in different quantum states $\kappa$ with binding energies $E_{j,\kappa}$ and statistical weights $g_{j,\kappa}$. Then the number densities $n_{j,\kappa}$ of ion species are given by the set of equations

$$\frac{n_{j,\kappa}}{n_{j',\kappa'}} = \left(\frac{n_e \lambda_e^3}{2}\right)^{j'-j} \left(\frac{m_j}{m_{j'}}\right)^{3/2} \frac{g_{j,\kappa}}{g_{j',\kappa'}} \frac{w_{j,\kappa}}{w_{j',\kappa'}} \exp\left(\frac{E_{j,\kappa} - E_{j',\kappa'}}{k_B T}\right), \tag{2.170}$$

which differ from the usual (ideal-gas) equations of chemical equilibrium by a factor $w_{j',\kappa'}/w_{j,\kappa}$. Thus the occupation probability $w_{j,\kappa}$ is a multiplicative factor to $g_{j,\kappa}$. The ionization equilibrium equation (for $n_j \equiv \sum_\kappa n_{j,\kappa}$) is immediately obtained from Eq. (2.170). Neglecting the difference between $m_j$ and $m_{j'}$, we get

$$\frac{n_j}{n_{j+1}} = \frac{n_e \lambda_e^3}{2} \frac{\mathcal{Z}_{\mathrm{int},j}}{\mathcal{Z}_{\mathrm{int},j+1}} \exp\left(\frac{E_{j,\mathrm{ion}}}{k_B T}\right), \tag{2.171}$$

where

$$\mathcal{Z}_{\mathrm{int},j} = \sum_\kappa g_{j,\kappa}\, w_{j,\kappa} \exp\left(\frac{E_{j,\kappa} - E_{j,\mathrm{gr-st}}}{k_B T}\right) \tag{2.172}$$

is the internal partition function, $E_{j,\mathrm{gr-st}}$ is the ground-state binding energy for an $j$th ionization stage, and $E_{j,\mathrm{ion}} = E_{j,\mathrm{gr-st}} - E_{j+1,\mathrm{gr-st}}$ is the ionization energy. Equation (2.171) reproduces the usual Saha equation, but the internal partition function is now *generalized* through the inclusion of the occupation probabilities $w_{j,\kappa}$. As shown by Fermi (1924), $w_{j,\kappa}$ cannot be arbitrary but should be consistent with the adopted form of $F_{\mathrm{ex}}$. Equation (2.34) implies (Potekhin, 1996b)

$$w_{j,\kappa} = \exp\{[(Z-j)\,\Delta\mu_e + \Delta\mu_Z - \Delta\mu_{j,\kappa}]/k_B T\}, \tag{2.173}$$

where

$$\Delta\mu_e = \partial F_{\mathrm{ex}}/\partial N_e, \quad \Delta\mu_Z = \partial F_{\mathrm{ex}}/\partial \mu_Z, \quad \text{and} \quad \Delta\mu_{j,\kappa} = \partial F_{\mathrm{ex}}/\partial \mu_{j,\kappa}$$

are non-ideal corrections to respective chemical potentials (Fontaine *et al.*, 1977).

Many attempts have been made to construct physically reasonable occupation probabilities. Some of them were discussed by Hummer & Mihalas (1988). For instance, one often uses the Planck-Larkin partition function, which is constructed by replacing the Boltzmann factors $e^{E/k_B T}$ by $(e^{E/k_B T} - E/k_B T - 1)$, in order to achieve the convergence of the internal partition function for Rydberg bound states. This approach yields a correct EOS at low densities where the virial expansion holds (Ebeling *et al.*, 1977). It requires, however, a redefinition of the electron continuum states in a manner consistent with the modification of the Boltzmann factors (Rogers, 1986). Therefore, the Planck-Larkin formalism fails at higher densities, where the continuum is perturbed so strongly that an exact solution for a free electron cannot be incorporated in the chemical picture of the plasma (see, e.g., Däppen *et al.* 1987).

Hummer & Mihalas (1988) proposed a model in which the atoms are assumed to be destroyed by plasma microfields that exceed a critical value for the Stark merging of resonant lines in an absorption spectrum of the plasma, according to the criterion of Inglis & Teller (1939). Microscopically, it looks as if an atom disintegrated as soon as a plasma ion falls inside a sphere whose radius is several times larger than the quantum-mechanical radius of the atom. It was found (Mihalas *et al.*, 1988) that this model does not provide proper pressure ionization at high densities. In order to force the ionization, Mihalas *et al.* (1988) introduced an artificial "pressure ionization term" into the free energy. This term ensures a reasonable EOS at $\rho \lesssim 10^{-2}$ g cm$^{-3}$ but produces an unrealistically stiff EOS at higher $\rho$, as noted by Saumon *et al.* (1995).

The origin of this shortcoming can be seen from the following consideration (Potekhin, 1996b). Let us assume that an electron becomes "unbound" when another plasma electron falls inside the quantum-mechanical volume of the first electron and shields the attraction to the nucleus. On the other hand, positively charged ions approaching the atom do not make the electron unbound, but tend to form the cluster (hopping) states mentioned above. Indeed, the ion-atom interaction yields a negative contribution to the energy, whereas unbinding implies this contribution to be positive. Thus in the chemical picture, the contribution of the cluster states to the free energy may be replaced by the contribution of atoms. This is closer to reality than the replacement of such states by a collection of free electrons and nuclei.

While adopting the above approach, one should keep in mind, however, that optical properties of clusters strongly differ from those of isolated atoms. Therefore, some additional *optical* occupation probabilities have to be introduced, which affect only the spectrum, but not thermodynamics. Thus the optical continuum lies lower than the thermodynamic one and overlaps some loosely bound states which actually belong to clusters, rather than to isolated atoms. This dichotomy of the optical and thermodynamic continua was first emphasized by Rogers (1986), who utilized it in the physical picture of plasmas. In the framework of the chemical picture it was realized in the model described in §2.5.2.

Models based on the chemical picture of the plasma have the unsurpassed advantage that they provide detailed information about occupation numbers of various species in the plasma, which is of paramount importance in astrophysics since it enables one to calculate radiative opacities. A simple example of such a model is given below.

## 2.5.2 An example: Partially ionized hydrogen plasma

The EOS of hydrogen in various physical conditions has been studied in many papers. Here we focus on the partially ionized atomic hydrogen at $T \gtrsim 3 \times 10^4$ K, that is relevant to neutron star atmospheres. Let us employ the chemical picture of the plasma. A detailed model in the framework of the chemical picture of a plasma consisting of protons, electrons, H atoms, and $H_2$ molecules (in all quantum-mechanical states) was developed by Saumon *et al.* (1995). A completely analytic version of the associated free-energy model, applicable to the case of weakly degenerate electrons in the absence of molecules, was presented by Potekhin (1996b) and extended to the case of arbitrary electron degeneracy by Potekhin *et al.* (1999b). In this model,

$$F_{\rm id} = F_{\rm id}^{(e)} + F_{\rm id}^{(p)} + F_{\rm id}^{\rm neu} + F_{\rm rad} \tag{2.174}$$

is the sum of the ideal-gas free energies of electrons (2.49), protons (2.71), neutral species, and photons (thermal radiation), respectively, and

$$F_{\rm ex} = F_{\rm ii} + F_{\rm xc} + F_{\rm ie} + F_{\rm ex}^{\rm neu} \tag{2.175}$$

includes the contributions (2.87), (2.150), and (2.154) associated with interactions of protons and electrons, while $F_{\rm ex}^{\rm neu}$ is responsible for interactions of neutral species with electrons, protons, and other neutral species.

The radiation term (which can be important only at low $\rho$ or very high $T$) reads

$$F_{\rm rad} = -(4\sigma_{\rm SB}/3c)\, VT^4, \tag{2.176}$$

where $\sigma_{\rm SB}$ is the Stefan-Boltzmann constant.

The ideal-gas contribution of atoms reads

$$F_{\rm id}^{\rm neu} = k_{\rm B}T \sum_\kappa N_\kappa \left[\ln(n_\kappa \lambda_{\rm H}^3/g_\kappa) - 1 - E_\kappa/k_{\rm B}T\right], \tag{2.177}$$

where $N_\kappa$ is the number of atoms in a quantum state $|\kappa\rangle$ with the statistical weight $g_\kappa$ and the binding energy $E_\kappa$. Following Saumon *et al.* (1995), we use a simplified treatment; the states with the same principal quantum number but different orbital and magnetic quantum numbers are not considered separately. In this case $\kappa$ is set equal to the principal quantum number (usually denoted by $n$), the binding energy is $E_\kappa = e^2/(2a_0\kappa^2)$, and the statistical weight $g_\kappa = \kappa^2$.

It should be noted that, although the non-ideality effects are not included in $F_{\rm id}$ explicitly, they do affect $F_{\rm id}$ through the particle numbers. In particular, the distribution of $N_\kappa$ in Eq. (2.177) is *not* assumed to obey the ideal-gas Boltzmann law.

The *excluded-volume* effect, arising from finite atomic sizes (§ 2.5.1) is treated in the "hard-sphere approximation." One assumes that a pair of atoms in states $\kappa$ and $\kappa'$ cannot be found closer to each other than at a distance

$d_{\kappa\kappa'} = l_\kappa + l_{\kappa'}$, where $l_\kappa$ is the root-mean-square proton-electron distance in an isolated atom. In the same way, an atom and a proton cannot be treated separately if the distance between the bound and free protons is smaller than $d_{\kappa p} = l_\kappa$. This gives rise to the excluded-volume ($F_{\rm ex-vol}$) contribution to $F_{\rm ex}^{\rm neu}$.

Another contribution comes from long-range interactions of atoms with atoms and protons via effective potentials (considered, e.g., by Saumon *et al.* 1995). However, as demonstrated by Potekhin (1996b) and Potekhin *et al.* (1999b), this contribution has almost no effect on the EOS of hydrogen at $T \gtrsim 3 \times 10^4$ K. Therefore, it will be neglected below.

The hard-sphere contribution was calculated by Saumon *et al.* (1995) using the analytic formula by Mansoori *et al.* (1971), derived for pure hard-sphere mixtures (without Coulomb interactions). Since the hard-sphere picture is already a very rough approximation, one can employ a simpler treatment using of the so-called van der Waals one-fluid model. In this model, one uses the formula for a system of identical hard spheres (Carnahan & Starling, 1969):

$$\frac{F_{ex-vol}}{N_{\rm tot} k_{\rm B} T} = \frac{4\eta_{\rm pack} - 3\eta_{\rm pack}^2}{(1-\eta_{\rm pack})^2}, \tag{2.178}$$

where $N_{\rm tot} = \sum_\alpha N_\alpha$ is the total number of particles,

$$\eta_{\rm pack} = \frac{\pi}{6 N_{\rm tot} V} \sum_{\alpha\alpha'} N_\alpha N_{\alpha'} d_{\alpha\alpha'}^3 \tag{2.179}$$

is the effective packing fraction, and $d_{\alpha\alpha'}$ are hard-sphere interaction diameters. In our case, the subscript $\alpha$ enumerates atomic quantum states described by quantum numbers $\kappa$ and plane-wave states of free protons. For interactions among charged particles, we set $d_{\alpha\alpha'} = 0$, because the repulsive interaction is already included into the Coulomb part of the free energy.

The Saha equation (2.171) corrected for the electron degeneracy may be written as

$$n_{\rm H} = n_p \frac{n_e \lambda_e^3}{2} \mathcal{Z}_{\rm int} \exp\left(\frac{E_{\rm gr-st} + \Delta\mu_{\rm deg}}{k_{\rm B} T}\right), \tag{2.180}$$

where

$$\Delta\mu_{\rm deg} = \partial F_{\rm id}^{(e)} / \partial N_e - \ln(n_e \lambda_e^3/2) \tag{2.181}$$

is the correction to the chemical potential due to the electron degeneracy ($\Delta\mu_{\rm deg} \to 0$ at $\theta \gg 1$),

$$\mathcal{Z}_{\rm int} = \sum_\kappa g_\kappa w_\kappa \exp\left(\frac{E_\kappa - E_{\rm gr-st}}{k_{\rm B} T}\right), \tag{2.182}$$

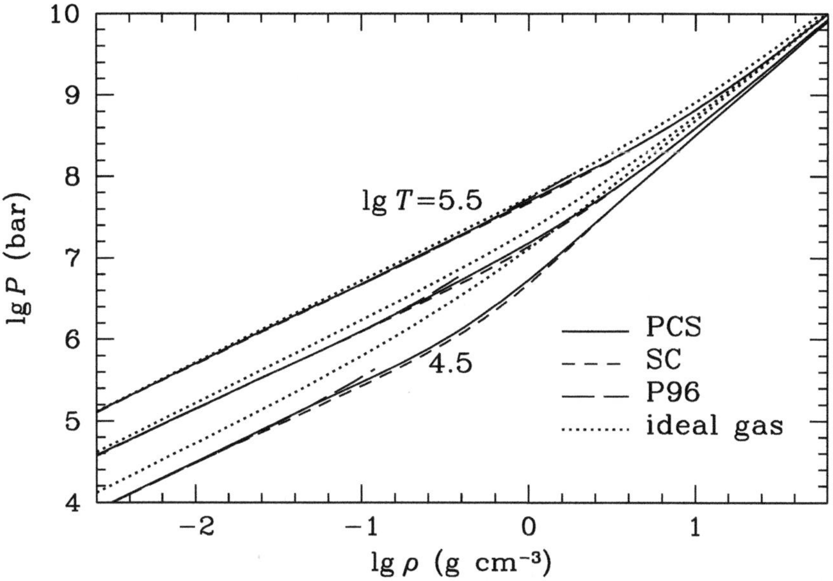


*Figure 2.8.* Pressure isotherms ($\lg T$ [K] $= 4.5$, 5.0, and 5.5) of partially ionized hydrogen given by three theoretical models in the chemical picture of the plasma (see text), compared with the EOS of ideal fully ionized gas (shown by dotted lines).

and $w_\kappa$ is the occupation probability (2.173). Since $F$ is analytic in our model, $w_\kappa$ can also be written in an analytic form (Potekhin, 1996b). The Boltzmann distribution of the atoms, corrected for non-ideality, reads

$$n_\kappa = \mathcal{Z}_{\rm int}^{-1}\, n_{\rm H}\, g_\kappa\, w_\kappa\, {\rm e}^{(E_\kappa - E_{\rm gr-st})/k_{\rm B}T}. \tag{2.183}$$

The minimum of the free energy can be sought by solving Eqs. (2.180)–(2.183) iteratively. Once the equilibrium distribution of plasma particles is found, the thermodynamic quantities are obtained using the relations of § 2.2.

Figure 2.8 demonstrates the EOS obtained using the described model. In the region of weak degeneracy, it coincides with the model labeled P96 (Potekhin, 1996b). Moreover, there is a good agreement with the more detailed SC model (Saumon *et al.*, 1995) and PCS model (Potekhin *et al.*, 1999b). Small differences occur only at those relatively low $T$, where the latter model predicts an appreciable amount of molecules. However, these three models give much lower pressure than the pressure of the ideal fully ionized hydrogen plasma (also shown in the figure), because both the Coulomb non-ideality and atomic recombination give negative contributions.

The adiabatic temperature gradient (2.40) is shown in Fig. 2.9. Note that $\nabla_{\rm ad} = 0.4$ for the ideal gas. The formation of bound species and the non-ideality effects are seen to cause deviations from this value. In its validity region

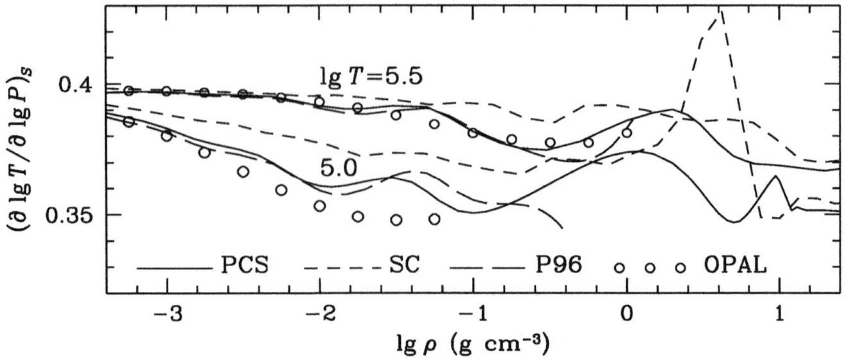


*Figure 2.9.* Comparison of the adiabatic gradient isotherms ($\lg T$ [K] $= 5.0$ and 5.5) for the three EOSs shown in Fig. 2.8 and the OPAL data (Rogers *et al.*, 1996), obtained using the physical picture of the plasma.

(i.e., at low densities), the model of Potekhin (1996b) is in good agreement with the present one. The differences from the predictions of Saumon *et al.* (1995) are somewhat larger. In all the models, the isotherms "wiggle" in the region of consecutive pressure destruction of excited atomic states. However, such wiggles are absent in the OPAL data (Rogers *et al.*, 1996), based on the physical picture of the plasma and also shown in Fig. 2.9. This clearly shows that second-order thermodynamic quantities are more sensitive to the details of a thermodynamic model than first-order ones.

The example of the hydrogen EOS described here is rather simple. More elaborated theoretical models, as well as numerical Monte Carlo simulations (e.g., Bezkrovniy *et al.*, 2004, and references therein) are available in the literature. Nevertheless, the relative simplicity of the described model makes it suitable for astrophysical modeling, and for including the effects of quantizing magnetic fields, which will be considered in Chapter 4.

# Chapter 3

# STRUCTURE AND EQUATION OF STATE OF NEUTRON STAR CRUSTS

After discussing the Coulomb properties of neutron star envelopes in Chapter 2, we will mainly focus on the nuclear part of the problem. Following a widespread convention we will often call an entire neutron star envelope as a crust (see a discussion on this point in the introductory part of Chapter 2).

Up to the density $\rho \sim 10^{11}$ g cm$^{-3}$, the atomic nuclei in the outer crust are expected to be those studied in the laboratory. However, for $\rho \gtrsim 10^{10}$ g cm$^{-3}$ they are beta-unstable in laboratory, with the maximum neutron excess $\delta = (N-Z)/A \simeq 0.3$, where $N$, $Z$, and $A$ are, respectively, the neutron number, proton number, and the total nucleon (mass) number of an atomic nucleus. With increasing density the neutron excess increases and for $\delta > 0.3$ the nuclei in the crust become so neutron rich that their highest neutron energy levels are unbound (§ 1.3.1). As a result, the nuclei are immersed in a gas of free neutrons. Such nuclei cannot exist in laboratory, because they are beyond the neutron drip line under terrestrial conditions; they would disintegrate via neutron emission on timescales $\sim 10^{-20}$ s. With increasing density and neutron excess, nuclear properties are more and more modified by the external neutron liquid. Our knowledge of the properties of matter under the conditions characteristic for the inner crust, with $10^{11}$ g cm$^{-3}$ $\lesssim \rho \lesssim 10^{14}$ g cm$^{-3}$ and $0.3 \lesssim \delta \lesssim 0.8$, is based on theoretical models. It should be stressed, that the advances in solving the nuclear many-body problem enabled one to calculate reliably the properties of many-nucleon systems not available to experiment (pure neutron matter at $\rho < \rho_0$, where $\rho_0 = 2.8 \times 10^{14}$ g cm$^{-3}$ is the saturation density of symmetric nuclear matter). This is very helpful for theoretical studies of the properties of matter near the bottom of the inner crust. Needless to say, these developments in the nuclear many-body problem have been stimulated and motivated by observations of neutron stars.

This chapter is devoted to the composition, equation of state (EOS), and the structure of the neutron star crust. In § 3.1 we briefly describe the formation of the crust of a newly born neutron star. We will restrict ourselves to the case of the crust in nuclear equilibrium (catalyzed matter). The properties of the outer crust in this approximation are described in § 3.2. Theoretical models of the inner crust at $\rho \lesssim 10^{14}$ g cm$^{-3}$ are presented in § 3.3. Section 3.4 is devoted to theoretical models of the ground state in the bottom layers of the inner crust, at $\rho \gtrsim 10^{14}$ g cm$^{-3}$, and to the determination of the bottom edge of the crust. In § 3.6 we describe the EOS of the crust. Elastic properties of the crust are discussed in § 3.7. Possible deviations from idealized crust model studied in the preceding sections are reviewed in § 3.8. Appendix A presents the tabulated equation of state.

## 3.1. The formation of the crust in a newly born neutron star

A neutron star formed in gravitational collapse of a stellar core (§ 1.4.2) is initially very hot, with the internal temperature $\sim 10^{11}$ K. At such high temperatures, the composition and equation of state of the envelope of the newly-born star ($\rho \lesssim 10^{14}$ g cm$^{-3}$, $n_{\rm b} \lesssim 0.1$ fm$^{-3}$) is different from that of the older star. This envelope will eventually become the neutron star crust.

In what follows, we will restrict ourselves to the case in which matter is transparent to neutrinos; this condition is satisfied for $T \lesssim 10^{10}$ K ($k_{\rm B}T \lesssim 1$ MeV). The hot envelope is then a mixture of heavy and light atomic nuclei (mostly $\alpha$-particles, because of their large binding energy of 28.3 MeV), neutrons, protons, electrons, positrons, and photons. At high densities and temperatures the density of nucleons outside nuclei can be large, and a consistent treatment of both nuclei and nucleons is required. The nuclei and the outside nucleons should be described using the same nucleon interaction (nucleon Hamiltonian). Modifications of the nuclear surface properties and pressure exerted by the nucleons on the nuclei have to be calculated in a consistent way. At high densities, where the distance between the nuclei becomes comparable to the nuclear size, one should also modify the nuclear Coulomb energy. Another important complication is that, at the temperatures under consideration, excited states of the nuclei become populated and must therefore be taken into account while calculating thermodynamic quantities.

All these effects have been incorporated in models of dense, hot matter using three different approaches. First, full Hartree-Fock calculations with an effective nucleon-nucleon (NN) interaction, for unit cells of matter containing one nucleus, were performed by Bonche & Vautherin (1981) and Wolff (1983). Second, finite temperature Thomas-Fermi calculations have been done by Marcos *et al.* (1982) and Ogasawara & Sato (1983). Third, calculations in which the nuclei have been described using the finite temperature compressible liquid-drop model have been performed by Lattimer *et al.* (1985).

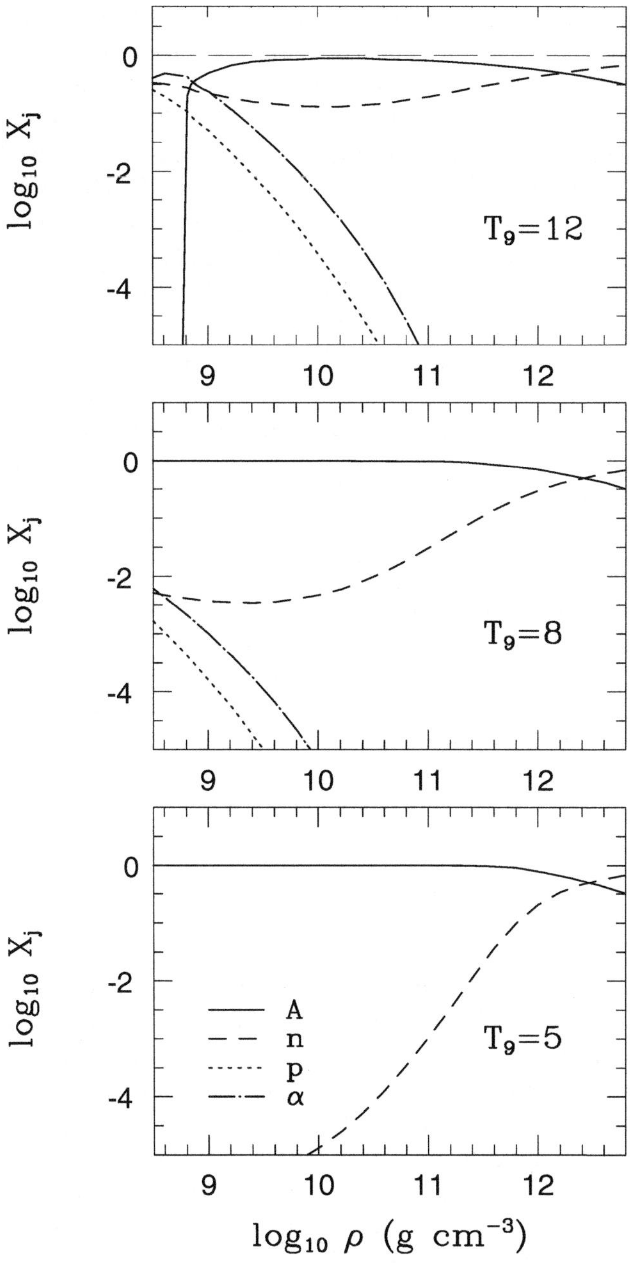


*Figure 3.1.* Mass fractions of different constituents of the outer envelope of a newly born neutron star versus matter density in beta equilibrium at different temperatures $T_9 = T/(10^9 \text{ K})$ (after Haensel *et al.* 1996). Calculations are performed for the Lattimer & Swesty (1991) model of nucleon matter with a specific choice ($K_0 = 220$ MeV) of the incompressibility of cold symmetric nuclear matter at the saturation density.

We will summarize the results using a more recent version of the compressible liquid–drop model, formulated and developed by Lattimer & Swesty (1991), selecting a specific value of the incompressibility of symmetric nuclear matter at saturation (equilibrium) density, $K_0 = 220$ MeV (for the definition of $K_0$, see § 5.4). We assume nuclear equilibrium as well as beta equilibrium of the matter. The assumption of nuclear equilibrium is justified by high temperature. Beta equilibrium is adopted for simplicity; a very rapid cooling of matter at highest temperatures can produce deviations from beta equilibrium.

In Fig. 3.1 we show the composition of the hot matter of the neutron star envelope for $T = 5 \times 10^9$ K, $8 \times 10^9$ K, and $1.2 \times 10^{10}$ K. We restrict ourselves to $\rho \lesssim 10^{13}$ g cm$^{-3}$, because at higher densities the thermal effects on matter composition are negligible. At $T \gtrsim 5 \times 10^9$ K, the shell and pairing effects, so visible in the $T = 0$ (ground state) approximation, particularly through jumps in the density dependence of various quantities (see § 3.2), are washed out by the thermal effects.

At $T = 1.2 \times 10^{10}$K, the nuclei evaporate completely for $\rho \lesssim 10^9$ g cm$^{-3}$. This can be understood within the compressible liquid-drop model; the nuclei are then considered as droplets of nuclear matter. At $\rho \lesssim 10^{11}$ g cm$^{-3}$, these droplets have to coexist with a vapor of neutrons, protons and $\alpha$-particles. However, the coexistence of two different nucleon phases (denser – nuclear liquid, less dense – vapor of nucleons and $\alpha$–particles) is possible only at $T$ lower than some critical temperature at given a density, $T_{\rm crit}(\rho)$. For $\rho \lesssim 10^9$ g cm$^{-3}$, one has $T_{\rm crit}(\rho) \lesssim 1.2 \times 10^{10}$ K.

With decreasing temperature, the mass fraction of evaporated nucleons and $\alpha$-particles decreases. For $T = 8 \times 10^9$ K, $\alpha$-particles are present at $\rho \lesssim 10^{10}$ g cm$^{-3}$, while free protons appear at even lower $\rho$. Free neutrons are present at all densities, but their fraction does not exceed one percent for $\rho \lesssim 10^{11}$ g cm$^{-3}$.

At $T = 5 \times 10^9$ K the thermal effects are weak and imply mainly the appearance of a small fraction of free neutrons ("neutron vapor") below zero-temperature neutron drip density $\rho_{\rm ND}$; this fraction falls below $10^{-5}$ at $\rho = 10^{10}$ g cm$^{-3}$. Further decrease of $T$ leads to the disappearance of neutrons at $\rho < \rho_{\rm ND}$, and to switching-on of shell effects. Another important effect will be the onset of superfluidity of neutrons (both inside and outside the nuclei) and protons. The composition freezes and does not change with further decrease of the temperature. An initially fluid element of the matter solidifies if its temperature falls below the melting temperature $T_{\rm m}$ that depends on local density and composition (§ 2.3.3).

## 3.2. Ground state of the outer crust

The ground state of the matter at densities and pressures, at which all neutrons are bound in nuclei (i.e., before the neutron drip) can be described by a model formulated in the classical paper of Baym *et al.* (1971b), hereafter referred to as BPS. An essential input for this model are the ground-state masses of atomic nuclei confined in crystalline lattice sites. At lowest densities, the relevant nuclei are those whose ground-state masses are determined with high precision by laboratory measurements. However, at higher densities the nuclei in the ground state of matter become more neutron rich. When the BPS paper was written, the last experimentally studied nucleus, present in the ground state of dense matter, was $^{84}$Se (with $Z/A = 0.405$). This nucleus is unstable in laboratory, and its beta-decay half-life time is 3.1 min. The maximum density, at which this nucleus is present, was found to be $8.2 \times 10^9$ g cm$^{-3}$.

During three subsequent decades significant progress has been made in the experimental nuclear physics techniques, and the masses of many new neutron-rich isotopes have been measured.[1] As for this writing (2006), the most neutron-rich nucleus, produced in laboratory is $^{78}$Ni (for its first experimental identification, see Engelmann *et al.* 1995). However, the most neutron rich nucleus with measured mass and expected to be present in the ground state of the crust (at $\sim 5 \times 10^{10}$ g cm$^{-3}$) is $^{80}$Zn; its beta-decay half-time in laboratory is 0.54 s.

We assume that the matter is in its ground state (complete thermodynamic equilibrium – cold catalyzed matter) and forms a perfect crystal of a single nuclear species, $(A,\ Z)$. Deviations from this rule will be discussed in § 3.8. At a given baryon density $n_{\rm b}$, the ground state corresponds to the minimum energy per nucleon $E = \mathcal{E}/n_{\rm b}$. However, $n_{\rm b}$ (or $\rho$) may be not a good variable for stellar matter because it suffers jumps (discontinuities) at some values of the pressure. On the contrary, the pressure is strictly continuous in the stellar interior and increases monotonically with decreasing distance from the star center. Therefore, it is convenient to find the ground state of cold ($T = 0$) matter at any given pressure $P$. This corresponds to minimizing the zero-temperature Gibbs free energy per nucleon, $h = (\mathcal{E} + P)/n_{\rm b}$.

Let us start with $P = 0$, i.e., $h = E = \mathcal{E}/n_{\rm b}$. In this case the minimum energy per nucleon is reached for a body-centered-cubic (bcc) lattice of $^{56}$Fe, and is $E(^{56}{\rm Fe}) = 930.4$ MeV. It corresponds to $\rho = 7.86$ g cm$^{-3}$ and $n_{\rm b} = 4.73 \times 10^{24}\ {\rm cm}^{-3} = 4.73 \times 10^{-15}\ {\rm fm}^{-3}$. It is worth to mention that $^{56}$Fe is not the most tightly bound free atomic nucleus. The maximum binding energy per nucleon $b \equiv [(A - Z)m_n c^2 + Z m_p c^2 - M(A, Z)c^2]/A$ in a nucleus with the ground-state mass $M(A, Z)$ is reached for $^{62}$Ni, $b(^{62}{\rm Ni}) = 8.7945$ MeV, to be compared with $b(^{56}{\rm Fe}) = 8.7902$ MeV. Let us notice that $b(^{58}{\rm Fe}) = 8.7921$ MeV is also higher than $b(^{56}{\rm Fe})$.

The bcc $^{56}$Fe crystal remains the ground state of cold matter up to pressures $\sim 10^{30}$ dyn cm$^{-2}$, at which the matter is compressed to $\sim 10^6$ g cm$^{-3}$ (Salpeter

[1]The results can be found at `http://nucleardata.nuclear.lu.se/database/masses/`; a recent review is given by Lunney *et al.* (2003).

1961, BPS). At such a high density, the matter is a plasma of nuclei and electrons which form a nearly uniform Fermi gas. At a given pressure, the values of the average electron density $n_e$ and the number density of the nuclei $n_{\mathcal{N}}$ are determined from the relations

$$n_e = Z n_{\mathcal{N}} \; , \qquad P = P_e(n_e, Z) + P_{\rm L}(n_{\mathcal{N}}, Z) \; , \tag{3.1}$$

where $P_e$ is the electron pressure and $P_{\rm L}$ is the "lattice" contribution (also called the electrostatic correction) resulting from the Coulomb interactions (see § 2.3.5 for details).

Let us divide the system into electrically neutral unit (Wigner-Seitz) cells containing one nucleus. The number density of the nuclei is $n_{\mathcal{N}} = n_{\rm b}/A$, and the volume of any cell $V_{\rm c} = 1/n_{\mathcal{N}}$, with the equivalent cell radius $r_{\rm c} = (4\pi n_{\mathcal{N}}/3)^{-1/3}$ (in Chapter 2 the radius $r_{\rm c}$ was denoted by $a_{\rm i}$ and called the ion-sphere radius; see § 2.1.3). For a given $(A, Z)$ nuclide, the Gibbs free energy per one unit cell is

$$G_{\rm cell}(A, Z) = W_{\mathcal{N}}(A, Z) + W_{\rm L}(Z, n_{\mathcal{N}}) + [\mathcal{E}_e(n_e, Z) + P]/n_{\mathcal{N}} \; , \tag{3.2}$$

where $W_{\mathcal{N}}$ is the energy of the nucleus (including rest energy of nucleons), $W_L$ is the lattice energy per cell (BPS), and $\mathcal{E}_e$ is the mean electron energy density. Neglecting quantum and thermal corrections and the nonuniformity of the electron gas (§ 2.3.3), we have

$$W_{\rm L} = -C_{\rm M} \, Z^2 e^2/r_{\rm c}, \quad C_{\rm M} \approx 0.9. \tag{3.3}$$

The lattice contribution to the pressure, Eq.(3.1), is thus $P_{\rm L} = \frac{1}{3} W_{\rm L} n_{\mathcal{N}}$.

The Gibbs free energy per nucleon $h = G_{\rm cell}/A$ is just the baryon chemical potential $\mu_{\rm b}(A, Z)$ for a given nuclide. To determine the ground state at a given $P$, one has to minimize $\mu_{\rm b}(A, Z)$ with respect to $A$ and $Z$.

For densities below the neutron-drip density ($\rho < \rho_{\rm ND}$) the lattice correction to $P$ and $\mu_{\rm b}$ is small. One can then easily see the reason for matter neutronization using the approximation $\mu_{\rm b}(A, Z) \simeq W_{\mathcal{N}}(A, Z)/A + Z\mu_e/A$, $P \simeq P_e$. Notice that for $\rho \gg 10^6$ g cm$^{-3}$, the electrons are ultra-relativistic and, therefore, $\mu_e \propto P^{1/4}$. With increasing $P$, it is energetically advantageous to replace $(A, Z)$ by $(A', Z')$ with higher $W_{\mathcal{N}}$ but smaller $Z'/A'$, because the increase in $W_{\mathcal{N}}/A$ is more than compensated by the decrease of the $Z\mu_e/A$ term.

We will follow the determination of the ground state of cold dense matter by Haensel & Pichon (1994) (hereafter referred to as HP). There are small differences between the approximations used by HP and BPS. HP obtained the values of $W_{\mathcal{N}}$ from the atomic masses by subtracting not only the electron rest energies, but removing also the atomic electron binding energies. On the contrary, BPS kept the atomic binding energies in the definition of $W_{\mathcal{N}}$, to simulate the electron screening effects in dense matter. Also, HP used a better approximation for the electron screening effects in dense matter. Their expression for

$\mathcal{E}_e$ takes into account a weak nonuniformity of the electron density resulting from the electron screening effects. HP included also a small exchange term in $\mathcal{E}_e$, which had been neglected in BPS. For $\rho \gg 10^6$ g cm$^{-3}$ the HP formula becomes $\mathcal{E}_e(n_e, Z) = (1.00116 - 1.78 \times 10^{-5} Z^{4/3})\, \mathcal{E}_e^{\rm FG}(n_e)$, where $\mathcal{E}_e^{\rm FG}(n_e)$ is the energy density of a uniform, free Fermi gas (§ 2.3.1).

To a very good approximation, a density jump, at which optimal values $(A, Z)$ change into $(A', Z')$, is given by

$$\frac{\Delta\rho}{\rho} \approx \frac{\Delta n_{\rm b}}{n_{\rm b}} \approx \frac{Z}{A}\frac{A'}{Z'} - 1 \,. \tag{3.4}$$

This equation follows from the continuity of the pressure $P \simeq P_e$.

A sharp discontinuity in $\rho$ and $n_{\rm b}$ is a consequence of the assumed one-component plasma model. Detailed calculations of the ground state of dense matter by Jog & Smith (1982) show, that actually the transition between $(A,\ Z)$ and $(A',\ Z')$ shells takes places through a very thin layer of a *mixed state* of these two species. However, since the pressure interval, where this mixed phase exists, is $\sim 10^{-4} P$, the approximation of a sharp density jump is quite adequate.

HP took experimental masses of nuclei from the tables of Audi (1992, 1993 – private communication).[2] Because of pairing effect, only even-even nuclei are relevant for the ground-state problem. For the remaining isotopes, up to the last one stable with respect to the emission of a neutron pair, HP used theoretical masses obtained from the mass formula of Möller (1992, private communication); the formalism is described by Möller & Nix (1988)).

The baseline HP calculation was done using the (central) experimental atomic masses (without considering experimental errors). As discussed by HP, these errors may become substantial for short-lived nuclei with largest neutron excess. The equilibrium nuclides present in the cold catalyzed matter are listed in Table 3.1. In the fifth column one finds the maximum density $\rho_{\rm max}$ at which a given nuclide survives. The sixth column gives the electron chemical potential $\mu_e$ at $\rho = \rho_{\rm max}$. The transition to the next nuclide has a character of a first-order phase transition. The corresponding fractional density jump $\Delta\rho/\rho$ is given in the last column. The last row above the horizontal line, which divides the table into two parts, corresponds to the maximum density, at which the ground state of dense matter contains a nucleus with mass measured in laboratory. Notice, that in the original HP table the "laboratory region" included also $^{78}$Ni nucleus with the mass evaluated through a semi-empirical method. Here, we restrict ourselves to masses measured in laboratory. Experimental uncertainties in masses of short-lived nuclei with very large neutron excess, as well as uncertainties given

[2] Some masses of unstable nuclei in these tables are actually semi-empirical evaluations based on the knowledge of masses of neighboring isotopes. More recent evaluations of nuclear masses are given by Audi *et al.* (1997) and Audi *et al.* (2003).

*Table 3.1.* Nuclei in the ground state of cold dense matter (after Haensel & Pichon 1994, with a modification explained in the text). Upper part is obtained with experimentally measured nuclear masses. Lower part: from mass formula of Möller. The last line corresponds to the neutron drip point.

| element | $Z$ | $N$ | $Z/A$ | $\rho_{\max}$ (g cm$^{-3}$) | $\mu_e$ (MeV) | $\Delta\rho/\rho$ (%) |
|---|---|---|---|---|---|---|
| $^{56}$Fe | 26 | 30 | 0.4643 | $7.96 \times 10^6$ | 0.95 | 2.9 |
| $^{62}$Ni | 28 | 34 | 0.4516 | $2.71 \times 10^8$ | 2.61 | 3.1 |
| $^{64}$Ni | 28 | 36 | 0.4375 | $1.30 \times 10^9$ | 4.31 | 3.1 |
| $^{66}$Ni | 28 | 38 | 0.4242 | $1.48 \times 10^9$ | 4.45 | 2.0 |
| $^{86}$Kr | 36 | 50 | 0.4186 | $3.12 \times 10^9$ | 5.66 | 3.3 |
| $^{84}$Se | 34 | 50 | 0.4048 | $1.10 \times 10^{10}$ | 8.49 | 3.6 |
| $^{82}$Ge | 32 | 50 | 0.3902 | $2.80 \times 10^{10}$ | 11.4 | 3.9 |
| $^{80}$Zn | 30 | 50 | 0.3750 | $5.44 \times 10^{10}$ | 14.1 | 4.3 |
| $^{78}$Ni | 28 | 50 | 0.3590 | $9.64 \times 10^{10}$ | 16.8 | 4.0 |
| $^{126}$Ru | 44 | 82 | 0.3492 | $1.29 \times 10^{11}$ | 18.3 | 3.0 |
| $^{124}$Mo | 42 | 82 | 0.3387 | $1.88 \times 10^{11}$ | 20.6 | 3.2 |
| $^{122}$Zr | 40 | 82 | 0.3279 | $2.67 \times 10^{11}$ | 22.9 | 3.4 |
| $^{120}$Sr | 38 | 82 | 0.3167 | $3.79 \times 10^{11}$ | 25.4 | 3.6 |
| $^{118}$Kr | 36 | 82 | 0.3051 | $(4.32 \times 10^{11})$ | (26.2 ) | |

by the mass formula beyond the experimental limits are discussed in HP. The last row of Table 3.1 corresponds to the neutron drip point which is determined theoretically.

A rough estimate of the neutron drip density can be obtained using a simplified version of the nuclear mass formula. Neglecting Coulomb, surface and all other finite-size terms and keeping only quadratic term in $\delta = (N - Z)/A$, one can write the energy per nucleon in an atomic nucleus (subtracting the rest energy and neglecting neutron-proton mass difference) as

$$E_{\mathcal{N}}(A, Z)/A \simeq E_0 + S_0\, \delta^2 \,, \tag{3.5}$$

where $E_0$ is the energy per nucleon in the symmetric nuclear matter and $S_0$ is the symmetry energy, both calculated at saturation density (see § 5.4). In this approximation, the neutron and proton chemical potentials (without rest energy contribution) are

$$\mu_n' = E_0 + (2\delta + \delta^2)\, S_0 \,, \quad \mu_p' = E_0 + (-2\delta + \delta^2)\, S_0 \,. \tag{3.6}$$

The value of $\delta$ corresponding to $\rho_{\rm ND}$ can be calculated from the condition $\mu_n' = 0$,

$$\delta_{\rm ND} = \sqrt{1 - (E_0/S_0)} - 1 \ . \tag{3.7}$$

Putting the experimental values $E_0 = -16$ MeV and $S_0 = 32$ MeV (§ 5.4) we get $\delta_{\rm ND} = 0.225$. On the other hand, the beta equilibrium condition reads

$$\mu_e = \mu_n - \mu_p \simeq 4S_0\,\delta \tag{3.8}$$

(to be compared with the consideration in Appendix D). Using the formula $\mu_e = 0.516\ (\rho_6 Z/A)^{1/3}$ MeV, we get a rough estimate

$$\text{bulk approximation :} \qquad \rho_{\rm ND} \simeq 2.2 \times 10^{11}\ {\rm g\ cm^{-3}}, \tag{3.9}$$

which is quite close to the refined theoretical value.

In order to calculate $\rho_{\rm ND}$ precisely, one has to consider the matter composed of nuclei and an outside dilute neutron gas. The density of the neutron gas in the vicinity of the neutron drip point is so small that its influence on the properties of the nuclei can be neglected. In order to determine the ground state of matter in this dripped phase, it is convenient to perform the minimization procedure at a fixed baryon number density $n_{\rm b}$. The quantity to be minimized is the mean energy density,

$$\mathcal{E}^{(\rm d)} = n_{\mathcal N}[W_{\mathcal N}(A,Z) + W_{\rm L}(n_{\mathcal N},Z)] + (1 - V_{\mathcal N} n_{\mathcal N})\,\mathcal{E}_n(n_n) + \mathcal{E}_e(n_e)\,, \tag{3.10}$$

where $\mathcal{E}_n$ is the the energy density of the neutron gas (including rest energy) and $V_{\mathcal N}$ is the volume of the nucleus. To a good approximation, the electron gas is spatially uniform and fills all the volume of the system. Close to the neutron drip point, the nuclear volume can be approximated by the volume of the nucleus in vacuum, $V_{\mathcal N} \simeq \frac{4\pi}{3} R_{\mathcal N}^3$, with the nuclear radius given by $R_{\mathcal N} \simeq 1.2A^{1/3}$ (see, e.g., Preston & Bhaduri 1975). Let us notice, that at $\rho \sim \rho_{\rm ND}$ the fraction of the volume occupied by nuclei is $V_{\mathcal N}/V_{\rm c} \sim \rho_{\rm ND}/\rho_0 \sim 10^{-4}$. At a fixed $n_{\rm b}$, one can express $n_n$ and $n_{\mathcal N}$ in terms of $n_e$,

$$n_{\mathcal N} = \frac{n_e}{Z}\,, \qquad n_n = \frac{n_{\rm b} - n_{\mathcal N} A}{1 - V_{\mathcal N} n_{\mathcal N}}\,, \tag{3.11}$$

and minimize $\mathcal{E}^{(\rm d)}$ as a function of $n_e$. The resulting quantity $\mathcal{E}^{(\rm d)}(A,Z)$ can then be minimized with respect to $A$ and $Z$. The pressure in the dripped phase (which is frequently called the $\mathcal{N}ne$ or the $Ane$ phase) can be obtained from the formula

$$P = P_e + P_{\rm L} + P_n\ . \tag{3.12}$$

At a given $n_{\rm b}$, the dripped phase is the actual ground state of the cold catalyzed matter if $\mathcal{E}^{(\rm d)} < \mathcal{E}$, where $\mathcal{E}$ is the energy density of the $Ae$ phase (containing only electrons and nuclei). Notice, that because of pairing phenomenon, the stability of nuclei with respect to neutron drip in vacuum corresponds to the stability of even-even nuclei with respect to two-neutron emission. This condition, as well as the condition of non-absorption of the outside neutrons by the nuclei is automatically satisfied, provided the minimization is performed and the state with a dripped neutron gas *is energetically preferred* over the state without free neutrons. Calculation shows, that the values of $A, Z$ do not change across the neutron drip point, and therefore $n_n(P \to P_{\rm ND}) = 0$.

Single-particle energy levels in nuclei are discreet, with large energy gaps between "major shells." The local maxima in the binding energies of the nuclei with "magic numbers" $Z = 28$ and $N = 50,\ 82$ are associated with filling up these major shells (see, e.g., Preston & Bhaduri 1975). The effect of closed proton and neutron shells on the composition of the ground-state matter is very strong; except for the $^{56}$Fe nucleus, present in the ground state at lowest densities, all the nuclides are those with a closed proton or neutron shell (Table 3.1). A sequence of three increasingly neutron-rich isotopes of nickel ($Z = 28$) is followed by a sequence of $N = 50$ nuclides with decreasing $Z$, ending at the last experimentally identified $^{78}$Ni. This last nuclide is doubly magic ($N = 50,\ Z = 28$).

In the density range $10^{11}$ g cm$^{-3}$ $\lesssim \rho < \rho_{\rm ND}$, HP get a sequence of $N = 82$ nuclides, with decreasing proton number, from $Z = 44$ down to $Z = 36$; they obtain the neutron drip at $\rho_{\rm ND} = 4.3 \times 10^{11}$ g cm$^{-3}$ (Table 3.1). As shown by HP, the results obtained using the mass formula of Pearson *et al.* (1992) are quite similar to those obtained using the mass formula of Möller.

While the persistence of the $N = 50$ and/or $Z = 28$ nuclei in the ground state of the outer crust may be treated as an *experimental fact*, the strong effect of the $N = 82$ shell, dominating at $10^{11}$ g cm$^{-3}$ $\lesssim \rho < \rho_{\rm ND}$, might be in principle an artifact of the extrapolation via the semiempirical mass formulae. It should be mentioned, that according to some many-body calculations of masses of very neutron rich nuclei the effect of the closed $N = 82$ shell might be much weaker, and could even be replaced by the strong effect of the closure of the $Z = 40$ subshell (Haensel *et al.*, 1989). Recent theoretical calculations of the composition in the bottom layers of the outer crust also give model-dependent results (Rüster *et al.*, 2006). Clearly, there is a need for a better understanding of the shell effects in nuclei close to the neutron drip.

The ground-state composition at a given pressure corresponds to the absolute minimum of the baryon chemical potential $\mu_{\rm b}(N, Z)$ in the $N - Z$ plane. Typically, there is only one well distinguished minimum. A well pronounced second minimum appears only close to the transition pressure between two nucleus species (this is the origin of a thin shell with a mixed phase found by Jog & Smith 1982 and mentioned above). With increasing pressure it becomes a new absolute minimum. Absolute minima lie in a valley, which may be called the "beta stability valley" in dense matter. With increasing pressure, the valley shifts in the $N$-direction, with a slight change of the inclination angle (HP).

## 3.3. Ground state of the inner crust at $\rho \lesssim 10^{14}$ g cm$^{-3}$

The existence of the inner neutron star crust, where very neutron rich nuclei are immersed in a gas of dripped neutrons, has been realized long before the discovery of pulsars (Harrison *et al.*, 1958). The first approach to describe the inner crust consisted in employing a semiempirical mass formula to calculate (or rather estimate) the masses of nuclei; this formula was combined with an expression for the energy of a neutron gas (Harrison *et al.*, 1958, 1965; Tsuruta & Cameron, 1965; Langer *et al.*, 1969; Bethe *et al.*, 1970). It is worth to mention that as early as in 1965 Tsuruta and Cameron estimated the neutron drip density and the density at the crust-core interface as $\rho_{\rm ND} \simeq 3 \times 10^{11}$ g cm$^{-3}$ and $\rho_{\rm cc} \simeq 8 \times 10^{13}$ g cm$^{-3}$, surprisingly close to the presently accepted values. Further work concentrated on a consistent description of ambient neutron gas and the nuclear matter inside neutron rich nuclei, using a unified expression for the energy density of the nuclear matter as a function of neutron and proton number densities and density gradients (Baym *et al.* 1971a, hereafter referred to as BBP; Buchler & Barkat 1971a,b; Arponen 1972). The most ambitious early attempt to calculate the ground state of the inner crust was the Hartree-Fock calculation of Negele & Vautherin (1973). Later calculations focused on a consistent description of the bottom layers of the crust including an updated treatment of pure neutron matter and effective NN interaction (Hashimoto *et al.*, 1984; Oyamatsu *et al.*, 1984; Oyamatsu, 1993; Lorenz *et al.*, 1993; Cheng *et al.*, 1997; Douchin *et al.*, 2000; Douchin & Haensel, 2000).

It should be stressed that many-body calculations of properties of the inner crust, employing the realistic interaction, are beyond the reach of the modern nuclear theory. They would require a precise treatment of a system of up to thousand nucleons in a unit cell, interacting via complicated *bare NN interaction* (determined from NN scattering data and deuteron properties). Moreover, it would be necessary to include three-nucleon forces (§ 5.5.2), required to describe correctly the properties of light nuclei ($^3$H, $^4$He) and the saturation parameters of nuclear matter.

The best calculations performed for the inner crust cannot go beyond calculations for heavy laboratory nuclei. They are based on phenomenological *effective NN interactions*, describing the interaction of nucleons in nuclear matter. They can be divided into three groups, according to the many-body technique used. These groups include the Hartree-Fock approximation employing an effective NN interaction; the semi-classical Extended Thomas-Fermi (ETF) approximation; and the compressible liquid drop model (CLDM).

### 3.3.1 Hartree-Fock calculations with effective nucleon-nucleon interaction

Let us divide matter into unit cells. Let a unit cell contain $N$ neutrons and $Z$ protons. The effective nuclear Hamiltonian for such a system of $A = N + Z$ nucleons is

$$\hat{H}_{\rm N}^{\rm eff} = \sum_{j=1}^{A} \hat{t}_j + \sum_{k<j\leq A} \hat{v}_{jk}^{\rm eff} , \tag{3.13}$$

where $\hat{t}_j$ is the kinetic energy operator of $j$-th nucleon, while $\hat{v}_{jk}^{\rm eff}$ is an operator of an effective two-body interaction between a $jk$ nucleon pair. Usually, $\hat{v}^{\rm eff}$ contains a component which is an effective two-body representation of three-body forces, important in dense nucleon medium (§ 5.5.2).

The effective nuclear Hamiltonian $\hat{H}_{\rm N}^{\rm eff}$ has to reproduce – as accurate as possible within the Hartree-Fock approximation – relevant properties of the ground state of a many-nucleon system, particularly, the ground state energy $E_0$. This last condition can be written as $\langle\Phi_0|\hat{H}_{\rm N}^{\rm eff}|\Phi_0\rangle \simeq \langle\Psi_0|\hat{H}_{\rm N}|\Psi_0\rangle$, where $\Phi_0$ and $\Psi_0$ are, respectively, the Hartree-Fock and exact wave functions, and $\hat{H}_{\rm N}$ is the exact nuclear Hamiltonian.

The basic assumption which justifies a Skyrme-type effective NN interaction (Skyrme, 1956) is that its range is small as compared with the internucleon distances. This means, that in momentum representation the effective NN interaction $\hat{\widetilde{v}}^{\rm eff}(\boldsymbol{k},\boldsymbol{k}')$ can be approximated by a momentum independent term plus terms quadratic in the initial and final relative momenta of an interacting nucleon pair, $\boldsymbol{k}$ and $\boldsymbol{k}'$, with the appropriate spin dependence. Its standard parametrization is

$$\begin{aligned}\hat{\widetilde{v}}_{12}^{\rm eff}(\boldsymbol{k},\boldsymbol{k}') &= t_0\,(1+x_0\hat{P}_\sigma) + \frac{1}{2}\,t_1\,(1+x_1\hat{P}_\sigma)\left(\boldsymbol{k}^2+\boldsymbol{k}'^2\right) \\ &\quad + t_2\,(1+x_2\hat{P}_\sigma)(\boldsymbol{k}\cdot\boldsymbol{k}') + t_3\,(1+x_3\hat{P}_\sigma)\,n_{\rm b}^\gamma\,, \end{aligned} \tag{3.14}$$

where $\hat{P}_\sigma = \frac{1}{2}(1-\boldsymbol{\sigma}_1\cdot\boldsymbol{\sigma}_2)$ is the spin exchange operator. Numerical values of the parameters $t_0,\ x_0,\ \ldots,\ \gamma$ are to be determined from fitting masses of laboratory nuclei in ground states and low-lying excited states. In coordinate space, the first and the last terms of Eq. (3.14) generate zero-range (contact) component of $\hat{v}_{12}^{\rm eff}(\boldsymbol{r})$, proportional to $\delta(\boldsymbol{r})$, where $\boldsymbol{r} = \boldsymbol{r}_1 - \boldsymbol{r}_2$. The remaining terms contain spatial gradient operators. The basic features of the Skyrme model of $\hat{v}^{\rm eff}$ (zero range plus quadratic gradient terms) represent an enormous practical advantage.

The complete Hamiltonian of the unit cell is $\hat{H}_{\rm cell}^{\rm eff} = \hat{H}_{\rm N}^{\rm eff} + V_{\rm Coul} + \hat{H}_e$, where $V_{\rm Coul}$ describes Coulomb interaction between protons and electrons, and $\hat{H}_e$ corresponds to a uniform electron gas. The Hartree-Fock approximation for the many-body nucleon wave function is

$$\Phi_{NZ} = \mathcal{C}_{NZ} \det\left[\varphi^{(p)}_{\alpha_i}(\xi_k)\right] \det\left[\varphi^{(n)}_{\beta_j}(\zeta_l)\right] , \tag{3.15}$$

where $\varphi^{(n)}_{\beta_j}(\zeta_l)$ and $\varphi^{(p)}_{\alpha_i}(\xi_k)$ are single-particle wave functions (orbitals) for neutrons ($j, l = 1, \ldots, N$) and protons ($i, k = 1, \ldots, Z$), respectively, and $\mathcal{C}_{NZ}$ is a normalization constant. The space and spin coordinates of a $k$-th proton are represented by $\xi_k$; $\zeta_l$ is the same for an $l$-th neutron; $\{\alpha_i\}$ and $\{\beta_j\}$ are sets of quantum numbers of occupied single-particle states for protons and neutrons, respectively.

Further approximation used by Negele & Vautherin (1973) consisted in imposing spherical symmetry; the unit cell was approximated by a sphere of radius $r_{\rm c}$. The Hartree-Fock equations for $\varphi^{(p)}$ and $\varphi^{(n)}$ were derived by minimizing the Hartree-Fock energy functional at a fixed volume $V_{\rm c}$ of the unit cell,

$$E_{\rm cell}\left[\varphi^{(p)}_{\alpha}, \varphi^{(n)}_{\beta}\right] = \langle \Phi_{NZ}\Phi_e \,|\hat{H}^{\rm eff}_{\rm cell}|\, \Phi_{NZ}\Phi_e\rangle = \text{minimum} , \tag{3.16}$$

where $\Phi_e$ is the plane-wave Slater determinant for an ultra-relativistic electron gas of constant density $n_e = Z/V_{\rm c}$. The minimization was performed at fixed average neutron and proton densities, $\overline{n}_n = N/V_{\rm c}$, $\overline{n}_p = Z/V_{\rm c} = n_{_N} Z$. For details the reader is referred to the original paper (Negele & Vautherin, 1973).

The boundary conditions for unbound neutrons at the edge of the unit cell used by Negele & Vautherin (1973) deserve a separate comment. The authors argue, that because unbound neutrons are described by a statistical theory, one is free to choose "any convenient boundary condition which is consistent with a statistical neutron density distribution at the cell edge". Their actual boundary condition was to require that at $r = r_{\rm c}$ all single-particle neutron wave functions of even parity (even $\ell$, orbital angular momentum) vanish, and the radial derivatives of odd-parity single-particle functions vanish as well. The arbitrariness of this boundary condition, together with the simplifying assumption of spherical symmetry, may be worrying at $\sim 10^{14}$ g cm$^{-3}$, where the nuclear radius becomes comparable to $r_{\rm c}$.

Once the Hartree-Fock orbitals $\varphi^{(n)}_{\beta}$ and $\varphi^{(p)}_{\alpha}$ are determined, one finds the minimum (ground state) value of $E_{\rm cell}(N, Z)$, filling the lowest $N$ neutron and $Z$ proton states. Then, the absolute ground state configuration is found by minimizing $E_{\rm cell}(N, Z)$ at a fixed $A = N + Z$. Let us notice, that $\alpha_Z$ and $\beta_N$ correspond to "Fermi levels" for protons and neutrons, respectively. In terms of the single-nucleon orbitals, the neutron drip point corresponds to the threshold density at which the neutron Fermi level becomes *unbound*, i.e., $\varphi^{(n)}_{\beta_N}$ extends over the entire unit cell.

A sample of neutron and proton density profiles is shown in Fig. 3.2. One notices wiggles in density profiles within the nuclei; they result from specific radial quantum oscillations of the neutron and proton orbitals. Even at highest densities considered, no proton drip occurs. With increasing density of the

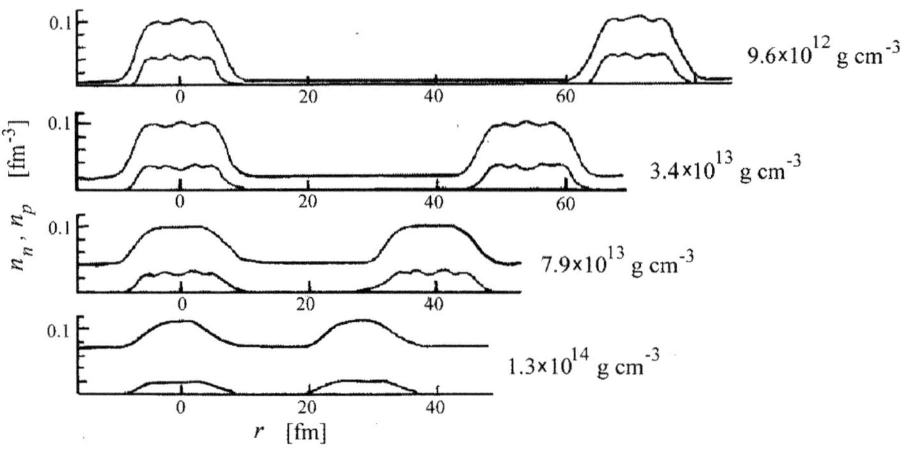


*Figure 3.2.* Density profiles of neutron and protons, at several average densities $\rho$, along a line joining the centers of two adjacent unit cells. Based on Fig. 3 of Negele & Vautherin (1973) with the permission of the authors.

matter, the number density of neutrons outside the nuclei grows up and the number density of protons within the nuclei goes down. As Negele & Vautherin (1973) find, at $\rho \gtrsim 8 \times 10^{13}$ g cm$^{-3}$ the differences in energy between various local minima of $E_{\text{cell}}(N, Z)$ become so small, that it is not meaningful to proceed with their calculational scheme to higher densities.

One of the most interesting results of Negele & Vautherin (1973) was the prediction of the strong shell effect for protons. It manifests itself by the persistence of $Z = 40$ (closed proton subshell) from the neutron drip point to $\rho \sim 3 \times 10^{12}$ g cm$^{-3}$, and $Z = 50$ (closed major proton shell) for $3 \times 10^{12}$ g cm$^{-3} \lesssim \rho \lesssim 3 \times 10^{13}$ g cm$^{-3}$, Fig. 3.3.

Negele & Vautherin (1973) derived suitable analytical formulae, which smoothly reproduced their EOS for the inner crust. However, we should warn the reader that the authors based their expression for the effective NN interaction on the microscopic calculations of asymmetric nuclear and pure neutron matter by Siemens (1970) and Siemens & Pandharipande (1971), in which the Reid soft-core bare NN interaction (Reid, 1968) was used. Since then, the knowledge of the bare NN interaction and of the EOS for nuclear matter and pure neutron matter at $\rho < \rho_0$ has become much more precise. Fortunately, the results of Siemens & Pandharipande (1971) for pure neutron matter at subnuclear densities were found to be in nice agreement with more advanced microscopic calculations (Pethick *et al.*, 1995).

Alas, after the monumental work of Negele & Vautherin (1973) no attempts of a Hartree-Fock calculation of nuclear structures in the ground state of the

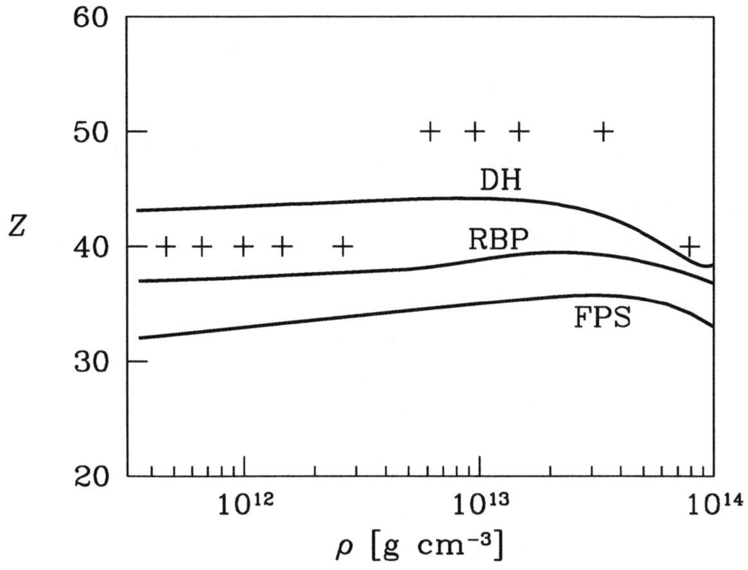


*Figure 3.3.* Numbers of protons per nucleus in the ground state of the inner crust obtained by various authors. Solid lines: RBP – Ravenhall *et al.* (1972); FPS – as quoted in Pethick & Ravenhall (1995); DH – Douchin & Haensel (2000). Crosses – Negele & Vautherin (1973).

inner crust were carried out in the next thirty years. This might result from an unsolved problem of correct boundary conditions at the spherical unit cell edge, supplemented by the difficulties in finding an absolute minimum of the Hartree-Fock energy functional. These problems did not prevent carrying out Hartree-Fock unit-cell calculations of nuclear structures in hot dense matter, relevant for collapsing stellar cores (Bonche & Vautherin, 1981; Wolff, 1983). For the hot matter at $k_{\rm B}T \gtrsim 1$ MeV, the thermal averaging and a much smaller contribution of the nucleon gas outside the nuclei (in the relevant case of entropy per nucleon $\sim (1-2)\, k_{\rm B}$) make the results much less sensitive to the boundary conditions.

Let us mention also the problem of pairing correlations between nucleons. Pairing may be treated by extending the Hartree-Fock scheme to the Hartree-Fock-Bogoliubov framework (see, e.g., Ring & Schuck 1980). Superfluid properties of the inner neutron star crust (pairing in nuclei and in a free neutron liquid, § 1.3.4) were studied by many authors (see, e.g., Lombardo & Schulze, 2001; Sandulescu *et al.*, 2004, and references therein) but the results are drastically different and very model dependent. Luckily, superfluidity has almost no effect on the EOS of dense matter. However, it might influence the composition of the inner crust (Baldo *et al.*, 2005). Clearly, this problem requires further studies.

In principle, the boundary conditions in the spherical Wigner-Seitz cell approximation are incompatible with periodic symmetry of the crystal. In order to respect crystal symmetry, one has to use the appropriate Bloch single particle functions (see Carter *et al.*, 2005, for a discussion of this point).

### 3.3.2 Extended Thomas-Fermi (ETF) calculations

In the neutron drip regime, the number of nucleons in the unit cell grows rapidly with increasing density. At $\rho \sim 10^{13}$ g cm$^{-3}$ one has $A_{\rm cell} \sim 1000$ (Negele & Vautherin, 1973), and the implementation of the self-consistent Hartree-Fock scheme would require an enormous amount of work and computer time. Large sizes of the nuclei suggest further simplifications of the Hartree-Fock model via *semiclassical approximation*, in which the relevant quantities are represented "on the average", with quantum fluctuations (oscillations) being averaged out. The energy of the unit cell is a sum of the nuclear energy $E_{\rm N}$ (which includes nucleon rest energies), the Coulomb energy $E_{\rm Coul}$, and the energy of the electron gas, $E_e$. In the ETF approximation (e.g., Ring & Schuck 1980, Chapter 13; Brack *et al.* 1985), the nuclear energy of a unit cell is

$$\begin{aligned} E_{\rm N} &= \int_{\rm cell} \{\mathcal{E}_{\rm N}\left[n_n(\boldsymbol{r}), n_p(\boldsymbol{r}), \nabla n_n(\boldsymbol{r}), \nabla n_p(\boldsymbol{r})\right] + m_n c^2 n_n(\boldsymbol{r}) \\ &\quad + m_p c^2 n_p(\boldsymbol{r})\}\, {\rm d}^3 r\ . \end{aligned} \tag{3.17}$$

The nuclear energy density functional $\mathcal{E}_{\rm N}$ is nonlocal, as it depends on density gradients. For the ETF approximation to be valid, characteristic length-scales for variations of $n_n(\boldsymbol{r})$ and $n_p(\boldsymbol{r})$ have to be much larger than the internucleon distance. Then one can keep only quadratic gradient terms in $\mathcal{E}_{\rm N}$. To a very good approximation, the electron gas is uniform,[3] with $n_e = Z/V_{\rm c}$, and the Coulomb energy of the unit cell is given by

$$E_{\rm Coul} = \frac{1}{2}\int_{\rm cell} e\left[n_p(\boldsymbol{r}) - n_e\right]\phi(\boldsymbol{r})\, {\rm d}^3 r\ , \tag{3.18}$$

where $\phi(\boldsymbol{r})$ is the electrostatic potential to be calculated from the Poisson equation,

$$\nabla^2\phi(\boldsymbol{r}) = 4\pi e\left[n_p(\boldsymbol{r}) - n_e\right]\ . \tag{3.19}$$

To determine the ground state at a given $n_{\rm b}$, one has to find $n_n(\boldsymbol{r})$ and $n_p(\boldsymbol{r})$, which minimize $E_{\rm cell}/V_{\rm c}$ under the constraints

$$V_{\rm c} n_{\rm b} = \int_{\rm cell}\left[n_n(\boldsymbol{r}) + n_p(\boldsymbol{r})\right] {\rm d}^3 r\ , \qquad \int_{\rm cell}\left[n_p(\boldsymbol{r}) - n_e\right] {\rm d}^3 r = 0\ . \tag{3.20}$$

[3] This approximation can be relaxed and one can treat $n_e(\boldsymbol{r})$ within the ETF model as a quantity to be determined from the condition $E_{\rm cell}/V_{\rm c} = \min$. However, the effect of the electron density nonuniformity is small; the electron screening length is much larger than $r_{\rm c}$ (Chapter 2).

The problem is simplified assuming spherical symmetry; in this case the unit cell is approximated by a sphere of the radius $r_c = (3V_c/4\pi)^{1/3}$. The boundary conditions are such that far from the nucleus surface the nucleon densities are uniform. This requires the nuclear radius to be significantly smaller than $r_c$.

The ETF method was first applied for studying the matter in the inner crust by Buchler & Barkat (1971a); see also Buchler & Barkat (1971b) and Barkat *et al.* (1972). According to these authors, the presence of the outer neutron fluid strongly alters the properties of the nuclei, which casts doubts on the validity of the previous calculations based on the extrapolation of semi-empirical mass formulae (Harrison *et al.*, 1965; Langer *et al.*, 1969; Bethe *et al.*, 1970). The dependence of $\mathcal{E}_N$ on $n_n(\boldsymbol{r})$ and $n_p(\boldsymbol{r})$ was taken form the results of Brueckner *et al.* (1968, 1969) for asymmetric nuclear matter and pure neutron matter. The coefficient of the quadratic gradient-term was determined by fitting masses of laboratory nuclei. The minimization problem was reduced to solving the appropriate Euler-Lagrange equations with required boundary conditions. The neutron drip was found at $\rho_{ND} \approx 4 \times 10^{11}$ g cm$^{-3}$, and the crust bottom was at $\rho_{cc} \sim 10^{14}$ g cm$^{-3}$. The Buchler-Barkat values of $Z$ stayed roughly constant $Z \sim 30$ for $10^{11}$ g cm$^{-3}$ $\lesssim \rho \lesssim 5 \times 10^{13}$ g cm$^{-3}$, and then decreased to $Z \sim 20$ at $\rho \simeq 10^{14}$ g cm$^{-3}$.

In the 1980s the main effort was focused on dense and hot matter, relevant for the gravitational collapse of stellar cores and type II supernova explosions. An exception from this rule was the paper by Ogasawara & Sato (1983), who devoted their §3.1 to cold catalyzed matter. Their calculational scheme was similar to that used by Barkat *et al.* (1972). However, Ogasawara & Sato (1983) used different models for the potential energy of asymmetric nuclear matter. They obtained the neutron drip density $\rho_{ND} \approx (3-4) \times 10^{11}$ g cm$^{-3}$ and the values $Z = 35 - 45$, higher than those of Barkat *et al.* (1972). The results of Ogasawara and Sato are in good agreement with the Hartree-Fock results of Negele & Vautherin (1973).

Significant progress in the 1980s was achieved in calculations of the properties of asymmetric nuclear matter and pure neutron matter with realistic bare NN interactions (see Friedman & Pandharipande 1981; Wiringa *et al.* 1988). On the other hand, calculations based on the HF method and its semi-classical simplifications, with new models of effective NN interaction, reached a high degree of precision in reproducing the properties of atomic nuclei. The ETF calculation in the 1980s focused on the possible appearance of nonspherical nuclei in the crust bottom, as will be described in § 3.4. Oyamatsu (1993) studied the ground state of the inner crust within the ETF approximation using four different energy density functionals $\mathcal{E}_N$. These functionals were constructed in such a way as to reproduce the gross properties of laboratory nuclei and to match the EOS of pure neutron matter obtained by Friedman & Pandharipande (1981) for a realistic bare NN interaction. Oyamatsu performed explicit mini-

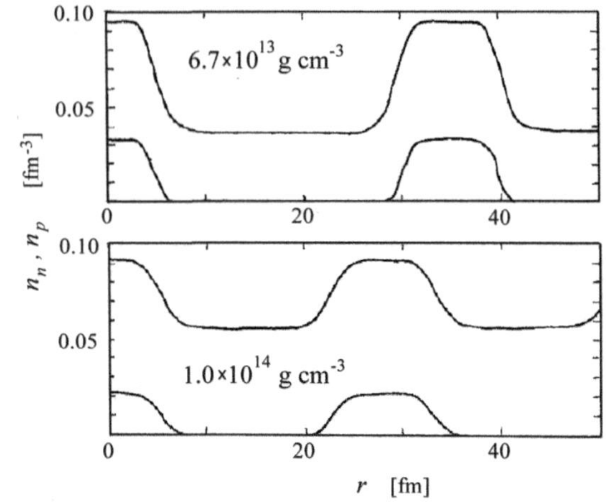


*Figure 3.4.* Neutron and proton density profiles at two average mass densities along a line joining the centers of adjacent unit cells. Based on Fig. 5 of Cheng *et al.* (1997), with the permission of the authors.

mization of the TF energy functional within a family of *parameterized* $n_n(\boldsymbol{r})$ and $n_p(\boldsymbol{r})$ density profiles [Eq. (B.7) of Appendix B]. At a given average nucleon density $n_\mathrm{b}$, the ground state configuration was obtained by minimizing $E_\mathrm{cell}/V_\mathrm{c}$ with respect to the parameters of the nucleon density profiles. In the density interval from the neutron drip to $10^{14}$ g cm$^{-3}$ Oyamatsu gets $Z \simeq 40$ for all four models, in good agreement with the Hartree-Fock calculations of Negele & Vautherin (1973).

Simultaneously with the application of the relativistic Brueckner-Hartree-Fock (RBHF, see § 5.9.1, page 240) approach to neutron star matter at supranuclear densities, the semi-classical ETF approximation based on the RBHF model was applied to the inner crust. Starting from the RBHF results for the bulk asymmetric nuclear matter, Sumiyoshi *et al.* (1998) employed the ETF scheme of Oyamatsu (1993), with his parameterization of the nucleon density profiles. The quadratic gradient term in the energy density functional was determined by fitting the properties of terrestrial nuclei. The authors found the neutron drip at $\rho_\mathrm{ND} \approx 2.4 \times 10^{11}$ g cm$^{-3}$. Their values of $Z$ in the inner crust are systematically lower than obtained in previous works, with $Z \simeq 35$ near the neutron drip, decreasing down to about 20 at $\rho \simeq 10^{14}$ g cm$^{-3}$. This may be attributed to a relatively large Coulomb energy of nuclei in their model owing to smaller nuclear radii. Notice that their RBHF value $0.185$ fm$^{-3}$ for the saturation density of the symmetric nuclear matter is significantly larger than the experimental value of $0.16$ fm$^{-3}$ which may explain the compactness of their nuclei.

The relativistic Hartree energy functional, calculated in the nonlinear relativistic mean field model of dense nucleon matter, can be simplified using the relativistic extended Thomas-Fermi (RETF) approximation proposed by Centelles *et al.* (1992, 1993). In the RETF approximation, one gets the $\mathcal{E}_\mathrm{N}$ functional containing the terms quadratic in $\nabla n_n$ and $\nabla n_p$, which are completely specified within the model. The RETF model was applied by Cheng *et al.* (1997) to determine the ground state of the inner crust, using the nonlinear $\sigma$–$\omega$–$\rho$ model Lagrangian by Boguta & Bodmer (1977). Three sets of the Lagrangian parameters were used. Cheng *et al.* (1997) exactly solved the Euler-Lagrange equations for $n_n(r)$ and $n_p(r)$ in a spherical unit cell. A sample of their density profiles at $\rho \lesssim 10^{14}$ g cm$^{-3}$ is shown in Fig. 3.4. Notice the absence of any wiggles in the neutron density distributions within the nuclei, characteristic for the Hartree-Fock results of Negele & Vautherin (1973); cf. Fig. 3.2. The wiggles are absent because of the semi-classical character of the ETF approximation. Cheng *et al.* (1997) did not publish their values of $Z$ as a function of the matter density. However, an analysis of their figures and tables indicates that, similarly to Sumiyoshi *et al.* (1998), they got the nuclei which are relatively small, with the values of $Z$ significantly lower than those obtained in nonrelativistic calculations. Shen (2002) used the combination of a Relativistic Mean Field model for the bulk energy and the parameterization of the surface energy proposed by Oyamatsu (1993). He obtained the inner crust extending from $\rho_\mathrm{ND} \approx 4 \times 10^{11}$ g cm$^{-3}$ to $\rho_\mathrm{cc} \approx 10^{14}$ g cm$^{-3}$.

### 3.3.3 Compressible liquid drop model (CLDM)

The nature of the Hartree-Fock and ETF calculations does not allow one to study separate physical contributions and effects, whose interplay leads to a particular structure of the crustal matter. The compressible liquid drop model (CLDM) enables one to separate various terms in $E_\mathrm{cell}$ and identify their contribution.

There are also other practical advantages of the CLDM. On the one hand, it can be considered as a suitable and economical parameterization of the results of microscopic calculations of the Hartree-Fock or ETF type. On the other hand, the CLDM avoids technical complications related to the choice of the boundary conditions in the Hartree-Fock approach (§ 3.3.1). Finally, the CLDM allows for thermodynamically consistent and systematic treatment of bulk and finite-size effects and is particularly convenient for studying phase transitions in dense matter (see § 3.4). In particular, the CLDM treats two major effects of the outer neutron gas on the nuclei: 1) the decrease of the surface tension with growing density under the action of free neutrons; 2) the compression of the nuclear matter within the nuclei by the pressure of the ambient neutron gas. However, all these attractive features of the CLDM are meaningful only when finite-size contributions are calculated using a microscopic Hartree-Fock or ETF approach

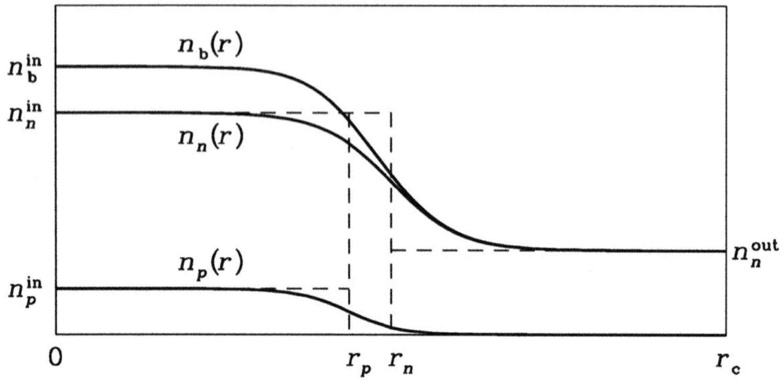


*Figure 3.5.* Local proton and neutron number densities within a spherical unit cell in the inner crust. Solid lines are actual density profiles, dashed lines correspond to the compressible liquid drop model.

from the same effective nucleon Hamiltonian as used for the calculation of the bulk (volume) terms. In particular, only then the decrease of the surface tension due to the presence of the ambient neutron gas is treated in a correct way.

In the CLDM, one divides the nuclear contribution $E_{\rm N}$ (without the Coulomb interactions) into the bulk and surface terms, $E_{\rm N,bulk}$ and $E_{\rm N,surf}$. Let $E_{\rm Coul}$ be the Coulomb contribution to the energy of the unit cell. The electrons are assumed to form an uniform Fermi gas and give the rest and kinetic energy contribution denoted by $E_e$ (§ 2.3.1). The total energy of the unit cell is, therefore,

$$E_{\rm cell} = E_{\rm N,bulk} + E_{\rm N,surf} + E_{\rm Coul} + E_e \ . \tag{3.21}$$

Contrary to $E_{\rm N,bulk}$, $E_{\rm N,surf}$ and $E_{\rm Coul}$ depend on sizes and shapes of nuclear structures.

In the CLDM the nucleons are distributed between the three subsystems: the denser nucleon fluid to be labeled by "i"; the less dense neutron fluid labeled by "o"; and the nuclear surface ("i-o" phase interface) labeled by "s". One requires mechanical and chemical equilibrium between these subsystems. Far from the nuclear surface, the nucleon densities are constant, being equal to $n_{n,\rm i}$ and $n_{p,\rm i}$ in the "i" phase, and $n_{n,\rm o}$ and $n_{p,\rm o} = 0$ in the "o" gas. The definition of the surface term is ambiguous. For the spherical nuclei in the inner crust it is convenient to define it by the *reference proton radius* $r_p$, such that $4\pi r_p^3 n_{p,\rm i}/3 = Z$. This definition is convenient because of the presence of the Coulomb term in $E_{\rm cell}$ determined solely by the proton density distribution. Similarly, the neutron radius $r_n$ is defined by $\frac{4\pi}{3}[r_n^3(n_{n,\rm i} - n_{n,\rm o}) + r_{\rm c}^3 n_{n,\rm o}] = N_{\rm cell}$ (see Fig. 3.5). In view of a significant neutron excess, the interface includes the *neutron skin* of the thickness $s_n = r_n - r_p$ formed by the neutrons adsorbed on the nucleus surface.

The bulk energy of the nucleons in the cell is

$$E_{\rm N,bulk} = V_{\rm c}\left[w\,\mathcal{E}_{\rm N,i} + (1-w)\,\mathcal{E}_{n,\rm o}\right] , \tag{3.22}$$

where $\mathcal{E}_{\rm N,i}$ is the energy density in the nucleus far from the surface and $\mathcal{E}_{n,\rm o}$ is the energy density of the ambient neutron gas. The volume fraction occupied by the phase "i" (i.e., protons) is $w = V_p/V_{\rm c} = (r_p/r_{\rm c})^3$.

Fundamentals of the thermodynamics of surfaces can be found in §§ 154, 156, 157 of Landau & Lifshitz (1993). The nuclear surface-energy term, $E_{\rm N,surf}$, gives the contribution of the interface between the neutron gas and nuclear matter; it includes the contribution from the neutron skin (Lattimer *et al.*, 1985; Pethick & Ravenhall, 1995; Lorenz, 1991),

$$E_{\rm N,surf} = \mathcal{A}\,\sigma + N_{\rm s}\,\mu_{n,\rm s} , \tag{3.23}$$

where $\sigma$ is the *surface thermodynamic potential per unit area*, $\mathcal{A}$ is the area of the nuclear surface ($\mathcal{A} = 4\pi r_p^2$ for spherical nuclei), $N_{\rm s}$ is the number of neutrons in the neutron skin, and $\mu_{n,\rm s}$ is the chemical potential of neutrons adsorbed on the nucleus surface. In the simplest approximation, in which the curvature corrections in $E_{\rm N,surf}$, proportional to $\mathcal{A}/r_p$, are neglected, $\sigma$ is approximated by the *surface tension* $\sigma_{\rm s}$, and $N_{\rm s} = \nu_{\rm s}\,\mathcal{A} \simeq (n_{n,\rm i} - n_{n,\rm o})s_n\mathcal{A}$, where $\nu_{\rm s} = N_{\rm s}/\mathcal{A}$ is the surface density of adsorbed neutrons. A more precise expression for $E_{\rm N,surf}$ can be obtained by including the curvature corrections (Lorenz, 1991; Douchin, 1999). In view of the possibility of nuclear structures with infinite volumes (§ 3.4) it is convenient to introduce the contribution of neutrons from the neutron skin to the total (average) nucleon density, $n_{\rm s}$ (Lorenz, 1991). For spherical nuclei, the number of neutrons in the neutron skin per unit volume is $n_{\rm s} = N_{\rm s}/V_{\rm c}$ (Lorenz, 1991).

In order to calculate $E_{\rm Coul}$, one uses the Wigner-Seitz approximation. Neglecting the diffuseness of the proton distribution one gets

$$E_{\rm Coul} = \frac{16\pi^2}{15}\,(n_{p,\rm i}e)^2\,r_p^5 f_3(w) , \qquad f_3(w) = 1 - \frac{3}{2}\,w^{1/3} + \frac{1}{2}\,w . \tag{3.24}$$

At $T = 0$ the equilibrium can be determined by minimizing the total energy density, $\mathcal{E} = E_{\rm cell}/V_{\rm c}$, at a fixed value of $n_{\rm b}$. The quantity $\mathcal{E}$ is a function of seven independent variables. A convenient set of variables is: $n_{n,\rm i}$, $n_{p,\rm i}$, $n_{\rm s}$, $n_{n,\rm o}$, $r_p$, $r_n$, $r_{\rm c}$; in this way all independent variables will be finite even for infinite nuclear structures considered in § 3.4. Fixing $n_{\rm b}$ and requiring charge neutrality of the cell, we reduce the number of independent variables to five. Therefore, there will be five conditions of equilibrium resulting from the stationarity of $\mathcal{E}$ with respect to variations of thermodynamic variables. All these conditions has well defined physical meaning. The first condition requires

$$(\partial \mathcal{E}/\partial n_{n,\mathrm{i}})_{[n_{p,\mathrm{i}}][n_{\mathrm{s}}][r_p][r_{\mathrm{c}}]} = 0 \ , \tag{3.25}$$

where quantities in square brackets are kept constant. Neglecting the curvature corrections, it implies the equality of the neutron chemical potentials in the bulk phases of nucleon matter, $\mu_{n,\mathrm{i}}^{\mathrm{bulk}} = \mu_{n,\mathrm{o}}^{\mathrm{bulk}}$. The second equation results from minimization with respect to the number of protons,

$$(\partial \mathcal{E}/\partial n_{p,\mathrm{i}})_{[n_{\mathrm{s}}][n_{n,\mathrm{o}}][r_p][r_{\mathrm{c}}]} = 0 \ , \tag{3.26}$$

and yields the beta equilibrium condition between neutrons, protons, and electrons. Neglecting the curvature corrections, it reads

$$\mu_{n,\mathrm{i}}^{\mathrm{bulk}} - \mu_{p,\mathrm{i}}^{\mathrm{bulk}} - \mu_e = \frac{8\pi}{5}\, e^2\, n_{p,\mathrm{i}}\, r_p^2\, f_3(w) \ . \tag{3.27}$$

We also need the condition on the number of surface neutrons. It results from the requirement of stationarity with respect to transfer of a neutron from the nucleus interior to the surface, all other particle numbers being fixed,

$$(\partial \mathcal{E}/\partial n_{\mathrm{s}})_{[n_{p,\mathrm{i}}][w n_{n,\mathrm{i}}+n_{\mathrm{s}}][r_p][r_{\mathrm{c}}]} = 0 \ . \tag{3.28}$$

Neglecting the curvature corrections, this condition implies that the chemical potential of the surface neutrons is equal to the bulk chemical potentials in both phases, $\mu_{n,\mathrm{s}} = \mu_{n,\mathrm{i}}^{\mathrm{bulk}} = \mu_{n,\mathrm{o}}^{\mathrm{bulk}}$.

To these three conditions, expressing chemical equilibrium within the system, we have to add two equations of mechanical equilibrium. The condition number four results from the requirement of stationarity with respect to variations of $r_p$. It expresses the equality of the pressure inside and outside the nucleus,

$$(\partial \mathcal{E}/\partial r_p)_{[n_p][n_{\mathrm{s}}][n_n][r_{\mathrm{c}}]} = 0 \ , \tag{3.29}$$

where the constancy of $n_n = N_{\mathrm{cell}}/V_{\mathrm{c}}$ and $n_p = Z/V_{\mathrm{c}}$ is equivalent to keeping constant numbers of protons and neutrons in a fixed $V_{\mathrm{c}}$. Neglecting the curvature corrections, this condition reads

$$P_{\mathrm{i}}^{\mathrm{bulk}} - P_{\mathrm{o}}^{\mathrm{bulk}} = \frac{2\sigma_{\mathrm{s}}}{r_p} - \frac{4\pi}{15}\, e^2\, n_{p,\mathrm{i}}^2\, r_p^2\, (1-w) \ , \tag{3.30}$$

where $P_{\mathrm{j}}^{\mathrm{bulk}} = n_{\mathrm{j}}^2\, \partial(\mathcal{E}_{\mathrm{j}}^{\mathrm{bulk}}/n_{\mathrm{j}})/\partial n_{\mathrm{j}}$.

Finally, the fifth equation determines the equilibrium size of the cell. It results from the condition of stationarity with respect to variations of the cell radius, while $w$ and all the densities including $n_{\mathrm{s}}$ are kept constant,

$$(\partial \mathcal{E}/\partial r_{\mathrm{c}})_{[n_{n,\mathrm{i}}][n_{p,\mathrm{i}}][n_{\mathrm{s}}][w]} = 0 \ . \tag{3.31}$$

Because $w$ is constant here, the condition involves only the Coulomb and surface terms in $E_{\rm cell}$. In our approximation it can be written as

$$\mathcal{E}_{\rm N,surf} = 2\,\mathcal{E}_{\rm Coul}\ . \tag{3.32}$$

This is the "virial theorem" of the simplified CLDM with no curvature corrections (BBP); it enables one to express $r_p$ in terms of the remaining variables. Generalization of the "virial theorem" to the case of nonstandard nuclear shapes will be discussed in § 3.4.

Let us write an explicit expression for the nuclear component of the energy density, neglecting, for simplicity, the curvature corrections in $\mathcal{E}_{\rm N,surf}$. Both the surface tension $\sigma \simeq \sigma_{\rm s}$ and the neutron skin thickness $s_n = r_n - r_p$ are calculated under the condition of thermodynamic and mechanical equilibrium of the semi-infinite "i" and "o" phases, separated by a plane interface. Therefore, $\sigma$ and $s_n$ depend only on one thermodynamic variable, e.g., on the proton fraction in the bulk "i" phase, $x_{\rm i} = n_{p,{\rm i}}/n_{\rm i}$, where $n_{\rm i} = n_{n,{\rm i}} + n_{p,{\rm i}}$. The formula for the energy density $\mathcal{E}_{\rm N}$ reads then

$$\mathcal{E}_{\rm N} = w\,\mathcal{E}_{\rm N,i} + (1-w)\,\mathcal{E}_{n,{\rm o}} + \frac{3w}{r_p}\left[\sigma_{\rm s} + (n_{n,{\rm i}} - n_{n,{\rm o}})\,s_n\mu_n\right]\ . \tag{3.33}$$

Let us remind, that according to Eq. (3.28) the chemical potential of neutrons adsorbed on the nuclear surface is equal to the value of $\mu_n$ common for both bulk phases.

Historically, the first CLDM calculations for the ground state of the inner crust were performed in the classical paper by BBP, who laid the foundation of subsequent CLDM calculations. Unfortunately, BBP used simplified qualitative expressions, based on dimensionality arguments, to describe the reduction of $\sigma$ with increasing density. This resulted in a too rapid increase of $Z$ with increasing density, which was corrected by Ravenhall *et al.* (1972). Later CLDM calculations were performed by Lorenz (1991); Douchin *et al.* (2000), and Douchin & Haensel (2000). They were based on the effective NN interactions particularly suitable for strongly asymmetric nuclear systems. Lorenz used the FPS model (**F**riedman **P**andharipande **S**kyrme, derived by Pandharipande & Ravenhall 1989) consistent with the results of many-body calculations of dense asymmetric nuclear matter (see Chapter 5 and Friedman & Pandharipande 1981) with a realistic bare NN interaction and a phenomenological three-nucleon force. Douchin and Haensel used the SLy (**S**kyrme **Ly**on, Chabanat *et al.* 1997, 1998) effective forces, adjusted to the properties of neutron-rich nuclei, and adjusted also, at $\rho > \rho_0$, to the results of many-body calculations of the properties of dense asymmetric nuclear matter by Wiringa *et al.* (1988) (based on the bare two-nucleon interaction AV14 and the phenomenological UVII three-nucleon interaction).

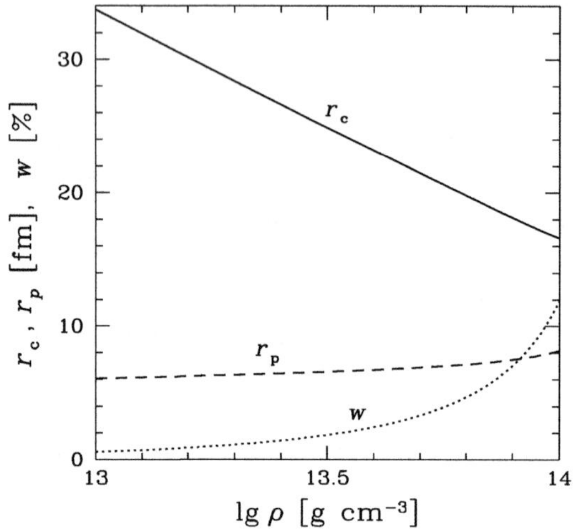


*Figure 3.6.* Radius of the spherical unit cell $r_{\rm c}$, the proton radius of spherical nuclei $r_p$, and the fraction of volume $w$ filled by protons (in percent) versus matter density $\rho$. Based on the results of Douchin & Haensel (2000).

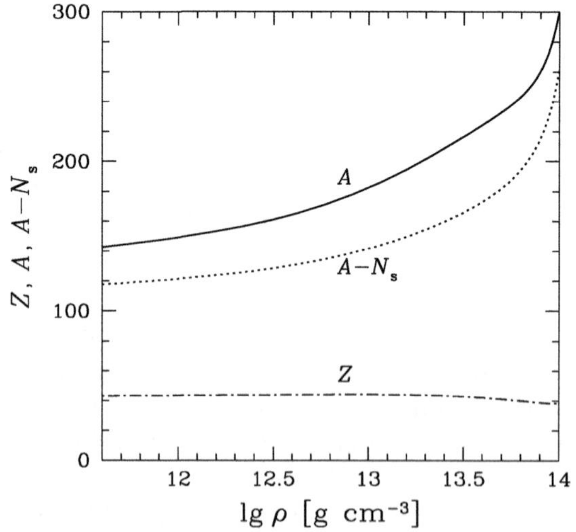


*Figure 3.7.* Mass number of spherical nuclei $A$ and their proton number $Z$ versus the average matter density $\rho$. The dotted line gives number of the nucleons after subtracting the neutrons belonging to neutron skin. Based on the results of Douchin & Haensel (2000).

For illustration, let us discuss the CLDM results for spherical nuclei obtained by Douchin & Haensel (2000). Geometrical parameters of the nuclei in the inner crust, up to $\rho = 10^{14}$ g cm$^{-3}$, are shown in Fig. 3.6. More detailed information can be found in Fig. 3.7. The number of nucleons in a nucleus, $A$, grows monotonically and reaches $\sim$300 at $\rho = 10^{14}$ g cm$^{-3}$, where $A_{\rm cell} \simeq 1000$. However, the number of protons changes rather weakly, $Z \simeq 40$. The values of $Z$ for spherical nuclei are quite similar to those obtained by Oyamatsu (1993) and Ravenhall *et al.* (1972), but they are somewhat higher than those obtained using relativistic mean-field models (Sumiyoshi *et al.*, 1998; Cheng *et al.*, 1997). An interesting quantity is the number of neutrons forming the neutron skin, $N_{\rm s}$. As seen from Fig. 3.7, for $\rho \gtrsim \frac{1}{3}\rho_0$ the value of $N_{\rm s}$ decreases with increasing density; this happens because $n_{n,{\rm i}}$ and $n_{n,{\rm o}}$ become more and more alike.

Since for $\rho \simeq 10^{14}$ g cm$^{-3}$ the spherical nuclei are very heavy, with $A \simeq 300$, the problem of their stability with respect to deformation and fission arises.

Originally, the Bohr-Wheeler condition for fission (e.g., Preston & Bhaduri 1975) was derived for isolated nuclei which were treated as drops of incompressible and charged nuclear matter. Let $E_{\rm Coul}$ and $E_{\rm surf}$ denote the Coulomb and surface energies of a spherical nucleus. The Bohr-Wheeler condition, which applies to nuclei in vacuum, states that for $E_{\rm Coul}^{(0)} \geq 2E_{\rm surf}$ a spherical nucleus is unstable with respect to small quadrupolar deformations. It is, therefore, expected to to undergo spontaneous deformation and fission into smaller fragments. Here, $E_{\rm Coul}^{(0)}$ is the Coulomb energy in vacuum, to be distinguished from $E_{\rm Coul}$ in the inner crust. In the crust, one has to correct the Bohr-Wheeler condition for the presence of electron background and other nuclei. Such corrections were calculated by Brandt (1985; as quoted by Pethick & Ravenhall 1995). The leading corrections were found to be of the order of $(r_p/r_{\rm c})^3$, in contrast to the corrections to $E_{\rm Coul}$, where the leading correction term is linear in $r_p/r_{\rm c}$ [see Eq. (3.24)]. Keeping only the leading correction to the Coulomb energy, one can rewrite the equilibrium condition (3.32) in the approximate form

$$E_{\rm surf} \simeq 2E_{\rm Coul}^{(0)}\left(1 - \frac{3}{2}\frac{r_p}{r_{\rm c}}\right) . \tag{3.34}$$

The Coulomb energy $E_{\rm Coul}^{(0)}$ in vacuum is larger than the total unit cell energy, equal to one half of $E_{\rm surf}$. As the density increases, $E_{\rm Coul}^{(0)}$ can become sufficiently large for the Bohr-Wheeler condition to be satisfied. Within the linear approximation, this would happen for $r_p/r_{\rm c} > 1/2$, i.e., when the nuclei fill more than $(1/2)^3 = 1/8$ of space. As seen from Fig. 3.6, this does not happen for spherical nuclei at $\rho < 10^{14}$ g cm$^{-3}$ for the Douchin & Haensel (2000) model of the inner crust. However, at $\rho \sim 10^{14}$ g cm$^{-3}$ the value of $(r_p/r_{\rm c})^3$ is rather close to the critical value of $1/8$.

## 3.4. Ground state of the bottom layers of the crust

For $\rho \lesssim 10^{13.5}$ g cm$^{-3}$ the ground state of the inner crust contains spherical nuclei. We will see that this structure is stable with respect to the transition into different nuclear shapes or into a uniform $npe$ matter. Of course, we *expect* the nuclei to be spherical (or quasispherical) as long as $r_p \ll r_c$. This is particularly clear within the CLDM, where it is the spherical shape which minimizes the shape-dependent (finite-size) contribution $E_{\rm N,surf} + E_{\rm Coul}$ for $r_p \ll r_c$. However, the situation is far from being obvious at $\rho \gtrsim 10^{13.5}$ g cm$^{-3}$, where $r_p/r_c \gtrsim 0.5$.

In order to get familiar with the state of matter near the bottom of the crust, in § 3.4.1 we discuss the simplest bulk approximation. Then in § 3.4.2 we study possible shapes of nuclear structures. Finally, in § 3.5 we present an alternative (with respect to the CLDM model) method of determining the density $\rho_{\rm cc}$ of the crust-core interface.

### 3.4.1 The bulk approximation

The bulk approximation consists in neglecting the Coulomb and surface effects. At a given mean nucleon density $n_{\rm b}$, nucleons are generally present in both of the coexisting "i" and "o" fluids, characterized by the constant local number densities $n_{n,{\rm i}}$, $n_{p,{\rm i}}$, $n_{n,{\rm o}}$, and $n_{p,{\rm o}}$ (the latter number density is nonzero in the proton drip regime, at $n_{\rm b} > n_{\rm PD}$). The equilibrium between the "i" and "o" fluids, ensured by the (strong) NN interaction, implies the equality of the chemical potentials of nucleons and the equality of the nucleon pressures,

$$\mu_{n,{\rm i}} = \mu_{n,{\rm o}}, \qquad \mu_{p,{\rm i}} = \mu_{p,{\rm o}} \; , \qquad P_{\rm N,i} = P_{\rm N,o} \; , \tag{3.35}$$

where the label N indicates the nucleon contribution to the pressure, and the condition on the proton chemical potentials applies only at $n_{\rm b} > n_{\rm PD}$. At a given $n_{\rm b}$, Eqs. (3.35) enable one to determine $n_{n,{\rm i}}$, $n_{p,{\rm i}}$, $n_{n,{\rm o}}$, and $n_{p,{\rm o}}$.

As before, let $w$ be the volume fraction occupied by the "i" phase. At a given mean nucleon number density $n_{\rm b} = w\, n_{\rm i} + (1-w)\, n_{\rm o}$, the total energy density is

$$\mathcal{E} = w\, \mathcal{E}_{\rm N,i} + (1-w)\, \mathcal{E}_{\rm N,o} + \mathcal{E}_e \; , \tag{3.36}$$

where $\mathcal{E}_e$ is the electron energy density. Beta equilibrium implies the relation between the chemical potentials of neutrons, protons and electrons,

$$\mu_n = \mu_p + \mu_e \; , \tag{3.37}$$

while the requirement of the overall charge neutrality leads to

$$n_e = w\, n_{p,{\rm i}} + (1-w)\, n_{p,{\rm o}} \; . \tag{3.38}$$

The set of Eqs. (3.35)–(3.38) determines completely the equilibrium state at a given $n_{\rm b}$.

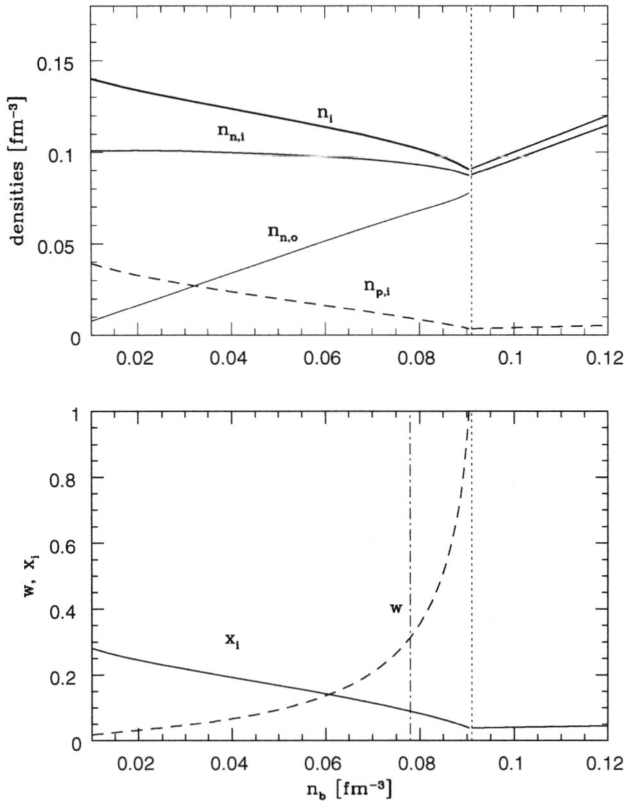


*Figure 3.8.* Ground state of the inner crust in the bulk approximation for the SLy4 effective NN interaction (Douchin & Haensel, 2000). Two-fluid phase to the left, one-fluid (uniform) phase to the right of the vertical dotted line (at $n_{2\leftrightarrow1} = 0.091\ \mathrm{fm}^{-3}$); $n_{p,\mathrm{o}}$ is not shown, because it is too small. Proton drip occurs at $0.085\ \mathrm{fm}^{-3}$ and the maximum value of $n_{p,\mathrm{o}}$ (reached at $n_{2\leftrightarrow1}$) is only $\sim 5 \times 10^{-4}\ \mathrm{fm}^{-3}$. In the lower panel, the thin dash-and-dot vertical line shows the crust-core transition in the CLDM. At the actual transition $n_\mathrm{b} = n_\mathrm{cc} = 0.078\ \mathrm{fm}^{-3}$ the bulk approximation gives $w \simeq 0.3$, in good agreement with the CLDM.

The two-fluid phase is stable at those $n_\mathrm{b}$ at which the energy per nucleon is lower than in the one-fluid phase (which is the uniform $npe$ matter). This stability condition breaks down at a density $n_{2\leftrightarrow1}$. As seen from Fig. 3.8, while $n_\mathrm{b}$ approaches $n_{2\leftrightarrow1}$ from the lower density (two-fluid) side, the volume fraction $w \to 1$ and the (denser) fluid "i" fills the entire volume (cf. Pethick *et al.* 1995). Therefore, the $2 \leftrightarrow 1$ transition is continuous, without any density jump. These properties are characteristic to the bulk approximation. Because the surface and Coulomb contributions increase the energy density, the real transition to a uniform $npe$ matter occurs at a baryon number density $n_\mathrm{cc}$ lower

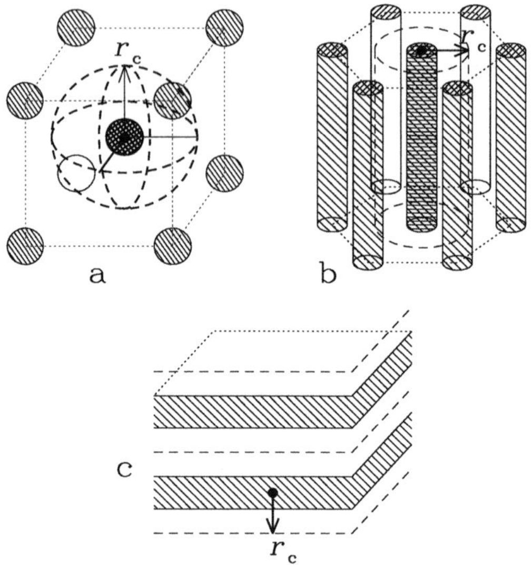


*Figure 3.9.* Unit cells for a set of nuclear shapes (spheres, rods, plates) in the inner crust. The size of the unit cell is $r_c$. Hatched regions show nuclear matter, while blank regions show neutron gas. In "bubbular phases" (tubes, spherical bubbles) one has to exchange the blank and hatched regions.

than $n_{2\leftrightarrow1}$. Thus, $n_{2\leftrightarrow1}$ is a robust *upper bound* to $n_{cc}$. For the SLy4 effective NN interaction one gets the proton drip density $n_{PD} = 0.085\,\mathrm{fm}^{-3}$ and $n_{2\leftrightarrow1} = 0.091\ \mathrm{fm}^{-3}$ (Douchin & Haensel, 2000).

### 3.4.2 Unusual nuclear shapes

As pointed out long ago by BBP, the nuclei will turn "inside-out" when the fraction of the volume occupied by the nuclear matter exceeds 50%. Then spherical bubbles of the neutron gas in the nuclear matter will become energetically preferable (BBP). This result was obtained within the CLDM neglecting the curvature contribution to the surface energy. Generally, while minimizing the energy, the nuclear shape has to be treated as a thermodynamic variable. The actual shape of nuclei in the ground state corresponds to the minimum of $\mathcal{E}$ at a given $n_b$.

Historically, the first studies of such a type were connected with the structure of matter in gravitational collapse of massive stellar cores. Calculations performed within the CLDM for dense hot matter, with $T \gtrsim 10^{10}$ K and the entropy per nucleon $(1-2)\ k_B$, indicated that before the transition into the uniform plasma the matter undergoes a series of phase transitions accompanied by changes of nuclear shapes. It was demonstrated by Ravenhall *et al.* (1983) who considered a basic set of spherical, cylindrical, and planar geometries, corresponding to *dimensionality* $d = 3,\ 2,\ 1$. For each $d$, they restricted themselves to the simplest shapes with a single curvature radius (maximal symmetry). For $d = 3$, Ravenhall *et al.* (1983) considered spherical nuclei in a nucleon gas and spherical bubbles in a denser nuclear matter, referred to as 3N and 3B, respectively. For $d = 2$, the nuclear structures were cylindrical nuclei (rods, 2N) and cylindrical holes in the nuclear matter filled with the nucleon gas (tubes, 2B). Finally, for $d = 1$ the authors considered parallel plates of the nuclear matter separated by the nucleon gas; in that case the "bubbular" and "nuclear" phases coincided and were denoted by 1N.[4] With increasing density, Ravenhall *et al.* (1983) found a sequence of phase transitions 3N $\to$ 2N $\to$ 1N $\to$ 2B $\to$ 3B which preceded the transition to the uniform matter. These transitions were accompanied by the increase of the fraction of volume occupied by denser (nuclear matter) phase. These results were confirmed in later calculations (see Watanabe *et al.* 2001, and references therein).

One of the virtues of the CLDM is its flexibility as far as the shape of nuclei is concerned. The terms $\mathcal{E}_{\rm N,bulk}$ and $\mathcal{E}_e$ are shape independent. The surface and Coulomb terms do depend on the shape, but they can easily be calculated neglecting the curvature corrections. In what follows we present the formulae for $\mathcal{E}_{\rm N,surf}$ and $\mathcal{E}_{\rm Coul}$ in this simple approximation. For the sake of completeness, we include also the case of spherical nuclei (phase 3N) already considered in §3.3.3. Using elementary considerations, one can show that the general formula for the surface energy is

$$\mathcal{E}_{\rm N,surf} = \frac{w\,d}{r_p}\left[(n_{n,\rm i} - n_{n,\rm o})\,\mu_n s_n + \sigma_{\rm s}\right] \ , \tag{3.39}$$

where $d$ is the dimensionality and $w = (r_p/r_{\rm c})^d$ is the filling factor.

The Coulomb contribution is more complicated, but the result can also be represented by a universal expression (Ravenhall *et al.*, 1983). The calculation is based on the Wigner-Seitz approximation. The unit cells for the 3N, 2N and 1N phases are shown in Fig. 3.9.[5] In the case of rods, the unit cell is approximated by a cylinder, coaxial with a rod, of radius $r_{\rm c}$. The number of rods per unit area of the plane perpendicular to the rods is $1/(\pi r_{\rm c})^2$. In the case of plates, the boundary of the unit cell consists of two planes parallel to a slab

---

[4]For obvious reasons, culinary terms are also frequently used to denote various phases; 3B, 2N, and 1N are frequently referred to as *Swiss cheese*, *spaghetti*, and *lasagna* phases, respectively.

[5]The exact shape of the unit cells is such as to fill the entire space. Therefore, in 3N and 3B phases, which form bcc lattices, the exact cell is rhombic dodecahedron, while rods and tubes would require unit cells in the form of regular hexagonal prism. Replacing the exact unit cells by spheres (3N, 3B) and cylinders (2N, 2B) simplifies enormously the calculations and turns out to be precise within better than one percent in the relevant range of $w$ (e.g., Pethick & Potekhin 1998).

of nuclear matter, at a distance $r_c$ from the slab symmetry plane. For the phases of spherical nuclei (3N), nuclear matter rods (2N), and plates (1N) one obtains

$$\mathcal{E}_{\rm Coul} = \frac{4\pi}{5}\,(n_{p,\rm i}\,e\,r_p)^2\,f_d(w)\;, \tag{3.40}$$

where

$$f_d(w) = \frac{5}{(d+2)}\left[\frac{1}{d-2}\left(1-\frac{1}{2}\,w^{1-2/d}\,d\right)+\frac{1}{2}\,w\right]\;. \tag{3.41}$$

In the case of $d = 2$ (rods) one has to take the limit of $d \to 2$ which gives

$$f_2(w) = \frac{5}{8}\left(\ln\frac{1}{w} - 1 + w\right)\;. \tag{3.42}$$

These formulae hold also for neutron gas tubes (2B) and neutron gas bubbles (3B) but in these cases one has to replace $w \to 1 - w$, and $r_p$ stands for the radius of tubes or bubbles.

The virial theorem, which states that $\mathcal{E}_{\rm N,surf} = 2\mathcal{E}_{\rm Coul}$ in equilibrium, is valid for any phase. It is a consequence of scaling of the Coulomb and surface energies with respect to $r_p$, $\mathcal{E}_{\rm N,surf} \propto r_p^{-1}$, $\mathcal{E}_{\rm Coul} \propto r_p^2$. The pressure equilibrium condition for spheres, rods and plates reads

$$P_{\rm i}^{\rm bulk} - P_{\rm o}^{\rm bulk} = (d-1)\,\frac{\sigma_{\rm s}}{r_p} + \frac{4\pi}{5}\,e^2\,n_{p,\rm i}^2\,r_p^2\,f_d(w)\left[\frac{2}{d} + w\,\frac{f_d'(w)}{4f_d(w)} - 1\right]. \tag{3.43}$$

For bubbular phases (bubbles, tubes), one has to put minus in front of the surface term, and replace $w$ by $1 - w$.

The beautiful simplicity of the formulae is lost when one introduces the "curvature corrections" to the finite-size terms. In the case of the surface terms, they result from the dependence of the nuclear surface energy on the surface curvature $\kappa$. In the case of the five nuclear shapes under consideration the curvature is given by $\kappa = (d-1)/r_p$ for the 3N and 2N phases, and $\kappa = -(d-1)/r_p$ for the 3B and 2B phases. The surface thermodynamic potential, calculated including the lowest-order curvature correction, is then given by $\sigma = \sigma_{\rm s} + \kappa\sigma_{\rm c}$. It should be stressed, that in contrast to the surface tension $\sigma_{\rm s}$, the *curvature tension* $\sigma_{\rm c}$ does depend on the choice of the "reference surface," which in our case is taken at $r = r_p$ (e.g., Kolehmainen *et al.* 1985; Douchin *et al.* 2000). The curvature corrections to the Coulomb energy appear when we include the diffuseness of the proton surface. These corrections were studied in detail by Lorenz (1991) (also see Douchin 1999).

One should stress important differences in physical conditions prevailing in a collapsing stellar core (the case considered by Ravenhall *et al.* 1983) and in the cold catalyzed matter. First, in contrast to the cold catalyzed matter, the

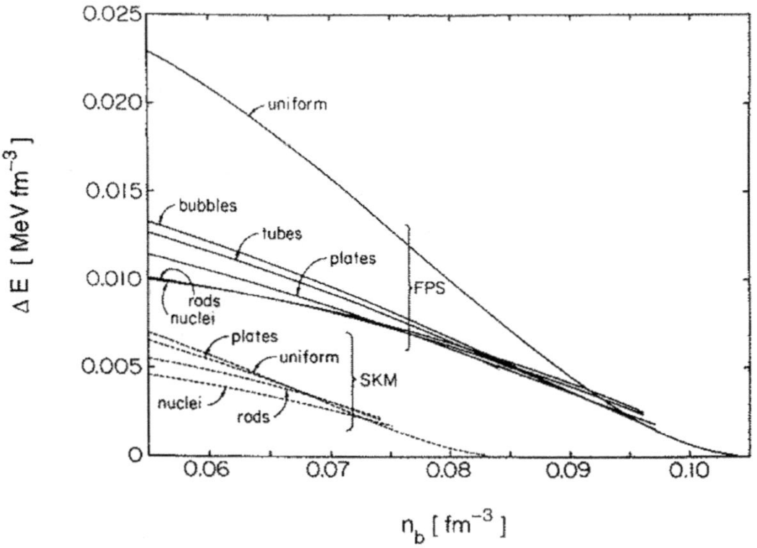


*Figure 3.10.* Energy density of a given phase of the inner crust minus the bulk energy density versus the average baryon density $n_\mathrm{b}$ for the FPS and SkM effective NN interactions (from Lorenz *et al.* 1993, with the permission of the authors). Label "uniform" refers to the uniform $npe$ matter.

majority of nucleons in the collapsing matter belong to the denser (nuclear) phase even at $\rho \sim 10^{14}$ g cm$^{-3}$; moreover, the nucleon gas in the collapsing matter contains not only neutrons but also some protons. Second, because of neutrino trapping (see § 1.4.2), the average proton fraction in the collapsing matter is about 30%, to be contrasted with a few percent fraction in the neutron star matter at $\rho \sim 10^{14}$ g cm$^{-3}$.

The first detailed study of the inner crust at $\rho \gtrsim 10^{14}$ g cm$^{-3}$, performed by Lorenz *et al.* (1993) within the CLDM, indicated that the presence of unusual nuclear shapes before the transition to the uniform $npe$ matter depends on the model for the effective NN interaction. For the FPS model, they found a sequence of phase transitions 3N $\to$ 2N $\to$ 1N $\to$ 2B $\to$ 3B, which started at $n_\mathrm{b} \approx 0.064$ fm$^{-3}$ $\simeq \frac{1}{3}n_0$ ($1.1 \times 10^{14}$ g cm$^{-3}$) and ended at the crust-core interface ($n_\mathrm{cc} = 0.096$ fm$^{-3}$, $\rho_\mathrm{cc} = 1.6 \times 10^{14}$ g cm$^{-3}$) with the transition from the 3B phase to the uniform $npe$ matter. It should be stressed that in the indicated density range the differences between $\mathcal{E}(n_\mathrm{b})$ for various shapes are very small, $\lesssim$ keV fm$^{-3}$. This can be compared with typical energy density differences between the crust and the uniform phases, $\simeq (10-20)$ keV fm$^{-3}$ in the same density range (see Fig. 3.10).

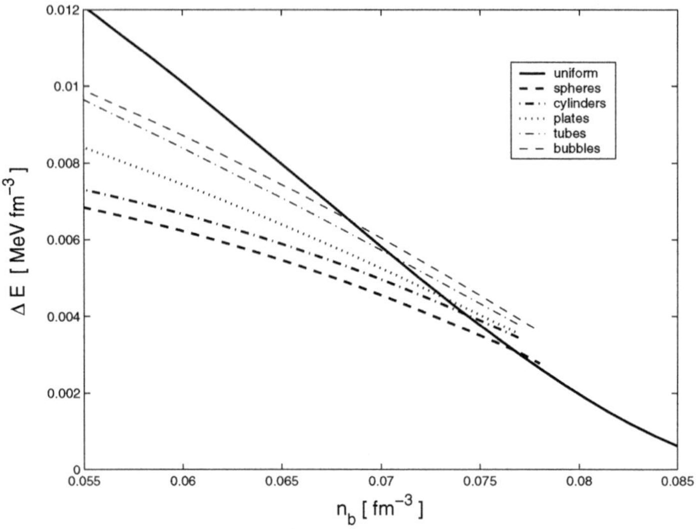


*Figure 3.11.* Energy density of a given phase of inner crust minus the bulk energy density as a function of the average baryon density $n_b$ for the SLy4 effective NN interaction. Label "uniform" refers to the uniform $npe$ matter. Based on Douchin & Haensel (2000).

As shown by Lorenz *et al.* (1993), the very presence of unusual shapes depends on the assumed model of $v_{\rm NN}^{\rm eff}$. In the case of the SkM force (used by Bonche & Vautherin 1981, 1982 in their studies of dense and hot matter) spherical nuclei were energetically preferred to the very bottom of the crust, which was found at significantly lower density $n_{\rm cc} = 0.074~{\rm fm}^{-3}$ ($\rho_{\rm cc} = 1.2 \times 10^{14}~{\rm g~cm}^{-3}$). Further calculations confirmed this unfortunate ambiguity. Using parametrized density profiles in the ETF energy density functional, Oyamatsu (1993) found a complete sequence of phase transitions in the density range $(1.0 - 1.5) \times 10^{14}~{\rm g~cm}^{-3}$. A similar sequence was found by Sumiyoshi *et al.* (1998), who obtained however a much narrower density range of unusual nuclear shapes $(0.050 - 0.058)~{\rm fm}^{-3}$ before the final transition to the uniform $npe$ matter. On the contrary, Cheng *et al.* (1997), using an unconstrained relativistic ETF approach, found that spherical nuclei persist in the ground state to the crust edge, which ranged from $0.058~{\rm fm}^{-3}$ to $0.073~{\rm fm}^{-3}$, depending on the parameters of their $\sigma$–$\omega$–$\rho$ Lagrangian. Similarly, calculations by Douchin & Haensel (2000) with the SLy4 effective NN force indicated the absence of unusual nuclear shapes (see Fig. 3.11); the transition to the uniform matter took place at $n_{\rm cc} = 0.078~{\rm fm}^{-3}$ ($\rho_{\rm cc} = 1.3 \times 10^{14}~{\rm g~cm}^{-3}$).

The existing calculations enable one to make some generic statements on unusual nuclear shapes at $\rho \lesssim \rho_{\rm cc}$. The very phenomenon of phase transitions between the various shapes results from the interplay of three quantities: finite-size (surface and Coulomb) term in $E_{\rm cell}$, the dominating bulk energy term, and the volume fraction $w$ of the denser nucleon fluid. If the finite-size terms are small, then $\rho_{\rm cc}$ is reached at a relatively low value of $w$. However, unusual shapes become energetically advantageous only at sufficiently large values of $w$. Therefore, a small surface tension may prohibit the appearance of unusual shapes before $\rho_{\rm cc}$ is reached (as in the case of the SLy4 and SkM forces, Figs. 3.10 and 3.11). It should be stressed, however, that the phase transitions themselves result from very small energy differences (see Figures 3.10 and 3.11) of the energy densities; the finite-size terms are very small compared to $\mathcal{E}_{\rm bulk}$.

In the case of the CLDM, one should stress the importance of the curvature term in $E_{\rm N,surf}$. This term should be included in any calculation of the crust-core transition. As mentioned above, introducing the curvature corrections in the finite-size terms complicates the problem. According to Oyamatsu *et al.* (1984), in the absence of the curvature correction to $E_{\rm N,surf}$ the 3N$\rightarrow$2N transition occurs at $w = 0.2$. However, in the presence of this correction the 3N phase can persist at larger $w$.

According to the CLDM or ETF calculations, a change of the nuclear shape is accompanied by a very small (less than one percent) density jump leading to a weak first-order phase transition (Lorenz *et al.*, 1993; Oyamatsu, 1993; Sumiyoshi *et al.*, 1995). The EOS in the vicinities of such transitions is derived using the Maxwell construction. Notice, however, that these results have been obtained in the Wigner-Seitz approximation for $\mathcal{E}_{\rm Coul}$, whose precision at $\sim 10^{14}$ g cm$^{-3}$ has to be carefully checked.

### 3.4.3 Quantum corrections to the CLDM energy

The CLDM model is *par excellence* classical. Also, the ETF scheme is a semiclassical approximation to a quantum-mechanical many-body problem. As the differences of the energy densities between phases with different nuclear shapes are very small, one may worry about possible importance of neglected quantum effects, which result from a discrete character of quantum single-particle states and from the interference of single-particle wave functions. In terrestrial nuclear physics, there exists a systematic procedure of adding quantum (shell) corrections to the smooth CLDM energies of nuclei (the Strutinsky method, described , e.g., by Preston & Bhaduri 1975).

As we have seen in § 3.3.1, the shell effects for protons in spherical nuclei are strong, which leads to the persistence of $Z = 40$ and $Z = 50$ in the inner crust. The energy correction for nonspherical nuclei, resulting from the shell effects of protons, has been calculated by Oyamatsu & Yamada (1994). It has been found that the inclusion of the proton shell effects shifts the changes of nuclear shapes to higher densities. In other words, the proton shell effects make a phase with a given nuclear shape more stable. However, deviations from the results of the CLDM turn out to be small. The shell effect for neutrons bound in nuclei is expected to be smaller than for protons, because neutrons are much more numerous and occupy mostly unbound (continuum) states. In their calculations performed for the 1N (slab) geometry, Oyamatsu & Yamada (1994) found the shell-energy correction for neutrons an order of magnitude smaller than for protons. Proton shell effects for spherical nuclei in the inner crust were calculated by Dutta *et al.* (2004).

The shell-energy correction for dripped neutrons was studied in detail in a series of papers by Magierski, Bulgac, and Heenen (see Magierski *et al.* 2001 and references therein). The shell effects for unbound neutrons result from the possibility of forming resonant quantum states owing to neutron-nucleus scattering. This scattering leads to an effective interaction between nuclei immersed in a neutron gas (neglecting phase shifts of neutron wave functions owing to neutron collisions at $T \neq 0$). Such a phenomenon is similar to the Casimir effect, familiar in the quantum field theory and in the condensed matter physics (e.g., Itzykson & Zuber 1980, pp. 138–142). As a result of the "fermionic Casimir effect" for dripped neutrons in the presence of nuclei, the energy density acquires a small term $\mathcal{E}_{n,\text{shell}}$, which fluctuates rapidly as a function of $\rho$. At $\rho \sim 10^{14}$ g cm$^{-3}$ the amplitude of oscillations of $\mathcal{E}_{n,\text{shell}}$ turns out to be of the same order of magnitude as the energy difference $\Delta\mathcal{E}$ between the various phases of nuclear shapes, displayed in Figs. 3.10 and 3.11. This could lead to complicated phase transition patterns at $T \lesssim \mathcal{E}_{n,\text{shell}}/n_\text{b}k_\text{B} \sim 10^9$ K (while at higher $T$ the shell effects are washed out by thermal fluctuations). As Magierski & Heenen (2002) argue, their result suggests, that at $T \lesssim 10^9$ K and $\rho \sim 10^{14}$ g cm$^{-3}$ the number of phase transitions could significantly increase. In particular, the order of transitions predicted by the CLDM could be reversed. Moreover, several phases could coexist at a given density. Finally, one cannot neglect the possibility of a mixed, disordered state of the crust.

Another quantum effect neglected in the CLDM or ETF model is pairing of nucleons. It seems to be unimportant for phase transitions. As shown by Magierski & Heenen (2002), neutron pairing does not change the values and density dependence of $\mathcal{E}_{n,\text{shell}}$. On the other hand, Baldo *et al.* (2005) find a rather strong effect of nucleon pairing on the values of $Z$ in the inner crust. Clearly, the effect of nucleon pairing on the structure of the inner crust requires a further study.

## 3.5. Reaching the crust-core interface from the core side

The determination of the bottom edge of the crust based on the CLDM of nuclei requires a highly precise calculation of the finite-size contribution, $E_\text{N,surf} + E_\text{Coul}$, in $E_\text{cell}$. One has to construct a CLDM representation of the ground state of the inner crust and then find the crust-core transition from the condition of thermodynamic phase equilibrium. One should use the same nuclear Hamiltonian for the crust and the core phases. One needs also a very precise many-body method to describe nuclear structures at the crust bottom, which is a very difficult task (see Figs. 3.10 and 3.11). Luckily, the crust-core phase transition is *very weakly first-order* (see above). Therefore, one can locate this transition using a completely different method well known in the theory of phase transitions in condensed matter. This will serve as an independent test for the precision of the CLDM calculation of $\rho_{cc}$, described above. We will locate the crust edge by checking the stability of the uniform $npe$ matter. We start from high densities, where the homogeneous phase is certainly stable with respect to the formation of spatial inhomogeneities (BBP, Pethick *et al.* 1995). By lowering the density, we will eventually find the threshold density, at which the uniform $npe$ matter becomes unstable. This threshold gives a very good approximation of the actual $\rho_{cc}$.

For a given $n_b$, the ground state of the uniform $npe$ matter minimizes the energy density $\mathcal{E}(n_n, n_p, n_e) = \mathcal{E}_0$ at a fixed baryon number density ($n_n + n_p = n_b$) and under the condition of electric charge neutrality ($n_e = n_p$). This implies beta equilibrium between the matter constituents and ensures vanishing of the first variation of $\mathcal{E}$ due to small perturbations $\delta n_j(\boldsymbol{r})$ (where $j = n, p, e$) of the equilibrium solution (at a fixed total nucleon number and a global charge neutrality of the system). However, this does not guarantee the stability of the spatially homogeneous state of the $npe$ matter; the stability requires the second variation of $\mathcal{E}$, quadratic in $\delta n_j$, be positive.

The energy functional of a slightly nonuniform matter can be calculated using the semi-classical ETF treatment of the kinetic and the spin-gradient terms in the nucleon contribution to $\mathcal{E}$ (Brack *et al.*, 1985). Assuming small spatial gradients, we keep only the quadratic gradient terms in the ETF expressions. This approximation is justified because characteristic wavelengths of periodic perturbations are indeed much larger than internucleon distances. In this case the change of the energy (per unit volume) induced by density perturbations can be expressed (BBP, Pethick *et al.* 1995) as

$$\mathcal{E} - \mathcal{E}_0 = \frac{1}{2} \int \frac{\mathrm{d}\boldsymbol{q}}{(2\pi)^3} \sum_{j,k} F_{jk}(\boldsymbol{q})\, \delta n_j(\boldsymbol{q})\, \delta n_k(\boldsymbol{q})^* \,, \tag{3.44}$$

where we use the Fourier representation

$$\delta n_j(\boldsymbol{r}) = \int \frac{\mathrm{d}\boldsymbol{q}}{(2\pi)^3}\, \delta n_j(\boldsymbol{q})\, \mathrm{e}^{\mathrm{i}\boldsymbol{q}\boldsymbol{r}} \,. \tag{3.45}$$

The Hermitian matrix $F_{ik}(\boldsymbol{q})$ determines the stability of the uniform $npe$ matter with respect to spatially periodic perturbations with a wave vector $\boldsymbol{q}$. Due to the isotropy of the uniform $npe$ matter, $F_{ik}$ depends only on $q = |\boldsymbol{q}|$. The matrix elements $F_{ik}$ can be calculated from the second variation of the microscopic energy functional $\mathcal{E}[n_j(r), \nabla n_j(r)]$ (BBP, Pethick *et al.* 1995).

The condition for the $F_{ij}$ matrix to be positive-definite is equivalent to the requirement that the matrix determinant is positive (Pethick *et al.*, 1995). At any density $n_{\rm b}$, one has thus to check whether $\det[F_{ij}(q)] > 0$. Let us start with some $n_{\rm b}$, at which $\det[F_{ij}(q)] > 0$ for any $q$. By decreasing $n_{\rm b}$, we find eventually a wave-number $Q$ at which the stability condition is violated for the first time; let it happen at a density $n_Q$. For $n_{\rm b} < n_Q$ the homogeneous state ceases be the ground state of the $npe$ matter.

Calculations performed using several effective nuclear Hamiltonians indicate that $n_Q \simeq n_{\rm cc}$, within a percent or better (Pethick *et al.*, 1995; Douchin & Haensel, 2000). For the ETF approximation to be correct, the wavelength $\lambda_Q = 2\pi/Q$ of critical density perturbations must be significantly larger than the mean inter-nucleon distance. The critical wave numbers $Q$ are typically $\sim 0.3\ {\rm fm}^{-1}$ so that $\lambda_Q \sim 20$ fm. At $n_Q \sim 0.1\ {\rm fm}^{-3}$ the fraction of protons is only about 3–4% but $\lambda_Q \sim 20$ fm is still 2–3 times larger than an inter-proton distance. The ratio of $\lambda_Q$ to the inter-neutron distance is typically about eight. Calculations of $n_Q$ show that the actual precision of the ETF approximation is much better than guessed from the ratio of $\lambda_Q$ to the inter-nucleon distance. This feature is well known from the ETF calculations of the energy of terrestrial atomic nuclei (Ring & Schuck, 1980).

The instability at $n_Q$ signals a phase transition with a loss of translational symmetry of the $npe$ matter, producing nuclear structures. The agreement of $n_Q$ and $n_{\rm cc}$ is a good test for the precision of calculated values of $n_{\rm cc}$. It indicates that the approximation of the spherical unit cell for 3N or 3B phases remains valid at $\rho \approx \rho_{\rm cc}$. It means also that the linear approximation of the curvature correction in $\sigma$ is sufficiently precise. Finally, it is a convincing argument for the validity of the CLDM at very large neutron excesses.

## 3.6. The equation of state of the neutron star crust

In the present section we discuss the EOS of the outer and inner crusts, built of cold catalyzed matter.

The EOS in the outer crust is rather well established. We suggest to use the HP EOS given in Table A.1 of Appendix A. It is similar to the older BPS EOS. In some pressure intervals the two EOSs give a few percent difference in $\rho$, resulting from the difference in nuclides present at those pressures.

As soon as one leaves the region of experimentally known nuclei, the EOS of cold catalyzed matter becomes uncertain. The uncertainty increases at $\rho > \rho_{\rm ND}$. The properties of nuclei are affected by the ambient neutron gas which contributes more and more to the total pressure. Therefore, the problem of correct modelling of the EOS of pure neutron gas at subnuclear densities becomes important. The real EOS of cold catalyzed matter stems from the real nucleon Hamiltonian, which is expected to describe nucleon interactions at $\rho \lesssim 2\rho_0$ (while at higher $\rho$ non-nucleon degrees of freedom can be important, Chapter

5). In practice, in order to make the solution of the many-body problem feasible, the task was reduced to finding an *effective nucleon Hamiltonian*, which would enable one to calculate reliably the EOS for $10^{11}$ g cm$^{-3} \lesssim \rho \lesssim \rho_0$, including the crust-core transition.

Usually, an effective nuclear interaction is constructed in such a way as to reproduce the properties of laboratory nuclei. A minimal condition is to reproduce experimental saturation parameters of the symmetric nuclear matter (§ 5.4). In the crust at $\rho \gtrsim 10^{13}$ g cm$^{-3}$, the pressure is mostly supplied by free neutrons. Therefore, an effective nuclear force has to give an accurate description of the EOS of the neutron gas at subnuclear densities. We have no direct experimental access to the EOS of the neutron matter. However, we have precise numerical calculations of the ground-state properties of the neutron matter at $n_n \lesssim n_0$, based on the modern many-body theories with best *bare* nucleon interactions; they be used in lieu of experimental data (Pethick *et al.*, 1995; Pethick & Ravenhall, 1995).[6] In Fig. 3.12 we plot the energy per neutron, $\mathcal{E}/n_n$, in the neutron matter versus $n_n$ for several effective nuclear interactions. One notices good overall agreement of the SLy curve with the microscopic many-body results of Wiringa *et al.* (1988) and Akmal *et al.* (1998); for the remaining interactions the agreement is worse. Let us stress that the SLy force was constructed to fit the results of Wiringa *et al.* (1988) at $n_\mathrm{b} \gtrsim n_0$.

Of course, for $\rho \lesssim 4 \times 10^{11}$ g cm$^{-3}$ one can use an EOS based on experimental or semi-empirical nuclear masses, but it is reassuring to check that this EOS is in nice agreement with a "theoretical EOS", based on the effective NN FPS and SLy interactions. As seen from Fig. 3.13, significant differences between the SLy and FPS EOSs occur in the density interval from $4 \times 10^{11}$ g cm$^{-3}$ to $4 \times 10^{12}$ g cm$^{-3}$. They result mainly from the difference of the neutron drip thresholds, $\rho_\mathrm{ND}(\mathrm{SLy}) \simeq 4 \times 10^{11}$ g cm$^{-3}$ (in good agreement with the "empirical EOS" of HP) versus $\rho_\mathrm{ND}(\mathrm{FPS}) \simeq 6 \times 10^{11}$ g cm$^{-3}$. For $4 \times 10^{12}$ g cm$^{-3} \lesssim \rho \lesssim 10^{14}$ g cm$^{-3}$ the SLy and FPS EOSs are very similar. The behavior of the two EOSs near the crust-core transition is shown in Fig. 3.14. The FPS EOS is softer there.

In the SLy EOS the crust-core transition takes place as a very weak first-order phase transition, with the relative density jump $\sim 1\%$. Let us remind that for this model spherical nuclei persist to the very crust bottom. As seen from Fig. 3.14, the crust-core transition is accompanied by a noticeable jump of the slope (stiffening) of the EOS. For the FPS EOS, the crust-core transition is preceded by a sequence of phase transitions with changes of nuclear shapes. They make the crust-core transition smoother than in the SLy case, with a gradual increase of the stiffness, which nevertheless undergoes a visible jump

[6]These statements may be over-optimistic. Monte Carlo calculations by Sarsa *et al.* (2003) give an EOS of neutron matter which is noticeably stiffer than the EOSs of Wiringa *et al.* (1988) and Akmal *et al.* (1998).

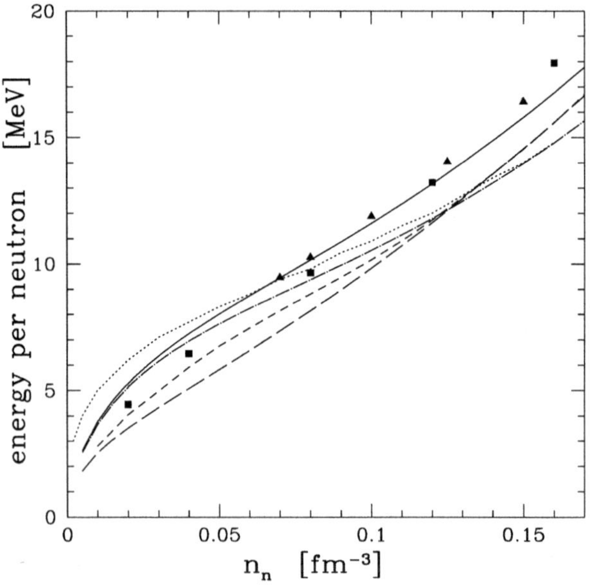


*Figure 3.12.* Energy per neutron versus neutron number density for pure neutron matter. We show the results of Wiringa *et al.* (1988) (the UV14+UVIII interaction; filled triangles); Akmal *et al.* (1998) (the UV18+$\delta v$+UIX* interaction; filled squares); Pethick *et al.* (1995) (the FPS effective nuclear interaction; the dotted line); the SLy model (the solid line); the SkM* model (the long-dashed line); the Sk1′ model (the long dashed-and-dot line); Sumiyoshi et al. (Sumiyoshi *et al.* 1995 and Sumiyoshi 2000, private communication; the Relativistic Mean Field Theory; the dashed line).

at the bottom of the bubble-layer edge. All in all, while exotic nuclear shapes may strongly affect the transport phenomena, neutrino emission, and elastic properties of the matter, their effect on the EOS is small.

The overall SLy EOS of the crust, calculated including adjacent segments of the core and the outer crust, is shown in Fig. 3.15. In the outer crust, the SLy EOS cannot be visually distinguished from the HP EOS (based on experimental nuclear masses). An important stiffness parameter of the EOS is the local adiabatic index, $\gamma = (n_\mathrm{b}/P)\,\mathrm{d}P/\,\mathrm{d}n_\mathrm{b}$, which can be approximated by $\gamma \simeq (\rho/P)\,\mathrm{d}P/\,\mathrm{d}\rho$ at subnuclear densities. In Fig. 3.16 we plot $\gamma$ versus $\rho$ for the SLy EOS at densities $\rho > \rho_\mathrm{ND}$. At $\rho < \rho_\mathrm{ND}$, we show two plots of $\gamma$. The solid line corresponds to the precise calculation of $\gamma$ from the HP model, but neglecting density discontinuities associated with density jumps owing to changes of ground-state nuclides. To a very good approximation, this adiabatic index is $4/3$ (see also BPS). This is because before the neutron drip the pressure is very well approximated by the sum of the pressure $P_e$ of

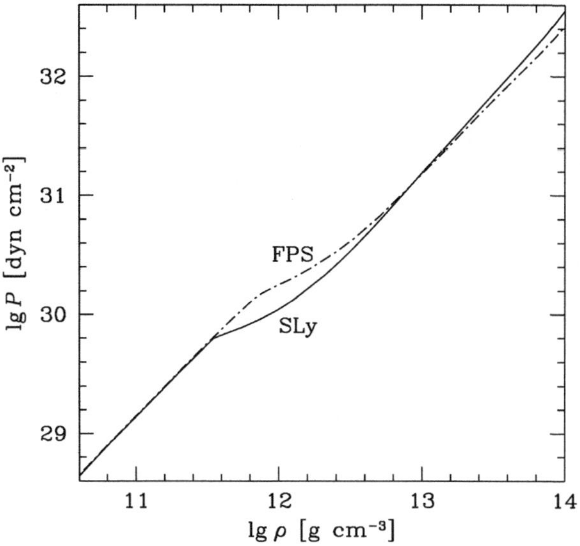


*Figure 3.13.* Comparison of the SLy and FPS EOSs.

ultra-relativistic electrons and the lattice contribution $P_L$; both are proportional to $\rho^{4/3}$ (see §§ 2.3.5 and 3.2). Averaging the pressure over density jumps performed implicitly in the CLDM softens the EOS and yields lower value of $\gamma$. After a dramatic softening in the density region following the neutron drip point, the EOS stiffens gradually, with a visible increase of $\gamma$ near $\rho_{cc}$. Then follows a jump in $\gamma$ associated with the disappearance of nuclei, and subsequent stiffening (due to NN interaction) in the uniform *npe* liquid.

The tabulated EOSs discussed above are given in Appendix A; their analytical approximations are presented in Appendix C.

## 3.7. Elastic properties of neutron star crust

In contrast to the liquid core, the solid crust can sustain an *elastic strain.* As neutron stars are relativistic objects, a relativistic theory of elastic media in a curved space-time should be used. Such a theory of elasticity was developed by Carter & Quintana (1972) who applied it to rotating neutron stars (Carter & Quintana, 1975a,b). However, for the sake of simplicity, we will restrict ourselves to the Newtonian version of the theory of elasticity (Landau & Lifshitz, 1984). This approximation can be used to obtain the elastic moduli of the crust, which are the same as those appearing in the general relativistic formulation.

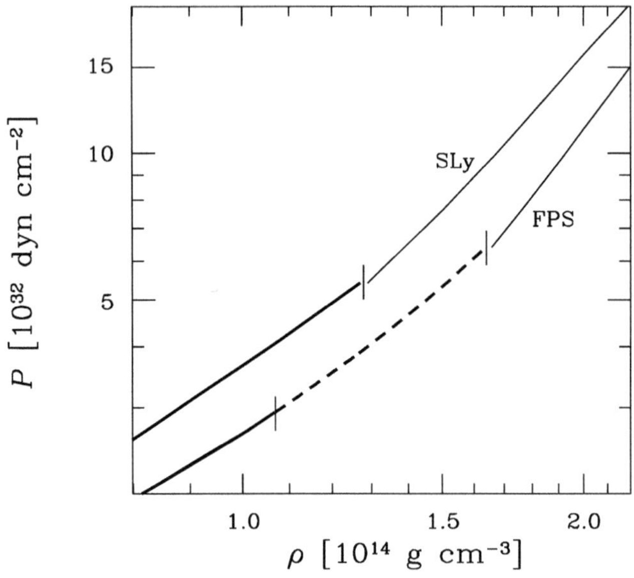


*Figure 3.14.* Comparison of the SLy and FPS EOSs near the crust-core transition. Thick solid lines refer to the inner crust with spherical nuclei. The dashed line is for "exotic nuclear shapes". Thin solid lines refer to the uniform *npe* matter.

The state of thermodynamic equilibrium of a matter element corresponds to *equilibrium positions* of nuclei, which will be denoted by $\boldsymbol{r}$. Neutron star evolution (e.g., spin-down, accretion) or some outer effects may *deform* the crust. In what follows, we neglect the thermal contributions to thermodynamic quantities.

A deformation of a crust element implies a *displacement* of the nuclei into their new positions $\boldsymbol{r}' = \boldsymbol{r} + \boldsymbol{u}$, where $\boldsymbol{u} = \boldsymbol{u}(\boldsymbol{r})$ is the displacement vector. In the continuum-medium limit, relevant for macroscopic phenomena, both $\boldsymbol{r}$ and $\boldsymbol{u}$ are treated as continuous fields. The displacement $\boldsymbol{u}$ produces an elastic strain (i.e., a force which tends to return the matter element to the equilibrium state with the minimum energy $\mathcal{E}_0$),[7] and determines the *deformation energy* $\mathcal{E}_{\rm def} = \mathcal{E}(\boldsymbol{u}) - \mathcal{E}_0$. A uniform translation does not contribute to $\mathcal{E}_{\rm def}$. The deformation is described by the symmetric *strain tensor*

$$u_{ik} = u_{ki} = \frac{1}{2}\left(\frac{\partial u_i}{\partial x_k} + \frac{\partial u_k}{\partial x_i}\right) , \tag{3.46}$$

[7] In this section, by "energy" we mean energy per unit volume (energy density)

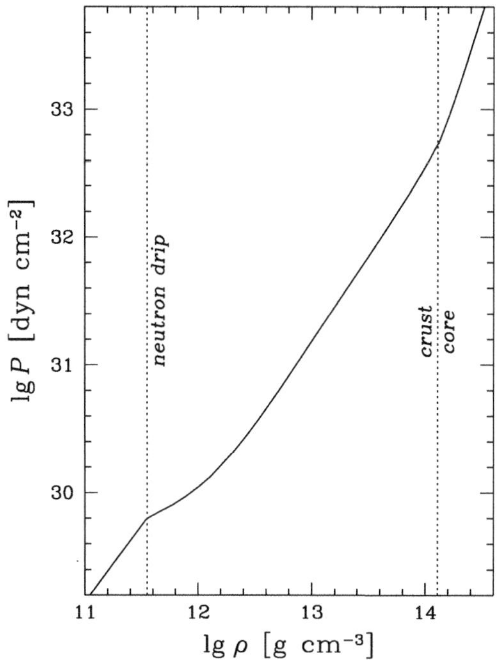


*Figure 3.15.* The SLy EOS. Dotted vertical lines show the neutron drip and crust-core interface.

where $i, j = 1, 2$, and 3. This expression is valid as long as the displacement $\boldsymbol{u}$ is small and terms quadratic in $\boldsymbol{u}$ can be neglected (Landau & Lifshitz, 1984).

Any deformation can be decomposed into the *compression* and *shear* components,

$$u_{ik} = u_{ik}^{\text{comp}} + u_{ik}^{\text{shear}} , \tag{3.47}$$

where $u_{ik}^{\text{comp}} = \frac{1}{3}\,\delta_{ik}\;\nabla\cdot\boldsymbol{u}$ and $u_{ik}^{\text{shear}} = u_{ik} - u_{ik}^{\text{comp}}$. The deformation changes the volume of a matter element according to $\mathrm{d}V' = (1+\nabla\cdot\boldsymbol{u})\;\mathrm{d}V$. A pure compression, which does not affect the shape of the matter element, is described by $u_{ik} = a\,\delta_{ik}$. A pure shear deformation keeps the volume of the matter element constant, so that $\nabla\cdot\boldsymbol{u} = 0$.

To the lowest order, the deformation energy is quadratic in the deformation tensor,

$$\mathcal{E}_{\text{def}} = \frac{1}{2}\,\lambda_{iklm}\,u_{ik}\,u_{lm} , \tag{3.48}$$

where summation is over repeated indices. Since $\mathcal{E}_{\text{def}}$ is a scalar, $\lambda_{iklm}$ are components of a rank-four tensor. While the total number of these components is 81, general symmetry relations reduce the maximum number of linearly independent components (elastic moduli) to 21. The number of independent elastic

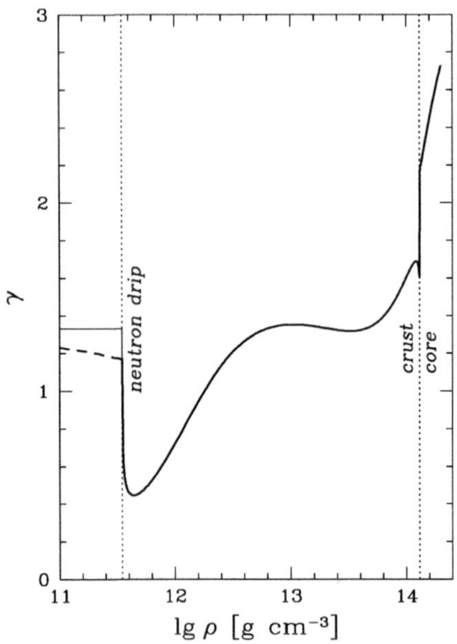


*Figure 3.16.* Adiabatic index $\gamma$ for the ground state of the neutron star matter for the HP model before neutron drip and the CLDM of Douchin & Haensel (2000, 2001) at higher $\rho$. Dotted vertical lines indicate the neutron drip and crust-core interface. The dashed line before the neutron drip shows $\gamma$ obtained using the Douchin-Haensel model.

moduli decreases with increasing symmetry of elastic medium and becomes as small as three for a bcc crystal, and two for an isotropic solid. The elastic stress tensor $\sigma_{ik}$ is derived from the deformation energy via $\sigma_{ik} = \partial\mathcal{E}_{\rm def}/\partial u_{ik}$.

### 3.7.1 From bcc lattice to isotropic solid

Although microscopically the ground state of the neutron star crust at $\rho \lesssim 10^{14}$ g cm$^{-3}$ corresponds to a bcc lattice, one usually assumes that its macroscopic properties, relevant for neutron star models, are those of an isotropic bcc polycrystal. This simplifying assumption is made, because it is likely that the crustal matter is better approximated by a polycrystal than a monocrystal. The elastic properties of an isotropic solid are described by two elastic moduli,

$$\mathcal{E}_{\rm def} = \frac{1}{2} K \, (\nabla \cdot \boldsymbol{u})^2 + \mu \left( u_{ik} - \frac{1}{3} \, \delta_{ik} \, \nabla \cdot \boldsymbol{u} \right)^2 . \tag{3.49}$$

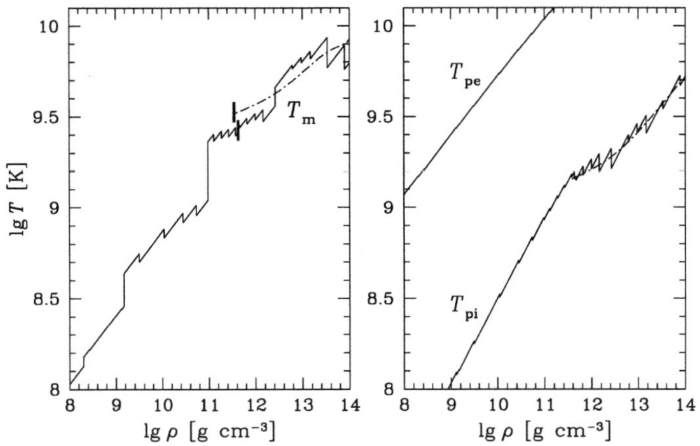


*Figure 3.17.* Melting temperature (*left*) and electron and ion plasma temperatures (*right*) of the ground-state matter in the crust. Solid lines are based on the models of Haensel & Pichon (1994) and Negele & Vautherin (1973) for the outer and inner crusts, respectively. Jumps are associated with changes of nuclides. Dot-and-dashed lines are based on the CLDM of Douchin & Haensel (2000); their smooth behavior is an approximation inherent to the model. Thick vertical dashes indicate the neutron drip.

Here, $\mu$ is the *shear modulus* and $K$ is the *compression modulus*. Then the stress tensor is

$$\sigma_{ik} = \frac{\partial \mathcal{E}_{\rm def}}{\partial u_{ik}} = K\delta_{ik}\nabla \cdot \boldsymbol{u} + 2\mu\left(u_{ik} - \frac{1}{3}\,\delta_{ik}\,\nabla \cdot \boldsymbol{u}\right) . \tag{3.50}$$

For a pure uniform compression

$$K = n_{\rm b}\,(\partial P/\partial n_{\rm b}) = \gamma P \;, \tag{3.51}$$

where $\gamma$ is the adiabatic index.

Detailed calculations of the direction-averaged effective shear modulus of a bcc Coulomb solid, appropriate for the polycrystalline crusts, were performed by Ogata & Ichimaru (1990). These authors considered a one component bcc Coulomb crystal, neglecting screening by degenerate electrons, as well as quantum zero-point motion of ions.

For an ideal bcc lattice there are only three independent elastic moduli denoted traditionally as $c_{11}$, $c_{12}$, and $c_{44}$ (Chapter 6 of Kittel 1986). A compressional deformation ($\nabla \cdot \boldsymbol{u} = 0$) is determined only by two independent elastic moduli, because

$$\mathcal{E}_{\rm def} = b_{11}\left(u_{xx}^2 + u_{yy}^2 + u_{zz}^2\right) + 2c_{44}\left(u_{xy}^2 + u_{xz}^2 + u_{yz}^2\right) \;, \tag{3.52}$$

where $b_{11} = \frac{1}{2}(c_{11} - c_{12})$. For $T = 0$, Ogata & Ichimaru (1990) find $b_{11} = 0.0245\,n_{\mathcal{N}}\,(Ze)^2/r_{\rm c}$, $c_{44} = 0.1827\,n_{\mathcal{N}}\,(Ze)^2/r_{\rm c}$. These values agree with the

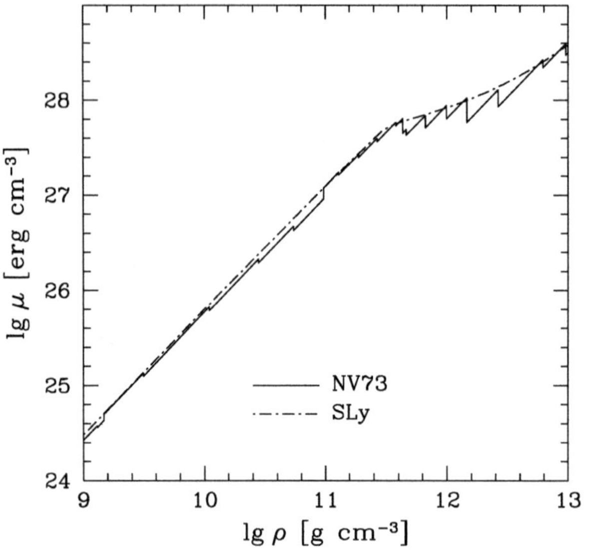


*Figure 3.18.* Effective shear modulus $\mu$ versus density at $T = 0$ for bcc lattice. The solid line is for the models of Haensel & Pichon (1994) and Negele & Vautherin (1973) (in the outer and inner crust, respectively; §§ 3.2 and § 3.3). The dot-and-dashed line is for the model of Douchin & Haensel (2000) (§ 3.3).

classical result of Fuchs (1936). A significant difference between $b_{11}$ and $c_{44}$ indicates a strong elastic anisotropy of an ideal bcc monocrystal.

While treating the crust as an isotropic solid is a reasonable approximation (we most probably deal with a bcc polycrystal), the choice of an "effective" shear modulus deserves a comment. In numerous papers treating the elastic aspects of neutron star dynamics, a standard choice was $\mu = c_{44}$ (Baym & Pines 1971; Pandharipande *et al.* 1976; McDermott *et al.* 1988b, and references therein). It is clear, that replacing $\mu$ by a single *maximal* elastic modulus of a strongly anisotropic lattice is not accurate. The correct effective value of $\mu$ was calculated by Ogata & Ichimaru (1990) by direct averaging over rotations of Cartesian axes. For $T = 0$ their result is

$$\mu = \frac{1}{5}\,(2b_{11} + 3c_{44}) = 0.1194\,\frac{n_N\,(Ze)^2}{r_c}\,, \tag{3.53}$$

about 1.5 times smaller than $\mu = c_{44}$ used in previous papers.

Now return to the isotropic solid. Equation (3.53) can be rewritten as

$$\mu = 0.0159\,(Z/26)^{2/3}\,P_e\,, \tag{3.54}$$

where $P_e$ is the pressure of ultra-relativistic degenerate electrons. Therefore,

$$\mu/K = 0.016\ (Z/26)^{2/3}\ (P_{\rm e}/\gamma P) \ll 1\ . \tag{3.55}$$

The crust is much more susceptible to shear than to compression; its Poisson coefficient $\sigma \simeq 1/2$, while its Young modulus $E \simeq 3\mu$.

Strictly speaking, the above formulae hold for the outer crust, where $r_{\mathcal{N}} \ll r_{\rm c}$ and $P \simeq P_{\rm e}$. In the inner crust they are only approximate.

### 3.7.2 Exotic nuclei

Some models predict the existence of unusual nuclei (rods, plates, tubes, bubbles) in the crust bottom, at $\rho \gtrsim 10^{14}$ g cm$^{-3}$ (§ 3.4). Let us concentrate on two unusual shapes, rods and plates, which are expected to fill most of the bottom layer. The elastic moduli for the tube-phase can be obtained from those for rods by replacing $w \rightarrow 1-w$. The properties of matter containing rods and plates are intermediate between those of solids and liquids. For example, any displacement of a plate along its surface or a rod along its axis is not opposed by restoring forces, which is a typical property for a liquid. However, an elastic strain opposes any bending of planes or rods, a property specific for a solid. These intermediate kinds of matter are usually referred to as *mesomorphic* phases, or *liquid crystals* (see, e.g., Landau & Lifshitz 1984; de Gennes & Prost 1993). The elastic properties of rod and plate phases were studied by Pethick & Potekhin (1998) (also see Pethick & Ravenhall 1995).

As stressed by Pethick & Potekhin (1998), the physical reasons for forming mesomorphic phases in neutron star matter are very different from those in laboratory. For a laboratory liquid crystal, it is the interaction between very nonspherical molecules which drives them to form rods or plates. In the neutron star interior one deals with *spontaneous symmetry breaking* resulting from the competition between the Coulomb energy and nuclear surface energy.

In our discussion of the elastic properties of rod and plate phases in neutron star matter we closely follow Pethick & Potekhin (1998). The energies of mesomorphic phases are calculated using the generalized Liquid Drop Model (§ 3.4). It is assumed that the average (total) nucleon density remains constant. Also, it is assumed that the fraction of the total volume occupied by nuclear matter, and the densities of the nuclear matter and the neutron gas, remain constant. Thus, only the Coulomb and surface energies are altered by deformations.

The plate phase has rotational symmetry about any axis perpendicular to the plates. Therefore, it is similar to the *smectics A* phase in the nomenclature of liquid crystals physics (de Gennes & Prost, 1993). Let the $z$-axis coincide with the symmetry axis of the equilibrium (nondeformed, or relaxed) configuration. Only a displacement in the $z$-direction can be opposed by a restoring force, so that we can set $\boldsymbol{u} = (0, 0, u)$. The deformation energy of a unit volume can

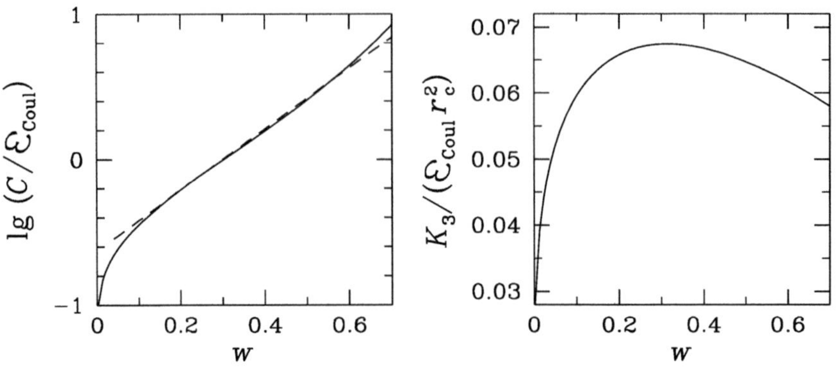


*Figure 3.19.* Shear coefficient $C$ (in units of $\mathcal{E}_{\rm Coul}$) and bend coefficient $K_3$ (in units of $\mathcal{E}_{\rm Coul} r_{\rm c}^2$) of the rod (columnar) phase of neutron star matter versus the filling factor $w = (r_p/r_{\rm c})^2$. The dashed line in the left panel is the approximation given by Eq. (3.62). Based on Pethick & Potekhin (1998).

then be written as (de Gennes & Prost, 1993)

$$\mathcal{E}_{\rm def} = \frac{1}{2} B \left[ \frac{\partial u}{\partial z} - \frac{1}{2} (\nabla_\perp u)^2 \right]^2 + \frac{1}{2} K_1 \left( \nabla_\perp^2 u \right)^2 , \qquad (3.56)$$

where $\nabla_\perp \equiv (\partial/\partial x,\ \partial/\partial y,\ 0)$. By considering appropriate types of deformations and calculating corresponding $\mathcal{E}_{\rm def}$ within the generalized Liquid Drop Model, Pethick & Potekhin (1998) found

$$B = 6\,\mathcal{E}_{\rm Coul} , \qquad K_1 = \frac{2}{15} r_{\rm c}^2\, \mathcal{E}_{\rm Coul} \left(1 + 2w - 2w^2\right) . \qquad (3.57)$$

Here, $\mathcal{E}_{\rm Coul}$ is the Coulomb energy in equilibrium. According to Eq. (3.40),

$$\text{plates} - \text{equilibrium}: \ \mathcal{E}_{\rm Coul} = \frac{2\pi}{3} (e n_{p,{\rm i}} r_{\rm c})^2 (1-w)^2 w^2 , \qquad (3.58)$$

where $r_{\rm c}$ is the half-distance between the plates (Fig. 3.9) and $n_{p,{\rm i}}$ is the proton number density in the nuclear matter, while $w$ is the volume fraction occupied by the nuclear matter. All quantities are calculated for the relaxed system. The corrections to the above expressions for $B$ and $K_1$, resulting from the curvature term in the nuclear surface energy and from the changes in the plate thickness, measured along the $z$-axis, are small (Pethick & Potekhin, 1998).

Now let us consider the rod phase (which can also be called *columnar phase*, see de Gennes & Prost 1993). The basic parameters describing the ground (relaxed) state are the rod radius $r_p$ and the unit cell radius $r_{\rm c}$ (Fig. 3.9). The number of rods per unit area perpendicular to them is $1/\pi r_{\rm c}^2$. The ground-state configuration forms a two-dimensional triangular lattice. The Coulomb energy in equilibrium (3.42) reads

$$\text{rods} - \text{equilibrium}: \ \mathcal{E}_{\text{Coul}} = \frac{\pi}{2} (en_{p,\text{i}} r_{\text{c}})^2 w^2 \left[ \ln\left(\frac{1}{w}\right) - 1 + w \right] . \quad (3.59)$$

The displacements which affect the elastic energy are of the form $\boldsymbol{u} = (u_x, u_y, 0)$. Then

$$\mathcal{E}_{\text{def}} = \frac{1}{2} B \left( \frac{\partial u_x}{\partial x} + \frac{\partial u_y}{\partial y} \right)^2 + \frac{1}{2} C \left[ \left( \frac{\partial u_x}{\partial x} - \frac{\partial u_y}{\partial y} \right)^2 + \left( \frac{\partial u_x}{\partial y} + \frac{\partial u_y}{\partial x} \right)^2 \right]$$
$$+ \frac{1}{2} K_3 \left( \frac{\partial^2 \boldsymbol{u}}{\partial x^2} \right)^2 + B' \left( \frac{\partial u_x}{\partial x} + \frac{\partial u_y}{\partial y} \right) \left( \frac{\partial \boldsymbol{u}}{\partial z} \right)^2 + \frac{1}{2} B'' \left( \frac{\partial \boldsymbol{u}}{\partial z} \right)^4 . \quad (3.60)$$

The elasticity constant $B$ is associated with a uniform transverse compression or dilatation, while the elasticity moduli $C$ and $K_3$ are associated with transverse shearing and bending of the rod lattice, respectively. The second line of Eq. (3.60) contains higher-order terms, which may be important for large longitudinal deformations. All three terms containing $B$, $B'$, and $B''$ result from the change of the the unit cell radius induced by the deformation. The calculation of $\mathcal{E}_{\text{def}}$ within the generalized Liquid Drop Model yields (Pethick & Potekhin, 1998)

$$B = \frac{3}{2} \mathcal{E}_{\text{Coul}} , \quad B' = -\frac{3}{4} \mathcal{E}_{\text{Coul}} , \quad B'' = \frac{3}{8} \mathcal{E}_{\text{Coul}} . \quad (3.61)$$

The elasticity constant $C$ turns out to be a sensitive function of the filling factor $w$, as shown in Fig. 3.19. Within the relevant range of filling factors, $0.2 \lesssim w \lesssim 0.4$, it can be very well fitted by (Pethick & Potekhin, 1998)

$$C \simeq 10^{2.1(w-0.3)} \mathcal{E}_{\text{Coul}} . \quad (3.62)$$

The "bend constant" $K_3$ determines the deformation energy associated with bending of rods. Its dependence on $w$ and $r_{\text{c}}$ can be understood from Fig. 3.19. At $0.2 \lesssim w \lesssim 0.4$ one can use the approximation $K_3 \simeq 0.06\ \mathcal{E}_{\text{Coul}}\, r_{\text{c}}^2$.

## 3.8. Deviations from an idealized ground state model

The ground-state crust, described in the preceding sections, is built of an idealized one-component plasma of atomic nuclei and assumes full thermodynamic equilibrium. The real crust may deviate from these idealized models. The knowledge of "imperfections" of the crust is particularly important for its transport properties (§ 1.3.6). On the other hand, a crust which is out of nuclear equilibrium, constitutes a reservoir of energy which could be released and influence neutron star evolution.

### 3.8.1 Crust formation in a newly-born neutron star

Let us consider a newly born neutron star. Crystallization during its cooling is first-order phase transition in a Coulomb plasma. Just after the temperature falls below the local melting temperature $T_{\rm m}(\rho)$ at a given density $\rho$ (Fig. 3.17), the matter becomes an overcooled liquid. This state is metastable, and the crystallization is expected to start at $T < T_{\rm m}(\rho)$ via the formation of nucleation seeds and subsequent crystal growth. A possible scenario of crystallization at $\rho = 8 \times 10^{13}$ g cm$^{-3}$ was studied by De Blasio (1995).

The initial temperature of a newly born neutron star exceeds $10^{10}$ K. The composition of the hot matter is characterized by a statistical distribution of $(A, Z)$ nuclei. An initial spread of $(A, Z)$ is wide (Burrows & Lattimer, 1984) but thermodynamic functions deviate little from a single-nucleus approximation. With decreasing temperature, the $(A, Z)$ peak becomes narrower and the single-nucleus approximation is more precise (Burrows & Lattimer, 1984). After the solidification of the crust its composition is expected to be almost frozen, reflecting the situation at the crystallization point rather than in the ground state. For instance, consider the outer crust with the melting temperature ranging from $T_{\rm m} \sim 10^8$ K at $\rho = 10^8$ g cm$^{-3}$ to $T_{\rm m} \sim 10^9$ K at $\rho = 10^{11}$ g cm$^{-3}$ (Fig. 3.17). In contrast to the ground-state composition at $T = 0$, transitions between shells $(A_1, Z_1)$ and $(A_2, Z_2)$ at $T \simeq T_{\rm m}$ will be continuous. A one-component plasma can exist only sufficiently far from transition layers. Two-component transition layers were studied by De Blasio (1998, 2000).

Higher temperatures are characterized by larger fraction of evaporated nucleons. The most sensitive region is around the neutron drip point in the cold catalyzed matter, $\rho_{\rm ND} \simeq 4 \times 10^{11}$ g cm$^{-3}$. At $T \simeq 5 \times 10^9$ K, there is a non-negligible fraction of free neutrons for $10^{11}$ g cm$^{-3}$ $\lesssim \rho \lesssim \rho_{\rm ND}$ (see the lower panel of Fig. 3.1). In general, one notices a significant excess of free neutrons for the densities $10^{11}$ g cm$^{-3}$ $\lesssim \rho \lesssim 10^{12}$ g cm$^{-3}$ as compared to the matter at $T = 0$. With further cooling, there will be a tendency to absorb these excess neutrons by nuclei, which in turn will increase their $A$ and modify their $Z$ due to weak-interaction processes. At $T \lesssim 10^9$ K and $\rho \sim 10^{11}$ g cm$^{-3}$ the fraction of dripped neutrons is negligibly small (Fig. 3.1). However, the temperature may be too low to reach full nuclear equilibrium, mainly because of high Coulomb barriers and the lack of free protons and $\alpha$-particles. Therefore, one may expect deviations from the ground-state composition in the cooled crust at $10^{11}$ g cm$^{-3}$ $\lesssim \rho \lesssim 10^{12}$ g cm$^{-3}$ (Bisnovatyi-Kogan & Chechetkin, 1979). However, theoretical estimates of this effect depend on the nuclear properties at $(N - Z)/A \simeq 0.3$ and rely on extrapolations of semiempirical mass formulae. It would be desirable to reconsider this problem using modern description of very neutron-rich nuclei.

Now let us outline a general problem of thermal fluctuations of $Z$ and $N_{\rm cell}$ at $T_{\rm m} \gtrsim 10^9$ K (Jones, 2001c). We neglect a small fraction of dripped protons

(Fig. 3.1) and set $Z_{\rm cell} \simeq Z$. Because thermal fluctuations are accompanied by fluctuations of volume of an electrically neutral cell, the calculation has to be done at fixed $P$ and $T$. First consider the case of $T = 0$. At a given pressure, the enthalpy per nucleon, $h$, is a function of $Z$ and $N_{\rm cell}$. Some examples of the $h(Z, N_{\rm cell})$ maps for the outer crust, where $N_{\rm cell} = N$, are presented by Haensel & Pichon (1994). Unfortunately, in the inner crust, only the maps of the energy per nucleon $e(Z, N_{\rm cell})$ at a given $n_{\rm b}$ are available (Negele & Vautherin, 1973). At $n_{\rm b} = 5.76 \times 10^{-3}$ fm$^{-3}$ ($\rho = 9.6 \times 10^{12}$ g cm$^{-3}$) the difference between the absolute and secondary minima of $e(Z, N_{\rm cell})$ is $\Delta e \sim 10$ keV. Similar differences are expected between the values of the free enthalpy per nucleon $h = e + P/n_{\rm b}$ at a given pressure. Let $Z_{\rm min}$ and $N_{\rm min}$ denote the values of $Z$ and $N_{\rm cell}$ which correspond to the absolute minimum. The properties of matter in the inner crust at $T = 0$ suggest, that at $T = T_{\rm m} \gtrsim 10^9$ K thermal fluctuations of $Z$ and $N_{\rm cell}$ may be large. The reason is that the formation enthalpy of impurities with $Z_{\rm imp} = Z_{\rm min} + \Delta Z$, $N_{\rm cell,imp} = N_{\rm min} + \Delta N_{\rm cell}$ corresponding to the secondary minimum is expected to be smaller than $k_{\rm B}T_{\rm m}$ (these two minima are associated with proton shell closure at $Z = 40$ and $Z = 50$). Because the minima of $h(Z, N_{\rm cell})$ are rather flat, one may obtain substantial concentrations of nuclei with $Z \neq Z_{\rm min}$ and $N_{\rm cell} \neq N_{\rm min}$ during the solidification. Jones (1999) calculated the formation enthalpy of impurities, using the CLDM and adding the shell and pairing effects. His results indicate a high heterogeneity of $Z$ which becomes frozen after the temperature falls below $T_{\rm m}$, with a substantial population of two closed $Z$-shells. Let us mention that high values of $T_{\rm m}$ (high thermal energies) and large number densities of neutrons (predominantly unbound) in the inner crust are favorable for impurity fractions higher than those in the outer crust.

These calculations of impurity fractions should be regarded as preliminary. Unfortunately, kinetics of phase transitions is notoriously difficult for theoretical modeling, especially if the approximations cannot be tested in laboratory. The existing models can be improved by a more careful treatment of thermal contributions at $T \simeq 10^9$ K; the use of the $T = 0$ approximation may be questionable for some thermodynamic quantities. Also, the importance of the shell and pairing effects during solidification should be clarified. Fortunately, the EOS in the crust is rather insensitive to deviations from the one-nucleus model.

Investigations of nuclear structures in the bottom crustal layers, formed during one minute in which a proto neutron star with $T \gtrsim 10^{11}$ K transforms to a hot neutron star with $T \lesssim 10^{10}$ K (§ 1.4.2), are very complicated. One of the approaches is based on the quantum molecular dynamics simulations in a box containing from a few thousand to a few tens of thousand nucleons (see, e.g., Watanabe *et al.*, 2004, and references therein). In addition to "nuclear pasta" phases with a well defined geometry (rods, spheres, tubes, etc.) these authors find also what they call "spongelike" phase with complicated geometry, representing a mixture of several different geometries; it appears at $T \gtrsim 3 \times 10^{10}$ MeV. Similar intermediate (mixed) phases were obtained, at subnuclear densities, from the quantum molecular dynamics simulations for *cold dense matter* with an effective NN interaction by Watanabe *et al.* (2003). Together with the Hartree-Fock calculations for the cold matter at subnuclear densities (Magierski *et al.*, 2003, and references therein) these results indicate that the structure of the bottom layers might be heterogeneous. However, even in this case the impact of the heterogeneity on the EOS would be small.

### 3.8.2 Accreted crust

While a newly born neutron star is made of hot matter in nuclear equilibrium, its subsequent evolution can lead to the formation of regions, where the matter is out of nuclear equilibrium. This may happen in a neutron star crust, where reshuffling of nucleons necessary for the formation of large nuclei (present in the cold catalyzed matter) may be prohibited by high Coulomb barriers. This is the case of an old accreting neutron star. For an accretion rate $\sim 10^{-10} M_{\odot}\ \mathrm{yr}^{-1}$, the typical temperature in the neutron star interior is $\sim 10^8$K (Fujimoto *et al.* 1984; Miralda-Escudé *et al.* 1990).

Let us consider a standard scenario of the evolution of accreted matter. Explosive burning of the helium layer leads to the formation of matter consisting mainly of $^{56}$Ni, which transforms into $^{56}$Fe by electron captures. The growing layer of the processed accreted matter pushes down the original crust. The original catalyzed (ground-state) outer crust is replaced by a new, noncatalyzed one in $\sim 10^5$ years. In view of low temperature ($T \lesssim 10^8$ K), the only processes which can take place when the accreted matter sinks inwards are electron captures and beta decays, neutron emission or absorption and, at sufficiently high densities, pycnonuclear fusion. A detailed study of these processes was done by Sato (1979), who considered several scenarios with different initial compositions of the matter, and by Haensel & Zdunik (1990a) (also see Bisnovatyi-Kogan & Chechetkin 1979 and references therein).

A noncatalyzed crust represents a reservoir of energy. The energy release takes place owing to the nonequilibrium processes. Some aspects of this problem were first considered by Vartanyan & Ovakimova (1976). Later nonequilibrium processes and resulting crustal heating were studied in detail by Haensel & Zdunik (1990a). These processes lead to the appearance of very thin layers where heat is produced at a rate proportional to the accretion rate. As shown by Haensel & Zdunik (1990a), the associated total heat release – *deep crustal heating* – in the crust can be larger than the original inward heat flow resulting from the steady hydrogen burning between the helium flashes (Fujimoto *et al.*, 1984). The total heat release per one accreted nucleon, $\sim 1.5$ MeV, depends rather weakly on the initial composition of ashes produced by X-ray bursts (Haensel & Zdunik, 2003).

In the scenario of Haensel & Zdunik (1990a), nuclei in the inner crust have $Z \lesssim 20$, to be compared with $Z \approx 40$–$50$ in the catalyzed matter. The nuclei in the inner accreted crust appear to be much lighter than in the catalyzed matter. Very recently Jones (2005) argued that the deviation of the inner accreted crust from nuclear equilibrium could be much smaller than in the scenarios of Sato (1979) and Haensel & Zdunik (1990a), which would result in a much weaker deep crustal heating. Clearly, the problem requires further studies.

Many neutron stars in close X-ray binaries are transient accretors (transients); § 1.4.6. They exhibit X-ray bursts separated by long periods (months or even years) of quiescence. It is believed that the quiescence corresponds to a low-level, or even halted, accretion onto the neutron star. During high-state accretion episodes, the heat is deposited by nonequilibrium processes in the deep layers ($10^{12} - 10^{13}$ g cm$^{-3}$) of the crust. This deep crustal heating can maintain the temperature of the neutron star interior at a sufficiently high level to explain a persistent thermal X-ray radiation in quiescence (Brown *et al.*, 1998).

The problem of the detailed outcome of time-dependent nucleosynthesis during X-ray bursts is very complicated and is not completely solved (see, e.g., Rembges *et al.* 1997; Schatz *et al.* 1999, 2001). The nature of the unstable thermonuclear burning at higher accretion rates $10^{-8}\ M_\odot\ \mathrm{yr}^{-1} \lesssim \dot{M} \lesssim 10^{-9}\ M_\odot\ \mathrm{yr}^{-1}$, is not well understood. The ashes from such a burning might contain some admixture of nuclei beyond the iron group, with $A \simeq 60-100$ (Schatz *et al.*, 1999, 2001).

The case of thermally stable burning of hydrogen and helium at sufficiently high accretion rates should be considered separately. This regime corresponds to most of the X-ray pulsars – magnetized accreting neutron stars (surface magnetic field $B \gtrsim 10^{12}$ G), where the local accretion rate in the polar cap region is thought to be large enough for a stable burning. A similar situation is encountered at very high rates of accretion ($\dot{M} \gtrsim 10^{-8}\ M_\odot\ \mathrm{yr}^{-1}$) on weakly magnetized ($B \ll 10^{11}$ G) neutron stars. At high temperatures, corresponding to the high accretion rates, the hydrogen burns via the rapid proton capture producing a mix of elements beyond the iron group. It is expected that the compression of this heterogeneous matter will produce an impure solid crust.

If the starting composition is a mix with significant fractions of different nuclides, its further evolution may keep heterogeneity of the matter. The thermal and electrical conductivities of such a crust could be lower than in a perfect crystal. The distribution of nuclides would be rather smooth, in contrast to the extreme case of a one-nucleus model with noticeable density jumps. The average values of $Z$ and $A$ will still be lower than in the cold catalyzed matter.

Chapter 4

# ENVELOPES WITH STRONG MAGNETIC FIELDS

Magnetic fields $B \gtrsim 10^{12}$ G, typical for isolated neutron stars (§ 1.3.8), drastically modify many physical properties of the matter. Motion of free electrons and ions perpendicular to the field lines is quantized into Landau orbitals with a characteristic transverse scale equal to the *magnetic length* $a_\mathrm{m} = (\hbar c/eB)^{1/2}$. This brings to the scene an atomic field-strength parameter $\gamma = (a_0/a_\mathrm{m})^2$, where $a_0$ is the Bohr radius. If this parameter is large, the Lorentz force acting on valence electrons in atoms exceeds the Coulomb force. The Landau energy levels of electrons are modified by relativistic effects if the field strength in the relativistic units,

$$b = \hbar\omega_\mathrm{c}/(m_e c^2) = B/B_\mathrm{r}, \tag{4.1}$$

becomes $b \gtrsim 1$. Here, $\omega_\mathrm{c} = eB/(m_e c)$ is the electron cyclotron frequency and $B_\mathrm{r} = m_e^2 c^3/(e\hbar) = 4.414 \times 10^{13}$ G is often called the relativistic magnetic field.

Introducing the notation $B_{12} = B/10^{12}$ G, we have

$$\gamma = 425.44\, B_{12}, \quad b = \alpha_\mathrm{f}^2 \gamma = B_{12}/44.14. \tag{4.2}$$

A magnetic field will be called *strong* if $\gamma \gg 1$ (which is typical for radio pulsars) and *superstrong* if $b \gg 1$ (which occurs in *magnetars*; see § 1.4, particularly, §§ 1.4.4 and 1.4.5).

In § 4.1 we review the main magnetic effects in a fully ionized plasma of charged pointlike particles (electrons and ions). In the domain of partial ionization, one should take into account quantum-mechanical effects of the magnetic field on bound species. These effects are outlined in § 4.2. In § 4.3 we briefly discuss thermodynamics of partially ionized layers of magnetic neutron-star envelopes and consider the best studied hydrogen atmosphere in more detail.

## 4.1. Fully ionized plasmas

### 4.1.1 Free electron in a magnetic field

Let us consider a uniform magnetic field $\boldsymbol{B}$ directed along the $z$-axis. In this case, the vector potential in the Landau gauge reads $\boldsymbol{A} = (-By, 0, 0)$. Quantum states of a free electron in the magnetic field form a complete orthogonal basis. The basic states can be labeled by $(p_x, p_z, n, s)$, where $n = 0, 1, 2, \ldots$ is the Landau quantum number, $s$ is the spin quantum number, $p_z$ is the $z$-projection of the electron momentum, while its $x$-projection $p_x$ determines the $y$-coordinate of the guiding center of electronic motion, $y_c = p_x/(m_e\omega_c)$. An explicit solution of the Dirac equation reads (e.g., Akhiezer & Berestetskiĭ 1965; Sokolov & Ternov 1974)

$$\Psi_{p_x,p_z,n,s}(\boldsymbol{r}) = \frac{\exp[\mathrm{i}(p_x x + p_z z)/\hbar]}{(L_x L_z)^{1/2}}\, \psi_{ns}(p_z, y - y_c), \tag{4.3}$$

where $L_x$ and $L_z$ are normalization lengths. The ground Landau level $n = 0$ is nondegenerate with respect to spin (spin is antiparallel to $\boldsymbol{B}$, $s = -1$; the statistical weight $g_0 = 1$), whereas the levels $n > 0$ are double degenerate ($s = \pm 1$, $g_n = 2$). The latter degeneracy allows different choices of the electron basic bispinors $\psi_{ns}$. The simplest choice which is often the most convenient one is

$$\psi_{n,1}(p_z, y) = \frac{1}{\sqrt{2\tilde\epsilon(\tilde\epsilon + 1)a_\mathrm{m}}} \begin{pmatrix} (\tilde\epsilon + 1)\,\mathcal{H}_{n-1}(y/a_\mathrm{m}) \\ 0 \\ \tilde p_z\,\mathcal{H}_{n-1}(y/a_\mathrm{m}) \\ -\sqrt{2bn}\,\mathcal{H}_n(y/a_\mathrm{m}) \end{pmatrix}, \tag{4.4a}$$

$$\psi_{n,-1}(p_z, y) = \frac{1}{\sqrt{2\tilde\epsilon(\tilde\epsilon + 1)a_\mathrm{m}}} \begin{pmatrix} 0 \\ (\tilde\epsilon + 1)\,\mathcal{H}_n(y/a_\mathrm{m}) \\ -\sqrt{2bn}\,\mathcal{H}_{n-1}(y/a_\mathrm{m}) \\ -\tilde p_z\,\mathcal{H}_n(y/a_\mathrm{m}) \end{pmatrix}. \tag{4.4b}$$

Here, $\tilde\epsilon = \epsilon/(m_e c^2)$ and $\tilde p_z = p_z/(m_e c)$ are, respectively, the electron energy and longitudinal momentum in the relativistic units,

$$\mathcal{H}_n(\xi) = \frac{\exp(-\xi^2/2)}{\pi^{1/4}(2^n n!)^{1/2}} H_n(\xi) \tag{4.5}$$

is a harmonic-oscillator function, and $H_n(\xi) = (-1)^n \mathrm{e}^{\xi^2} \mathrm{d}^n \mathrm{e}^{-\xi^2}/\mathrm{d}\xi^n$ is a Hermite polynomial. The electron energy and the absolute value of $p_z$ are inter-related as

$$\epsilon = \epsilon_n(p_z) = c\,\left(m_e^2 c^2 + 2\hbar\omega_\mathrm{c} m_e n + p_z^2\right)^{1/2}, \tag{4.6}$$

$$|p_z| = p_n(\epsilon) = [(\epsilon/c)^2 - (m_e c)^2 - 2m_e\hbar\omega_\mathrm{c} n]^{1/2}, \quad n \le n_\mathrm{max}, \tag{4.7}$$

where $n_{\rm max}$ is the maximum Landau number for a given $\epsilon$. The latter number is an integral part of the variable $\Upsilon$

$$n_{\rm max}(\epsilon) = {\rm Int}(\Upsilon), \quad \Upsilon = \frac{1}{\hbar\omega_{\rm c}} \frac{p_0^2(\epsilon)}{2m_e}. \tag{4.8}$$

Instead of Eq. (4.4), one can use any other basis of the type

$$\psi'_{ns} = \psi_{\,ns} \cos\varphi - s\,\psi_{\,n,-s} \sin\varphi\,. \tag{4.9}$$

It is sufficient to assume that $0 \le \varphi \le \pi/2$; $\varphi$ may depend on $n$ but should be zero for $n = 0$.

In particular, by choosing $\varphi = \arcsin\sqrt{(1 - p_z/p_0)/2}$ we obtain the basis of states with fixed electron helicities (i.e., spin projections on the canonical momentum). In this case, $s$ defines the helicity sign.

The spin magnetic moment of the electron contains a small anomalous part whose relative magnitude is determined by the difference of the electron gyromagnetic factor $g_e = 1.00116$ (the ratio of the actual magnetic moment to the Bohr magneton) from 1. The anomalous magnetic moment splits the energy levels $n \ge 1$ and, strictly speaking, removes the spin degeneracy. In neutron star envelopes, however, this splitting is typically negligible, because $\delta\epsilon$ is smaller than either the thermal width $\sim k_{\rm B}T$ of the Fermi level or the collisional width of the Landau levels (see, e.g., Kaminker & Yakovlev 1981).

**Non-relativistic limit.** In the non-relativistic limit, the basis of bispinors (4.4) is often most convenient, because it corresponds to fixed spin projections ($s\hbar/2$) on the $z$-axis (two lower components of bispinors $\psi_{ns}$ are negligible in this case). Then the coordinate part of the wave function is formally given by Eq. (4.3) with

$$\psi_{n,1} = a_{\rm m}^{-1/2}\,\mathcal{H}_{n-1}(y/a_{\rm m}), \quad \psi_{n,-1} = a_{\rm m}^{-1/2}\,\mathcal{H}_n(y/a_{\rm m}), \tag{4.10}$$

Let us also mention that in the cylindrical gauge of the vector potential, $\boldsymbol{A} = (-By/2, Bx/2, 0)$, $p_x$ is not a good quantum number; the magnetic quantum number $m$ (i.e., the $z$-projection of the angular momentum in units of $\hbar$) takes its place. At any $n$, one has $m = n,\ n-1,\ n-2\ldots$ In the non-relativistic limit, the coordinate parts of the basic wave functions do not depend on $s$ (but one should not forget different statistical weights of the Landau levels with $n = 0$ and $n > 0$). These coordinate parts are

$$\Psi'_{p_z,n,m,s}(\boldsymbol{r}) = \frac{{\rm e}^{{\rm i}p_z z/\hbar}}{L_z^{1/2}}\,\Phi_{n,-m}(\boldsymbol{r}_\perp). \tag{4.11}$$

Here $\boldsymbol{r}_\perp = (x, y) = (r_\perp\cos\phi, r_\perp\sin\phi)$,

$$\Phi_{n,-m}(r_\perp) = \frac{{\rm e}^{im\phi}}{\sqrt{2\pi}\,a_{\rm m}}\,I_{n-m,n}(r_\perp^2/2a_{\rm m}^2) \tag{4.12}$$

is a Landau function, $L_z$ is the normalization length, and $I_{nn'}(u) = (-1)^{n'-n} I_{n'n}(u)$ is a Laguerre function (Sokolov & Ternov, 1974). Assuming $n' \geq n$, one has

$$I_{n'n}(u) = \mathrm{e}^{-u/2} u^{(n'-n)/2} \sum_{k=0}^{n} (-1)^k \frac{\sqrt{n'!n!}}{k!(n-k)!(n'-n+k)!} u^k. \tag{4.13}$$

A wave function of the relativistic electron in a magnetic field in the cylindrical gauge can be also expressed in terms of $\Phi_{n,-m}(r_\perp)$ (e.g., Sokolov & Ternov 1974).

### 4.1.2 Thermodynamic functions of the ideal electron gas

Thermodynamic functions of the electron gas in a magnetic field are easily derived from the first principles (Landau & Lifshitz, 1993). The number of quantum states per longitudinal momentum interval $\Delta p_z$ for an electron with given spin, magnetic, and Landau quantum numbers $(s, m, n)$ in a volume $V$ equals $V\Delta p_z/(4\pi^2 a_\mathrm{m}^2 \hbar)$ (Landau & Lifshitz, 1976). This allows one to express the electron number density $n_e$ and the thermodynamic potential $\Omega = -PV$ as

$$n_e = \frac{1}{(2\pi a_\mathrm{m})^2 \hbar} \sum_{n=0}^{\infty} g_n \int_{-\infty}^{\infty} f^{(0)}(\epsilon - \mu, T)\, \mathrm{d}p_z, \tag{4.14}$$

$$\Omega = -\frac{V k_\mathrm{B} T}{2\pi^2 a_\mathrm{m}^2 \hbar} \sum_{n=0}^{\infty} g_n \int_0^{\infty} \ln\left(1 + \exp\left[\frac{\mu - \epsilon}{k_\mathrm{B} T}\right]\right) \mathrm{d}p_z, \tag{4.15}$$

where $f^{(0)}$ is the Fermi-Dirac function (2.46) and $\epsilon$ is given by Eq. (4.6). Integrating Eq. (4.14) by parts, we obtain

$$n_e = \int_{m_e c^2}^{\infty} \mathcal{N}_B(\epsilon) \left(-\frac{\partial f^{(0)}}{\partial \epsilon}\right) \mathrm{d}\epsilon, \tag{4.16}$$

where

$$\mathcal{N}_B(\epsilon) = \frac{1}{2\pi^2 a_\mathrm{m}^2 \hbar} \sum_{n=0}^{n_\mathrm{max}} g_n p_n(\epsilon), \tag{4.17}$$

and $n_\mathrm{max}$ is defined by Eq. (4.8). If $n_\mathrm{max} \gg 1$, the sum in Eq. (4.17) can be approximated by an integral, which gives the classical result [cf. Eq. (2.3)]

$$\mathcal{N}_0(\epsilon) = p_0^3(\epsilon)/(3\pi^2 \hbar^3). \tag{4.18}$$

In particular, using Eq. (4.16) we can write the squared Thomas-Fermi wave number (2.13) as

$$k_{\rm TF}^2 = 4\pi e^2 \int_{m_e c^2}^{\infty} \frac{\partial \mathcal{N}_B}{\partial \epsilon} \left( -\frac{\partial f^{(0)}}{\partial \epsilon} \right) \mathrm{d}\epsilon$$

$$= \frac{2\alpha_{\rm f}}{\pi a_{\rm m}^2} g_n \int_{m_e c^2}^{\infty} \sum_{n=0}^{n_{\rm max}(\epsilon)} \frac{\epsilon}{c p_n} \left( -\frac{\partial f^{(0)}}{\partial \epsilon} \right) \mathrm{d}\epsilon. \tag{4.19}$$

It is convenient to rewrite Eq. (4.16) as

$$n_e = \frac{\tau_0}{2\pi^2 a_{\rm m}^2 \lambda\!\!\!^{-}_{\rm C}} \sum_{n=0}^{\infty} g_n F(\chi_n, \tau_n^{-1}), \tag{4.20}$$

$$F(x,y) = \int_0^{\infty} \frac{\mathrm{e}^{t-x}}{(\mathrm{e}^{t-x}+1)^2} \sqrt{t(t+2y)}\, \mathrm{d}t, \tag{4.21}$$

$$\chi_n = \frac{\mu - \epsilon_n(0)}{k_{\rm B} T}, \qquad \tau_n = \frac{t_{\rm r}}{\sqrt{1+2bn}}, \tag{4.22}$$

and use the approximation (Potekhin, 1996a)

$$F(x,y) \approx \ln(1+\mathrm{e}^x)\, \frac{1+y+\xi+c(y)\, a(\xi)\, \sqrt{\xi+2y}}{1+\xi+c(y)\, b(\xi)}, \tag{4.23}$$

where $\xi = \ln\{1+\exp[x - x_0(y)]\}$, $x_0(y) = 1/(1+0.623\, y^{1.603})$,

$$c(y) = 0.9422\, y^{1.7262},$$
$$a(\xi) = \sqrt{\pi}/2 + (0.103 + 0.043\, \xi^2)\sqrt{\xi},$$
$$b(\xi) = 1 + 0.0802\sqrt{\xi} + 0.2944\, \xi + 0.043\, \xi^3.$$

This approximation reproduces correct asymptotes at small and large $x$ and $y$, and remains accurate within 0.6% at any $x$ and $y$. The chemical potential $\mu$ at a given density $n_e$ can be found by the numerical inversion of Eq. (4.14) with the use of the fit (4.23).

The $x$-derivative of the right-hand side of Eq. (4.23) reproduces the exact derivative $\partial F(x,y)/\partial x$ with a maximum relative error of 2% . Using this derivative in Eq. (4.19) and replacing $\partial f^{(0)}/\partial\epsilon \to -\delta(\epsilon-\mu)$ we get the electron screening wave number

$$k_{\rm TF}^2 = \frac{2\alpha_{\rm f}}{\pi a_{\rm m}^2} \sum_{n=0}^{n_{\rm max}} g_n \left( \frac{\partial F(x, \tau_n^{-1})}{\partial x} \right)_{x=\chi_n}. \tag{4.24}$$

Integrating Eq. (4.15) by parts, we obtain

$$P = \int_{m_e c^2}^{\infty} \mathcal{N}_B(\epsilon)\, f^{(0)}\, \mathrm{d}\epsilon = P_{\rm r} \frac{b\tau_0^{3/2}}{\sqrt{2}\pi^2} \sum_{n=0}^{\infty} g_n (1+2bn)^{1/4} I_{1/2}(\chi_n, \tau_n), \tag{4.25}$$

where $P_\mathrm{r} = m_e c^2/\lambda\!\!\!^{-}_\mathrm{C}{}^3$ is the relativistic unit of pressure introduced in § 2.3.1. The Fermi-Dirac integral $I_{1/2}(\chi, \tau)$ is readily evaluated using Eqs. (2.54)–(2.56).

Let us comment, in passing, that the *kinetic* pressure of an electron gas, calculated as the quantum-mechanical average $n_e\langle p_\alpha v_\alpha\rangle$, is anisotropic in quantizing magnetic fields. For instance, the kinetic pressure in the transverse direction, $P_\perp = n_e\langle p_x v_x\rangle$, is much smaller than $P = n_e\,\langle p_z v_z\rangle$, if the field is strongly quantizing. However, the kinetic pressure is only one part of the total pressure in the magnetized plasma. As proven by Blandford & Hernquist (1982), a deficit of the kinetic pressure in the transverse direction is exactly balanced by the pressure excess caused by magnetization currents. Thus, the total actually thermodynamic pressure is *isotropic* at any field strength, and Eq. (4.25) is always valid.

**Strongly quantizing magnetic field.** Let $p_\mathrm{F0} = \hbar k_\mathrm{F0}$, $\epsilon_\mathrm{F0}$, and $T_\mathrm{F0}$ denote, respectively, the *non-magnetic* Fermi momentum, energy, and temperature at a given density (§ 2.1.2). We reserve the notations $p_\mathrm{F} = \hbar k_\mathrm{F}$, $\epsilon_\mathrm{F}$, and $T_\mathrm{F}$ for the same quantities in a magnetic field. *We keep the parameters* $x_\mathrm{r}$, $\gamma_\mathrm{r}$, *and* $\beta_\mathrm{r}$ *expressed through* $p_\mathrm{F0}$, as in Chapter 2. For instance, $x_\mathrm{r} = p_\mathrm{F0}/m_e c$ is a convenient measure of the density regardless the magnetic field strength.

At $T \ll T_\mathrm{F}$, one can replace $(-\partial f^{(0)}/\partial\epsilon)$ in Eq. (4.16) by the delta function $\delta(\epsilon - \epsilon_\mathrm{F})$:

$$n_e = \mathcal{N}_B(\epsilon_\mathrm{F}), \tag{4.26}$$

$$\begin{aligned} P &= \int_{m_e c^2}^{\epsilon_\mathrm{F}} \mathcal{N}_B(\epsilon)\,\mathrm{d}\epsilon \\ &= \frac{P_\mathrm{r}}{4\pi^2}\, b \sum_{n=0}^{n_\mathrm{max}(\epsilon_\mathrm{F})} g_n\,(1+2bn)\,[x_n\sqrt{1+x_n^2} - \ln(x_n + \sqrt{1+x_n^2})], \end{aligned} \tag{4.27}$$

where $x_n = c p_n(\epsilon_\mathrm{F})/\epsilon_n(0)$, and $P_\mathrm{r}$ is the same as in Eq. (2.66). The Fermi energy $\epsilon_\mathrm{F}$ at a given $n_e$ is found by the inversion of Eq. (4.26).

A magnetic field is called *strongly quantizing*, if it confines most of the electrons to the ground Landau level. This occurs at sufficiently low temperatures and densities. In this case, from Eq. (4.26) one obtains $\epsilon_\mathrm{F} = m_e c^2\,\sqrt{1+x_B^2}$, where

$$x_B \equiv \lambda\!\!\!^{-}_\mathrm{C} k_\mathrm{F} = 2\pi^2 a_\mathrm{m}^2 \lambda\!\!\!^{-}_\mathrm{C} n_e = \frac{2x_\mathrm{r}^3}{3b} \approx 30.2\,\frac{\langle Z\rangle}{A'}\,\frac{\rho_6}{B_{12}}, \tag{4.28}$$

while $A'$ and $\langle Z\rangle$ are the mean effective atomic mass and charge numbers, respectively (see § 2.1.1); $x_\mathrm{r}$ and $\rho_6$ are introduced in § 2.1.2. With increasing density at a fixed $B$, the electron number density $n_e$ reaches some critical value

$n_B$, at which $\epsilon_F = \epsilon_1(0)$ and degenerate electrons start to populate the first excited Landau level. From Eq. (4.28) we see that $n_B = 1/(\pi^2\sqrt{2}\,a_m^3)$. Hence the strongly quantizing regime occurs at $T \ll T_{\rm cycl}$ and $\rho < \rho_B$, where

$$T_{\rm cycl} = \hbar\omega_c/k_B \approx 1.343 \times 10^8\, B_{12}\ {\rm K}, \tag{4.29}$$

$$\rho_B = \frac{A' m_u}{\langle Z\rangle} n_B \approx 7045 \frac{A'}{\langle Z\rangle} B_{12}^{3/2}\ {\rm g\ cm^{-3}}. \tag{4.30}$$

Comparing Eqs. (4.28) and (2.3), we see that $k_F = (4/3)^{1/3}(\rho/\rho_B)^{2/3}\, k_{F0}$ in this regime. Therefore, $T_F$ is strongly reduced at $\rho \ll \rho_B$, compared to its non-magnetic value $T_{F0}$:

$$T_F = \frac{\gamma_B - 1}{\gamma_r - 1} T_{F0}, \quad x_B = \left(\frac{4\rho^2}{3\rho_B^2}\right)^{1/3} x_r, \quad \gamma_B = \sqrt{1 + x_B^2}. \tag{4.31}$$

The nondegenerate electron gas obeys classical statistics. According to the Bohr-van Leeuwen theorem (see footnote 1 on page 54), the magnetic field in this case does not affect the EOS. On the contrary, the EOS is changed drastically, if the electron gas is strongly degenerate and the magnetic field is strongly quantizing. In that case only the $n = 0$ term survives in Eq. (4.27), and the EOS can be presented in the form

$$P = \frac{P_r\, b}{(2\pi)^2}\left[x_B\,\gamma_B - \ln(x_B + \gamma_B)\right] = \frac{P_r\, b}{2\pi^2}\frac{x_B^{\gamma_{\rm ad}}}{\gamma_{\rm ad}} \propto \frac{\rho^{\gamma_{\rm ad}}}{B^{\gamma_{\rm ad}-1}}, \tag{4.32}$$

where $x_B$ and $\gamma_B$ are given by Eq. (4.31). In Eq. (4.32) we have introduced a quasi-adiabatic index $\gamma_{\rm ad}$ which, in general, depends on $x_B$, but takes on the constant values 3 and 2 in the non-relativistic ($x_B \ll 1$) and ultrarelativistic ($x_B \gg 1$) limits, respectively. Compared with the non-magnetic case, Eq. (2.70), $\gamma_{\rm ad}$ is higher (the density dependence of $P$ is steeper), but the numerical value of $P$ is lower everywhere except in the vicinity of the first Landau threshold. This means that a strongly quantizing magnetic field softens the EOS of degenerate electrons.

**Non-quantizing magnetic field.** If the temperature or density is high enough, the electron distribution is smeared over many Landau levels, and one can replace $\mathcal{N}_B(\epsilon)$ by $\mathcal{N}_0(\epsilon)$. Then the field is *non-quantizing*. This happens either at $\rho \gg \rho_B$ or at $T \gg T_B$, where

$$T_B = \begin{cases} T_{\rm cycl}, & \text{if } \rho < \rho_B, \\ T_{\rm cycl}/\sqrt{\gamma_r}, & \text{if } \rho > \rho_B \end{cases} \tag{4.33}$$

(in this chapter we assume that $T \ll T_r$). In the relativistic regime at $\rho > \rho_B$, $T_B$ is smaller than $T_{\rm cycl}$, because the distance between *excited* Landau levels

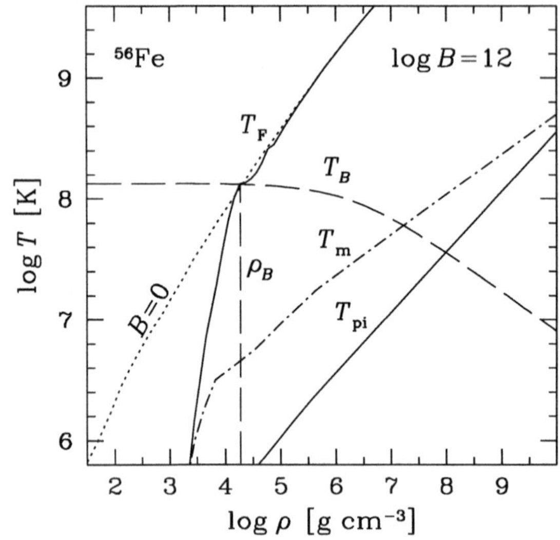


*Figure 4.1.* Characteristic parameter domains in the $\rho$–$T$ plane for iron matter at $B = 10^{12}$ G. Solid lines show $T_\mathrm{F}$ and $T_\mathrm{pi}$; the dot-dashed line presents $T_\mathrm{m}$ ($\Gamma = 175$). The dotted line display $T_\mathrm{F}$ at $B = 0$ (Fig. 2.2). Long-dashed lines show $T_B$ and $\rho_B$ and separate the regions of strong (the lower left sector) and weak (the lower right sector) magnetic quantization, and the domain of the non-quantizing field ($T \gtrsim T_B$).

for the electrons with $\epsilon = \epsilon_\mathrm{F}$ is $\approx \hbar\omega_\mathrm{g} < \hbar\omega_\mathrm{c}$, where $\omega_\mathrm{g}(\epsilon) = eBc/\epsilon$ is the electron gyrofrequency.

In the non-quantizing magnetic field, many Landau levels contribute to sums over $n$ in Eqs. (4.16) and (4.25). In this case, the summation can be approximately replaced by the integration. Then, integrating by parts, we can reduce Eqs. (4.16) and (4.25) to Eqs. (2.51) and (2.50), respectively.

If $\rho > \rho_B$ and $T \lesssim T_B$, the Landau quantization can remain important for a phenomenon under study. In this case the field is called *weakly quantizing*. Usually it happens if only a few Landau levels are populated. Higher-order thermodynamic quantities (such as the electron heat capacity, entropy, magnetization) are much stronger affected by magnetic fields in this regime than the bulk quantities (for instance, the electron energy density, chemical potential, pressure).

**A density-temperature diagram.** Characteristic $\rho$–$T$ domains for the outer neutron-star envelope composed of iron are shown in Fig. 4.1 for $B = 0$ and $10^{12}$ G. Partial ionization is taken into account in the mean-ion approximation. The electrons are degenerate below $T_\mathrm{F}$; the ions are classical above $T_\mathrm{pi}$. The mean-ion charge number $Z_\mathrm{eff}$ has been evaluated assuming that the pressure created by free electrons and by free ions with this $Z_{\rm eff}$ equals the pressure given by the finite-temperature Thomas-Fermi model of Thorolfsson *et al.* (1998) (discussed below in §4.3). For comparison, the dotted line reproduces non-magnetic $T_{\rm F}$ from Fig. 2.2. Finally, the long-dashed lines separate three regions, where the magnetic field is strongly quantizing (to the left of $\rho_B$ and considerably below $T_B$), weakly quantizing (to the right of $\rho_B$ at $T \lesssim T_B$), or classical (above $T_B$).

**Non-relativistic limit.** Thermodynamic functions of the ideal electron gas in a magnetic field simplify in the non-relativistic limit ($p_{\rm F} \ll m_e c$, $T \ll T_{\rm r}$). In this case the electron pressure and number density are given by

$$P_e = \frac{k_{\rm B}T}{\pi^{3/2} a_{\rm m}^2 \lambda_e} \sum_{ns} I_{1/2}(\chi_n), \quad n_e = \frac{1}{2\pi^{3/2} a_{\rm m}^2 \lambda_e} \sum_{ns} I_{-1/2}(\chi_n). \tag{4.34}$$

Here, $\lambda_e$ is the electron thermal wavelength given by Eq. (2.27). In the nondegenerate regime ($T \gg T_{\rm F}$), one has $I_\nu(\chi) \approx {\rm e}^\chi\, \Gamma(\nu+1)$, were $\Gamma(\nu+1)$ is the gamma-function. Therefore, Eq. (4.34) yields $P_e = n_e k_{\rm B} T$ and

$$\chi_0 = \ln(n_e \lambda_e^3/2) - \ln \zeta_e + \ln(\tanh \zeta_e), \tag{4.35}$$

where

$$\zeta_e \equiv \frac{\hbar\omega_{\rm c}}{2k_{\rm B}T} = \frac{T_{\rm cycl}}{2T}. \tag{4.36}$$

This provides an explicit analytical form of the Helmholtz free energy $F_{\rm id}^{(e)} = (\chi_0 - 1)\, N_e k_{\rm B} T$ (in this chapter we do not include the rest energy $m_e\, c^2$ into the free energy). In the non-quantizing field ($\zeta_e \ll 1$), the last two terms in Eq. (4.35) cancel out and the classical non-magnetic result is recovered, $F_{\rm id}^{(e)} = N_e k_{\rm B} T\, [\ln(n_e \lambda_e^3/2) - 1]$. In the strongly quantizing, nondegenerate regime ($\rho < \rho_B$ and $T_{\rm F} \ll T \ll T_{\rm cycl}$), the last term of Eq. (4.35) vanishes, which yields

$$F_{\rm id}^{(e)} = N_e k_{\rm B} T \left[\ln(2\pi a_{\rm m}^2 \lambda_e n_e) - 1\right]. \tag{4.37}$$

### 4.1.3 Magnetic oscillations

In the equations of §4.1.2, the summation over discrete Landau numbers $n$ reflects consecutive population of new Landau levels with growing density, which leads to magnetic quantum oscillations of thermodynamic and kinetic functions (see, e.g., Lifshitz & Pitaevskiĭ 1980). When the field is weakly quantizing, these quantities oscillate, as a rule, around their values obtained neglecting the magnetic quantization. For first-order (bulk) thermodynamic quantities ($P$, $U$, $\mu$), the oscillations are relatively weak, whereas for second-order quantities ($C_V$, $C_P$, $k_{\rm TF}$) they are more pronounced. For example, the oscillations of the density exponent $\chi_\rho$ defined by Eq. (2.38), will be shown in Fig. 4.11 on p. 204. The oscillations are smoothed by the thermal broadening of the Fermi distribution function and by the quantum broadening of the Landau levels (particularly, owing to electron collisions; see Yakovlev & Kaminker 1994, for references).

For example, we can mention the well known de Haas-van Alphen effect — oscillations of magnetic susceptibility (e.g., Landau & Lifshitz 1993). In quantizing magnetic fields, these oscillations can be large, resulting in high magnetization. Canuto & Chiu (1971) suggested that this effect could lead to a spontaneous magnetization of the electron gas in neutron star envelopes, the so called Landau orbital ferromagnetism – LOFER. They found that the oscillations of the kinetic pressure due to the Landau quantization of degenerate electron gas may result in a state of permanent quasistable macroscopic magnetism, which is stronger for higher pressure. Neglecting the broadening of the Landau levels, the authors obtained a density dependence of the maximum LOFER field strength, which implied that in the outer neutron star crust (at $\rho \lesssim 10^{11.5}$ g cm$^{-3}$) LOFER could produce $B$ up to $\sim 10^{12}$ G (and still larger $B$ in the inner crust). However, the thermal and quantum broadening of oscillations prevents the spontaneous magnetization. For instance, Schmid-Burgk (1973) showed that LOFER is smeared out by the thermal broadening in the outer crust of a neutron star at $T \gtrsim 10^4$ K.

The de Haas-van Alphen effect may also violate the condition for thermodynamic equilibrium which states that the field strength should increase with the growth of the magnetic induction. This instability leads to the formation of domains with alternating magnetization (Lifshitz & Pitaevskiĭ, 1980). Blandford *et al.* (1983) showed that in a neutron star envelope with $B = 10^{12}$ G this instability may develop at $T$ up to $10^7$ K. However, since the magnetization is weak (a few percent of the field strength), this effect can hardly have any observable consequences.

### 4.1.4 The effects of the magnetic field on plasma ions

The effects of the magnetic field on plasma ions are twofold. First, the magnetic field acts on the ions directly; second, it affects them through the electrons. The direct influence becomes appreciable when the ion cyclotron frequency $\omega_{\rm ci} = Z\,(m_{\rm e}/m_{\rm i})\,\omega_{\rm c}$ exceeds the ion plasma frequency $\omega_{\rm pi}$ [Eq. (2.30)] and the ion cyclotron energy $\hbar\omega_{\rm ci}$ exceeds the thermal energy $k_{\rm B}T$. This happens at $B_{12} \gtrsim 100\,\sqrt{\rho_6}$ and $B_{12} \gtrsim T/10^7$ K. In this case, transverse motion of the ions is quantized in the Landau orbitals with the energy of an elementary excitation equal to $\hbar\omega_{\rm ci}$. In contrast to the case of electrons, the spin degeneracy of the Landau levels is taken off completely because of relatively large abnormal magnetic moments of the nuclei.

The effects of the second type are caused by the electron polarization in quantizing magnetic fields. For a weakly coupled Coulomb plasma which composes, for instance, warm neutron star atmospheres, these effects will be discussed in § 4.1.5.

Below we consider the direct effect of a quantizing magnetic field on the free energy of the ideal ion gas. The opposite case of the ion crystal will be analyzed in § 4.1.6.

**Ion gas in a strongly magnetized outer envelope.** Under realistic conditions in outer neutron-star envelopes, the ions are nondegenerate and non-relativistic. Then the ideal part of the free energy can be written explicitly. In its essence, it differs from Eq. (4.37) by the inclusion of excited Landau levels, which can be summed up explicitly using the Boltzmann distribution over these levels. The result reads

$$\frac{F_{\rm id}^{(\rm i)}}{N_{\mathcal{N}} k_{\rm B} T} = \ln(2\pi a_{\rm m}^2 \lambda_{\rm i} n_{\mathcal{N}}) + \ln\left[1 - \exp\left(-\frac{\hbar\omega_{\rm ci}}{k_{\rm B}T}\right)\right] - 1 + \frac{\Delta F}{N_{\mathcal{N}} k_{\rm B} T}, \quad (4.38)$$

where $\lambda_{\rm i}$ is the ion thermal wavelength, Eq. (2.27). As in Eq. (2.71), the ion rest energy is excluded. The term $\Delta F$ arises from the zero-point energy, $\frac{1}{2}\hbar\omega_{\rm ci}$, and the spin energy. This contribution is the same for free and bound atomic nuclei. Therefore, it affects neither ionization equilibrium nor pressure, but it does affect the internal energy and specific heat. For example, for protons

$$\Delta F = N_{\mathcal{N}} \left\{ \frac{1}{2}\,\hbar\omega_{cp} - k_{\rm B} T \ln\left[2\cosh\left(\frac{g_p \hbar\omega_{cp}}{4 k_{\rm B} T}\right)\right]\right\}, \quad (4.39)$$

where $g_p = 5.5857$ is the proton spin gyromagnetic factor (the doubled ratio of the proton magnetic moment to the nuclear magneton, Landau & Lifshitz 1976), and $\hbar\omega_{cp}$ the proton cyclotron energy.

Equations (4.38) and (4.39) provide a good approximation to the ionic part of the free energy in not too cold neutron-star atmospheres, where the ions are nearly ideal.

**Ideal-gas model for the ultra-magnetized inner crust.** At densities $\rho \gtrsim$ a few $\times\ 10^{11}$ g cm$^{-3}$, typical for the inner crust of a neutron star, non-relativistic magnetic fields $B \lesssim B_{\rm r}$ are nonquantizing (see Eq. (4.1)). Therefore, they cannot directly affect the EOS of the inner crust, which is mainly determined by degenerate, weakly quantized electrons and free neutrons. A superstrong field, however, can strongly quantize particle motion and thus affect the EOS. In the inner crust, the strong quantization implies $B \gtrsim 10^{17}$ G. This effect was studied, for instance, by Suh & Mathews (2001) using the simplest model in which the crust matter was approximated by a gas of degenerate noninteracting neutrons, protons and electrons. The authors found that superstrong magnetic fields shift beta-equilibrium and increase the proton fraction. In addition, for $B \gtrsim 10^{18}$ G the softening of the EOS caused by the Landau quantization is overridden by the stiffening associated with the anomalous magnetic moments of the nucleons. The EOS of matter composed of strongly degenerate nucleons and electrons in strong magnetic fields was studied in more detail by Broderick *et al.* (2000) for the conditions prevailing in neutron star cores; they considered not only ideal $npe$ gas but also the mean-field model. The latter study shows that the results of Suh & Mathews (2001) are qualitatively correct.

It is also possible that a superstrong magnetic field $B \gtrsim 10^{16}$ G can change the nuclear shell energies and nuclear magic numbers and, consequently, affect the nuclear composition and the EOS of the inner crust of a neutron star (Kondratyev, 2002).

Note, however, that the field strengths $B \gtrsim 10^{18}$ G are unrealistic from the point of view of neutron star physics (see § 1.3.8). Moreover, there is no observational indication on the existence of fields $B > 10^{16}$ G in neutron star crusts (see § 1.4).

### 4.1.5 Weakly non-ideal Coulomb plasma

Theoretical studies of thermodynamics of interacting charged particles in strong magnetic fields have a long history, but the results are much less comprehensive than in the field-free case. Only some limiting cases have been considered which we review briefly in this section.

According to the Bohr-van Leeuwen theorem (p. 54) the magnetic field does not affect thermodynamics of classical charged particles. Thus the excess free energy $F_{\rm ii}(\Gamma)$ for a classical OCP of ions is independent of $B$ at any ion coupling parameter $\Gamma$ (defined by Eq. (2.22)). The classical regime for an electron-ion plasma corresponds to $r_s \gg 1$ and $\Gamma \ll 1$ in the absence of electron degeneracy. In this case the excess Coulomb free energy is given by the Debye-Hückel formula $F_{\rm ex}^{\rm C} = -N_e k_{\rm B} T \, Z \, \sqrt{(1+Z)/3} \, \Gamma^{3/2}$ (compare to Eq. (2.73)). Indeed, it is easy to check that for classical plasma particles this law holds independently of $B$ (Abrahams & Shapiro, 1991a; Cornu, 1998).

The magnetic field, however, affects quantum contributions to $F_{\rm ex}^{\rm C}$. These effects have been studied either in the regime of low $T$ and high $\rho$ (considered in the following section), or at low densities. In the latter case, a general power-series expansion for the free energy of a Coulomb plasma in an arbitrary magnetic field (up to the terms $\propto \rho^{5/2}$) was derived by Cornu (1998). The coefficients of this expansion are not given by analytical expressions but require numerical evaluation. For the OCP, a Wigner-Kirkwood-type expansion in powers of $\hbar$ is available (Cornu, 1998). Its lowest-order term (the quantum diffraction term $\propto \hbar^2$) was first obtained by Alastuey & Jancovici (1980). It generalizes Eq. (2.85) to the case of a quantizing magnetic field. For the nondegenerate electrons, this term can be written as

$$F_{\rm diff}^{(e)} = N_e k_{\rm B} T \, \frac{\Gamma_e^2}{8r_s} \left( \frac{2}{\zeta_e \, \tanh \zeta_e} - \frac{2}{\zeta_e^2} + \frac{1}{3} \right), \tag{4.40}$$

with $\zeta_e$ defined by Eq. (4.36). The brackct in Eq. (4.40) goes to 1 at $\zeta_e \to 0$, recovering the well known zero-field result, and to $1/3$ at $\zeta_e \gg 1$, indicating that two of the three degrees of freedom of electron motion are frozen out in a strongly quantizing field. Equation (4.40) is valid in the low-density regime (one can show that the validity conditions imply $r_s \gg \max(\Gamma, \Gamma^{-1})$). In this regime, the correction (4.40) is smaller than the classical OCP corrections to the Debye-Hückel formula owing to ion correlations. In the electron-ion plasmas, $F_{\rm diff}^{(e)}$ is exactly canceled because of the local neutrality relation (Cornu, 1998). Thus Eq. (4.40), although elegant by itself, can hardly be useful for neutron star modeling.

If $(\Gamma r_s)^{-1} \approx 3.167\, T_6 \gtrsim 1$ and $\Gamma / r_s^2 \approx 0.118\, \rho \, \langle Z \rangle / (A' \, T_6 \text{ g cm}^{-3}) \lesssim 1$, a high-temperature expansion of the free energy in powers of $e^2$ is applicable. These conditions are often fulfilled in the atmospheres of neutron stars. The expansion is in powers of two small parameters, $s_1 = \sqrt{\Gamma}/r_s$ and $s_2 = \sqrt{\Gamma r_s}$. In the field-free case, the expansion terms up to $\rho^{5/2}$ were obtained by DeWitt *et al.* (1995). The lowest-order term $\propto e^2$ is also known in an arbitrary magnetic field (Steinberg *et al.*, 1998):

$$\frac{F_{\rm HF}}{N_e k_{\rm B} T} = -\frac{3\Gamma^2}{8r_s} f_1(\zeta_e), \tag{4.41}$$

where the function

$$f_1(\zeta_e) = \frac{\cosh(2\zeta_e)}{\cosh^2 \zeta_e} \, \frac{\tanh \zeta_e}{\zeta_e} \, \frac{\operatorname{arctanh}(\sqrt{1 - \zeta_e^{-1} \tanh \zeta_e})}{\sqrt{1 - \zeta_e^{-1} \tanh \zeta_e}} \tag{4.42}$$

goes to 1 at small $\zeta_e$ and decreases as $\ln(4\zeta_e)/\zeta_e$ at very large $\zeta_e$.

Steinberg *et al.* (1998) calculated also the corrections $\propto e^4$ (the Montroll–Ward and exchange terms) in the magnetic field. For the electron gas, these corrections can be written in the form

$$\frac{F_4}{N_e k_{\rm B} T} = \frac{3\sqrt{\pi}}{16} \, \frac{\Gamma^{5/2}}{\sqrt{r_s}} \, [f_2^{ee}(\zeta_e) + f_3^{ee}(\zeta_e) \ln 2], \tag{4.43}$$

where $f_2^{ee}(\zeta_e)$ and $f_3^{ee}(\zeta_e)$ go to 1 at $\zeta_e \to 0$, reproducing the field-free result (DeWitt *et al.*, 1995), and decrease at large $\zeta_e$.

For the electron-ion plasma, the polarization screening contribution $F_{\rm ie}$ should be taken into account. At $B = 0$, it has been calculated in a number of papers and fitted by analytical expressions (see §2.4.4). However, the magnetic field effect on $F_{\rm ie}$ is known in a more restricted range of $\rho$, $T$, and $B$, than in the OCP of ions considered above.

In the regime of low electron degeneracy and weak Coulomb coupling, the lowest order contribution of the low-density expansion was obtained by Steinberg *et al.* (1998):

$$\frac{F_{\rm ie}^{(4)}}{N_{\!\mathcal{N}} k_{\rm B}T} = \frac{\pi}{4}\left(1+\frac{m_e}{m_{\rm i}}\right)^{1/2}\left(\frac{Ze^2}{k_{\rm B}T}\right)^2 \lambda_e n_e f_2^{\rm ie}(\zeta_e,\zeta_{\rm i}), \tag{4.44}$$

where $\zeta_{\rm i} = \hbar\omega_{\rm ci}/2k_{\rm B}T = \zeta_e\, Z\, m_e/m_{\rm i}$, and the function $f_2^{\rm ie}$ is known in an integral form. At $\zeta_e \to 0$, this function tends to 1, reproducing the well known zero-field result (e.g., DeWitt *et al.* 1995). Because $m_{\rm i} \gg m_e$, one can use an analytical approximation (Potekhin *et al.*, 1999b) accurate to 0.5%:

$$f_2^{\rm ie} = \frac{1}{2} + t^{0.9}\,\frac{\operatorname{arctanh}\left[(1-t)^{0.6}\right]}{2\,(1-t)^{0.6}}, \tag{4.45}$$

where $t \equiv (0.4\zeta_e)^{-1}\tanh(0.4\zeta_e)$. Integral representations for coefficients of the low-density expansion have been obtained by Cornu (1998).

### 4.1.6 Strongly coupled Coulomb plasma

The effects of strong magnetic fields on a strongly coupled Coulomb plasma can be important for the physics of magnetars. Let us summarize available results.

#### 4.1.6 a Ground-state energy

The ground-state exchange energy of the electron gas (per one electron) in a strongly quantizing field (Danz & Glasser, 1971; Fushiki *et al.*, 1989) behaves as

$$-2.25\,[\ln(\gamma r_s^2) - 0.457 + \ldots]\,(\gamma r_s^3)^{-1}\, e^2/a_0,$$

compared with

$$-0.75\,(9\pi/4)^{1/3}(\pi r_s)^{-1}\, e^2/a_0$$

in the non-magnetic case (e.g., Perrot & Dharma-wardana 1984). Thus, the magnetic field suppresses the exchange energy at $T \ll T_{\rm F}$ by a factor of $0.2036\,\gamma r_s^2/[\ln(\gamma r_s^2) - 0.457]$. Note that the condition for the strong magnetic quantization, $\rho < \rho_B$, requires that $\gamma r_s^2 > 2.23$, so that the suppression factor is greater than 1. This result is applicable at $T \ll T_{\rm F}$ and $\rho < \rho_B$, that is in the outer envelopes of very cold or ultra-magnetized neutron stars.

Using the linear response theory in the Thomas-Fermi limit, Fushiki *et al.* (1989) have analytically evaluated the electron polarization energy $F_{\rm ie}$ for a dense plasma in a strongly quantizing magnetic field at zero temperature. A comparison with the analogous zero-field result (2.158) shows that the strongly quantizing magnetic field ($\gamma r_s^2 > 2.23$) increases $F_{\rm ie}$ at high densities ($r_s \ll 1$) by a factor of $0.8846\,\gamma^2 r_s^4$.

In order to incorporate these results and the results reviewed in § 4.1.5 into the analytical free-energy model, Potekhin *et al.* (1999a) suggested an interpolation which reproduces the known limiting cases of $r_s \gg 1$ (the classical electron-ion plasma), $\gamma r_s^2 \ll 1$ (the non-quantizing regime), $\gamma r_s^2 \gg 1$ and $T \ll T_{\rm F}$ (the strongly quantizing degenerate regime), and $\gamma r_s^2 \gg 1$ and $T \gg T_{\rm F}$ (the strongly quantizing nondegenerate regime). The interpolation reproduces also Eq. (4.41) in its validity range. Nevertheless, the accuracy of the interpolation remains uncertain in a large range of $\rho$, $T$, and $B$, where none of these limiting cases can apply.

#### 4.1.6 b Coulomb crystal in a superstrong magnetic field

The magnetic field affects phonon spectrum of Coulomb crystals and respective ion thermodynamic properties. This happens at sufficiently high $B$ (to change the phonon spectrum) and low $T$ for the field to become quantizing. The magnetic field changes also polarizability of the electron gas and affects the ion thermodynamics in this way. As a rule, magnetic fields of ordinary pulsars cannot affect noticeably the ion thermodynamics in the degenerate layers of the envelope, but superstrong magnetic fields $B \gg 10^{14}$ G of magnetars can.

In a superstrong magnetic field, thermodynamic functions of a Coulomb crystal become dependent on the magnetic field strength, crystal orientation with regard to $\boldsymbol{B}$, and the crystal type. Kaplan & Glasser (1972) argued that a superstrong magnetic field could increase the stability of a quantum crystal of charged fermions against melting at high densities and suggested that the ground state in this case would be a hexagonal lattice. Nagai & Fukuyama (1982; 1983) compared the energies of zero-point vibrations for body-centered cubic (bcc), face-centered cubic (fcp), and hexagonal closed-packed (hcp) Coulomb lattices at zero temperature and found that the hcp lattice becomes more stable than the bcc one, if $R_S \lesssim 10^4$, where $R_S$ is the ion density parameter defined by Eq. (2.32).

Usov *et al.* (1980) derived dispersion equations for a Coulomb crystal in a quantizing magnetic field, qualitatively studied the phonon spectrum, and obtained an asymptotic dependence of the heat capacity in the limit of ultra-high $B$ and low $T$. Their method was further used by Baiko (2000), who performed a more detailed study of vibration and thermodynamic properties of strongly magnetized Coulomb crystals with a rigid electron background.

The results of Baiko (2000) extend those of § 2.3.3 to the case of strong magnetic fields. For simplicity, he focused on bcc crystals with the magnetic field directed from one ion to a closest neighbor (which minimizes the Coulomb energy). The dependence of the free energy on the crystal type (bcc or fcc) has turned out to be weak, as well as the dependence on the magnetic field

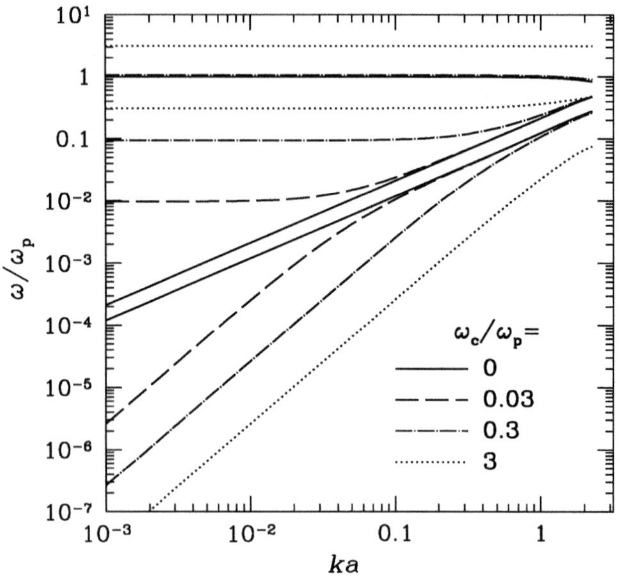


*Figure 4.2.* Vibration spectrum of a bcc crystal as a function of $ka_{\rm i}$ in a magnetic field for a fixed direction of the wave vector, $\boldsymbol{k}/k$=(0.720, 0.615, 0.323), at several values of $\omega_{\rm ci}/\omega_{\rm pi}$. From Baiko (2000) with the kind permission of the author.

orientation. The magnetic field strongly affects lattice thermodynamics as long as $\omega_{\rm ci} \gtrsim \omega_{\rm pi}$, which translates to $\rho \lesssim B^2/(4\pi c^2)$. This happens in the outer layers of neutron star crusts (at $\rho \lesssim 10^8$ and $10^{10}$ g cm$^{-3}$ for $B = 10^{15}$ and $10^{16}$ G, respectively).

The crystal vibration spectrum in a magnetic field $\boldsymbol{B}$ is calculated in the same manner as is in the field-free case but taking into account the Lorentz force acting on vibrating ions. The dispersion equation is rather simple. For a bcc crystal it gives three vibration branches, $\omega_1(\boldsymbol{k}) \le \omega_2(\boldsymbol{k}) \le \omega_3(\boldsymbol{k})$. An example is shown in Fig. 4.2. The vibration frequencies are seen to be noticeably affected by the magnetic field at $ka_{\rm i} \lesssim \omega_{\rm ci}/\omega_{\rm pi}$. Their behaviour at $k \to 0$ is of special interest. Let us recall that in the field-free case (§ 2.3.3) we have two acoustic vibration modes, $\omega_{1,2}(\boldsymbol{k}) \propto k$, and one optical mode $\omega_3(\boldsymbol{k}) \to \omega_{\rm pi}$. The magnetic field violates this simplicity. The vibration mode $\omega_1(\boldsymbol{k})$ remains acoustic ($\omega_1(\boldsymbol{k}) \propto k$) only if $\boldsymbol{k}$ is parallel to $\boldsymbol{B}$. In other cases one has $\omega_1 \propto k^2$ at sufficiently small $k$ (which means the "softening" of this mode). The modes $\omega_2(\boldsymbol{k})$ and $\omega_3(\boldsymbol{k})$ appear to be optical ($\omega_{2,3}(\boldsymbol{k}) \to$ const as $k \to 0$).

Baiko (2000) numerically realized a procedure suggested by Usov *et al.* (1980) to quantize ion vibrations (which is much more sophisticated than at $\boldsymbol{B} = 0$ because of Larmor motion). This enabled him to calculate (by the same technique as in the field-free case) the phonon entropy, heat capacity, and

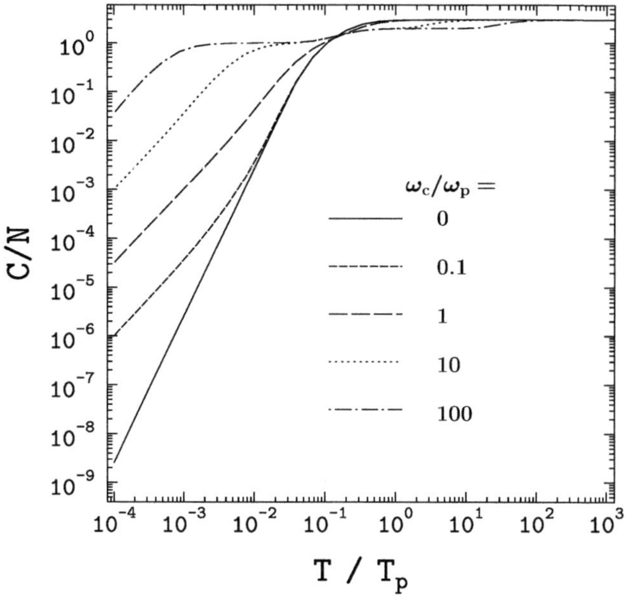


*Figure 4.3.* Harmonic lattice heat capacity (per one ion, divided by $k_B$) of bcc crystal as a function of $T/T_{pi}$ for different magnetic fields. From Baiko (2000) with the kind permission of the author.

pressure for a wide range of plasma parameters. For example, Fig. 4.3 shows the temperature dependence of the harmonic-lattice heat capacity $C = C_i$ per one ion (normalized by $k_B$) at several values of the magnetic field (compare with the results of § 2.4.6 for $B = 0$). At high enough $T$, as expected, the magnetic field is non-quantizing and its effect disappears. If $\omega_{ci}/\omega_{pi} \gg 1$, we can notice a substantial reduction of $C_i$ at $T_{pi} \lesssim T \ll T_{Bi}$, where $k_B T_{Bi} = \hbar\omega_{ci}$. At low $T \ll \min(T_{Bi}, T_{pi})$, the magnetic field increases the heat capacity by orders of magnitude and changes its temperature dependence, in agreement with the prediction of Usov *et al.* (1980). In this case, one has $C_i \propto T^{3/2}$ (instead of $C_i \propto T^3$ at $B = 0$). The effect is explained by the appearance of the soft vibration mode ($\omega_1(\boldsymbol{k}) \propto k^2$ at $k \to 0$).

An additional contribution to the specific heat comes from the free energy correction (4.39) produced by magnetic moments of atomic nuclei. One can easily show that this contribution has a noticeable maximum $\sim k_B$ per ion at $k_B T \sim g_i \hbar\omega_{cp}$ but becomes insignificant at much low and higher $T$ ($g_i$ being a gyromagnetic factor of a nucleus).

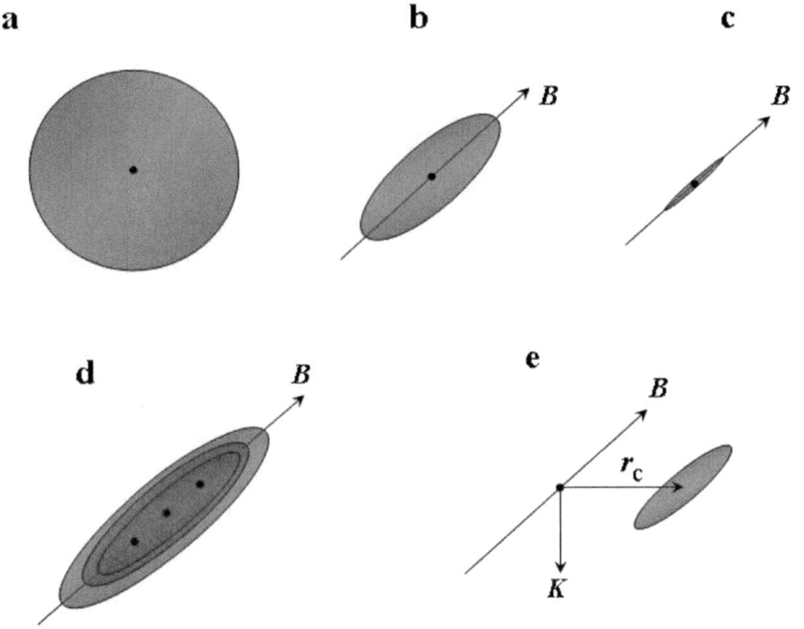


*Figure 4.4.* Schematic view of the effects of a strong magnetic field on atoms and molecules. (**a–c**) The H atom in the ground state becomes compressed and elongated with increasing the field strength from (**a**) $B \ll 10^9$ G, to (**b**) $B \sim 10^{10}$ G, and to (**c**) $B \sim 10^{12}$ G. (**d**) The field stabilizes molecular chains ($H_3$, for example). (**e**) An atom moving across the field $\boldsymbol{B}$ becomes decentered (with the relative guiding center $\boldsymbol{r}_c$ perpendicular to $\boldsymbol{B}$ and to the atomic pseudomomentum $\boldsymbol{K}$). The grey areas are ellipsoids, where the probability to find an electron exceeds $e^{-1}$; the solid dots show protons. The radius of the grey sphere in case (**a**) is $\approx 1.08\, a_0$. Coaxial ellipsoids in **d** correspond to electron orbitals with zero Landau quantum number and the smallest consecutive magnetic quantum numbers, which form the ground state of the molecular chain in the strong magnetic field.

## 4.2. Bound species in strong magnetic fields

In this section, we outline the main properties of bound species at the conditions typical for magnetized neutron star atmospheres. For a more detailed review see, e.g., Lai (2001).

### 4.2.1 Atoms

#### 4.2.1 a Hydrogen atom

The effects of a strong magnetic field on bound species are spectacular. Figure 4.4 schematically shows some of them for the simplest case of hydrogen in the ground state. The H atom becomes increasingly compressed and elongated with the growth of $B$. In a sufficiently strong field, molecular chains $H_N$ can be formed. An atomic center-of-mass motion across the field shifts the electron cloud off the nucleus. These effects will be briefly discussed below.

Atoms in strong magnetic fields have been intensively studied by many authors (see, e.g., Ruder *et al.* 1994, for review). At $\gamma \gg 1$, an electron cloud bound to a nucleus acquires the shape of a cylinder with the radius $\sim a_\mathrm{m}$ and much larger length. The binding energies of the ground and some excited states of the H atom increase as $(\ln \gamma)^2$ (such atomic states are called *tightly bound*). The energies of other (*hydrogenlike*) states remain restricted within $\sim 1$ Ry, where Ry $= m_e e^4/2\hbar^2 = 13.605\,692$ eV.

Quantum-mechanical characteristics of strongly magnetized hydrogen atoms are obtained by solving numerically the Schrödinger equation (Canuto & Ventura, 1977; Rösner *et al.*, 1984; Forster *et al.*, 1984; Potekhin *et al.*, 1997b) or the Dirac equation (Lindgren & Virtamo, 1979; Chen & Goldman, 1992). At $\gamma \gtrsim 0.1$, an atomic state can be characterized by the electron Landau quantum number $n$, the projection $\hbar m$ of the relative electron-to-proton angular momentum on the field direction ($m \leq n$), and by the number of nodes $\nu$ of a wave function in the field direction. At $\gamma \gtrsim 1$, only the states with $n = 0$ remain bound; other discrete states are quasibound (can decay through autoionization channels). The atomic binding energy can be written as

$$E_\mathrm{bind} = -E_{nm\nu}, \quad E_{nm\nu} = E^{\|}_{nm\nu} - m\hbar\omega_\mathrm{cp} + n\hbar(\omega_\mathrm{c} + \omega_\mathrm{cp}), \tag{4.46}$$

where $E^{\|}_{nm\nu}$ is the "longitudinal" energy (which is negative for discrete spectrum states). The proton spin energy in a magnetic field is dropped from Eq. (4.46), because proton spin-flip processes (forbidden in the electric dipole approximation) are so inefficient that the subsystems with proton spin "up" and "down" can be treated as independent in most of the applications where bound species are involved.

As already mentioned, an electron cloud at $\gamma \gtrsim 1$ is elongated. Its sizes transverse to the magnetic field are those of the Landau function $\Phi_{n,-m}$ given by Eq. (4.12), that is $l_x = l_y \approx \sqrt{2n - m + 1}\, a_\mathrm{m} \sim a_0/\sqrt{\gamma}$. The longitudinal size is much larger: $l_z \sim a_0/\ln\gamma$ for the tightly bound states ($\nu = 0$) and $l_z \sim a_0\nu^2$ for the hydrogenlike states ($\nu \geq 1$). At $\gamma \gg 1$, an atomic wave function is well described by the *adiabatic approximation*:

$$\psi_{nm\nu}(\boldsymbol{r}) \approx \Phi_{n,-m}(\boldsymbol{r}_\perp)\, g_{nm\nu}(z). \tag{4.47}$$

This approximation was widely used in early papers (e.g., Canuto & Ventura 1977 and references therein). Accurate wave functions and binding energies were obtained by Rösner *et al.* (1984) using the expansion

$$\psi_{nm\nu}(\boldsymbol{r}) = \sum_{n'=0}^{\infty} \Phi_{n',-m}(\boldsymbol{r}_\perp)\, g_{n',nm\nu}(z), \tag{4.48}$$

*Table 4.1.* Parameters $a_i$ and maximum fractional error of Eq. (4.49)

| state | $a_1$ | $a_2$ | $a_3$ | $a_4$ | $a_5$ | $a_6$ | err. (%) |
|---|---|---|---|---|---|---|---|
| $1s_0/(000)$ | 1 | 0.09016 | 0.13966 | 0.02328 | 0.017538 | 0.0008506 | 0.12 |
| $2p_{-1}/(010)$ | 5 | 0.2603 | 0.11135 | 0.010359 | 0.010278 | 0.0005253 | 0.23 |
| $3d_{-2}/(020)$ | 12 | 0.5924 | 0.13558 | 0.008498 | 0.008176 | 0.0006525 | 0.52 |
| $4f_{-3}/(030)$ | 21 | 1.1523 | 0.17860 | 0.007172 | 0.002976 | 0.0008771 | 0.87 |

where $g_{n',nm\nu}(z)$ were found from a system of ordinary differential equations which arises after substituting Eq. (4.48) into the Schrödinger equation. This solution was used by Forster *et al.* (1984) to calculate oscillator strengths of radiative transitions. Potekhin *et al.* (1997b) extended this method to continuum states and calculated the photoionization cross sections. In particular, they found that the coupling of different $(n, m)$-channels of electron scattering leads to the appearance of resonances in the photoionization cross sections (Beutler–Fano type resonances).

The longitudinal energies of tightly bound states can be approximated by (Ho *et al.*, 2003)

$$E^{\|}_{0m0} = -\frac{(|m|+1)^{-2} + (|m|+1)\,x/a_1 + a_3\,x^3 + a_4\,x^4 + a_6\,x^6}{1 + a_2\,x^2 + a_5\,x^3 + a_6\,x^4}\ \text{Ry}, \tag{4.49}$$

where $x = \ln(1 + a_1\gamma)$ and the parameters $a_1$–$a_6$ are given in Table 4.1 for three values of $m$. The last column of the table gives the maximum fractional error of the fit in the interval $0 < \gamma < 10^8$. The fit reproduces also the exact asymptotes

$$E^{\|}_{0m0} = -\left[\frac{1}{(|m|+1)^2} + (|m|+1)\,\gamma + O(\gamma^2)\right]\ \text{Ry} \quad \text{at} \quad \gamma \ll 1, \tag{4.50}$$

$$E^{\|}_{0m0} \simeq -(\ln\gamma)^2\ \text{Ry} \quad \text{at} \quad \gamma \to \infty. \tag{4.51}$$

Another fit for $E^{\|}_{0m0}$ with $-7 \le m \le 0$, accurate for $0.1 < \gamma < 10^4$ (typical for radio pulsars), is given by Eq. (10) of Potekhin (1998).

The longitudinal energies of hydrogenlike states tend to the Rydberg series,

$$E^{\|}_{nm\nu} = -\frac{1\ \text{Ry}}{(\ell+\delta)^2}, \quad \text{where} \quad \begin{cases} \ell = (\nu+1)/2, \ \ \delta \sim \gamma^{-1} & \text{(odd } \nu) \\ \ell = \nu/2, \ \ \delta \sim (\ln\gamma)^{-1} & \text{(even } \nu). \end{cases} \tag{4.52}$$

The quantum defect $\delta$ is different for even- and odd-parity states. For odd $\nu$, one has

$$\delta \approx (a_\nu + b_\nu\sqrt{\gamma} + 0.077\gamma)^{-1}, \tag{4.53}$$

*Table 4.2.* Parameters of the analytical approximations (4.52)–(4.54) for the energies of hydrogenlike states $|00\nu\rangle$ at $1 \leq \gamma \leq 10^4$.

| $\nu$ | 1 | 2 | 3 | 4 | 5 | 6 |
|---|---|---|---|---|---|---|
| $a_\nu$ | 0.785 | 0.578 | 0.901 | 0.631 | 0.970 | 0.660 |
| $b_\nu$ | 1.724 | 0.765 | 1.847 | 0.717 | 1.866 | 0.693 |

with $a_\nu \approx 1$ and $b_\nu \approx 2$. For even $\nu$,

$$\delta \approx \left[a_\nu + 1.28 \ln(1 + b_\nu \gamma^{1/3})\right]^{-1}, \tag{4.54}$$

with $a_\nu \approx b_\nu \approx \frac{2}{3}$. Accurate values of $a_\nu$ and $b_\nu$ are given in Table 4.2 (for $n = 0$, after Potekhin 1998). At $\gamma \geq 1$, typical errors of Eqs. (4.53) and (4.54) lie within $10^{-3}$.

Finally, note that binding energies of a non-relativistic, non-moving H atom in an arbitrary magnetic field can be calculated exactly (Kravchenko *et al.*, 1997).

### 4.2.1 b Other atoms and ions

Some calculations of the binding energies of multi-electron atoms and ions in strong magnetic fields were done in the Thomas-Fermi approximation (e.g., Rögnvaldsson *et al.* 1993) and using the DFT – the density functional theory (Jones, 1985; Relovsky & Ruder, 1996; Medin & Lai, 2006a). Since these methods are statistical, they are expected to be adequate when the number of electrons is large.

For several atoms from He to Fe, the Hartree-Fock approximation was used in combination with the adiabatic approximation [cf. Eq. (4.47)], employing basis functions which contain free-electron transverse parts (Neuhauser *et al.*, 1987; Miller & Neuhauser, 1991) [cf. Eq. (4.47)]. If the field is sufficiently strong, this adiabatic Hartree-Fock approximation is reasonably accurate for not too large charge $Z_{\rm nuc}$ of the atomic nucleus.

With growing $Z_{\rm nuc}$, inner atomic shells become progressively distorted by the Coulomb attraction, making the adiabatic approximation less accurate. In the 1990s, a more accurate "two-dimensional" Hartree-Fock approximation was used, which allowed one to vary radial parts of the basis functions. It was applied for calculating binding energies of various quantum states of He atom (Thurner *et al.*, 1993; Ivanov, 1994; Jones *et al.*, 1996), $H^-$ ion (Jones *et al.*, 1996), B atom and $B^+$ ion (Ivanov & Schmelcher, 2001), and also the ground state of atoms and their singly positive ions up to Ne (Ivanov & Schmelcher, 2000). A comparison with the "adiabatic Hartree-Fock" results shows that the latter are accurate within 10% as long as, for instance, $B_{12} > 0.1$ and $Z_{\rm nuc} \le 6$, or $B_{12} > 0.5$ and $Z_{\rm nuc} \le 10$. For $B_{12} \ge 5$ and $Z_{\rm nuc} \le 10$, the inaccuracy in the ground-state energy does not exceed 1.3%.

The dipole oscillator strengths, required for modelling of strongly magnetized neutron star atmospheres, were calculated for the He atom in the two-dimensional Hartree-Fock approximation by Thurner *et al.* (1993) and Jones *et al.* (1998) and for the Fe atom in the adiabatic approximation by Miller & Neuhauser (1991).

All calculations of multi-electron atoms and ions mentioned in this section used the approximation of an infinitely massive nucleus (whose position was fixed). This is an essential simplification. For instance, the characteristics of hydrogenlike ions with fixed nuclei obey a scaling with $Z_{\rm nuc}$ (Wunner *et al.*, 1981). In particular, the binding energies scale as $E(Z_{\rm nuc}, B) = Z_{\rm nuc}^2 E(1, B/Z_{\rm nuc}^2)$. As we shall see in § 4.2.3, motion across a strong magnetic field can qualitatively modify atomic states and violate this scaling.

### 4.2.2 Molecules and chains

**Neutral molecules.** The properties and the very existence of various types of molecules in strong magnetic fields were debated during decades and still remain a subject of investigation.

For a non-moving diatomic molecule whose axis is directed along the field the three-dimensional problem reduces to a two-dimensional one because of the cylindrical symmetry. It is, therefore, natural that this *parallel configuration* has been best studied.

For obvious reasons, the $H_2$ molecule is the best explored one (see, e.g., Demeur *et al.* 1994; Lai & Salpeter 1996; Detmer *et al.* 1998). Fit formulae for dissociation energies in the parallel configuration at $\gamma \gtrsim 10^3$ are given by Lai & Salpeter (1996, 1997). At such fields, the dissociation energy grows as $(\ln\gamma)^2$, approximately at the same rate as the atomic ground-state energy. For example, at $B = 10^{12}$ G, the cohesive energy of the $H_2$ molecule (that is the difference between the ground-state energies of two atoms and the molecule) is 45.5 eV and the adiabatic dissociation energy is 91 eV (compare with 4.48 eV at $B = 0$). The equilibrium internuclear distance decreases as $1/\ln\gamma$, becoming as small as $\frac{1}{4}a_0$ at $B = 10^{12}$ G, again roughly proportional to the longitudinal size of the H atom.

However, according to Detmer *et al.* (1998), the ground state of the $H_2$ molecule is unbound at weaker fields, $0.18 < \gamma < 12.3$, most typical for millisecond pulsars.

There are relatively few results on heavier molecules in strong magnetic fields. Some of them have been reviewed by Lai (2001). More recently, Medin & Lai (2006a) applied DFT for calculating binding energies of different hydrogen, helium, carbon, and iron molecules at several field strengths between $10^{12}$ G and $2 \times 10^{15}$ G (see also references therein to earlier works).

**Molecular ions.** One-electron molecular ions in strong magnetic fields have been studied thoroughly. Detailed papers by Wille (1986) and Kappes *et al.* (1994) were devoted to the $H_2{}^+$ molecular ion. The authors considered various orientations of the molecular axis with respect to the field. In particular, Wille (1986) showed that low-lying rotational states of the ion transform into vibrational ones as the field increases. This happens because the lowest energy in the transverse configuration (i.e., for the molecular axis perpendicular to the field lines) is higher than the ground-state energy in the parallel configuration. Thus the angular dependence of the energy creates a potential barrier to rotation. If the field is strong enough, the barrier becomes higher than the lowest rotational levels. Then the lowest rotational states correspond to oscillations of the molecular axis about the field direction. Kappes & Schmelcher (1996) calculated potential surfaces of the $H_2{}^+$ ion as a function of an internuclear distance and an angle between the ion axis and the magnetic field direction for various electron-vibrational-rotational states at $B \leq 10^{10}$ G. Wille (1987) considered also the $(\text{H-He})^{2+}$ system at $B \leq 10^8$ G.

For $B > 10^{10}$ G, the parallel configuration of the $H_2{}^+$ ion has been analyzed especially thoroughly (e.g., Kravchenko & Liberman 1997 and references therein). Non-parallel configurations were studied by Turbiner & López Vieyra (2003, 2004). In particular, the latter authors found that at $B \gtrsim 10^{11}$ G large inclination angles lead to the decay $H_2{}^+ \to \text{H} + p$.

The $H_3{}^{++}$ ion in a strong magnetic field was studied by López Vieyra & Turbiner (2002) and Turbiner *et al.* (2005) using a variational method. They found that $H_3{}^{++}$ with the protons in a line along the field is stable at $B > 10^{10}$ G. They also found that a configuration of protons forming an equilateral triangle perpendicular to the magnetic line is quasi-bound (metastable) at $10^8 < B < 10^{11}$ G and decays into the H atom and two protons.

**Molecular chains.** Strong magnetic fields stabilize polymer chains aligned with the fields, as first suggested by Ruderman (1971). In the 1970s, it was commonly accepted that such chains in the strong fields can be formed of any atoms. This belief was refuted by Müller (1984), who performed variational calculations and demonstrated that a molecular chain composed of iron is unbound at $1 \leq B_{12} \leq 5$. Later density-functional (Jones, 1985) and Hartree-Fock (Neuhauser *et al.*, 1987; Lai *et al.*, 1992; Demeur *et al.*, 1994) calculations confirmed this result and showed that, at $B \sim 10^{12} - 10^{14}$ G, infinite chains are stable only for the elements with atomic numbers $Z_{\text{nuc}} \lesssim 4 - 6$, while heavier elements do not form a stable molecular bond in the strong fields. On the contrary, Thomas-Fermi-like models, such as the Thomas-Fermi-Dirac-Weizsäcker model of Abrahams & Shapiro (1991b) still yield a strong binding. Recent DFT calculations by Medin & Lai (2006b) show that hydrogen, helium, carbon and iron infinite chains all are bound relative to individual atoms for magnetic fields $B \gtrsim 10^{12}$ G (and may be lower), but iron chains are not significantly bound as long as $B \lesssim 10^{14}$ G. Since all these studies are approximate and give different results, it remains unclear, whether the chains of heavy atoms can be really bound in the strong fields.

## 4.2.3 Effects of motion

Motion across the magnetic field (non-negligible in warm neutron star atmospheres) breaks down the cylindrical symmetry of an atom. The quantum-mechanical operator that generates velocity boosts is the pseudomomentum $\boldsymbol{K}$ (a pedagogical and insightful introduction of this quantity is given by Johnson *et al.* 1983). Therefore, $\boldsymbol{K}$ is relevant to describe atomic motion in a magnetic field. The pseudomomentum is collinear with the atomic velocity, but differs from the canonical center-of-mass momentum. With increasing $K_\perp$ (the transverse component of $\boldsymbol{K}$), the binding energy decreases, while the atomic size increases, producing a constant dipole moment perpendicular to $\boldsymbol{B}$ and $\boldsymbol{K}$. At $\gamma \gg 1$, the effects of collective motion (Avron *et al.*, 1978; Johnson *et al.*, 1983; Vincke & Baye, 1988) become especially pronounced. In particular, the so-called decentered states (with an electron localized mostly in the "magnetic well" displaced from the Coulomb center; see Fig. 4.4) are likely to be populated. For the hydrogen atom, these exotic states were first predicted by Burkova *et al.* (1976).

Quantum-mechanical effects of motion of hydrogenlike ions in strong magnetic fields were analyzed by Bezchastnov *et al.* (1998). For multi-electron atoms, ions, and molecules, such effects remain unexplored.

Now let us consider in more detail the simplest but important example of a hydrogen atom moving in a strong magnetic field.

### 4.2.3 a Hydrogen atom moving in a strong magnetic field

The first numerical solutions of the Schrödinger equation for an atom moving arbitrarily in strong magnetic fields were presented by Vincke *et al.* (1992). At superstrong fields, the binding energies were calculated by Lai & Salpeter (1995) in the non-relativistic approximation. Potekhin (1994) used the expansion of a wave function analogous to Eq. (4.48), but supplemented the sum over $n$ by the sum over magnetic quantum numbers $m$ (because $m$ is not a good quantum number for a moving atom). He numerically solved the corresponding system of coupled-channel equations and calculated binding energies, wave functions, and radiative transition rates. Analogous system of equations for continuum states was solved by Potekhin & Pavlov (1997). They extended the results of Potekhin *et al.* (1997b), mentioned above, to the case of a moving atom, and adapted the $R$-matrix formalism (Wigner & Eisenbud, 1947) to this case.

According to these studies, an atom moving across the field lines acquires a constant electric dipole moment in the direction opposite to its *guiding center* $\boldsymbol{r}_c = c\,(eB^2)^{-1}\boldsymbol{B}\times\boldsymbol{K}$. When $K_\perp$ is small enough, the dipole moment is also small, and the energy $E^\|(K_\perp)$ increases by $K_\perp^2/(2m_{nm\nu}^\perp)$. Here, $m_{nm\nu}^\perp$ is the so-called *effective transverse mass*, which exceeds the atomic mass $m_\mathrm{H}$ and grows with the field strength. In this case, the average transverse velocity is $v_\perp = K_\perp/m_{nm\nu}^\perp$. When $K_\perp$ reaches some critical value $K_c \approx \sqrt{2m_\mathrm{H}E^\|(0)} \sim 10^2\hbar/a_0$, the atom becomes *decentered*: $v_\perp$ attains its maximum and starts to decrease, while the electron-proton separation approaches $r_c$. Note that $r_c$ is proportional to $K_\perp$ and inversely proportional to $B$; an accurate fit of $K_c$ as a function of $B$ is given by Potekhin (1998). Thus, for decentered states, the transverse pseudomomentum $K_\perp$ characterizes the electron-proton separation, rather than the velocity.

In the limiting case where $K_\perp \gg \gamma(\nu+\frac{1}{2})^2\hbar/a_0$, all longitudinal energies approach the asymptote $E^\| \sim -e^2/r_c$. In this case all the states with $m \neq 0$ become unbound. Indeed, since $E^\|$ is small for large $K_\perp$, the binding energy (4.46) becomes negative at $m < 0$, However, at $m = 0$ and arbitrarily large $K_\perp$, there still remains an infinite series of truly bound states (enumerated by $\nu$).

Since $r_c = a_0^2 K_\perp/\gamma\hbar$, the decentered states at $\gamma \lesssim 1$ have huge sizes. Hence, they are likely destroyed by collisions with surrounding particles in laboratory and in atmospheres of magnetic white dwarfs. However, in neutron star atmospheres at $\gamma \gtrsim 10^3$ the decentered states may be significantly populated, as will be shown below.

Photoionization of an H atom moving in a strong magnetic field was studied by Bezchastnov & Potekhin (1994) and Kopidakis *et al.* (1996), using different modifications of the adiabatic approximation. A complete numerical treatment beyond the adiabatic approximation has been developed by Potekhin & Pavlov (1997). These authors showed that none of the versions of the adiabatic approximation can provide accurate photoionization cross sections for all values of $K_\perp$ and for any photon polarization, particularly because the continuum-channel coupling strongly affects the absorption of circularly polarized photons at sufficiently large $K_\perp$.

For astrophysical applications, it is useful to have analytical fits to the binding energies, quantum-mechanical sizes, and main oscillator strengths of moving H atoms. Such fits have been constructed by Potekhin (1998) for $7\times10^{11}$ G $\leq B \lesssim 3\times10^{13}$ G. Figure 4.5 demonstrates the dependence of binding energies of the hydrogen atom on $K_\perp$.

For field strengths outside the above range, reliable fitting formulae are absent. At stronger magnetic fields, tables of binding energies and all relevant

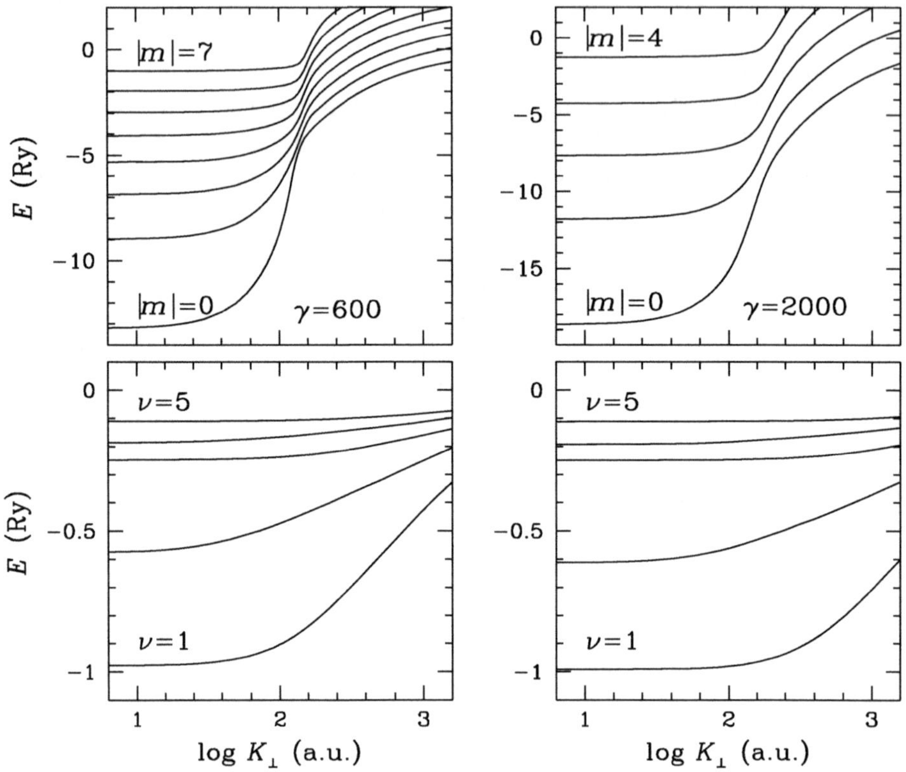


*Figure 4.5.* Energy spectrum of the hydrogen atom moving across a strong magnetic field as a function of the transverse component of pseudomomentum in atomic units (a.u.).

bound-bound oscillator strengths and bound-free cross sections were calculated by Potekhin & Chabrier (2004). For weaker magnetic fields, the binding energies and bound-bound transition rates of a moving H atom were first calculated by Lozovik & Volkov (2004).

Finally, let us mention that the coupling of center-of-mass and relative motions of an electron and a nucleus in a strong magnetic field affects also the probabilities of free-free transitions and corresponding opacities, especially if the radiation frequency is $\omega \lesssim \omega_{\rm ci}$, where $\omega_{\rm ci}$ is the ion cyclotron frequency (see, e.g., Potekhin & Chabrier 2003).

### 4.2.4 Magnetic condensation

Ruderman (1971) pointed out that polymer chains aligned with the strong magnetic fields should attract one another because of the quadrupole-quadrupole interactions, and eventually form a solid. The magnitude of such an interaction for hydrogen chains was estimated by Lai & Salpeter (1997) who concluded that hydrogen may form a solid stellar surface at superstrong fields ($B \gg 10^{13}$ G).

For iron and other heavy elements, the situation is very uncertain. Hartree-Fock studies cited above, as well as earlier DFT calculations (Jones, 1986) indicated that the stellar surface may be diffuse even at zero temperature, contrary to Thomas-Fermi-like models with a gradient correction (Abrahams & Shapiro, 1991a; Fushiki *et al.*, 1989; Rögnvaldsson *et al.*, 1993). However, finite-temperature corrections, introduced in these models, transform the solid surface into a diffuse atmosphere (Abrahams & Shapiro, 1991a; Thorolfsson *et al.*, 1998).

Lai (2001) formulated approximate conditions for the magnetic surface condensation. According to his study, the critical condensation temperature increases with the growth of the magnetic field strength, and can be as high as $10^6$ K (for a Fe surface at $B \sim 10^{13}$ G or an H surface at $B \sim$ a few $\times\, 10^{14}$ G). This means, in particular, that the thermal radiation of a neutron star can directly emerge from the degenerate metallic condensed surface, without passing through a gaseous atmosphere. According to Lai (2001), the zero-pressure density of the condensed matter can be estimated as

$$\rho_s \simeq 560\,\eta\,A\,Z^{-3/5}\,B_{12}^{6/5}\ \mathrm{g\ cm^{-3}}, \tag{4.55}$$

where $\eta$ is an unknown correction factor of the order of unity ($\eta = 1$ corresponds to the uniform electron gas model in the Wigner-Seitz approximation). The estimate Eq. (4.55) agrees, within a factor of a few, with the typical density of the plasma phase transition for strongly magnetized hydrogen (Potekhin *et al.*, 1999b; Potekhin & Chabrier, 2004). Medin & Lai (2006b) performed DFT calculations of the cohesive energies and work functions for zero-pressure condensed hydrogen, helium, carbon, and iron at $10^{12}$ G $\leq B \lesssim 10^{15}$ G. For instance, they found that the cohesive energy per carbon atom ranges from $\sim 50$ eV at $B = 10^{12}$ G to 20 keV at $10^{15}$ G. The cohesive energy per iron atom varies from $\sim 0.8$ keV at $B = 10^{13}$ G to 33 keV at $10^{15}$ G.

Calculations of the dielectric tensor and the thermal radiation spectrum of a strongly magnetized condensed surface were attempted previously by several research groups (e.g., Brinkmann 1980; Turolla *et al.* 2004). The most accurate calculations were done by van Adelsberg *et al.* (2005) and Pérez-Azorín *et al.* (2005).

## 4.3. Models of strongly magnetized outer envelopes of neutron stars

There are three main effects of a strong magnetic field on outer envelopes of neutron stars.

First, the bottom density of the photosphere greatly increases due to the reduction of radiative opacities. Since the same magnetic field that reduces the opacities also suppresses the electron degeneracy, the atmosphere remains typically nondegenerate in spite of this density increase. Consequently, at a given density, the pressure in the magnetized atmosphere is considerably lower than in the non-magnetic one.

Second, the increase of atomic binding energies tends to lower the ionization degree, as first guessed by Cohen *et al.* (1970). Hence, there can be a significant amount of bound species in a highly magnetized atmosphere, even if it was negligibly small at the same temperature in the field-free case.

Third, the quantum-mechanical effects of atomic and ionic thermal motion across the field (§ 4.2.3) have a great impact on the EOS. In general, the motion lowers the binding energies but increases the statistical weight of the atoms. The net effect on the ionization degree results from a balance of these two factors.

Let us mention that a strong magnetic field affects convection in neutron star envelopes. The super-adiabatic gradient (2.41) is a necessary but not sufficient condition for the convective instability. Miralles *et al.* (1997) argue that magnetic fields $B \sim 10^{11}$–$10^{13}$ G stabilize the atmosphere against convection. The influence of a low magnetic field on the convection remains questionable. Rajagopal & Romani (1996) suggest that even the lowest known neutron-star magnetic field ($\sim 10^8$ G) should strongly suppress convection, while according to Miralles *et al.* (1997) there may be rapid convective flows in neutron star envelopes at $B \lesssim 10^9$ G. Moreover, according to Urpin (2004, 2005), even a strongly magnetized neutron star ocean can be unstable, if the temperature varies along the surface.

## 4.3.1 Strongly magnetized iron envelopes

In the low-density regime, where the electrons are nondegenerate, a model of a magnetized iron atmosphere was developed by Rajagopal *et al.* (1997) who generalized the equation of ionization equilibrium given by Khersonskii (1987a):

$$\frac{n_j}{n_{j+1}} = \frac{n_e \lambda_e^3}{2} \frac{\sinh \zeta_j}{\zeta_j} \frac{\zeta_{j+1}}{\sinh \zeta_{j+1}} \frac{\tanh \zeta_e}{\zeta_e} \frac{\mathcal{Z}_{\mathrm{int},j}}{\mathcal{Z}_{\mathrm{int},j+1}} \exp\left(\frac{E_{j,\mathrm{ion}}}{k_\mathrm{B} T}\right). \quad (4.56)$$

This equation differs from the non-magnetic Saha equation (2.171) by the three ratios containing sinh and tanh. In these factors, $\zeta_j \equiv \hbar\omega_{cj}/2k_\mathrm{B}T$, $\zeta_e \equiv \hbar\omega_\mathrm{c}/2k_\mathrm{B}T$, and $\omega_{cj} = jeB/m_j c$ is the ion cyclotron frequency of an $j$th-ionized atom. These factors naturally come from partition functions of free ions and electrons whose transverse motion is quantized in the equidistant Landau levels. Obviously, for a neutral atom one should set $\sinh \zeta_0/\zeta_0 = 1$. The difference of these factors for the ions and electrons (sinh *versus* tanh) is caused by different treatments of their spin energies in the magnetic field. The electron spin energy $\pm\hbar\omega_\mathrm{c}/2$ provides an additional factor $(e^{\zeta_e} + e^{-\zeta_e})$ to the electron partition function, while analogous factors due to the nuclear spin energies at ionization stages $j$ and $j+1$ are equal and cancel each other in Eq. (4.56). However, the contribution of the nuclear spin energy must be taken into account, for instance, in calculating the specific heat (cf. § 4.1.4).

Constructing the models of magnetized iron atmospheres, Rajagopal *et al.* (1997) took into account the effects of atomic motion in an approximate way, based on the perturbation theory (Pavlov & Mészáros, 1993). Because of the high mass of iron atoms, this can be a reasonable approximation in relatively cold iron atmospheres. Based on the non-magnetic results for the Coulomb non-ideality, Rajagopal *et al.* (1997) estimated the influence of this non-ideality on the EOS and found it to be small (up to $\sim 10\%$) throughout the atmosphere.

The main uncertainty in the EOS of Rajagopal *et al.* (1997) comes from neglecting the effects of plasma non-ideality on bound species (particularly, on the pressure destruction). In order to achieve a convergence of the internal partition functions $\mathcal{Z}_j$, the authors arbitrarily replaced high-lying levels by a series of equidistant levels. This may be severely inaccurate if excited levels are significantly populated.

The EOS of subphotospheric stellar layers composed of heavy elements was calculated by a number of authors in the Thomas-Fermi approximation with gradient corrections. Fushiki *et al.* (1989) performed the first calculations of this kind and showed that the magnetic field strongly affects the EOS of a cold plasma. Using the approximation of zero temperature, they obtained a solid boundary (at $\rho \sim 10^3$ g cm$^{-3}$ for $B \sim 10^{12}$ G). Abrahams & Shapiro (1991a) showed that the boundary can be smeared away by the thermal effects. Rögnvaldsson *et al.* (1993) included into consideration the population of excited Landau levels. Both effects, of finite temperatures and higher Landau levels, were incorporated by Thorolfsson *et al.* (1998). The latter authors showed that the pressure $P_{\rm TF}$, calculated in the Thomas-Fermi approximation, is noticeably suppressed relative to the pressure $P_{\rm id}^{(e)}$ of the ideal electron gas as long as the field is strongly quantizing. However, $P_{\rm TF}$ rapidly approaches $P_{\rm id}^{(e)}$ when higher Landau levels are populated with increasing $\rho$ or $T$.

A comparison of different versions of Thomas-Fermi-like models presented by Abrahams & Shapiro (1991a) demonstrates a large uncertainty of the EOS for outer atmospheric layers, indicating that a more detailed model is required.

Nevertheless, we expect that the existing models give correct order-of-magnitude estimates of the magnetic-field effects. Solid lines in Fig. 4.6 show the finite-temperature Thomas-Fermi EOS of Thorolfsson *et al.* (1998) at two magnetic field strengths and two temperatures. For comparison, by dot-dashed lines we also plot the EOS of fully ionized iron in the same magnetic fields (taking into account the electron degeneracy but neglecting non-ideality), and by dotted lines we plot the EOS of fully ionized iron at $B = 0$. All the effects discussed above are clearly seen. In the strongly quantizing field (i.e., at $\rho \ll 10^4$ g cm$^{-3}$ and $\rho \ll 10^{5.5}$ g cm$^{-3}$ for $B = 10^{12}$ G and $10^{13}$ G, re-

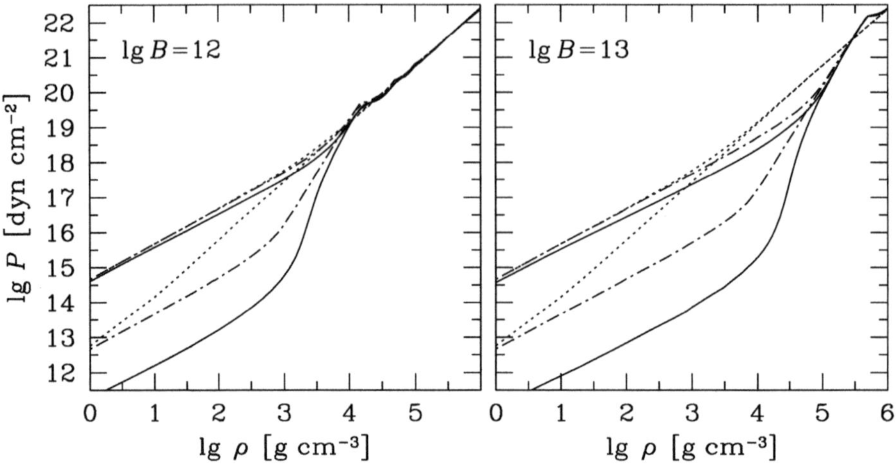


*Figure 4.6.* EOS of iron matter at $B = 10^{12}$ G (*left*) and $10^{13}$ G (*right*). Solid lines refer to the Thomas-Fermi model (Thorolfsson *et al.*, 1998), dot-dashed lines to fully ionized gas, dotted lines to non-magnetic fully ionized gas. A lower line of each type corresponds to $k_{\rm B}T = 10$ eV, ($T = 1.16 \times 10^5$ K) and an upper one to 1 keV ($1.16 \times 10^7$ K)

spectively), the magnetic effects are very pronounced at the lower temperature. The EOS of fully ionized and magnetized matter is much softer than its non-magnetic counterpart because of the delayed onset of the electron degeneracy (see Eq. (4.32) and a discussion therein). Even lower pressure is provided by the Thomas-Fermi EOS, which takes into account electron-ion attraction and, in an approximate way, partial ionization. After $\rho$ reaches and exceeds $\rho_B$, all three types of lines converge. In this case the EOS only slightly differs from the EOS of the ideal non-magnetic electron gas (§ 2.3.1 e). Some oscillations of the pressure around this non-magnetic value are still visible. They reflect consecutive filling of excited Landau states. At such high densities, the field is weakly quantizing.

The higher-$T$ curves reveal the same features but less pronounced, because higher temperatures partly remove the electron degeneracy and destroy bound states. That is why, for example, the upper curves of all three types on the left panel ($B = 10^{12}$ G) nearly coincide.

### 4.3.2 Strongly magnetized hydrogen atmosphere

The effects of strong magnetic fields on the thermodynamics of the atmospheric plasma have been studied most thoroughly for hydrogen atmospheres. The equation of ionization equilibrium was first derived by Gnedin *et al.* (1974), who neglected the quantization of proton motion and and the effects of atomic motion on the internal structure of hydrogen atoms (§ 4.2.3). The next step was made by Khersonskii (1987a). He took into account the quantization of proton motion (the terms containing the proton cyclotron frequency $\omega_{cp}$ in Eq. (4.46)), but neglected the effects of motion on atomic states. Khersonskii (1987b) analyzed also dissociation equilibrium of the $H_2{}^+$ molecular ion (ignoring the effects of motion on its states) and showed that its abundance in a neutron star atmosphere is typically very small.

The importance of effects of motion on atomic quantum states for the ionization equilibrium was fully recognized by Ventura *et al.* (1992) who, however, did not include them in calculations. Lai & Salpeter (1995, 1997) were the first who quantitatively evaluated these effects. Nevertheless, their treatment has a limited applicability since it was based on crude approximations for binding energies of moving atoms and for the pressure ionization (as discussed by Potekhin *et al.* 1999b). A more detailed model will be considered in § 4.3.2 b.

In § 4.3.2 a, following Potekhin *et al.* (1999b), we describe an EOS of partially ionized, strongly magnetized hydrogen plasma constructed in the framework of the chemical picture.

### 4.3.2 a Free energy model

Consider a plasma consisting of $N_e$ electrons, $N_p = N_e$ protons, and $N_{\rm H}$ hydrogen atoms in a volume $V$. The free energy model is a straightforward generalization of the non-magnetic model presented in § 2.5.2,

$$F = F_{\rm id}^{(e)} + F_{\rm id}^{(p)} + F_{\rm id}^{\rm neu} + F_{\rm rad} + F_{\rm ex}^{\rm C} + F_{\rm ex}^{\rm neu}. \qquad (4.57)$$

The free energies of the ideal electron and proton gases, $F_{\rm id}^{(e)}$ and $F_{\rm id}^{(p)}$, are given in §§ 4.1.1 and 4.1.4, respectively. The Coulomb part $F_{\rm ex}^{\rm C}$ has been discussed in § 4.1.5. The standard radiation term $F_{\rm rad}$, Eq. (2.176), can be important only at low $\rho$ or very high $T$. Here we focus on the ideal and non-ideal contributions, $F_{\rm id}^{\rm neu}$ and $F_{\rm ex}^{\rm neu}$, produced by bound species.

The coupling between atomic motion and binding energies precludes the separation of the translational free energy from the internal one [which is quite standard in the non-magnetic case, Eq. (2.35)]. Because quantum-mechanical characteristics of an atom in a strong magnetic field depend on the transverse pseudomomentum in a non-trivial way, the distribution of atoms over $K_\perp$ cannot be written in a closed form. It is only the distribution over $K_z$ that remains Maxwellian. Let us consider bound states at $\gamma \gg 1$. In this case the Landau number $n = 0$, and we do not write it explicitly. Let $p_{m\nu}(K_\perp)\,{\rm d}^2K_\perp$ be the probability to find an atom with given $(m, \nu)$ in an element ${\rm d}^2K_\perp$ near a point $\boldsymbol{K}_\perp$ of the transverse pseudomomentum plane. For the Maxwell distribution, we would have $p_{m\nu}(K_\perp) = (2\pi\hbar)^{-2}\lambda_{\rm H}^2 \exp[-K_\perp{}^2/(2m_{\rm H})]$. Here, $\lambda_{\rm H}$ is the thermal wavelength of the H atom, given by Eq. (2.27). Generally, the number of atoms in an element $\mathrm{d}^3K$ of pseudomomentum space is

$$\mathrm{d}N(\boldsymbol{K}) = N_{m\nu}\,\frac{\lambda_\mathrm{H}}{2\pi\hbar}\,\exp\left(-\frac{1}{k_\mathrm{B}T}\,\frac{K_z^2}{2m_\mathrm{H}}\right)\,p_{m\nu}(K_\perp)\,\mathrm{d}^3K, \tag{4.58}$$

where $N_{m\nu} = \int \mathrm{d}N_{m\nu}(\boldsymbol{K})$ is the total number of atoms with specified discrete quantum numbers. The distribution $N_{m\nu}p_{m\nu}(K_\perp)$ is not given in advance but should be calculated self-consistently via minimization of the total free energy, including the non-ideal terms.

It is convenient to introduce the deviations from the Maxwell-Boltzmann distribution through the occupation probabilities $w_{m\nu}(K_\perp)$:

$$p_{m\nu}(K_\perp) = \left(\frac{\lambda_\mathrm{H}}{2\pi\hbar}\right)^2 \frac{w_{m\nu}(K_\perp)}{\mathcal{Z}_{m\nu}}\,\exp\left(\frac{E_{m\nu}(K_\perp) - E_{m\nu}(0)}{k_\mathrm{B}T}\right), \tag{4.59}$$

$$\frac{N_{m\nu}}{N_\mathrm{H}} = \frac{\mathcal{Z}_{m\nu}}{\mathcal{Z}_\mathrm{int}}\,\exp\left(\frac{E_{m\nu}(0) - E_\mathrm{gr-st}}{k_\mathrm{B}T}\right), \tag{4.60}$$

where

$$\mathcal{Z}_{m\nu} = \frac{\lambda_\mathrm{H}^2}{2\pi\hbar^2}\int_0^\infty w_{m\nu}(K_\perp)\,\exp\left(\frac{E_{m\nu}(K_\perp) - E_{m\nu}(0)}{k_\mathrm{B}T}\right) K_\perp\,\mathrm{d}K_\perp, \tag{4.61}$$

$$\mathcal{Z}_\mathrm{int} = \sum_{m\nu} \mathcal{Z}_{m\nu}\,\exp\left(\frac{E_{m\nu}(0) - E_\mathrm{gr-st}}{k_\mathrm{B}T}\right), \tag{4.62}$$

and $E_\mathrm{gr-st} \equiv E_{00}(0)$.

The number of atoms per a unit phase-space cell equals (e.g., Landau & Lifshitz 1976) $[\mathrm{d}N(\boldsymbol{K})/\,\mathrm{d}^3K]\,(2\pi\hbar)^3/V$. The calculation of $F = U - TS$ for this distribution gives

$$\begin{aligned} F_\mathrm{id}^\mathrm{H} &= \sum_{m\nu} N_{m\nu}\left\{k_\mathrm{B}T\int \ln\left[n_{m\nu}\lambda_\mathrm{H}^3\,w_{m\nu}(K_\perp)\right]p_{m\nu}(K_\perp)\,\mathrm{d}^2K_\perp \right.\\ &\qquad \left. - k_\mathrm{B}T\,(\ln\mathcal{Z}_{m\nu} + 1) + E_\mathrm{gr-st} - E_{m\nu}(0)\right\}. \end{aligned} \tag{4.63}$$

To this expression we should add the contribution of molecules. This has been done by Potekhin *et al.* (1999b) for ground-state $H_2$ molecules, in order to estimate the validity range of the model where molecules are neglected. A more realistic treatment of molecules would require extensive quantum-mechanical calculations (with allowance for the center-of-mass motion and rotation).

The contribution of atoms to the non-ideal part $F_\mathrm{ex}$ of the free energy can be calculated in the hard-sphere approximation using the van der Waals one-fluid model according to Eq. (2.178). An obvious generalization of the sum in Eq. (2.179) includes $\int p_{m\nu}(K_\perp)\,\mathrm{d}^2K_\perp$; the root-mean-square proton-electron distance $l_\kappa = l_{m\nu}(K_\perp)$ was approximated by analytic formulae by Potekhin (1998).

### 4.3.2 b Equilibrium conditions

Minimization of the free energy yields the ionization equilibrium (generalized Saha) equation:

$$\frac{n_{\rm H}}{n_p} = n_e \frac{\lambda_p \lambda_e}{\lambda_{\rm H}^3} (2\pi a_{\rm m}^2)^2 \left[1 - \exp\left(-\frac{\hbar\omega_{cp}}{k_{\rm B}T}\right)\right] \times \mathcal{Z}_{\rm int} \exp\left(\frac{E_{\rm gr-st} + \Delta\mu_{\rm deg}}{k_{\rm B}T}\right), \qquad (4.64)$$

where

$$\Delta\mu_{\rm deg} = \mu_e + \frac{\partial \mu_e}{\partial \ln n_e} - \frac{\partial P_e}{\partial n_e} - k_{\rm B}T \ln(2\pi a_{\rm m}^2 \lambda_e n_e)$$

takes into account the effects of electron degeneracy and population of excited Landau levels. Here, $\mu_e$ and $P_e$ refer to the ideal electron gas (§ 4.1.2), and we have excluded the rest energy $m_e c^2$ from $\mu_e$ in this non-relativistic treatment.

The distributions of atoms over discrete quantum numbers and over transverse pseudomomenta are given by Eqs. (4.60) and (4.59), respectively. We do not present the $K$-dependent occupation probabilities $w_{m\nu}(K_\perp)$ in the latter equations (they are given in the original paper by Potekhin *et al.* 1999b), but discuss the results.

### 4.3.2 c Occupation numbers

Figure 4.7 displays the distribution of atoms over two quantum numbers, $m$ and $\nu$, for $B = 10^{12}$ G at two relatively low densities, $\rho = 0.001$ g cm$^{-3}$ and 0.1 g cm$^{-3}$. The left panel shows the relative occupation numbers for the tightly bound states $\nu = 0$ with different quantum numbers $m$. The distribution is broader for higher density. This apparently surprising feature is easily explained by the presence of the third quantum parameter $K_\perp$. At low densities, the majority of atoms reside in states with large values of $K_\perp$ because of the large statistical weight of such states, which all have $m = 0$ (§ 4.2.1). At higher densities, these strongly decentered states are removed by the excluded-volume (pressure-ionization) effects, and the distribution over $m$ grows broader. Conversely, on the right panel we observe a narrower $\nu$-distribution at higher density, because the excluded-volume effects eliminate hydrogenlike states. Ultimately, at still higher densities, only the ground centered state survives ($m = \nu = 0$, $K_\perp < K_c$).

Figure 4.8 shows the ionization curves at $B = 10^{12}$ G and $T = 10^6$ K. The heavy solid line represents the total fraction of atoms $x_{\rm H} = n_{\rm H}/n_0$ in all quantum states, calculated from Eq. (4.64); $n_0 = n_{\rm H} + n_p$. The thin solid line shows the fraction of atoms in the ground state ($m = \nu = 0$, but any $K_\perp$), and the dashed line shows the fraction of atoms in the centered states ($K_\perp < K_c$, any $m$ and $\nu$). For reference, triangles display the atomic fraction at $B = 0$.

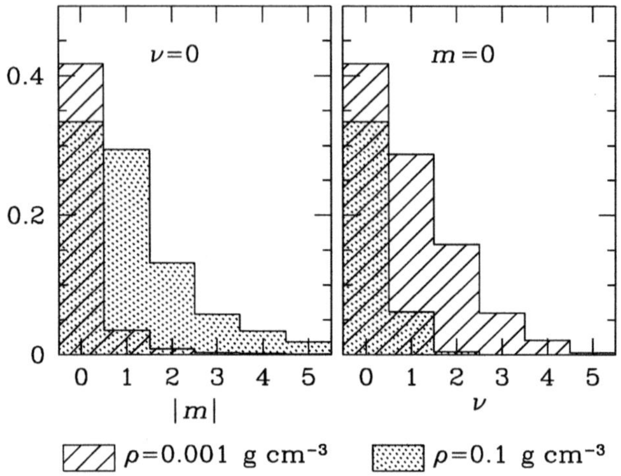


*Figure 4.7.* Distribution of atomic occupation numbers at $T = 10^6$ K for $B = 10^{12}$ G. The distribution over the quantum number $m$ at $\nu = 0$ and over $\nu$ at $m = 0$ is shown for two densities, $\rho = 0.001$ g cm$^{-3}$ (hatched histograms) and 0.1 g cm$^{-3}$ (shaded histograms).

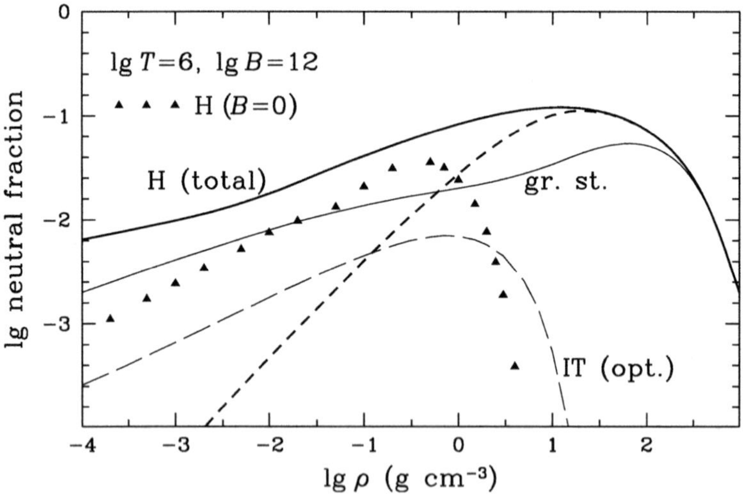


*Figure 4.8.* Ionization isotherms at $B = 10^{12}$ G and $T = 10^6$ K: The total fraction of atoms $x_\mathrm{H} = n_\mathrm{H}/n_0$ (the thick solid line) and the fractions of ground-state (the thin solid line), centered (short dashes), and optically identifiable (Inglis-Teller; long dashes) atoms. For comparison, triangles give $x_\mathrm{H}$ at $B = 0$.

We see that a strong magnetic field generally increases the neutral fraction. The excited and decentered atoms contribute significantly at low densities. Since the effective size of the decentered atoms is proportional to $K_\perp$, the integration in Eq. (4.61) gives roughly $\mathcal{Z}_{m\nu} \propto n_0^{-2/3}$, and, therefore, $x_{00}$ decreases asymptotically as $n_0^{1/3}$. Because of the broadening of the $\nu$-distribution (roughly, $\max \nu \propto n_0^{-1/6}$), the low-density wing of the curve for the total neutral fraction has a slope $x_\mathrm{H} \propto n_0^{1/6}$, which is very moderate compared to $x_\mathrm{H} \propto n_0^{1/2}$ in the non-magnetic case (triangles).

The centered atoms, whose pseudomomentum is bounded from above by the critical value $K_c$, have nearly density-independent internal partition function at low $\rho$. Therefore, their fraction behaves as $x_\mathrm{cen} \propto n_0$, and they disappear much faster at low $\rho$.

At high densities, on the contrary, the decentered states become depleted due to the excluded-volume effects, so that the dashed line in the figure merges with the solid one at $\rho \gtrsim 10$ g cm$^{-3}$. At still higher densities, $\rho \gtrsim 10^2$ g cm$^{-3}$, all excited states disappear, and only the state $m = \nu = 0$ survives. Because of the reduced atomic volume, the pressure ionization occurs at densities $\rho \sim 10^2 - 10^3$ g cm$^{-3}$, orders of magnitude larger than for the field-free case.

Not all neutral atoms that contribute to the EOS may be identified spectroscopically (cf. § 2.5.1). The interaction of plasma species produces a significant fraction of clusters. Such clusters contribute to the EOS similarly to atoms, lowering the pressure, but their radiation-absorption properties are clearly different from those of an isolated atom. Therefore, they should be removed from the neutral fraction used for the calculation of bound-free and bound-bound opacities. Analogously, at low $\rho$ we should not include those highly excited states that disobey the Inglis & Teller (1939) criterion for the dissolution of spectral lines under the Stark effect. The "dissolved" lines form an "optical continuum." The fraction of atoms that are not strongly perturbed by microfields and, therefore, do exhibit spectral characteristics of isolated atoms can be estimated by generalizing Hummer-Mihalas's occupation probabilities (Hummer & Mihalas, 1988) to the case of strong magnetic fields. At every $s$, $\nu$, and $K_\perp$, we calculate the "optical" occupation probability $w^\mathrm{o}_{\nu s}(K_\perp)$, replacing the Inglis-Teller criterion by an approximate criterion based on the average atomic size [Eq. (14) of Pavlov & Potekhin 1995]. Weakly perturbed atoms, which contribute to the bound-bound and bound-free opacities, constitute a fraction $w^\mathrm{o}_{\nu s}(K_\perp)/w^\mathrm{t}_{\nu s}(K_\perp) < 1$ of the total number of atoms. Here, $w^\mathrm{t}_{\nu s}(K_\perp)$ is the "thermodynamic" occupation probability derived from the free energy. In Fig. 4.8 this "Inglis-Teller" fraction is shown by the long-dashed line. Its rapid decrease at high densities indicates that atomic spectral features disappear at $\rho \sim 10$ g cm$^{-3}$, long before the pressure ionization (cf. § 2.5.2).

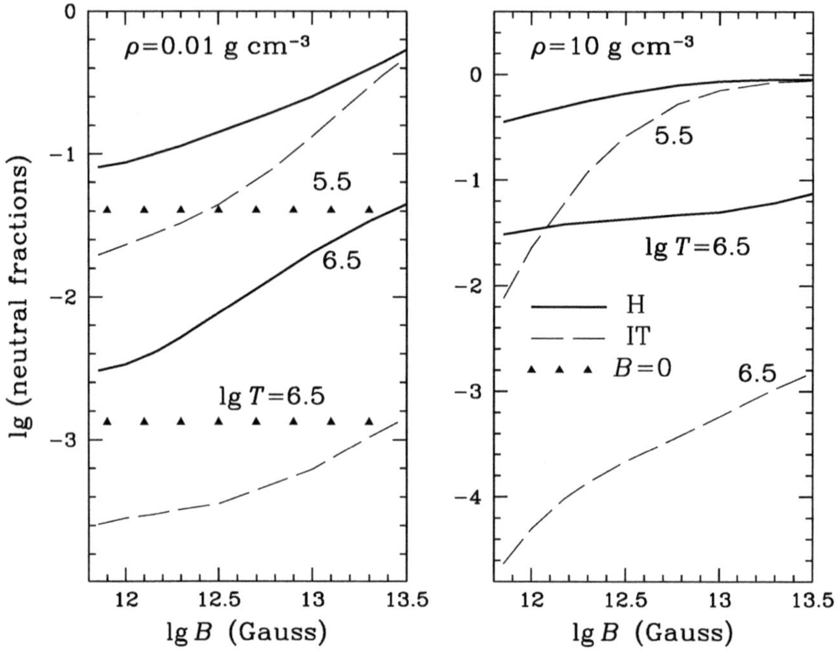


*Figure 4.9.* Dependence of the atomic fraction $x_{\rm H}$ (solid lines) and the fraction of optically identifiable atoms (dashed lines) on the magnetic field strength for $\rho = 0.01$ g cm$^{-3}$ *(left)*, $\rho = 10$ g cm$^{-3}$ *(right)*, $T = 10^{5.5}$ K (upper curves), and $10^{6.5}$ K (lower ones). Triangles show $x_{\rm H}$ at $B = 0$.

The $B$-dependence of the atomic fraction at two values of $T$ and two values of $\rho$ is shown in Fig. 4.9. The total $x_{\rm H}$ is drawn by solid lines, and the "optical" (Inglis-Teller) fraction by dashed lines. Triangles on the left panel show the total fraction of atoms at $B = 0$ (this fraction is negligible at $\rho = 10$ g cm$^{-3}$ on the right panel).

It was stated in early papers (Gnedin *et al.*, 1974; Khersonskii, 1987a; Miller & Neuhauser, 1991) that the ionization degree decreases with growing $B$ above $\sim 10^{12}$ G only at $T \lesssim 5 \times 10^5$ K but increases at higher $T$. However, Fig. 4.9 does not reveal such an increase. This is an effect of moving atoms, neglected in the early papers. First, the growth of $B$ increases the effective mass $m^\perp$ and thus the statistical weight of the centered atoms. Second, at low densities the internal partition function grows further under the effect of decentered states.

### 4.3.2 d Equation of state

Figure 4.10 presents four pressure isotherms derived from the free energy model which was described in §4.3.2 a. For comparison, we also show the

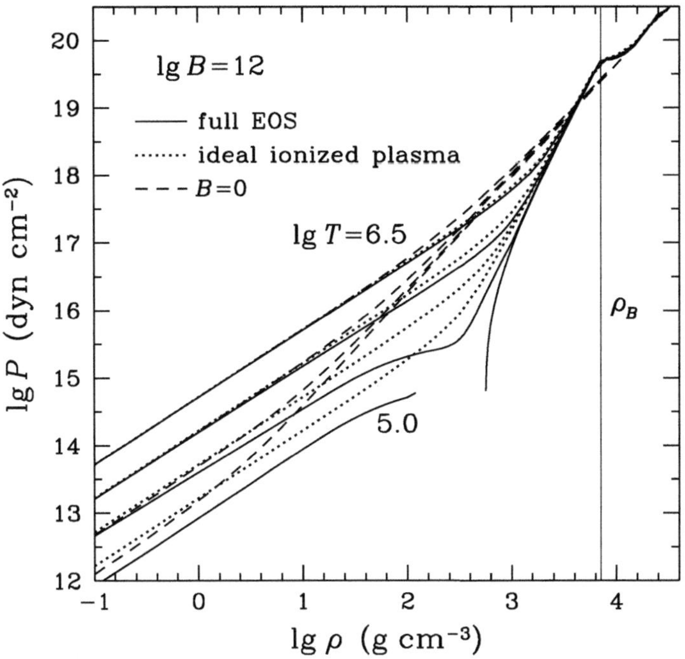


*Figure 4.10.* EOS of partially ionized atomic hydrogen at $B = 10^{12}$ G (solid lines) compared to the EOS of fully ionized ideal electron-proton plasma at the same $B$ (dotted lines) and the EOS of partially ionized hydrogen at $B = 0$ (dashed lines). The temperature logarithms are (from top to bottom) $\log T[\mathrm{K}] = 6.5, 6.0, 5.5$, and $5.0$. The vertical line shows the density $\rho_B$, above which excited Landau levels become populated.

EOS of magnetized, fully ionized ideal gas (§ 4.1.2) and the non-magnetic EOS (§ 2.5.2). The vertical line bounds the region $\rho < \rho_B$. Let us first discuss the low-density regime, $\rho \lesssim 10$ g cm$^{-3}$. At $T \gtrsim 10^6$ K, all three EOSs converge to $P = (n_0 + n_e)\, k_B T = 2 n_0 k_B T$. At lower temperatures, the pressure deviates from this law because of partial recombination of atoms. As discussed in the previous section, the strong magnetic field increases the neutral fraction. Therefore, the pressure is reduced more strongly compared to the field-free case.

In the intermediate density range, $10$ g cm$^{-3} \lesssim \rho \lesssim \rho_B$, the differences among the three considered cases are most important. For $B = 0$, the plasma is fully ionized in this region, and the electrons become partially degenerate, making the EOS stiffer. In the strong magnetic field, the electron degeneracy is reduced (§ 4.1). Hence the ideal-gas EOS is softer than at $B = 0$, except for the densities approaching $\rho_B$, where the degeneracy sets in, and the pressure grows rapidly. The partial recombination and Coulomb non-ideality further decrease

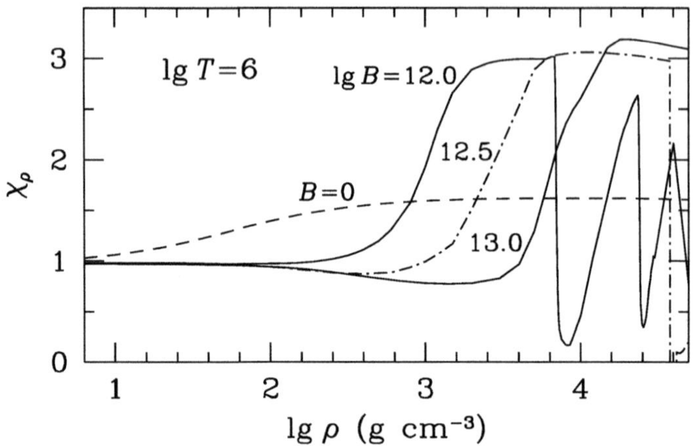


*Figure 4.11.* Density exponent $\chi_\rho = (\partial \ln P/\partial \ln \rho)_T$ at $T = 10^6$ K without magnetic field (dashed line) and in strong magnetic fields of various strengths (dot-dashed and solid lines). After Potekhin *et al.* (1999b).

$P$. The pressure ionization discussed above has two opposite effects. One is the positive contribution of free electrons, produced in the course of the ionization, and the positive non-ideal pressure of neutral species. The opposite effect is the negative Coulomb contribution. At low $T$, these effects may cause the thermodynamic instability ($\partial P/\partial \rho < 0$) leading to a phase transition, which we observe on the isotherm $T = 10^5$ K. The second lowest isotherm in the figure is slightly over the critical one. This is a complete analogue to the plasma phase transition, which is a first-order phase transition from the low-ionization state to a high-ionization state of the matter. In our model, it is caused by a strong Coulomb attraction between electrons and ions. The attraction produces a negative contribution to the pressure, that cannot be compensated at low temperatures until the degeneracy sets in. One should not forget, however, that the plasma phase transition is model-dependent and its reality has not been proven.[1]

At higher densities $\rho \gtrsim \rho_B$, excited Landau levels become populated due to the increase of the Fermi energy. Eventually, at $\rho \gg \rho_B$, the non-magnetic EOS is recovered.

Figure 4.11 demonstrates the effects of strong magnetic fields on the density exponent $\chi_\rho = (\partial \ln P/\partial \ln \rho)_T$. Although the pressure approaches the

[1]At $B = 0$, the plasma phase transition was first suggested independently by Wigner & Huntington (1935) and Landau & Zeldovitch (1943). Later it was predicted by several theoretical models (Norman & Starostin, 1968; Ebeling & Richert, 1985; Saumon *et al.*, 1995), but never confirmed in experiment.

non-magnetic value at $\rho > \rho_B$, the effects of magnetic quantization remain quite prominent for the derivative $\chi_\rho$, as shown by the curve corresponding to $B = 10^{12}$ G. A sudden fall of $\chi_\rho$ occurs at $\rho = \rho_B$. At higher densities, the field becomes weakly quantizing; it slightly modifies the pressure but strongly affects its derivative. Consecutive population of excited Landau levels causes oscillations of $\chi_\rho$ around the non-magnetic value. This is an example of magnetic oscillations described in §4.1.3.

We have discussed the main effects of strongly quantizing magnetic fields on a partially ionized hydrogen plasma. Other thermodynamic quantities, obtained in the framework of the present model, experience similar dramatic modifications. Certainly, such effects are important for theoretical modeling of neutron star atmospheres.

These new atmosphere models have been constructed by Ho *et al.* (2003) and Potekhin *et al.* (2004). The modelling confirms that partial ionization may strongly affect spectra of thermal radiation of magnetized neutron stars. For hydrogen atmospheres, the effects of atomic motion on atomic structure are crucial and cannot be neglected.

Chapter 5

# NEUTRON STAR CORES: NUCLEONS AND HYPERONS

## 5.1. Introduction

As we have seen in § 3.5, nuclei cannot exist at densities exceeding $\sim (1.5 - 2.0) \times 10^{14}$ g cm$^{-3}$. At these densities the matter becomes a uniform plasma of neutrons, protons, and electrons. Up to $\rho \sim 2\rho_0$ its properties can be calculated in a rather reliable way using the methods of the nuclear many-body theory, which have been applied with some success for the microscopic description of ordinary nuclear structure. At such densities the only baryons present in the ground state of the matter are nucleons which are in beta equilibrium with electrons (and muons if the electron Fermi energy exceeds $m_\mu c^2 = 105.7$ MeV). The shell of the neutron-star core with densities $\rho \lesssim 2\rho_0$ is called the *outer core*. Its matter will be called the $npe\mu$ matter. At higher densities our assumption of the $npe\mu$ composition of dense matter can be invalid because the matter may contain hyperons. The derivation of the equation of state (EOS) of the matter composed of nucleons, hyperons, electrons, and muons will be the topic of the present chapter.

As soon as the density significantly exceeds the normal nuclear density, the structure and composition of a neutron-star core becomes more uncertain. Several exotic phases of the matter have been suggested on theoretical grounds. For instance, the matter can be inhomogeneous, containing a condensate of pions or kaons, and acquiring a periodic structure. One can also contemplate a transition to a deconfined quark plasma. For all these phase transitions, the ground state of the matter can be either in a single (pure) phase at a given pressure or in a mixed-phase state, in which the thermodynamic equilibrium is realized by two coexisting phases, their proportion depending on the pressure. There is a plethora of theoretical possibilities concerning the structure of the inner part of neutron star with $\rho \gtrsim 2\rho_0$; this part is called the *inner core*. The hypothetical exotic phases of dense matter and their effect on the EOS of neutron-star core will be studied in Chapter 7.

The EOS of the core at $\rho \gtrsim (10^{14} - 10^{15})$ g cm$^{-3}$ is necessary to construct the models of neutron stars and calculate their maximum mass $M_{\rm max}$. In its greatest generality, the problem can be formulated as follows: we are seeking for the ground state of matter composed of hadrons and leptons. Two charges are conserved and can be fixed while minimizing the energy: the total electric charge, set to zero, and the baryonic charge, $A_{\rm b}$. After fixing the Hamiltonian and the volume of the system, we have to calculate a state with minimum energy per baryon. This state depends only on the average baryon density $n_{\rm b} = A_{\rm b}/\text{volume}$.

Calculation of the EOS at $\rho \gtrsim (10^{14} - 10^{15})$ g cm$^{-3}$ is an intellectual challenge. As soon as the density exceeds the normal nuclear density, a theorist enters an area where the theoretical calculations can be tested exclusively by neutron star observations. The difficulties are legion: one needs a precise solution of a many body-problem for a very dense system with strong and complicated interactions, which are often poorly known. The progress in the calculation of the EOS is intimately related to the development of the many-body theory of nuclear matter. In the next two sections we briefly describe the development of ideas and theories which led to the present models of the EOS of neutron star cores built of nucleons, hyperons, and leptons.

## 5.2. Before the discovery of pulsars: 1932–1967

In his classical paper, written before the discovery of the neutron, Landau (1932) gave an independent derivation of the maximum (Chandrasekhar) mass limit of white dwarfs and considered the final fate of more massive stars. Landau argued that such stars cannot have a stable equilibrium configuration of a white dwarf type and consequently collapse to much higher densities, reaching the densities beyond those of atomic nuclei (see § 1.2). Slightly later, a theoretical study of the composition and EOS of dense stellar matter was presented by Sterne (1933). He studied thermodynamic equilibrium of the matter composed of neutrons, protons, electrons, and atomic nuclei.[1] Using the today terminology, we can say that Sterne tried to determine the EOS of catalyzed matter. He concluded, that in the case of low temperatures, the low-density catalyzed matter is composed of $^{56}$Fe, while at $\rho \gtrsim 2 \times 10^{10}$ g cm$^{-3}$ the matter consists of neutrons only.[2] In this way, it has been realized, on theoretical

[1]Sterne included only a few nuclei, $^{4}$He, $^{12}$C, $^{14}$N, $^{16}$O, $^{56}$Fe, $^{214}$Pb (this last nuclide was called then $^{214}$Ra B).

[2]The most neutron-rich nucleus known was $^{214}$Pb, with $Z/A = 0.383$, so that at $\rho > 2.3 \times 10^{10}$ g cm$^{-3}$ it was energetically advantageous to replace nuclei by free neutrons.

grounds, that the compression of the matter during stellar evolution will induce neutronization. As concluded by Sterne (1933) "... matter at these high densities would be literally squeezed together into a form of neutrons." Because Sterne assumed thermodynamic equilibrium between matter constituents, he could avoid a discussion of the mechanism of nuclear transformations (the Fermi theory of weak interactions, required for studying beta-transformations, was formulated one year later, in 1934).

While Chandrasekhar's studies were restricted to white dwarfs, in one of his papers on this subject he mentions (Chandrasekhar 1935) an unpublished result of J. von Neumann, who "... has shown that the *very* ultimate EOS for matter should *always* be $P = \frac{1}{3}\mathcal{E}$ ...". Now we know that the limiting EOS of nucleonic matter is $P = \mathcal{E}$ (see § 5.15 below), but von Neumann's result is perfectly valid as a strict limit at $\rho \longrightarrow \infty$, just due to the asymptotic freedom of the QCD (§ 7.5). The status of the theory of dense matter by the mid 1930s was reviewed by Hund (1936). In his review, written after Baade & Zwicky (1934a) introduced neutron stars in the realm of astronomy, Hund considered the densities, at which the EOS could be affected by nuclear forces between nucleons. At that time, the range of nuclear forces was estimated as $\sim$ 1–10 fm. Therefore, Hund argued, the nuclear forces become important at such $\rho$, where the distance between nucleons is comparable to the range of nuclear forces. This implies $(m_n/\rho)^{1/3} \sim (10^{-13}–10^{-12})$ cm. Thus, the nuclear forces were expected to become important at $\rho \sim 10^{12} - 10^{15}$ g cm$^{-3}$, a very reasonable estimate from the today perspective.

Oppenheimer & Volkoff (1939) calculated first neutron star models using the exact form of the equations of hydrostatic equilibrium in General Relativity which they derived simultaneously with Tolman (1939; see §§ 1.2 and 6.1) from the Einstein equations. They assumed the EOS of free Fermi gas of neutrons at $T = 0$, getting the maximum mass of neutron stars $0.71\ M_\odot$. This limiting mass was only one half of the Chandrasekhar mass limit for white dwarfs. Therefore, they remarked: "Since neutron star core will not tend to form by collapse of ordinary matter for masses under 1.5 $\mathrm{M}_\odot$ (Landau limit),[3] it seems unlikely that static neutron star core can play any great part in stellar evolution ..." They realized, however, that the actual value of $M_{\rm max}$ might be much higher than $0.7\ M_\odot$, the reason being that "... the EOS we have used so far fails to describe behavior of highly condensed matter." The authors tried to estimate the maximum effect of possible repulsive nuclear forces at $\rho > 10^{15}$ g cm$^{-3}$ respecting the inequality $P < \frac{1}{3}\mathcal{E}$, which at that time was thought to be valid for all physically sound EOSs. The extreme case was $P = \frac{1}{3}\mathcal{E}$ but this did not significantly increase $M_{\rm max}$. Oppenheimer and Volkoff understood that if

[3]That is how they called the Chandrasekhar limit.

$P$ were noticeably larger than $\frac{1}{3}\mathcal{E}$ then $M_{\rm max}$ could be noticeably larger than $0.7\ M_{\odot}$. However, the lack of knowledge of nuclear forces in 1939 prevented them from evaluating the effect of nuclear interaction on the EOS.

Any further progress necessitated better knowledge of nuclear structure and nuclear interactions. A breakthrough in this domain occurred in two decades following the Oppenheimer-Volkoff paper, but it had no impact on the neutron star theory, just because the topic was widely considered as "academic".[4] The studies of the matter at high density were revived by J.A. Wheeler and his collaborators in the mid-1950s. They considered the end-point of the evolution of massive stars, the stability of neutron stars, and the gravitational collapse of massive stellar cores into black holes. Their results were summarized by Harrison *et al.* (1958) who introduced the concept of *cold catalyzed matter* – a state which realizes the minimum energy per baryon at a given $n_{\rm b}$. Their results relevant for neutron-star crusts were mentioned in § 3.3. For the EOS in a neutron-star core, they used the approximation of a non-interacting $npe$ gas. Therefore, the only progress in two decades after the classical work of Oppenheimer & Volkoff (1939) was the inclusion of beta equilibrium. It slightly softened the EOS at $\rho = (10^{14} - 10^{16})$ g cm$^{-3}$, yielding $M_{\rm max} = 0.70\ M_{\odot}$ instead of $0.71\ M_{\odot}$ obtained by Oppenheimer and Volkoff for the model of pure neutron gas. Harrison *et al.* (1958) considered also the density range $\rho = (10^{18}–10^{22}$ ) g cm$^{-3}$, where the nucleons should become ultra-relativistic and $n_p/n_n \longrightarrow 1/8$. Today this result is only of historic interest because we know that nucleons dissolve into a quark-gluon plasma well before such huge densities are reached (§ 7.5).

It was A.G.W. Cameron, a young nuclear theorist at that time, who pointed out the crucial importance of nuclear forces and evaluated their effect on the EOS (Cameron, 1959). Cameron used an effective nucleon-nucleon (NN) interaction constructed by Skyrme (1959).[5] A neutron star matter was approximated by a gas of neutrons interacting via an effective Skyrme-type potential. This interaction was of paramount importance as it increased $M_{\rm max}$ from $0.7\ M_{\odot}$ to $2.0\ M_{\odot}$. This last value seems quite reasonable today, more than four decades later! Cameron (1959) discussed also the qualitative effects of the appearance

[4] An exception was F. Zwicky. A.G.W. Cameron recalling his postdoc academic year 1959-1960 at the California Institute of Technology (Caltech) reminds (Cameron, 1999) that "For years Fritz (Zwicky) had been pushing his ideas about neutron stars to anyone who would listen and had been universally ignored. I believe that part of the problem was his personality, which implied strongly that people were idiots if they did not believe in neutron stars." Cameron was one of a few persons who listened.

[5] The idea of Skyrme was to approximate the effective nuclear interaction in nuclear matter by zero-range (contact) two-body and three-body forces, with additional momentum-dependent terms representing the finite range of the interaction. The numerical parameters were to be determined from the properties of atomic nuclei. These "Skyrme forces" played central role in the development of the many-body theory of nuclear structure after powerful computers became available in early 1970s. Application of this approach to a neutron-star crust is discussed in § 3.3.1.

of muons and hyperons. One can only agree with his conclusion that "... it has been very gratifying to find that the upper limit on the mass of neutron stars is sufficiently large that no difficulty should arise in their formation by sudden condensation in supernova explosions." The difficulty pointed out by Oppenheimer & Volkoff (1939) was therefore removed.

As soon as nuclear forces, tested experimentally in nuclear physics, are introduced into the EOS, the question whether the model keeps its validity at $\rho \gtrsim 10^{15}$ g cm$^{-3}$ should be raised. Is this EOS consistent with basic physical conditions? By 1960, a widely accepted opinion was that an EOS has to satisfy the condition $3P \geq \mathcal{E}$ stemming, as it was thought, from the non-negative character of the trace of the stress-energy tensor $T_i^i = \mathcal{E} - 3P \geq 0$.[6] It was thought that $P$ reaches its upper limit of $\frac{1}{3}\mathcal{E}$ at $\rho \longrightarrow \infty$. The limit corresponds to any ultra-relativistic free gas. In this limit the speed of sound is $v_{\rm s} = c/\sqrt{3}$. It is easy to see that the interaction contribution to the EOS of Cameron (1959) increases so rapidly with growing density that at sufficiently high $\rho$ the inequality $P \leq \mathcal{E}/3$ is violated. However, this could be attributed to a non-relativistic character of the Skyrme forces.

The situation was clarified by Zeldovich (1961) who presented an interaction model which was compatible with special relativity but led to an EOS that violated the inequality $P < \mathcal{E}/3$. In his model, one has $P \simeq \mathcal{E}$ at sufficiently high $n_{\rm b}$ (see § 5.15 below). Simultaneously, $v_{\rm s} \simeq c$.[7]

Cameron (1959) was the first who pointed out that hyperons could be present in neutron star cores. Salpeter (1960) presented the arguments why $\Sigma^-$ should appear at lower density than the less massive hyperon $\Lambda^0$, a feature confirmed later by detailed calculations (see § 5.14). Ambartsumyan & Saakyan (1960) performed the first detailed calculation of the composition and the EOS of a mixture of free degenerate Fermi gases of baryons, mesons, electrons, and muons in full thermodynamic equilibrium. They included all baryons and mesons known at that time. They showed that with the increase of density new hyperons appear successively, increasing the hyperon fraction in dense matter. Hyperons are unstable under terrestrial conditions, but they are stable in a sufficiently dense matter because of the Pauli exclusion principle, which prohibits their decay into nucleons. The lowest threshold density, $0.64$ fm$^{-3}$ ($1.1 \times 10^{15}$ g cm$^{-3}$), was obtained for the $\Sigma^-$ hyperon. According to Ambartsumyan & Saakyan (1960), the core of a massive neutron star consists of an inner hyperon core and an outer nucleon shell. Some effects of strong baryon-baryon interactions on

---

[6]See, e.g., §35 of Landau & Lifshitz (1999) and §27 of Landau & Lifshitz (1993). Notice, however, wise footnotes to these sections stating that "... there is as yet no general proof of this inequality valid for all types of interaction (not only electromagnetic) which exist between particles in Nature."

[7]The Zeldovich model of 1961 ignored the quark structure of hadrons. Contemporary theory of dense hadronic matter predicts the deconfinement of quarks and the asymptotic free-quark gas behavior ($v_{\rm s} \simeq c/\sqrt{3}$) at $\rho \longrightarrow \infty$, see § 7.5.

the EOS of dense matter were considered by Sahakian & Vartanyan (1963). In their model, baryons of species $j$ moved in density-dependent potential wells $U_j$. After deducing the values of $U_j$ from nuclear matter calculations, and making rough estimates of $U_j$ for hyperons, they found a strong effect of $U_j$ on the threshold densities of hyperon creation. For example, $\Sigma^-$ appeared at $n_b = 0.21 \text{ fm}^{-3}$ in their model, to be compared with $0.64 \text{ fm}^{-3}$ in the case of the free Fermi gas.

Tsuruta & Cameron (1965) studied the effect of nuclear interactions on the EOS of dense matter using two phenomenological interaction models, based on the Levinger & Simmons (1961) nucleon-nucleon (NN) potential. The first hyperon ($\Sigma^-$) appeared at $10^{15}$ g cm$^{-3}$. The dependence of the EOS on the assumed baryon-baryon (BB) interaction model was very strong. For a more repulsive potential $V_\alpha$ they got a rather stiff EOS yielding $M_{\max} = 2\,M_\odot$, while for a softer $V_\beta$ potential they obtained $M_{\max} = 1\,M_\odot$. They stated that this difference measured "...the basic uncertainty in the properties of the neutron-star models due to our lack of knowledge of nuclear forces." It is interesting to note an obvious scepticism of Harrison *et al.* (1965) with respect to models of nuclear interactions and introduction of hyperons. At $\rho > 10^{14}$ g cm$^{-3}$, the Harrison-Wheeler EOS (Chapter 10 of Harrison *et al.* 1965) assumed a mixture of free Fermi gases of neutrons, protons, and electrons. In this way, their EOS predicted $M_{\max} = 0.7\,M_\odot$. They wrote: "...to estimate the pressure-density relation in detail is out of the question at this time when it is difficult even to give a theoretical account good to 50 percent of the absolute binding energy of nuclear matter at normal density!" This illustrates their real opinion on the potential of nuclear theory at that time.

It is fair to say that before the discovery of pulsars neutron stars were mostly considered as exotic (very questionable) stellar objects connected with fundamental unsolved problems of nuclear and particle physics and posed by the need to formulate the theory of matter at $\rho = (10^{14} - 10^{16})$ g cm$^{-3}$.[8] It is therefore not surprising that during the long period from 1932 to late 1950s the number of researchers who worked on neutron stars was small; the number of papers was not much larger than the number of years elapsed since the discovery of the neutron. The interest in neutron stars began to raise in the mid 1960s with the advent of the X-ray astronomy.

[8] An example is a fundamental problem of relevance of the quark structure of baryons to the structure of neutron star cores. It was first raised by Ivanenko & Kurdgelaidze (1965, 1969) who considered quark cores of neutron stars. Using thermodynamic (energy) arguments they studied a transition from the matter composed of the ultra-relativistic baryons to the matter built of massive non-relativistic quarks.

## 5.3. After the discovery of pulsars: 1968–2005

The discovery of pulsars and their identification as neutron stars (see § 1.2) changed the problem of the EOS at $\rho = (10^{14} - 10^{16})$ g cm$^{-3}$ from academic to urgent. The dense matter and neutron stars immediately became timely and attractive albeit difficult areas of research. The number of papers increased enormously. Therefore, the review given below is but a very brief account of the progress made in the period from 1968 to 2005. The studies can be subdivided in various ways, but the two basic subdivisions are: according to the physical model and according to the method of solving the many-body problem.

The methods of calculating the ground state of matter at $\rho > 10^{14}$ g cm$^{-3}$ are described in § 5.9. The main methods are based on: perturbation expansion within the Brueckner-Bethe-Goldstone theory (BBG, § 5.9.1), perturbation expansion within the Green's-function theory (GFT, § 5.9.2), variational method (§ 5.9.3), relativistic mean-field model (RMF, § 5.9.4), and effective energy-density functionals (§ 5.9.5).

In the present section, as in the whole chapter, we restrict ourselves to two physical models of neutron star matter. We will start with the simplest ("minimal") model where the basic hadronic constituents are well known from terrestrial physics: neutrons and protons. They form *nuclear matter* of atomic nuclei. Then, we will review models allowing for the presence of hyperons. These baryons are also studied in laboratory. The lightest hyperons form, together with nucleons, *hypernuclear matter* of hypernuclei.

### 5.3.1 Nucleons

At $\rho \sim \rho_0$ the proton and electron fractions are much smaller than 10% (§ 5.12). In the first approximation one can replace neutron-star matter by a purely neutron matter. This EOS has been calculated by many authors. The BBG theory was used by Siemens & Pandharipande (1971) starting from the Reid soft-core NN potential (Reid, 1968) and by Buchler & Ingber (1971) with the Ingber NN potential (Ingber, 1968). A lowest-order (with additional constrains) variational calculation of Pandharipande (1971a), as well as an improved variational scheme of Pandharipande & Bethe (1973), gave systematically higher energy per nucleon than the BBG theory. By the end of 1970s, it became clear that the difference between the BBG and variational results was due to unjustified approximations used in both theories; the results converged after adding most important higher-order terms in the energy per nucleon. The remaining uncertainties resulted mainly from the poorly known three-body (NNN) forces in the nuclear Hamiltonian (see Friedman & Pandharipande 1981; Wiringa *et al.* 1988). The impressive paper of Wiringa *et al.* (1988) and their results on the EOS of neutron matter has remained a standard for other calculations of neutron matter for many years. An unpleasant feature of their EOS was that at $\rho = 3\rho_0$ more than half of the calculated energy per nucleon resulted from poorly known and phenomenologically treated NNN forces (see § 5.5.2).

The next step – the calculation of the EOS of the $npe\mu$ matter in beta equilibrium – could be done easily, using the results obtained for the ground-state energy of asymmetric nuclear matter (uniform neutron-rich mixture of nucleons). A strong dependence on a nucleon-nucleon interaction potential, clearly seen in the beginning of 1970s, significantly diminished as soon as more realistic NN potentials have become available reproducing with high precision more accurate NN scattering data. In the late 1980s the most sophisticated and apparently most reliable calculations of the ground state of the $npe\mu$ matter, based on realistic NN and NNN interactions, implied a rather stiff EOS, with $M_{\max} \simeq (1.9 - 2.1)\ M_\odot$. The 1990s witnessed the progress in two important aspects. First, new NN interactions, which satisfactorily reproduce new very precise and complete NN scattering data, have become available (Stoks *et al.*, 1994; Wiringa *et al.*, 1995; Machleidt *et al.*, 1996). Second, a huge increase of the computational power has enabled one to produce much more precise BBG and variational calculations. The BBG calculations included the effect of the NNN forces, and also allowed one to evaluate the higher-order contributions (three-hole-line diagrams etc., — see, e.g., Baldo & Burgio 2001, and references therein). Moreover, the BBG calculations were frequently performed within a relativistic formulation, with a direct use of the field-theoretic meson-exchange NN interactions. The variational calculations included the leading relativistic effects (boost corrections to the NN interaction) which reduced the effect of the NNN forces on the EOS. The progress consisted also in a more precise treatment of correlations and higher-order terms in the ground-state energy. The advance made in the decade after Wiringa *et al.* (1988) was summarized in the monumental paper of Akmal *et al.* (1998).

Since the 1970s, only a few calculations of the EOS have used the Green's function formalism of the many-body theory (§ 5.9.2). This theory is of great beauty, and has an advantage of a natural relativistic formulation. Therefore, it can also be used with field-theoretic meson-exchange models of strong interactions. However, actual calculations of the EOS using this method have been based on very rough approximations (ladder approximation, § 5.9.2), and have not reached the precision and completeness of the most sophisticated BBG and variational calculations, especially as far as the evaluation of important higher-order corrections is concerned.

A different kind of many-body theory, the RMF model, was formulated in the classical paper of Walecka (1974), and further developed and extended by other authors (e.g., Serot 1977; Boguta & Bodmer 1977; Glendenning *et al.* 1983a,b; see § 5.9.4). Since the 1970s the various versions of the RMF model have been applied to calculations of the EOS. The basic object of the RMF model is an *effective Lagrangian* describing the nucleon interactions via coupling to meson fields. The most complete RMF calculations, based on models consistent with experimentally measured properties of nuclear matter (§ 5.9.4), lead to stiff EOSs, giving typical $M_{max}$ above 2 $M_{\odot}$.

As for the time of this writing (2006), the most precise many-body calculations of the $npe\mu$ matter in beta equilibrium, starting from realistic NN and NNN interactions, give a rather stiff EOS. It is very comforting that the EOSs obtained within the BBG theory and variational theories (basically different many-body theories, employing different up-to-date NN interactions) do not differ much. The largest differences occur at highest $\rho \gtrsim 10^{15}$ g cm$^{-3}$, and result in a scatter of $M_{max} = (1.9 - 2.2)\ M_{\odot}$.

### 5.3.2 Nucleons and hyperons

The inclusion of hyperons has been a great challenge since the very beginning. The first difficulty resulted from the complicated multi-component character of the strongly interacting baryon matter. In principle, it could be overcome using sufficiently powerful computers. The second difficulty was much harder: a poor knowledge of the nucleon-hyperon (NH) interaction and a lack of knowledge of the hyperon-hyperon (HH) interaction. Some researchers have been trying to overcome this problem by exploiting approximate symmetries of strong interactions. Langer & Rosen (1970) used an effective baryon Hamiltonian. They approximated the NH and HH interactions by the NN ones (the same as in Tsuruta & Cameron 1966b) in the same spin and isospin channels. The ground state was calculated in the Hartree-Fock approximation. The calculations included nucleons, $\Sigma$, $\Lambda$, $\Xi$ hyperons, and $\Delta$ resonances; all of them were found present in neutron-star matter at $\rho \gtrsim 1.2 \times 10^{15}$ g cm$^{-3}$. The $\Delta$ resonances were treated as individual particles.[9] $\Sigma^-$ hyperons appeared at an unrealistically low density $\rho \sim \rho_0$.

The first application of the variational method to the multi-component baryonic matter was done by Pandharipande (1971b), who obtained a very soft EOS yielding $M_{max} = 1.4\ M_{\odot}$. Generally, the results were very sensitive to the (very poorly known) hyperonic (NH and HH) interactions and to the calculational scheme (in particular, to the type of correlations included in variational trial wave functions). For example, Pandharipande & Garde (1972) included tensor correlations in trial wave functions and found no hyperons even at $n_b \gtrsim 1$ fm$^{-3}$!

---

[9]The inclusion of $\Delta$ resonances as real constituents of dense matter is not evident. In vacuum, $\Delta$ decays via strong interactions and is a resonant state with spin 3/2 and isospin 3/2, produced in the nucleon-pion scattering (its energy half-width is 120 MeV, about 10% of the mass-energy). In contrast, $\Delta$ resonances may naturally appear as *virtual* particles in intermediate states, contributing to NN and NNN interactions (§§ 5.6 and 5.5.2; particularly, Figs. 5.3 and 5.2). Sawyer (1972a) found a large positive energy shift of the $\Delta$ isobars in dense matter, due to the presence of filled virtual isobar states. According to his calculations, this shift prohibits the appearance of $\Delta$ isobars at $\rho \lesssim 10^{16}$ g cm$^{-3}$.

To a large extent, the differences between the various EOSs of hyperonic matter were due to different NN, NH and HH interactions employed. Using the variational method Bethe & Johnson (1974) studied this interaction-dependence within a set of their five phenomenological baryon-baryon (BB) potentials. They used meson-exchange type models to construct various components of their BB potentials and required that hyperonic potentials should be consistent with experimental data on the properties of hypernuclei. The first hyperon ($\Sigma^-$) appeared at $\rho \sim 3\rho_0$. As demonstrated by Malone *et al.* (1975), the scatter of the stiffness of the Bethe-Johnson EOSs results in the scatter of $M_{\max} = (1.65 - 1.85)\, M_\odot$.

Because of the computational complexity, full self-consistent BBG calculations of the hyperonic matter were not feasible in the 1970s. Moszkowski (1974) used density-dependent $G$-matrices of the BBG theory of asymmetric nuclear matter. Employing the quark model of baryons, he expressed $G_{HN}$ and $G_{HH}$ in terms of $G_{NN}$.[10] Independently of the specific form of his model, $\Sigma^-$ hyperons appeared at a lowest density. However, the composition of the matter at $\rho \sim 10^{15}$ g cm$^{-3}$ was very model dependent. Balberg & Gal (1997) used a phenomenological approach and constructed an energy-density functional for the baryon matter of nucleons, $\Sigma$, $\Lambda$, and $\Xi$ hyperons. The parameters of their model were adjusted to experimental nuclear matter parameters and experimental data on hyperons and their interactions (in hypernuclei). Using their energy-density functionals, Balberg and Gal calculated several EOSs of dense matter. Some of their models included a strongly repulsive component of the $\Sigma$N interaction, suggested by experiments in the 1990s. In these models, $\Sigma^-$ hyperons did not appear in neutron star cores.

Extensive calculations in the framework of the RMF model (§ 5.9.4) were done by Glendenning (1985). The parameters of the nucleon component of the effective Lagrangian were fixed by fitting the experimental parameters of nuclear matter at the saturation point. However, the parameters of the hyperonic component were not strongly constrained by experiments. Therefore, the choice of several sets of phenomenological constants describing the coupling of hyperons to mesons was, to a large extent, based on arguments of strong-interaction symmetry. First hyperons appeared at $\rho \sim 2\rho_0$; at $\rho \sim 5\rho_0$ the hyperons constituted a sizable fraction of baryons and the matter became a "baryon soup". An inclusion of hyperons softened the EOS compared to that containing nucleons alone. Further calculations using various versions of the relativistic mean field model were performed by many authors (see, e.g., Glendenning *et al.* 1992, Knorren *et al.* 1995, Schaffner & Mishustin 1996; § 5.9.4).

[10]Our convention is that Roman symbols N and H are abbreviations for "nucleon" and "hyperon", while italic symbols $N$ and $H$ label baryon species: $N = n, p$ and $H = \Lambda, \Sigma^-, \Xi^0, \ldots$

In the simplest version of the Green's function theory (the ladder approximation, see § 5.9.2) the calculations of the ground-state energy of the multi-component baryonic matter are a straightforward generalization of the nucleon matter case. Calculations of such an EOS were done, e.g., by Weber & Weigel (1989a,b).

In the 1990s, computational power became sufficient to perform *self-consistent* BBG calculations of the EOS for baryon matter containing nucleons and hyperons. Preliminary calculations with two different baryon-baryon potentials were carried out by Schulze *et al.* (1998); they were mostly focused on conditions relevant for terrestrial hypernuclear physics. Calculations of the EOS of baryonic matter in beta equilibrium at densities appropriate for a neutron-star core were done by Baldo *et al.* (2001), Vidaña *et al.* (2000a,b), and Nishizaki *et al.* (2002). Uncertainties of such EOSs resulted mainly from poor knowledge of NH and HH interactions (these interactions were constrained by a limited experimental information on HN scattering and on the physics of hypernuclei). The lack of data on three-body forces involving hyperons contributed additionally to the uncertainty in the EOS.

The best calculations performed by the time of this writing (2006) indicate that the presence of hyperons leads to a significant softening of the EOS of a neutron-star core (with respect to the $npe\mu$ case). This softening implies the lowering of the maximum neutron-star mass $M_{\rm max}$ by about $(0.3-0.5)\ M_\odot$, compared to neutron-star models with purely nucleonic cores. Therefore, we expect, on theoretical grounds, that neutron stars with hyperonic cores have $M_{\rm max} \lesssim (1.8-1.9)\ M_\odot$.

## 5.4. The properties of nuclear matter

Any many-body theory of neutron-star matter has to reproduce empirical data on bulk nuclear matter. Roughly speaking, the nuclear matter is what the heavy atomic nuclei are built of. Strictly speaking, it is an idealized infinite uniform system of nucleons, where the Coulomb interaction is switched off. The notion of the nuclear matter appears naturally within the Liquid Drop Model of nuclei, if we put $E_{\rm Coul} = 0$ and pass to the limit of $A \longrightarrow \infty$. In this limit, the energy per nucleon, $E$, depends only on the neutron and proton densities. It is convenient to express this dependence in terms of the nucleon density $n_{\rm b}$ and the asymmetry parameter $\delta = (n_n - n_p)/n_{\rm b}$, so that $n_n = (1+\delta)n_{\rm b}/2$, $n_p = (1-\delta)n_{\rm b}/2$. Charge symmetry of nuclear forces (see, e.g., Preston & Bhaduri 1975) implies that $E(n_{\rm b}, \delta) = E(n_{\rm b}, -\delta)$, i.e., $E$ does not change if protons are replaced by neutrons and vice versa. The case of $\delta = 0$ corresponds to *symmetric nuclear matter*, while for $\delta = 1$ we are dealing with *neutron matter*. The case of the symmetric nuclear matter is especially simple: in view of charge symmetry of nuclear forces this matter can be treated as a many-body system composed of one kind of particles – nucleons. Small effects of charge-symmetry breaking, like neutron-proton mass difference or charge-symmetry breaking terms in the NN interaction can be neglected (see Haensel 1977, for a detailed discussion of these effects). The symmetric nuclear matter is the simplest approximation of the bulk nuclear matter in heavy atomic nuclei. The effects of small $\delta > 0$ (quadratic in $\delta$ because of charge symmetry of nuclear forces) can be considered as corrections to the leading $\delta = 0$ term. Typically, one has $\delta^2 \lesssim 0.04$, for terrestrial nuclei, and the symmetric nuclear matter is then a reasonable approximation. On the other hand, a pure neutron matter, which is a one-component system, is the simplest approximation of the matter in a neutron star core. However, in general case one should deal with a two-component system.

Since the end of the 1950s, nuclear matter calculations represent the testing ground of nuclear many-body theories (the present status of these theories is described in Baldo & Burgio 2001). Calculations yield the energy per nucleon, $E$, versus the nucleon number density $n_{\rm b}$. In this section, we will not include the nucleon rest energy into $E$. Some examples of $E(n_{\rm b})$ are shown in Fig. 5.1. They are calculated for a specific model of nuclear matter, but their qualitative features are generic.

The minimum of the $E(n_{\rm b})$ curve for symmetric nuclear matter ($\delta = 0$) corresponds to a bound equilibrium state at zero pressure. The values of $E$ and $n_{\rm b}$ at this minimum will be denoted by $E_0$ and $n_0$. Since $P = n_{\rm b}^2 \, {\rm d}E/\,{\rm d}n_{\rm b}$, the dotted segment corresponds to negative pressure and is therefore not interesting. The solid segment gives $E(n_{\rm b})$ for symmetric nuclear matter compressed to a density $n_{\rm b} > n_0$. As clear from Fig. 5.1, $B_0 = -E_0$ is the maximum binding energy per nucleon in nuclear matter. The binding energy per nucleon $B(A, \delta)$ in a self-bound (i.e., bound under zero pressure) system of $A$ nucleons with a nonzero neutron excess parameter $\delta$ will be smaller than $B_0$. The value of $B(A, \delta)$ will tend to $B_0$ from below, if $A \longrightarrow \infty$, $\delta \longrightarrow 0$, and the Coulomb forces are switched off. Simultaneously, the mean number density of the system will tend to $n_0$. This property, resulting from the interplay of the short-distance repulsion and the long-distance attraction in the NN interaction, is called *saturation*; $B_0 = -E_0$ is called the binding energy at saturation, and $n_0$ is the saturation density.

First let us consider the case of small $\delta$ and small $(n_{\rm b} - n_0)/n_0$, characteristic of terrestrial nuclei. Keeping only the quadratic terms,[11] we get

$$E(n_{\rm b}, \delta) \simeq E_0 + S_0 \, \delta^2 + \frac{K_0}{9} \left( \frac{n_{\rm b} - n_0}{n_0} \right)^2 , \tag{5.1}$$

[11]The linear term $E_{\rm a}\delta$ resulting from charge-symmetry breaking in NN interaction can be neglected because of the smallness of $E_{\rm a}$ (Haensel 1977)

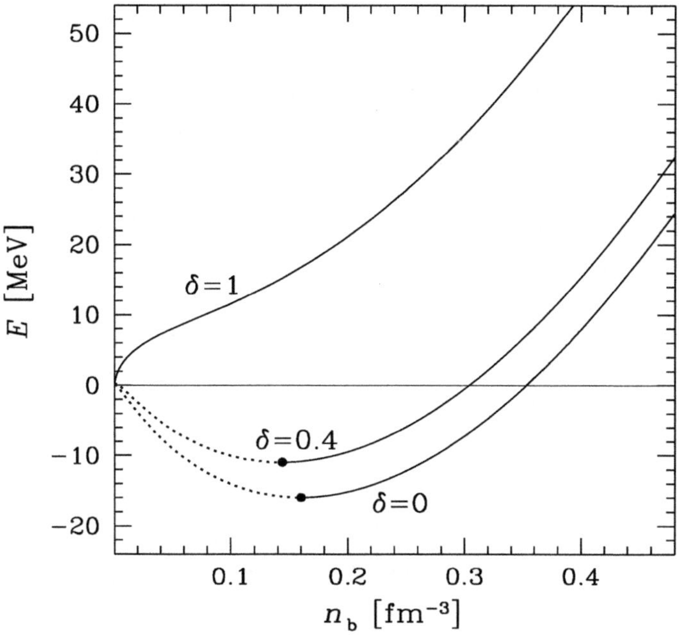


*Figure 5.1.* Energy per nucleon versus baryon number density for symmetric nuclear matter ($\delta = 0$), asymmetric nuclear matter with $\delta = 0.4$ (such an asymmetry corresponds to the neutron-drip point in a neutron star crust and to a central core of a newly born protoneutron star), and pure neutron matter ($\delta = 1$). Minima of the $E(n_b)$ curves are indicated by filled dots. Dotted segments correspond to negative pressure. Calculations are performed for the SLy4 model of effective nuclear Hamiltonian, which was used to calculate the SLy EOS by Douchin & Haensel (2001). It yields $n_0 = 0.16\ \mathrm{fm}^{-3}$ and $E_0 = -16.0$ MeV.

where $S_0$ and $K_0$ are, respectively, the nuclear *symmetry energy* and *incompressibility* at the saturation point,[12]

$$S_0 = \frac{1}{2}\left(\frac{\partial^2 E}{\partial \delta^2}\right)_{n_b = n_0,\ \delta=0}, \qquad K_0 = 9\left(n_b^2 \frac{\partial^2 E}{\partial n_b^2}\right)_{n_b = n_0,\ \delta=0}. \tag{5.2}$$

The symmetry energy $S_0$ determines the increase in the energy per nucleon due to a small asymmetry $\delta$; the incompressibility $K_0$ gives the curvature of the

[12]A traditional factor of nine in the definition of $K_0$ is introduced for historical reasons. In the original definition of $K_0$ the energy per nucleon in the symmetric nuclear matter was treated as a function of a common Fermi momentum (in units of $\hbar$) for neutrons and protons, $k_F$, related to $n_b$ via $n_b = 2k_F^3/(3\pi^2)$. This resulted in $K_0 \equiv (k_F^2 \mathrm{d}E/\mathrm{d}k_F^2)_{k_F = k_{F0}}$ and produced a factor of nine while replacing the derivative with respect to $k_F$ by the derivative with respect to $n_b$.

$E(n_{\rm b})$ curve at $n_{\rm b} = n_0$ and the associated increase of the energy per nucleon of the symmetric nuclear matter due to a small compression or rarefaction.

The values of $B_0$, $n_0$, $S_0$, and $K_0$ can be extracted from experimentally measured nuclear masses. The uncertainties in the empirical values of these parameters result from the ambiguities of semi-empirical mass formulae, used to reproduce thousands of nuclear masses (e.g, Myers 1976; Groote *et al.* 1976; Seeger & Havard 1976; Bauer 1976; Jänecke 1976). One obtains $n_0 = 0.16 \pm 0.01$ fm$^{-3}$ and $B_0 = 16.0 \pm 1.0$ MeV. It is more difficult to extract $S_0$, because of the ambiguity in separating the bulk and surface symmetry terms in binding energies. Thus, the uncertainty of the empirical value is rather large, $S_0 = 32 \pm 6$ MeV. The extraction of $K_0$ from experimental data is even more complicated. Analyses of isoscalar giant monopole modes in heavy nuclei, summarized by Blaizot (1980), suggested $K_0 = 210 \pm 30$ MeV. A more recent analysis, based on precise measurements of the properties of giant monopole resonances in $^{90}$Zr, $^{116}$Sn, $^{144}$Sm, and $^{208}$Pb (excited by inelastic scattering of $\alpha$ particles), yields $K_0 = 231 \pm 5$ MeV (Cavedon *et al.*, 1987; Youngblood *et al.*, 1999). This result is consistent with another recent determination, $K_0 \approx 234$ MeV, by Myers (1998) using a phenomenological Thomas-Fermi model fitted to measured nuclear masses and diffusenesses of the nuclear surface.

The last parameter discussed in this section is the nucleon effective mass, $m^*$, calculated at the Fermi surface in saturated symmetric nuclear matter. It can be evaluated theoretically from the momentum dependence of the nucleon quasiparticle energy $\varepsilon_k$ via

$$(\mathrm{d}\varepsilon_k/\,\mathrm{d}k)_{k=k_{\rm F}} = \hbar^2 k_{\rm F}/m^* \; . \tag{5.3}$$

The notion of a nucleon quasiparticle is naturally introduced while considering low-lying excitations of nuclear matter. Such excitations can be treated in terms of quasiparticles (with number density $\ll n_{\rm b}$) of energy $\varepsilon_k$, where $\hbar k$ is the quasiparticle momentum. The value of $m^*$ enters the density of quasiparticle states (per unit energy and volume) at the Fermi surface (taking into account spin and isospin degeneracies),

$$(\mathrm{d}N_\varepsilon/\mathrm{d}\varepsilon)_{\varepsilon=\varepsilon_{\rm F}} = 2m^* k_{\rm F}/(\pi^2\hbar^2) \tag{5.4}$$

(where $N_\varepsilon$ is the number of states with the energies below $\varepsilon$).

A determination of $m^*$ from nuclear physics experiments is a complicated task, because of the coupling of quasiparticle states with nuclear surface vibration modes. The nuclear matter value of $m^*$ at saturation is evaluated as $m^* \simeq 0.8m$ (see, e.g., Onsi & Pearson 2002 and references therein).

## 5.5. The Hamiltonian

The Hamiltonian should contain contributions of separate constituents of dense matter and their interactions. In principle, it should be given by the quantum chromodynamics (QCD). The electromagnetic interaction is negligible for the EOS, and the weak interaction enters the problem only indirectly by opening some channels for reaching the ground state of the matter. Assuming zero temperature, we should determine the ground state of a system; this is a basic quantum-mechanical problem. In a language of statistical physics we should find the equilibrium (most stable) state at a given $n_{\rm b}$ and $T = 0$.

Unfortunately, this project cannot be realized explicitly. We have to use an effective theory, where quark degrees of freedom are not treated explicitly but are replaced by hadrons – baryons and mesons – in which quarks are confined. As the hadronic Hamiltonian cannot be presently derived from the QCD, we have to use phenomenological models of strong (hadronic) interaction, based partly on mesonic theories, where strong interaction between hadrons is modeled by the exchange of mesons. Most refined and complete phenomenological models have been constructed for the NN interactions. They have been tested using thousands of experimental data on NN scattering cross sections supplemented with experimental deuteron ($^2$H) properties. Experimental information on the NH and HH interactions is restricted to the lowest-mass hyperons $\Lambda$ and $\Sigma$. It is mainly obtained from studies of hypernuclei. As for other hadrons, their interaction models are incomplete and plagued by uncertainties due to scarcity (or non-existence) of experimental data.

As we know now, two-body hadronic interactions yield only a part of the hadronic Hamiltonian of dense matter. At $\rho \sim 10^{15}$ g cm$^{-3}$, interactions involving three and more hadrons are important. Our experimental knowledge of three-body interaction is restricted to nucleons. The three-nucleon (NNN) force is necessary to reproduce properties of $^3$H and $^4$He and to obtain correct parameters of symmetric nuclear matter at saturation.

In view of such a high degree of our ignorance, it seems reasonable to start with a model which is the simplest, and not obviously wrong. Such a "minimalistic" approach consists in extending the $npe\mu$ model to $\rho \gtrsim 2\rho_0$. The calculated EOS has to be confronted with observations, to see whether it is *sufficient* to explain observational data. After fulfilling this minimal program, we can try richer models, including hyperons and exotic phases of hadronic matter. Whatever model of dense matter we assume, we should calculate its ground state as a function of density.

### 5.5.1 Phenomenological nucleon-nucleon interaction

Since the dawn of nuclear physics the determination of forces which bind atomic nuclei has been a central problem for experimentalists and theoreticians.

In its most basic formulation, the problem consists in determining the nucleon-nucleon potential which would explain the nucleon-nucleon scattering data and the properties of $^2$H. It has turned out to be a very difficult task. Bethe (1953) estimated that during the preceding 25 years more hours of human work had been devoted to this problem than to any other scientific problem.

Present phenomenological NN potentials fit very precisely a few thousand of NN scattering data in the energy range up to 350 MeV (in laboratory reference frame). At higher energies, non-elastic processes of pion production switch on and the potential model represented by a Hermitian operator becomes meaningless.

The complexity of the NN interaction was realized very early. In 1939 Rabi and collaborators discovered that $^2$H has a non-zero electric quadrupole moment. It indicated the presence of tensor forces which couple nucleon spins and relative spatial coordinates (Kellogg *et al.*, 1939). Now we know that the NN potential depends on many quantities characterizing a two-nucleon system. These quantities enter via operator invariants consistent with the symmetries of strong interactions. For an $ij$ pair of interacting nucleons these quantities are represented by the following operators: the relative position vector $\boldsymbol{r}_{ij} = \boldsymbol{r}_i - \boldsymbol{r}_j$; spins $\boldsymbol{\sigma}_i$ and $\boldsymbol{\sigma}_j$ (in the units of $\hbar/2$); isospins $\boldsymbol{\tau}_i$ and $\boldsymbol{\tau}_j$ (in the units of 1/2); the relative momentum $\hat{\boldsymbol{p}}_{ij} = \hat{\boldsymbol{p}}_i - \hat{\boldsymbol{p}}_j$; the total orbital angular momentum $\hat{\boldsymbol{L}} = \boldsymbol{r}_{ij} \times \hat{\boldsymbol{p}}_{ij}$ and its square $\hat{L}^2$ in the center-of-mass system. Let us introduce also the operators of the total spin $\hat{\boldsymbol{S}} = \frac{1}{2}(\boldsymbol{\sigma}_i + \boldsymbol{\sigma}_j)$ (in units of $\hbar$) and the total isospin $\hat{\boldsymbol{T}} = \frac{1}{2}(\boldsymbol{\tau}_i + \boldsymbol{\tau}_j)$, which act in spin and isospin spaces, respectively. The tensor coupling enters via the tensor operator

$$\hat{S}_{ij} = 3(\boldsymbol{\sigma}_i \cdot \boldsymbol{n}_{ij})(\boldsymbol{\sigma}_j \cdot \boldsymbol{n}_{ij}) - \boldsymbol{\sigma}_i \cdot \boldsymbol{\sigma}_j \ , \tag{5.5}$$

where $\boldsymbol{n}_{ij} = \boldsymbol{r}_{ij}/r_{ij}$, and the spin-orbit coupling enters via $\hat{\boldsymbol{L}} \cdot \hat{\boldsymbol{S}}$. Both couplings are necessary for explaining experimental data.

The NN potential acting between a nucleon pair $ij$ is a Hermitian operator $\hat{v}_{ij}$ in coordinate, spin, and isospin spaces. The operator $\hat{v}_{ij}$ commutes with $\hat{\boldsymbol{J}} = \hat{\boldsymbol{L}} + \hat{\boldsymbol{S}}$, $\hat{T}^2$, and $\hat{S}^2$, which leads to vanishing matrix elements of $\hat{v}_{ij}$ between states with different $(JST)$. However, because of the tensor force, the $S = 1$ (spin triplet) states with different $L = J \pm 1$ can mix. The Pauli exclusion principle allows only for two-nucleon states with an odd value of the sum $L + S + T$.

The form of $\hat{v}_{ij}$, which is sufficiently general to reproduce the wealth of NN scattering data, is

$$\hat{v}_{ij} = \sum_{u=1}^{18} v_u(r_{ij}) \hat{O}_{ij}^u \ , \tag{5.6}$$

where the first fourteen operators are charge-independent, i.e., invariant with respect to rotation in the isospin space:

$$\begin{aligned}\hat{O}_{ij}^{u=1,\ldots,14} &= 1, \boldsymbol{\tau}_i \cdot \boldsymbol{\tau}_j \,, \boldsymbol{\sigma}_i \cdot \boldsymbol{\sigma}_j \,, (\boldsymbol{\sigma}_i \cdot \boldsymbol{\sigma}_j)(\boldsymbol{\tau}_i \cdot \boldsymbol{\tau}_j) \,, \hat{S}_{ij} \,, \hat{S}_{ij}(\boldsymbol{\tau}_i \cdot \boldsymbol{\tau}_j) \,, \\ & \hat{\boldsymbol{L}} \cdot \hat{\boldsymbol{S}} \,, \hat{\boldsymbol{L}} \cdot \hat{\boldsymbol{S}}(\boldsymbol{\tau}_i \cdot \boldsymbol{\tau}_j) \,, \hat{L}^2 \,, \hat{L}^2(\boldsymbol{\tau}_i \cdot \boldsymbol{\tau}_j) \,, \hat{L}^2(\boldsymbol{\sigma}_i \cdot \boldsymbol{\sigma}_j) \,, \\ & \hat{L}^2(\boldsymbol{\sigma}_i \cdot \boldsymbol{\sigma}_j)(\boldsymbol{\tau}_i \cdot \boldsymbol{\tau}_j) \,, (\hat{\boldsymbol{L}} \cdot \hat{\boldsymbol{S}})^2 \,, (\hat{\boldsymbol{L}} \cdot \hat{\boldsymbol{S}})^2(\boldsymbol{\tau}_i \cdot \boldsymbol{\tau}_j) \,. \end{aligned} \tag{5.7}$$

The form of $\hat{v}_{ij}$, Eq. (5.6), is still quite restrictive, because apart from the angular-momentum dependent terms, the interaction is *local*, i.e., depends only on $\boldsymbol{r}_{ij}$. Non-local NN interactions derived within meson-exchange models will be discussed in § 5.6.

The terms with $\hat{O}_{ij}^{u=15,\ldots,18}$ are small and break charge independence; they are not invariant with respect to a rotation in the isospin space. The charge independence corresponds to $v_{np}(T=1) = v_{nn} = v_{pp}$, while the charge symmetry implies only that $v_{nn} = v_{pp}$. Modern fits to very precise nucleon scattering data indicate the existence of charge-independence breaking. However, the effect of forces, which break charge-independence, on the energy of nucleon matter is much smaller than uncertainties of many-body calculations, and such forces can be neglected while constructing the EOS.[13]

It is well known that the long-range part of $\hat{v}_{ij}$ results from one-pion exchange and has a range of $\hbar/m_\pi \simeq 1.4$ fm. Therefore, one can write

$$\hat{v}_{ij} = \hat{v}_{ij}^{\pi} + \hat{v}_{ij}^{\rm IS} \,, \tag{5.8}$$

where $\hat{v}_{ij}^{\pi}$ is a one-pion exchange part and $\hat{v}_{ij}^{\rm IS}$ is a phenomenological intermediate- and short-range (IS) component. For a given pair of nucleons ($ij = nn$, $pp$, or $np$), $\hat{v}_{ij}^{\pi}$ involves only operators $\boldsymbol{\sigma}_i \cdot \boldsymbol{\sigma}_j$ and $\hat{S}_{ij}$ (see, e.g., Preston & Bhaduri 1975, §5–11 ).

### 5.5.2 Three-body interaction

Calculations show that the two-body interactions which satisfactorily reproduce NN scattering and $^2$H properties, give the binding energies of $^3$H and $^4$He systematically lower than experimental ones. This indicates the necessity of introducing three-body interaction into the nuclear Hamiltonian. The underbinding of light nuclei can be corrected by introducing three-body forces. As we will see later, four and more-body interactions are not needed.

---

[13] Some charge dependence of nuclear force is already seen on the one-pion exchange level, resulting from the difference between charged- and neutral-pion masses, $m_{\pi^\pm}c^2 = 139.57$ MeV, $m_{\pi^0}c^2 = 134.97$ MeV. Other theoretical predictions of charge asymmetry are related to intermediate- and short-range parts of nuclear interaction and are based, e.g., on the mixing of heavy mesons, like the $\rho - \omega$-mixing (see, e.g., Langacker & Sparrow 1982; Coon & Barrett 1994; Coon & Scadron 1994).

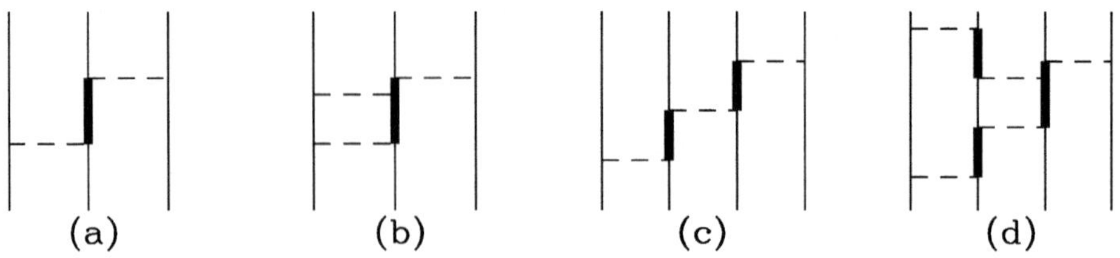


*Figure 5.2.* Some meson-exchange Feynman diagrams describing processes contributing to NNN and four-nucleon interactions. Time goes upwards. Thin vertical solid lines: nucleon states. Thick vertical segments: $\Delta$ resonance in intermediate states. Dashed horizontal lines: exchanged mesons.

On the other hand, it is known that realistic two-body forces saturate the nuclear matter at a too high density. In other words, they produce the minimum in the energy per nucleon versus nucleon density at a density significantly higher than $0.16\,\mathrm{fm}^{-3}$. The three-body interaction can correct this, provided it supplies a repulsion which increases rapidly with the growth of nucleon density.

The three body interaction $\hat{V}_{ijk}$ depends on spatial, spin, and isospin coordinates of three nucleons and cannot be reduced to a sum of two-body interactions involving these coordinates. In the meson theory of nuclear forces, a three body interaction appears naturally at the level of two-pion exchange (§ 5.6.1). This suggests a decomposition used in constructing the so called Urbana models UVII and UIX of the NNN interaction (Schiavilla *et al.*, 1986; Pudliner *et al.*, 1995). In those models, in analogy with the two-body interaction, whose long-range part is given by the one-pion exchange, the NNN interaction between three nucleons $ijk$ is written as

$$\hat{V}_{ijk} = \hat{V}_{ijk}^{2\pi} + \hat{V}_{ijk}^{\mathrm{IS}} \,. \tag{5.9}$$

The two-pion exchange part corresponds to the diagram (a) in Fig. 5.2, while the IS (**I**ntermediate and **S**hort range) part is phenomenological. The two-pion exchange part dominates at low densities and provides an additional binding for the $^3$H and $^4$He nuclei. The IS part is repulsive and supplies a repulsion at higher densities needed to saturate the symmetric nuclear matter at lower density; this part is insignificant in low-density systems such as light nuclei. Phenomenological constants are chosen to give the best overall fit to the binding energies of $^3$H and $^4$He, and to saturation properties of nuclear matter. Older model of the NNN force of Lagaris & Pandharipande (1981b), called TNI (**T**hree **N**ucleon **I**nteraction), did not separate the two-pion exchange part. It was fully phenomenological and explicitly depended on nucleon density and neutron excess. Its parameters were calculated by fitting experimental saturation properties of nuclear matter. The phenomenological character of the IS component of the NNN interaction has practical advantage: in fact, this term mimics additionally the contribution of four-nucleon and higher-order forces in the nuclear Hamiltonian, which turn out to be important at high densities. More sophisticated models of the NNN interaction based on the boson-exchange theory of nuclear forces were constructed by Coon *et al.* (1984); Coelho *et al.* (1983); Coon & Peña (1993). They were used in microscopic calculations of $^3$H and $^4$He. However, because of their purely three-body character and the lack of phenomenological parameters they are not very useful for modelling the high-density nucleon matter.

Some examples of Feynman diagrams (diagrams (c) and (d)), which represent meson-exchange contributions to the four-nucleon interaction, are shown in Fig. 5.2. By analogy with the NNN force, the magnitude of the four-nucleon force can be evaluated by studying $A = 4$ nuclei. Let us consider the interaction part of the nuclear Hamiltonian consisting of the two-body part $\hat{V}^{(2)}$ and the three-body part $\hat{V}^{(3)}$ and assume that it reproduces NN data (NN scattering and deuteron properties) as well as the properties of $A = 3$ nuclei. A natural strategy is to check how well does $\hat{V}^{(2)} + \hat{V}^{(3)}$ reproduce the binding energy of $^4$He. The difference between the calculated and experimental binding energies of $^4$He, $B_{\rm calc}$ and $B_{\rm exp}$, should then be attributed to the four-nucleon force, absent in the adopted Hamiltonian. Such a program was carried out using precise solutions of the four-nucleon problem. The difference $B_{\rm calc} - B_{\rm exp}$ turned out to be negligibly small. For example, for the Argonne AV18 two-nucleon and Urbana UIX three-nucleon forces one gets $|B_{\rm calc} - B_{\rm exp}|/B_{\rm exp} < 1\%$ (Nogga *et al.*, 2000), which means that there is no need to introduce the four-nucleon force on the phenomenological level. This can be contrasted with the NNN force which is clearly *needed* to correct parameters of $A = 3$ nuclei calculated using the NN force alone. Four-body forces in dense matter are indirectly included through a density dependence of the IS component of the NNN force.

### 5.5.3 Relativistic corrections and three-body interaction

The NN potential is non-relativistic. It was fitted to NN scattering data, reduced to the center-of-mass reference frame, and it was constructed in that particular frame with vanishing total momentum of an $ij$ pair, $\boldsymbol{P} = 0$. The interaction in any other frame, where $\boldsymbol{P} \neq 0$, will differ from that in the center-of-mass frame by the so-called *boost interaction* $\delta\hat{v}_{ij}(\boldsymbol{P})$. The relativistic corrections are obtained via expansion in powers of $v/c$, where $v$ is a nucleon velocity. The boost interaction is well known for the Coulomb potential. By including terms $\sim (v/c)^2$, it yields the Breit-Coulomb formula for the electromagnetic interaction of two charged particles (see, e.g., §83 of Berestetskiĭ *et al.* 1982). A general formula for $\delta\hat{v}$, quadratic in $v/c$, was derived by Friar

(1975):

$$\delta\hat{v} = -\frac{\hat{\boldsymbol{P}}^2}{8m^2c^2}\hat{v} + \frac{1}{8m^2c^2}\left[(\hat{\boldsymbol{P}}\cdot\boldsymbol{r})\,(\hat{\boldsymbol{P}}\cdot\nabla),\hat{v}\right] + \frac{1}{8m^2c^2}\left[\left((\boldsymbol{\sigma}_i-\boldsymbol{\sigma}_j)\times\hat{\boldsymbol{P}}\right)\cdot\nabla,\hat{v}\right], \tag{5.10}$$

where $\nabla \equiv \nabla_i + \nabla_j$, $\hat{\boldsymbol{P}} = -i\hbar\nabla$ is the total (nucleon pair) momentum operator, and $[\hat{a},\hat{b}] \equiv \hat{a}\hat{b} - \hat{b}\hat{a}$. Putting the boost term into the nucleon Hamiltonian $\hat{H}_{\rm N}$, one gets a new Hamiltonian $\hat{H}_{\rm N}^*$ which includes all the terms quadratic in the nucleon velocities. However, parameters of the NNN interaction should be refitted to experimental binding energies of $^3$H and $^4$He, and to nuclear matter data. Actually, only the IS part of the NNN interaction is modified by the presence of the boost interaction. The "boost-corrected" $\hat{V}_{ijk}^{\rm IS*}$ turns out to be significantly weaker than $\hat{V}_{ijk}^{\rm IS}$. For example, the strength of the IS component of the UIX* is by $\sim$40% smaller than the strength of the UIX NNN interaction (Akmal *et al.*, 1998). Consequently, the introduction of the boost interaction reduces the contribution of repulsive many-body forces in the EOS at high densities.

## 5.6. Meson-exchange nucleon-nucleon interaction

On the hadronic level, strong NN interaction results from the exchange of mesons between the nucleons. The origin of this description goes back to 1935, when Yukawa (1935) proposed that nucleons interact via exchange of virtual massive particles of the Compton wavelength $\sim 10^{-13}$ cm. In the language of the field theory, strong interactions result from the coupling of the nucleon fields to the meson fields.

Of course, the fundamental theory of strong interactions between hadrons is the QCD, where the fundamental fields are those of quarks and gluons. From the today's perspective, the Meson Exchange Model (MEM) of strong interactions is an *effective theory*, where quarks and gluons do not appear explicitly, and the building blocks are mesons, nucleons, and their resonances (like $\Delta$ isobars). The MEM is successful in describing NN scattering data (at laboratory energies $\lesssim$350 MeV), the $^2$H properties, and properties of "dilute" nucleon systems. The MEM operates with the nucleon and meson fields, $\psi$ and $\varphi$. It is sufficient to include only mesons of rest mass below 1 GeV$/c^2$. Therefore, the MEM does not describe very short-range NN interactions at distances $\lesssim$ 0.2 fm. Meson-nucleon couplings are described by corresponding Lagrangian densities, depending on the symmetry behavior of a meson field under rotations and reflections. As far as the symmetry is concerned, the selected mesons are:

pseudoscalar (field $\varphi^{(\mathrm{ps})}$, mass $m_{\mathrm{ps}}$), scalar (field $\varphi^{(\mathrm{s})}$, mass $m_{\mathrm{s}}$), and vector (field $\varphi_\mu^{(\mathrm{v})}$, $\mu = 0, \ldots, 3$, mass $m_{\mathrm{v}}$).[14]

In a short-hand notation, in which the isospin structure is not indicated, we have the following meson-nucleon coupling Lagrangian densities:

pseudoscalar (ps) mesons $\pi$, $\eta$ $(J^{\mathcal{P}} = 0^-)$ :

$$\mathcal{L}_{\mathrm{ps}} = \frac{f_{\mathrm{ps}}}{m_{\mathrm{ps}}} \overline{\psi} \gamma^5 \psi \varphi^{(\mathrm{ps})} \; ; \tag{5.11}$$

scalar (s) mesons $\sigma$, $\delta$ $(J^{\mathcal{P}} = 0^+)$ : $\mathcal{L}_{\mathrm{s}} = g_{\mathrm{s}} \overline{\psi} \psi \varphi^{(\mathrm{s})}$ ; (5.12)

vector (v) mesons $\rho$, $\omega$ $(J^{\mathcal{P}} = 1^+)$ :

$$\mathcal{L}_{\mathrm{v}} = g_{\mathrm{v}} \overline{\psi} \gamma^\mu \psi \varphi_\mu^{(\mathrm{v})} + \frac{f_{\mathrm{v}}}{4m} \overline{\psi} \sigma^{\mu\nu} \psi \left( \partial_\mu \varphi_\nu^{(\mathrm{v})} - \partial_\nu \varphi_\mu^{(\mathrm{v})} \right) ; \tag{5.13}$$

where $\partial_\mu F \equiv \partial F / \partial x^\mu$, $\sigma^{\mu\nu} = [\gamma^\mu, \gamma^\nu]/2$, $m$ is the nucleon mass, while $J^{\mathcal{P}}$ denotes the meson spin $J$ and parity $\mathcal{P}$.[15] The experimentally measured meson masses are: $m_\pi c^2 = 138$ MeV, $m_\eta c^2 = 548$ MeV, $m_\rho c^2 = 769$ MeV, $m_\omega c^2 = 783$ MeV, and $m_\delta c^2 = 983$ MeV. The scalar $\sigma$ meson plays a special role: it represents a scalar state of an exchanged pion pair, and its mass is found from fitting the MEM to NN scattering data (in this way, one gets $m_\sigma c^2 = 550$ MeV, see below). For the sake of simplicity, we neglect charge splitting of meson masses. Apart from experimental meson masses, the MEM contains coupling constants $f$ and $g$ determined by fitting experimental data. Finally, in order to account for finite sizes of interacting hadrons, one has to introduce form-factors at every meson-nucleon vertex. These form-factors, parameterized in terms of momentum transfer in a meson exchange, are determined by fitting experimental NN data. The form-factors describe the effect of shortest-range strong interactions, which depend on the quark structure of baryons and are not calculable within the MEM.

One-meson exchange processes can be visualized as lowest order (second order in the meson-nucleon coupling constant) Feynman diagrams, e.g., diagram (a) in Fig. 5.3. One-$\pi$ and one-$\omega$ exchange contributions explain two basic features of the NN interaction. The longest range ($\hbar/m_\pi c \approx 1.4$ fm) one-$\pi$ exchange yields a long-range tensor force, while one-$\omega$ exchange produces a

---

[14]Our notations follow Berestetskiĭ *et al.* (1982). We use Greek indices $\mu = 0, 1, 2, 3$ to label time-space components, and Latin indices $k = 1, 2, 3$ to label spatial components. The $2 \times 2$ Pauli matrices are denoted by $\sigma^k$, and the $2 \times 2$ unit matrix is denoted by $I$. The spacetime metric tensor is $g_{\mu\nu} = \mathrm{diag}(+1, -1, -1, -1)$ and the $4 \times 4$ Dirac matrices are: $\gamma^0 = \begin{pmatrix} I & 0 \\ 0 & -I \end{pmatrix}$, $\gamma^k = \begin{pmatrix} 0 & \sigma^k \\ -\sigma^k & 0 \end{pmatrix}$, $\gamma_5 = \gamma^5 = -\mathrm{i}\gamma^0\gamma^1\gamma^2\gamma^3 = \begin{pmatrix} 0 & I \\ I & 0 \end{pmatrix}$. The Dirac conjugate of $\psi$ is $\overline{\psi} \equiv \psi^\dagger \gamma^0$.

[15]Detailed description of the history and the modern version of the MEM is given by Machleidt (1989). However, he uses different from ours definitions of $\gamma_5$ and $\sigma^{\mu\nu}$.

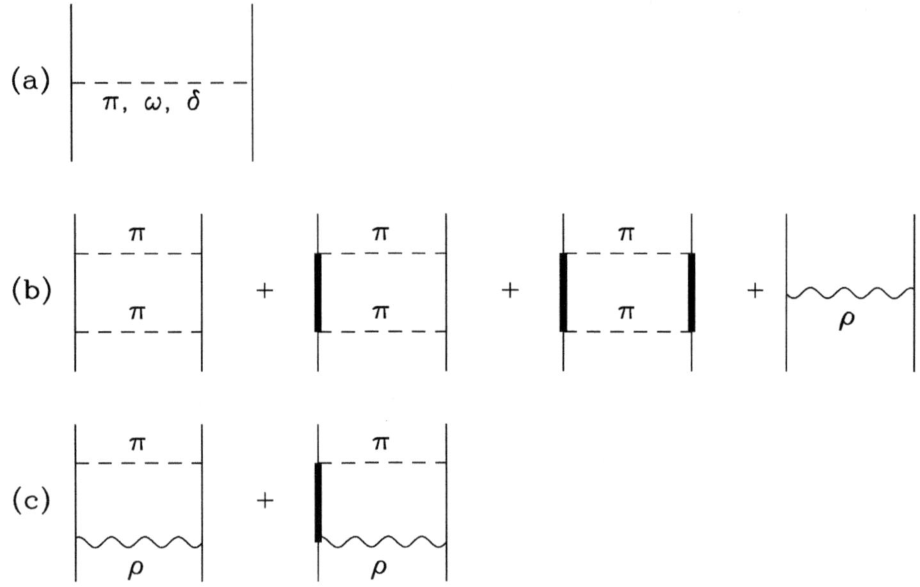


*Figure 5.3.* Some Feynman diagrams describing the most important meson-exchange processes which contribute to the NN interaction. Time goes upwards. Thin vertical lines: nucleons. Thick vertical segments: $\Delta$ resonance in an intermediate state.

short-range ($\hbar/m_\omega c \approx 0.25$ fm) repulsion and a sizable spin-orbit term. The contribution of one-$\delta$ exchange ($\hbar/m_\delta c \approx 0.2$ fm) is relatively small, but is included in the model. The contribution of one-$\eta$ exchange is usually neglected because of its weak coupling to nucleons.

An important intermediate-range attractive component of the NN interaction is produced by two-pion exchange processes ($\hbar/2m_\pi c \approx 0.7$ fm). Some Feynman diagrams contributing to this component in the MEM are shown in Fig. 5.3 (diagrams (b)). Intermediate-state nucleons can be in an excited $\Delta$-resonance state with $m_\Delta c^2 = 1232$ MeV. This is the lowest excited nucleon state (spin 3/2, isospin 3/2). The first three diagrams represent the so called *uncorrelated two-pion exchange* processes in which pions do not interact strongly between themselves. In addition, one has to include diagrams where exchanged pions interact strongly with themselves, forming, e.g., hadronic resonances. An example is given by a right-most diagram (b) in Fig. 5.3, where exchanged pair of interacting (*correlated*) pions forms a vector ($J^{\mathcal{P}} = 1^+$) $\rho$-meson.

Because the meson-nucleon coupling is strong, one should check the contribution of many higher-order processes for their importance in the NN interaction. It turns out that the most important are those involving simultaneous exchange of a pion and a $\rho$-meson. Two Feynman diagrams (diagrams (c)) describing these processes are shown in Fig. 5.3.

The summation of Feynman diagrams yields the scattering matrix which has to reproduce NN scattering data as well as the $^2$H properties. Using a fitting procedure, one determines with a high precision both the meson-nucleon coupling constants and the form-factors.

However, for the convenience of many-body calculation one is interested in representing the NN interaction in the form of a "potential" which would be equivalent to the NN field-theoretic MEM. This is usually done in the frame of the one-boson-exchange (OBE) model, where many-pion exchange contributions are modeled by the exchange of a scalar $\sigma$-meson ($m_\sigma c^2 \approx 550\,\mathrm{MeV}$) with an appropriate $\sigma N$ coupling constant. This fictitious $\sigma$ meson of the OBE model reproduces a very important intermediate-range ($\hbar/m_\sigma c \approx 0.4\,\mathrm{fm}$) component of the NN interaction.

In the OBE approximation one can represent the NN interaction potential as a sum of one-boson exchange contributions,

$$\hat{v}_{ij}^{\mathrm{OBE}} = \sum_{\alpha=\pi,\rho,\eta,\omega,\delta,\sigma} \hat{v}_{ij,\alpha}^{\mathrm{OBE}} . \tag{5.14}$$

This potential can be written explicitly in a relativistically covariant way in momentum space,

$$\left(\boldsymbol{q}'|\hat{v}_{ij}^{\mathrm{OBE}}|\boldsymbol{q}\right) , \quad \boldsymbol{q} = \boldsymbol{p}_i - \boldsymbol{p}_j , \quad \boldsymbol{q}' = \boldsymbol{p}_i' - \boldsymbol{p}_j' . \tag{5.15}$$

The OBE potential is *nonlocal* in coordinate space. Acting on a NN wave function $\Psi$, it gives

$$\left(\boldsymbol{r}|\hat{v}_{ij}^{\mathrm{OBE}}|\Psi\right) = \int \mathrm{d}\boldsymbol{r}' \left(\boldsymbol{r}|\hat{v}_{ij}^{\mathrm{OBE}}|\boldsymbol{r}'\right) \Psi(\boldsymbol{r}') , \tag{5.16}$$

where $\boldsymbol{r} = \boldsymbol{r}_i - \boldsymbol{r}_j$ and $\boldsymbol{r}' = \boldsymbol{r}_i' - \boldsymbol{r}_j'$. The interaction term depends not only on the relative distance between nucleons, but involves its neighborhood and a two-body wave function in this neighborhood.[16]

We may still represent it in an equivalent form of a formally local but momentum-dependent operator $\hat{\mathcal{V}}_{ij}^{\mathrm{OBE}}(\boldsymbol{r}, \hat{\boldsymbol{q}})$, where $\hat{\boldsymbol{q}} = -\mathrm{i}\hbar\nabla_{\boldsymbol{r}}$. Using the well known property of the space-translation operator (see, e.g., §26 of Schiff 1968)

$$\mathrm{e}^{\mathrm{i}(\boldsymbol{r}'-\boldsymbol{r})\cdot\hat{\boldsymbol{q}}/\hbar}\Psi(\boldsymbol{r}) = \Psi(\boldsymbol{r}') , \tag{5.17}$$

[16]To be contrasted with the standard case of a local interaction $\hat{v}_{ij}^{\mathrm{loc}}$, where $(\boldsymbol{r}|\hat{v}_{ij}^{\mathrm{loc}}|\boldsymbol{r}') = v^{\mathrm{loc}}(\boldsymbol{r})\delta(\boldsymbol{r}'-\boldsymbol{r})$, so that $(\boldsymbol{r}|\hat{v}_{ij}^{\mathrm{loc}}|\Psi) = v_{ij}^{\mathrm{loc}}(\boldsymbol{r})\Psi(\boldsymbol{r})$ .

we can rewrite the $\hat{v}_{ij}^{\rm OBE}\Psi$ term in the coordinate representation as

$$\begin{aligned}(\boldsymbol{r}|\hat{v}_{ij}^{\rm OBE}|\Psi) &= \hat{\mathcal{V}}_{ij}^{\rm OBE}(\boldsymbol{r},\hat{\boldsymbol{q}})\Psi(\boldsymbol{r})\;,\\ \hat{\mathcal{V}}_{ij}^{\rm OBE}(\boldsymbol{r},\hat{\boldsymbol{q}}) &= \int {\rm d}\boldsymbol{r}'\hat{v}_{ij}^{\rm OBE}(\boldsymbol{r}'){\rm e}^{{\rm i}(\boldsymbol{r}'-\boldsymbol{r})\cdot\hat{\boldsymbol{q}}/\hbar}\;.\end{aligned} \tag{5.18}$$

We see, that the non-local $\hat{v}_{ij}^{\rm OBE}$ is equivalent to a formally local momentum-dependent $\hat{\mathcal{V}}_{ij}^{\rm OBE}(\boldsymbol{r},\hat{\boldsymbol{q}})$. In the non-relativistic approximation, only the terms quadratic in momenta are retained. In this case, the spin and momentum structure takes a very familiar form, expressed via the operator invariants considered already in § 5.5.1 while constructing a phenomenological NN interaction. The pseudoscalar-meson component $\hat{\mathcal{V}}_{\pi,\eta}$ is a static (momentum-independent) operator involving the $\boldsymbol{\sigma}_i\cdot\boldsymbol{\sigma}_j$, $\hat{S}_{ij}$ operators alone. The scalar-meson contribution $\hat{\mathcal{V}}_{\sigma,\delta}^{\rm OBE}$ contains $\hat{\boldsymbol{q}}^2$ and spin-orbit $\hat{\boldsymbol{L}}\cdot\hat{\boldsymbol{S}}$ operators. The richest spin and momentum structure results from the exchange of $\rho$ and $\omega$ vector mesons. This $\hat{\mathcal{V}}_{\rho,\omega}^{\rm OBE}$ contains all previously listed operators. However, one should keep in mind that the OBE potential is par excellence a relativistic model. Therefore, its exact momentum dependence, that reflects the non-locality in coordinate space, is actually much more complicated.

An OBE model can very well reproduce existing NN data. Recent OBE models fit $\sim 4300$ $pp$ and $np$ scattering cross sections at collision energies below 350 MeV (in laboratory frame). The very high quality of fitting is similar to that reached for NN-interaction potentials constructed by the Nijmegen group (Stoks *et al.* 1994, $\chi^2/{\rm datum} = 1.03$) and the Argonne group (Wiringa *et al.* 1995, $\chi^2/{\rm datum} = 1.09$).

### 5.6.1 Three- and more-nucleon interactions

Many-body interactions arise naturally in the meson-exchange models: they are represented by Feynman diagrams which cannot be reduced to a sequence of NN interactions.

For instance, in diagram (a) of Fig. 5.2 the first meson exchange transforms a nucleon into a $\Delta$ resonance. This diagram does not describe an NN interaction which should have a nucleon pair in the final state. The second pion-exchange process starts with an $N\Delta$ pair,[17] so that it is clearly not an $NN \longrightarrow NN$ process. The presence of the third nucleon is necessary for the whole process to occur, and therefore the diagram describes a genuine NNN interaction. Another example of a Feynman diagram contributing to the NNN interaction is shown in Fig. 5.2 (diagram (b)).

[17] We remind that $N = n$ or $p$, while N is an abbreviation for the word "nucleon".

Diagrams representing the three-body interaction contain an exchange of *at least* two mesons. The NNN interaction resulting from two-pion exchange becomes important in systems where three nucleons can be localized simultaneously within a range $\hbar/m_\pi c \approx 1.4$ fm. Therefore, NNN forces can be significant in nuclear matter at normal nuclear density. Their effect is also visible in the ground-state energy of much less dense few-nucleon systems like $^3$H and $^4$He.

Diagram (a) in Fig. 5.2 leads to a long-range component $V_{ijk}^{2\pi}$ of the NNN potential of § 5.5.2. One can also consider the four-body interaction, generated, e.g., by meson-exchange processes described by Feynman diagrams (c) and (d) of Fig. 5.2. The longest-range component corresponds to the exchange of three pions (diagram (c)).

In practice, one theoretically calculates the long-range two-pion exchange component of the NNN force, but adjusts its strength to reproduce properties of $A = 3$ and $A = 4$ nuclei. In this way one gets the final form of $\hat{V}_{ijk}^{2\pi}$. The intermediate and short-range component of the NNN force is treated phenomenologically, as described in § 5.5.2.

Some examples of Feynman diagrams which represent meson-exchange processes and contribute to the four-nucleon interaction are shown in Fig. 5.2 (diagrams (c) and (d)). As we have discussed in § 5.5.2, there in no need to introduce the four-body force on the phenomenological level to describe the $A = 4$ nuclei. In contrast, the NNN force is clearly *needed* to correct for the underbinding of the $A = 3$ nuclei by the NN force alone and to make theoretical saturation parameters of nuclear matter consistent with experimental ones.

## 5.7. The hyperon interactions

Experimental data on nucleon-hyperon (NH) and hyperon-hyperon (HH) interactions are scarce and rather imprecise, in sharp contrast with highly precise and complete NN data. Only a few points of NH scattering cross sections are available, while HH scattering data are absent. Many bound-state energies of hypernuclei containing a single $\Lambda$ hyperon have been measured. These data are important for studying the $\Lambda$N interaction. Particularly useful are the data on the $A = 3$ and the $A = 4$ hypernuclei, because they allow one to check precise solutions of the three- and four-body problem. The data on ${}^{3}_{\Lambda}$H, ${}^{4}_{\Lambda}$H, and ${}^{4}_{\Lambda}$He hypernuclei can be used to test the N$\Lambda$ interaction *in vacuum*. Models of *in medium* (effective) $\Lambda\Lambda$ interaction can be tested in double-$\Lambda$ hypernuclei, such as ${}^{6}_{\Lambda\Lambda}$He, ${}^{10}_{\Lambda\Lambda}$Be, and ${}^{13}_{\Lambda\Lambda}$B. As we will see, the $\Lambda$N interaction in hypernuclei is responsible for an important effect of the so called $\Lambda$–$\Sigma$ conversion. Consequently, data on hypernuclei yield also an indirect information on the N$\Sigma$ and $\Lambda\Sigma$ interactions. Studies of ${}^{4}_{\Sigma}$He give a valuable information on the N$\Sigma$ interaction. All in all, one can experimentally test models of BB interaction for the baryon-pair strangeness $S = 0, -1, -2$.

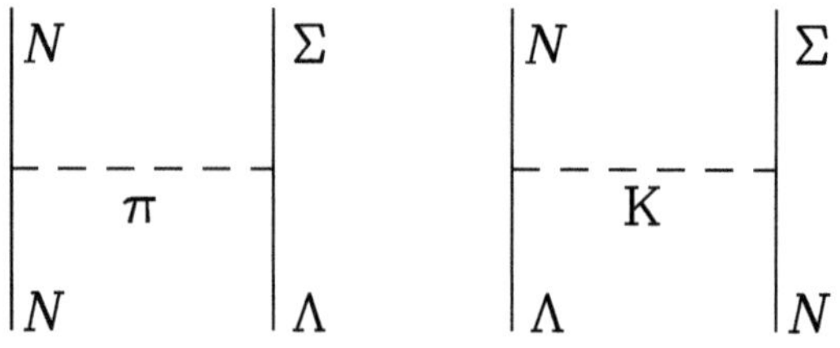


*Figure 5.4.* Two Feynman diagrams describing strong-interaction one-meson-exchange processes accompanied by $\Lambda$–$\Sigma$ conversion. Notice that $K$ meson in the right-hand-side diagram transfers strangeness.

Unfortunately, there are no data on $\Xi$ N, $\Xi\Lambda$, $\Xi\Sigma$, and $\Xi\Xi$ scattering. Even worse, there are no data on double-hypernuclei containing $\Xi\Lambda$ or $\Xi\Xi$ pairs, which could supply information on interactions of baryon pairs with $S = -3$ and $S = -4$. Under such unfavorable experimental situation, one has no choice but to extend BB interaction models from the experimentally available $S = 0, -1, -2$ sectors to the $S = -3, -4$ sectors using the SU(3) symmetry within the $J^{\mathcal{P}} = \frac{1}{2}^+$ baryon octet. This symmetry is experimentally known to be inexact. In order to construct realistic models one has to introduce its breaking. The program along these lines was initiated in the late 1970s (Nagels *et al.*, 1977, 1979). The BB interaction is based on the OBE model (see § 5.6). It was necessary to enlarge the set of exchanged mesons by including kaons (Maessen *et al.*, 1989; Rijken *et al.*, 1999; Reuber *et al.*, 1997). At the quark level, the SU(3) breaking makes the $s$-quark mass much larger than the masses of $u$ and $d$ quarks. At the hadronic level, the SU(3)-breaking effect in BB interaction is significant ($\sim$ 20% of the interaction strength).

Above an energy threshold, BB interaction opens the possibility of baryon conversion, for instance,

$$\Lambda + p \longrightarrow \Sigma^+ + n, \qquad \Lambda + p \longrightarrow \Sigma^0 + p. \tag{5.19}$$

Some examples of one-boson-exchange Feynman diagrams corresponding to conversion channels are shown in Fig. 5.4. The BB interaction becomes multichannel; it can be represented by a matrix $(B_1 B_2 | \hat{v}_{ij} | B_3 B_4)$. The importance of the $\Lambda$–$\Sigma$ conversion results from a relatively small mass difference $(m_\Sigma - m_\Lambda)c^2 \approx 74$ MeV which is much smaller than $(m_\Delta - m_N)c^2 \approx$ 293 MeV.

Calculations of the energy difference between the $0^+$ ground state and $1^+$ excited state in ${}^4_\Lambda$H and ${}^4_\Lambda$He with modern BB potentials lead to a systematic underbinding (Akaishi *et al.*, 2000; Nogga *et al.*, 2002). This could be attributed to an attractive $\Lambda$NN force (Akaishi *et al.*, 2000).

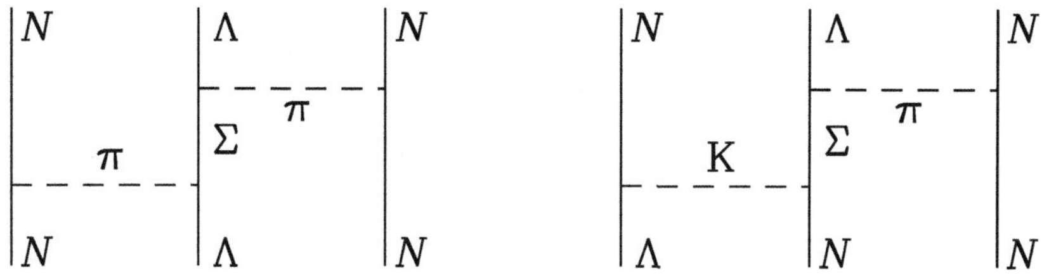


*Figure 5.5.* Feynman diagrams which describe second-order three-body correlations involving two-body interactions in dense matter; they do not contribute to genuine three-body ΛNN interaction.

Let us remind, that lowest-order Feynman diagrams, which generate an NNN force, involve $\Delta$ resonance in an intermediate state (Fig. 5.2). At the first glance, the lowest order $\Lambda$NN force should be represented by a diagram obtained from Fig. 5.2a by replacing $\Delta$ resonance by $\Sigma$ hyperon. This would lead to diagrams represented in Fig. 5.5. However, these diagrams describe two consecutive acts of two-body interaction with $\Lambda$–$\Sigma$ conversion. Therefore, Fig. 5.5 displays second-order three-body correlations in dense matter resulting from two-body interactions.

## 5.8. Solving the many-body problem – an overview

The basic formula for the ground-state energy per baryon of a system of $A_{\rm b}$ baryons is

$$E_{\rm B} = \frac{\left(\Psi_0|\hat{H}_{\rm B}|\Psi_0\right)}{A_{\rm b}\left(\Psi_0|\Psi_0\right)}\ , \tag{5.20}$$

where $\hat{H}_{\rm B}$ is a baryon (B) Hamiltonian operator and $\Psi_0$ is a ground-state wave function of the system. In our case $E_{\rm B}$ should be calculated in the thermodynamic limit ($A_{\rm b} \longrightarrow \infty$, volume of the system $\longrightarrow \infty$). In the simplest case of nucleon matter (B=N), the calculation yields $E_{\rm N}$ as a function of $n_n$ and $n_p$. The knowledge of $E_{\rm N}(n_n, n_p)$ is sufficient for calculating the EOS of matter consisting of nucleons and leptons (the so called $npe\mu$ matter). In a more general case of hyperonic matter, one needs $E_{\rm B}$ as a function of all baryon densities $n_B$ ($B = n, p, \Sigma^-, \Lambda, \ldots$). In § 5.9 we will review the methods of calculating $E_{\rm B}$. These methods can be divided into several groups.

(*1*) The methods of the first group are the most ambitious: they attempt to determine $E_{\rm B}$ from nuclear interactions in vacuum. Some of these methods are based on summations of infinite classes of diagrams which describe a perturbation expansion of $E_{\rm B}$, Eq. (5.20). Because the interaction is strong, standard expansions in powers of interaction strength are useless. On the contrary, some selected diagrams can be summed up to infinite order. The oldest and the most widely used method is the Brueckner-Bethe-Goldstone theory, devised initially for the nuclear matter (§ 5.9.1). The approach based on the Green's functions technique (§ 5.9.2) is less popular. Another approach employs the variational method which is is well known in quantum mechanics (see, e.g., §32 of Schiff 1968) and gives

$$E_{\rm B}^{(\rm var)} = \frac{\left(\Psi_{\rm var}|\hat{H}_{\rm B}|\Psi_{\rm var}\right)}{A_{\rm b}\left(\Psi_{\rm var}|\Psi_{\rm var}\right)} \geq E_{\rm B}^{(\rm exact)}, \tag{5.21}$$

where $\Psi_{\rm var}$ is a trial wave function. This method consists in minimizing the energy functional $E_{\rm B}^{(\rm var)}$ within a set of trial wave functions, which should be sufficiently rich in their structure, reflecting the structure of $\hat{H}_{\rm B}$ (§ 5.9.3).[18]

(2) The methods of the second group employ approximate solutions of many-body problem based on effective interactions. The starting point is a phenomenological interaction *in a dense baryon medium*, which strongly differs from that in vacuum. In particular, many-body correlations are assumed to be absorbed in the effective Hamiltonian $\hat{H}_{\rm B}^{\rm eff}$. This Hamiltonian may well reproduce the true value of $E_{\rm B}$ within the Hartree-Fock approximation for an $A_{\rm b}$-body wave function,

$$E_{\rm B}^{(\rm HF)} = \frac{\left(\Psi_{\rm HF}|\hat{H}_{\rm B}^{\rm eff}|\Psi_{\rm HF}\right)}{A_{\rm b}\left(\Psi_{\rm HF}|\Psi_{\rm HF}\right)} \simeq E_{\rm B}^{(\rm exact)}. \tag{5.22}$$

This method is of the mean-field character. It is well suited for studying a multi-component baryonic matter. It is particularly popular in its relativistic formulation, being called by proud names of Relativistic Mean Field Theory or Relativistic Hadrodynamics. It is based on a relativistic effective Lagrangian which describes hadronic interactions in baryon matter. The Lagrangian $\mathcal{L}_{\rm B}^{\rm eff}$ reminds the bare (in vacuum) Lagrangian of strong interactions in its structure but involves different phenomenological coupling constants and additional terms, which are adjusted to fit the experimental properties of the nuclear matter at saturation. The effective Hamiltonian $\hat{H}_{\rm B}^{\rm eff}$ is then derived from $\mathcal{L}_{\rm B}^{\rm eff}$ using standard methods of the field theory. The ground-state energy is obtained by solving the mean-field (Hartree) equations for the ground state (see, e.g., Glendenning 2000). The virtues of this approach are: formal elegance, relativistic

[18]Recently Sarsa *et al.* (2003) carried out Monte Carlo calculations of the ground state of neutron matter with realistic models of NN and NNN interactions. Such calculations would be exact (for an adopted interaction model) in the limit of very many particles included into Monte Carlo runs. However, the computational effort grows exponentially with the number of neutrons put into a periodic box (see Sarsa *et al.* 2003 and references therein)

invariance, and numerical simplicity even in the case of a multi-component hyperonic matter (§ 5.9.4).

The mean-field method exists also in the non-relativistic form, where $\hat{H}_{\rm B}^{\rm eff}$ includes two-body and three-body in-medium interactions. Such $\hat{H}_{\rm B}^{\rm eff}$ has been used very successfully, starting from the early 1970s, in the Hartree-Fock calculations of nuclear structure. A very early example of $\hat{H}_{\rm B}^{\rm eff}$ was proposed by Cameron (1959) who calculated the EOS of a neutron-star core employing an effective interaction constructed by Skyrme (1959). Starting from the early 1970s, the Skyrme-type effective interactions have been used very successfully in the Hartree-Fock calculations of the structure of atomic nuclei. The parameters of $\hat{H}_{\rm N}^{\rm eff}$ can be fixed by the requirement to reproduce the properties of atomic nuclei and the EOS of pure neutron matter obtained in a more fundamental approach. Both relativistic and non-relativistic mean-field approaches can be used to calculate unified EOSs of neutron star interiors (crust and core) from one effective Hamiltonian.

## 5.9. Energy of nucleon matter

Consider a system of $A_{\rm b}$ nucleons ($N$ neutrons and $Z$ protons) in a large box of volume $V$ in the limit $V \longrightarrow \infty$, $A \longrightarrow \infty$, $Z/A_{\rm b}$ =const. This (thermodynamic) limit corresponds to a homogeneous nucleon matter with $n_n = N/V$, $n_p = Z/V$, and $n_{\rm b} = A_{\rm b}/V = n_n + n_p$. We need to calculate the energy per nucleon $E(n_n, n_p)$ in the ground state of the system. We neglect Coulomb interaction and the difference between neutron and proton masses; the nucleon mass in vacuum will be denoted by $m$. Let us review several methods developed to carry out this task.

### 5.9.1 Brueckner-Bethe-Goldstone (BBG) theory

This theory, formulated in 1954–1965, is an example of a successful application of field-theoretic methods to strongly interacting many-body systems. Brueckner proposed a general theory of nuclear matter valid for strong and even singular NN interactions (Brueckner *et al.*, 1954; Brueckner, 1954, 1955; Brueckner & Levinson, 1955). He re-formulated the perturbation expansion in terms of diagrams involving the *reaction matrix* G. The G-matrix itself (actually Brueckner called it the K-matrix) summed two-body NN correlations up to the infinite order in the NN interaction and self-consistently included dispersive effects of a dense medium. Goldstone (1957) proved the crucial "linked cluster theorem" which enabled one to select those infinite classes of diagrams which contributed to the ground state energy of nuclear matter. Bethe & Goldstone (1957) derived the equation for a wave function (the Bethe-Goldstone equation) of a pair of interacting nucleons moving in nuclear medium. It clarified the importance of the Pauli exclusion principle for two-body correlations in nuclear matter.[19] Finally, Bethe (1965) made an important step beyond the lowest-order theory in the G-matrix (i.e., beyond the two-body correlations). He summed infinite series of important diagrams of the second and higher order in G (three-body clusters), which turned out to be crucial for the convergence of the BBG theory. A comprehensive review of the BBG theory is given by Baldo (1999). Modern BBG calculations of nuclear matter and neutron-star matter are reviewed by Baldo *et al.* (2001).

Let us start from the non-relativistic formulation of the BBG theory. The Hamiltonian is

$$\hat{H} = \hat{H}_{\rm kin} + \hat{H}_{\rm int} , \tag{5.23}$$

where

$$\hat{H}_{\rm kin} = -\sum_{i=1}^{A} \frac{\hbar^2 \nabla_i^2}{2m} . \tag{5.24}$$

and $\hat{H}_{\rm int}$ is the nuclear interaction part. Let us initially restrict ourselves to two-body nuclear forces, so that

$$\hat{H}_{\rm int} = \hat{V}^{(2)} = \frac{1}{2} \sum_{i,j=1}^{A}{}' \hat{v}_{ij} , \tag{5.25}$$

where the prime over the sum indicates that terms with $i = j$ are excluded.

It is crucial to split $\hat{H}$ into an unperturbed single-particle part and a perturbation:

$$\hat{H} = \hat{H}_0 + \hat{H}_1 , \tag{5.26}$$

where

$$\hat{H}_0 = \hat{H}_{\rm kin} + \hat{U} , \quad \hat{H}_1 = \hat{H}_{\rm int} - \hat{U} , \quad \hat{U} = \sum_{j=1}^{A} \hat{U}_j , \tag{5.27}$$

$\hat{U}_j$ being a single-particle potential acting on a nucleon $j$. The definition of $\hat{U}_j$ is important for the convergence of the BBG expansion series, where $\hat{H}_1$ is treated as a perturbation. As the system is spatially uniform, $\hat{U}_j$ is constant in space, and unperturbed nucleon states are plane waves $|\boldsymbol{p}\rangle$. For simplicity, we omit spin indices and express nucleon momenta $\boldsymbol{p}$ in units of $\hbar$. In momentum representation, the unperturbed single-particle energy is

$$e_N(p) = \frac{\hbar^2 p^2}{2m} + U_N(p) , \tag{5.28}$$

[19]The linked cluster theorem, formulated in terms of diagrams, which were latter named "Goldstone diagrams", and the equation governing the wave function of a nucleon pair in nuclear matter (Bethe-Goldstone equation) were all derived during the graduate studies of Goldstone at Trinity College of the Cambridge University (Cambridge, England).

where $N = n$ or $p$. Because of the isotropy of the nuclear matter, which is assumed to be spin-unpolarized, single-particle energies are independent of momentum direction and nucleon spin. However, owing to a neutron excess, the single-particle potentials for neutrons and protons are different and the G-matrix is no longer charge symmetric ($G_{nn} \neq G_{pp}$). As we will see below, $U_N$ can be expressed in terms of the G-matrix, the central quantity of the BBG theory. Its calculation is equivalent to the summation of the "ladder diagrams" of the BBG expansion series. It is performed by solving the integral equation, which can be written in the operator form:

$$\hat{G}_{NN'} = \hat{v}_{NN'} + \hat{v}_{NN'} \, \frac{\hat{Q}_{NN'}}{z - \hat{h}_{NN'}} \, \hat{G}_{NN'} \, . \tag{5.29}$$

Here, $\hat{Q}_{NN'}$ is the two-particle exclusion-principle operator, which projects particle states outside the Fermi surface, and $\hat{h}_{NN'}$ is the Hamiltonian operator acting on uncorrelated two-particle states,

$$\hat{Q}_{NN'}|\boldsymbol{p}_1\boldsymbol{p}_2) = \Theta(p_1 - p_{\mathrm{F}N})\Theta(p_2 - p_{\mathrm{F}N'})|\boldsymbol{p}_1\boldsymbol{p}_2) \, , \tag{5.30}$$

$$\hat{h}_{NN'}|\boldsymbol{p}_1\boldsymbol{p}_2) = \left[e_N(p_1) + e_{N'}(p_2)\right]|\boldsymbol{p}_1\boldsymbol{p}_2) \, , \tag{5.31}$$

while $z$ is the *starting energy* parameter. In the low-density limit we get

$$\hat{h}_{NN'}|\boldsymbol{p}_1\boldsymbol{p}_2) \longrightarrow \frac{\hbar^2}{2m}(p_1^2 + p_2^2)|\boldsymbol{p}_1\boldsymbol{p}_2) \, , \quad \hat{Q}_{NN'} \longrightarrow 1 \, . \tag{5.32}$$

In this case the G-matrix equation transforms into the well known equation for the scattering T-matrix, which describes the NN scattering in vacuum (see, e.g., Messiah 1961, vol. II, Chapter XIX, §14). Passing to the momentum representation, we get

$$\begin{aligned}(\boldsymbol{p}_1'\boldsymbol{p}_2'|G_{NN'}(z)|\boldsymbol{p}_1\boldsymbol{p}_2) &= (\boldsymbol{p}_1'\boldsymbol{p}_2'|v_{NN'}|\boldsymbol{p}_1\boldsymbol{p}_2) \\ &+ \int \frac{d\boldsymbol{k}_1}{(2\pi)^3}\frac{d\boldsymbol{k}_2}{(2\pi)^3}(\boldsymbol{p}_1'\boldsymbol{p}_2'|v_{NN'}|\boldsymbol{k}_1\boldsymbol{k}_2) \\ &\times \frac{Q_{NN'}(\boldsymbol{k}_1,\boldsymbol{k}_2)}{z - e_N(k_1) - e_{N'}(k_2)}(\boldsymbol{k}_1\boldsymbol{k}_2|G_{NN'}(z)|\boldsymbol{p}_1\boldsymbol{p}_2) \, .\end{aligned} \tag{5.33}$$

In G-matrix elements relevant for calculating the ground-state energy, the starting energy is the sum of single-particle energies of the initial $|\boldsymbol{p}_1\boldsymbol{p}_2)$ state:

$$z = e_N(p_1) + e_{N'}(p_2) \, , \tag{5.34}$$

with single-particle potentials given in terms of the G-matrix via a Hartree-Fock expression

$$U_n(p_1) = 2\int_p \mathrm{d}\boldsymbol{p}_3(\boldsymbol{p}_1\boldsymbol{p}_3|G_{np}(e_n(p_1)+e_p(p_3))|\boldsymbol{p}_1\boldsymbol{p}_3)$$
$$+2\int_n \mathrm{d}\boldsymbol{p}_3(\boldsymbol{p}_1\boldsymbol{p}_3|G_{nn}(e_n(p_1)+e_n(p_3))|\boldsymbol{p}_1\boldsymbol{p}_3)_\mathrm{a}\,, \tag{5.35a}$$

$$U_p(p_1) = 2\int_n \mathrm{d}\boldsymbol{p}_3(\boldsymbol{p}_1\boldsymbol{p}_3|G_{pn}(e_p(p_1)+e_n(p_3))|\boldsymbol{p}_1\boldsymbol{p}_3)$$
$$+2\int_p \mathrm{d}\boldsymbol{p}_3(\boldsymbol{p}_1\boldsymbol{p}_3|G_{pp}(e_p(p_1)+e_p(p_3))|\boldsymbol{p}_1\boldsymbol{p}_3)_\mathrm{a}\,. \tag{5.35b}$$

Here, for the sake of compactness, we use the notations

$$|\boldsymbol{p}_1\boldsymbol{p}_2)_\mathrm{a} \equiv |\boldsymbol{p}_1\boldsymbol{p}_2) - |\boldsymbol{p}_2\boldsymbol{p}_1)\,,\quad \int_N \mathrm{d}\boldsymbol{p} \equiv \int \frac{\mathrm{d}\boldsymbol{p}}{(2\pi)^3}\Theta(k_{\mathrm{F}N}-p)\,. \tag{5.36}$$

As we are dealing with the spin-unpolarized system, we can use the spin-averaged G-matrix. The spin degeneracy gives a factor of two in front of the integrals.

The auxiliary single-particle potential term $U(p)$, Eq. (5.27), crucial for the convergence of the linked-cluster expansion, deserves an additional comment. The choice of $U(p)$ for states above the Fermi surface ($p > p_\mathrm{F}$) has been a subject of a long debate since the formulation of the BBG theory. Eventually, the so called *continuous prescription* for $U$ has been regarded as the most advantageous. According to this prescription, no energy gap is introduced between the energies of occupied ($p < p_\mathrm{F}$) and empty ($p > p_\mathrm{F}$) momentum states.[20] Such a choice turns out to be particularly suitable in view of the rapid convergence of the BBG expansion (see, e.g., Baldo *et al.* 2000, 2001). The lowest-order BBG approximation for the energy density $\mathcal{E}$ (without nucleon rest energy contribution) is given by the Hartree-Fock expression, where the G-matrix acts as an effective interaction. This justifies the name "Brueckner-Hartree-Fock" (BHF) approximation, used by many authors. The BHF expression for $\mathcal{E}$ reads

$$\begin{aligned}\mathcal{E} &= \mathcal{E}_\mathrm{FFG}(n_n,n_p) + \frac{1}{2}\int_p \mathrm{d}\boldsymbol{p}_1\int_p \mathrm{d}\boldsymbol{p}_2(\boldsymbol{p}_1\boldsymbol{p}_2|G_{pp}(e_p(p_1)+e_p(p_2))|\boldsymbol{p}_1\boldsymbol{p}_2)_\mathrm{a}\\ &+ \frac{1}{2}\int_n \mathrm{d}\boldsymbol{p}_1\int_n \mathrm{d}\boldsymbol{p}_2(\boldsymbol{p}_1\boldsymbol{p}_2|G_{nn}(e_n(p_1)+e_n(p_2))|\boldsymbol{p}_1\boldsymbol{p}_2)_\mathrm{a}\\ &+ \int_n \mathrm{d}\boldsymbol{p}_1\int_p \mathrm{d}\boldsymbol{p}_2(\boldsymbol{p}_1\boldsymbol{p}_2|G_{np}(e_n(p_1)+e_p(p_2))|\boldsymbol{p}_1\boldsymbol{p}_2)\,,\end{aligned} \tag{5.37}$$

[20]This introduces singularity in the kernel of the G-matrix equation at $p_1 = p_2 = k_1 = k_2 = k_\mathrm{F}$; (Cauchy) principal values of the integrals are to be taken.

where $\mathcal{E}_{\rm FFG}(n_n, n_p)$ is the energy density of a free Fermi gas.

In order to check the convergence of the BBG theory one has to go beyond the BHF approximation. The next step is to include contributions $\mathcal{E}^{(3)}$ from *irreducible three-body correlations*, which cannot be reduced to a sequence of two-body correlations.

The calculation of $\mathcal{E}^{(3)}$ requires a generalization of the two-body in-medium scattering formalism to the three-body one. It took a long time to formulate correctly the three-body (in vacuum) scattering problem in quantum mechanics. It was done by Faddeev (Faddeev 1960; also see Faddeev 1965 and Chapter 2 of Sitenko & Tartakovskii 1997). Bethe (1965) used the Faddeev formalism to evaluate $\mathcal{E}^{(3)}$ in symmetric nuclear matter. The central quantity in the Bethe theory is a three-body scattering matrix $\mathcal{T}$, to be obtained from the Bethe-Faddeev equation. The three-body scattering matrix in medium is decomposed into a sum $\mathcal{T}^{(1)} + \mathcal{T}^{(2)} + \mathcal{T}^{(3)}$, where $\mathcal{T}^{(1)}$, $\mathcal{T}^{(2)}$, and $\mathcal{T}^{(3)}$ satisfy a system of coupled scattering equations containing the G-matrix. These equations involve the three-particle exclusion operator,

$$\hat{Q}^{(3)}|\boldsymbol{p}_1\boldsymbol{p}_2\boldsymbol{p}_3) = \Theta(p_1 - p_{\rm F})\Theta(p_2 - p_{\rm F})\Theta(p_3 - p_{\rm F})|\boldsymbol{p}_1\boldsymbol{p}_2\boldsymbol{p}_3) \,, \tag{5.38}$$

and the three-particle mean-field Hamiltonian operator $\hat{h}^{(3)}$ which appears in the denominator of the three-particle propagator,

$$\hat{h}^{(3)}|\boldsymbol{p}_1\boldsymbol{p}_2\boldsymbol{p}_3) = [e(p_1) + e(p_2) + e(p_3)]\,|\boldsymbol{p}_1\boldsymbol{p}_2\boldsymbol{p}_3) \,. \tag{5.39}$$

The $\mathcal{T}^{(j)}$ matrices sum all contributions of those three-body correlation diagrams, where an $j$-th nucleon is a spectator (i.e., does not change its momentum state) in the last scattering event.

Calculations based on a realistic two-body NN interaction fail to reproduce experimental parameters of symmetric nuclear matter at saturation. Realistic NN forces are also known to underbind $^3$H and $^4$He (§ 5.5.2). These deficiencies can be corrected by adding a three-body (NNN) interaction to the nuclear Hamiltonian,

$$\hat{V}^{(3)} = \frac{1}{6}\sum_{i,j,k}^{A}{}'' \hat{V}_{ijk} \,, \tag{5.40}$$

where the double prime indicates that terms with two or three equal indices $i$, $j$, $k$ are omitted. In calculations, the NNN force can be replaced by an effective NN force. This can be done by averaging over positions of the third nucleon taking into account pair correlations of this nucleon with two remaining ones (Lejeune *et al.*, 1986). Such a procedure gives a density dependent effective NN force to be added to $\hat{V}^{(2)}$ (Baldo *et al.*, 1997).

The BBG theory presented above is non-relativistic. Relativistic effects grow with increasing density. For a model of free neutron Fermi gas, the Fermi velocity in the units of $c$ is

$$\frac{v_{\rm F}}{c} = \frac{\hbar p_{\rm F}}{m} = 0.353 \left(\frac{n_{\rm b}}{n_0}\right)^{1/3} , \qquad (5.41)$$

and becomes significant in the center of a massive neutron star, where $n_{\rm b}/n_0 \gtrsim 4-6$. A relativistic extension of the BBG theory was worked out in the 1980s (see ter Haar & Malfiet 1987; Fuchs 2004, and references therein). The essential point is to use the Dirac equation for single-particle states in nuclear matter. The majority of calculations adopted OBE interactions. The relativistic extension of the BHF approximation is called the relativistic Brueckner-Hartree-Fock (RBHF) or Dirac-Brueckner-Hartree-Fock theory.

### 5.9.2 Green's Function Theory

The ground-state energy per baryon, Eq. (5.20), can also be calculated using other perturbative expansions. A very elegant method is based on the Martin-Schwinger hierarchy of Green's functions (Martin & Schwinger 1958; also see Chapters 5–12 of Weber 1999 for applications to the EOS of dense matter). The basic quantities in the Martin-Schwinger Green's function theory (GFT) are n-particle (2n-point) Green's functions $g_{\rm n}$ ($\rm n = 1, 2, 3, \ldots$):

$$g_1(1;1') = {\rm i}\,(\Psi_0|\hat{\boldsymbol{T}}\{\hat{\psi}(1)\hat{\overline{\psi}}(1')\}|\Psi_0) , \qquad (5.42a)$$

$$g_2(1,2;1',2') = -(\Psi_0|\hat{\boldsymbol{T}}\{\hat{\psi}(1)\hat{\psi}(2)\hat{\overline{\psi}}(2')\hat{\overline{\psi}}(1')\}|\Psi_0) , \qquad (5.42b)$$

$$g_3(1,2,3;1',2',3') = -{\rm i}\,(\Psi_0|\hat{\boldsymbol{T}}\{\hat{\psi}(1)\hat{\psi}(2)\hat{\psi}(3)\hat{\overline{\psi}}(3')\hat{\overline{\psi}}(2')\hat{\overline{\psi}}(1')\}|\Psi_0) \ldots , \qquad (5.42c)$$

where $\Psi_0$ is the ground state of nuclear matter, $\hat{\psi}$ is the nucleon-field annihilation operator, $\hat{\overline{\psi}} \equiv \hat{\psi}^\dagger\gamma^0$, and $\hat{\boldsymbol{T}}$ is the time-ordering operator (e.g., Berestetskiĭ *et al.*, 1982). For the sake of compactness, numbers 1, $1'$ and etc. denote a set of space-time and spin-isospin coordinates of a nucleon. The Martin-Schwinger hierarchy of Green's functions is a chain of the integro-differential equations which couple $g_{\rm n}$ with $g_{\rm n+1}$. An equation for $g_1$ involves $g_2$, an equation for $g_2$ involves $g_1$ and $g_3$, and so on. The chain of these equations can be terminated provided some approximation of many-body correlations is made. For example, if two-body correlations are neglected, then we get the mean-field (Hartree-Fock) approximation, where all $g_{\rm n}$ are expressed in terms of one-particle propagators $g_1$ in an external self-consistent Hartree-Fock field. A more realistic case of strongly interacting many-body system includes two-body correlations (but to an infinite order in two-body interaction). In this case $g_3$ can be approximated by a properly symmetrized sum of products of $g_2$ and $g_1$. This enables one to sum, to infinite order, two-body correlations in the nucleon matter. This approximation is frequently called the "independent pair approximation" (because it exactly treats the two-body problem including medium effects on the mean-field level), or "ladder approximation" (because diagrams, which are summed up, are ladder-like).

An important quantity of the GFT is the nucleon *mass operator* (or the so called *self-energy*), denoted by $\Sigma_N$. It takes different values for neutrons and protons because of the neutron excess. In a uniform and spin-unpolarized nucleon matter $\Sigma_N$ depends on two parameters of a single-particle state: momentum $p = |\boldsymbol{p}|$ and energy $\omega$. Many ground-state properties of nucleon matter, particularly the ground-state energy, are derived from the real part of $\Sigma_N(p, \omega)$, denoted by $V_N(p, \omega)$. We will review main formulae below using a standard *quasiparticle approximation*. The single-particle energy $e_N(p)$ is obtained from $V_N(p, \omega)$ via the *dispersion equation*

$$e_N(p) = \frac{\hbar^2 p^2}{2m} + V_N\left(p, e_N(p)\right) . \tag{5.43}$$

As we see, single-particle energies of neutrons and protons are determined by the real part of the self-energy calculated *on the energy shell*, where $\omega = e_N(p)$. In contrast to the BBG theory, the actual momentum distribution of nucleons, $f_N(p)$, is used instead of the free Fermi gas one, $f_N^{(0)}(p) = \Theta(k_{\mathrm{F}N} - p)$. The Fermi momentum $k_{\mathrm{F}N}$ satisfies the condition that the Fermi energy is equal to the chemical potential:

$$\mu_N = \frac{\partial \mathcal{E}}{\partial n_N} = \frac{\hbar^2 k_{\mathrm{F}N}^2}{2m} + V_N(k_{\mathrm{F}N}, \mu_N) . \tag{5.44}$$

The momentum distribution of nucleons, $f_N(p)$, is calculated in a self-consistent way from $V_N(p, \omega)$,

$$f_N(p) = \left[1 - \frac{\partial}{\partial \omega} V_N(p, \omega)\right]^{-1}_{\omega = e_N(p)} . \tag{5.45}$$

At $p = k_{\mathrm{F}N}$, which coincides with the Fermi momentum of a free Fermi gas of the same number density $n_N$, the momentum distribution suffers a discontinuity (a drop). It is a general property of normal Fermi liquids (see, e.g., §10 of Lifshitz & Pitaevskiĭ 1980). We have

$$n_N = \frac{p_{FN}^3}{3\pi^2} = \frac{2}{(2\pi)^3} \int \mathrm{d}\boldsymbol{p}\, f_N(p) . \tag{5.46}$$

The ground-state energy density is expressed in terms of $e_N(p)$ and $f_N(p)$ by

$$\mathcal{E} = \sum_{N=n,p} \int_N \mathrm{d}\boldsymbol{p}\, f_N(p) \left[e_N(p) + \frac{\hbar^2 p^2}{2m}\right] , \tag{5.47}$$

where we use the short-handed notation, given by Eq. (5.36), for the integration over momentum states below the Fermi surface.

The central quantity of the independent pair approximation is the in-medium T-matrix, which describes two-nucleon scattering in a nucleon medium. In the momentum representation, the T-matrix for an $NN'$ nucleon pair satisfies the integral equation

$$\begin{aligned}(\boldsymbol{p}_1'\boldsymbol{p}_2'|T_{NN'}(\Omega+\mathrm{i}\eta)|\boldsymbol{p}_1\boldsymbol{p}_2) &= (\boldsymbol{p}_1'\boldsymbol{p}_2'|v_{NN'}|\boldsymbol{p}_1\boldsymbol{p}_2) \\ &+ \int \frac{\mathrm{d}\boldsymbol{k}_1}{(2\pi)^3}\frac{\mathrm{d}\boldsymbol{k}_2}{(2\pi)^3}(\boldsymbol{p}_1'\boldsymbol{p}_2'|v_{NN'}|\boldsymbol{k}_1\boldsymbol{k}_2) \\ &\times \Lambda_{NN'}(\boldsymbol{k}_1,\boldsymbol{k}_2,\Omega+\mathrm{i}\eta)\,(\boldsymbol{k}_1\boldsymbol{k}_2|T_{NN'}(\Omega+\mathrm{i}\eta)|\boldsymbol{p}_1\boldsymbol{p}_2)\,, \quad (5.48)\end{aligned}$$

where $\Lambda_{NN'}$ is a two-body Green's function in the momentum representation. The parameter $\Omega$ is real; it is analogous to the starting energy of the BBG theory; $\eta$ is infinitesimally small and positive, $\eta \longrightarrow +0$.

In the independent-pair approximation, $V_N(p,\omega)$ can be expressed in terms of the in-medium T-matrix, which is analogous to the G-matrix in the BBG theory. For example, the real part of the self-energy for neutrons and protons is given by

$$\begin{aligned}V_n(p,\omega) &= 2\int_n \mathrm{d}\boldsymbol{p}'\, f_n(p')\,\mathrm{Re}(\boldsymbol{p}\boldsymbol{p}'|T_{nn}\left(\omega+\mathrm{i}\eta+e_n(p')\right)|\boldsymbol{p}\boldsymbol{p}')_{\mathrm{a}} \\ &+2\int_p \mathrm{d}\boldsymbol{p}'\, f_p(p')\,\mathrm{Re}(\boldsymbol{p}\boldsymbol{p}'|T_{np}\left(\omega+\mathrm{i}\eta+e_p(p')\right)|\boldsymbol{p}\boldsymbol{p}')\,, \quad (5.49\mathrm{a})\end{aligned}$$

$$\begin{aligned}V_p(p,\omega) &= 2\int_p \mathrm{d}\boldsymbol{p}'\, f_p(p')\,\mathrm{Re}(\boldsymbol{p}\boldsymbol{p}'|T_{pp}\left(\omega+\mathrm{i}\eta+e_p(p')\right)|\boldsymbol{p}\boldsymbol{p}')_{\mathrm{a}} \\ &+2\int_n \mathrm{d}\boldsymbol{p}'\, f_n(p')\,\mathrm{Re}(\boldsymbol{p}\boldsymbol{p}'|T_{pn}\left(\omega+\mathrm{i}\eta+e_n(p')\right)|\boldsymbol{p}\boldsymbol{p}')\,. \quad (5.49\mathrm{b})\end{aligned}$$

The T-matrix can be calculated from Eq. (5.48) using several approximations called $\Lambda^{00}$, $\Lambda^{10}$, and $\Lambda^{11}$ and described below. In what follows, for the sake of simplicity, we omit spin and isospin variables. At the first step we discuss various approximations of the Green's functions in the space-time representation. The simplest approximation consists in replacing a two-particle Green's function by a product of two free one-particle ones:

$$g_2(1,2;3,4) \approx \mathrm{i}\, g_1^0(1,3)\, g_1^0(2,4)\,. \quad (5.50)$$

This is the so called $\Lambda^{00}$ approximation, which is a popular GFT-scheme (see, e.g., Weber 1999 and references therein). Alas, the only justification of this approximation is the simplicity of the model. As no medium effects are included, one calculates just the T-matrix in vacuum – a rather standard problem in the two-body scattering theory.

More ambitious, but more complicated, are the so called $\Lambda^{01}$ and $\Lambda^{11}$ approximations, which are very naturally defined by

$$\begin{aligned}
\Lambda(1,2;3,4) \simeq \Lambda^{10}(1,2;3,4) &\equiv \frac{\mathrm{i}}{2}\left[g_1(1,3)\,g_1^0(2,4) + g_1^0(1,3)\,g_1(3,4)\right] \ , \\
\Lambda(1,2;3,4) \simeq \Lambda^{11}(1,2;3,4) &\equiv \mathrm{i}\,g_1(1,3)\,g_1(2,4) \ . \qquad (5.51)
\end{aligned}$$

Now let us consider the two-particle propagator in momentum space; it enters the equation for the T-matrix. In the $\Lambda^{00}$ approximation both particles in intermediate states are free, i.e., the dispersion and exclusion-principle effects are neglected. This leads to

$$\Lambda^{00}(k_1, k_2, \Omega + \mathrm{i}\eta) = \left[\Omega - \varepsilon_{k_1} - \varepsilon_{k_2} + \mathrm{i}\eta\right]^{-1} \ . \qquad (5.52)$$

The $\Lambda^{10}$ approximation partly incorporates both the exclusion principle and dispersion effects for intermediate states. Finally, the $\Lambda^{11}$ approximation incorporates both effects for states above the Fermi surface (the expressions in the momentum representation are given, e.g., by Weber & Weigel 1985).

One of the technical difficulties of the GFT is a complicated character of its basic quantities. The theory can easily be formulated in a relativistic way (see, e.g., Wilets 1979). Notice that the main trick used to terminate and close the chain of the GFT equations is based on non-perturbative arguments.

### 5.9.3 Variational method

The method is based on the minimization of the expectation value of the Hamiltonian in the space of trial many-body wave functions $\{\Psi_{\mathrm{var}}\}$. Let us describe a general method developed by Pandharipande and his collaborators (Akmal *et al.* 1998 and references therein).

A trial wave function is constructed as

$$\Psi_{\mathrm{var}} = \hat{F}\Phi \ , \qquad (5.53)$$

where the operator $\hat{F}$ describes correlations and $\Phi$ is a ground-state wavefunction of $A_{\mathrm{b}}$ non-interacting nucleons in a large volume $V$. We will use the isospin formalism, where single-nucleon states are labeled by $\alpha_i = \{\boldsymbol{p}_i, s_i, t_i\}$ and nucleon variables in coordinate-spin-isospin spaces are $\xi_j = \{\boldsymbol{r}_j, \sigma_j, \tau_j\}$. A non-interacting state is given by the Slater determinant

$$\Phi = (A_{\mathrm{b}}!)^{-1/2} \det\{\phi_{\alpha_i}(\xi_j)\} \ , \qquad (5.54)$$

where a single-nucleon state $\phi_{\alpha_i}$ is a plane wave multiplied by spin and isospin spinors. It is sufficient to consider the thermodynamic limit $A_{\mathrm{b}} \longrightarrow \infty$ and $V \longrightarrow \infty$ with $A_{\mathrm{b}}/V = n_{\mathrm{b}}$. As both $\Phi$ and $\Psi_{\mathrm{var}}$ are antisymmetric with respect to any transposition of variables of two nucleons, the correlation operator $\hat{F}$ should be symmetric with respect to these transpositions.

Correlations are induced by the NN interaction, represented by an NN potential, Eq. (5.25). They are also induced by many-nucleon interactions, represented by an NNN potential, Eq. (5.9). However, even in the most advanced calculations $\hat{F}$ includes only two-body correlations,

$$\hat{F} = \hat{\mathcal{S}} \prod_{i<j}^{A_{\mathrm{b}}} \hat{f}_{ij} \ , \tag{5.55}$$

where $\hat{f}_{ij}$ is a two-body correlation operator and $\hat{\mathcal{S}}$ is the symmetrization operator. The operator $\hat{f}_{ij}$ is given by the expression analogous to Eq. (5.6) for the NN potential:

$$\hat{f}_{ij} = \sum_u f_u(r_{ij}) \hat{O}^u_{ij} \ . \tag{5.56}$$

Our system is spin-unpolarized but isospin-polarized: it contains an excess of neutrons. The procedure to determine the functions $f_u$ was developed by Lagaris & Pandharipande (1980, 1981a,b); it is briefly described below.

The functional to be minimized is the expectation value

$$E_{\mathrm{var}}\,[f] = \frac{(\Phi|\hat{F}^\dagger \hat{H} \hat{F}|\Phi)}{A_{\mathrm{b}}(\Phi|\hat{F}^\dagger \hat{F}|\Phi)} \ . \tag{5.57}$$

The minimization of $E_{\mathrm{var}}$ is performed in two steps. At the first step, $f_u$ functions, called $f_u^{(2)}$, are calculated by minimizing the functional $E_{\mathrm{var}}^{(2)}$ obtained from Eq. (5.57) by removing three-body interaction from the Hamiltonian, $\hat{H} \longrightarrow \hat{H}^{(2)}$. The partial-differential equations for $f_u^{(2)}$ are the Euler-Lagrange equations derived from the minimization of the functional $E_{\mathrm{var}}^{(2)}$ on the set of the functions $\{f_u^{(2)}\}$.

In order to make the task feasible, the variation within the set $\{f_u^{(2)}\}$ is *constrained.* Let us start with the central correlations, where the direction of $\boldsymbol{r}_{ij}$ is not coupled to the spin and/or relative momentum of a nucleon pair $(1 \leq u \leq 4)$. The physical condition, which leads then to a constraint on $f_u^{(2)}$ $(1 \leq u \leq 4)$, is the requirement, that sufficiently distant nucleons are not correlated. This condition can be expressed via

$$\text{central} \ \ u = 1, \ldots, 4: \qquad f_u^{(2)}(r_{ij} \geq d_u) = 1. \tag{5.58}$$

It is called the *healing constraint* on the central correlation functions, and $d_u$ (which is a variational parameter) is called the *healing distance*. For $r_{ij} > d_u$ the central component of two-nucleon wave function is no longer perturbed ("wounded") by the short-range interaction, being the same as for a non-interacting pair.

The situation with noncentral correlations, expressed in terms of the $f_u^{(2)} \hat{O}_{ij}^u$ components with $u = 5, \ldots, 14$ , is different. The noncentral components in a free-pair wave function do not exist. Therefore, the correlations $f_u^{(2)}$ with $u = 5, \ldots, 14$ should vanish at large $r_{ij}$:

$$\text{noncentral } u = 5, \ldots, 14 : \qquad f_u^{(2)}(r_{ij} \geq d_u) = 0 \, , \tag{5.59}$$

where $d_u$ is again a variational parameter.

For the sake of simplicity let us consider a purely neutron matter. As our system is spin-unpolarized, the functions $f^{(2)}$ for the spin-singlet and spin-triplet states can be considered separately. The tensor and spin-orbit terms vanish in the spin-singlet channel. Neglecting the term involving $(\hat{\boldsymbol{L}})^2_{ij}$, we find that $\hat{f}_{ij}^{(2)}$ reduces to a single function of $r \equiv r_{ij}$. For the spin-singlet states the Euler-Lagrange equation then reads

$$-\frac{\hbar^2}{m}\left[\varphi \nabla^2 f^{(2)} + 2\nabla\varphi \cdot \nabla f^{(2)}\right] + (v_{ij} - \lambda)\, \varphi f^{(2)} = 0 \, , \tag{5.60}$$

where $\varphi = 2\left[l(k_{\rm F} r)\right]^{1/2}$, $v_{ij}$ is the central component of the $nn$ interaction in the spin-singlet state, and $l(x) = 3\,(\sin x - x\cos x)/x^3$ .

Equation (5.60) should be solved under the constraint given by Eq. (5.58), supplemented by an additional condition on the radial derivative of $f^{(2)}$,

$$\left[\frac{\mathrm{d}}{\mathrm{d}r} f^{(2)}(r)\right]_{r=d} = 0 \, . \tag{5.61}$$

Trial values of the healing distance $d$ must be chosen in such a way that, on average, there is only one particle within a distance $d$ from a given particle. A precise value of $d$ is determined at the final step of the minimization of the $E_{\rm var}$ functional (see below). At a given $d$, a solution of Eq. (5.60) exists only for a specific value of $\lambda$, which can be found via an iterative procedure.

In the spin-triplet channel, the operator $\hat{f}_{ij}^{(2)}$ is

$$\hat{f}_{ij}^{(2)} = f_1 + f_2\,(\hat{S}_{12})_{ij} + f_3\,(\hat{\boldsymbol{L}} \cdot \hat{\boldsymbol{S}})_{ij} + f_4\,(\hat{\boldsymbol{L}})^2_{ij} + f_5\,(\hat{\boldsymbol{L}} \cdot \hat{\boldsymbol{S}})^2_{ij} \, . \tag{5.62}$$

The Euler-Lagrange equations constitute a system of five coupled partial differential equations for the functions $f_u^{(2)}$, with the healing constraints given by Eqs. (5.58) and (5.59). These constraints specify the constants $\lambda_u$. Combining the spin-singlet and spin-triplet functions $f_u^{(2)}$, one gets an operator $\hat{f}^{(2)}$ which depends on healing distances $\{d\}$.

At the next step one must calculate the expectation value of the Hamiltonian, Eq. (5.57), in the limit of $A_{\rm b} \longrightarrow \infty$. Of course, the explicit integration over $3A_{\rm b}$ coordinates is impossible. However, because the correlations are short-ranged, one may use a diagrammatic technique, developed in the theory of strongly interacting quantum liquids, where the contributions from various classes of correlations are summed to the infinite order via *variational chain summation* (see Akmal *et al.* 1998 and references therein). In its original form this technique was developed for dense Fermi-liquids with central correlations. In the case of nucleon matter, noncentral correlations are important, and new techniques for summing contributions from noncentral interaction and correlation components were developed. The calculations are particularly complicated because of the non-commuting character of the tensor and momentum-dependent operators in the Hamiltonian and correlations.

The operator $\hat{F} = \hat{F}^{(2)}$ which minimizes $E^{(2)}_{\rm var}$ does not minimize the complete $E_{\rm var}$ that contains the NNN interaction. In order to come as close as possible to the minimum of $E_{\rm var}$, one recalculates $E_{\rm var}$ replacing $\hat{f}^{(2)}_{ij}$ by

$$\hat{f}_{ij} = \sum_u \beta_u f_u \, \hat{O}^u_{ij} \; . \tag{5.63}$$

In this way one gets $E_{\rm var}$ which depends on a set of variational parameters $\{d\}$ and $\{\beta\}$. At the next step one must minimize $E_{\rm var}$ with respect to $\{d\}$ and $\{\beta\}$.

An actual variational calculation of $E_0 \simeq (E_{\rm var})_{\rm min}$ contains many additional simplifications and approximations, but still, viewing the complexity of the problem, it represents a monumental computational achievement (Wiringa *et al.*, 1988; Akmal *et al.*, 1998). Further improved variational calculations for pure neutron matter were done by Morales *et al.* (2002)

## 5.9.4 Relativistic mean-field model

### 5.9.4 a Historical overview

It is well known that the main properties of atomic nuclei can be explained in terms of two basic and seemingly contradictory models, the shell model and the liquid-drop model. In the shell model, the $A$-body problem reduces to a one-body problem of a nucleon moving in a mean-field created by its interaction with remaining nucleons. In the liquid-drop model, nucleons are so strongly coupled between themselves that their motion is fully collective. In this case a nucleus is treated as a drop of strongly coupled nucleons.

In the 1950s, nuclear theoreticians started to explain these different facets of nuclei in the language of the field theory, where the strong interaction between nucleons originates from their coupling to various meson fields. From the very beginning the main strategy was in proposing simple solvable models. This approach was initiated by Johnson & Teller (1955) in the paper "Classical Field Theory of Nuclear Forces". The authors proposed to describe a motion of nucleons in heavy nuclei and the gross (liquid-drop) nuclear properties using a solvable model in which nucleons are coupled to a scalar and a vector meson fields. The scalar meson mediated long- and intermediate-range attraction needed to explain the binding of nucleons in nuclei, as well as the depth of the shell-model potential well. The vector meson mediated a short-range repulsion between nucleons to achieve the saturation of nuclear matter. The parameters of the model were adjusted by fitting basic experimental nuclear data.

Johnson & Teller (1955) proposed actually a "minimal nuclear mean-field model", which involved two electrically neutral meson fields; we will call them $\sigma$ and $\omega$. The energy of nuclei was calculated using the mean-field approximation (in the lowest-order in coupling constants). The authors stressed, that their scalar meson "... need not be an elementary particle in any sense of this word. It may be a virtual state composed of other mesons." They got $m_\sigma c^2 \sim 600$ MeV, remarkably close to the $\sigma$-meson mass of the contemporary models of NN interaction (§ 5.6). The $\omega$ field was vector and, therefore, led to a momentum-dependent single-particle potential and generated a density-dependent nucleon effective mass $m^* < m$. The $\omega$-meson contribution to the single-particle potential was repulsive and increased with growing density. Accordingly, the density dependence of the energy per nucleon for the symmetric nuclear matter exhibited a minimum, which coincided with saturation parameters of nuclear matter after the adjustment of meson masses and couplings. Two main deficiencies of the model were: its non-relativistic character and the presence of a peculiar bound state of nuclear matter at very high density. Moreover, the meson fields were classical (non-quantized).

The Johnson-Teller model was improved by Duerr (1956), who developed a relativistically invariant version of the $\sigma$–$\omega$ model. The Johnson-Teller model can be reproduced as the non-relativistic limit of the Duerr model. Duerr discussed also the validity of the mean-field approximation. In his model, nucleons are point sources of meson fields, represented by pointlike density distributions,

$$n_{\rm b}(\boldsymbol{r}) = \sum_{i=1}^{A} \delta(\boldsymbol{r} - \boldsymbol{r}_i) . \tag{5.64}$$

The mean-field approximation consists in replacing these distributions by a mean constant density. It is a good approximation if the distance scale of the exchanged meson $\phi$ is much larger than the internucleon distance,

$$\frac{\hbar}{m_\phi c} \gg n_{\rm b}^{-1/3} \Longrightarrow \frac{n_{\rm b}}{n_0} \gg 102 \left( \frac{m_\phi c^2}{500 \text{ MeV}} \right)^3 . \tag{5.65}$$

It is clear that this condition cannot be satisfied, because nucleons should transform into quarks at a density an order of magnitude lower. Moreover, the main approximation of replacing all fields by their expectation values resulted from the Hartree approximation for a many-body wave function of nucleon matter. This implies the absence of the exchange (Fock) term in the Hartree-Fock approximation for $\mathcal{E}$. However, it is well known, that for a fermion system with two-body interaction of range $a$ the Fock terms in $\mathcal{E}$ become negligible in the limit of $k_{\rm F}a \longrightarrow \infty$. This limit is similar to Eq. (5.65), and it is never reached in a neutron-star core. All in all, the model is phenomenological; it cannot be justified from first principles.

Disregarding this basic difficulty, we can just employ the mean field approximation as the basis of a phenomenological model. This procedure was carried out by Duerr. Below we briefly describe the solution of the relativistic mean-field model of Duerr, which is the simplest version of the $\sigma$–$\omega$ model in the nuclear theory. We will use the formulation of the $\sigma$–$\omega$ theory, developed in detail in a seminal paper of Walecka (1974).

### 5.9.4 b Solution of an RMF model

In the rest of this section we will use, in order to simplify the formulae, a system of units, where $\hbar = c = 1$. The mean-field approximation for a spatially uniform ground state of nucleon matter corresponds to nucleons filling momentum states within the Fermi spheres of radii $k_{{\rm F}n}$ and $k_{{\rm F}p}$; nucleon eigenfunctions are plane-wave solutions of the Dirac equations. The nucleon energy spectrum is then given by eigenvalues of the Dirac equations for nucleons moving in constant $\sigma$ and $\omega$ fields (with $\omega_0 \neq 0$ and $\omega_1 = \omega_2 = \omega_3 = 0$). These energy eigenvalues are charge and spin degenerate,

$$e_n(\boldsymbol{k}) = e_p(\boldsymbol{k}) = \sqrt{k^2 + (m - g_\sigma\sigma)^2} + g_\omega\omega_0 \;, \tag{5.66}$$

where *constant* meson field values are determined from the field equations reduced to

$$g_\sigma\sigma = \left(\frac{g_\sigma}{m_\sigma}\right)^2 \frac{1}{\pi^2} \sum_{N=n,p} \int_0^{k_{{\rm F}N}} {\rm d}k \, \frac{k^2(m_N - g_\sigma\sigma)}{\sqrt{k^2 + (m_N - g_\sigma\sigma)^2}} \;, \tag{5.67}$$

$$g_\omega\omega_0 = \left(\frac{g_\omega}{m_\omega}\right)^2 n_{\rm b} \;. \tag{5.68}$$

The energy density and the pressure are then calculated as the ground-state expectation values of the components of the energy-momentum tensor $T^{\mu\nu}$,[21]

[21] Detailed derivations of the formulae, based on the standard methods of the field theory, can be found in the monograph of Glendenning (2000).

$$\mathcal{E}(n_n, n_p) = \langle \Psi_0 | T^{00} | \Psi_0 \rangle = \frac{1}{\pi^2} \sum_{N=n,p} \int_0^{k_{\mathrm{F}N}} \mathrm{d}k \, \frac{k^2 \, (m_N - g_\sigma \sigma)}{\sqrt{k^2 + (m_N - g_\sigma \sigma)^2}}$$

$$+ \frac{1}{2} \left( \frac{m_\sigma}{g_\sigma} \right)^2 (g_\sigma \sigma)^2 + \frac{1}{2} \left( \frac{m_\sigma}{g_\sigma} \right)^2 (g_\omega \omega_0)^2 \,, \tag{5.69}$$

$$P(n_n, n_p) = -\langle \Psi_0 | T^{11} | \Psi_0 \rangle = \frac{1}{3\pi^2} \sum_{N=n,p} \int_0^{k_{\mathrm{F}N}} \mathrm{d}k \, \frac{k^4}{\sqrt{k^2 + (m_N - g_\sigma \sigma)^2}}$$

$$- \frac{1}{2} \left( \frac{m_\sigma}{g_\sigma} \right)^2 (g_\sigma \sigma)^2 + \frac{1}{2} \left( \frac{m_\omega}{g_\omega} \right)^2 (g_\omega \omega_0)^2 \,. \tag{5.70}$$

One can easily see, that the model contains only two free parameters, $g_\sigma / m_\sigma$ and $g_\omega / m_\omega$. They can be adjusted to reproduce two selected experimental parameters of the saturated nuclear matter,[22] for instance, the binding energy per nucleon and the saturated nucleon density, $B_0$ and $n_0$. No wonder that in this case the values of remaining parameters of the saturated nuclear matter, such as the incompressibility $K_0$ and the symmetry energy $S_0$, strongly deviate from the experimental values (§ 5.4): $K_0 \simeq 2\, K_0^{\mathrm{exp}}$ and $S_0 \simeq 0.5\, S_0^{\mathrm{exp}}$. Moreover, the EOS of pure neutron matter predicts a non-existent bound state at $n_\mathrm{b} \simeq n_0$.

The problem of incorrect incompressibility and fictitious bound state was eliminated by Boguta & Bodmer (1977) who extended the $\sigma$–$\omega$ model by including self-interactions of the $\sigma$-field. These interactions were introduced via the term

$$U(\sigma) = \frac{1}{3} \, b \, m_N (g_\sigma \sigma)^3 + \frac{1}{4} \, c \, (g_\sigma \sigma)^4 \,, \tag{5.71}$$

where $b$ and $c$ are constants; $U(\sigma)$ is added to the Hamiltonian density (and subtracted from the Lagrangian density). The argument for including $U(\sigma)$ is that it makes the quantum version of the model *renormalizable* (Boguta & Bodmer, 1977). Two additional free parameters $b$ and $c$ can be adjusted to reproduce correct values of $K_0$ and $m^*$ (§ 5.4). Still there remains the problem of the correct value of $S_0$. It is solved by introducing the charge-triplet of the vector $\rho$-meson fields, which couple to the neutron-proton number density difference and, therefore, contribute to the symmetry energy. In this way one gets a modern version of the RMF model, used in neutron-star calculations since the mid-1980s and described in detail by Glendenning (2000).

The RMF model involves the nucleon bispinor fields $\psi_n$ and $\psi_p$, and five meson fields. The meson fields are: the neutral scalar $\sigma$, the neutral vector $\omega_\mu$,

---

[22] Interestingly, Duerr (1956) demonstrated the existence of a strong spin-orbit coupling of nucleons moving in a nuclear potential well. This important feature is typical of *relativistic* mean field models, where the strong spin-orbit term is not postulated separately (in contrast to the non-relativistic mean-field theories) but is rigorously derived. However, the spin-orbit contribution to the energy averages out to zero in the mean-field approximation for a spatially uniform nuclear matter.

and the charge-triplet vector $\rho^q_\mu$ ($q = -1, 0, +1$). Each meson field $\phi$ of mass $m_\phi$ is coupled to an appropriate nucleon current (with a coupling constant $g_\phi$) and yields a scalar interaction term in the Lagrangian $\mathcal{L}_{N\phi}$. Within the $\rho$ triplet, $g_\rho$ does not depend on the meson charge, which reflects the charge independence of the strong interactions. The crucial assumption which makes the RMF model so easy in use (and therefore so attractive) is that nucleon currents are treated as uniform (constant) in space and that the Fock terms in the energy expectation value are neglected. These assumptions do not come from first principles; they are strictly valid only in the limit of $n_{\rm b} \longrightarrow \infty$, as we have already mentioned in the context of the Duerr (1956) paper [see Eq. (5.65)], so that the model is *essentially phenomenological.*

Within the RMF model, all three spatial components of nucleon currents and meson fields, as well as of charged $\rho$-fields, vanish in the ground state $\Psi_0$. Non-vanishing time-like components of vector fields describe the meson potentials in which nucleons move. These time-like components are constant in space, as a result of the assumed constancy of field sources. The ground-state momentum distribution for nucleons is that of a free Fermi gas.

The *constant* values of the meson fields are determined from the Euler-Lagrange equations derived for the RMF model Lagrangian. The solutions for the $\omega$ and $\rho$ fields are particularly simple, because the Lagrangian is linear in these fields,

$$g_\omega \omega_0 = \left(\frac{g_\omega}{m_\omega}\right)^2 (n_n + n_p) \,, \quad g_\rho \rho^0_0 = \frac{1}{2}\left(\frac{g_\rho}{m_\rho}\right)^2 (n_p - n_n) \,. \tag{5.72}$$

The equation for the sigma field is much more complicated and reads

$$\begin{aligned} g_\sigma \sigma \;&=\; \left(\frac{g_\sigma}{m_\sigma}\right)^2 \frac{1}{\pi^2} \sum_{N=n,p} \int_0^{k_{\mathrm{F}N}} \mathrm{d}k \, \frac{k^2 (m - g_\sigma \sigma)}{\sqrt{k^2 + (m - g_\sigma \sigma)^2}} \\ &\quad - \left(\frac{g_\sigma}{m_\sigma}\right)^2 \left[ b\, m (g_\sigma \sigma)^2 + c (g_\sigma \sigma)^3 \right] \,. \end{aligned} \tag{5.73}$$

The nucleon energy spectrum is obtained from the Dirac equations for nucleon bispinors $\psi_n$ and $\psi_p$:

$$e_n(\boldsymbol{k}) \;=\; \sqrt{k^2 + (m_N - g_\sigma \sigma)^2} + g_\omega \omega_0 - g_\rho \rho^0_0 \,, \tag{5.74a}$$

$$e_p(\boldsymbol{k}) \;=\; \sqrt{k^2 + (m_N - g_\sigma \sigma)^2} + g_\omega \omega_0 + g_\rho \rho^0_0 \,. \tag{5.74b}$$

While the charge independence of $e(\boldsymbol{k})$ is broken, the spin degeneracy still remains, because the matter is spin-unpolarized. We can now express the energy density in the ground state, $\mathcal{E} = \langle \Psi_0 | T^{00} | \Psi_0 \rangle$, by summing energies of nucleon states up to the Fermi level and adding energies of meson fields:

$$\mathcal{E}_{\rm N}(n_n,n_p) = \frac{1}{\pi^2}\sum_{N=n,p}\int_0^{k_{{\rm F}N}} {\rm d}k\, k^2\sqrt{k^2+(m_N-g_\sigma\sigma)^2}$$
$$+\frac{1}{2}\left(\frac{m_\omega}{g_\omega}\right)^2 (g_\omega\omega^0)^2 + \frac{1}{2}\left(\frac{m_\rho}{g_\rho}\right)^2 (g_\rho\rho_0^0)^2 + \frac{1}{2}\left(\frac{m_\sigma}{g_\sigma}\right)^2 (g_\sigma\sigma)^2$$
$$+\frac{1}{3}b\, m(g_\sigma\sigma)^3 + \frac{1}{4}c\,(g_\sigma\sigma)^4 . \tag{5.75}$$

The pressure is calculated as $P = -\langle\Psi_0|T^{11}|\Psi_0\rangle$:

$$P_{\rm N}(n_n,n_p) = \frac{1}{3\pi^2}\sum_{N=n,p}\int_0^{k_{{\rm F}N}} {\rm d}k\, \frac{k^4}{\sqrt{k^2+(m-g_\sigma\sigma)^2}}$$
$$+\frac{1}{2}\left(\frac{m_\omega}{g_\omega}\right)^2 (g_\omega\omega^0)^2 + \frac{1}{2}\left(\frac{m_\rho}{g_\rho}\right)^2 (g_\rho\rho_0^0)^2 - \frac{1}{2}\left(\frac{m_\sigma}{g_\sigma}\right)^2 (g_\sigma\sigma)^2$$
$$-\frac{1}{3}b\, m(g_\sigma\sigma)^3 - \frac{1}{4}c\,(g_\sigma\sigma)^4 . \tag{5.76}$$

The time-like components of meson fields should be expressed through $n_n$ and $n_p$ by solving Eqs. (5.72) and (5.73).

The RMF model described above contains five free parameters: $g_\sigma/m_\sigma$, $g_\omega/m_\omega$, $g_\rho/m_\rho$, $b$, and $c$. These parameters can be adjusted to reproduce the five experimental parameters of nuclear matter at the saturation point: $n_0$, $B_0$, $S_0$, $K_0$, and the nucleon effective mass $m^*$ at the Fermi surface (§ 5.4).

The effective mass $m^*$ in the RMF model deserves an additional comment. The effective mass can be defined in various ways. Two important definitions are: (1) the *Dirac effective mass* relevant for the nucleon Dirac equation in the RMF model, $m_{\rm D}^* \equiv m - g_\sigma\sigma$; it comes from the nucleon kinetic energy $e_{\rm kin}(k) = \sqrt{(m_{\rm D}^*)^2+k^2}$; and (2) the *Landau effective mass* $m_{\rm L}^* \equiv (1/k_{\rm F})(\partial e(k)/\partial k)^{-1}_{k=k_{\rm F}}$. It is easy to show that these effective masses are related by $(m_{\rm L}^*)^2 = (m_{\rm D}^*)^2 + k_{\rm F}^2$. An effective mass in nuclear matter, measured experimentally (see § 5.4), should be identified with $m_{\rm L}^*$.

As we have mentioned before, the $\rho$-field is crucial for fitting the experimental nuclear symmetry energy at the saturation, $S_0^{\rm exp} \simeq 30$ MeV. To show this, let us consider a weakly asymmetric nuclear matter. This can be done by introducing the familiar variables $n_{\rm b} = n_n + n_p$ and $\delta = (n_n - n_p)/n_{\rm b}$ in the limit of $\delta \ll 1$. At a fixed $n_{\rm b}$, we have

$$\left[\mathcal{E}(n_{\rm b},\delta) - \mathcal{E}(n_{\rm b},0)\right]/n_{\rm b} \simeq S(n_{\rm b})\delta^2 , \tag{5.77a}$$

$$S(n_{\rm b}) = \frac{1}{6}\left(\frac{g_\rho}{m_\rho}\right)^2 n_{\rm b} + \frac{k_{\rm F}^2}{6\sqrt{k_{\rm F}^2+(m_N-g_\sigma\sigma)^2}} , \tag{5.77b}$$

*Table 5.1.* Examples of parameter sets of the RMF model adjusted to different sets of parameters of nuclear matter at the saturation point (Glendenning & Moszkowski, 1991; Glendenning, 2000).

| $n_0$ | $B_0$ | $m_D^*$ | $S_0$ | $K_0$ | $g_\sigma^2/m_\sigma^2$ | $g_\omega^2/m_\omega^2$ | $g_\rho^2/m_\rho^2$ | $10^3 b$ | $10^3 c$ |
|---|---|---|---|---|---|---|---|---|---|
| (fm$^{-3}$) | (MeV) | ($m_N$) | (MeV) | (MeV) | (fm$^2$) | (fm$^2$) | (fm$^2$) | | |
| 0.153 | 16.3 | 0.7 | 32.5 | 300 | 11.79 | 7.149 | 4.411 | 2.947 | −1.070 |
| 0.153 | 16.3 | 0.78 | 32.5 | 240 | 9.927 | 4.820 | 4.791 | 8.659 | −2.421 |
| 0.153 | 16.3 | 0.75 | 32.5 | 200 | 11.299 | 5.696 | 4.656 | 8.784 | −1.0098 |

where $k_F = (3\pi^2 n_b/2)^{1/3}$ is the nucleon Fermi momentum in the symmetric nuclear matter. The coupling of $\rho$ mesons to nucleons gives, therefore, a contribution to the symmetry energy which *grows linearly* with baryon density. As will be discussed in § 5.13, this property has an important impact on the high-density behavior of the proton fraction in the $npe\mu$ matter. At $n_b = n_0$, the $\rho$-field term contributes about half of the symmetry energy!

Specific values of the five parameters of the RMF model depend on values of experimental saturation parameters of nuclear matter. Some examples of parameter sets are given in Table 5.1. Notice that the coefficient of the quartic term in $U(\sigma)$, Eq. (5.71), is negative. Taken at its face value, this means that the energy-density functional is not bounded from below! However this feature does not seem to be dangerous, because the values of $g_\sigma \sigma$ are bounded by the bare nucleon mass. Consider Eq. (5.73) for the $\sigma$-field. One can show that $g_\sigma \sigma$ grows monotonically with increasing $n_b$ tending to $m$ from below as $n_b \longrightarrow \infty$. This property remains valid in the presence of the self-interaction potential $U(\sigma)$.

### 5.9.4 c Concluding remarks

The RMF model has many attractive features which make it a popular tool in the neutron-star physics. It is computationally simple even for a multicomponent dense baryonic matter consisting of nucleons and hyperons (§ 5.10.2). Its Lorentz invariance guarantees that the sound speed in dense matter does not exceed the speed of light at any density. However, one should not forget the underlying approximations and assumptions.

Let us start with the basic assumption of the spatial constancy of the meson-field sources. It is valid if the mean distance between nucleons $r_{NN}$ is much smaller than the spatial range of virtual mesons $r_\phi = \hbar/m_\phi c$. As seen from Eq. (5.65), this condition holds at $n_b \gg 100\, n_0$! Clearly, the assumption $r_{NN} \ll r_\phi$ is not satisfied in neutron-star cores. Moreover, at $n_b = n_0$, where the RMF model is adjusted to experimental data, the opposite is true: $r_{\rm NN} \gg r_\phi$! As we have already mentioned, the Fock terms in the ground-state energy are neglected, which is strictly valid in the limit of $n_{\rm b} \longrightarrow \infty$.

Except for the self-coupling term $U(\sigma)$, the interaction part $\mathcal{L}_{\rm int}^{\rm RMF}$ of the RMF Lagrangian density is identical to the analogous component of the one-boson-exchange Lagrangian, $\mathcal{L}_{\rm int}^{\rm OBE}$, used to describe NN data *in vacuum* (§ 5.6). However, applying a mean-field approximation to $\mathcal{L}^{\rm OBE}$ gives a symmetric nuclear matter which is strongly unbound at $n_{\rm b} = n_0$ (Akmal *et al.*, 1998). Clearly, $\mathcal{L}^{\rm RMF}$ is not directly related to $\mathcal{L}^{\rm OBE}$.

### 5.9.5 Effective energy density functionals

According to the Kohn-Sham density functional theory (Kohn & Sham, 1965), developed starting from the ideas of Hohenberg & Kohn (1964), the energy of a many-Fermion system is a unique functional of one-particle densities. The ground state energy is determined by the minimization of the energy-density functional with respect to one-particle densities (number density, spin density, and kinetic energy density). The density functional theory (DFT) has been widely applied in condensed matter physics (see, e.g., the monograph of Dreizler & Gross 1990) and in nuclear physics (see, e.g., the monograph of Petkov & Stoitsov 1991). Its application for describing the matter in a neutron-star crust is outlined in § 3.3.1. While the existence of a unique density functional representing the energy of a fermionic system is predicted by the DFT, its specific functional form is still to be determined either by confronting DFT predictions with experiments or by calculating the functional using a more fundamental many-body theory.

The energy-density functional becomes particularly simple for a uniform, spin-unpolarized nucleon matter, where it reduces to a function of $n_n$ and $n_p$, $\mathcal{E}_{\rm N}(n_n, n_p)$. The dependence of $\mathcal{E}_{\rm N}$ on nucleon number densities is usually presented in a parameterized form. Numerical coefficients are obtained by adjusting the model to empirical properties of nuclear matter at saturation and also to best *ab initio* many-body calculations of pure neutron matter (to the most advanced results based on realistic nucleon-nucleon interaction). Remaining parameters of the energy-density functional are obtained by fitting the properties of laboratory nuclei calculated in the Hartree-Fock approximation with account for nucleon pairing effects. In this way some Skyrme-type energy-density functionals were constructed (see, e.g., Chabanat *et al.* 1997, 1998). A detailed study of a very large set of Skyrme energy-density functionals and of their application to neutron star EOS and neutron star models was presented by Rikovska-Stone *et al.* (2003). Other energy-density functionals were based exclusively on many-body calculations with realistic nucleon-nucleon interactions (e.g., the FPS energy-density functional). Several selected $\mathcal{E}_{\rm N}(n_n, n_p)$ are presented in the Appendix D.

## 5.10. Energy of hyperon-nucleon matter

With increasing baryon density, hyperons may appear in neutron-star cores replacing highly energetic neutrons. Instead of the nucleon energy density $\mathcal{E}_{\rm N}(n_n, n_p)$, one needs a more general baryon-energy functional $\mathcal{E}_{\rm B}(\{n_B\})$, where $\{n_B\}$ is a set of number densities of baryon species $\{B\}$,

$$\sum_B n_B = n_{\rm b} \,. \tag{5.78}$$

At baryon densities $n_{\rm b} \lesssim 10\, n_0$, relevant for neutron-star cores, it is sufficient to consider the octet of lightest baryons. The octet includes the nucleon charge-doublet $N$, the charge-singlet $\Lambda$, the charge-triplet $\Sigma$, and the charge-doublet $\Xi$. The parameters of these baryons are given in Table 5.2. Baryon fields $\psi_B$ have well defined properties in the *isospin space*: nucleons and $\Xi$-hyperons are represented by an iso-spinor field (isospin $I = 1/2$), $\Lambda$ by an iso-scalar field ($I = 0$), and $\Sigma$-hyperons by an iso-vector field ($I = 1$). Each baryon field is an eigenstate of the third isospin component $I_3$, related to its electric charge $Q$ and strangeness $S$ by (see, e.g., Perkins 2000)

$$I_3 = Q - (S+1)/2 \,. \tag{5.79}$$

The electric charge density and the strangeness per baryon are given by

$$q_{\rm b} = \sum_B n_B Q_B \,, \quad s_{\rm b} = \sum_B n_B S_B / n_{\rm b} \,. \tag{5.80}$$

### 5.10.1 Brueckner-Bethe-Goldstone theory

At the lowest-order (BHF) level the generalization to the multicomponent baryon matter should take into account the multichannel nature of the BB interaction resulting from the $\Lambda$–$\Sigma$ conversion. Consider the set of $N$, $\Lambda$ and $\Sigma$. One can introduce G-matrices $G_{N\Lambda,N\Lambda}$, $G_{N\Lambda,N\Sigma}$, $G_{N\Sigma,N\Lambda}$, $G_{N\Sigma,N\Sigma}$, which are coupled by $(N\Sigma|\hat{v}_{ij}|N\Lambda)$ matrix elements of the interaction operator $\hat{v}_{ij}$ in the generalized G-matrix equations.[23] Moreover, each G-matrix is coupled to remaining ones via the self-consistent single-particle potentials

$$\begin{aligned} U_B(p_1) &= 2\sum_{B'\neq B} \int_{B'} \mathrm{d}\boldsymbol{p}_3 \, (\boldsymbol{p}_1\boldsymbol{p}_3|G_{BB',BB'}(e_B(p_1)+e_{B'}(p_3))|\boldsymbol{p}_1\boldsymbol{p}_3) \\ &+ 2\int_B \mathrm{d}\boldsymbol{p}_3 \, (\boldsymbol{p}_1\boldsymbol{p}_3|G_{BB,BB}(e_B(p_1)+e_B(p_3))|\boldsymbol{p}_1\boldsymbol{p}_3)_{\rm a} \,. \end{aligned} \tag{5.81}$$

[23]For example, $(p\Sigma^-|\hat{v}_{ij}|n\Lambda)$ means strong interaction with $n\Lambda$ in an initial state and with $p\Sigma^-$ in a final state; similar notation is used for G-matrix elements.

*Table 5.2.* Masses, electric charges, strangeness, and e-folding (mean) lifetimes of the baryon octet, measured in laboratory. The baryon number, spin, and parity of all these baryons are 1, $1/2$, and $+1$, respectively.

| baryon name | $mc^2$ (MeV) | $Q$ $(e)$ | $S$ | $\tau$ (s) |
|---|---|---|---|---|
| $p$ | 938.27 | 1 | 0 | $> 10^{32}$ |
| $n$ | 939.56 | 0 | 0 | 886 |
| $\Lambda^0$ | 1115.7 | 0 | $-1$ | $2.6 \times 10^{-10}$ |
| $\Sigma^+$ | 1189.4 | 1 | $-1$ | $0.80 \times 10^{-10}$ |
| $\Sigma^0$ | 1192.6 | 0 | $-1$ | $7.4 \times 10^{-20}$ |
| $\Sigma^-$ | 1197.4 | $-1$ | $-1$ | $1.5 \times 10^{-10}$ |
| $\Xi^0$ | 1314.8 | 0 | $-2$ | $2.9 \times 10^{-10}$ |
| $\Xi^-$ | 1321.3 | $-1$ | $-2$ | $1.6 \times 10^{-10}$ |

Here, we have returned to physical units and short-hand notations, Eq. (5.36), used in §§ 5.9.1 and 5.9.2. The BHF approximation for the energy density reads

$$\begin{aligned}\mathcal{E}(\{n_B\}) &= \mathcal{E}_{\rm FFG}(\{n_B\}) \\ &+ \frac{1}{2}\sum_{B'}\int_{B'}\mathrm{d}\boldsymbol{p}_1\int_{B'}\mathrm{d}\boldsymbol{p}_2\,(\boldsymbol{p}_1\boldsymbol{p}_2|G_{B'B',B'B'}(e_{B'}(p_1)+e_{B'}(p_2))|\boldsymbol{p}_1\boldsymbol{p}_2)_{\rm a} \\ &+ \frac{1}{2}\sum_{B''\neq B'}\int_{B''}\mathrm{d}\boldsymbol{p}_1\int_{B'}\mathrm{d}\boldsymbol{p}_2\,(\boldsymbol{p}_1\boldsymbol{p}_2|G_{B''B',B''B'}(e_{B''}(p_1)+e_{B'}(p_2))|\boldsymbol{p}_1\boldsymbol{p}_2)\,.\end{aligned} \tag{5.82}$$

A self-consistent solution of the BBG equations is a formidable numerical task, because of non-linear coupling of integral equations for G-matrices. No wonder that such calculations became feasible only in the late 1990s (Schulze *et al.* 1998; Vidaña *et al.* 2000a,b; Nishizaki *et al.* 2002; also see the review paper by Baldo & Burgio 2001 and references therein). Because of a tremendous complexity, an evaluation of the three-body correlation contribution *à la* Bethe-Faddeev has not been carried out up to now (till 2006). However, in view of uncertainties of the NH and HH interactions in dense matter, even the calculation of the EOS at the BHF level is so model-dependent, that the unsolved problem of higher-order correlations seems of second priority. Additional large uncertainties result from our ignorance of three-body interaction involving hyperons. As visualized by the BBG calculations of Nishizaki *et al.* (2002), three body interactions can play a decisive role in the EOS of hyperonic matter (see § 6.5.5). The EOS based on the two-body interactions only is usually too soft to support $1.44\ \mathrm{M}_\odot$ of the Hulse-Taylor pulsar. Inclusion of realistic three-body interactions involving hyperons is therefore a major task for the many-body calculations of the EOS.

### 5.10.2 Relativistic mean-field model

The mean-field method is particularly suitable for calculating the EOS of a uniform multi-component dense matter. Strong interactions of hadrons are invariant with respect to rotation in the isospin space, and therefore $\mathcal{L}$ is an iso-scalar. In particular, the coupling constants $g_{\phi B}$ do not depend on $I_{3B}$. The generalization of the RMF model equations to the full baryon octet is then straightforward: the total source of a meson field is a sum of contributions from all baryon fields. For the sake of generality, we will write all the equations in the form which enables one to include baryons beyond the lowest-mass baryon octet. In this section we will use units in which $\hbar = c = 1$.

The equations for the non-vanishing components of the $\omega$ and $\rho$ fields are:

$$\omega_0 = \sum_B \frac{g_{\omega B}}{m_\omega^2} n_B \,, \qquad \rho_0^0 = \sum_B \frac{g_{\rho B}}{m_\rho^2} I_{3B} n_B \,. \tag{5.83}$$

In the equation for the $\sigma$ field, the nucleon term has to replaced by a sum over all baryon terms, taking due account of the spin degeneracy,

$$\begin{aligned} m_\sigma^2 \sigma &= \sum_B \frac{2J_B + 1}{2\pi^2} g_{\sigma B} \int_0^{k_{\mathrm{F}B}} \mathrm{d}k \, \frac{k^2 (m_B - g_{\sigma B}\sigma)}{\sqrt{k^2 + (m_B - g_{\sigma B}\sigma)^2}} \\ &\quad - g_{\sigma N} \left[ b\, m_N (g_\sigma \sigma)^2 + c\, (g_\sigma \sigma)^3 \right] \,, \end{aligned} \tag{5.84}$$

where $J_\mathrm{B}$ is the baryon spin. The Fermi momenta of baryons $B$ are related to their number density $n_B$ by

$$n_B = \frac{2J_B + 1}{6\pi^2} k_{\mathrm{F}B}^3 \,. \tag{5.85}$$

In order to calculate the EOS of the ground-state matter, one needs the energy spectrum of baryons. It is obtained from the Dirac equations for bispinors $\psi_B$:

$$e_B(\boldsymbol{k}) = \sqrt{k^2 + (m_B - g_{\sigma B}\sigma)^2} + g_{\omega B}\omega_0 + g_{\rho B} I_{3B} \rho_0^0 \,. \tag{5.86}$$

The only modification in the expressions for the energy density and pressure consists in replacing nucleon contributions by sums over all baryons,

$$\mathcal{E}_{\rm B}(\{n_B\}) = \sum_B \frac{2J_B+1}{2\pi^2}\int_0^{k_{{\rm F}B}} {\rm d}k\, k^2\sqrt{k^2+(m_B-g_{\sigma B}\sigma)^2} + \frac{1}{2}m_\sigma^2\sigma^2$$
$$+\frac{1}{2}m_\rho^2(\rho_0^0)^2 + \frac{1}{2}m_\omega\omega_0^2 + \frac{1}{3}\,b\,m_N(g_{\sigma N}\sigma)^3 + \frac{1}{4}\,c\,(g_{\sigma N}\sigma)^4\,, \qquad (5.87)$$

$$P_{\rm B}(\{n_B\}) = \frac{1}{3}\sum_B \frac{2J_B+1}{2\pi^2}\int_0^{k_{{\rm F}B}} {\rm d}k\,\frac{k^4}{\sqrt{k^2+(m_B-g_{\sigma B}\sigma)^2}} - \frac{1}{2}m_\sigma^2\sigma^2$$
$$+\frac{1}{2}m_\rho^2(\rho_0^0)^2 + \frac{1}{2}m_\omega\omega_0^2 - \frac{1}{3}\,b\,m_N(g_{\sigma N}\sigma)^3 - \frac{1}{4}\,c\,(g_{\sigma N}\sigma)^4\,, \qquad (5.88)$$

where the values of time-like components of meson fields are expressed via $\{n_B\}$ by solving Eqs. (5.83) and (5.84).

The free parameters of the model are $g_{\sigma B}/m_\sigma$, $g_{\omega B}/m_\omega$, $g_{\rho B}/m_\rho$, $b$, and $c$. Experimental values of $m_B$ are used.

Due to the mean-field character of the model and the charge independence of strong interactions, the case of the full baryon octet is not much more complicated than the minimal case of the nucleon matter. Also, going beyond the baryon octet by including $\Delta$ resonances is not a very difficult task within the RMF model.

## 5.11. The equation of state of the outer core

The theoretical description of the matter at $\rho \lesssim 2\rho_0$ is within the reach of the modern nuclear theory. The nuclear Hamiltonian, albeit very complicated, is known reasonably well. The calculation of the ground state of nucleon matter requires big computing resources but can be carried out with a reasonable accuracy.

### 5.11.1 Calculating the equation of state

Consider the matter composed of nucleons, electrons, and possibly muons. Nucleons form a strongly interacting Fermi liquid, while electrons and muons constitute nearly ideal Fermi gases. The energy per unit volume is

$$\mathcal{E}(n_n,n_p,n_e,n_\mu) = \mathcal{E}_{\rm N}(n_n,n_p) + \mathcal{E}_e(n_e) + \mathcal{E}_\mu(n_\mu)\,, \qquad (5.89)$$

where $\mathcal{E}_{\rm N}$ is the nucleon contribution. In what follows, we will assume full thermodynamic equilibrium. The pressure and energy-density depend on a single parameter; its best choice is usually the baryon density $n_{\rm b}$. The equilibrium at given $n_{\rm b}$ corresponds to the minimum of $\mathcal{E}$ under the condition of electrical neutrality.

We will derive the equilibrium equations using a general method which can also be used in a multi-component matter containing, e.g., hyperons. Let us employ the method of Lagrange multipliers, particularly suitable for calculating the minimum of a function of many variables under additional constraints. In our case, the variables are the number densities $n_j$, $j = n, p, e, \mu$, and the constraints are

$$\text{fixed baryon density:} \qquad n_n + n_p - n_{\rm b} = 0 \,, \tag{5.90a}$$

$$\text{electrical neutrality:} \qquad n_e + n_\mu - n_p = 0 \,. \tag{5.90b}$$

Let us introduce the auxiliary function $\widetilde{\mathcal{E}}$, defined by

$$\widetilde{\mathcal{E}} = \mathcal{E} + \lambda_1(n_e + n_\mu - n_p) + \lambda_2(n_n + n_p - n_{\rm b}) \,. \tag{5.91}$$

In this case $\lambda_i$ are Lagrange multipliers to be determined from the unconstrained minimization of $\widetilde{\mathcal{E}}$ by requiring $\partial\widetilde{\mathcal{E}}/\partial n_j = 0$ for all $j$:

$$\partial\widetilde{\mathcal{E}}/\partial n_n = \mu_n + \lambda_2 = 0 \,, \tag{5.92a}$$

$$\partial\widetilde{\mathcal{E}}/\partial n_p = \mu_p - \lambda_1 + \lambda_2 = 0 \,, \tag{5.92b}$$

$$\partial\widetilde{\mathcal{E}}/\partial n_e = \mu_e + \lambda_1 = 0 \,, \tag{5.92c}$$

$$\partial\widetilde{\mathcal{E}}/\partial n_\mu = \mu_\mu + \lambda_1 = 0 \,, \tag{5.92d}$$

with $\partial\mathcal{E}/\partial n_j = \mu_j$. Eliminating the Lagrange multipliers from Eqs. (5.92) one gets the relation between the chemical potentials

$$\mu_n = \mu_p + \mu_e \,, \qquad \mu_\mu = \mu_e \,, \tag{5.93}$$

which expresses the equilibrium with respect to the weak-interaction processes

$$n \longrightarrow p + e + \overline{\nu}_e \,, \qquad p + e \longrightarrow n + \nu_e \,, \tag{5.94a}$$

$$n \longrightarrow p + \mu + \overline{\nu}_\mu \,, \qquad p + \mu \longrightarrow n + \nu_\mu \,. \tag{5.94b}$$

We consider a neutron-star core transparent for neutrinos (which occurs, typically, as soon as $T \lesssim 10^9 - 10^{10}$ K). In this case neutrinos do not affect the matter thermodynamics, and we can put $\mu_{\nu_e} = \mu_{\overline{\nu}_e} = \mu_{\nu_\mu} = \mu_{\overline{\nu}_\mu} = 0$.

Equations (5.93) supplemented by the constraints (5.90) form a closed system of equations which determine the equilibrium composition of the $npe\mu$ matter.

Electrons are ultra-relativistic, so that (in physical units, in which the electron Fermi momentum is $\hbar p_{{\rm F}e}$)

$$\mu_e = \hbar c p_{{\rm F}e} \approx 122.1\,(n_e/0.05 n_0)^{1/3}\ {\rm MeV} \,, \tag{5.95}$$

while muons are mildly relativistic

$$\mu_\mu = m_\mu c^2 \sqrt{1 + (\hbar p_{{\rm F}\mu}/m_\mu c)^2} \,. \tag{5.96}$$

Muons are present only if $\mu_e > m_\mu c^2 = 105.65$ MeV; in the opposite case we are dealing with the $npe$ matter.

Once the equilibrium is determined, the pressure is calculated from the first law of thermodynamics (at $T = 0$):

$$P = n_{\rm b}^2 \frac{{\rm d}(\mathcal{E}/n_{\rm b})}{{\rm d}n_{\rm b}} \ . \tag{5.97}$$

The derivative is taken at the equilibrium composition.

One may worry about density-dependent particle composition. Generally, the density dependence of particle fractions $x_j \equiv n_j/n_{\rm b}$ gives a non-vanishing contribution to the density derivative of the energy per nucleon. Let us treat $\mathcal{E}$ as a function of $n_{\rm b}$, $x_p$, $x_e$, and $x_\mu$. Then

$$P = n_{\rm b}^2 \left(\frac{\partial(\mathcal{E}/n_{\rm b})}{\partial n_{\rm b}}\right)_{\rm eq} + \frac{1}{n_{\rm b}} \sum_{j=p,e,\mu} \left(\frac{\partial \mathcal{E}}{\partial x_j}\right)_{\rm eq} \left(\frac{{\rm d}x_j}{{\rm d}n_{\rm b}}\right)_{\rm eq} \ , \tag{5.98}$$

where derivatives are taken at equilibrium. However, using Eqs. (5.90) and (5.93) one can see that the second term on the right-hand-side of Eq. (5.98) vanishes, i.e., both formulae for $P$ give the same result.

### 5.11.2 The nuclear symmetry energy and the proton fraction

Many-body calculations of the energy per nucleon in an asymmetric nuclear matter with realistic nucleon-nucleon interactions show that, to a very good approximation, the dependence on the neutron excess $\delta = 1 - 2x_p$ is quadratic (see, e.g., Lagaris & Pandharipande 1981c; Wiringa *et al.* 1988; Akmal *et al.* 1998):

$$E_{\rm N}(n_{\rm b}, \delta) \simeq E_0(n_{\rm b}) + S(n_{\rm b})\, \delta^2 \ . \tag{5.99}$$

Here, $E_0(n_{\rm b})$ refers to the symmetric nuclear matter and $S(n_{\rm b})$ is the symmetry energy. A very high precision of this formula, even for $\delta \simeq 1$, indicates that the higher-order terms of the expansion in $\delta$ are small.

In this context it is instructive to consider the free-Fermi gas (FFG) model of the nuclear matter, where the energy per baryon is

$$E^{\rm FFG}(n_{\rm b}, \delta) = \frac{3}{10}\, \epsilon_{\rm F}(n_{\rm b}) \left[(1+\delta)^{5/3} + (1-\delta)^{5/3}\right] \ . \tag{5.100}$$

Here, $m \equiv (m_n + m_p)c^2/2 = 938.93$ MeV is the mean nucleon mass and $\epsilon_{\rm F}$ is the Fermi energy in the symmetric nuclear matter at a given $n_{\rm b}$,

$$\epsilon_{\rm F} = \frac{\hbar^2}{2m} \left(\frac{3}{2}\pi^2 n_{\rm b}\right)^{2/3} \approx 36.8 \left(\frac{n_{\rm b}}{n_0}\right)^{2/3} \ {\rm MeV} \ . \tag{5.101}$$

The small-$\delta$ expansion of $E^{\rm FFG}$ reads then

$$E^{\rm FFG}(n_{\rm b},\delta) = \frac{3}{5}\,\epsilon_{\rm F}(n_{\rm b}) + \frac{1}{3}\,\epsilon_{\rm F}(n_{\rm b})\,\delta^2\,, \tag{5.102}$$

which gives the symmetry energy for the free Fermi gas model in the form

$$S^{\rm FFG} \approx 12.3\left(\frac{n_{\rm b}}{n_0}\right)^{2/3}\ {\rm MeV}\,. \tag{5.103}$$

It is easy to check that the quadratic approximation, Eq. (5.99), is very precise even at $\delta = 1$. From Eq. (5.100) applied to a pure neutron matter we obtain $E^{\rm FFG}(n_{\rm b},1) = (3/5)\,2^{2/3}\,\epsilon_{\rm F}(n_{\rm b}) \approx 0.9524\,\epsilon_{\rm F}(n_{\rm b})$, while Eq. (5.102) gives $14\epsilon_{\rm F}(n_{\rm b})/15 \approx 0.9333\,\epsilon_{\rm F}(n_{\rm b})$ which is only 2% smaller!

The simple form of the dependence of $E_{\rm N}$ on $x_p$ enables us to clarify the relation between the symmetry energy and the composition of the $npe$ matter at beta-equilibrium. Using Eq. (5.99) we can easily calculate the difference between the chemical potentials of neutrons and protons,

$$\mu_n - \mu_p = 4(1-2x_p)S(n_{\rm b})\,. \tag{5.104}$$

The beta-equilibrium in the $npe$ matter (where $x_e = x_p$) implies, therefore,

$$\frac{x_p^{1/3}}{1-2x_p} = \frac{4S(n_{\rm b})}{\hbar c\,(3\pi^2 n_{\rm b})^{1/3}}\,. \tag{5.105}$$

Accordingly, the proton fraction at a given $n_{\rm b}$ is determined by the symmetry energy. Under typical conditions, the proton fraction is small, $x_p \ll 1$, and

$$x_p(n_0) \approx \frac{64\,[S(n_{\rm b})]^3}{3\pi^2(\hbar c)^3\,n_{\rm b}} \approx 4.75\times10^{-2}\left(\frac{n_0}{n_{\rm b}}\right)\left(\frac{S(n_{\rm b})}{30\ {\rm MeV}}\right)^3\,, \tag{5.106}$$

i.e., the dependence of $x_p$ on the symmetry energy is very strong.

The proton fraction at the normal nuclear density $n_{\rm b} = n_0$ is directly determined by the symmetry energy $S_{\rm b}(n_0) = S_0$ at the saturation point (§ 5.4). As the experimental value of $S_0$ is $S^{\rm exp} \simeq 30$ MeV, the proton fraction in the neutron-star matter at the normal nuclear density should be $x_p(n_0) \simeq 5\%$, independently of any specific EOS of dense matter. On the other hand, Eq. (5.106) tells us that the actual value of $x_p(n_0)$ for a given model of dense matter is very sensitive to the value of $S_0$ of that model. In particular, the free Fermi-gas model yields a very small $S_0$, Eq. (5.103), and gives an unrealistically low value $x_p^{\rm FFG}(n_0) \approx 0.0033$.

## 5.12. Equation of state and composition

In the rectangular insert in Fig. 5.6 we show EOSs of the outer core calculated using five models of nucleon matter. They are briefly described in Table 5.3.

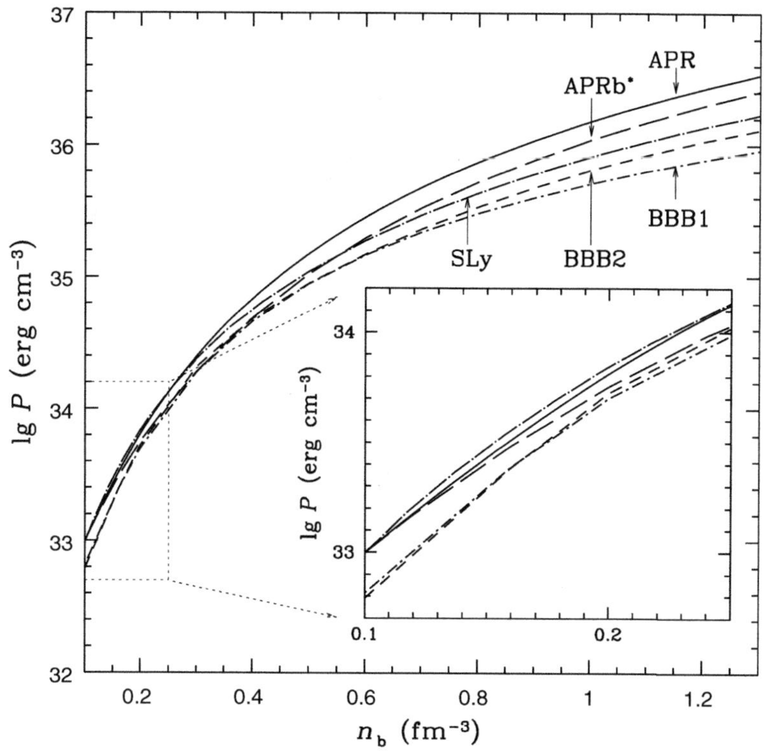


*Figure 5.6.* Pressure versus baryon number density for several EOSs of the $npe\mu$ matter in beta equilibrium. Labels are the same as in Table 5.3. The tables of the APR and APRb* EOSs have been kindly provided by J.M. Lattimer.

Two of them, BBB1 and BBB2, are obtained in the framework of the BBG theory assuming different realistic NN potentials and a model of the NNN interaction. Although the NN potentials are essentially different (the local U14 Urbana potential and the non-local momentum-dependent Paris potentials) they fit equally well NN data and give very similar EOSs of the outer neutron-star core.

The APR and APRb* EOSs are calculated using the variational method. The center-of-mass NN potential is the same (Argonne A18). However, the APRb* EOS includes a repulsive boost NN-interaction and uses the Urbana NNN force UIX* readjusted to experimental data (see § 5.5.3). The repulsive component of the UIX* NNN force is much weaker than that of the UIX NNN force in the APR EOS. However, these differences in the Hamiltonian slightly affect the EOS of the outer neutron-star core. Finally, the SLy EOS is based on a Skyrme-type energy density functional; it is very similar to the APR one.

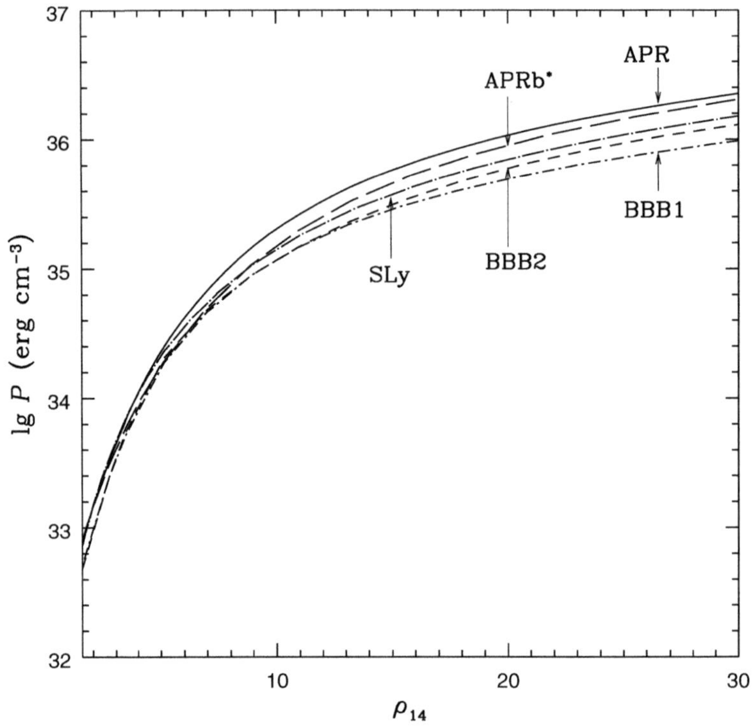


*Figure 5.7.* Pressure versus mass density for several EOSs of the $npe\mu$ matter. Labels are the same as in Table 5.3.

As we have shown in § 5.11.2, the proton fraction depends strongly on the symmetry energy $S(n_b)$. While the expression for the pressure contains $S(n_b)$ and $S'(n_b)$, the proton fraction is roughly proportional to the cube of $S(n_b)$. In this way, $x_p(n_b)$ "amplifies" the differences in $S(n_b)$ given by various nuclear-matter theories. This feature is clearly seen in the rectangular insert in Fig. 5.8.

The APR and APRb* $x_p(n_b)$ curves exhibit a kink at $n_b \sim 0.2\ \mathrm{fm}^{-3}$ while other curves are monotonous and smooth. According to Akmal *et al.* (1998), the kink is associated with a phase transition. Within the variational approach, the phase transition is reflected by a sudden change of the correlation functions, and specifically by a sudden and strong change of some healing distances $d_u$ (§ 5.9.3). Akmal *et al.* (1998) identify this phase transition with a $\pi^0$ condensation (see § 7.3). It should be stressed, however, that variational wave-functions incorporate only short-range correlations and cannot accurately describe a phase transition connected with long-range correlations (Akmal & Pandharipande, 1997; Akmal *et al.*, 1998). The actual phase transition is ac-

*Table 5.3.* Selected EOSs of neutron star cores

| EOS | model | reference |
|---|---|---|
| BPAL12 | $npe\mu$ energy density functional | Bombaci (1995) |
| BGN1H1 | $np\Lambda\Xi e\mu$ energy density functional | Balberg & Gal (1997) |
| FPS | $npe\mu$ energy density functional | Pandharipande & Ravenhall (1989) |
| BGN2H1 | $np\Lambda\Xi e\mu$ energy density functional | Balberg & Gal (1997) |
| BGN1 | $npe\mu$ energy density functional | Balberg & Gal (1997) |
| BBB2 | $npe\mu$ Brueckner theory, Paris NN plus Urbana UVII NNN potentials | Baldo *et al.* (1997) |
| BBB1 | $npe\mu$ Brueckner theory, Argonne A14 NN plus Urbana UVII NNN potentials | Baldo *et al.* (1997) |
| SLy | $npe\mu$ energy density functional | Douchin & Haensel (2001) |
| APR | $npe\mu$ variational theory, Argonne A18 NN plus Urbana UIX NNN potentials | Akmal *et al.* (1998) |
| APRb* | $npe\mu$ variational theory, Argonne A18 NN with boost correction plus adjusted Urbana UIX* NNN potentials | Akmal *et al.* (1998) |
| BGN2 | $npe\mu$ effective nucleon energy functional | Balberg & Gal (1997) |

companied by a density jump at the phase interface. It is incorporated in the EOS through the mixed-phase construction in the region $n_1 < n_b < n_2$, where the fraction of the denser phase increases monotonically from zero at $n_b = n_1$ to one at $n_b = n_2$ (we refer the reader to §7.2.3 for a general discussion of mixed phases). However, the effects of phase transitions on the EOS are rather weak. They are invisible in the APR and APRb* curves in Figs. 5.6 and 5.7, where interpolation between tabulated EOS points is made.

It is interesting to compare the values of $x_p$ given by different theories at $n_b = n_0$. These values range from 0.035 for the U14+UVII model to 0.06 for the APRb* one. A difference by a factor $\sim 2$ stems from using $S_0 = 28$ MeV and $S_0 = 35$ MeV, respectively; see Eq. (5.106). These values of the symmetry energy are still within extreme experimental values of $S_0$.

## 5.13. Inner core: minimal model – nucleons and leptons

### 5.13.1 Equation of state of the $npe\mu$ matter

The assumption that the matter at $\rho \gtrsim 2\rho_0$ has the same composition as the outer core leads to the "minimal model" of the inner neutron-star core. In Figs. 5.6 and 5.7 we present several selected EOSs calculated under this assump-

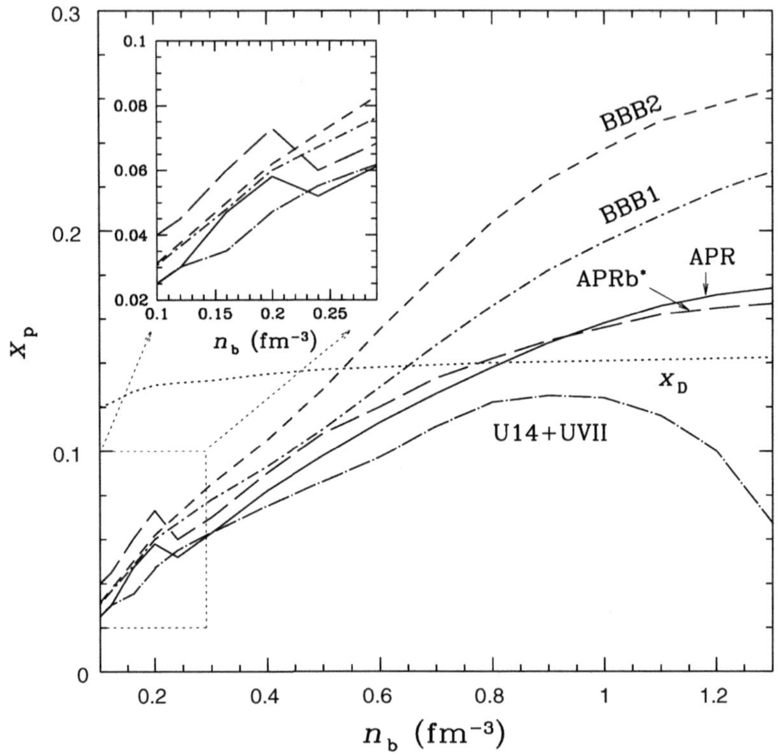


*Figure 5.8.* Proton fraction in $npe\mu$ matter at beta equilibrium for different EOSs (labeled as in Table 5.3). Characteristic negative-slope segments for the APR and APRb* models correspond to a narrow mixed-phase region. The dotted line gives the threshold value of $x_p$ above which the direct Urca process is allowed.

tion. They employ different nuclear Hamiltonians and many-body theories as described in § 5.12. The comparison of the APR and APRb* curves shows that a consistent inclusion of lowest-order relativistic effects in the interaction part of the two-body Hamiltonian combined with a simultaneous readjustment of the three-body force weakly affects the EOS even at $n_b = 1.2\ \mathrm{fm}^{-3}$. To some extent, the difference between the more modern APR and APRb* EOSs (based on the up-to-date NN and NNN Hamiltonians) and the BBB1 and BBB2 EOSs (based on older interactions constructed in the 1980s) results from the differences in the Hamiltonians themselves. The APR and APRb* EOSs are stiffer than those which employ older versions of NN and NNN interactions. The SLy EOS uses a phenomenological energy-density functional; its stiffness is intermediate between the EOSs obtained from the BBG theory and the variational method.

The overall divergence of $P(\rho)$ curves at higher densities is smaller than the divergence of $P(n_b)$ curves. This is natural because a higher stiffness implies also a steeper increase of the interaction contribution to $\mathcal{E} \equiv \rho c^2$. The hydrostatic equilibrium of a star is determined by the $P(\rho)$ curve. At $\rho = 10^{15}$ g cm$^{-3}$ pressures given by the different EOSs in Fig. 5.7 range within $(1.2-1.8) \times 10^{35}$ dyn cm$^{-2}$.[24] As we will see in § 6.5.5, these EOSs lead to the maximum masses of neutron stars $M_{\max} = (1.9 - 2.2)\ M_{\odot}$.

## 5.13.2 Proton fraction and direct Urca threshold

The behaviour of the proton fraction $x_p(n_b)$ is crucial for the cooling of neutron stars. A neutron star is formed as a hot compact object with an internal temperature $T \sim 10^{11}$ K in a gravitational collapse of a degenerate stellar core (see § 1.4.2). During the initial $10^5$–$10^6$ years of its life a star cools via neutrino emission from its core. The most efficient cooling channel is due to the so called direct Urca processes, mentioned in § 1.3.7,

$$n \longrightarrow p + e + \overline{\nu}_e\ , \qquad p + e \longrightarrow n + \nu_e\ , \tag{5.107a}$$

$$n \longrightarrow p + \mu + \overline{\nu}_\mu\ , \qquad p + \mu \longrightarrow n + \nu_\mu\ . \tag{5.107b}$$

Let us first consider the electron Urca processes in the upper line. These reactions are allowed only at rather high $n_b$ at which $x_p(n_b)$ exceeds a threshold value $x_D(n_b) \approx 0.11 - 0.14$ (Lattimer *et al.*, 1991). The threshold condition can be obtained as follows. Because neutrons, protons, and electrons form degenerate Fermi liquids, only the states close to the Fermi surfaces (within a shell of the thickness $\sim k_B T$ around the Fermi energy) are involved in the processes (5.107). Therefore, the momenta of neutrons, protons, and electrons can be approximated by their Fermi momenta $p_{Fj}$ ($j = n, p, e, \mu$), while the neutrino momentum $p_\nu \sim k_B T/c \ll p_{Fj}$. Neglecting small corrections $k_B T/c \ll p_{Fj}$ one sees that momentum conservation imposes the triangle rule:

$$p_{Fn} < p_{Fp} + p_{Fe}\ , \tag{5.108}$$

which is satisfied for $x_p > x_D$. In the absence of muons, $x_D = 1/9$; their presence slightly increases $x_D$ above $1/9$, and $x_D$ may become as large as 0.14 (see Fig. 5.8). Replacing electrons by muons in Eqs. (5.108) one can get the threshold proton fraction which opens the muon direct Urca process. This process becomes allowed at a slightly higher density than the electron one (Lattimer *et al.*, 1991).

The threshold density depends on an EOS model. For some EOSs, $x_p$ never reaches $x_D$ and direct Urca processes remain forbidden in neutron-star cores.

[24] Generalization of this statement, which refers to a restricted set of selected EOSs, should be taken with a grain of salt (see Sarsa *et al.* 2003; Morales *et al.* 2002)

An example of such a dense-matter model is U14+UVII of Wiringa *et al.* (1988), Fig. 5.8.[25] The neutrino emission proceeds then via the so called modified Urca processes with an additional nucleon in initial and final states of Eq. (5.107). This additional nucleon does not participate in beta processes but only opens it via a momentum transfer mediated by strong interactions. This strongly suppresses the neutrino emission rate. If direct Urca processes operate, then a non-superfluid neutron-star core cools to $10^9$ K in a minute, and to $10^8$ K in a year. If they are not allowed, the timescales will be one year and $10^5$ years, respectively.

### 5.13.3 Adiabatic index

An important parameter which characterizes the stiffness of the EOS with respect to density perturbations is the *adiabatic index*, defined by

$$\gamma = \frac{n_{\rm b}}{P}\,\frac{{\rm d}P}{{\rm d}n_{\rm b}} = \frac{P+\mathcal{E}}{P}\,\frac{{\rm d}P}{{\rm d}\mathcal{E}}\ . \tag{5.109}$$

It is calculated assuming full equilibrium of matter constituents and is generally a function of density. The function $\gamma(n_{\rm b})$ is plotted in Fig. 5.9 for some EOSs of the $npe\mu$ matter. For most of EOSs, $\gamma(n_{\rm b})$ varies from 2 to 4. However, for the BGN1, SLy and BG2 EOSs the adiabatic index is a weak function of density. These EOSs are special, because they are based on the Skyrme-type energy-density functionals,with a very specific smooth dependence of the pressure on neutron and proton number densities. The remaining EOSs are constructed in the framework of many-body theories based on realistic nuclear Hamiltonians. They have a complicated density dependence, which reflects different contributions from various components of nuclear interaction, particularly from the NNN component.

The composition of the $npe\mu$ matter is density dependent. Therefore, a change in density disturbs beta equilibrium. This in turn initiates weak interaction processes which move the matter toward the equilibrium on a characteristic relaxation timescale. Neutron star pulsations have characteristic periods $\sim 10^{-3}$ s, which are much shorter than the timescales of relaxation towards beta equilibrium. Thus, sound waves inside a neutron star core propagate at a constant composition, and the corresponding adiabatic index should be calculated at fixed (frozen) $x_j$. Such an adiabatic index will be denoted by $\gamma_{\rm fr}$; it is larger than the index $\gamma$ calculated from Eq. (5.109) assuming full equilibrium.

[25]The model U14+UVII of Wiringa *et al.* (1988) belongs to a class of models, which were developed in the 1980s and showed a non-monotonous $S(n_{\rm b})$. For these models, $S(n_{\rm b})$ starts to decrease with increasing $n_{\rm b}$ above some density, and can even vanish at a very large $n_{\rm b} > 1\ {\rm fm}^{-3}$. Other models of Wiringa *et al.* (1988), for instance A14+UVII, also show a non-monotonous behavior of $S(n_{\rm b})$ but their symmetry energy never vanishes in a neutron-star core. For more recent EOSs, developed in the 1990s, both $x_p$ and $S(n_{\rm b})$ increase monotonously with increasing $n_{\rm b}$ in the stellar core.

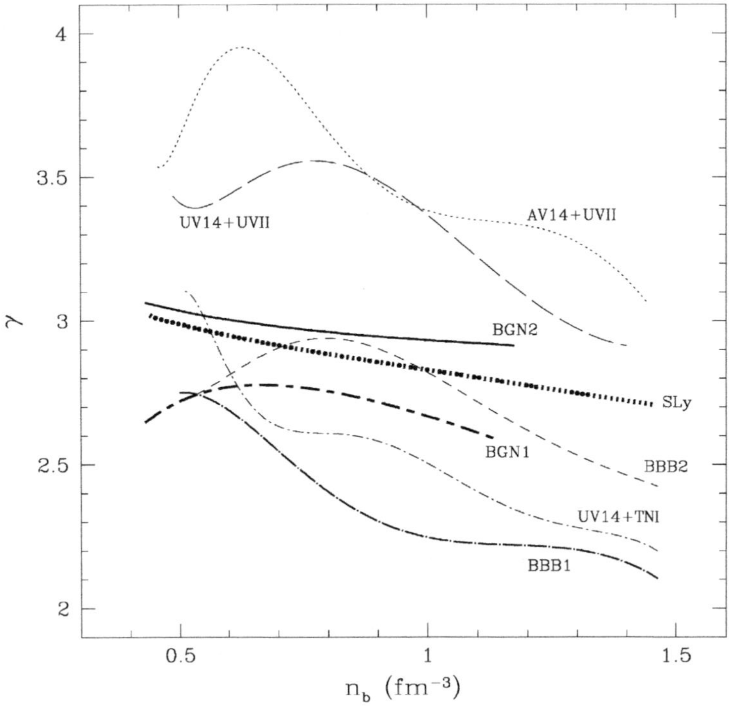


*Figure 5.9.* Adiabatic index of the $npe\mu$ matter in an inner neutron-star core versus baryon density for selected EOSs. All but three EOSs are from Table 5.3, the remaining ones are the EOSs of Wiringa *et al.* (1988). The index is calculated using analytic fits of $P(n_\mathrm{b})$ to tabulated EOSs.

It is $\gamma_\mathrm{fr}$ (not $\gamma$), which enters the expression for the local velocity of sound $v_\mathrm{s}$:

$$\frac{v_\mathrm{s}}{c} = \left(\frac{\mathrm{d}P}{\mathrm{d}\mathcal{E}}\right)^{1/2}_\mathrm{fr} = \left(\frac{\gamma_\mathrm{fr}P}{\mathcal{E}+P}\right)^{1/2} . \tag{5.110}$$

## 5.14. Hyperons in the inner core

A presupernova core at the brink of a collapse contains atomic nuclei (of the Fe-Ni group), alpha particles, free nucleons, electrons and positrons, but not hyperons. However, a huge gravitational compression can initiate the transformation of nucleons into hyperons, as soon as such transformation lowers the energy density at a given $n_\mathrm{b}$. This process is mediated by the strangeness-changing weak interaction and may become possible at $\rho \gtrsim 2\rho_0$.

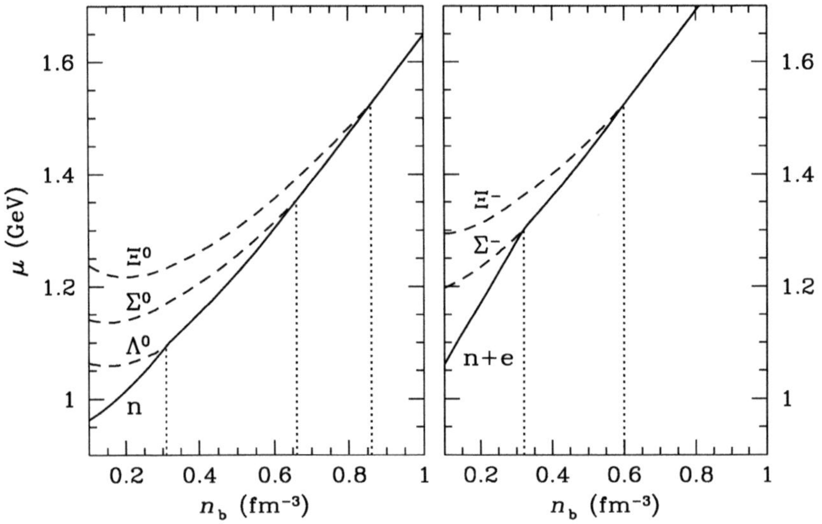


*Figure 5.10.* Threshold chemical potentials of neutral hyperons and neutron (*left*) and of negatively charged hyperons and the sum $\mu_e + \mu_n$ (*right*) versus baryon number density for model C of Glendenning (1985). Vertical dotted lines mark the thresholds for the creation of new hyperons; dashed lines show minimum enthalpies $\mu_H^0$ of unstable hyperons before the thresholds.

## 5.14.1 Hyperonic composition

Let us consider an electrically neutral matter composed of baryons $B$ (nucleons and hyperons) and leptons $\ell$ (electron and muons) at a given baryon number density $n_{\rm b}$. The baryon density is

$$\sum_B n_B = n_{\rm b} \, , \tag{5.111}$$

while the electric charge neutrality implies

$$\sum_B n_B Q_B - \sum_{\ell=e,\mu} n_\ell = 0 \, , \tag{5.112}$$

where $Q_B$ is the electric charge of a baryon $B$ in units of $e$. The energy density depends on the number densities of baryons $\{n_B\}$ and leptons $(n_e, n_\mu)$, $\mathcal{E} = \mathcal{E}(\{n_B\}, n_e, n_\mu)$. The equilibrium state has to be determined by minimizing $\mathcal{E}$ under the constraints given by Eqs. (5.111) and (5.112). To this aim, we will use the method of Lagrange multipliers described in § 5.11.1. In analogy with Eq. (5.91) we define the auxiliary energy density $\widetilde{\mathcal{E}}$

$$\widetilde{\mathcal{E}} = \mathcal{E} + \lambda_{\rm b} \left( \sum_B n_B - n_{\rm b} \right) + \lambda_{\rm q} \left( \sum_B Q_B n_B - \sum_{\ell=e,\mu} n_\ell \right) . \tag{5.113}$$

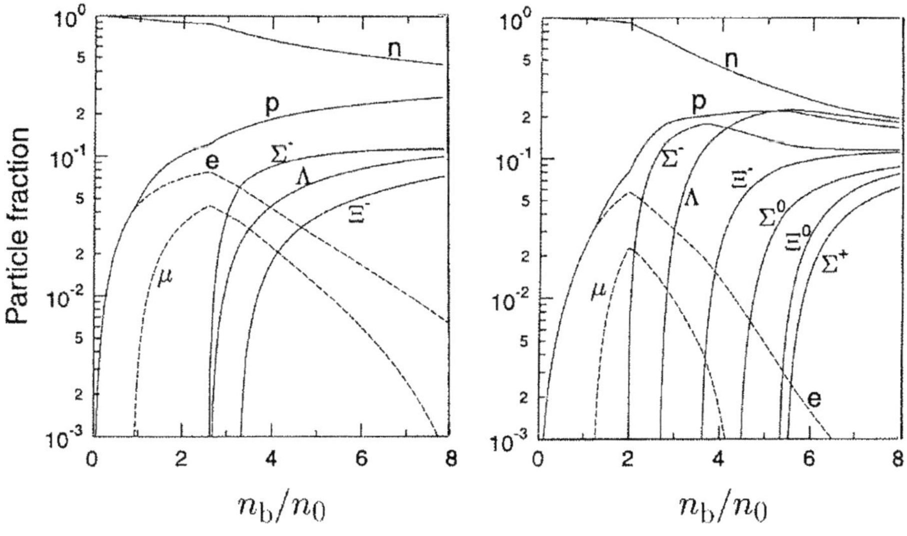


*Figure 5.11.* Fractions of particles $x_j = n_j/n_\mathrm{b}$ versus baryon number density $n_\mathrm{b}$ (in units of $n_0 = 0.16\ \mathrm{fm}^{-3}$) calculated by Hanauske *et al.* (2000) for two relativistic models of baryonic interactions (with the kind permission of the author). *Left*: Effective chiral model of Hanauske *et al.* (2000). *Right*: Relativistic mean field model TM1 of Sugahara & Toki (1971).

Let $N_\mathrm{B}$ be the number of the baryon species. Minimizing $\widetilde{\mathcal{E}}$, we get a set of $N_\mathrm{B} + 2$ equations

$$\partial\widetilde{\mathcal{E}}/\partial n_B = \mu_B + \lambda_\mathrm{b} + \lambda_\mathrm{q}\, Q_B = 0 \qquad (B = 1, \ldots, N_\mathrm{B}), \quad (5.114a)$$

$$\partial\widetilde{\mathcal{E}}/\partial n_\ell = \mu_\ell - \lambda_\mathrm{q} = 0 \qquad (\ell = e,\ \mu), \quad (5.114b)$$

where $\lambda_\mathrm{b}$ and $\lambda_\mathrm{q}$ are Lagrange multipliers and $\mu_j = \partial\mathcal{E}/\partial n_j$.

Eliminating Lagrange multipliers, we get a system of $N_\mathrm{B}$ relations for $N_\mathrm{B} + 2$ chemical potentials. We have two additional relations, Eqs. (5.111) and (5.112), so that the total number of equations is equal to $N_\mathrm{B} + 2$. The relations involving the chemical potentials of nucleons and leptons are equivalent to Eqs. (5.93) obtained for the $npe\mu$ matter: $\mu_e = \mu_\mu$, $\mu_n = \mu_p + \mu_e$. However, we have now additional equations which describe equilibrium with respect to weak interactions. The equilibrium equations depend on $Q_B$. In our case $Q_B = -1, 0, 1$:

$$Q_B = -1 \ : \quad \mu_{B^-} = \mu_n + \mu_e\,, \quad (5.115a)$$

$$Q_B = 0 \ : \quad \mu_{B^0} = \mu_n\,, \quad (5.115b)$$

$$Q_B = +1 \ : \quad \mu_{B^+} = \mu_n - \mu_e\,. \quad (5.115c)$$

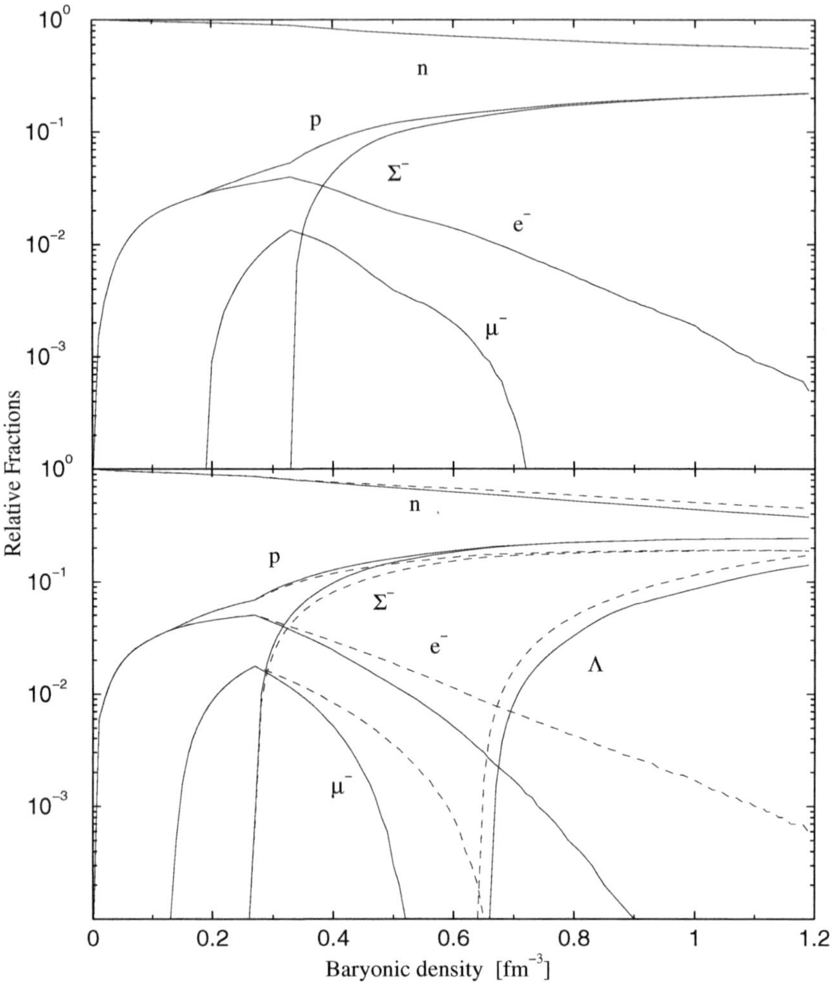


*Figure 5.12.* Particle fractions $x_j = n_j/n_\mathrm{b}$ versus $n_\mathrm{b}$ as calculated by Vidaña *et al.* (2000b) in the BHF approximation for two models of baryonic interactions (with the kind permission of the authors). *Upper panel*: Nijmegen model E of Rijken *et al.* (1999): only $\Sigma^-$ hyperon is present in neutron-star cores. *Lower panel*: APR model for the nucleon sector (Table 5.3, Akmal *et al.* 1998) and Nijmegen model E of Rijken *et al.* (1999) for NH and HH interactions. In contrast to the upper panel, $\Lambda$ is present in dense matter. Solid lines in both panels: all baryon-baryon (NN, NH, HH) interactions are included. Dashed lines in the lower panel: HH interaction is (artificially) switched off.

The lightest baryons form an octet, containing nucleons and $\Lambda$, $\Sigma$, $\Xi$ hyperons (§ 5.10.2, Table 5.2).

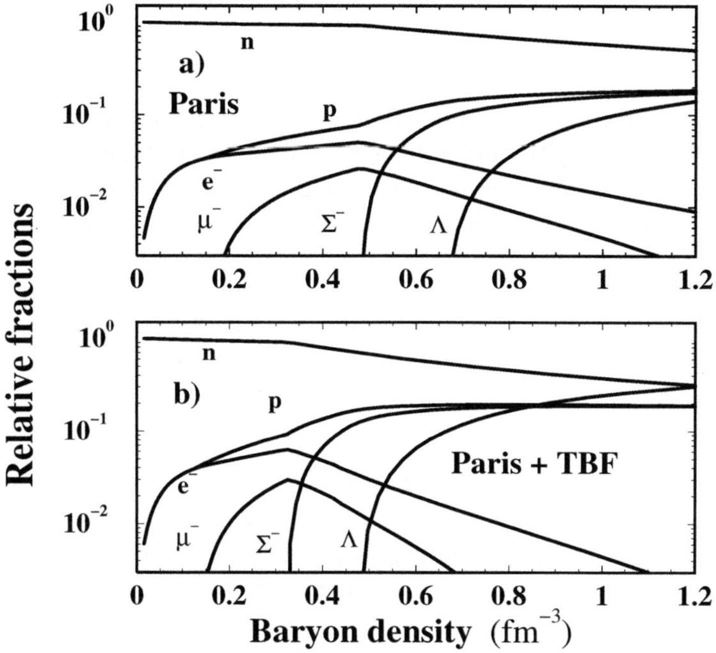


*Figure 5.13.* Effect of the three-body forces between nucleons on particle fractions $x_j = n_j/n_{\rm b}$ in dense matter; from Baldo & Burgio (2001) with the kind permission of the authors (see text for details).

One can calculate the threshold densities $\rho_H$ of hyperons ($H$) by checking the threshold condition at various $\rho$. One can start at $\rho = 0.5\,\rho_0$, where hyperons are certainly absent. Nevertheless, one can always calculate the minimum increase of the energy of the matter produced by adding a single hyperon $H$ at a fixed pressure $P$. This can be done by considering the energy of the matter with an admixture of given hyperons and by calculating numerically the limit of the derivative

$$\lim_{n_H \longrightarrow 0} (\partial\mathcal{E}/\partial n_H)_{\rm eq} \equiv \mu_H^0 \; . \tag{5.116}$$

To be specific, consider the lightest $\Lambda$ hyperon. As long as $\mu_\Lambda^0 > \mu_n$, this hyperon cannot survive because the system will lower its energy via an exothermic reaction $\Lambda + N \longrightarrow n + N$. However, $\mu_n$ increases with growing $n_{\rm b}$ and the functions $\mu_\Lambda^0(n_{\rm b})$ and $\mu_n(n_{\rm b})$ intersect at some $n_{\rm b} = n_{\rm c}^\Lambda$ (the left panel in Fig. 5.10). For $n_{\rm b} > n_{\rm c}^\Lambda$, the $\Lambda$ hyperons become stable in dense matter: their decay is blocked by the Pauli principle.

As we have already mentioned, usually $\Lambda$ is not the first one to appear in a neutron-star core, because $\Sigma^-$ appears at lower density (§ 5.2). This is visualized in Figs. 5.11, 5.12, and 5.13. The threshold condition for the $\Sigma^-$ hyperon creation is

$$\mu^0_{\Sigma^-} = \mu_n + \mu_e \ . \tag{5.117}$$

The electron Fermi energy $\mu_e$ adds to $\mu_n$ and the threshold condition is usually satisfied at lower density (and at lower $\mu_n$) than for $\Lambda$. However, this is not a strict rule, as one can see by comparing the left and right panels of Fig. 5.10. The figure shows also examples of the appearance of other hyperons (for $Q = 0$ and $Q = -1$).

Large values of $\mu_e$ may prohibit the appearance of $Q = +1$ hyperons in neutron-star cores. For example, consider $\Sigma^+$, the lightest positively charged hyperon. As follows from the bottom line of Eqs. (5.115), the condition of its appearance is

$$\mu^0_{\Sigma^+} = \mu_n - \mu_e \ . \tag{5.118}$$

The subtraction of $\mu_e$ can easily lead to $\mu^0_{\Sigma^+} > \mu_n - \mu_e$ in dense matter, making $\Sigma^+$ unstable (because the process $\Sigma^+ + e \to n + \nu_e$ is exothermic). Accordingly, $\Sigma^+$ hyperons do not appear in neutron-star cores in some models (see, e.g., Weber & Weigel 1989a; Vidaña *et al.* 2000b). However, there is no strict rule to forbid their presence. First, a rapid increase of the $\Sigma^-$ fraction at $\rho > \rho_{\rm c}^{\Sigma^-}$ leads to a strong decrease of the electron and muon fractions with increasing density: it is energetically advantageous to replace negatively charged and highly energetic electrons by slow and massive $\Sigma^-$ hyperons with the same charge (see Figs. 5.11–5.13). The removal of electrons lowers $\mu_e$ and allows $\Sigma^+$ to appear at $n_{\rm b} \gtrsim 5\, n_0$ in many models (see the right panel of Fig. 5.11; also see, e.g., Glendenning 1985; Hanauske *et al.* 2000; Balberg & Gal 1997).

The lack of the precise knowledge of NH interaction and *a fortiori* of HH interaction strongly affects the predicted hyperonic composition of neutron-star cores. Another uncertainty is introduced by many-body theories as illustrated in Fig. 5.11. Both baryon-baryon interaction models fit equally well empirical parameters of nuclear matter at saturation. However, their hyperonic sectors and, consequently, dense-matter parameters are different. The left panel of Fig. 5.11 refers to baryonic matter composed mostly of neutrons, with $\sim$20% of protons and only a few percent of hyperons even at $n = 6\, n_0 \simeq 1$ fm$^{-3}$. The right panel shows a drastically different situation. At $n_{\rm b} \simeq 1$ fm$^{-3}$, neutrons constitute less than one third of baryons: we are dealing with a "baryon soup," where six hyperon species are present. At $n_{\rm b} = 8\, n_0$, we have $n_\Lambda \simeq n_n$! While $\Sigma^0$, $\Xi^0$, and $\Sigma^+$ hyperons are absent in the left panel of Fig. 5.11, they are all present at $n_{\rm b} \gtrsim 6\, n_0$ in the right panel.

The dependence of the matter composition on the baryon-baryon interaction is further illustrated in Fig. 5.12. The curves are calculated using the BHF theory (Vidaña *et al.*, 2000b). The solid curves in both panels are obtained assuming the same NH and HH interactions (model E of Rijken *et al.* 1999) but different versions of NN interaction. Specifically, in the upper panel one employs model E of NN interaction of Rijken *et al.* (1999), while in the lower panel one uses the A18 model of NN interaction supplemented with the UIX model of the NNN force of Akmal *et al.* (1998). The EOS model of Akmal *et al.* (1998) with its three nucleon force is much stiffer than that of Rijken *et al.* (1999). Clearly, the softening of nucleon EOS shifts the hyperon appearance to higher densities; for instance, we have $n_{\rm c}^{\Lambda} > 1.2~{\rm fm}^{-3}$ in the upper panel! On the other hand, the comparison of the solid and dotted curves in the lower panel of Fig. 5.12 shows that HH interaction removes leptons from high-density matter; switching off the HH forces strongly increases the lepton fraction via substantial lowering of the $\Sigma^-$ fraction.

Figure 5.13 illustrates the effect of the NNN interaction on the composition of the matter (from Baldo & Burgio 2001). The upper panel corresponds to purely two-body interactions, while the lower panel is obtained including NNN forces. The latter forces stiffen the nucleon component of the EOS, lower the thresholds of the hyperon appearance and increase hyperon fractions at high densities. One has $n_{\Sigma^-} \simeq n_p$ at $n_{\rm b} = 0.8~{\rm fm}^{-3}$ and $n_\Lambda \simeq n_n$ at $n_{\rm b} = 1.2~{\rm fm}^{-3}$. A sizable abundance of $\Sigma^-$ (the lower panel) dramatically decreases the electron fraction at high $\rho$.

In spite of a strong model dependence, one can notice generic features of hyperonic and leptonic compositions. The $\Sigma^-$ hyperons appear usually at lowest densities which depend mostly on the stiffness of the EOS of the $npe\mu$ matter: the stiffer the EOS, the lower the threshold.[26] Hyperonization implies a deleptonization: negatively charged $e$ and $\mu$ are replaced by $\Sigma^-$ and other negatively charged hyperons. Finally, $\Sigma^+$ is usually the only $Q = +1$ hyperon present in neutron-star cores, but its presence is model-dependent.

All Figs. 5.11–5.13 demonstrate a generic shape of $x_e(n_{\rm b})$ and $x_\mu(n_{\rm b})$ curves. Muons appear at $n_{\rm b} \sim n_0$. Both lepton fractions increase while the density grows up to $n_{\rm c}^{\Sigma^-}$ and then decrease since leptons are replaced by $\Sigma^-$ hyperons. Models with high hyperon fractions show an evanescence of leptons at $n_{\rm b} \sim 1~{\rm fm}^{-3}$ producing a "baryon soup" at $n_{\rm b} \sim 1~{\rm fm}^{-3}$ with high strangeness per baryon (close to $-1$ in "hyperon stars" with $M \simeq M_{\rm max}$).

### 5.14.2 Hyperon softening of equation of state

The formation of hyperons softens the EOS because high-energy neutrons are replaced by more massive low-energy hyperons (producing lower pressure).

[26]If $\Sigma^-$ interaction with nucleon matter has a strong repulsive component, then $\Sigma^-$ might be absent in dense matter. This is so in the BGN1H1 and BGN2H1 models of Balberg & Gal (1997), who introduced a repulsive component of interaction of $\Sigma^-$ with nuclear matter, as suggested by some analyses of $\Sigma^-$ atoms (Batty *et al.*, 1994; Mareš *et al.*, 1995).

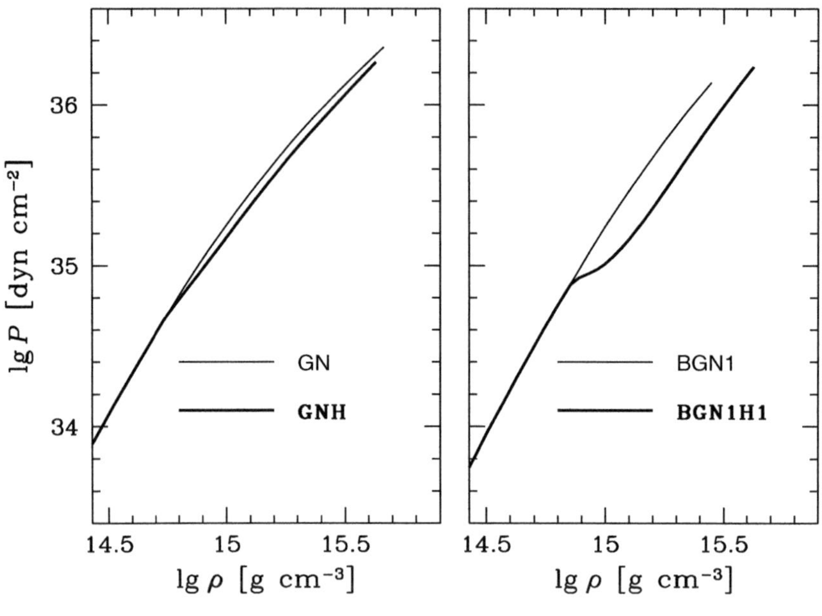


*Figure 5.14.* Softening of EOSs by the presence of hyperons. Each panel shows an EOS with (thick line) and without (thin line) hyperons. *Left*: model EOSs of Glendenning (1985). *Right*: the BGN1 and BGN1H1 EOSs of Balberg & Gal (1997) (labeled as in Table 5.3)

The softening is a generic effect independent of models of NH and HH interactions, as illustrated in Fig. 5.14, but its magnitude is model-dependent.

The softening is reflected in the density dependence of the adiabatic index $\gamma$, Eq. (5.109), as demonstrated in Fig. 5.15. The index drops at each hyperon threshold $\rho_H$. At densities slightly higher than $\rho_H$, the number density of hyperon species $H$ is small, i.e., these hyperons mainly interact with nucleons. Thus, the softening and associated drop of $\gamma$ strongly depend on the NH interaction. For example, these effects are much stronger for the BGN1H1 EOS of Balberg & Gal (1997) than for the GNH EOS of Glendenning (1985).

Let us stress that the adiabatic index considered here is calculated for dense matter in full equilibrium and can be used for static neutron stars. The adiabatic index relevant for neutron-star pulsations has to be calculated without any assumption of equilibrium with respect to weak interactions and may be quite different from the equilibrium one (Haensel *et al.*, 2002b).

The EOSs at $n_b \gtrsim 5\,n_0$ become especially model-dependent. In particular, a possible contribution of three-body forces involving hyperons is almost unknown. A very preliminary study of the contribution of three-baryon interaction, carried out within the BBG theory, was presented by Nishizaki *et al.* (2002). Assuming universal three-body repulsion acting between baryons,

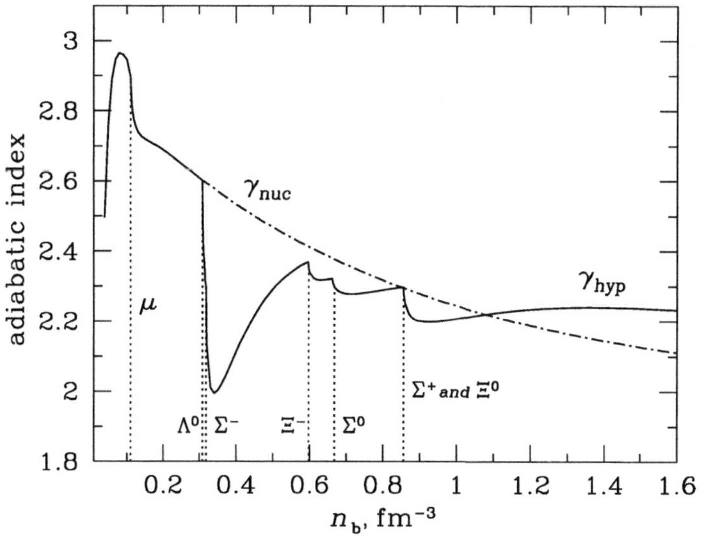


*Figure 5.15.* The adiabatic index $\gamma$ versus $n_b$ in a neutron-star core. Calculations are performed for the model EOS of Glendenning (1985). The solid line ($\gamma_{hyp}$) is for the hyperonic matter (vertical dotted lines indicate thresholds for the appearance of muons and hyperons); the dot-and-dashed line ($\gamma_{nuc}$) corresponds to the case in which the appearance of hyperons is artificially forbidden.

Nishizaki *et al.* (2002) found a strong stiffening of the high-density EOS, resulting in a significant increase of $M_{max}$ (see § 6.5.5).

## 5.15. Superluminal and ultrabaric equations of state

Since the EOS at $n_b \gg n_0$ is very uncertain, it important to impose model-independent bounds. The basic requirement is that any EOS should respect *Lorentz invariance* and *causality*. These requirements can be formulated in different (and not equivalent) ways. For instance, it has been claimed that a physically correct EOS can be neither *superluminal* nor *ultrabaric*, where

$$\text{ultrabaric EOS:} \qquad P > \mathcal{E}\ ; \tag{5.119}$$

$$\text{superluminal EOS:} \qquad \mathrm{d}P/\,\mathrm{d}\mathcal{E} > 1\ . \tag{5.120}$$

Some of EOSs in Fig. 5.16 are superluminal beyond a point indicated by an asterisk (BGN2, SLy). For the SLy EOS, the superluminality occurs at densities which are not realized in stable neutron stars. However, for the BGN2 EOS the most massive neutron stars have a superluminal central core with $v_s > c$. Both EOSs are derived within a non-relativistic many-body theory. Therefore one may attribute their superluminality to the lack of Lorentz invariance in non-relativistic theories.

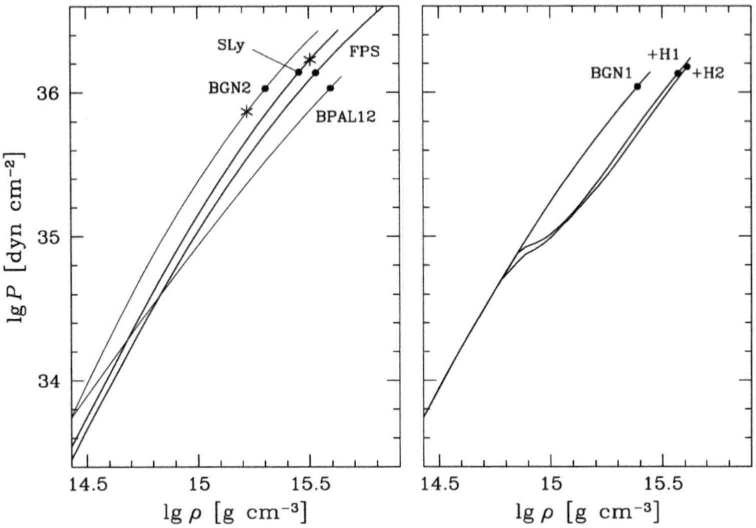


*Figure 5.16.* Selected model EOSs of neutron-star cores (denoted as in Table 5.3). Filled circles show maximum densities in stable neutron stars, while asterisks indicate the densities above which EOSs are superluminal ($v_{\rm s} > c$).

Still, there is a need to clarify the relation between the widely used condition of $v_{\rm s} < c$, Lorentz invariance, and causality. For this purpose we formulate two "questions of principle": (*a*) can a Lorentz-invariant theory give a superluminal or ultrabaric EOS? (*b*) does a superluminal EOS with $\mathrm{d}P/\,\mathrm{d}\mathcal{E} > 1$ necessarily contradict causality? These problems have been analyzed by many authors (Bludman & Ruderman, 1968, 1970; Ruderman, 1968; Caporaso & Brecher, 1979; Olson , 2000). Let us start with the simplest solvable model of dense matter, an ideal gas of (noninteracting) particles:

$$\mathcal{E} = mn\,\frac{1}{A}\left\langle \sum_{i=1}^{A} c^2/\sqrt{1-v_i^2/c^2} \right\rangle , \tag{5.121a}$$

$$P = \frac{1}{3}\,mn\,\frac{1}{A}\left\langle \sum_{i=1}^{A} v_i^2/\sqrt{1-v_i^2/c^2} \right\rangle , \tag{5.121b}$$

where $A$ is the number of particles, $n$ and $m$ is their number density and mass, $v_i$ is a particle velocity, and $\langle \ldots \rangle$ means averaging over a particle distribution. Since $v_i^2 \le c^2$ one gets $P \le \frac{1}{3}\mathcal{E}$ for an ideal gas model, and the equality $P = \frac{1}{3}\mathcal{E}$ is achieved in the ultra-relativistic limit.

The upper bound $P \le \frac{1}{3}\mathcal{E}$, valid for an ideal gas, can be broken in the presence of strong short-range repulsive interactions. Long ago, Zeldovich

(1961) presented a classical Lorentz-invariant model, where a strong repulsive interaction between nucleons was produced by the exchange of a neutral vector (spin $J = 1$) meson of mass $m_{\rm v}$.[27] The interaction potential was taken in the Yukawa form

$$V_{ij} = g_{\rm v}^2 \, \frac{{\rm e}^{-\mu_{\rm v} r_{ij}}}{r_{ij}} \, , \tag{5.122}$$

where $g_{\rm v}$ is the coupling constant and $\mu_{\rm v} = m_{\rm v}c/\hbar$. The energy density is

$$\mathcal{E} = \mathcal{E}_{\rm FFG} + \mathcal{E}_{\rm int} \, , \tag{5.123}$$

where $\mathcal{E}_{\rm FFG}$ is the free Fermi-gas (FFG) contribution. The interaction term is

$$\mathcal{E}_{\rm int} = \frac{1}{2} \, g_{\rm v}^2 n_{\rm b} \sum_{i \neq j}^{A} \frac{{\rm e}^{-\mu_{\rm v} r_{ij}}}{r_{ij}} \, . \tag{5.124}$$

Let us define a characteristic internucleon distance $a$ by $a^{-3} = 4\pi n_{\rm b}/3$. In the limit of $a \ll 1/\mu_{\rm v}$, one can replace summation by integration[28]

$$\frac{1}{A} \sum_{i \neq j}^{A} f(r_{ij}) \longrightarrow n_{\rm b} \int {\rm d}\boldsymbol{r} \, f(r) \, , \tag{5.125}$$

which gives $\mathcal{E}_{\rm int} = \beta \, n_{\rm b}^2$, with $\beta = 2\pi g_{\rm v}^2/\mu_{\rm v}^2$, and coincides with the familiar expression obtained in the quantum theory using the Hartree approximation for the ground-state of nucleon matter. The complete EOS reads

$$\mathcal{E} = \mathcal{E}_{\rm FFG} + \beta n_{\rm b}^2 \, , \quad P = P_{\rm FFG} + \beta n_{\rm b}^2 \, . \tag{5.126}$$

For sufficiently large $n_{\rm b}$ the interaction term will imply $P > \mathcal{E}/3$ and for $n_{\rm b} \longrightarrow \infty$ one gets $P \longrightarrow \mathcal{E}$. The Zeldovich example shows that a Lorentz-invariant dense-matter model can lead to the asymptotic behavior $P/\mathcal{E} \longrightarrow 1$ and ${\rm d}P/\,{\rm d}\mathcal{E} \longrightarrow 1$ (from below) for $n_{\rm b} \longrightarrow \infty$, but the EOS never becomes superluminal or ultrabaric.

*Is a superluminal EOS incompatible with Lorentz invariance?* Bludman & Ruderman (1968) gave a negative answer to this question. They presented two models which are Lorentz-invariant and still lead to $v_{\rm s} > c$ and $P > \mathcal{E}$ for sufficiently large $n_{\rm b}$. Let us consider a one-parameter EOS $\mathcal{E} = \mathcal{E}(n_{\rm b})$ and $P = P(n_{\rm b})$. It is easy to show the sound speed can be expressed as

$$v_{\rm s}/c = \left( n_{\rm b} \mathcal{E}''/\mathcal{E}' \right)^{1/2} \, , \tag{5.127}$$

[27] It corresponds to $\omega$ meson of the relativistic mean-field model of § 5.9.4.

[28] This limit is reached at densities $n_{\rm b} \gg 3\mu_{\rm v}^3/4\pi$. For $m_{\rm v}c^2 = 500$ MeV we have $1/\mu_{\rm v} \approx 0.4$ fm which corresponds to $n_{\rm b} \gg 24\, n_0$.

where $\mathcal{E}' = \mathrm{d}\mathcal{E}/\,\mathrm{d}n_\mathrm{b}$ and $\mathcal{E}'' = \mathrm{d}^2\mathcal{E}/\,\mathrm{d}^2 n_\mathrm{b}$. In this case

$$n_\mathrm{b}\mathcal{E}'' > \mathcal{E}' \Longrightarrow v_\mathrm{s} > c\,. \tag{5.128}$$

Assuming $\mathcal{E} \propto n_\mathrm{b}^{2+\eta}$ in the limit of $n_\mathrm{b} \longrightarrow \infty$, we have

$$v_\mathrm{s}/c = \sqrt{1+\eta} > 1 \quad \text{if} \quad \eta > 0\,. \tag{5.129}$$

The Zeldovich model employed the Hartree approximation, which neglects both quantum effects and correlations between nucleons due to the strong short-range repulsion. The energy can be lowered by including (classical) correlations if a distance between neighboring nucleons becomes comparable or smaller than the range of the repulsive interaction, $r_{ij} \lesssim 1/\mu_\mathrm{v}$. At a fixed $n_\mathrm{b}$, the correlations decrease $\mathcal{E}$ and increase $P$, compared to the values obtained in the Zeldovich (1961) model. Therefore, the correlations stiffen the EOS, which corresponds to $\eta > 0$ and $v_\mathrm{s} > c$ for $n_\mathrm{b} \longrightarrow \infty$. Moreover, an EOS which includes the correlations can even become ultrabaric provided the interaction energy in a particle-cell volume $1/n_\mathrm{b}$ exceeds $mc^2$. All in all, the examples presented by Bludman & Ruderman (1968) show that the Lorentz invariance does not necessarily imply $v_\mathrm{s} < c$ and $P < \mathcal{E}$ at $n_\mathrm{b} \longrightarrow \infty$.

*Is a superluminal EOS inconsistent with causality?* The inequality $v_\mathrm{s} > c$ *can* reflect the lack of causality, as a signal (a density perturbation) propagates faster than light. However, $v_\mathrm{s} > c$ can also result from an interference of various spectral components of a sound wave, with amplification at high-frequency components and damping (destructive interference) of low-frequency ones. This may happen if the medium is not in its true ground state.

Up to this point, $\mathcal{E}_\mathrm{int}$ was calculated in the Hartree approximation, i.e., neglecting quantum correlations which result from the exclusion principle. This approach can be improved by using the Hartree-Fock approximation. Additional (exchange) terms in $\mathcal{E}_\mathrm{int}$ can also lead to $\eta > 0$, and they can make the nucleon matter with a neutral vector-meson repulsion superluminal and ultrabaric at $n_\mathrm{b} \longrightarrow \infty$ (Bludman & Ruderman, 1970). However, the most important quantum effect are connected with a possible formation of nucleon-antinucleon ($\mathrm{N\overline{N}}$) pairs. A strong NN repulsion corresponds to a strong $\mathrm{N\overline{N}}$ attraction. If the NN repulsion makes the matter superluminal and ultrabaric, then the matter becomes unstable with respect to the $\mathrm{N\overline{N}}$ formation: the matter described by such an EOS is in a highly excited state (Bludman & Ruderman, 1970)!

Let us summarize the attempts to answer the questions of principle. As we have seen, Lorentz invariance and causality do not exclude the possibility of $v_\mathrm{s} > c$. However, "counter examples", where the inequality $v_\mathrm{s} < c$ is violated, seem to be irrelevant to neutron-star physics. Moreover, the densities at which EOSs based on these "counter-example models" become superluminal are much higher than the density of quark deconfinement. And the Asymptotic Freedom property of the QCD implies that $v_{\rm s} \longrightarrow c/\sqrt{3}$ for $n_{\rm b} \longrightarrow \infty$ in quark matter.

## 5.16. Effect of baryon superfluidity on equation of state

Baryons are most probably superfluid in neutron star cores. Since typical critical temperatures are $T_{\rm c} \sim 10^9 - 10^{10}$ K, superfluidity occurs soon after the neutron star birth. Let us consider the simplest model of a superfluid neutron gas and use the BCS model with an isotropic ${}^1S_0$ neutron pairing (see, e.g., §51 of Fetter & Walecka 1971). For temperatures $T \ll T_{\rm c}$, the superfluid gap can be approximated by its zero-temperature value, $\Delta(T) \simeq \Delta(0) \equiv \Delta_0$, where $\Delta_0 \approx 1.76\, k_{\rm B}\, T_{\rm c}$. Moreover, let us approximate the normal phase by a free Fermi gas. At a given baryon density $n_{\rm b}$, the difference of the energy densities between the normal (n) and the superfluid (s) phases can be shown to be given by

$$\mathcal{E}_{\rm n} - \mathcal{E}_{\rm s} = \frac{m k_{\rm F}}{4\pi^2 \hbar^2}\, \Delta_0^2\,, \tag{5.130}$$

where the energy gap is assumed to be much smaller than the neutron Fermi energy, so that the lowest power of $\Delta_0$ is retained. The relative change in $\mathcal{E}$ resulting from superfluidity is therefore

$$\frac{\mathcal{E}_{\rm n} - \mathcal{E}_{\rm s}}{\mathcal{E}_{\rm n}} = \frac{5}{8} \left(\frac{\Delta_0}{\epsilon_{\rm F}}\right)^2 \approx 1.83 \times 10^{-4} \left(\frac{\Delta_0}{\rm MeV}\right)^2 \left(\frac{n_0}{n_{\rm b}}\right)^{2/3}\,, \tag{5.131}$$

where $\epsilon_{\rm F}$ is the neutron Fermi energy. One can also show, within our simple model, that the effect of superfluid transition on the pressure is $P_{\rm n} - P_{\rm s} = -\mathcal{E}_{\rm n} + \mathcal{E}_{\rm s}$. These estimates can be extended to a more complicated case of non-isotropic pairing and a multicomponent baryon matter. All in all, they show that as long as superfluid gaps $\Delta \lesssim 1$ MeV, the relative effect of superfluidity on the EOS at supranuclear density is less than $10^{-4} - 10^{-2}$.

## 5.17. Effect of strong magnetic field on equation of state

As already mentioned in § 4.1.5, the magnetic field strongly affects the EOS if the quantization of particle motion across the field (Landau orbitals) and the interaction of particle magnetic moments with the field become important. These quantum effects are perceptible if particles occupy only a few magnetically induced quantum levels. Otherwise, the effects of the magnetic field on the EOS are minor.

These statements can be illustrated for the case of electrons considered in § 4.1.2. We have shown that Eq. (4.25) for the pressure in an arbitrary magnetic field reduces to the non-magnetic expression (2.50), if the field is nonquantizing (i.e., many Landau levels are occupied). Thus, the magnetic field significantly affects the pressure only if $T_B \gtrsim T$ and $\rho_B \gtrsim \rho$, where $T_B$ and $\rho_B$ are given by Eqs. (4.29), (4.33), and (4.30). The latter (density) restriction translates into

$$B \gtrsim (3.8 \times 10^{19}\ \mathrm{G})\,(x_e n_\mathrm{b}/\mathrm{fm}^{-3})^{2/3}, \tag{5.132}$$

where $x_e$ is the number of electrons per baryon. We have $n_\mathrm{b} \sim 0.1\ \mathrm{fm}^{-3}$ near the crust-core interface, and $x_e$ is typically a few percent throughout the core. Therefore, the electron pressure in the core is appreciably affected by the fields $B \gg 10^{17}$ G.

One can easily generalize Eq. (5.132) for other fermions ($\mu$-mesons, nucleons) in the ideal-gas model. In this case, $x_e$ should be replaced by the number of given particles per baryon, and the right-hand side should be multiplied by $m_\mu/m_e = 206.77$ for muons and $\sim 10^3$ (of the order of nucleon-to-electron mass and electron-to-nucleon magnetic moment ratios) for protons and neutrons. Accordingly, the partial proton and neutron pressures of the $npe\mu$ gas at neutron-star densities cannot be affected by a magnetic field unless $B \gtrsim 10^{20}$ G. However, the values of $B \gtrsim 10^{18}$ G are unrealistic (see § 1.3.8).

Suh & Mathews (2001) considered the effects of superstrong magnetic fields on muon production and pion condensation in the ideal $npe\mu^-\pi^-$ gas at neutron-star densities. They showed that the EOS and the threshold densities of the $\mu^-$ and $\pi^-$ appearance can be affected by the magnetic field $B \gg 10^{17}$ G.

More elaborated models of neutron-star matter in superstrong magnetic fields have been studied by Broderick *et al.* (2000), who considered not only the ideal $npe\mu$ gas, but also interacting matter in the framework of the RMF model. The magnetic field affects their EOS in a neutron-star core at $B \gtrsim 10^{18}$ G, which agrees with the simpler estimates cited above. Since this field is close to the upper limit on the magnetic field for dynamically stable stellar configurations, it is unlikely that a magnetic modification of the EOS could be important in the cores of neutron stars.

Chapter 6

# NEUTRON STAR STRUCTURE

## 6.1. Equations of hydrostatic equilibrium

Neutron stars are relativistic objects. Their structure and evolution should be studied using the General Theory of Relativity. The importance of relativistic effects for a star of mass $M$ and radius $R$ is characterized by the *compactness parameter* $r_g/R$ given by Eq. (1.4), where $r_g$ is the Schwarzschild radius. Typically, one has $r_g/R \sim 0.2 - 0.4$ for a neutron star, whereas $r_g/R \ll 1$ for all other stars. For instance, $r_g/R \sim 10^{-4}$ for white dwarfs and $r_g/R \sim 10^{-6}$ for main-sequence stars of $M \simeq M_\odot$.

Let us outline the derivation of relativistic equations of stellar structure for static spherically symmetric stellar objects. The starting point is the expression for the metric of a stationary, spherically symmetric space-time,

$$\mathrm{d}s^2 = c^2\,\mathrm{d}t^2\,\mathrm{e}^{2\Phi} - \mathrm{e}^{2\lambda}\,\mathrm{d}r^2 - r^2\,(\mathrm{d}\theta^2 + \sin^2\theta\,\mathrm{d}\phi^2), \tag{6.1}$$

where $t$ is a time-like coordinate, $r$ is a radial coordinate, $\theta$ and $\phi$ are the polar and azimuthal angles, respectively, whereas $\Phi = \Phi(r)$ and $\lambda = \lambda(r)$ are some functions of $r$. The angular geometry (with respect to $\theta$ and $\phi$) is the same as in the flat space-time because of spherical symmetry, but space-time is generally curved along "directions" of $r$ and $t$. In our case, space-time curvature is produced by a massive stellar body. For the flat space-time, we would have $\Phi(r) = \lambda(r) = 0$.

Let us introduce four coordinates $x^0 = ct$, $x^1 = r$, $x^2 = \theta$, and $x^3 = \phi$, and write $\mathrm{d}s^2 = g_{ik}\,\mathrm{d}x^i\,\mathrm{d}x^k$ ($i, k$ = 0, 1, 2, 3, implying sum over repeated indices). The metric tensor $g_{ik}$ of the space-time in question is diagonal with the diagonal components

$$g_{00} = \mathrm{e}^{2\Phi}, \quad g_{11} = -\mathrm{e}^{2\lambda}, \quad g_{22} = -r^2, \quad g_{33} = -r^2\sin^2\theta. \tag{6.2}$$

One can easily analyse this metric and understand the meaning of all variables. Let us remind that $r$ and $t$ are Schwarzschild coordinates, suitable for an observer at infinity. First, let us fix $t$, $r$, $\theta = \pi/2$ and vary $\phi$ from 0 to $2\pi$, producing circumference around the stellar center. We immediately obtain that the proper length of the circle (i.e., the length measured in a the local inertial reference frame) is $l = 2\pi r$. Thus $r$ can be called the *circumferential radius*. The proper area of the spherical surface at a given $r$ is $4\pi r^2$. Second, let us fix $t$, $\theta$, $\phi$ and vary $r$ from $r = 0$ to some value $r_1$. A proper length element in the radial direction (the radial distance as measured by a local observer) is $\mathrm{d}l = \mathrm{e}^{\lambda}\,\mathrm{d}r$. Thus, the proper radial length $l_1 = \int_0^{r_1} \mathrm{e}^{\lambda}\,\mathrm{d}r$ is generally non-equal to $r_1$ because space-time is curved. Therefore, $\lambda(r)$ determines the space-time curvature in the radial direction. Instead of $\lambda(r)$ it is more convenient to introduce a new function $m(r)$ related to $\lambda(r)$ through

$$\mathrm{e}^{-\lambda} = \sqrt{1 - 2Gm/(rc^2)}. \tag{6.3}$$

One can show (e.g., Landau & Lifshitz 1999) that $m(r)$ represents the *gravitational mass* confined inside a sphere with radial coordinate $r$. The gravitational mass of a neutron star is generally smaller than the baryon mass ("rest mass") due to gravitational mass defect (see §6.2). Using Eq. (6.3) we can write a proper radial length element $\mathrm{d}l$ and proper volume $\mathrm{d}V$ between close spherical shells with radial coordinates $r$ and $r + \mathrm{d}r$ as

$$\mathrm{d}l = \frac{\mathrm{d}r}{\sqrt{1 - 2Gm/(c^2 r)}}, \qquad \mathrm{d}V = \frac{4\pi r^2\,\mathrm{d}r}{\sqrt{1 - 2Gm/(c^2 r)}}. \tag{6.4}$$

Finally, let us fix $r$, $\theta$, $\phi$ and vary $t$. According to Eq. (6.1), a proper time interval (i.e., the interval measured in a local inertial reference frame) is

$$\mathrm{d}\tau = \mathrm{e}^{\Phi(r)}\,\mathrm{d}t. \tag{6.5}$$

Let us imagine a source that produces periodic signals of frequency $\nu(r) = \mathrm{d}N/\mathrm{d}\tau = \mathrm{e}^{-\Phi(r)}\mathrm{d}N/\mathrm{d}t$ in a point with a radial coordinate $r$, where $\mathrm{d}N$ is the number of signals over a time interval $\mathrm{d}\tau$. A distant observer will detect signals of frequency $\nu_\infty = \mathrm{d}N/\mathrm{d}t$ (because $\Phi(r) \to 0$ and space-time becomes asymptotically flat as $r \to \infty$). Therefore, $\nu_\infty = \nu(r)\,\mathrm{e}^{\Phi(r)}$ which means that $\Phi(r)$ determines the gravitational redshift of periodic signals, $z(r) \equiv \nu(r)/\nu_\infty - 1 = \mathrm{e}^{-\Phi(r)} - 1$.

The relationship between geometry of space-time and material properties of massive bodies is given by the Einstein equations

$$\mathcal{R}_{ik} - \frac{1}{2}\, g_{ik}\,\mathcal{R} = \frac{8\pi G}{c^4}\, T_{ik}, \tag{6.6}$$

where $\mathcal{R}_{ik}$ is the Ricci curvature tensor and $\mathcal{R} = \mathcal{R}_i^i$ is the scalar curvature, whereas $T_{ik}$ is the stress-energy tensor. The tensors $\mathcal{R}_{ik}$ and $\mathcal{R}$ are calculated from the metric tensor $g_{ik}$, Eq. (6.2), using standard equations of the Riemann geometry (e.g., Landau & Lifshitz 1999). Let us treat neutron star matter as a perfect fluid (non-viscous medium of total energy density $\mathcal{E}$, in which all stresses are zero except for an isotropic pressure $P$). This approximation is justified because shear stresses produced, for instance, by elastic strain in the solid crust (§ 6.11) or by strong magnetic field (§ 6.9.1) are generally negligible compared to the pressure.

In the perfect-fluid approximation we have $T_{ik} = (P + \mathcal{E})u_i u_k - P g_{ik}$ , where $u_i$ is 4-velocity of matter (Landau & Lifshitz, 1999). Here, $P$ and $\mathcal{E}$ are defined in a local inertial reference frame comoving with matter. The energy density includes all forms of the internal energy, like the rest energy and kinetic energy of matter constituents, and the energy of their interactions. However, it does not include contributions of gravitational forces. In the spirit of General Relativity, the effects of gravitational interactions are incorporated via space-time geometry. Local thermodynamic properties of the matter are not affected by general relativistic effects, because the space-time curvature is thought to be negligible on microscopic scales.

For a static neutron star, one has to determine four functions of $r$, which are the metric functions $\Phi(r)$ and $\lambda(r)$, as well as the pressure and mass-density profiles, $P(r)$ and $\rho(r) = \mathcal{E}(r)/c^2$. The problem is simplified by a strong degeneracy of neutron star interior where thermal contributions to $P$ and $\rho$ can be neglected. Then, the EOS involves only $P$ and $\rho$, but not the temperature (important exceptions are: neutron star atmospheres, newly born neutron stars, envelopes of exploding X-ray bursters).

Calculating $\mathcal{R}_{ik}$ for the metric (6.1) and using the Einstein equations (6.6), one comes to the three relativistic equations of hydrostatic equilibrium for a static spherically symmetric neutron star,

$$\frac{\mathrm{d}P}{\mathrm{d}r} = -\frac{G\rho m}{r^2}\left(1+\frac{P}{\rho c^2}\right)\left(1+\frac{4\pi P r^3}{mc^2}\right)\left(1-\frac{2Gm}{c^2 r}\right)^{-1}, \tag{6.7}$$

$$\frac{\mathrm{d}m}{\mathrm{d}r} = 4\pi r^2 \rho, \tag{6.8}$$

$$\frac{\mathrm{d}\Phi}{\mathrm{d}r} = -\frac{1}{\rho c^2}\frac{\mathrm{d}P}{\mathrm{d}r}\left(1+\frac{P}{\rho c^2}\right)^{-1}. \tag{6.9}$$

Equation (6.7) is called the Tolman-Oppenheimer-Volkoff equation of hydrostatic equilibrium (Tolman 1939; Oppenheimer & Volkoff 1939; also see § 1.2). Equation (6.8) describes mass balance; its apparently Newtonian form is illusive because the proper volume of a spherical shell, given by Eq. (6.4), is not simply $4\pi r^2\,\mathrm{d}r$. Finally, Eq. (6.9) is a relativistic equation for the metric function $\Phi(r)$. These equations should be supplemented by an EOS, $P = P(\rho)$. The above equations constitute a closed system of equations to be solved for obtaining $P(r)$, $\rho(r)$, $m(r)$, and $\Phi(r)$. Actually, Eqs. (6.7) and (6.8) do not contain $\Phi(r)$ and can be solved separately to determine $P(r)$, $\rho(r)$ and $m(r)$. The function $\Phi(r)$ can be found then from Eq. (6.9).

In the stellar layers where $P \ll \rho c^2$ and $Pr^3 \ll mc^2$, the Tolman-Oppenheimer-Volkoff equation can be rewritten in the quasi-Newtonian form:

$$\frac{\mathrm{d}P}{\mathrm{d}l} = -g\rho, \quad g(r) = \frac{Gm}{r^2\sqrt{1-2Gm/(c^2r)}}, \tag{6.10}$$

where $g$ can be called the *local gravitational acceleration* and $\mathrm{d}l$ is given by Eq. (6.4). This form will be used in § 6.9 for studying the structure of the outer neutron star envelope.

In the stellar interior, $P > 0$ and $\mathrm{d}P/\mathrm{d}r < 0$. The stellar radius is determined from the condition $P(R) = 0$. Outside the star (for $r > R$), we have $P = 0$ and $\rho = 0$. Then Eq. (6.8) gives $m(r > R) = M = \text{const}$. The latter quantity is called the *total gravitational mass of the star*; the total energy content of the star is $E = Mc^2$. Combining Eqs. (6.7) and (6.9) in the limit of $P \longrightarrow 0$, we obtain that $\mathrm{e}^{2\Phi} = 1 - r_\mathrm{g}/r$ outside the star. Therefore, for $r > R$, Eqs. (6.7)–(6.9) yield the well-known Schwarzschild metric,

$$\mathrm{d}s^2 = c^2\,\mathrm{d}t^2\left(1-\frac{r_\mathrm{g}}{r}\right) - \left(1-\frac{r_\mathrm{g}}{r}\right)^{-1}\mathrm{d}r^2 - r^2\left(\mathrm{d}\theta^2 + \sin^2\theta\,\mathrm{d}\phi^2\right), \tag{6.11}$$

which describes a curved space-time around any massive, spherically symmetric body (not necessarily static). Thus, the gravitational redshift of signals emitted from the neutron star surface ($r = R$) is

$$\nu_\infty = (1 - r_\mathrm{g}/R)^{1/2}\,\nu(R), \quad z_\mathrm{surf} = (1 - r_\mathrm{g}/R)^{-1/2} - 1\,. \tag{6.12}$$

At large distances from the star ($r \gg r_\mathrm{g}$) the Schwarzschild space-time (6.11) becomes asymptotically flat. Therefore, Schwarzschild time $t$ is a proper time for a distant observer.

Finally, a non-relativistic star with $P \ll \rho c^2$, $Pr^3 \ll mc^2$, and $r_\mathrm{g} \ll R$ creates a weak space-time curvature. In this case Eqs. (6.7)–(6.9) reduce to the familiar Newtonian equations of stellar equilibrium,

$$\frac{\mathrm{d}P}{\mathrm{d}r} = -\frac{Gm\rho}{r^2}, \qquad \frac{\mathrm{d}m}{\mathrm{d}r} = 4\pi r^2\rho, \qquad \frac{\mathrm{d}\Phi}{\mathrm{d}r} = \frac{Gm}{c^2r^2}\,. \tag{6.13}$$

We see that in the non-relativistic limit $\Phi(r)\,c^2$ becomes the Newtonian gravitational potential.

## 6.2. Baryon number, mass and chemical potential. Binding energy of neutron stars

Let us consider an element of matter of a proper volume $\mathrm{d}V$. The baryon number in this volume is $n_\mathrm{b}\,\mathrm{d}V$, where $n_\mathrm{b}$ is the baryon number density measured in a local reference frame. An isolated star has a fixed baryon number, which will be denoted by $A_{\rm b}$. Although some theories of fundamental interactions predict the breaking of baryon number conservation, the timescales involved are much longer than the Universe age $\sim 10^{10}$ years.[1]

In a static, spherically symmetric star the number of baryons confined within a sphere of radius $r$ is

$$a_{\rm b}(r) = 4\pi \int_0^r n_{\rm b}(r')\,{\rm e}^{\lambda(r')}\,r'^2\,{\rm d}r' \,. \tag{6.14}$$

One can rewrite this equation in differential form to be solved simultaneously with Eqs. (6.7) and (6.8),

$${\rm d}a_{\rm b}/{\rm d}r = 4\pi r^2\,n_{\rm b}/\sqrt{1-2Gm/(rc^2)}, \tag{6.15}$$

with the natural boundary condition $a_{\rm b}(0) = 0$. In view of baryon-number conservation, $a_{\rm b}$ can be treated as a *Lagrangian coordinate* in the star; it specifies the position of a matter element during spherically-symmetric evolution of the star (radial accretion, pulsations, collapse, etc.). The total baryon number of the star is

$$A_{\rm b} = a_{\rm b}(R) = 4\pi \int_0^R n_{\rm b}(r)\,{\rm e}^{\lambda(r)}\,r^2\,{\rm d}r \,. \tag{6.16}$$

The baryon number of the canonical neutron star ($M = 1.4\,M_\odot$, see § 1.3.3) can be estimated as $A_{\rm b} \simeq 1.4\,M_\odot/m_n = 1.7 \times 10^{57}$. It is convenient to introduce an equivalent mass of non-interacting baryons called the *baryon mass* (or the *rest mass*) of the star, $M_{\rm b} \equiv A_{\rm b} m_{\rm b}$ ($m_{\rm b} \approx m_n$ is the mass of one baryon). One of the possible choices of $m_{\rm b}$ is $m_{\rm b} = m_n$. Such a choice implies

$$M_{\rm b} = A_{\rm b} m_n \simeq 0.842\,A_{\rm b57}\,M_\odot \,, \tag{6.17}$$

where $A_{\rm b57} \equiv A_{\rm b}/10^{57}$. Other choices of $m_{\rm b}$ are also possible (see below). The values of $A_{\rm b}$ and $M_{\rm b}$ remain constant along an evolutionary path of an isolated star.

Actually, the value of $A_{\rm b}$ or $M_{\rm b}$ can be calculated directly for a static neutron star built of cold catalyzed matter. In this case it is instructive to introduce the *baryon chemical potential*

$$\mu_{\rm b} = {\rm d}\mathcal{E}/{\rm d}n_{\rm b} = (\rho c^2 + P)/n_{\rm b}. \tag{6.18}$$

Using Eq. (6.9) combined with the zero-temperature thermodynamic identity ${\rm d}P = n_{\rm b}\,{\rm d}\mu_{\rm b}$, one comes to the important conclusion (Harrison *et al.* 1965) that

$$\mu_{\rm b}(r)\,{\rm e}^{\Phi(r)} = {\rm constant} \tag{6.19}$$

[1]The fate of neutron stars on timescales $\sim 10^{35}$ years or longer was studied by Adams & Laughlin (1997) and Adams *et al.* (1998).

throughout the entire star.

To be specific, let $m_{\rm b} = m_0 = 1.6586 \times 10^{-24}$ g, the mass per baryon in $^{56}$Fe.[2] Then Eq. (6.19) implies

$$n_{\rm b}(r) = (1 + z_{\rm surf}) \left(\rho(r)c^2 + P(r)\right) {\rm e}^{\Phi(r)}/m_0 . \tag{6.20}$$

Therefore,

$$a_{\rm b}(r) = 4\pi \, \frac{1 + z_{\rm surf}}{m_0 c^2} \int_0^r {\rm e}^{\Phi+\lambda} \left(\rho c^2 + P\right) r'^2 {\rm d}\, r' \, , \tag{6.21}$$

and the total baryon number is $A_{\rm b} = a_{\rm b}(R)$. To the best of our knowledge, these formulae were first derived by Goldman (1989). They enable one to calculate $a_{\rm b}(r)$ and $A_{\rm b}$ from the solution of Eqs. (6.7)–(6.9).

In contrast to an atomic nucleus, a neutron star is bound by gravity. We define the binding energy of the star similarly to the binding energy of the nucleus, as the mass defect with respect to the dispersed configuration of its constituents, multiplied by $c^2$. However, in the case of the neutron star several definitions of "dispersed configuration" may be considered. First of all, consider a neutron star was formed in gravitational collapse of a degenerate iron core of a massive star at the endpoint of thermonuclear evolution. The mass of the iron core containing $A_{\rm b}$ nucleons is, to a good approximation, $M_{\rm Fe} = A_{\rm b} m_0$. Strictly speaking, $M_{\rm Fe}$ is the mass of a dispersed cloud of $^{56}$Fe dust. The binding energy of the neutron star with respect to presupernova (PSN) core is thus

$$E_{\rm bind}^{\rm (PSN)} \simeq E_{\rm bind}^{\rm (Fe)} = \left[A_{\rm b} m_0 - M(A_{\rm b})\right] c^2 \, , \tag{6.22}$$

where $M(A_{\rm b})$ is the gravitational mass of the neutron star of baryon number $A_{\rm b}$ (notice that when a neutron star forms some nucleons could transform into hyperons, but the baryon number $A_{\rm b}$ stays constant). This definition of the binding energy is particularly useful in the context of energy balance in type II supernova explosions (§ 1.4.2). In this case $E_{\rm bind}^{\rm (PSN)}$ constitutes an upper bound on the energy released in a type II supernova. As more than 99% of the energy is carried away by neutrinos, the total energy of the neutrino burst is expected to be nearly equal to $E_{\rm bind}^{\rm (Fe)}$.

Basically, the very initial dispersed, primordial configuration of dense matter is a hydrogen cloud. The binding energy of a neutron star with respect to this dispersed state is

$$E_{\rm bind}^{\rm (H)} = \left[A_{\rm b} m_{\rm H} - M(A_{\rm b})\right] c^2 \, , \tag{6.23}$$

[2] Let us remind that $m_0 c^2 = 930.4$ MeV is the energy per baryon in the ground state of matter at zero pressure, see § 3.2.

where $m_{\rm H} \simeq m_p + m_e = 1.6735 \times 10^{-24}$ g is the mass of hydrogen atom. The difference between these two definitions of $E_{\rm bind}$ is

$$E_{\rm bind}^{\rm (H)} - E_{\rm bind}^{\rm (Fe)} \simeq A_{\rm b} \left( m_{\rm H} - m_0 \right) c^2 = 1.33 \, A_{\rm b57} \times 10^{52} \, {\rm erg} \, . \qquad (6.24)$$

For a $1.4 \, M_\odot$ neutron star, $E_{\rm bind}^{\rm (H)}$ is typically $\sim$10% higher than $E_{\rm bind}^{\rm (Fe)}$ (see § 6.7).

## 6.3. Proper mass and gravitational energy

In the Newtonian theory, it is useful to split the total energy of a star into the internal energy $E_{\rm int}$, gravitational energy $E_{\rm grav}$, rotational energy, etc. As a rule, such a decomposition has no meaning in General Relativity, but it is meaningful for spherical static stars considered in the present chapter (Misner *et al.*, 1973). The proper internal energy density is defined as the excess of the proper energy density over the proper rest energy density $n_{\rm b} m_0 c^2$. Accordingly, the total internal energy of the star is defined as

$$E_{\rm int} = 4\pi \int_0^R \left( \mathcal{E} - n_{\rm b} m_0 c^2 \right) {\rm e}^{\lambda} \, r^2 \, {\rm d}r = \left( M_{\rm P} - M_{\rm b0} \right) c^2 \, , \qquad (6.25)$$

where $M_{\rm P}$ is the so called *proper mass* of the star and $M_{\rm b0}$ is its baryon (rest) mass calculated for $m_{\rm b} = m_0$. The proper mass is the total proper stellar energy divided by $c^2$. The gravitational energy of the star should then be defined as

$$E_{\rm grav} = \left( M - M_{\rm P} \right) c^2 \, . \qquad (6.26)$$

In the Newtonian approximation $E_{\rm grav}^{\rm Newt} = -4\pi G \int_0^R m \rho r \, {\rm d}r$.

The quantities $E_{\rm int}$ and $E_{\rm grav}$ are analogous to the corresponding quantities in the Newtonian theory. In particular, $E_{\rm int} > 0$ and $E_{\rm grav} < 0$, and $E_{\rm grav} + E_{\rm int} < 0$ for a self-bound star. Let us recall that in the Newtonian approximation the gravitational energy is related via the *virial theorem* to the pressure integrated over the equilibrium configuration: $E_{\rm grav}^{\rm Newt} = -12\pi \int_0^R P \, r^2 \, {\rm d}r$. The virial theorem in General Relativity is much more complicated (see Gourgoulhon & Bonazzola 1993 and references therein).

## 6.4. Constructing neutron star models

### 6.4.1 Introductory remarks

In order to construct a neutron star model one has to solve the set of Eqs. (6.7)–(6.9) supplemented by an EOS of dense matter. General physical conditions in the stellar interior require the pressure to be positive, continuous, and monotonically decreasing from the stellar center to the surface. The mass density and baryon density are not constrained to be continuous; discontinuities appear at those values of $P$ which correspond to first-order phase transitions (see Chapters 3 and 5). The functions of $r$ to be determined from Eqs. (6.7)–(6.9) and (6.15) are: $P = P(r)$, $m = m(r)$, $\Phi = \Phi(r)$ and $a_{\rm b} = a_{\rm b}(r)$; they are all continuous. The density profile is then determined from the EOS, $\rho = \rho(P)$.

Equations (6.7)–(6.8) and (6.15) are usually integrated from the stellar center with the boundary conditions: $P(0) = P_{\rm c}$, $m(0) = 0$, and $a_{\rm b}(0) = 0$. Notice that the pressure and density profiles in the star can be found by integrating Eqs. (6.7) and (6.8) alone. The stellar surface, $r = R$, is determined from the condition $P(R) = 0$ (which neglects the presence of a very thin atmosphere). It is very well justified in calculating global parameters of neutron stars.

The metric function $\Phi(r)$ can be found from Eq. (6.9) by matching the solution at $r = R$ with the outside Schwarzschild solution, $e^{\Phi(R)} = \sqrt{1 - r_{\rm g}/R}$ (§ 6.1). Once $P(r)$ and $\rho(r)$ are determined from Eqs. (6.7) and (6.8) at the preceding step, the right-hand side of Eq. (6.9) becomes a known function of $r$, which makes numerical calculation of $\Phi(r)$ straightforward. Calculation of $A_{\rm b}$ from Eq. (6.15) is equally straightforward.

In many cases, building neutron star models does not require special efforts. In particular, it is sufficient to use logarithmic interpolation between values of $P$ and $\rho$ tabulated on a restricted number of mesh points.

However, sometimes one needs to construct very precise neutron star models. Then a good test of the precision is supplied by Eq. (6.19). Strictly speaking, if Eq. (6.19) does not hold, the configuration *is not* in hydrostatic equilibrium (Harrison *et al.*, 1965). While deviations of $\mu_{\rm b}(r)\, e^{\Phi(r)}$ from constancy are usually small (e.g., a fraction of percent) they signal that either the neutron star model is unstable or at least one of the two quantities, $M$ or $A_{\rm b}$, is in error. This may occur independently of numerical precision of solving differential equations (6.7)–(6.9) and (6.15) by a mere use of the logarithmic interpolation of the EOS (see, e.g., Harrison *et al.* 1965, Baym *et al.* 1971b, Arnett & Bowers 1977). Resulting inconsistencies, which may seem minor as far as the values of global stellar parameters are concerned, may lead to serious problems if very high precision of *simultaneous* determination of $M$ and $A_{\rm b}$ is required.

For instance, let us follow the neutron star evolution at a constant $A_{\rm b}$ with a phase transition occurring in the stellar core (§ 7.9). Let the total energy release in the phase transition correspond to the decrease of the total mass by $\Delta M = 10^{-6} M$. It is obvious, that calculating $\Delta M$ with two significant figures requires better than $10^{-8}$ relative precision in determining $A_{\rm b}$ for stellar models before and after the phase transition. To get this precision one has to use such an interpolation procedure between the tabulated EOS points, in which Eq. (5.97), resulting from the first law of thermodynamics, is strictly fulfilled. Then, the constancy of $\mu_{\rm b}(r) e^{\Phi(r)}$ and the accuracy of the calculation of stellar parameters will be limited only by the numerical precision of the computer code for solving the differential equations of stellar structure (see, e.g., Haensel & Prószyński 1982). A particularly useful thermodynamically consistent interpolation of an EOS was developed by Swesty (1996). His method, based on the interpolation in terms of Hermite polynomials of third degree, fulfills, by construction, thermodynamic identities at any $r$ (except at phase transition points).

### 6.4.2 Equation of state

In order to construct neutron star models one needs the EOS of dense matter. Both the history and present status of the EOS problem have been described in Chapters 1, 3 and 5.

In this chapter we analyze the effect of EOS on neutron star structure using some illustrative examples. In the outer neutron star crust we will employ the EOS of Haensel & Pichon (1994) based on experimental masses of neutron-rich nuclei. The EOS of the outermost iron envelope of the crust will be taken from BPS. In the inner crust we use the SLy EOS, and in a few cases the FPS EOS (Lorenz, 1991). The SLy and FPS EOSs describe the core and the inner crust in a unified manner. These EOSs are discussed in Chapter 3. Usually, the crustal EOS has little importance for global parameters of neutron star models, which are mostly determined by the EOS of a stellar core.

In the present chapter we restrict ourselves to models of neutron star cores composed of nucleons and hyperons. Stellar models containing exotic hadronic phases, such as pion condensate, kaon condensate, and deconfined quark matter, will be considered in Chapter 7. Hypothetical strange (quark) stars will be studied in Chapter 8.

The EOS of a neutron star core is mainly determined by strong interactions between baryons (Chapter 5). The reliability of these EOSs decreases rapidly with increasing $\rho$. Our discussion of the impact of the EOS at $\rho \gtrsim \rho_0$ on the neutron star structure will be based on the limited, but hopefully representative set of model EOSs. The selected models are discussed in Chapter 5 and are listed in Table 5.3 (also see Table 6.1).

As we stressed in Chapter 5, the most important qualitative feature of an EOS at $\rho \gtrsim \rho_0$ – as far as neutron star structure is concerned – is its stiffness. However, the stiffness may vary with density (e.g., owing to hyperonization of matter). Therefore, it is useful to introduce the *effective stiffness* of the EOS based on the value of the maximum mass of neutron stars, $M_{\rm max}$ (§ 6.5.1), which is a *functional* of the EOS. The topmost EOS in Table 6.1 is the softest one, and the effective stiffness of the EOSs increases to the bottom of the table. Let us remind, that the stiffest/softest EOSs are extreme models, characterized by the compression modulus of symmetric nuclear matter, which is significantly higher/lower than the standard value of $K_0 = 220$ MeV (§ 5.4). We include these extreme EOSs to represent the stiffest/softest neutron star models.

*Table 6.1.* Maximum mass configurations for the EOSs of the neutron star core listed in Table 5.3. In all cases, except FPS, we use the SLy EOS of the crust. We present radius $R$, compactness $r_g/R$, central baryon density $n_c$, central mass density $\rho_c$, and binding energy $E_{\rm bind}^{\rm (Fe)}$ (see § 6.2).

| EOS | $M_{\rm max}$ | $R$ | $r_g/R$ | $n_c$ | $\rho_c$ | $E_{\rm bind}^{\rm (Fe)}$ |
|---|---|---|---|---|---|---|
| | $[M_\odot]$ | [km] | | $[{\rm fm}^{-3}]$ | $[10^{15}\ {\rm g\ cm}^{-3}]$ | $[10^{53}\ {\rm erg}]$ |
| BPAL12 | 1.46 | 9.04 | 0.478 | 1.76 | 3.94 | 3.19 |
| BGN1H1 | 1.64 | 9.42 | 0.516 | 1.59 | 3.71 | 3.82 |
| BBB1 | 1.79 | 9.67 | 0.547 | 1.37 | 3.09 | 5.26 |
| FPS | 1.80 | 9.27 | 0.572 | 1.46 | 3.40 | 5.37 |
| BGN2H1 | 1.82 | 9.54 | 0.564 | 1.46 | 3.51 | 4.83 |
| BBB2 | 1.92 | 9.50 | 0.596 | 1.35 | 3.20 | 6.17 |
| SLy | 2.05 | 9.99 | 0.605 | 1.21 | 2.86 | 6.79 |
| BGN1 | 2.18 | 10.9 | 0.591 | 1.05 | 2.46 | 7.28 |
| APR | 2.21 | 10.0 | 0.651 | 1.15 | 2.73 | 9.13 |
| BGN2 | 2.48 | 11.7 | 0.626 | 0.86 | 2.02 | 9.40 |

Two of the EOSs listed in Table 6.1 – SLy and FPS – are the unified ones and describe, within a single physical model, both the crust and the core of a star (§ 3.6). In other cases, an overall EOS is constructed by matching smoothly the EOSs of the crust and the core. If the crust and core are described by different nuclear models, the matching is thermodynamically inconsistent and numerically ambiguous. However, as the matching is restricted to a thin shell near the crust-core interface, this is usually of little importance.

Unified EOSs can be fitted by analytical and continuous functions (Haensel & Potekhin 2004, Appendix C). This enables one to avoid ambiguities and thermodynamic inconsistency plaguing tabulated EOSs.

## 6.5. Masses and stability of neutron stars

### 6.5.1 Stellar oscillations and stability

After calculating an equilibrium stellar model one should check its stability. Only models in *stable equilibrium* are of astrophysical interest. Below we briefly discuss the stability with respect to small perturbations. For simplicity, we restrict ourselves to non-rotating spherically symmetric equilibrium models.

The stability problem can be formulated in several ways (e.g., Shapiro & Teukolsky 1983). One of them is based on the analysis of stellar oscillations. One can introduce the field of small Lagrangian displacements $\xi^i(r, \theta, \phi, t)$

($i = 1, 2, 3$) of fluid elements in a perturbed star, which induces small changes $h_{ij}(r, \theta, \phi, t)$ in the metric. We consider linear approximation in $\xi^i$. The velocity field of a pulsating star is then $\mathrm{e}^{-\Phi}\partial\xi^i/\partial t$. A standard procedure consists in factoring out the dependence of $\xi^i$ on $\theta, \phi$ via spherical harmonics $Y_{\ell m}(\theta, \phi)$.

The study of small *radial* pulsations, with $\ell = m = 0$, $\xi^\theta = \xi^\phi = 0$, $\xi^r = \xi(r, t)$, is not much more complicated in General Relativity than in Newtonian theory (Chandrasekhar 1964; Harrison *et al.* 1965; Meltzer & Thorne 1966; Bardeen *et al.* 1966; Misner *et al.* 1973). In particular, gravitational field in vacuum outside the star is not affected by radial pulsations ($h_{ij} = 0$ outside the star). On the contrary, nonradial pulsations with $\ell > 0$ are more complicated for relativistic stars (Thorne & Campolattaro 1967).

First let us describe radial pulsations. After linearizing Einstein equations, one can get an equation governing the dynamics of radial pulsations,

$$\frac{\partial^2 \xi}{\partial t^2} = -\hat{\mathcal{H}}\xi \,, \tag{6.27}$$

where $\hat{\mathcal{H}}$ is a self-adjoint (Hermitian) linear second-order differential operator, independent of $t$, $\theta$, $\phi$, and determined by the equilibrium stellar model. Its solution can be sought in the form $\xi(r, t) = \mathrm{e}^{-\mathrm{i}\omega t}\,\zeta(r)$. Then Eq. (6.27) reduces to

$$\lambda\,\zeta = \hat{\mathcal{H}}\zeta, \qquad \lambda \equiv \omega^2. \tag{6.28}$$

This equation, supplemented by appropriate boundary conditions at $r = 0$ and $r = R$, constitutes a Sturm-Liouville problem for eigenfrequencies $\omega$ and eigenvectors $\zeta(r)$ of radial oscillations (analogous to a stationary quantum mechanical problem of discrete eigenstates in a quantum system with a Hamiltonian operator $\hat{\mathcal{H}}$). Generally, there is an infinite number of various stellar oscillation modes (studied by astroseismology; see, e.g., § 1.3.9). Radial oscillations can be labeled by the index $n = 0, 1, 2, \ldots$ which enumerates nodes of the radial function $\zeta(r)$ within the star. As in Quantum Mechanics, the *lowest* eigenvalue $\lambda_0 = \omega_0^2$ belongs to the *fundamental* mode of radial pulsations without any nodes ($n = 0$); higher modes correspond to larger $\lambda$ ($\lambda_0 \leq \lambda_1 \leq \lambda_2 \ldots$). As a rule, the fundamental mode is very simple ($\zeta(r)$ is nearly linear in $r$ and describes homologous contractions and rarefactions of the star). An order-of-magnitude estimate is $\omega_0 \sim \sqrt{G\bar{\rho}}$, where $\bar{\rho}$ is the mean density of the star. This estimate is universal (valid for all types of stars); the pulsation period for the fundamental mode, $2\pi/\omega_0$, can serve as an estimate of a hydrodynamical time-scale for a given star. We have $\omega_0 \sim 10^4$ s$^{-1}$ for neutron stars.

Now it is clear that the stability of the star with respect to small radial perturbations is determined by the eigenvalue $\lambda_0 = \omega_0^2$ for the fundamental mode. If $\lambda_0 > 0$ then $\lambda_n > 0$ for all other $n$ and all eigenfrequencies $\omega_n$ are real, meaning that the equilibrium stellar model is *stable* (undergoes small-amplitude harmonic oscillations). If $\lambda_0 < 0$ then at least the fundamental mode is *unstable* ($\omega_0$ becomes purely imaginary and small displacements in the fundamental mode exponentially grow with time). In this case the highest instability increment (the most rapid exponential growth) is provided by the fundamental mode. Finally, at $\lambda_0 = 0$ we have $\omega_0 = 0$ and the star has *neutral* stability (is neither stable nor unstable).

The stability criterion described above requires the calculation of the eigenfrequency of the fundamental mode. It can be reformulated to avoid such calculation and simplify the stability analysis. The simplest "practical" criterion can be obtained assuming that the adiabatic index in a pulsating star is the same as in a slowly deformed matter. One can show that in this case the stellar model is *stable* if its mass $M$ increases with growing central density,

$$\mathrm{d}M/\mathrm{d}\rho_\mathrm{c} > 0. \tag{6.29}$$

This so called *static stability criterion* (Harrison *et al.*, 1965; Zeldovich & Novikov, 1971) is widely used in the literature. This condition is *necessary* but not sufficient (as we discuss below). The opposite inequality $\mathrm{d}M/\mathrm{d}\rho_\mathrm{c} < 0$ always implies instability of stellar models.

The equations governing *nonradial* pulsations of neutron stars in General Relativity were derived by Thorne & Campolattaro (1967). The stability of these modes was studied by Detweiler & Ipser (1973). Non-radial pulsations can be labeled by indices $n$, $\ell$ and $m$ (see above) and correspond to $\ell > 0$. They are accompanied by metric perturbations outside the star. Pulsations with $\ell > 1$ and $m \neq 0$ are accompanied by the emission of gravitational waves which introduces dissipation into pulsations of stars built of perfect (inviscid) fluid. In this case the appropriate operator $\hat{\mathcal{H}}$ for an equivalent eigenproblem (an analog of Eq. (6.28)) becomes non-Hermitian, its eigenvalues $\lambda$ and pulsation eigenfrequencies $\omega$ are complex numbers. Particularly well studied are $\ell = 2$ (quadrupole) modes, which are damped by gravitational radiation in 0.1 – 0.3 s (e.g., Lindblom 1983). This damping is usually weak on oscillation time-scales. As stressed by several authors (e.g., Detweiler & Ipser 1973, and references therein), it is sufficient to neglect the damping in the first approximation for calculating pulsation frequencies and eigenmodes of realistic neutron star models. In this first approximation the operator $\hat{\mathcal{H}}$ becomes Hermitian and represents an extension of the operator for radial pulsations to nonradial case. The approximate real eigenvalues of $\lambda$ for nonradial pulsations can be inserted into the row of $\lambda$-values for radial pulsations (as in Quantum Mechanical problem for a spherically symmetric potential). In particular, $\lambda$ for nonradial pulsations should exceed $\lambda_0$. This means that the stability of a non-rotating star with respect to small radial pulsations ($\lambda_0 > 0$) implies actually the stability with respect to any small pulsations of the star.

Perturbations of neutron stars may be accompanied by reactions (nuclear transformations) in dense matter. A finite timescale of reactions affects the EOS (particularly, the adiabatic index) and complicates the dynamical analysis of the stability (Meltzer & Thorne 1966; Chanmugam & Gabriel 1971; Gourgoulhon *et al.* 1995; also see Chapter 5).

### 6.5.2 Stability criteria based on the mass-radius diagram

Let us consider a family of equilibrium stellar models parameterized by the central density. We have thus $M = M(\rho_c)$ and $R = R(\rho_c)$ and can construct the mass-radius diagram which depends on the EOS of stellar matter. The stability of stellar models with respect to *radial* oscillations is intimately related to the shape of the $M(R)$ curve.

In Fig. 6.1 we show two examples of the $M(R)$ curves calculated for two different EOSs (the EOS on the right panel being overall stiffer). Each curve has three extrema, which will be called critical points ($C_1$, $C_2$ and $C_3$). These points divide the curves into four segments. Configurations lying on the segments with $\mathrm{d}M/\mathrm{d}\rho_c < 0$ are unstable with respect to small deformation (§ 6.5.1). However, as we already stressed, the condition $\mathrm{d}M/\mathrm{d}\rho_c > 0$ is necessary but not sufficient. A method which enables one to determine precise number of unstable normal radial modes using the $M(R)$ curve was described by Harrison *et al.* (1965) (see also Bardeen *et al.* 1966; Meltzer & Thorne 1966 and Thorne 1967). Let us formulate the stability criteria and illustrate them in Fig. 6.1.

**A.** *Changing stability.* At each critical point of the $M(R)$ curve one and only one normal radial mode changes its stability (stable $\rightarrow$ unstable, or unstable $\rightarrow$ stable). There are no changes of stability associated with radial pulsations at other points of the $M(R)$ curves.

**B.** *Number of nodes for a mode which changes stability.* A mode with even number $n$ of radial nodes changes its stability if and only if $\mathrm{d}R/\mathrm{d}\rho_c > 0$ at the critical point. A mode with odd $n$ changes its stability if and only if $\mathrm{d}R/\mathrm{d}\rho_c < 0$.

**C.** *Bend at a critical point and the character of stability change.* One mode becomes unstable (stable) if and only if the $M(R)$ curve bends counterclockwise (clockwise) at the critical point.

Let us apply these rules to the $M(R)$ curves in Fig. 6.1. Let the lowest density segment be stable for all radial modes. At $C_1$ the fundamental mode becomes unstable on both curves. It regains its stability at $C_2$ in the right panel, because $\mathrm{d}R/\mathrm{d}\rho_c < 0$ there. However, on the left panel $(\mathrm{d}R/\mathrm{d}\rho_c)_{C_2} > 0$; therefore, the fundamental mode remains unstable and the additional $n = 1$ mode becomes unstable. At $C_3$ the $n = 0$ mode becomes unstable in the right

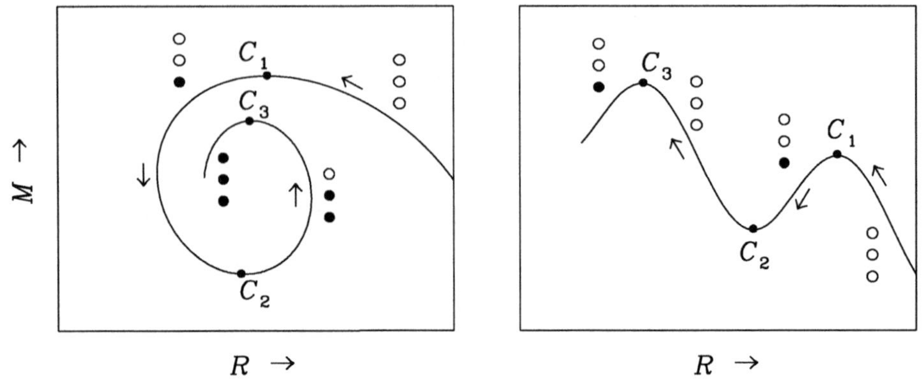


*Figure 6.1.* Two examples (*left* and *right*) of fragments of $M(R)$ curves for stellar models with different EOSs. Critical points are denoted by $C_1$, $C_2$ and $C_3$. Arrows near curves indicate the direction of increasing $\rho_c$. Three lowest modes of radial pulsations on a given segment are represented by a column of three circles, with unstable modes filled in black.

panel. In the left panel, the $n = 2$ mode becomes unstable, so that beyond $C_3$ all three lowest radial modes are unstable.

### 6.5.3 Neutron stars and white dwarfs

Having an EOS of cold dense matter one can easily calculate (§ 6.4) a family of stellar models parameterized by the central density $\rho_c$ and construct thus an $M - \rho_c$ diagram. A qualitative sketch of such a diagram is shown in Fig. 6.2. One can obtain a sufficiently smooth curve in a huge density interval ($\rho_c \lesssim 10^{16}$ g cm$^{-3}$). In the given case, two segments of the curve (plotted by the solid lines) correspond to $\mathrm{d}M/\mathrm{d}\rho_c > 0$ and refer to stable stellar configurations. They are two types of compact stars which are *white dwarf stars* ($\rho_c \lesssim 10^9$ g cm$^{-3}$) and *neutron stars* ($1.7 \times 10^{14}$ g cm$^{-3} \lesssim \rho_c \lesssim 3 \times 10^{15}$ g cm$^{-3}$). The filled dots denote (from left to right) the maximum-mass white dwarf, the minimum-mass neutron star and the maximum-mass neutron star, respectively. The dotted segments of the curve between the first and second filled dots and after the third filled dot correspond to unstable stellar models ($\mathrm{d}M/\mathrm{d}\rho_c < 0$). The instability of stellar configurations between the first and second filled dots occurs owing to the softening of the EOS due to electron captures and then the neutron drip (Chapter 3); the appropriate dotted segment of the curve is model-dependent.

White dwarfs are supported against gravitational contraction by the pressure of degenerate electron gas (Chapters 2 and 3); see, e.g., Shapiro & Teukolsky (1983). Massive white dwarfs loose their stability because of the softening of the electron pressure by relativistic effects and beta captures. The EOS of

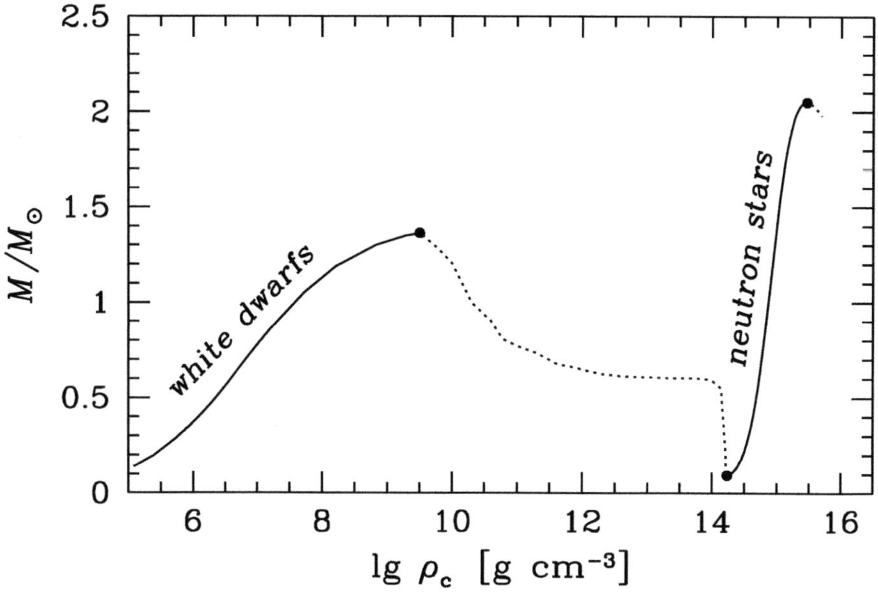


*Figure 6.2.* Schematic dependence of masses of equilibrium stellar models (built of cold dense matter) versus central density. Solid fragments show stable stars (either white dwarfs or neutron stars) while dotted fragments show unstable stars. Filled dots indicate maximum-mass and minimum-mass stars.

white dwarf matter is well known. The parameters of white dwarfs are reliably calculated and depend slightly on the composition of white dwarf cores.

Neutron stars are supported by the baryonic pressure associated with Fermi motion of baryons and their strong interactions (Chapter 5). The composition and EOS of the matter in neutron star cores is still uncertain which leads to large uncertainties in masses and radii of neutron stars as discussed below in this chapter.

One may continue the white-dwarf branch of the $M - \rho_c$ curve in Fig. 6.2 to lower $\rho_c$. White dwarfs occupy formally the part of this curve down to $M \sim 0.1\ M_\odot$. Lower-mass stars on this curve are called *brown dwarfs* (e.g., Chabrier & Baraffe 2000). The transition between white dwarfs and brown dwarfs is smooth, without any change of stability. A very low-mass extension of the same curve should reproduce planets. Hence, neutron stars, white dwarfs, brown dwarfs and planets are relatives (cold stellar objects). However, neutron stars and white dwarfs are final products of stellar evolution while brown dwarfs are different. They are thought to be created together with main sequence stars but undergo almost no nuclear evolution (their internal temperatures are too low to ignite efficient nuclear burning).

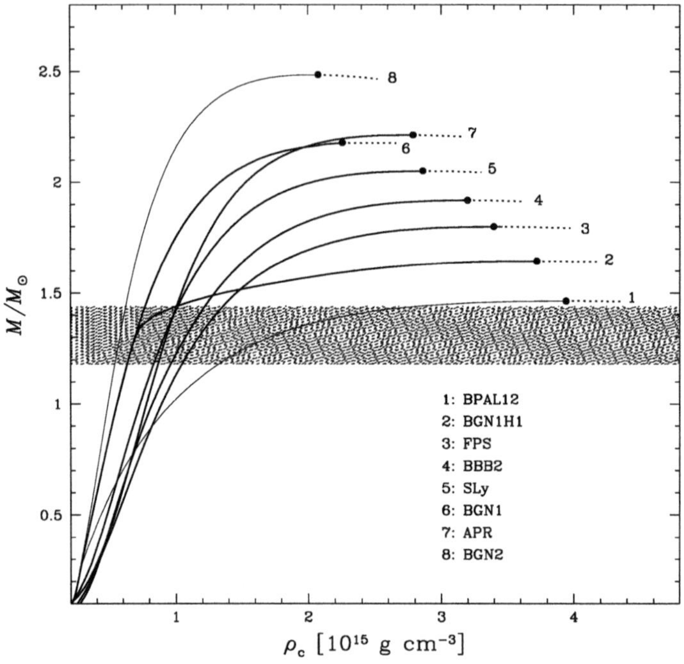


*Figure 6.3.* Gravitational mass $M$ versus central density $\rho_c$ of neutron star models for several EOSs. Filled circles show maxima on $M - \rho_c$ curves. Configurations to the right of the maxima (dotted segments) are unstable with respect to small radial perturbations. The shaded band is the range of precisely measured masses of binary radio pulsars (Chapter 9).

## 6.5.4 A variety of neutron star models

Now we focus on neutron star models. Once these models are computed for a given EOS, one can present them in the form of different diagrams discussed throughout this chapter.

In Fig. 6.3 we show the $M(\rho_c)$ diagram (at $\rho_c > 2.5 \times 10^{14}$ g cm$^{-3}$) for some EOSs from Table 6.1. On the higher-density side, $M(\rho_c)$ curves exhibit maxima $M = M_{\rm max}$ shown by filled dots.

On the lower-density side $M(\rho_c)$ curves exhibit minima $M_{\rm min} \simeq 0.1\ M_\odot$ (not shown in Fig. 6.3; a similar minimum is presented in Fig. 6.2). Properties of low-mass models with $M \sim M_{\rm min}$ will be discussed in § 6.5.8.

All points on $M(\rho_c)$ curves correspond to equilibrium stellar configurations. The configurations plotted by solid lines (between the minimum and maximum of $M$) correspond to $\mathrm{d}M/\mathrm{d}\rho_c > 0$ and appear to be stable. Configurations to the right of $M_{\max}$ (dotted segments in Fig. 6.3) and to the left of $M_{\min}$ are unstable. Non-rotating neutron stars with $M > M_{\max}$ will collapse into black holes. Numerical simulations of such events have been performed by several authors, particularly, by Gourgoulhon & Haensel (1993) and Baumgarte *et al.* (1996a,b). Neutron stars with $M < M_{\min}$ will explode (§ 6.5.8). Extreme configurations with $M = M_{\min}$ and $M = M_{\max}$ correspond to neutral equilibrium. In the presence of phase transitions in stellar interiors for more complicated EOSs, there may be two (generally, several) disconnected segments of stable neutron stars on an $M(\rho_c)$ curve. We will discuss such cases in Chapter 7. All in all, the theory provides a variety of different neutron star models.

### 6.5.5 Maximum masses of neutron stars

As seen from Table 6.1 and Fig. 6.3, the maximum mass $M_{\max}$ ranges from $1.46\ M_\odot$ to $2.48\ M_\odot$, for the selected EOSs. The central density in maximum-mass stars is even more uncertain. It ranges from $2 \times 10^{15}$ g cm$^{-3}$ $\sim 7\rho_0$ for the stiffest EOS to $4 \times 10^{15}$ g cm$^{-3}$ $\sim 15\rho_0$ for the softest one. Even $7\rho_0$ is far beyond the limits in which our EOSs can be considered as reliable.

Let us emphasize once more the dominant role of strong interactions for the structure of massive neutron stars (also see §§ 1.2 and 5.2). Replacing neutron-star matter by a free (non-interacting) Fermi gas of neutrons lowers the value of $M_{\max}$ to $0.72\ M_\odot$ (Oppenheimer & Volkoff, 1939). If one further allows for beta equilibrium between otherwise non-interacting nucleons, electrons, and muons, the EOS becomes even softer, with $M_{\max} = 0.70\ M_\odot$. Clearly, the values of $M_{\max}$ obtained neglecting nuclear interactions directly contradict precisely measured masses of neutron stars in compact binaries (§9.1.2). Turning the argument around, we may say that the very precisely measured mass $1.44\ M_\odot$ of the Hulse-Taylor pulsar implies that the nucleon-nucleon interaction is sufficiently repulsive at supranuclear densities to lift $M_{\max}$ by more than hundred percent over the Oppenheimer-Volkoff value.

In view of uncertainties in nucleon-hyperon (NH) and hyperon-hyperons (HH) interactions in dense matter, let us first consider a subset of EOSs with the N$e\mu$ composition (nucleons and leptons). Let us remove from this subset the extreme models BPAL12 and BGN2. Then the subset contains models of the N$e\mu$ matter which reproduce empirical saturation properties of nuclear matter (§ 5.4), with the narrow the range of theoretical maximum masses of neutron stars,

$$M_{\max}(\mathrm{N}e\mu) \simeq (1.8 - 2.2)\ M_\odot\ . \tag{6.30}$$

The appearance of hyperons softens an EOS and lowers $M_{\max}$. For the selected EOSs, the hyperonization at $\rho \simeq 2\rho_0$ lowers $M_{\max}$ even to a narrower range,

$$M_{\max}(\mathrm{NH}e\mu) \simeq (1.5 - 1.8)\ M_\odot\ . \tag{6.31}$$

Let us stress that this conclusion is obtained assuming that hyperons appear at $\rho \sim 2\rho_0$. However, the hyperonization threshold depends sensitively on the NH interaction in dense matter, which is poorly known for a neutron rich matter at $\rho \gtrsim \rho_0$. Therefore, the values of $M_{\max}$ in Eq. (6.31) have to be taken with caution. Actually, Eq. (6.31) reflects a typical effect of hyperons on $M_{\max}$, assuming that without hyperons $M_{\max}$ is given by Eq. (6.30). Alas, the lack of knowledge of the HH interaction implies a large uncertainty in the effect of hyperons on $M_{\max}$. For example, the EOS calculated by Vidaña *et al.* (2000b) shows especially strong softening by hyperonic interactions, with $M_{\max}(\text{N}e\mu) = 1.89\ M_\odot$ and $M_{\max}(\text{NH}e\mu) = 1.34\ M_\odot$. The latter value contradicts the measured mass of the Hulse-Taylor pulsar. When Vidaña *et al.* (2000b) remove (artificially) the HH interaction they get $M_{\max}(\text{NH}e\mu) = 1.47\ M_\odot$, which is marginally acceptable for explaining measured masses in double neutron star binaries (but not acceptable to explain masses of pulsars in binaries with white dwarfs, §9.1.3).

The problem of "too strong softening" of the EOS by hyperons might result from neglecting contribution from three-body interactions involving hyperons. As in the case of three-nucleon interaction, this contribution is strongly density-dependent and could stiffen the EOS at high densities. Its crucial role for increasing $M_{\max}$ was illustrated by Nishizaki *et al.* (2002). They considered two models of baryon-baryon interaction and obtained extremely low $M_{\max} = 1.08\ M_\odot$ and $1.10\ M_\odot$, respectively, when three-body interaction acted only between nucleons. Assuming that three-body force is universal and acts between all baryons, Nishizaki *et al.* (2002) obtained $M_{\max} = 1.52\ M_\odot$ and $1.82\ M_\odot$, respectively. The inclusion of realistic three-body interaction involving hyperons should be a priority task in forthcoming many-body calculations of the EOS.

### 6.5.6 The nature of the maximum mass of neutron stars

The existence of $M_{\max}$ can be explained using two kinds of physical arguments, related to the behavior of cold matter at ultrahigh densities and to General Relativity.

The arguments based on the behavior of the EOS at high densities were presented for Newtonian stars by Landau (1932) in a brief paper described in §1.2. These arguments are given in the first part of the paper, devoted to white dwarfs, but they can easily be rephrased for neutron stars (Harrison *et al.*, 1965). Consider a self-gravitating sphere of radius $R$ and total mass $M$ containing $A_{\rm b}$ neutrons. The Fermi gas of neutrons is squeezed with the increase of $M$ and becomes ultra-relativistic at sufficiently high densities ($\rho \gg 10^{15}$ g cm$^{-3}$). Let us assume that the interaction energy between neutrons can be neglected, compared to the kinetic energy. The internal (compression) energy of the star can then be estimated from the Fermi energy of ultra-relativistic neutrons,

$$\epsilon_{\rm F} \simeq \hbar c \left(A_{\rm b}/R^3\right)^{1/3} , \qquad E_{\rm int}(A_{\rm b}, R) \simeq A_{\rm b}\epsilon_{\rm F} \simeq (\hbar c/R)\, A_{\rm b}^{4/3} . \quad (6.32)$$

Let us suggest further that the gravitational energy is produced only by the rest mass of the matter, $M = A_{\rm b} m_n$. This gives

$$E_{\rm grav}(A_{\rm b}, R) \simeq -GM^2/R = -GA_{\rm b}^2 m_n^2/R . \quad (6.33)$$

For a fixed $A_{\rm b}$, a stable equilibrium is reached at the minimum of the total energy $E = E_{\rm int} + E_{\rm grav} = \alpha(A_{\rm b})/R$. The minimum can exist only if $\alpha > 0$. In this case $E$ will lower with increasing $R$, which will decrease $\epsilon_{\rm F}$. However non-relativistic effects will eventually stiffen the EOS and add the term $\propto R$ to $E_{\rm int}$. The latter term will increase $E$ after passing through a minimum corresponding to hydrostatic equilibrium. The condition $\alpha(A_{\rm b}) > 0$ is satisfied for

$$A_{\rm b} < A_{\rm max} \simeq \left(\hbar c/Gm_n^2\right)^{3/2} \approx 2.2 \times 10^{57} , \quad (6.34)$$
$$M < M_{\rm max}^{\rm Landau} = A_{\rm max} m_n \approx 1.8\, M_\odot . \quad (6.35)$$

Beautiful by its simplicity and astonishing by the realistic value of $M_{\rm max}$, the Landau-type derivation is based on two bold assumptions. First, one assumes that the matter becomes a free Fermi gas with increasing density (i.e., interactions between matter constituents become negligible). Second, one uses the Newtonian gravity and the baryon mass of the star. The crucial effect of space-time curvature will be discussed in the end of the present section, while now we focus on the behavior of the EOS at high densities.

A counterexample to the free gas behavior at very high densities was given by Zeldovich (1961). He showed (§ 5.15) that for a model of the neutron-neutron interaction mediated by the exchange of vector mesons, the interaction contribution dominates over the kinetic energy at high densities, so that

$$\mathcal{E} \simeq \mathcal{E}_{\rm pot} \propto n_{\rm b}^2 \propto A_{\rm b}^2/R^6 . \quad (6.36)$$

This would result in a compression (internal) energy of a cold star,

$$\text{Zeldovich model:} \qquad E_{\rm int} \simeq b\, A_{\rm b}^2/R^3 \propto n_{\rm b}^2 \propto A_{\rm b}^2/R^6 . \quad (6.37)$$

Then the minimum of $E = E_{\rm int} + E_{\rm grav}$ exists for any $A_{\rm b}$ and there is no upper bound on $M$ or $A_{\rm b}$ for Newtonian stars. In that case cold compact stars have a fixed radius, which does not depend on the stellar mass,[3]

$$\partial E/\partial R = 0 \quad \Longrightarrow \quad R = \sqrt{3b/Gm_0^2} . \quad (6.38)$$

[3]This feature is the well known *exact* result for Newtonian stars with a polytropic EOS, $P \propto \rho^2$ (the polytropic index $n = 1$).

In four decades following the Zeldovich paper our knowledge of dense matter has become much deeper. Now we know that at sufficiently high densities quarks become deconfined (albeit we are not sure whether such densities are reached in neutron star cores, see § 7.5). We know also that at still higher densities, owing to the asymptotic freedom property, quarks behave as nearly free fermions (§ 7.5). The internal energy of a cold star composed of free quarks is $E_{\rm int} \propto A_{\rm b}^{4/3}/R$. In this sense the Landau argument, based on the asymptotic high-density behavior of the EOS, remains valid.

Moreover, irrespectively of the EOS, the upper bound on $M$ is a consequence of General Relativity. Consider the right-hand-side of the Tolman-Oppenheimer-Volkoff equation of hydrostatic equilibrium, Eq. (6.7). It describes gravitational pull acting on a matter element of unit volume:

$$\text{Gravitational pull} = -\frac{Gm\rho}{r^2}\left(1+\frac{4\pi P}{mr^3}\right)\left(1+\frac{P}{\rho c^2}\right)\left(1-\frac{2Gm}{rc^2}\right)^{-1} . \tag{6.39}$$

This pull is given by a Newtonian-like term $-Gm\rho/r^2$ multiplied by three relativistic factors. With increasing $M$, all these factors amplify the pull compared to the Newtonian case, demanding higher pressure $P$ to keep hydrostatic equilibrium. Mathematically, the derivative $\mathrm{d}M/\mathrm{d}P_{\rm c}$ decreases with growing $M$ and makes stellar configurations less stable.

Let us illustrate this property with an *unphysical* case of incompressible fluid of constant density $\rho_{\rm inc}$. The total gravitational mass is then $M = (4\pi/3)\,\rho_{\rm inc}R^3$, and the pressure $P(r)$ in the star is determined analytically (the solution was obtained by Karl Schwarzschild in 1916; see, e.g., Box 23.2 of Misner *et al.* 1973). The central pressure $P_{\rm c}$ can be determined as a function of $r_{\rm g}/R$. As the mass increases, the pressure $P_{\rm c}$ and $r_{\rm g}/R$ grow up. In other words, both $r_{\rm g}/R$ and $M$ increase with increasing $P_{\rm c}$. For $P_{\rm c} \to \infty$, the radius tends to a finite value $R_{\rm max} = \frac{9}{8} r_{\rm g}$, and the mass tends to

$$M_{\rm max}^{\rm inc} = \frac{4\pi}{9}\frac{R_{\rm max}c^2}{G} = \frac{4c^3}{3^{5/2}\pi^{1/2}G^{3/2}\rho_{\rm inc}^{1/2}} \approx 5.09\,M_\odot \left(\frac{5\times 10^{14}\ {\rm g\ cm^{-3}}}{\rho_{\rm inc}}\right)^{\frac{1}{2}} . \tag{6.40}$$

This limit is the effect of General Relativity; there is no limit on the mass of incompressible-fluid stars in Newtonian gravitation. If $M_{\rm max}$ exists for an incompressible fluid, then it should *a fortiori* exist for *any* EOS of dense matter with finite compressibility.

Calculations show that the numerical value of $M_{\rm max}$ is mainly determined by the EOS at $\rho \gtrsim 2\rho_0$. This EOS is largely unknown leading to uncertainties in theoretical predictions of $M_{\rm max}$.

### 6.5.7 The upper bound on the maximum mass

Because $M_{\rm max}$ is uncertain, it is important to have a possibly firm upper bound on $M_{\rm max}$, based on general physical requirements. Let us assume that the EOS is known up to a certain density $\rho_{\rm u}$. This reliably known segment of the EOS ($0 < P \le P_{\rm u} = P_<(\rho_{\rm u})$) will be denoted by $P_<(\rho)$. Fixing $P_<(\rho)$, we can treat $M_{\rm max}$ as a *functional* of the EOS at $P > P_{\rm u}$; this *unknown* segment of the EOS will be denoted as $P_>(\rho)$. The inverse function $\rho_>(P)$ at $P > P_{\rm u}$ does not need to be continuous. General conditions imposed on $P_>(\rho)$ were discussed in § 5.15. As was shown there, apart from some very artificial situations (e.g., of highly excited medium), causality is in practice equivalent to the condition of subluminality of the EOS, $\mathrm{d}P/\mathrm{d}\rho \le c^2$. Additionally, the matter has to be stable, $\mathrm{d}P/\mathrm{d}\rho > 0$. The conditions imposed on $P_>(\rho)$ can be summarized as

$$0 < \mathrm{d}P_>/\mathrm{d}\rho \le c^2 \, . \tag{6.41}$$

Our task is to find the maximum of $M_{\rm max}$ on a set of EOSs $\{P_>(\rho)\}$ which satisfy these conditions. It turns out that the maximum $M_{\rm max} = M_{\rm max}^{\rm CL}$ is realized by the so called causality limit (CL) EOS,

$$P_>^{\rm CL}(\rho) = P_{\rm u} + (\rho - \rho_{\rm u})c^2 \, . \tag{6.42}$$

This is natural because the causal-limit EOS is the stiffest possible at $\rho > \rho_{\rm u}$. Calculations show that for $\rho_{\rm u} \lesssim 2\rho_0$ the effect of the outer layers with $\rho < \rho_{\rm u}$ on the value of $M_{\rm max}^{\rm CL}$ is weak.

In addition, one can introduce the *pure causal-limit EOS* of the form $P(\rho \ge \rho_{\rm s}) = (\rho - \rho_{\rm s})\, c^2$ and $P(\rho < \rho_{\rm s}) = 0$, where $\rho_{\rm s}$ is the stellar "surface" density. As shown in Appendix E, in this case one exactly gets $M_{\rm max} = 3.0\,(5 \times 10^{14}~{\rm g~cm^{-3}}/\rho_{\rm s})^{1/2}~M_\odot$. Putting formally $\rho_{\rm s} \sim 2\rho_0$, we would get $M_{\rm max}$ only slightly (by less than one percent) different than that obtained for initial causal-limit EOSs with $\rho_{\rm u} = \rho_{\rm s}$.

All in all, for $\rho_{\rm u} \lesssim 2\rho_0$ one gets (Rhoades & Ruffini 1974; Hartle 1978; Kalogera & Baym 1996; Glendenning 2000)

$$\begin{array}{l}\text{General Relati-}\\ \text{vity and } v_{\rm s} \le c\end{array} \Longrightarrow M_{\rm max} \le M_{\rm max}^{\rm CL} = 3.0 \sqrt{\frac{5 \times 10^{14}~{\rm g~cm^{-3}}}{\rho_{\rm u}}}~M_\odot . \tag{6.43}$$

The maximum mass for any EOS at $\rho > \rho_{\rm u}$ with a subluminal sound velocity is lower than this upper bound. The inequality (6.43) is widely accepted. Thus, it seems safe to state that the actual $M_{\rm max}$ of neutron stars built of baryonic matter is below $3~M_\odot$.[4] The value of $M_{\rm max}^{\rm CL}$ can be increased by rapid rotation, as discussed in §§ 6.12.7 and 6.12.5.

---

[4]For some very exotic hypothetical models of compact stars built, for instance, of a self-bound Q-matter, this upper bound may not apply as explained in § 8.20.

### 6.5.8 Low-mass neutron stars and the minimum mass

Calculations of the minimum mass $M_{\rm min}$ of neutron stars have a long history. The first correct estimate of $M_{\rm min}$ was obtained by Oppenheimer & Serber (1938). They showed, that the earlier estimate by Landau (1937) was much too small and based on incorrect arguments. Oppenheimer & Serber (1938) got $M_{\rm min} \simeq 0.17\ M_\odot$ neglecting nuclear interactions. They argued that the contribution of nucleon-nucleon interactions decreases $M_{\rm min}$ to $(0.03 - 0.10)\ M_\odot$, depending on the assumptions on these interactions (very poorly known at that time). Their stability criterion was based on energy arguments: a neutron star had to be stable with respect to dispersion into a gas of atomic nuclei (taken to be Ca as an example).

The next important step was made by Harrison *et al.* (1965) basing on the Harrison-Wheeler EOS, which described both the crust and the core of neutron stars. From the today perspective, their EOS in the neutron drip regime is unrealistic. They obtained $M_{\rm min} = 0.18\ M_\odot$ for the central density $\rho_{\rm c} \sim 3 \times 10^{13}$ g cm$^{-3}$. Using two versions of the Levinger & Simmons (1961) baryon-baryon potential, Tsuruta & Cameron (1966b) obtained $M_{\rm min} = (0.11 - 0.13)\ M_\odot$. Cohen & Cameron (1971) got $M_{\rm min} = 0.065\ M_\odot$, employing an EOS based on the updated version of the Levinger-Simmons nucleon-nucleon potential (Langer *et al.*, 1969). Finally, Baym *et al.* (1971b) in their classical paper (BPS) got $M_{\rm min} = 0.0925\ M_\odot$. They used their own (BPS) EOS of the outer crust, and the EOS of Baym *et al.* (1971a) (BBP) for the inner crust and outer layers of the core.

Neutron star models with $M \sim M_{\rm min}$ are sensitive to the EOS at subnuclear densities, especially at the crust-core interface. The EOS should be physically correct there; a brutal *ad hoc* matching of the crust and core EOSs are not sound for this purpose. A microscopic model, underlying the EOS, should be the same on both sides of the crust-core interface, basing on the same effective nuclear Hamiltonian. Of all EOSs in Table 6.1, only the SLy and FPS EOSs satisfy this condition. Let us summarize the results obtained by Haensel *et al.* (2002b) using these EOSs. The $M(\rho_{\rm c})$ curves near $M \approx M_{\rm min}$ are shown in Fig. 6.4. The minimum of an $M(\rho_{\rm c})$ curve is very sharp and the lower-density side of the curve is extremely steep. The value of $M_{\rm min}$ seems to be rather well established, $M_{\rm min} \simeq 0.1\ M_\odot$.

It is evident that the minimum mass of neutron stars can be strongly increased by rotation and thermal effects. As we will see in §6.12.3, rotation will also change the nature of the minimum-mass configuration. For newly born protoneutron stars, both thermal and neutrino-trapping effects are significant. They can increase the minimum mass to $(0.9 - 1.1)\ M_\odot$ (Goussard *et al.*, 1998; Strobel *et al.*, 1999). However, $M_{\rm min}$ obtained for cold neutron stars still remains the absolute lower bound on neutron star mass.

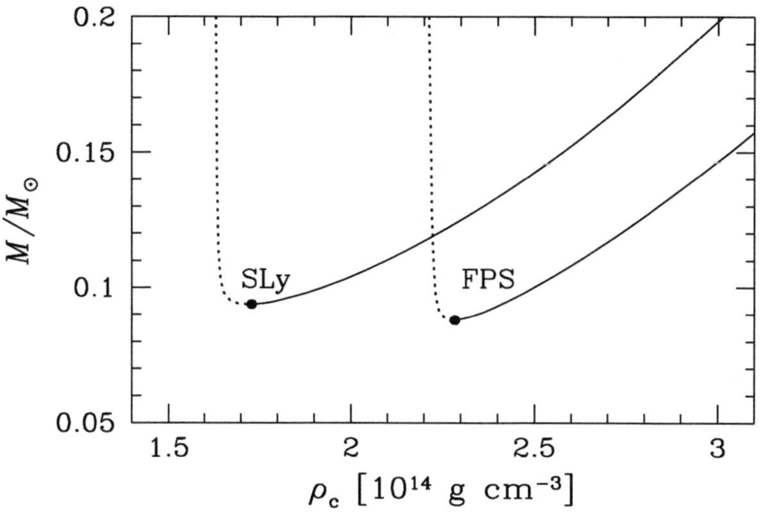


*Figure 6.4.* Gravitational mass $M$ versus central density $\rho_c$ of low-mass neutron star models near $M \approx M_{\rm min}$ for the FPS and SLy EOSs. The minima of $M(\rho_c)$ curves are indicated by filled circles. Configurations to the left of the minima (dotted lines) are unstable with respect to small radial perturbations.

Some astrophysical scenarios, in which a cold neutron star looses its matter and reaches the critical $M_{\rm min}$ value, were proposed by Blinnikov *et al.* (1984); Colpi *et al.* (1989, 1991); Sumiyoshi *et al.* (1998). These authors considered a neutron star in a compact binary with a more massive companion (another neutron star or a black hole). The companion may accrete matter from the neutron star. Accretion may be self-accelerating, because the decrease of neutron star mass increases its radius, making the star more susceptible to the mass loss. Once $M$ decreases below $M_{\rm min}$, no equilibrium configuration can be reached. Numerical simulations indicate that an unstable low-mass neutron star will explode.

## 6.6. Radii and surface redshifts

A small radius of neutron stars is one of their most specific features. While the expected masses of neutron stars $\sim (1{-}2)\, M_\odot$ are typical for main-sequence stars of solar-type, their radii are extremely small by ordinary stellar standards. A stellar object of $M \sim M_\odot$ and $R < 100$ km cannot be anything but a neutron star.[5] Moreover, neutron stars of $M \gtrsim M_\odot$ are expected to have sizable surface

[5]Exotic compact stellar objects, e.g., strange quark stars, can also fulfill this condition (Chapter 8), but by the "Occam's razor" principle we will consider them as hypothetical as long as non-ambiguous observational

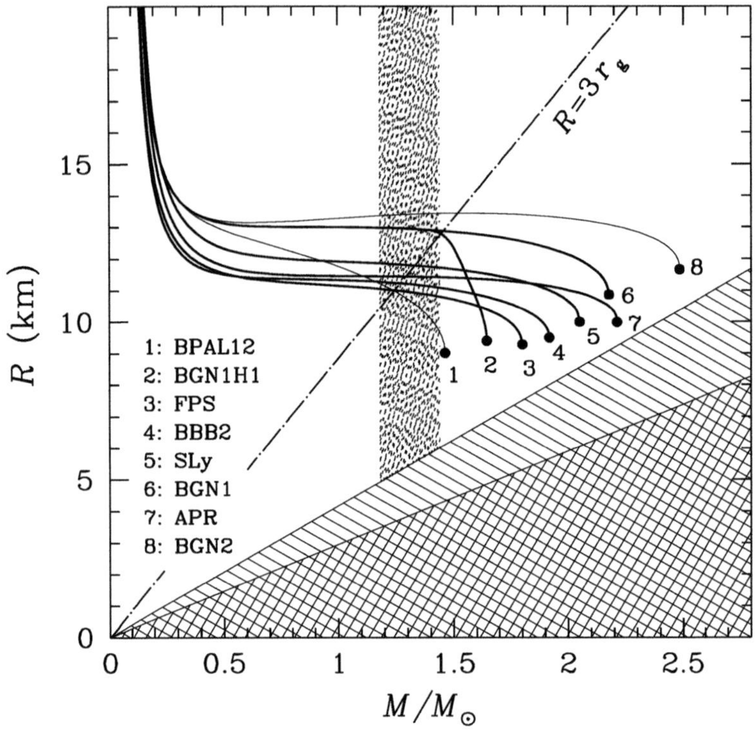


*Figure 6.5.* Circumferential neutron star radius $R$ versus gravitational mass $M$ for the selected EOSs of dense matter. The doubly hatched area ($R < r_g$) is prohibited by General Relativity. The entire hatched triangle is prohibited by General Relativity combined with the condition $v_s \leq c$. The shaded vertical band corresponds to the measured range of masses of double neutron star binaries (§9.1.2 c). The dot-and-dashed line $R = 3\,r_g$ shows the radius of the innermost stable orbit of a test particle rotating around a compact object with a given mass.

redshifts $\sim 0.2-0.6$ (another specific feature). Speaking of neutron star radius, one has to distinguish the "circumferential radius," that determines proper length of the stellar equator (§ 6.6.1), and the "apparent radius" (sometimes called the "radiation radius"), as measured by an observer at infinity (§ 6.6.6).

## 6.6.1 Circumferential radii

The radius of a neutron star of a given mass depends on the EOS of stellar matter. The lack of knowledge of the EOS implies the uncertainty in the $R(M)$

evidence for their existence is absent. The Occam's principle states that entities must not be multiplied beyond necessary limits. Its traditional interpretation is that phenomena should be explained in terms of the simplest possible causes. The principle was devised by William of Occam, a medieval English philosopher.

relation. Several $R(M)$ curves for the selected EOSs are displayed in Fig. 6.5. For all these EOSs, neutron stars with $M > 0.3\ M_\odot$ have $R < 14$ km. For a given $M$, the radius increases with the growth of the EOS stiffness. As seen from Fig. 6.5, General Relativity and subluminality of sound in dense matter ($v_s \leq c$) restrict the region of the $R - M$ plane, which neutron stars can access. Still, the uncertainty in theoretical predictions is large. For the canonical neutron star mass $M = 1.4\, M_\odot$, the radius ranges from 10 km to 14 km. The minimum radius is reached at $M = M_{\rm max}$ and ranges from 9 km to 12 km.

Let us notice a substantial difference between models of neutron stars containing nucleons alone and nucleons with hyperons. If we restrict ourselves to models consistent with empirical saturation parameters of nuclear matter, then purely nucleonic EOSs show very weak dependence of radius on mass for $M \gtrsim M_\odot$. Doubling $M$ decreases the radius by less than 20 percent.

The dependence of $R$ on $M$ has been thoroughly studied by Lattimer & Prakash (2001). Here we follow their arguments. Let us first approximate the EOS by the polytrope, $P = K n_b^\gamma$, with the adiabatic index $\gamma = 2$.[6] Then Newtonian stellar models have a constant radius, independent of $M$ and $\rho_c$. This radius is uniquely determined by the constant $K$ according to $R_{\rm Newt} \propto K^{1/2}$. Therefore, an accurate measurement of $R$ would allow one to determine $K$ and identify (constrain) the EOS. In General Relativity, the stellar radius weakly decreases with increasing mass provided $M$ is not too close to $M_{\rm max}$, and the dependence $R \propto K^{1/2}$ is approximate. Lattimer & Prakash (2001) looked for a more general correlation between the EOS and values of $R$. For a broad set of EOSs, they found an approximate relation

$$R(M) \simeq C(n_s)\ [P(n_s)]^{1/4}\ , \tag{6.44}$$

where $C$ is independent of the EOS and is remarkably independent of $M$ in the mass range $(1 - 1.4)\, M_\odot$, while $n_s$ is a selected baryon density in the range $(1 - 2)\, n_0$. For $n_s/n_0$=1, 1.5, and 2, they obtained $C(n_s) \simeq 9$, 7, and 6, respectively. Neglecting the weak dependence of $R$ on $M$, the "empirical relation" (6.44) enables one to determine the EOS in the density range $(1-2)n_0$ from a measured value of $R$, provided $M$ ranges from $M_\odot$ to $1.4\, M_\odot$ and is not too close to $M_{\rm max}$.

However, this "empirical relation" cannot be used for EOSs with too strong softening, for instance, owing to the appearance of hyperons. Neutron stars with hyperon cores exhibit different behavior. Their EOS beyond the hyperon threshold is so soft, that the stellar radius decreases rapidly with increasing

[6]This is only an approximation. As shown in Chapter 5, even for realistic EOSs of the N$e\mu$ matter we have density dependent $\gamma = 2-3$ (§ 5.13.3), while for the NH$e\mu$ models $\gamma = 1-3$ is a rapidly varying function of density after hyperons appear in dense matter (§ 5.14.2).

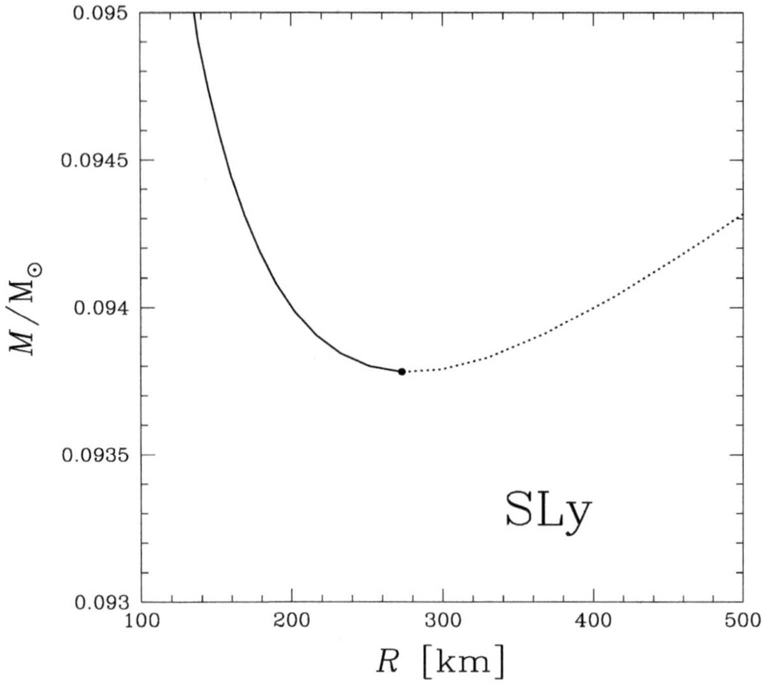


*Figure 6.6.* Gravitational mass versus radius at $M \sim M_{\rm min}$ for neutron stars with the SLy EOS. The solid line – stable configurations, the dotted line – configurations unstable with respect to small radial perturbations. The minimum-mass configuration is indicated with the filled circle.

mass. In a narrow range of masses close to $M_{\rm max}$ these stars may have very different radii.

### 6.6.2 Radii of low-mass neutron stars

The mass-radius relation for neutron stars with $M \sim M_{\rm min}$ deserves a separate discussion. For the SLy EOS it is shown in Fig. 6.6. In contrast to the $M(\rho_{\rm c})$ curve with a sharp minimum, the minimum in the $M(R)$ curve is flat.

Let $\rho_{\rm c,min}$ be the central density of the minimum-mass configuration. Configurations with $\rho_{\rm c} \simeq \rho_{\rm c,min}$ are very loosely bound by gravitational forces (see below). Their radii reach hundreds of kilometers, and a small difference in their mass is accompanied by a large difference in radius. For example, a decrease in mass by 0.001 $M_\odot$ (by ~one per cent) on the stable segment of the $M(R)$ curve in Fig. 6.6, implies an increase in radius by ~ hundred kilometers (by more than twice!).

This behavior is related to properties of the matter at subnuclear densities. Neutron stars with $M \sim M_{\rm min}$ are weakly bound, huge spheres of the solid

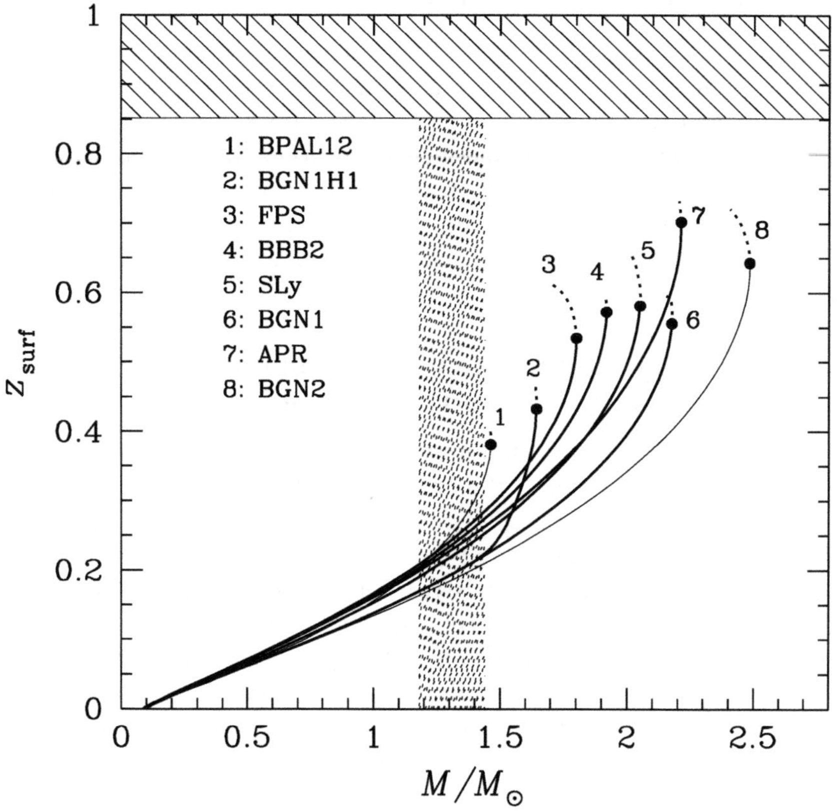


*Figure 6.7.* Surface gravitational redshift $z_{\rm surf}$ versus gravitational mass $M$ for several EOSs of dense matter. Full and dotted lines show stable and unstable configurations, respectively. The hatched area above the horizontal line is prohibited for subluminal EOSs ($v_{\rm s} \leq c$). The shaded vertical band is the range of precisely measured masses of double neutron star binaries (Chapter 9).

crust of hundreds kilometers thick, containing a tiny core. At $M = M_{\rm min}$ the core constitutes 2% of the stellar mass for the SLy EOS, and 3% for the FPS EOS; the core radius is 3.8 and 4.2 km for the SLy and FPS EOSs, respectively. As the SLy EOS is stiffer than the FPS EOS for $10^{13}\ {\rm g\ cm^{-3}} \lesssim \rho \lesssim \rho_0$ (see § 3.6), the SLy neutron stars with $M \simeq M_{\rm min}$ are less compact, and contain a smaller and less massive core. For $\rho_{\rm c} < \rho_{\rm c,min}$ , both the mass and radius of the core decrease very slowly with decreasing $\rho_{\rm c}$, while the total mass and radius of the star increase extremely steeply.

### 6.6.3 Gravitational redshifts

Gravitational redshift $z_{\rm surf}$ of frequencies of photons, emitted from a neutron-star surface, is given by Eq. (6.12). It is an important measure of the compactness parameter $x_{\rm GR} = r_{\rm g}/R$ (see § 1.3.3, Eq. (1.4)). A measured value of $z_{\rm surf}$ would give $x_{\rm GR}$ and, hence, the $M/R$-ratio. If, in addition, the mass $M$ were known, one could immediately determine the radius $R$. The plots of $z_{\rm surf}$ versus gravitational mass are given in Fig. 6.7. For the canonical mass of $1.4\ M_\odot$, the redshift decreases from 0.3 to 0.2 with increasing the stiffness of the EOS.

### 6.6.4 The upper bound on gravitational redshift and the lower bound on neutron star radius

For a given EOS, the maximum value $z_{\rm surf}^{\rm max}$ of stable neutron stars is reached at $M = M_{\rm max}$. On average, $z_{\rm surf}^{\rm max}$ increases with the growth of $M_{\rm max}$. However, this is not a strict rule because $z_{\rm surf}^{\rm max}$ depends on both $M_{\rm max}$ and $R_{M_{\rm max}}$. Taking a softer EOS lowers the value of $z_{\rm surf}^{\rm max}$. In particular, $z_{\rm surf}^{\rm max} < 0.4$ for the softest BPAL12 EOS. While the most massive star for this EOS is the most compact and dense stable stellar model, it is the "least relativistic" of all maximum-mass configurations owing to the low value of $M_{\rm max}$. In general, the EOS which maximizes $z_{\rm surf}^{\rm max}$ has to be relatively soft in the outer neutron-star layers (to reduce $R$) but stiff in the stellar core (to increase $M_{\rm max}$). Modern EOSs of the $npe\mu$ matter are of this type (Chapter 5) and give $z_{\rm surf}^{\rm max} \simeq 0.6$. The presence of hyperonic cores lowers the value of $z_{\rm surf}^{\rm max}$.

The maximum surface redshift of stable neutron stars is a functional of the EOS of dense matter. If one neglects the condition of subluminality, the absolute upper bound on $z_{\rm surf}$ is equal 2 (for the derivation of this result, see §6 of Chapter 11 of Weinberg 1972). If we restrict ourselves to subluminal EOSs with $0 < {\rm d}P/{\rm d}\rho \leq c^2$, then the value of $z_{\rm surf}^{\rm max}$ is maximized (Lindblom, 1984) for the pure causal-limit (CL) EOS with no normal neutron-star envelope: $P(\rho \geq \rho_{\rm s}) = (\rho - \rho_{\rm s})c^2$ and $P(\rho < \rho_{\rm s}) = 0$ (see § 6.5.7). The structure of the Tolman-Oppenheimer-Volkoff equation implies, that the value of $z_{\rm surf}^{\rm max}$ for this EOS does not depend on the "surface density" $\rho_{\rm s}$ (see Appendix E). This limiting redshift will be denoted as $z_{\rm surf}^{\rm CL}$.

The presence of a crust of normal matter around a causal-limit core decreases the value of $z_{\rm surf}^{\rm max}$. Let $P_{\rm b}$ and $\rho_{\rm b}$ be the pressure and density at the bottom of the normal crust. The case of $P_{\rm b} = 0$ corresponds to the pure casual-limit EOS without any crust. For $\rho_{\rm b} \lesssim \rho_0$ the effect of the normal crust on the maximum-mass configuration is small and can be treated using a perturbative approach. The relevant small parameter is $P_{\rm b}/\rho_{\rm b}c^2$. For $\rho_{\rm b} \sim \rho_0$ we have typically $P_{\rm b}/\rho_{\rm b}c^2 \sim 10^{-2}$ (see Chapter 3). Numerical calculations show that adding a crust around the causal-limit core increases $R_{M_{\rm max}}$ linearly in

$P_b/\rho_b c^2$. An associated increase in $M_{max}$ is quadratic in $P_b/\rho_b c^2$ and, therefore, negligible. Thus, the presence of the normal crust lowers the value of $z_{surf}^{max}$ as compared to the pure causal-limit EOS.

The precise upper bound on the surface redshift for subluminal EOSs is (Haensel *et al.*, 1999)[7]

$$\begin{array}{c}\text{General Relativity} \\ \text{and } v_s \le c\end{array} \Longrightarrow \begin{array}{l} z_{surf} \le z_{surf}^{CL} = 0.8509 \\ \text{and } r_g/R \le 0.7081 \,. \end{array} \tag{6.45}$$

Introducing some density discontinuities into the causal-limit EOS does not increase the value of $z_{surf}^{max}$ (Gondek & Zdunik, 1995). Modern EOSs give maximum surface redshifts of neutron stars typically 40% lower than $z_{surf}^{CL}$.

An upper bound on the surface redshift implies a lower bound on the radius of a neutron star of gravitational mass $M$

$$v_s \le c \quad \Longrightarrow \quad R \ge 1.412 \, r_g = 4.17 \, (M/M_\odot) \ \text{km} \,. \tag{6.46}$$

Accordingly, the circumferential radius of a neutron star of the canonical mass $1.4 \ M_\odot$ has to be larger than 5.83 km.

### 6.6.5 The upper bound on surface gravity

Gravitational acceleration on the stellar surface, usually called the surface gravity and denoted by $g_s$, is an important parameter of the theory of neutron star atmospheres (§ 1.3.1) and the theory relating internal and surface temperatures of neutron stars (e.g., Gudmundsson *et al.* 1983; see also Potekhin *et al.* 2003 and references therein). The expression for $g_s$, resulting from Eq. (6.10), includes the effect of space-time curvature and reads (also see § 1.3.3)

$$g_s = \frac{GM}{R^2\sqrt{1-x_{GR}}} = 15.25 \times 10^{14} \frac{x_{GR}^2}{\sqrt{1-x_{GR}}} \frac{M_\odot}{M} \ \text{cm s}^{-2}. \tag{6.47}$$

where, as usual, $x_{GR} = r_g/R$ and the effects of rotation are neglected. For a neutron star with $M = 1.4 \, M_\odot$ and $R = 10$ km we have $g_s = 2.43 \times 10^{14}$ cm s$^{-2}$. Thus it is convenient to measure $g_s$ in units of $10^{14}$ cm s$^{-2}$ and introduce $g_{s,14} \equiv g_s/(10^{14} \ \text{cm s}^{-2})$.

The surface gravity of neutron stars is many orders of magnitude larger than for other stars. It is $\sim 10^5$ times larger than for white dwarfs and $\sim 10^{10}$ times larger than at the solar surface. As shown by Bejger & Haensel (2004),

---

[7] Our value of $z_{surf}^{CL}$ is ≈5% lower than the value obtained via the extrapolation of redshifts from Table 1 of Lindblom (1984) to $P_b/\rho_b = 0$. Lindblom calculated the upper limit to $z_{surf}$ for stars with crusts. Notice that a precise determination of $M_{max}$ is rather easy, but the determination of $R_{M_{max}}$ and associated $z_{surf}^{max}$ with the same relative precision as $M_{max}$ is more difficult because of the flatness of $M(R)$ curves.

the value of $g_s$ depends very strongly on the largely unknown EOS in neutron star interiors. Let us discuss upper bounds on $g_s$ following these authors.

For subluminal EOSs of dense matter, a strict upper bound on $x_{GR}$ is $x_{GR}^{max} = 0.7081$ (see Eq. (6.45)). Using this value, from Eq. (6.47) we get an upper bound on the surface gravity of a non-rotating neutron star of mass $M$:

$$v_s \leq c \quad \Longrightarrow \quad g_{s,14} \leq g_{max,14}^{CL} = 14.1\,(M_\odot/M)\,. \tag{6.48}$$

For a given EOS, the maximum of $g_s$ is reached at $M = M_{max}$. The latter values should be higher than the highest measured neutron star mass, $M_{obs}^{max}$. Neutron star mass measurements are reviewed in § 9.1. Spin frequencies of observed neutron stars are insufficiently high to affect their structure, so that the effect of rotation can be neglected in this analysis (see § 6.12.2). Then we come to the upper bound

$$g_{s,14} < 14.1\,(M_\odot/M_{obs}^{max})\,. \tag{6.49}$$

The higher $M_{obs}^{max}$, the stronger we constrain the maximum $g_s$. A precisely measured mass of the Hulse-Taylor pulsar, $1.441\ M_\odot$ (§ 9.1), gives the upper bound of $9.79 \times 10^{14}$ cm s$^{-2}$. According to Clark *et al.* (2002), the neutron star in the high-mass X-ray binary 4U 1700–37 has the mass $M_{obs} > 1.9\ M_\odot$ at the $2\sigma$ confidence level. This would imply $g_s < 7.4 \times 10^{14}$ cm s$^{-2}$. However, this result has to used with caution because the compact object in 4U 1700-37 can be not a neutron star but a black hole (see § 9.1).

Bejger & Haensel (2004) calculated $g_s$ for neutron star models using more than thirty baryonic EOSs of dense matter, particularly those in Table 6.1, and constructed plots of $g_s(M)$ and $g_s(x_{GR})$. The $g_s(M)$ curves show a significant scatter. On the contrary, the $g_s(x_{GR})$ curves presented in Fig. 6.8 are much less sensitive to the EOS. They can be reproduced (with error $\lesssim$25%) by

$$g_{s,14} \simeq 5\,x_{GR}^{5/4}/\sqrt{1 - x_{GR}}\,. \tag{6.50}$$

Actually, the accuracy of this fit formula is higher if we exclude superluminal EOSs and unrealistically stiff ones (with too high incompressibility of nuclear matter at saturation, resulting in $M_{max} \gtrsim 2.5\ M_\odot$). Putting then the upper bound $x_{GR}^{max} = 0.708$ into Eq. (6.50) we get an approximate "realistic upper bound" $g_s < 6 \times 10^{14}$ cm s$^{-2}$. Subluminal EOSs involving only nucleons give $g_s^{max} \simeq (3-5) \times 10^{14}$ cm s$^{-2}$. The stiffer the EOS, the closer $g_s^{max}$ to the subluminal upper bound $g_s^{CL}(M_{max})$. Subluminal hyperonic EOSs give $g_s^{max}$ typically lower than $4 \times 10^{14}$ cm s$^{-2}$. For these EOSs, $g_s^{max}$ can be as small as one-fifth of the upper bound $g_s^{CL}(M_{max})$.

We close this section with brief comments on $g_s$ for EOSs involving meson condensates and quark matter, studied in Chapters 7 and 8. As shown by Bejger & Haensel (2004), subluminal EOSs with an exotic high-density phase have

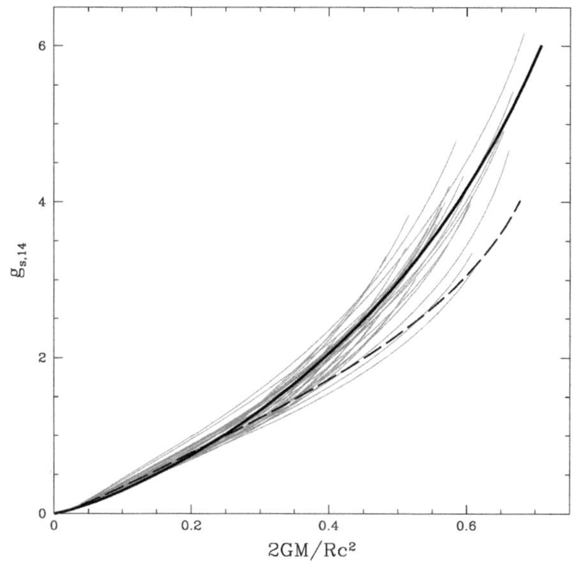


*Figure 6.8.* Plots of $g_{s,14}$ versus compactness parameter $r_g/R$. The thick solid line is the approximate formula, Eq. (6.50). The thick dashed line is obtained using the causal-limit EOS (§ 6.5.7) with $n_u = 0.3\ \mathrm{fm}^{-3}$. From Bejger & Haensel (2004).

relatively low $g_s^{\max}$. A phase transition softens the EOS, lowering the stellar radius. On the other hand, the softening decreases $M_{\max}$. The latter effect dominates over the former. The very special case of $g_s$ for self-bound strange quark stars will be studied in Chapter 8.

### 6.6.6 Apparent radii

For $M \gtrsim M_\odot$ circumferential radii $R$ of neutron stars can be as small as $R \sim 2\,r_g$. Here we discuss "apparent radii", $R_\infty$, which a distant observer would see if a telescope were able to resolve the star.

The apparent radius is related to the circumferential radius by (e.g., Thorne 1977)

$$R_\infty = R/\sqrt{1 - r_g/R}\,. \tag{6.51}$$

The dependences $R_\infty(M)$ and $R(M)$ are rather different. The difference reflects space-time curvature near the star and increases with the growth of $M$. Figure 6.9 shows $R_\infty(M)$ for some EOSs of dense matter. For stiff and moderately stiff EOSs ($M_{\max} \gtrsim 1.8\ M_\odot$) without any strong softening at highest densities, the apparent radius $R_\infty$ *increases* with growing $M$ at $M \gtrsim 0.5\ M_\odot$

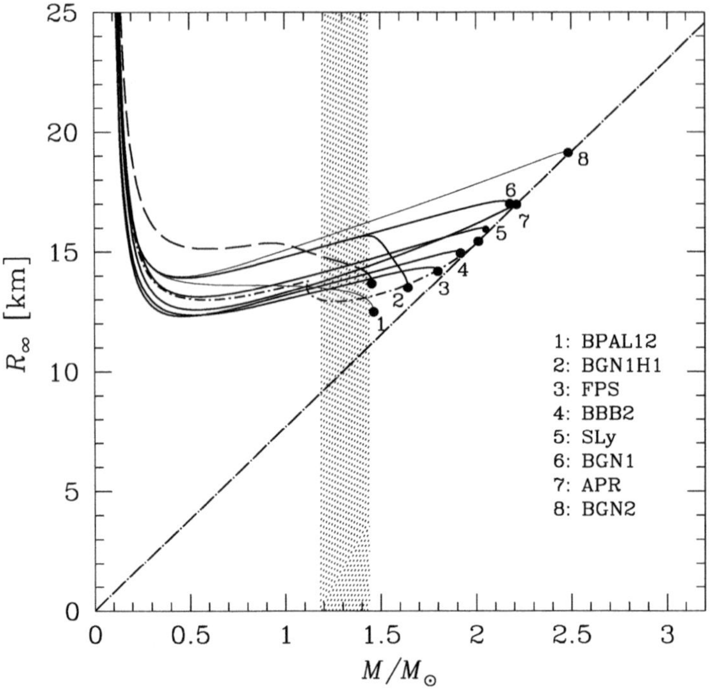


*Figure 6.9.* Apparent radius of neutron stars versus gravitational mass for selected EOSs of dense matter (labeled as in Table 6.1). The dashed curve is for hybrid neutron stars with a mixed baryon-quark phase (EOS from Table 9.1 of Glendenning 2000). The dashed-and-dot curve is for an EOS with first-order phase transition to a pure kaon-condensed phase (Kubis, 2001). The dash-and-dot straight line is the minimum value $R_\infty^{\rm min} = 7.66\ (M/M_\odot)$ km.

(except for a tiny region close to $M_{\rm max}$). In contrast, $R$ *decreases* in the same mass interval.

A strict lower bound on $R_\infty(M)$ results from the definition of $R_\infty$ (Lattimer & Prakash, 2001; Haensel, 2001). Specifically, Eq. (6.51) can be rewritten as $R_\infty/r_{\rm g} = x_{\rm GR}^{-1}(1-x_{\rm GR})^{-1/2}$, which means that $R_\infty/r_{\rm g}$ is a function of one parameter, $x_{\rm GR} = r_{\rm g}/R$. This function diverges at $x_{\rm GR} = 0$ and $x_{\rm GR} = 1$, and has a single minimum at $x_{\rm GR} = 2/3$. Therefore, the minimum value of $R_\infty(M)$ is $R_\infty^{\rm min}(M) = 7.66\ (M/M_\odot)$ km. This value is only 0.6% smaller than the apparent radius 7.71 $(M/M_\odot)$ km for the "true" maximum compactness $x_{\rm GR} = 0.7081$ consistent with $v_{\rm s} < c$ (§ 6.6.4).

While the subluminal upper bound on $x_{\rm GR}$ at a given $M$ is slightly larger than 2/3, *actual* maximum values $x_{\rm GR}(M_{\rm max})$ for various EOSs are lower than 2/3. However, if we restrict ourselves to medium stiff and stiff EOSs with

$M_{\rm max} \gtrsim 1.8\,M_\odot$, then $x_{\rm GR}(M_{\rm max}) \simeq 0.6$, which is only by 0.07 lower than 2/3. In such cases $R_\infty(M_{\rm max})$ is close to $R_\infty^{\rm min}(M_{\rm max})$ (Fig. 6.9).

According to Lattimer & Prakash (2001) and Haensel (2001) one can expect $R_\infty > 12$ km for any baryonic EOS, independently of neutron star mass. Our Fig. 6.9 confirms this "practical lower bound" on $R_\infty$.

## 6.7. Binding energy

A configuration with $M = M_{\rm max}$ has maximum baryon number $A_{\rm b}$ (and baryon mass $M_{\rm b}$), whereas a configuration with $M = M_{\rm min}$ has minimum baryon number (and baryon mass) (Zeldovich, 1962). To prove this statement let us consider two infinitesimally close configurations built of cold catalyzed matter. Let the first one consist of $A_{\rm b}$ baryons, and the second consist of $A_{\rm b} + {\rm d}A_{\rm b}$ baryons. The first configuration can be transformed into the second by adding ${\rm d}A_{\rm b}$ baryons on the stellar surface (and keeping the system in full equilibrium). The change of the total gravitational mass accompanying this transformation is given by the small-increment theorem which relates small increments ${\rm d}M$ and ${\rm d}A_{\rm b}$ (Zeldovich, 1962; Zeldovich & Novikov, 1971)

$$ {\rm d}M = m_0\sqrt{1 - x_{\rm GR}}\,{\rm d}A_{\rm b}\,. \tag{6.52}$$

Therefore, the extremum condition ${\rm d}M/{\rm d}\rho_{\rm c} = 0$ implies ${\rm d}A_{\rm b}/{\rm d}\rho_{\rm c} = 0$ (and ${\rm d}M_{\rm b}/{\rm d}\rho_{\rm c} = 0$), and vice versa. This means that the extremum of $M$ is reached simultaneously with the extremum of $A_{\rm b}$ (and $M_{\rm b}$).

An example of the $M_{\rm b}(\rho_{\rm c})$ dependence is displayed in Fig. 6.10. The configuration with $M = M_{\rm max}$ corresponds also to the maximum mass defect, which is about 15%. As it turns out, the maximum relative mass defect depends rather weakly on the EOS of baryon matter. Instead of the mass defect, it is convenient to use its energy equivalent, the binding energy $E_{\rm bind}$, defined in §6.2. Figure 6.11 shows $E_{\rm bind}$ as a function of $M$ for several EOSs. For $M = 1.4\,M_\odot$, the binding energy relative to the dispersion of the star into the gas of $^{56}$Fe ions is about $2 \times 10^{53}$ erg. The maximum binding energy increases with the growth of $M_{\rm max}$, and ranges from $\approx 3 \times 10^{53}$ erg for the softest to $\approx 10^{54}$ erg for the stiffest EOS of Table 6.1.

The behavior of $E_{\rm bind}$ at $M \approx M_{\rm max}$ deserves a comment. The spike at $M = M_{\rm max}$ (Fig. 6.11) becomes nonsingular, if considered as a function of $\rho_{\rm c}$ (rather than $M$). To show this, let us use the equation

$$ {\rm d}E_{\rm bind}/{\rm d}A_{\rm b} = \mu_0\left\{1 - \sqrt{1 - x_{\rm GR}}\right\} > 0\,, \tag{6.53}$$

where $\mu_0 = m_0\,c^2$. It is valid for configurations built of cold catalyzed matter, as a consequence of the relation (6.52). Therefore, ${\rm d}E_{\rm bind}/{\rm d}\rho_{\rm c} = ({\rm d}E_{\rm bind}/{\rm d}A_{\rm b})\,({\rm d}A_{\rm b}/{\rm d}\rho_{\rm c})$ is continuous at $\rho_{\rm c} \approx \rho_{\rm c,max}$ and vanishes at $\rho_{\rm c} = \rho_{\rm c,max}$. We have ${\rm d}E_{\rm bind}/{\rm d}\rho_{\rm c} > 0$ for $\rho_{\rm c} < \rho_{\rm c,max}$ and ${\rm d}E_{\rm bind}/{\rm d}\rho_{\rm c} < 0$ for $\rho_{\rm c} > \rho_{\rm c,max}$.

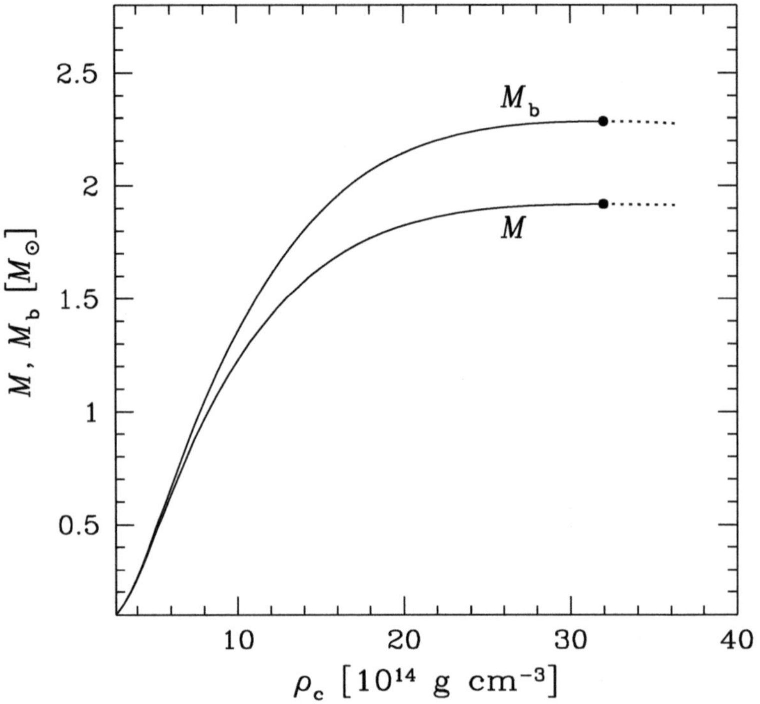


*Figure 6.10.* Baryon mass $M_b = Am_0$ and gravitational mass $M$ of neutron star models versus central density for the BBB2 EOS. The configuration with $M = M_{max}$ (filled circle) realizes the maximum of $M_b$. Configurations with higher $\rho_c$ (dotted segments) are unstable.

According to Fig. 6.11, of two configurations with $M \approx M_{max}$, containing the same number of baryons, the unstable one ($\rho_c > \rho_{c,max}$) is less bound than the stable one. This property is general and can be proved starting from basic Eq. (6.52). Using this equation, we get the following expression for the difference $\Delta E_{bind}(A_b) = E_{bind}(A_b) - E'_{bind}(A_b)$ in binding energies of stable and unstable configurations with $M \approx M_{max}$:

$$\Delta E_{bind}(A_b) = \mu_0 \int_{A_b}^{A_{max}} \left\{ \sqrt{1 - x_{GR}(\tilde{A}_b)} - \sqrt{1 - x'_{GR}(\tilde{A}_b)} \right\} d\tilde{A}_b, \tag{6.54}$$

where $x_{GR}(A_b) = 2GM(A_b)/R(A_b)c^2$, $x'_{GR} = 2GM'(A_b)/R'(A_b)c^2$, and primed quantities refer to unstable equilibrium configurations. At the same $A_b$, an unstable configuration is more compact than a stable one, $R'(A_b) < R(A_b)$. Moreover, variations in $M$ at $M \approx M_{max}$ are of the second order of smallness

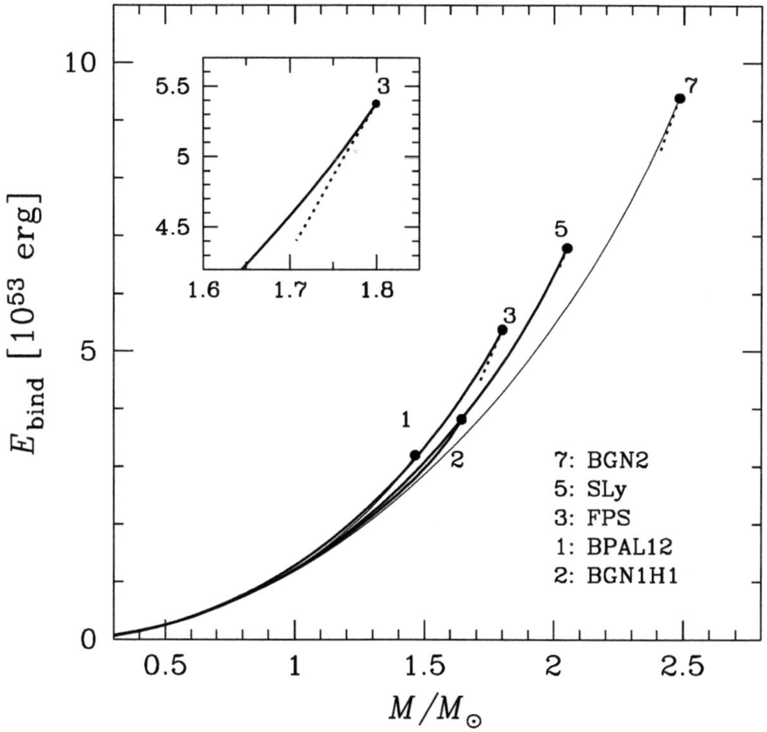


*Figure 6.11.* Binding energy (relative to the dispersion of a star into $^{56}$Fe gas) versus gravitational mass for several EOSs from Table 6.1. Maximum-mass configurations (filled circles) have maximum binding energy. Dotted segments show unstable configurations which are less bound than the stable ones. The inset shows a zoomed segment of the $E_{\rm bind}(M)$ curve for the FPS EOS at $M \approx M_{\rm max}$.

in $\rho_{\rm c} - \rho_{\rm c,max}$, and, therefore, are much smaller than variations in $R$. Thus, $R'(A_{\rm b}) < R(A_{\rm b})$ implies $E'_{\rm bind}(A_{\rm b}) < E_{\rm bind}(A_{\rm b})$.

### 6.7.1 Approximate formulae

Figure 6.11 shows that the binding energy of a stable neutron star correlates with its gravitational mass. A simple approximate "empirical formula" which describes the dependence $E_{\rm bind}(M)$ at $M > 0.5\ M_\odot$ for realistic EOSs reads (Lattimer & Yahil, 1989)

$$E_{\rm bind} \simeq 1.5 \times 10^{53}\ (M/M_\odot)^2\ {\rm erg}\ . \tag{6.55}$$

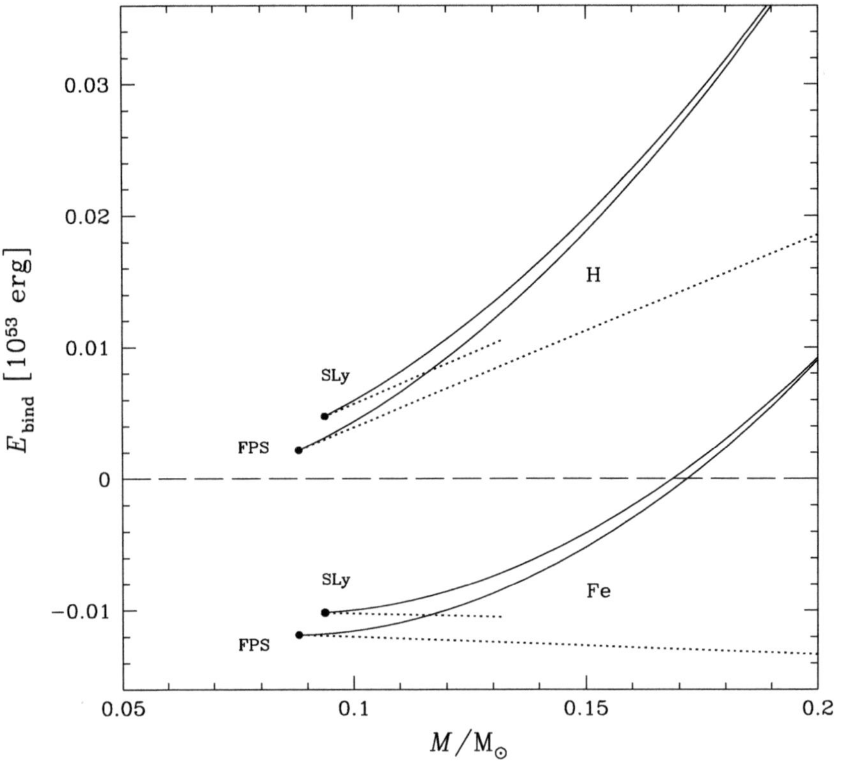


*Figure 6.12.* Binding energy (relative to the dispersion of a star into a gas of $^{56}$Fe and hydrogen) versus gravitational mass at $M \approx M_{\rm min}$ for the FPS and SLy EOSs. Solid lines – stable configurations; dotted lines – unstable configurations, filled dots – $M = M_{\rm min}$.

A more refined formula, containing the compactness parameter $x_{\rm GR} = r_{\rm g}/R$, was proposed by Lattimer & Prakash (2001):

$$E_{\rm bind} \simeq 1.6 \times 10^{53} \left(\frac{M}{M_\odot}\right) \left(\frac{x_{\rm GR}}{0.3}\right) \frac{1}{1 - 0.25\, x_{\rm GR}} \ {\rm erg}. \qquad (6.56)$$

It is quite precise for EOSs without any strong softening at high densities. It can be less accurate for massive neutron stars with hyperon cores.

Because $E_{\rm bind}$ reaches maximum at $M = M_{\rm max}$, Eq. (6.55) gives an approximate expression for the maximum binding energy $E_{\rm bind}^{\rm max}$,

$$E_{\rm bind}^{\rm max}({\rm EOS}) \simeq 1.5 \times 10^{53} \ (M_{\rm max}({\rm EOS})/M_\odot)^2 \ {\rm erg}\,. \qquad (6.57)$$

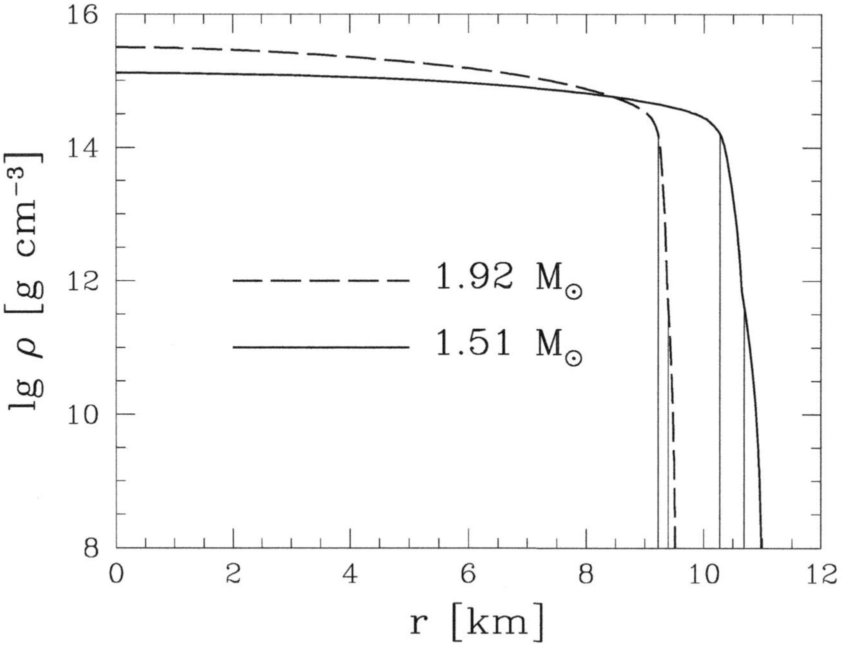


*Figure 6.13.* Density versus radial coordinate for neutron-star models with $M = 1.51\ M_\odot$ (the solid line) and $M = M_{\rm max} = 1.92\ M_\odot$ (the dashed line) employing the BBB2 EOS in stellar cores and the FPS EOS in the crusts. Thin vertical lines show the crust-core interface and the neutron-drip point.

This formula reproduces exact values of $E_{\rm bind}^{\rm max}$ typically better than within ten percent for our set of EOSs in Table 6.1.

## 6.7.2 Neutron star models with $M \approx M_{\rm min}$

The binding energy of neutron stars with $M \approx M_{\rm min}$ is shown Fig. 6.12. Let us focus on stable configurations (solid segments). At $M \approx M_{\rm min}$ the sign of $E_{\rm bind}$ depends on the choice of the dispersed reference system. Neutron stars with lowest masses are bound ($E_{\rm bind} > 0$) with respect to the hydrogen gas, but are unbound ($E_{\rm bind} < 0$) with respect to the dispersed $^{56}$Fe gas.

The case of $E_{\rm bind} < 0$ deserves a comment. Negative binding energies of neutron stars with $M < 0.17\ M_\odot$ indicate that these stars are actually *metastable* with respect to the transformation into a cloud of $^{56}$Fe dust. Their decompression (while keeping the matter in the ground state) would eventually *lower* the total energy, after overcoming an initial energy barrier (that results from the gravitational binding of the outer layers). Being metastable with respect to *large amplitude* decompression, these configurations are, however,

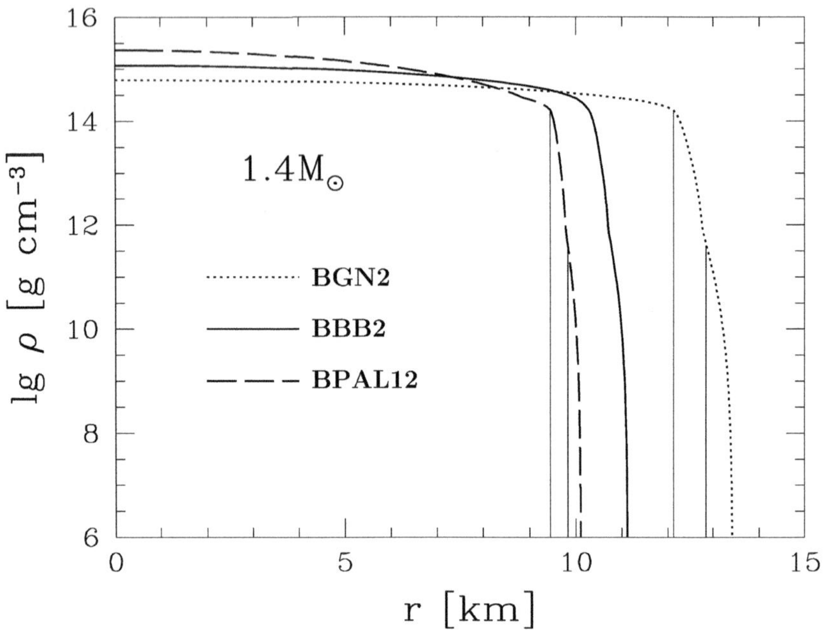


*Figure 6.14.* Density versus radial coordinate for neutron star models with $M = 1.4\ M_{\odot}$. Stellar cores are described either by BPAL12, or by BBB2, or by BGN2 EOSs (the dashed, solid or dotted lines, respectively). In all cases, the FPS EOS of the crust is used. Thin vertical lines indicate crust-core interface and neutron-drip point for the BPAL12 and BGN2 models.

*stable* with respect to *small amplitude* perturbations. This can be contrasted with the *instability* of configurations at $\rho_c < \rho_{c,min}$. The latter equilibrium configurations will explosively expand to huge white-dwarf like configurations with radii exceeding $10^4$ km under the effect of any small perturbation.

## 6.8. Basic internal structure of neutron stars

The distribution of matter within a neutron star depends on its mass and results from an interplay between the pressure and gravity. In Fig. 6.13 we show the mass density within a neutron star as a function of radial coordinate, calculated for the BBB2 EOS in the stellar core for two masses. The higher mass is equal to the maximum mass for this EOS. Calculations show that the density within the core is rather uniform for $M = 1.2 - 1.5\ M_{\odot}$. The pronounced density drop occurs near the crust-core interface. The density drop becomes steeper near the neutron drip point, as a result of the EOS softening due to the neutron drip. The growth of the stellar mass increases the gravitational pull within the crust, which makes the crustal density profile steeper and the crust thinner. The effect is most pronounced for stars with $M = M_{max}$. For the medium-stiff BBB2 EOS in the core, the crust contains $1.4\%$ and $0.8\%$ of the total gravitational mass of the stars with $M = 1.24\ M_\odot$ and $1.51\ M_\odot$, respectively. The crust thickness of these stars is $1.01$ km and $0.72$ km. For $M = M_{\rm max}$, the crust contains only $0.2\%$ of the stellar mass, and the crust thickness becomes as small as $0.29$ km. The crust mass and thickness are smaller for softer EOSs.

While the crustal EOS is rather well established (see Chapter 3), the crust structure results from the interplay of its EOS and the gravitational pull exerted by the core; the latter depends on the core compactness. Therefore, the uncertainty of the EOS in the stellar core introduces some uncertainty in the crust structure. This is visualized in Fig. 6.14, where we show the density profiles inside a $1.4\ M_\odot$ star calculated for the BPAL12, BBB2, and BGN2 EOSs in the core. The crust thickness ranges from 0.7 km for the softest EOS up to 1.3 km for the stiffest one. The dependence of the fraction of stellar mass contained in the crust on the core EOS is even more dramatic: this fraction ranges from $0.7\%$ for the softest BPAL12 EOS to $2.2\%$ (three times larger!) for the stiffest BGN2 EOS. The BBB2 model is typical for medium-stiff EOSs; the crust thickness is about 0.8 km and the crustal mass fraction is $1\%$.

## 6.9. Universal structure of the outer envelope

A solution to the equations of stellar structure is greatly simplified (Urpin & Yakovlev, 1979) in the outer neutron star envelope which is much thinner than $R$ and contains a small fraction of mass. In these layers, the density and pressure vary much more rapidly with $r$ than $m(r) \approx M$. Accordingly, the metric (6.1) reduces to the Schwarzschild metric (6.11), where $r$ can be safely replaced by $R$ in the brackets and in the last term,

$$\mathrm{d}s^2 = c^2\,\mathrm{d}t^2 \left(1 - \frac{r_{\rm g}}{R}\right) - \left(1 - \frac{r_{\rm g}}{R}\right)^{-1} \mathrm{d}r^2 - R^2\,(\mathrm{d}\theta^2 + \sin^2\theta\,\mathrm{d}\phi^2). \quad (6.58)$$

Introducing the local proper time $\tau$ and the local proper depth $z$ measured from the stellar surface:

$$\tau = (1 - r_{\rm g}/R)^{1/2}\,t, \quad z = (R - r)\,(1 - r_{\rm g}/R)^{-1/2}, \qquad (6.59)$$

we reduce the metric to the locally flat one, $\mathrm{d}s^2 = c^2\,\mathrm{d}\tau^2 - \mathrm{d}z^2 - R^2\,(\mathrm{d}\theta^2 + \sin^2\theta\,\mathrm{d}\phi^2)$.

For a low-density matter in the outer neutron star envelope, we have $P \ll \rho c^2$ and $Pr^3 \ll Mc^2$. With these simplifications, Eq. (6.7) can be rewritten as the familiar Newtonian equation for a plane-parallel atmosphere (also see Eq. (6.10)),

$$\mathrm{d}P/\mathrm{d}z = g_{\rm s}\,\rho, \qquad (6.60)$$

where $g_{\rm s}$ is the surface gravity given by Eq. (6.47) and discussed in § 6.6.5.

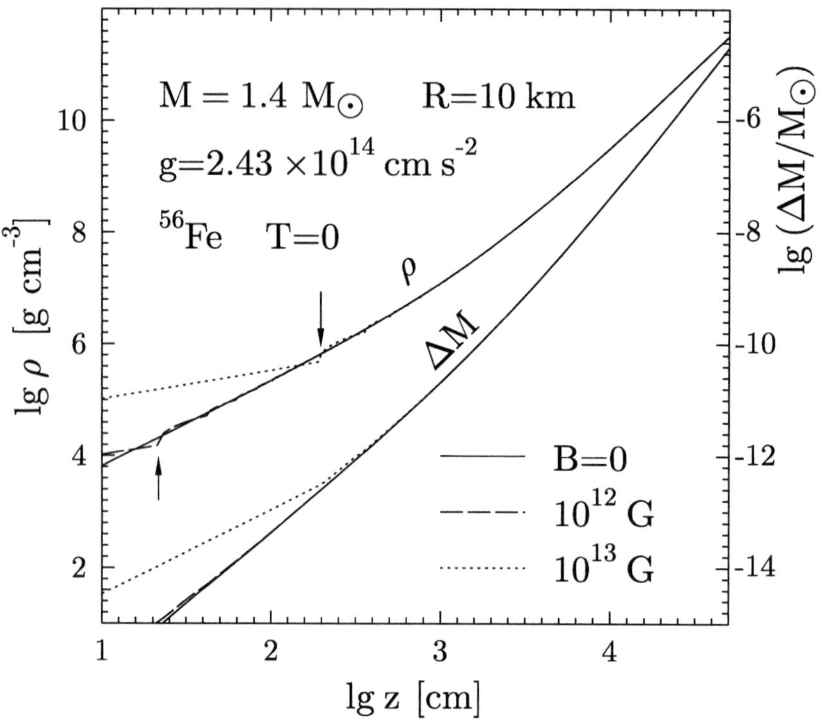


*Figure 6.15.* Density (left vertical scale) and surface mass $\Delta M$ (right scale) in a neutron star crust versus depth $z$ for $B = 0$, $10^{12}$ and $10^{13}$ G. Arrows indicate kinks of the density profiles of magnetic envelopes at which the first Landau level becomes populated with increasing $z$.

Because the main contribution to the pressure of the outer degenerate ($T \ll T_{\rm F}$) stellar layers comes from a free electron gas, we may set $P \approx P_e$ and use the zero-temperature thermodynamic relationship $\mathrm{d}P_e = n_e \, \mathrm{d}\mu_e$, where $\mu_e$ is the electron chemical potential (including the rest-mass term). On the other hand, the main contribution into the mass density comes from ions, and we have $\rho \approx n_{\rm i} m_{\rm i} \approx m_{\rm u} n_e A_{\rm i}/Z_{\rm i}$, where $m_{\rm u}$ is the atomic mass unit, while $A_{\rm i}$ and $Z_{\rm i}$ are the mass and charge numbers of the ions (nuclei). Inserting these relationships into Eq. (6.60), we get

$$\mathrm{d}\mu_e/\mathrm{d}z = g_{\rm s} m_{\rm u} A_{\rm i}/Z_{\rm i} \; . \tag{6.61}$$

Integrating this equation, we can obtain a solution in a closed form. It can be further simplified by assuming that $A_{\rm i}/Z_{\rm i} = \mathrm{constant}$ over the outer neutron star envelope. In this case by integrating Eq. (6.61) from the surface (where

$\mu_e = m_e c^2$) we obtain

$$\mu_e - m_e c^2 = g_s m_u z A_i / Z_i. \tag{6.62}$$

Because $\mu_e = m_e c^2 \sqrt{1 + x_r^2}$, this solution can be rewritten as

$$x_r^3 = \left[ \frac{z}{z_0} \left( 2 + \frac{z}{z_0} \right) \right]^{3/2}, \quad z_0 = \frac{m_e c^2 Z_i}{m_u g_s A_i} = \frac{49.3\, Z_i}{A_i\, g_{s14}} \text{ m}, \tag{6.63}$$

where $g_{s14}$ is $g_s$ in units of $10^{14}$ cm s$^{-2}$, $x_r$ is the electron relativity parameter defined by Eq. (2.2) and $z_0$ is a depth-scale at which degenerate electrons become relativistic. Typically, we have $g_{s14} \sim 2-3$, so that $z_0 \sim 10$ m. Because $x_r^3 \approx \rho_6 Z_i / A_i$, Eq. (6.63) describes the density growth within the degenerate layers of the outer neutron star envelope. For a non-magnetic neutron star with $M \sim (1-2) M_\odot$ the solution is valid at densities $\rho \lesssim 10^{11}$ g cm$^{-3}$ and geometrical depths $z$ about some hundred meters. Nevertheless, the solution (6.63) is invalid in a very thin surface layer, where the electron gas is non-degenerate or where the assumption that $P = P_e$ is violated (see § 2.5). Its thickness ranges from ~1 meter in a hot neutron star to several millimeters in a cold one. On the contrary, Eq. (6.60) remains valid everywhere near the surface, and that outermost meter of the envelope can be studied separately by combining Eq. (6.60) and the equations of thermal structure.

According to Eq. (6.63), we have $\rho \propto z^{3/2}$ in the very outer layer of non-relativistic degenerate electron gas ($\rho \ll 10^6$ g cm$^{-3}$, $z \ll z_0$), and $\rho \propto z^3$ in the deeper layers, where electrons are ultrarelativistic ($\rho \gg 10^6$ g cm$^{-3}$, $z \gg z_0$). These density profiles correspond to polytropic equations of state with the polytropic indices $n = 1.5$ and $n = 3$ (polytropic exponents $\gamma = 5/3$ and $\gamma = 4/3$), respectively (compare § 2.3.1 d).

Finally, integrating Eqs. (6.60) and (6.8) we can easily find the gravitational mass $\Delta M(z) = m(R) - m(r)$ contained in a layer from the surface to a given depth $z$:

$$\Delta M(z) = 4\pi R^2\, P(z)\, g_s^{-1} \sqrt{1 - r_g/R}. \tag{6.64}$$

This mass is solely determined by the pressure $P(z)$ at the bottom of the given layer. We see that the structure of the outer neutron star crust is *universal* (*selfsimilar*).

For instance, consider the neutron star with $M = 1.4\, M_\odot$, $R = 10$ km, and the outer envelope composed of $^{56}$Fe. The $\rho(z)$ and $\Delta M(z)$ profiles are depicted in Fig. 6.15. The density $\rho = 10^{10}$ g cm$^{-3}$ is reached at the depth $z \simeq 150$ m, and the mass of the envelope at this depth is $\Delta M \simeq 2 \times 10^{-7}\, M_\odot$. At the base of the outer crust, $\rho = 4.3 \times 10^{11}$ g cm$^{-3}$, we obtain $z \simeq 550$ m and $\Delta M \simeq 3 \times 10^{-5}\, M_\odot$. However, our solution can become inaccurate there because our initial assumptions that $z \ll R$ and $A_i/Z_i = \text{constant}$ can already be violated.

### 6.9.1 The structure of magnetic envelopes

The above consideration does not take into account the effects of strong magnetic fields. As explained in Chapter 4, these effects can be dramatic near the neutron star surface, but weaken with growing density. They are most important in the layers where the field is strongly quantizing – that is, at $\rho < \rho_B$ and $T \ll T_B$, where $\rho_B$ and $T_B$ are given by Eqs. (4.30) and (4.33), respectively. In these layers, the assumption of temperature-independent EOS ($T \ll T_{\rm F}$) becomes less accurate than in the nonmagnetic case. Indeed, according to Eq. (4.31), the Fermi temperature $T_{\rm F}$ in a strongly quantizing field $B$ is reduced by a factor $x_B/x_{\rm r} = (4/3)^{1/3}(\rho/\rho_B)^{2/3}$ [$x_B$ being given by Eq. (4.28)] for ultrarelativistic electrons and by the square of this factor for non-relativistic electrons.

If $T \ll T_F$, the electrons are nearly free (§4.1.2), and the magnetic field can be regarded as constant (and, hence, force-free) in a local part of the thin outer envelope, Eq. (6.62) is valid for any field strength $B$. Thus, the structure of the envelope is again described by a selfsimilar solution. However, the character of the solution at $\rho < \rho_B$ is different. In particular, in Eq. (6.63) one should replace the nonmagnetic electron relativity parameter $x_{\rm r}$ by the parameter $x_B$ that is appropriate for a strongly quantizing magnetic field. In this case, a given geometrical depth $z$ corresponds to a higher density $\rho$. This is illustrated by Fig. 6.15, where we plot the $\rho(z)$ and $\Delta M(z)$ profiles in the neutron star envelope for $B = 10^{12}$ G and $10^{13}$ G. For instance, the magnetic field $B = 10^{13}$ G strongly affects the density distribution in the layer $\rho < \rho_B \sim 10^6$ g cm$^{-3}$, located at $z \lesssim 3$ m. According to Eq. (4.30), the density at the bottom of this layer scales as $B^{3/2}$. Equation (4.32) shows that the EOS in this layer is polytropic, but the polytropic index differs from that at $B = 0$; now it is equal to $n = \frac{1}{2}$ ($\gamma = 3$) or $n = 1$ ($\gamma = 2$) for the non-relativistic or ultrarelativistic electron gases, respectively.

When $\rho$ exceeds $\rho_B$, higher Landau levels become populated, and the non-magnetic solution is approximately recovered. The exact solution, however, oscillates around the nonmagnetic one, as seen from Fig. 6.15. The kinks, especially pronounced in the density profiles, are associated with occupation of new Landau levels with increasing electron density (or geometrical depth). Arrows show the kinks produced by the occupation of the first excited Landau level. These kinks should be smeared out by thermal effects (by thermal broadening of the Landau levels). Moreover, the magnetic field in surface layers may be not force-free, and the magnetic force may further distort the structure of these layers.

## 6.10. Moment of inertia for slow rigid rotation

Many neutron stars show rapid rotation (§§ 1.4.4, 9.4). However, even the most rapidly spinning pulsar PSR J1748–2446ad, with the spin period of 1.396 ms, is distorted by rotation only slightly (provided its mass $M \gtrsim M_\odot$, see, e.g., Cook *et al.* 1994; Datta *et al.* 1998). Therefore, the effects of rotation on the structure of observed pulsars can be treated using the lowest-order perturbative approximation (Hartle, 1967), which is usually referred to as the approximation of *slow rotation*. A precise treatment of more rapid rotation requires two dimensional calculations, as will be discussed in § 6.12.

In this section, we restrict ourselves to a slow, uniform (rigid) rotation with an angular frequency $\Omega$. Strictly speaking, $\Omega$ is the spin frequency measured by a distant observer. The slow rotation implies, that centrifugal forces are small compared to the gravity, $R^3\Omega^2/(GM) \ll 1$. This condition is satisfied even for PSR J1748–2446ad, the fastest observed rotator ($\Omega = 4501 \text{ s}^{-1}$). Assuming $M = 1.4\ M_\odot$ and $R = 10$ km, we get $R^3\Omega^2/(GM) \simeq 0.11$ for this pulsar.

A rotation induces polar flattening of the star. A slowly rotating neutron star is axially symmetric. Because of time-reversal invariance, lowest-order rotational perturbations of the stellar structure are quadratic in $\Omega$. On the other hand, the leading contribution to the total angular momentum $J$ is proportional to $\Omega$. Therefore, to the lowest order in $\Omega$, the moment of inertia $I = J/\Omega$ is determined by the mass distribution and space-time curvature of a non-rotating star (Hartle, 1967) as discussed below. In this approximation, $I$ is given by Eq. (6.70).

### 6.10.1 Dragging of local inertial reference frames and moment of inertia

The centrifugal force acting on a fluid element in a star depends on the local spin frequency $\overline{\omega}$, as measured in a local inertial reference frame. Therefore, $\overline{\omega}$ is given by the difference between $\Omega$ and the spin frequency $\omega$ of the local inertial reference frame in the vicinity of the given fluid element, $\overline{\omega} = \Omega - \omega$. To the lowest order in $\Omega$, $\omega$ depends only on $r$. The quantity $\omega(r)$ describes *dragging of local inertial frames* in a rotating star; it vanishes in the Newtonian limit, where $\overline{\omega} = \Omega$. In order to calculate $J$, one has to know $\overline{\omega}$ everywhere in the star. Dragging of local inertial frames implies that $\overline{\omega} < \Omega$. The function $\overline{\omega}$ obeys the homogeneous, second-order linear differential equation (Hartle, 1967)

$$\frac{1}{r^4}\frac{\mathrm{d}}{\mathrm{d}r}\left(r^4\,\mathrm{e}^{-\Phi-\lambda}\,\frac{\mathrm{d}\overline{\omega}}{\mathrm{d}r}\right) + \frac{4}{r}\left(\frac{\mathrm{d}}{\mathrm{d}r}\,\mathrm{e}^{-\Phi-\lambda}\right)\overline{\omega} = 0\,, \tag{6.65}$$

where $\Phi$ and $\lambda$ are the metric functions of the non-rotating star (see Eq. (6.1)).

A solution of Eq. (6.65) is subject to two boundary conditions: regularity (finiteness) at $r = 0$ and vanishing of the dragging frequency at infinity ($\overline{\omega} \to \Omega$ for $r \to \infty$). In addition, $\overline{\omega}$ has to be continuous at the stellar surface.

For $r \geq R$, we have $\mathrm{e}^{-\Phi-\lambda} = 1$ (§ 6.1) which gives

$$\overline{\omega}(r \geq R) = \Omega - 2GJ/(c^2 r^3) . \tag{6.66}$$

Equation (6.65) can be replaced by two coupled first-order differential equations (Haensel & Prószyński, 1982),

$$\frac{\mathrm{d}j}{\mathrm{d}r} = \frac{8\pi}{3} r^4 \left( \rho + \frac{P}{c^2} \right) \mathrm{e}^{-\Phi-\lambda} \, \overline{\omega} \, , \tag{6.67}$$

$$\frac{\mathrm{d}\overline{\omega}}{\mathrm{d}r} = \frac{G \mathrm{e}^{\Phi+\lambda}}{c^2 r^4} \, j \, . \tag{6.68}$$

Here, $j(r)$ is the contribution to $J$ from a sphere of radius $r$. The total stellar angular momentum is $J = j(R)$. Equations (6.67) and (6.68) can be integrated simultaneously with Eqs. (6.7)–(6.9) of stellar structure from $r = 0$ to $r = R$, with the boundary conditions $j(0) = 0$ and $\overline{\omega}(0) = \overline{\omega}_0$. The value of $\overline{\omega}_0$ is arbitrary but sufficiently small (to be in the regime of slow rotation). The value of $\Omega$, corresponding to the selected value of $\overline{\omega}_0$, is obtained by matching the interior and exterior solutions, $\overline{\omega}(r \leq R)$ and $\overline{\omega}(r \geq R)$, at $r = R$:

$$\Omega = \overline{\omega}(R) + 2GJ/(c^2 R^3) . \tag{6.69}$$

The total moment of inertia of the star is then calculated as $I = J/\Omega$. Using Eq. (6.67), one gets

$$I = \frac{8\pi}{3} \int_0^R \mathrm{d}r \, r^4 \left( \rho + \frac{P}{c^2} \right) \frac{\overline{\omega}}{\Omega} \, \mathrm{e}^{-\lambda-\Phi} \, , \tag{6.70}$$

where the ratio $\overline{\omega}/\Omega$ is a function of $r$ independent of $\Omega$ (or $\omega_0$). General relativistic effects are incorporated in $(\overline{\omega}/\Omega)\, \mathrm{e}^{-\lambda-\Phi}$ and in $P/c^2$. Equation (6.70) explicitly takes into account space-time curvature ($\sim r_{\mathrm{g}}/R$) in the spherically symmetric approximation. Also, it takes into account dragging of inertial reference frames in the lowest order in $\Omega$; this gives a correction, linear in $r_{\mathrm{g}}/R$, contained in $\overline{\omega}/\Omega$. Nevertheless, Eq. (6.70) neglects the centrifugal corrections $\sim R^3\Omega^2/GM$. In this approximation $I$ can be calculated from a non-rotating stellar model.

In the Newtonian limit, $\overline{\omega} = \Omega$, $\lambda = \Phi = 0$, $P \ll \rho c^2$, and we get the familiar expression $I_{\mathrm{Newton}} = \frac{8}{3}\pi \int_0^R \mathrm{d}r \, r^4 \rho$.

## 6.10.2 Moment of inertia versus $M$

In Fig. 6.16 we show the dependence of $I$ on the neutron star mass $M$ for several EOSs of dense matter. A suitable unit for $I$ is $10^{45}$ g cm$^2$ (because

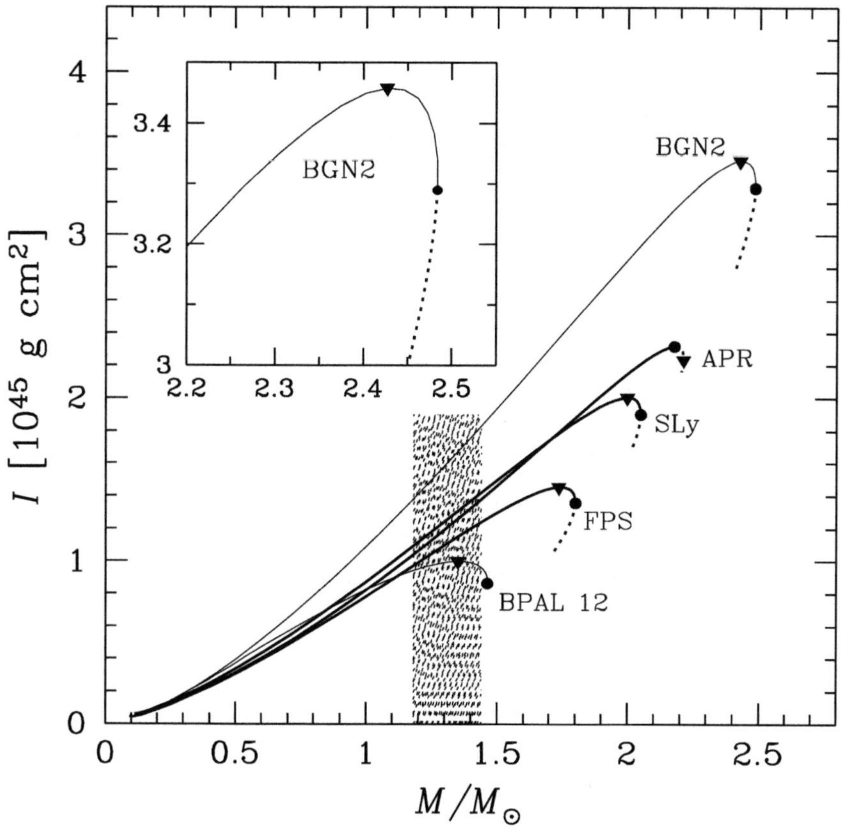


*Figure 6.16.* Moment of inertia $I$ of a slowly and rigidly spinning neutron star versus stellar mass $M$ for several EOSs of dense matter. Stellar configurations with maximum $I$ are indicated by filled triangles, and those with maximum $M$ by filled circles. The inset shows a zoomed segment of the $I(M)$ curve for the BGN2 EOS near the maximum of $I$. The shaded band shows the range of precisely measured masses of binary pulsars.

$M_\odot \times (10\,\mathrm{km})^2 \simeq 2.0 \times 10^{45}\,\mathrm{g\ cm^2}$); the value of $I$ in such units will be denoted by $I_{45}$. The $I(M)$ curve exhibits a maximum, $I_{\max}$, reached for a mass which is typically a few percent lower than $M_{\max}$.[8] However, because of the flatness of the peak of the $M(\rho_c)$ curve, the central density of the configuration with $I = I_{\max}$ is noticeably lower than $\rho_{c,\max}$ (by $\sim 30\%$ for very stiff EOSs and by a factor of $\sim 2$ for very soft ones).

[8]This property holds for the EOSs without any strong high-density softening due to hyperonization or phase transition to an exotic state. For an EOS with a strong high-density softening, $I_{\max}$ may be reached at $M$ significantly lower than $M_{\max}$. For examples and discussion see Bejger *et al.* (2005b)

According to Fig. 6.16, the value of $I_{45}$ at $M = 1.4\ M_\odot$ ranges from $I_{45} \sim 1$ for a very soft EOS to $\sim 2$ for a very stiff EOS. Even a more dramatic dependence on EOS is seen for the maximum moment of inertia $I_{45}^{\rm max}$, which ranges from $\sim 1$ for the softest EOS to nearly 4 for the stiffest one. Naturally, the $I(M)$ dependence is much more sensitive to the stiffness of the EOS, than the $R(M)$ dependence.

### 6.10.3 Crustal moment of inertia

The EOS of the neutron star crust is known much better than that of the core. Moreover, the crust is thin and contains a small fraction of stellar mass (§ 6.8). This allows us to derive an approximate formula for the crustal contribution to $I$, denoted by $I_{\rm crust}$. Using Eq. (6.70), we have

$$I_{\rm crust} = \frac{8\pi}{3} \int_{R_{\rm crust}}^{R} {\rm d}r\ r^4 \left(\rho + \frac{P}{c^2}\right) \frac{\bar{\omega}}{\Omega}\ {\rm e}^{-\lambda-\Phi} , \tag{6.71}$$

where $R_{\rm crust}$ is the radial coordinate of the crust bottom. In the crust we have $P/c^2 \ll \rho$. To the lowest order in $(R - R_{\rm crust})/R$, we get

$$I_{\rm crust} \simeq \frac{2}{3} M_{\rm crust} R^2\, \bar{\omega}(R)/\Omega\ . \tag{6.72}$$

Using Eq. (6.66), we come to the required expression,

$$I_{\rm crust} \simeq \frac{2}{3} M_{\rm crust} R^2 \left(1 - \frac{r_{\rm g}}{R} \frac{I}{MR^2}\right) \tag{6.73}$$

The factor $\frac{2}{3} M_{\rm crust} R^2$ is $I_{\rm crust}$ in the Newtonian approximation. The relativistic correction results from dragging of local inertial frames (the factor in brackets).

### 6.10.4 Moment of inertia versus $M$ and $R$

Now we present approximate formulae relating $I$ to stellar mass and radius (Ravenhall & Pethick, 1994; Lattimer & Prakash, 2001; Bejger & Haensel, 2002).

Ravenhall & Pethick (1994) considered neutron star models calculated for the FPS EOS. They showed that the ratio $I/MR^2$ depends mostly on the compactness parameter $x_{\rm GR} = r_{\rm g}/R$. Their formula can be rewritten as

$$I \simeq 0.21\ MR_\infty^2\ . \tag{6.74}$$

For a restricted range of (moderate) compactness, $x_{\rm GR} = 0.15 - 0.45$, it reproduces exact values of $I$ within $\sim 10\%$ for most of the EOSs considered by the authors.

Lattimer & Prakash (2001) analyzed in detail the behavior of $I/MR^2$ as a function of $x_{\rm GR}$ for a broad set of EOSs. They derived useful approximate

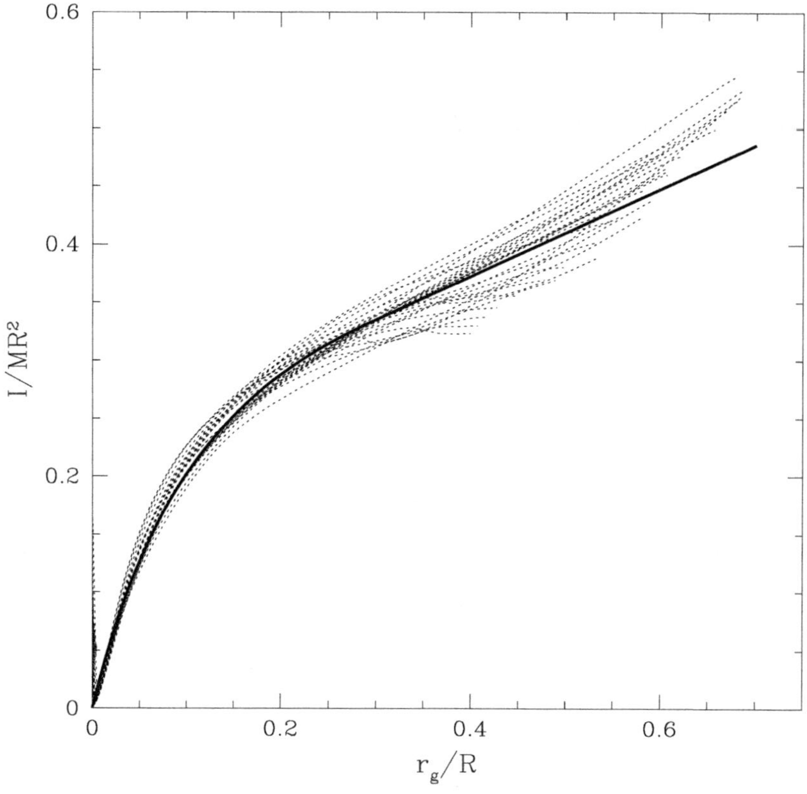


*Figure 6.17.* Dimensionless ratio $I/MR^2$ versus $r_g/R$ for thirty EOSs of dense baryon matter. The thick solid line corresponds to Eq. (6.75). Based on Bejger & Haensel (2002).

expressions for several analytic models of neutron-star structure, and studied the limit of low-mass neutron stars, $r_g/R \to 0$. In this limit Eq. (6.74) is invalid.

Bejger & Haensel (2002) studied a set of thirty EOSs, most of them constructed in the 1990s. They found that $I/MR^2$ is nearly universal function of $x_{\rm GR}$. Their result is displayed in Fig. 6.17, and their universal function reads

$$\frac{I}{MR^2} = \begin{cases} x_{\rm GR}/(0.295 + 2\,x_{\rm GR}) & \text{for } x_{\rm GR} \leq 0.3\,, \\ \frac{2}{9}\,(1 + 1.69\,x_{\rm GR}) & \text{for } x_{\rm GR} > 0.3\,. \end{cases} \tag{6.75}$$

As seen in Fig. 6.17, the precision of this approximation is typically better than 10%. The formula breaks down only for masses $M \lesssim 0.2\, M_\odot$, which correspond to $x_{\rm GR} < 0.04$.

### 6.10.5 Approximate formula for $I_{\rm max}$

As we stressed before, the value of $I_{\rm max}$ is more sensitive to the stiffness of the EOS, than $M_{\rm max}$ or $R_{M_{\rm max}}$. For a given EOS, one can write $I_{\rm max} = C_I M_{\rm max} R^2_{M_{\rm max}}$. It has been shown that, for a set of physically sound EOSs developed in the 1970s and 1980s, $C_I$ is only weakly dependent on the EOS (Haensel, 1990). As demonstrated by Bejger & Haensel (2002), $C_I$ is correlated with the compactness parameter $x_{\rm GR}^{\rm max}$ for maximum-mass stellar models. For the same set of EOSs that has been used to derive Eq. (6.75), the correlation can be described by the "empirical formula"

$$I_{45}^{\rm max} \simeq (2.414\, x_{\rm GR}^{\rm max} - 0.368)(M_{\rm max}/M_\odot)(R_{M_{\rm max}}/10\ {\rm km})^2\ . \qquad (6.76)$$

It reproduces exact values of $I_{\rm max}$ typically within $\sim 10\%$ and can be used to estimate $I_{\rm max}$ from known values of $M_{\rm max}$ and $R_{M_{\rm max}}$. Notice that Eq. (6.76) underestimates $I_{\rm max}$ for superluminal EOSs but overestimates $I_{\rm max}$ for EOSs with softening owing to the appearance of hyperons. Finally, its error is $\lesssim 5\%$ for "minimal composition EOSs" (nucleons and leptons).

One might argue that the validity of the simplest "empirical formula,"

$$I_{45}^{\rm max} \simeq (M/M_\odot)\,(R_{M_{\rm max}}/10\ {\rm km})^2 \qquad (6.77)$$

(Haensel, 1990; Bejger & Haensel, 2002), is a consequence of the fact that *realistic* EOSs are well approximated by polytropes. For a polytropic EOS $P = K n_{\rm b}^\gamma$ with a fixed $\gamma$, the relation

$$I_{45}^{\rm max} = C(\gamma)\,(M/M_\odot)\,(R_{M_{\rm max}}/10\ {\rm km})^2 \qquad (6.78)$$

is *exact*, with $C(\gamma)$ being independent of $K$. We get $C(2) = 0.83$ and $C(3) = 1.04$, so that the "best fit" value of $C$ for realistic EOSs lies roughly in the middle between $C(2)$ and $C(3)$. However, realistic EOSs at super-nuclear densities *are not polytropes*. One can introduce only a local, density dependent adiabatic index $\gamma(n_{\rm b})$ (see §§ 5.13.3 and 5.14.2). For example, for matter with hyperons the EOS softens considerably (and in a discontinuous way) at the thresholds of hyperon appearance, with $\gamma$ dropping below one (see § 5.14.2 and Balberg & Gal 1997). Clearly, the validity of the empirical formula for $I_{\rm max}$ is not due to the polytropic character of the EOS of dense matter.

## 6.11. Elastic shear strain, non-axial deformations, and gravitational radiation

A non-zero shear strain, which can build up in a neutron star crust, contributes to the stress tensor of crustal matter. Let us restrict ourselves, for the sake of simplicity, to the Newtonian approximation. The stress tensor is given by

$$\sigma_{ij} = -P\delta_{ij} + \sigma_{ij}^{\rm shear}\ . \qquad (6.79)$$

The shear strain can support *nonsphericity* of a non-rotating star and *non-axiality* of a spinning star. Let us follow a simplified Newtonian "one-parameter model" of Baym & Pines (1971) updated with new physics input. It can be reformulated in the framework of General Relativity (Carter & Quintana, 1975a).

Consider a slowly rotating and weakly deformed neutron star. The simplest deviation of the stellar shape from sphericity is caused by quadrupolar deformation. Let us assume that a stellar configuration is axially symmetric with respect to the $z$-axis. The amplitude of the quadrupolar deformation can be described by a single *oblateness parameter* $\epsilon$, related to the principal components of the moment of inertia tensor, $I_{xx} = I_{yy} \neq I_{zz}$, by

$$\epsilon = (I_{zz} - \overline{I})/\overline{I} \, , \tag{6.80}$$

where $\overline{I} = (I_{xx} + I_{yy} + I_{zz})/3$. Small deviations from spherical symmetry imply $\epsilon \ll 1$.

For a completely liquid star $\sigma_{ij}^{\rm shear} = 0$. The total energy of a non-spinning star (at a fixed total baryon number) is then minimized by a spherical shape, so that deviations from spherical symmetry can only increase the energy,

$$E^{\rm liq} = E_0 + \mathcal{A}\,\epsilon^2 \, . \tag{6.81}$$

However, the presence of $\sigma_{ij}^{\rm shear}$ results in an additional term in $E$, induced by a nonvanishing elastic shear strain. We will denote this term by $E^{\rm shear}$. In order to determine the dependence of $E^{\rm shear}$ on stellar deformation, we assume the existence of a reference (not necessarily static) *relaxed* configuration $\breve{\mathcal{C}}$ with zero elastic shear strain and the same number of baryons as the actual configuration. By construction, the reference configuration $\breve{\mathcal{C}}$ has such an oblateness parameter $\breve{\epsilon}$ that $E^{\rm shear}(\breve{\mathcal{C}}) = 0$. The actual *deformed* configuration with shear strain will be denoted by $\mathcal{C}$. Assume further, that $\mathcal{C}$ can be formed from $\breve{\mathcal{C}}$ by a continuous elastic deformation. Keeping only the terms quadratic in $\breve{\epsilon}$ and $\epsilon$, we approximate $E^{\rm shear} = \mathcal{B}(\epsilon - \breve{\epsilon})^2$. Then the total energy of configuration $\mathcal{C}$ is

$$E = E_0 + \mathcal{A}\epsilon^2 + \mathcal{B}\,(\epsilon - \breve{\epsilon})^2 \, . \tag{6.82}$$

For a purely liquid neutron star we have $\mathcal{B} = 0$ and, hence, $\epsilon = 0$. However, for $\mathcal{B} > 0$ and $\breve{\epsilon} \neq 0$, configuration $\mathcal{C}$ is no longer spherical, but has the oblateness parameter

$$\epsilon = \breve{\epsilon}\,\beta, \qquad \beta = \mathcal{B}/(\mathcal{A} + \mathcal{B}) \approx \mathcal{B}/\mathcal{A} \, , \tag{6.83}$$

where we take into account that $\mathcal{B} \ll \mathcal{A}$ (see below).

Early estimates of $\mathcal{A}$ for a neutron star with $M = 1.4\,M_\odot$ gave $\mathcal{A} \sim 10^{53}$ erg (Pandharipande *et al.*, 1976). Its dependence on $M$ was studied by Carlini & Treves (1989) who found that $\mathcal{A}$ decreases rapidly with decreasing $M$ (from $\mathcal{A} \gtrsim 10^{53}$ erg at $M = M_{\rm max}$ to $\mathcal{A} \sim 10^{50}$ erg at $M \simeq M_{\rm min} \simeq 0.1\,M_\odot$).

For a solid crust, $\mathcal{B}$ is determined by the crustal rigidity. The value of $\mathcal{B}$ can be calculated provided one knows the shear modulus of crustal matter and the distribution of elastic strain there. Early estimates for a neutron star with $M = 1.4\,M_\odot$ gave $\mathcal{B} \sim 10^{48}$ erg, which resulted in $\beta \sim 10^{-5}$ (Pandharipande *et al.*, 1976; Carlini & Treves, 1989). The dependence of $\mathcal{B}$ on $M$ was studied by Carlini & Treves (1989).

It should be stressed that the above estimates of $\mathcal{B}$ were based on old crustal models, with the crust-core interface at the baryon number density $n_{\rm cc} \simeq n_0 = 0.16\ {\rm fm}^{-3}$. Recent calculations give lower values $n_{\rm cc} = (0.06-0.10)\ {\rm fm}^{-3}$ (§§ 3.4.2 and 3.5). Another weakness of older estimates was a simplistic model of the distribution of elastic strain within a spinning-down neutron star. Using modern values of $n_{\rm cc}$ and a more realistic elastic strain field, Cutler *et al.* (2003) obtained much lower values $\beta \simeq (1.5-2.5) \times 10^{-7}$ unfavorable for detecting continuous gravitational waves from radio pulsars (§ 6.11.2).

### 6.11.1 Elastic strain and neutron stars with $M \approx M_{\rm min}$

Equilibrium configurations of low-mass neutron stars, studied in §§ 6.5.8 and 6.6.2, were calculated from Eqs. (6.7) and (6.8) which assume that dense matter is liquid. However, a neutron star with $M \sim 0.1\,M_\odot$ is nearly all solid, except for a small liquid core containing only $2-3\%$ of the stellar mass. In contrast to liquid, solidified matter can sustain shear strain, which changes the equations of hydrostatic equilibrium. For an isotropic Coulomb crystal of crustal matter, the elastic shear strain term in the stress tensor is determined by the shear modulus $\mu$ (§ 3.7). The effect of elastic strain on the stellar structure can roughly be estimated using the model of Baym & Pines (1971) discussed above. The shear strain makes a neutron star configuration nonspherical. Assuming, for simplicity, that the deviation from sphericity is quadrupolar, let us estimate a possible nonsphericity $\delta R/R$, where $\delta R$ is the absolute value of the difference between the polar and the equatorial radii. For a Coulomb crystal in the stellar crust, the maximum (breaking) strain does not exceed $\sim 10^{-2}$ (Ruderman, 1992). Then the Baym & Pines (1971) model predicts that $\delta R/R \lesssim 0.01\beta$, where $\beta \sim \int_{V_{\rm crust}} \mu\, {\rm d}V/|E_{\rm grav}|$ and $E_{\rm grav}$ is the gravitational energy of the star. For a neutron star with $M = 1.4\,M_\odot$ one has $\beta \sim 10^{-6}$, but for $M \simeq 0.1\,M_\odot$ one obtains $\beta \sim 10^{-3}$, due to a strong decrease of the factor $\mathcal{A}$ (see above). Therefore, we expect that the effect of elastic shear strain on the stellar radius at $M \sim 0.1\,M_\odot$ is $\delta R/R < 10^{-4}$.

### 6.11.2 Non-axial deformation of solitary pulsars and gravitational waves

Let us consider the evolution of a hot, young, rotating neutron star just after solidification of the crust. Such a star is very close to its liquid progenitor.

Therefore, its shear strain is negligibly small. This is a good model of $\breve{\mathcal{C}}$. Let us assume that it rotates rigidly with an angular frequency $\breve{\Omega}$ around the $z$-axis (coinciding with the body symmetry axis) which induces the rotational oblateness characterized by the parameter $\breve{\epsilon}$. The total energy of this star reads

$$E(\breve{\mathcal{C}}) = E_0 + \mathcal{A}\,\breve{\epsilon}^2 + J^2/(2I_{zz}) \,, \tag{6.84}$$

where $J$ is the stellar angular momentum. The value of $\breve{\epsilon}$ corresponds to the minimum of $E$ at a fixed $J$ and a fixed baryon number $A_{\rm b}$. Assuming $\breve{\epsilon} \ll 1$ one obtains

$$\breve{\epsilon} \simeq \breve{\Omega}^2 I_0/(4\mathcal{A}) \simeq 1.0 \times 10^{-3}\,(10\ {\rm ms}/\breve{P})^2 (I_{45}/\mathcal{A}_{53}) \,, \tag{6.85}$$

where $I_0$ is the moment of inertia of a spherical, non-rotating configuration $\mathcal{C}_0$, $I_{45} = I_0/(10^{45}\ {\rm g\ cm}^2)$, $\mathcal{A}_{53} = \mathcal{A}/(10^{53}\ {\rm erg})$, and $\breve{P}$ is the spin period of configuration $\breve{\mathcal{C}}$.

An elastic shear strain can be released during pulsar spindown by crust cracking or by plastic flow of crustal matter. Therefore, Eqs. (6.83) and (6.85) give an upper bound on the bulk oblateness of a spinning neutron star, supported by the elastic strain. We get $\epsilon_{\rm max}^{\rm shear} \sim 10^{-9}$–$10^{-8}$ for $\breve{P} = 10$–$3$ ms. Because both effects (rotation and elastic strain) on the neutron star structure are small, they are additive in the lowest order. Therefore, $\epsilon_{\rm max}^{\rm shear} \sim 10^{-9} - 10^{-8}$ is also the upper bound on the bulk non-axial asymmetry supported by the elastic shear strain,

$$\epsilon_{\rm asym} = (I_{xx} - I_{yy})/I \,, \tag{6.86}$$

for a neutron star rotating about the $z$-axis.

Our definition of the bulk non-axial asymmetry $\epsilon_{\rm asym}$ deserves a comment. In the literature, one frequently encounters a different measure of deviation from the axial symmetry, called ellipticity. This ellipticity is defined for a triaxial ellipsoid with semiaxes $a > b > c$ and $a - b \ll a$. The spin axis is assumed to coincide with $c$. The ellipticity is then introduced as (see, e.g., Shapiro & Teukolsky 1983)

$$\text{ellipticity} = \frac{a-b}{(a+b)/2} \,. \tag{6.87}$$

For a triaxial incompressible ellipsoid $I_{xx} - I_{yy} = \frac{1}{5}\,M\,(a^2 - b^2)$ and $I_{zz} = \frac{1}{5}\,M\,(a^2 + b^2)$. For $a - b \ll a$, one has $a^2 + b^2 \simeq (a+b)^2/2$, which means that the ellipticity given by Eq. (6.87) coincides with our $\epsilon_{\rm asym}$.

A non-zero value of $\epsilon_{\rm asym}$ induces a non-zero amplitude ($\sim \epsilon_{\rm asym} I$) of the quadrupole moment tensor of stellar mass distribution. A rotating neutron star (for instance, a pulsar) becomes then the source of persistent gravitational radiation at the frequency $\Omega_{\rm GW} = 2\Omega$, the doubled spin frequency $\Omega$ (see, e.g., Shapiro & Teukolsky 1983). The power of gravitational wave emission, averaged over the spin period $P = 2\pi/\Omega$, is

$$\dot{E}_{\rm GW} = \frac{32}{5}\frac{G}{c^5}\, I^2 \epsilon_{\rm asym}^2 \Omega^6 \; . \tag{6.88}$$

This formula can be used to constrain $\epsilon_{\rm asym}$ from pulsar timing. The emission of gravitational waves leads to the loss of the kinetic spin energy of the star, $\dot{E}_{\rm kin}^{\rm GW} = -\dot{E}_{\rm GW}$, and to the associated pulsar spindown $\dot{P}_{\rm GW}$. Of course, this spin-down is much slower than observed spin-down of pulsars ($\dot{E}_{\rm kin} = I\Omega\dot{\Omega}$, see § 1.4.4), associated with electromagnetic processes in pulsar magnetospheres. This means that $\dot{P}_{\rm GW} \ll \dot{P}_{\rm obs}$ and leads to an observational upper bound on the value of $\epsilon_{\rm asym}$ for a radio pulsar,

$$\epsilon_{\rm asym} \ll 4.3 \times 10^{-9}\, (P/1\;{\rm ms})^{3/2} \sqrt{\dot{P}/(10^{-19}\;{\rm s\,s^{-1}})} \; , \tag{6.89}$$

where we have set $I = 10^{45}$ g cm$^2$. The strongest bounds are obtained for millisecond pulsars, which have especially low $P$ and $\dot{P}$. For example, we obtain $\epsilon_{\rm asym} \ll 4 \times 10^{-8}$ for the nearby pulsar J0437–4715 ($P = 5.76$ ms, $\dot{P} = 5.7 \times 10^{-20}$, the distance $d$= 140 pc), and $\epsilon_{\rm asym} \ll 4 \times 10^{-9}$ for PSR B1957+20 ($P = 1.61$ ms, $\dot{P} = 1.7 \times 10^{-20}$, $d = 1.5$ kpc), which is the third among the fastest rotators (§ 9.4). These observational bounds are very stringent. For young pulsars, like the Crab pulsar, the observational bounds based on Eq. (6.89) are not useful, mostly because of relatively large $\dot{P}$.

Let us remind that gravitational waves are wave-like perturbations of space-time. Let the unperturbed (background) space be flat, with the metric $g_{ij}^{(0)} = {\rm diag}(1, -1, -1, -1)$. Then the metric of space with weak gravitational waves is

$$g_{ij} = g_{ij}^{(0)} + h_{ij} \; , \qquad |h_{ij}| \sim h \ll 1 \; , \tag{6.90}$$

where $h$ measures the amplitude of a gravitational wave. Gravitational waves are transverse, i.e., they produce an acceleration of free particles transverse to the wave propagation direction.

Consider a pair of test particles which are at rest in the absence of gravitational waves. Let $\boldsymbol{\xi}$ determine relative position of these particles. A gravitational wave induces their *relative acceleration* $\ddot{\boldsymbol{\xi}}$ which is a measurable quantity. A variation $\delta\boldsymbol{\xi}$ of $\boldsymbol{\xi}$ under the action of the gravitational wave is similar to that produced by a tidal force,

$$\delta\xi_j = h_{jk}^{TT}\, \xi_k/2 \; , \tag{6.91}$$

where $h_{jk}^{TT}$ is the relevant (traceless and transverse) part of $h_{jk}$. This equation leads to an estimate

$$\delta\xi/\xi \simeq h \; . \tag{6.92}$$

It is natural to call $h$ the relative gravitational wave strain. This is the major quantity which characterizes the detectability of gravitational waves. Typical values of $h$ at a distance $d$ from a pulsar with given $\epsilon_{\rm asym}$ are (e.g., Shapiro & Teukolsky 1983)

$$h = \frac{4G\Omega^2}{c^4 d} I\epsilon_{\rm asym} \simeq 4.2 \times 10^{-28} \left(\frac{\epsilon_{\rm asym}}{10^{-8}}\right) \left(\frac{10 \text{ ms}}{P}\right)^2 \left(\frac{1 \text{ kpc}}{d}\right) . \quad (6.93)$$

Nearby millisecond pulsars are the best potential sources of continuous periodic gravitational waves with a very stable frequency. These waves would be very different from bursting gravitational waves emitted from such astrophysical sources as coalescing stars or nonspherical collapses associated with type II supernova events. Periodic gravitational radiation from nearby millisecond pulsars could be searched using gravitational wave antennae (see, e.g., Pizzella *et al.* 2001) tuned to the doubled pulsar frequency. Observations of radio pulsars have advantage over observations of unpredictable, short bursting gravitational waves (New *et al.*, 1995). The most promising is the nearby millisecond pulsar B1957+20 mentioned above. Its expected gravitational wave strain is

$$\text{PSR B1957} + 20 \ : \qquad h \simeq 10^{-27} \left(\epsilon_{\rm asym}/10^{-9}\right) . \quad (6.94)$$

This amplitude is five orders of magnitude smaller than the recent observational limit of $h \sim 10^{-22}$ for the second fastest spinning pulsar B1937+21 ($P = 1.56$ ms), obtained in the first "science run" of the LIGO and GEO gravitational wave detectors (Allen & Woan, 2004). A dramatic improvement of detectability limits of the gravitational wave emission from 26 selected isolated radio pulsars was reached in the LIGO's second "science run" (Abbott *et al.* 2005). For some pulsars the limit become as low as $h \sim 10^{-24}$ and corresponded to $\epsilon_{\rm max} \sim 10^{-5}$.

The shear modulus for a hypothetical solid $\pi^0$-condensed matter in the neutron star core is estimated to be six orders of magnitude larger than in the inner crust (§ 7.7.2). A solid $\pi^0$-condensed core could support $\epsilon_{\rm asym}$ as high as $10^{-4}$. This value is excluded for some pulsars observed by Abbott *et al.* (2005). On the other hand, if PSR B1957+20 (not observed by Abbott *et al.* 2005) contains a $\pi^0$-condensed solid core of radius of a few kilometers, then the core elastic strain could be sufficient to produce asymmetry resulting in $h \sim 10^{-22}$, detectable in the LIGO observations. A recent discussion of asymmetries supported by elastic strain in exotic neutron star cores is presented in (Owen, 2005).

## 6.12. Rotating neutron stars

In this section we consider the effect rotation on neutron star structure beyond the slow-rotation regime of § 6.10. We limit ourselves to a stationary rigid (uniform) rotation, except for §6.12.5. This rotation is likely to be a good approximation for neutron stars older than one year. Differential rotation can be important for newly-born neutron stars formed in gravitational collapse or in coalescence of two neutron stars (see §6.12.5). For older neutron stars, deviations from rigid rotation connected, for instance, with crustal glitches, are expected to be small.[9]

For a stationary rigid rotation with the angular frequency $\Omega$, measured by a distant observer, is constant. We consider a star which is axially symmetric with respect to the spin axis. A stationary rotation of stellar bodies in General Relativity has been studied by many authors; see Stergioulas (2003) for a review.

### 6.12.1 Space-time metric and the equations of stationary motion

We use the familiar coordinates $t$, $r$, $\theta$, and $\phi$ and the axially-symmetric metric

$$\mathrm{d}s^2 = c^2 \mathrm{e}^{2\Phi}\,\mathrm{d}t^2 - \mathrm{e}^{2\lambda}\,r^2 \sin^2\!\theta\,(\mathrm{d}\phi - \omega\,\mathrm{d}t)^2 - \mathrm{e}^{2\alpha}\left(\mathrm{d}r^2 + r^2\,\mathrm{d}\theta^2\right), \quad (6.95)$$

where the metric functions $\Phi$, $\lambda$, $\omega$, and $\alpha$ depend solely on $r$ and $\theta$.

One has to distinguish between the inertial reference frame of a local observer $\mathcal{O}_0$ (at a point $(r, \theta, \phi)$ within the star) and an inertial reference frame of an observer at infinity $\mathcal{O}_1$. An important quantity is the fluid velocity in the $\phi$ direction, as measured in $\mathcal{O}_0$,

$$U = r\,\sin\theta\,(\Omega - \omega)\,\mathrm{e}^{\lambda - \Phi}\,. \quad (6.96)$$

The fact that $U = 0$ but $\Omega = \omega \neq 0$ for the fluid at rest with respect to $\mathcal{O}_0$ reflects the well known phenomenon of dragging of local inertial frames.

The Lorentz factor connecting measurements in $\mathcal{O}_0$ and $\mathcal{O}_1$ is

$$\Gamma = \left(1 - U^2/c^2\right)^{-\frac{1}{2}}\,. \quad (6.97)$$

In the perfect fluid approximation, equations of stationary axially symmetric motion reduce to

$$\frac{1}{\mathcal{E} + P}\frac{\partial P}{\partial x^i} + \frac{\partial \Phi}{\partial x^i} + \frac{\partial \ln\Gamma}{\partial x^i} = 0\,, \qquad x^i = r,\ \theta\,. \quad (6.98)$$

[9]Using the two-component model of glitches (§ 9.7), one can show that for pulsars exhibiting macroglitches the macroscopic spin frequency $\Omega_\mathrm{n}$ of neutron superfluid in the inner crust can be larger than the measured spin frequency $\Omega$ of the crust and the rest of the pulsar coupled to it by no more than one part in 10,000. In the strongest macroglitches, $\Delta\Omega/\Omega \sim 10^{-6}$. It is thought that $\Delta\Omega$ results from a transfer of an excess angular momentum of superfluid neutrons, $I_n \Delta\Omega_n$, to the crust-coupled component with the moment of inertia $I_\mathrm{c}$ (that is nearly equal to the total moment of inertia $I$). However, $I_n \sim 10^{-2}\,I_\mathrm{c} \simeq 10^{-2}\,I$. Therefore, angular momentum conservation implies $\Delta\Omega_n/\Omega \sim I_n\Delta\Omega/I\Omega \sim 10^{-4}$.

It is suitable to define a dimensionless *pseudo-enthalpy*

$$H(P) = \int_0^P \frac{\mathrm{d}P'}{\mathcal{E}(P') + P'} \,. \tag{6.99}$$

It vanishes at the stellar surface and increases monotonically towards the stellar center. It is a very useful variable in numerical calculations of rotating stars, because after defining $H$ one can easily obtain the first integral of motion

$$H(r,\theta) + \Phi(r,\theta) - \ln\Gamma(r,\theta) = H(0,0) + \Phi(0,0) = \text{constant} \,. \tag{6.100}$$

This is the General Relativistic version of the Bernoulli theorem for a stationary ideal fluid flow. Consequently, the most useful parameterization of EOSs for rotating stars is $\mathcal{E} = \mathcal{E}(H)$, $P = P(H)$. It is easy to see that $n_\mathrm{b}(H) = [\mathcal{E}(H) + P(H)]\,\mathrm{e}^{-H}/m_0c^2$, where we assume that the mass per baryon at the stellar surface equals $m_0$.

The quantity $H$ can be expressed in terms of the enthalpy per baryon

$$h(P) = (\mathcal{E} + P)/n_\mathrm{b} \,. \tag{6.101}$$

Using $\mathrm{d}P = n_\mathrm{b}\,\mathrm{d}h$ we get

$$H(P) = \int_{h(0)}^{h(P)} \frac{\mathrm{d}h}{h} = \ln\left(\frac{h(P)}{m_0c^2}\right) \,. \tag{6.102}$$

The calculation of the metric functions $\Phi$, $\lambda$, $\omega$, and $\alpha$ is reduced to solving a set of coupled second-order elliptic-type partial differential equations (see, e.g., Bonazzola *et al.* 1993) and will not be described here.

In order to reach high precision in the two-dimensional calculations of rotating configurations, it is preferable to use analytic representation of the EOS. Such representations, in the form of the functions $\rho = \mathcal{E}/c^2 = \rho(H)$, $P = P(H)$ for unified EOSs were obtained by Haensel & Potekhin (2004); they are presented in Appendix C.

### 6.12.2 Stationary configurations of rotating neutron stars

Rotating configurations form a two-parameter family and can be labeled, for instance, by the values of $\rho_\mathrm{c}$ and $\Omega$. Spinning neutron stars are flattened; their equatorial radius $r_\mathrm{eq}$ is larger than the polar radius, $r_\mathrm{pol}$. Here, $r_\mathrm{eq}$ and $r_\mathrm{pol}$ are the values of the radial coordinate $r$ in the metric (6.95) at the equator and pole, respectively. We will also use the circumferential radius, $R_\mathrm{eq}$, which is the proper length of the equator divided by $2\pi$. For non-rotating configurations $r_\mathrm{eq} = R_\mathrm{eq}$, but for rotating ones $R_\mathrm{eq} > r_\mathrm{eq}$. Configurations $\mathcal{C}(\rho_\mathrm{c}, \Omega)$ cover a region in the $M - R_\mathrm{eq}$ plane, which is shown in Fig. 6.18. A curve $M(R_\mathrm{eq};\Omega)$ for a fixed $\Omega$ is limited on the high-density and the low-density sides. On the

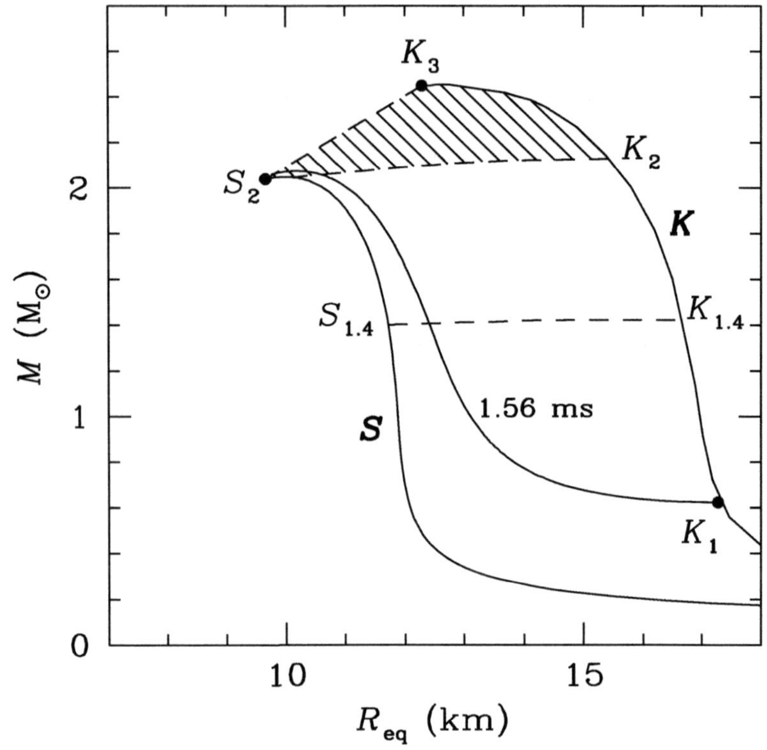


*Figure 6.18.* Families of stable stationary rotating neutron star configurations in the $M - R_{\rm eq}$ plane (for the SLy EOS). The central density increases upwards along each solid line. The curve "1.56 ms" corresponds to the spin period 1.56 ms. Other notations are explained in the text.

high-density side, it is limited by the stability with respect to small axially symmetric perturbations (the line segment $S_2K_3$ in Fig. 6.18).[10]

In order to check whether a configuration $\mathcal{C}(\rho_{\rm c1}, \Omega_1)$ with an angular momentum $J_1$ is stable, one has to consider a family of configurations $\mathcal{C}(\rho_{\rm c}, J_1)$ (in the neighborhood of $\mathcal{C}(\rho_{\rm c1}, \Omega_1)$, passing through $\mathcal{C}(\rho_{\rm c1}, \Omega_1)$) with the same angular momentum $J = J_1$. The configuration $\mathcal{C}(\rho_{\rm c1}, \Omega_1)$ is stable if

$$(\partial M/\partial \rho_{\rm c})_{\rho_{\rm c}=\rho_{\rm c1}} > 0\ , \tag{6.103}$$

where the derivative is calculated along the chosen $\{\mathcal{C}(\rho_{\rm c}, J_1)\}$ family. A line determined by $(\partial M/\partial \rho_{\rm c})_J = 0$ (e.g., the line segment $S_2K_3$ in Fig. 6.18) separates configurations stable and unstable with respect to small axially symmetric

[10]General stability criteria for rigidly rotating relativistic stars are given by Friedman *et al.* (1988)

perturbations. In the static limit, $J_1 = 0$, this condition coincides with the condition of stability for non-rotating stars, $\mathrm{d}M/\mathrm{d}\rho_c > 0$ (see § 6.5.1). Both conditions are necessary but may be not sufficient.

The low-density boundary is determined by the stability with respect to the mass-shedding from the equator. A necessary condition for the existence of a stationary rotating configuration requires the equatorial velocity of stellar matter $U(r = r_{\rm eq}, \theta = \pi/2)$ be smaller than the *Keplerian velocity* $U_{\rm K}(r_{\rm eq})$ of a test particle moving on a circular orbit of radius $r_{\rm eq}$ in the equatorial plane. The Keplerian velocity corresponds to the *Keplerian angular frequency* $\Omega_{\rm K}$, called also the *mass-shedding* angular frequency. A line on which $U(r = r_{\rm eq}, \theta = \pi/2) = U_{\rm K}(r_{\rm eq})$ determines the mass-shedding boundary (the curve **K** in Fig. 6.18). There are no stationary rotating configurations to the right of this line in the $M - R_{\rm eq}$ plane.

At a given baryon mass, rotation increases $R_{\rm eq}$. For illustration, we display an $M - R_{\rm eq}$ curve for the shortest observed pulsar period $P_{\rm min}^{\rm obs} = 1.56$ ms. The effect of such a rotation on the maximum-mass configuration is very small. For $M \simeq M_{\rm max}$ the period of 1.56 ms corresponds to the *slow rotation regime*. Therefore, $M_{\rm max}^{1.56\rm ms} - M_{\rm max}^{\rm stat}$ is quadratic in the small parameter $\overline{\Omega} = \Omega/\sqrt{GM/R^3}$ (see § 6.10). For example, in the case of the SLy EOS (Fig. 6.18), we obtain $\overline{\Omega}^2 = (\overline{\Omega}_{\rm max}^{\rm obs})^2 \simeq 0.06$. The maximum mass increase connected with rotation at $P = 1.56$ ms is $0.2\,(\overline{\Omega}_{\rm max}^{\rm obs})^2 \simeq 2\%$, which agrees very well with the exact numerical result (Douchin & Haensel, 2001). The effect will be smaller for a softer EOS (e.g., the FPS or BGN1H1 EOS) and larger for a stiffer EOS (e.g., the APR EOS). In contrast, the highest possible rotation frequency increases $M_{\rm max}$ by ~20% (§ 6.12.4).

At a given baryon mass $M_{\rm b} < M_{\rm b,max}^{\rm stat}$, rotation is limited by the mass-shedding limit (the line **K** in Fig. 6.18). The point $K_2$ corresponds to a Keplerian configuration with $M_{\rm b} = M_{\rm b,max}^{\rm stat}$. It can be reached by spinning-up the initial static maximum-mass configuration ($S_2$) to the Keplerian frequency. The inverse transition $K_2 \rightarrow S_2$ can be realized via angular momentum loss. The same is true for any pair of configurations $K$ and $S$ with the same baryon mass $M_{\rm b} < M_{\rm b,max}^{\rm stat}$. Solitary pulsars move along a part of such a track, as they spin-down during their lifetime. The track of the star with $M^{\rm stat} = 1.4\,M_\odot$ is denoted by $S_{1.4}K_{1.4}$.

On the contrary, rotating stationary configurations with $M_{\rm b} > M_{\rm b,max}$ cannot transform into the static ones via angular momentum loss. Such rotating configurations are called *supermassive*. They are doomed to collapse into black holes, just because the static configurations with $M_{\rm b} > M_{\rm b,max}^{\rm stat}$ do not exist. Consider a supermassive configuration $\mathcal{C}(M_{\rm b}, J)$, spinning down via angular momentum loss, so that $\dot{J} < 0$. There exists some value of $J = J_{\rm min}(M_{\rm b})$ that defines the last stable configuration, $\mathcal{C}(M_{\rm b}, J_{\rm min})$ (while for lower $J$ configurations collapse into black holes). In the limit of $J_{\rm min} \to 0$, the spin-down track reduces to one point, which is the static maximum-mass configuration $S_2$. Rotating supermassive configurations fill the region $S_2K_2K_3$ in the $M - R_{\rm eq}$ plane. The line segment $S_2K_3$ corresponds to the supermassive configurations with the smallest allowed $J$ at a given $M_{\rm b}$. This line separates rotating configurations stable with respect to the axisymmetric perturbations from the unstable ones (which cannot exist in a stationary state). The configuration $K_3$ has the maximum mass $M_{\rm max}^{\rm rot}$ (among all stable spinning configurations). At $K_3$, the line of Keplerian configurations, corresponding to the mass-shedding limit, joins the instability line. This configuration has also the maximum baryon mass, $M_{\rm b,max}^{\rm rot}$.

It should be mentioned, that $K_3$ slightly differs from the configuration with the maximum possible angular frequency $\Omega_{\rm max}$, which we call the *maximally rotating* configuration (Cook *et al.*, 1994). However, the mass of the maximally rotating configuration is only $\sim 0.1\%$ lower than $M_{\rm max}^{\rm rot}$, while the value of $\Omega$ at $K_4$ is by $\sim 1\%$ smaller than $\Omega_{\rm max}$ (Cook *et al.*, 1994). Therefore, the difference between these configurations is really small; it is actually of the same order of magnitude as an error (uncertainty) stemming from the ambiguity of the interpolation of the EOS between mesh points.

At a given $M_{\rm b}$ the centrifugal forces increase $r_{\rm eq}$ and decrease $r_{\rm pol}$. The effect is more pronounced for stiffer EOS and decreases with growing $M_{\rm b}$. The flattening of a spinning star can be described by the ratio $r_{\rm pol}/r_{\rm eq}$. In what follows, we will denote the circumferential radius of a non-rotating star by $R_0$. In the slow-rotation regime we get $r_{\rm eq} - R_0 \propto \overline{\Omega}^2$ and $R_0 - r_{\rm pol} \propto \overline{\Omega}^2$. The behavior of $r_{\rm pol}/r_{\rm eq}$, $r_{\rm eq}$, and $r_{\rm pol}$ as a function of the (cyclic) spin frequency $f = \Omega/2\pi$ for several baryon masses is displayed in Fig. 6.19. One can see a rapid increase of the flattening as $f$ approaches the Keplerian limit $f_{\rm K}$.[11] The centrifugal force produces the strongest deformation in the equatorial layer of the outer crust. This is clearly seen in Fig. 6.20.

Another effect of the centrifugal force is that, at a given $M_{\rm b}$, the central density decreases with increasing $f$. This effect is illustrated in Fig. 6.21. For a slow rotation, the effect is small and quadratic in $f$. It becomes stronger for stiffer EOS and weaker with increasing $M_{\rm b}$. Notice, that the increase of the central density due to the decrease of $f$, say, from $f_1$ to $f_2$ can be calculated from

$$\rho_{\rm c}(f_2) = \rho_{\rm c}(f_1) + \int_{f_2}^{f_1} |{\rm d}\rho_{\rm c}/{\rm d}f|\,{\rm d}f\ . \tag{6.104}$$

[11]Notice that for a neutron star with $M^{\rm stat} = 1.4\ M_\odot$, the BGN1H1 EOS is stiffer than the SLy one, see Figs. 6.3 and 6.5.

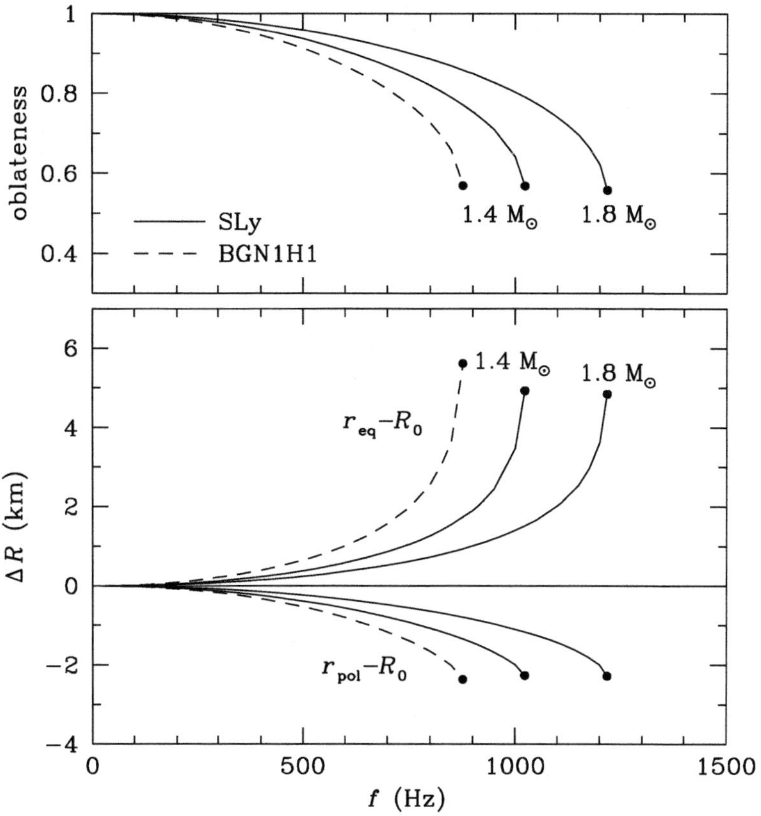


*Figure 6.19.* Effect of rotation on the neutron star shape; $R_0$ is the circumferential radius of a non-rotating star. *Top*: $r_{pol}/r_{eq}$ versus spin frequency for two baryon masses, equal to those for the static models with $M = 1.4$ and $1.8\ M_\odot$. Calculations are performed for the SLy and BGN1H1 EOSs. For the latter EOS, only lower-mass family is shown, because $M_{max}^{stat} < 1.8\,M_\odot$ for this EOS. *Bottom*: Increase of $r_{eq}$ and decrease of $r_{pol}$ due to stellar rotation.

An important parameter of the theory of rotating Newtonian stars is the ratio of the rotational (kinetic) energy to the gravitational energy, $\beta \equiv E_{rot}/|E_{grav}|$. A similar quantity can be defined for rotating relativistic stars, albeit not without ambiguity.

The kinetic spin energy is defined, by analogy with the Newtonian case, as

$$E_{kin} = \frac{1}{2}\,J\Omega\,. \tag{6.105}$$

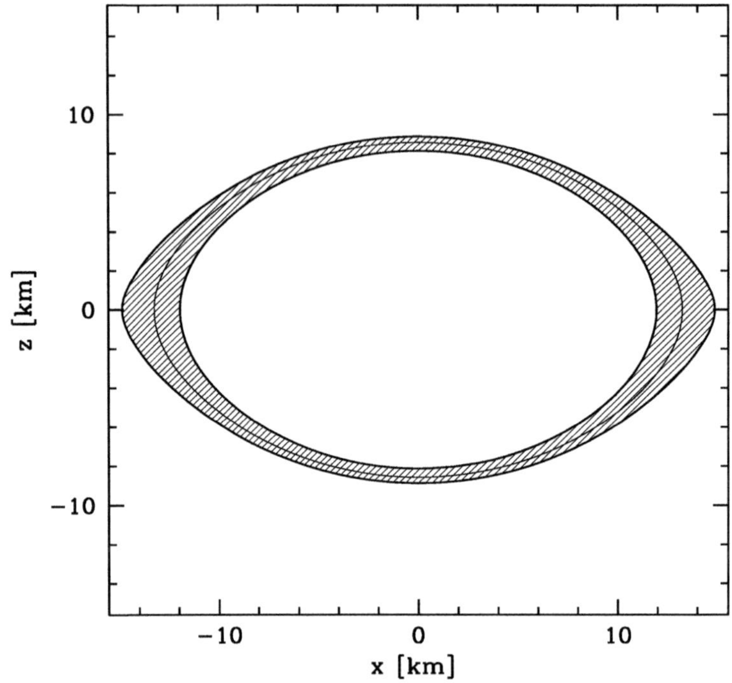


*Figure 6.20.* Cross section in the plane passing through the rotational axis of a neutron star with the SLy EOS spinning at $f$=1200 Hz. The baryon mass is $M_{\rm b} = 2.035\,M_\odot$, the gravitational mass $M = 1.82\;M_\odot$, the central baryon density $n_{\rm c} = 0.565\;{\rm fm}^{-3}$. The coordinates $x$ and $z$ are defined as $x = r\sin\theta\cos\phi$ and $z = r\cos\theta$, where $r$, $\theta$, and $\phi$ as the same as in Eq. (6.95). The contours are the lines of constant density. The inner contour is the crust-core interface; the outer one is the stellar surface, and the intermediate contour corresponds to the neutron drip.

The definition of $E_{\rm grav}$ for non-rotating stars was given in §6.3. We generalize it here for rotating stars,

$$E_{\rm grav} = Mc^2 - M_{\rm P}c^2 - E_{\rm kin}\;, \tag{6.106}$$

where $M_{\rm P}$ is the proper mass, defined as in the static case (§6.3) as an integral of the proper internal energy density over the stellar volume.

Critical values of $\beta$, above which a rotating star becomes unstable with respect to triaxial deformations, will be discussed in §6.12.10. For a fixed baryon mass, $\beta$ increases with increasing $f$. This dependence is shown in Fig. 6.21 for two values of $M_{\rm b}$. At a given $f$, the value of $\beta$ strongly depends on the baryon mass and on the stiffness of the EOS. However, the maximum value

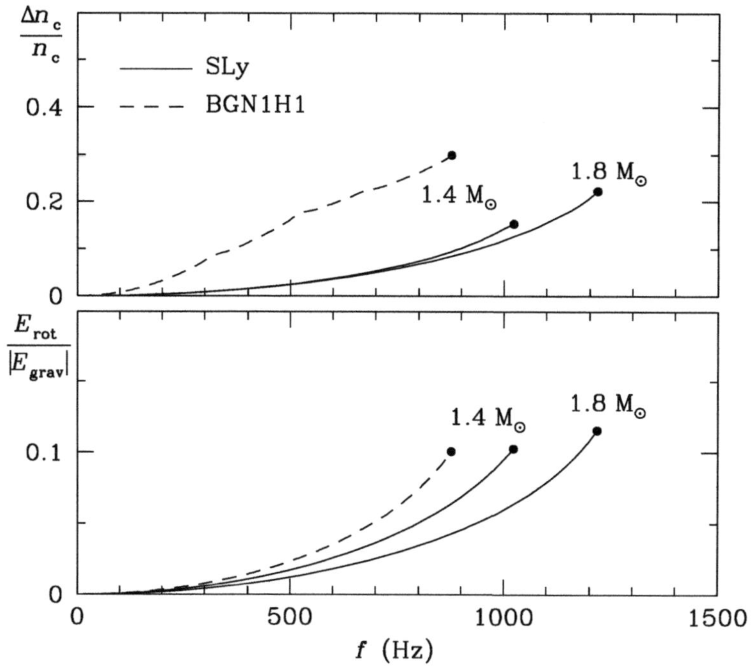


*Figure 6.21.* *Top*: Fractional decrease of the central density by rotation for neutron stars of two baryon masses with the SLy and BGN1H1 EOSs (the same as in Fig. 6.19). *Bottom*: Ratio of rotational to gravitational energy at fixed baryon masses versus spin frequency.

$\beta = \beta_{\rm max} \simeq 0.11 - 0.12$ for a given $M_{\rm b}$, reached at $f = f_{\rm K}$, depends rather weakly on the EOS and increases weakly with growing $M_{\rm b}$.

### 6.12.3 Spinning low-mass neutron stars

The shortest observed pulsar spin period $P_{\rm min}^{\rm obs} = 1.396$ ms corresponds to the spin frequency $f_{\rm max}^{\rm obs} = 716$ Hz. For neutron stars with $M \gtrsim 1.3\ M_\odot$ the rotation at such frequency has a small effect on neutron star structure and can be described using the *slow rotation approximation* (§6.10). The leading effects of rotation on the neutron star structure are then quadratic in $f$; they are small as long as the gravitational pull $\sim GM/R^2$, acting on a unit mass element at the equator, is much larger than the centrifugal force $\sim 4\pi^2 R f^2$. However, these effects get larger with increasing equatorial radius and decreasing mass; they become decisive for $f^2 \sim GM/(4\pi^2 R^3)$.

Spinning low-mass neutron stars were studied by Haensel *et al.* (2002b), who performed exact two-dimensional calculations of stellar models for the (unified) SLy and FPS EOSs. The rotational effects are stronger for the stiffer

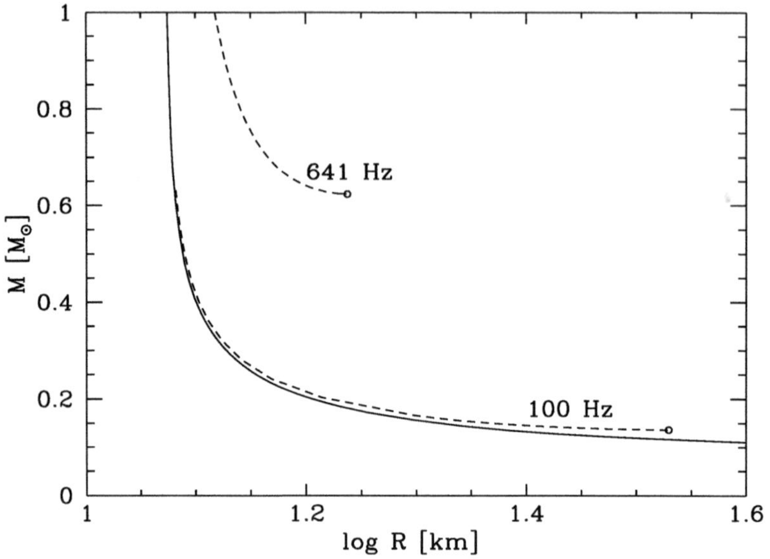


*Figure 6.22.* Gravitational mass $M$ versus equatorial radius for low-mass non-rotating (solid lines) and rotating (dashed lines) neutron stars with the SLy EOS. The upper dashed line is for the second shortest observed pulsar spin period 1.56 ms, while the lower dashed line is for the spin period 10 ms. The curves for rotating stars terminate at the Keplerian (mass shedding) configurations denoted by open circles. Based on Haensel *et al.* (2002b).

SLy EOS, which is most natural. Till January 2006, the shortest measured pulsar period was 1.56 ms. Therefore, Haensel *et al.* (2002b) used $f_{\rm max}^{\rm obs} = 641$ Hz. At $f$ =641 Hz and $M \lesssim 0.8\ M_\odot$, the effects of rotation would become huge. The lowest-mass configuration corresponds to the mass-shedding limit. One gets $M_{\rm min}^{\rm SLy}(641\ {\rm Hz}) = 0.61\ M_\odot$. For the softer FPS EOS, the effect is weaker ($M_{\rm min}^{\rm FPS}(641\ {\rm Hz}) = 0.54\ M_\odot$), because such low-mass neutron stars are more compact. Even rotation at $f = 100$ Hz affects noticeably low-mass stars. The minimum mass $M_{\rm min}^{\rm SLy}(100\ {\rm Hz}) = 0.13\ M_\odot$ reached at the mass-shedding limit is by nearly 40% larger than that for non-rotating neutron stars.

For the SLy minimum-mass configuration rotating at $f = 641$ Hz, the equatorial and polar radial coordinates of the stellar core are $r_{\rm eq}$=11 km and $r_{\rm pol}$=9 km, respectively. However, the crust is much stronger deformed than the core. The equatorial crust thickness is $\sim 6$ km, four times larger than the polar one, leading to a significant flattening, $r_{\rm pole}/r_{\rm eq} \approx 0.6$. The mass is still concentrated in the core which it contains 93% of the total mass.

The minimum-mass configuration, calculated for the SLy EOS at $f =$ 100 Hz, has a nearly spherical core containing half of the total mass. The equatorial radius of the core is only $\approx 6$ km, to be compared with the 28 km equatorial thickness of the crust. The flattening of the minimum-mass configuration is large, $r_{\rm pole}/r_{\rm eq} \simeq 0.7$.

Let us point out the basic difference between low-mass rotating and non-rotating configurations. Let $\rho_{\rm c,M_{\rm min}}$ denote the central density of the minimum-mass configuration. In the non-rotating case, the equilibrium configurations with $\rho_{\rm c} < \rho_{\rm c,M_{\rm min}}$ are secularly unstable with respect to small perturbations, whereas stationary rotating configurations with $\rho_{\rm c} < \rho_{\rm c,M_{\rm min}}$ just do not exist.

### 6.12.4 Maximum mass and minimum rotation period

Rotation increases $M_{\rm max}$ because the centrifugal forces oppose the gravity. For $\Omega \ll \Omega_{\rm K}$, this increase is small and quadratic in $\Omega/\Omega_{\rm K}$. More generally, $M_{\rm max}(\Omega) - M_{\rm max}^{\rm stat}$ is an even function of $\Omega$, because it does not depend on the orientation of the spin axis (neglecting the effects of strong magnetic fields).

Let us mention two important extreme configurations, with the maximum mass $M_{\rm max}^{\rm rot}$ and with the minimum period $P_{\rm min}$, which do not coincide but are very close to each other (see above). Depending on the EOS, the *maximally rotating configuration* with the rotation period $P_{\rm min}$ can have the central density higher or lower than the maximum-mass configuration, but the difference is a few per cent at most (Cook *et al.*, 1994). As a rule, the mass of the maximally rotating configuration is lower than $M_{\rm max}^{\rm rot}$ by much less than one per cent (Cook *et al.*, 1994).

It is useful to note that for realistic baryonic and subluminal EOSs $M_{\rm max}^{\rm rot}$ is (to a very good approximation, typically within 3%) proportional to the maximum mass of non-rotating configurations (Lasota *et al.*, 1996),

$$M_{\rm max}^{\rm rot} \simeq 1.18 M_{\rm max}^{\rm stat} \ . \tag{6.107}$$

However, the above formula is invalid for the EOSs of a self-bound matter, of the generic form $P = a(\rho - \rho_{\rm s})$ (in particular, for strange quark stars which will be considered in § 8.21). Let us remind that, in contrast to "ordinary neutron stars", such exotic stars are bound not only by gravity but also by strong interactions.

### 6.12.5 Rapid differential rotation and maximum mass

Up to now we have restricted ourselves to stationary rigidly rotating neutron stars. Rigid rotation was limited by the mass-shedding limit at the equator. As soon as we allow for a differential rotation, in which $\Omega$ within the star depends on $\varpi \equiv \sqrt{x^2 + y^2} = r \sin\theta$, we may contemplate a rotating configuration with such $\Omega(\varpi)$, that $\Omega_{\rm ax} = \Omega(0)$ (i.e., on the spin axis) is significantly larger than $\Omega_{\rm K}$ and simultaneously $\Omega(r_{\rm eq}) < \Omega_{\rm K}$. This dynamically stable hydrostatic equilibrium will be always secularly unstable, because viscous forces will tend to redistribute the angular momentum leading eventually to the uniform $\Omega$. During the redistribution the star may be driven into a critical configuration beyond which the pressure, centrifugal, and gravitational forces can no longer be balanced. For instance, the star may start to eject some matter or collapse into a black hole.

As we have seen in § 6.12.4, rigid rotation can increase $M_{\rm max}$ from $M_{\rm max}^{\rm stat}$ to $M_{\rm max}^{\rm rot(R)}$ ("R" for "rigid"), by about 20%. Clearly, a differential rotation with $\Omega_{\rm ax}$ significantly larger than $\Omega_{\rm eq}$ can lead to a much larger increase of $M_{\rm max}$. Calculations performed for realistic modern EOSs of dense matter show that the maximum baryon mass of dynamically stable configurations, $M_{\rm b,max}^{\rm rot(D)}$ ("D" for "differential"), can be $\sim 50\%$ larger than $M_{\rm b,max}^{\rm stat}$ (see Morrison *et al.*, 2004a, and references therein).

Introducing three critical masses, we obtain several types of dynamically stable neutron stars. Non-rotating hydrostatic configurations exist for baryon masses up to $M_{\rm b,max}^{\rm stat}$. Rigid rotation pushes the maximum baryon mass up to $M_{\rm b,max}^{\rm rot(R)}$. Rigidly rotating neutron stars with baryon masses below $M_{\rm b,max}^{\rm stat}$ are called *normal*; by loosing angular momentum they can finally reach a non-rotating state. Rigidly rotating configurations with baryon masses $M_{\rm b,max}^{\rm stat} < M_{\rm b} < M_{\rm b,max}^{\rm rot(R)}$ are called *supermassive*; they exist only as rotating stars and will collapse into black holes during angular momentum loss. Finally, configurations with $M_{\rm b,max}^{\rm rot(R)} < M_{\rm b} < M_{\rm b,max}^{\rm rot(D)}$ can exist only in a state of rapid differential rotation; they are called *hypermassive*. The viscous redistribution of angular momentum will erase the difference between $\Omega_{\rm ax}$ and $\Omega_{\rm eq}$ and lead eventually to the dynamical instability and collapse into a black hole.

Let us stress that the value $M_{\rm b,max}^{\rm rot(D)}$ is sensitive to the dependence of $\Omega$ on $\varpi$. The values $M_{\rm b,max}^{\rm rot(D)} \simeq 1.5\ M_{\rm b,max}^{\rm stat}$, reported for some realistic EOSs by Morrison *et al.* (2004a), were actually obtained within some class of functions $\Omega(\varpi)$ and for some moderate values of $\Omega_{\rm ax}/\Omega_{\rm eq}$. They cannot be regarded as absolute upper bounds for differentially rotating neutron stars.

A gravitational collapse of a massive stellar core during a type II supernova explosion or a coalescence of two neutron stars in a binary system can produce a differentially rotating neutron star with the baryon mass exceeding $M_{\rm b,max}^{\rm rot(R)}$. However, the lifetime of such a hypermassive configuration is short, determined by the angular momentum transport timescale. This timescale can be less than a second (Duez *et al.*, 2006). A hypermassive neutron star ends its short life by collapsing into a black hole. Supermassive neutron stars, with $M_{\rm b,max}^{\rm stat} < M_{\rm b} < M_{\rm b,max}^{\rm rot(R)}$, formed in type II supernova events, or in coalescence of two neutron stars, or by accretion spin-up in close binaries, can live for a very long time (billions of years in the case of accretion spin-up). However, they will also collapse into black holes after spinning down to some critical rotation frequency. Precise values of three critical masses $M_{\rm b,max}^{\rm stat}$, $M_{\rm b,max}^{\rm rot(R)}$, and $M_{\rm b,max}^{\rm rot(D)}$ depend on the poorly known EOS of stellar matter at densities exceeding $10^{15}$ g cm$^{-3}$. However, we have rather robust theoretical prediction that $M_{\rm b,max}^{\rm rot(R)} \simeq 1.2 M_{\rm b,max}^{\rm stat}$ and $M_{\rm b,max}^{\rm rot(D)} \gtrsim 1.5\ M_{\rm b,max}^{\rm stat}$.

### 6.12.6 Approximate formulae for $P_{\rm min}$

A precise two-dimensional calculation of the structure of rotating stars is much more difficult than for non-rotating stars. Therefore, it is very fortunate that the value of $P_{\rm min}$ can be expressed, with a surprisingly good precision, in terms of the mass and radius of the *non-rotating* configuration with the maximum mass (Haensel & Zdunik 1989; correlations between these three quantities were also pointed out by Friedman *et al.* 1989 and Shapiro *et al.* 1989). The updated formula, based on numerical results obtained for a broad set of the *subluminal* ($v_{\rm sound} < c$) EOSs, reads (Haensel *et al.*, 1995)

$$P_{\rm min} \simeq 0.82 \left(\frac{M_\odot}{M_{\rm max}^{\rm stat}}\right)^{\frac{1}{2}} \left(\frac{R_{M_{\rm max}}^{\rm stat}}{10\ {\rm km}}\right)^{\frac{3}{2}} \ {\rm ms}. \tag{6.108}$$

The precision of this formula (for subluminal EOSs) is within 5%. It is valid not only for baryonic EOSs, but also for strange quark stars and other exotic stars (§ 8.21). It is, therefore, more universal than Eq. (6.107).

One can propose a more precise formula which takes into account an additional linear correlation between the ratio $(M_{\rm max}^{\rm stat})^{1/2}/P_{\rm min}(R_{M_{\rm max}}^{\rm stat})^{3/2}$ and the compactness parameter of the static maximum-mass configuration, $x_{\rm GR}^{\rm max} \equiv 2GM_{\rm max}^{\rm stat}/R_{M_{\rm max}}^{\rm stat}c^2$ (Lasota *et al.*, 1996),

$$P_{\rm min} \simeq \frac{0.187}{(x_{\rm GR}^{\rm max})^{3/2}\left(1 + 0.808\, x_{\rm GR}^{\rm max}\right)} \frac{M_{\rm max}^{\rm stat}}{M_\odot} \ {\rm ms}\ . \tag{6.109}$$

This formula will be useful for deriving an absolute lower bound on the shortest period $P_{\rm min}$ of rigid rotation from the upper bound on the surface redshift of non-rotating neutron stars (§ 6.12.9).

### 6.12.7 The causal upper bound on the mass of spinning neutron stars

In § 6.5.7 we derived the absolute upper bound on the static neutron star mass based on the knowledge of the EOS at $\rho < \rho_{\rm u} \sim 2\rho_0$ under the constraint of $v_{\rm sound} \leq c$. Let $M_{\rm max}^{\rm CL,stat}$ be the upper bound for non-rotating neutron stars. Rotation will increase it, $M_{\rm max}^{\rm CL}(\Omega) > M_{\rm max}^{\rm CL,stat}$. The upper bound $M_{\rm max}^{\rm CL,rot}$ for rotating stars has been obtained for the same causal-limit EOS as for the non-rotating models; it is reached at $\Omega$ very close to $\Omega_{\rm max}$. Its precise value is (Koranda *et al.*, 1997)

$$v_{\rm sound} \leq c\ :\quad M_{\rm max}^{\rm CL,rot} = 3.89\ M_\odot \left(5 \times 10^{14}\ {\rm g\ cm^{-3}}/\rho_{\rm u}\right)^{\frac{1}{2}}\ . \tag{6.110}$$

For a given $\rho_u$ it is $\sim$30% larger than $M_{max}^{CL,stat}$.

### 6.12.8 The lower bound on rotation period

The minimum rotation period is realized for an EOS which leads to most compact and dense stars, especially stable with respect to the mass shedding from the equator. We know from § 6.6.4 that the most compact non-rotating stars are obtained for the pure causality limit EOS $P = c^2(\rho - \rho_s)$. The same is true for rigidly rotating configurations (Koranda *et al.* 1997). For such an EOS, both the minimum rotation period and the maximum mass configurations exhibit *exact* scaling with respect to variation of $\rho_s$: $P_{min}^{CL} \propto \rho_s^{-1/2}$, $M_{max}^{CL,stat} \propto \rho_s^{-1/2}$ (see Appendix E). In other words, for this EOS one has $P_{min}^{CL} \propto M_{max}^{CL,stat}$. Precise numerical calculations of Koranda *et al.* (1997) yield

$$P_{min}^{CL} = 0.196\ (M_{max}^{CL,stat}/M_\odot)\ \text{ms}. \tag{6.111}$$

Pulsars with precisely measured masses rotate at $\Omega \ll \Omega_K$ (see § 9.1). Therefore, the effects of rotation on their structure can be neglected and their EOS has to fulfill the inequality $M_{max}^{obs} < M_{max}^{stat}(\text{EOS})$. Equation (6.111) implies then

$$v_s \leq c \quad \Longrightarrow \quad P > P_{min}^{CL} = 0.196\ (M_{max}^{obs}/M_\odot)\ \text{ms}\ . \tag{6.112}$$

At the time of this writing, the most massive neutron star with the accurately measured mass is the Hulse-Taylor pulsar, $M_{max}^{obs} = 1.442\ M_\odot$, leading to $P > 0.283$ ms. A future precise measurements of masses of heavier pulsars with $\Omega \ll \Omega_K$ will increase this *absolute lower limit* on $P$. For example, if the mass of PSR J0751+1807 in binary with a white dwarf were confirmed to be $\simeq 2\ M_\odot$ (see § 9.1.3) then we would have $P > 0.39$ ms!

One can slightly increase the value of the lower limit on $P$ assuming (as in the maximum mass problem), that we know the EOS at $\rho < \rho_u$, and that at $\rho > \rho_u$ ($n_b > n_u$) the EOS has $v_{sound} = c$. For example, using the FPS EOS at $n_b < 0.1\ \text{fm}^{-3}$, Koranda *et al.* (1997) obtain

$$P > 0.2045\ (M_{max}^{obs}/M_\odot)\ \text{ms}\ , \tag{6.113}$$

which gives $P > 0.295$ ms for $M_{max}^{obs} = 1.442\ M_\odot$.

### 6.12.9 From the upper bound on $z_{surf}$ to the lower bound on the rigid-rotation period

Following Haensel *et al.* (1999), let us remark that the improved empirical formula (6.109) for $P_{min}$ gives

$$P_{min} \simeq f(x_{GR}^{max})\ M_{max}^{stat}/M_\odot\ , \quad f(x) = 0.187\ x^{-3/2}\ (1 + 0.808x)^{-1}\ . \tag{6.114}$$

The function $f(x_{\rm GR}^{\rm max})$ decreases monotonically with increasing $x_{\rm GR}^{\rm max}$ and reaches its minimum value at the upper bound of $x_{\rm GR}^{\rm max}$. General Relativity and the condition $v_{\rm sound} \leq c$ yield the upper bound $z_{\rm surf} < z_{\rm surf}^{\rm CL} = 0.8504$ (§ 6.6.4), which results in the upper bound $x_{\rm max} < x_{\rm max}^{\rm CL} = 0.7081$. Therefore, we obtain $f_{\rm min} = f(x_{\rm max}^{\rm CL}) = 0.200$. On the other hand, we have an observational constraint $M_{\rm max}^{\rm stat} > M_{\rm max}^{\rm obs}$. This gives

$$z_{\rm surf} < z_{\rm surf}^{\rm CL} = 0.8504 \quad \Longrightarrow \quad P_{\rm min}^{\rm CL} \simeq 0.200\,(M_{\rm max}^{\rm obs}/M_\odot)\ {\rm ms}, \tag{6.115}$$

and $P > 0.288$ ms for $M_{\rm max}^{\rm obs} = 1.442\ M_\odot$. This approximate lower bound on $P$ is only 2% higher than the result of extensive and precise two-dimensional calculations of Koranda *et al.* (1997), mentioned in the preceding section.

### 6.12.10 Rapid rotation and gravitational-radiation instabilities

A rotating neutron star is susceptible to various secular non-axially symmetric instabilities, which are growing on timescales much longer than the dynamical timescale $\tau_{\rm dyn} \sim 1$ ms. These instabilities break the axial symmetry of the star and induce angular momentum loss owing to the emission of gravitational radiation. Therefore, if the critical angular frequency above which an instability sets in is $\Omega_{\rm crit} < \Omega_{\rm K}$, then stable rotation is limited by $\Omega_{\rm crit}$ (rather than by $\Omega_{\rm K}$).

Within the Newtonian gravitation, at sufficiently high ratio of the kinetic-to-gravitational energy $\beta \equiv E_{\rm rot}/|E_{\rm grav}| > \beta_{\rm crit}^{\rm (dyn)}$ a rotating star becomes dynamically unstable with respect to the triaxial deformation ("bar instability"). For an incompressible fluid, $\beta_{\rm crit}^{\rm (dyn)} = 0.27$ (see, e.g., Shapiro & Teukolsky 1983 and references therein). The same is expected to occur in General Relativity, although the definitions of $E_{\rm rot}$ and $E_{\rm grav}$ become ambiguous. In what follows we use the widely accepted definition given in § 6.12.2. The value of $\beta_{\rm crit}^{\rm (dyn)}$ for relativistic stars turns out to be higher than in the Newtonian case (see Gondek-Rosińska & Gourgoulhon 2002 and references therein). A rotating star which is dynamically unstable with respect to the bar instability would lose angular momentum via gravitational radiation, so that a stable rotation with $\beta > \beta_{\rm crit}^{\rm (dyn)}$ would be impossible. However, as we have seen in the preceding section, for realistic EOSs one gets $\beta \simeq 0.1$ even at $\Omega = \Omega_{\rm K}$. Therefore, rigidly rotating neutron stars are expected to be always stable with respect to the dynamical bar-instability.

The earliest known secular non-axially symmetric instability is that driven by viscosity (Roberts & Stewartson 1963 and references therein). It is connected with the transition to a lower-energy triaxial state owing to the viscous angular momentum redistribution. This *viscosity driven* instability proceeds on a viscous timescale and is absent in a rotating perfect-fluid star. The crucial parameter is again $\beta \equiv E_{\rm rot}/|E_{\rm grav}|$. For incompressible-fluid Newtonian stars $\beta_{\rm crit}^{\rm (vis)} = 0.14$ (Shapiro & Teukolsky 1983 and references therein). For realistic Newtonian stars $\beta_{\rm crit}^{\rm (vis)}$ is slightly lower (Bonazzola *et al.*, 1996). Relativistic effects significantly increase $\beta_{\rm crit}^{\rm (vis)}$ (Bonazzola *et al.*, 1998a; Gondek-Rosińska & Gourgoulhon, 2002).

A very different type of instabilities was discovered by Chandrasekhar (1970) for spinning stars built of the perfect incompressible fluid stars. It is driven by the back-reaction of the gravitational radiation. These modes grow on a timescale $\tau_{\rm GRR}$ resulting from the coupling of the gravitational radiation to the stellar matter (gravitational radiation reaction). If a non-axially symmetric perturbation propagates in the direction opposite to the stellar rotation in the star reference frame and is observed at infinity as corotating with the star (which happens at a sufficiently rapid rotation), then such a perturbation will grow owing to the gravitational radiation. A simple argument, valid for a small perturbation (linear regime), is that the star looses its angular momentum by increasing the counter-rotating flow; the perturbation amplitude will then grow with time. As shown by Friedman & Schutz (1978), such an instability is generic and present in all rotating perfect-fluid stars. For a perturbation with a given angular frequency $\omega$ and azimuthal dependence $e^{{\rm i}m\phi}$ $(m > 0)$, the instability appears when $\Omega > \Omega_{\rm crit} = \omega/m$. However, in real neutron stars the Chandrasekhar-Friedman-Schutz (CFS) instability is damped by the viscosity provided $\tau_{\rm GRR} > \tau_{\rm vis}$, where $\tau_{\rm vis}$ is determined by all viscous dissipative mechanisms,

$$\frac{1}{\tau_{\rm vis}} = \frac{1}{\tau_\eta} + \frac{1}{\tau_\zeta} + \dots\ ; \tag{6.116}$$

all timescales have to be calculated using eigenfunctions of a given CFS-mode. Here, we have included explicitly the damping resulting from the shear ($\tau_\eta$) and bulk ($\tau_\zeta$) viscosities of stellar matter. Notice that the value of $\tau_\eta$ is quadratic in the velocity gradients and decreases rapidly with growing $m$. On the contrary, the perfect-fluid threshold $\Omega_{\rm crit} \propto 1/m$ decreases for larger $m$, simplifying the instability development in a star built of a perfect fluid. Numerical calculations show a compromise between these two tendencies in a viscous star; the most unstable CFS modes are those with $m = 4-5$. In general, the viscous damping of the CFS instabilities studied before 1998 was found to be very strong. This damping led to a small (a few per cent) decrease of the maximum spin frequency below the Keplerian value.

In 1998, N. Andersson showed that a special class of the CFS instabilities (connected with the so called Rossby ($r$) modes) leads to an exceptionally strong effect on rotating neutron stars. The $r$-mode instability was predicted to grow on timescales $\tau_r$ orders of magnitude shorter than other CFS instabilities (as reviewed by Andersson & Kokkotas 2001). In perfect-fluid stars, $r$-modes are unstable at any $\Omega > 0$, in contrast to the previously discussed CFS modes, which are unstable at $\Omega > \Omega_{\rm crit} = \omega/m$. Therefore, the growth rate $1/\tau_r$ can be large even at $\Omega$ significantly smaller than $\Omega_{\rm K}$. In the presence of the viscosity, the condition $\tau_r < \tau_{\rm vis}$ shifts the threshold value of $\Omega$ from zero to some finite value $\Omega_{\rm crit}$. An $r$-mode is completely damped by viscosity if $\Omega_{\rm crit} > \Omega_{\rm K}$. If an unstable $r$-mode could grow to a large amplitude, then rotating neutron stars would be promising sources of gravitational radiation. However, although $r$-modes do not produce large density variations, they would be damped by the bulk viscosity if hyperons are present in the neutron star cores (Jones, 2001a,b; Haensel *et al.*, 2002b; Lindblom & Owen, 2002). A possible existence and a size of the window in the $\rho_{\rm c} - T$ plane in which a necessary condition for the $r$-mode instability could be satisfied is a subject of vigorous debates.

Chapter 7

# NEUTRON STARS WITH EXOTIC CORES

## 7.1. Introduction

It has been recognized since the 1960s that with increasing density $\rho$ above the standard nuclear-matter density $\rho_0$ the matter can undergo *phase transitions* to states qualitatively different from the state at $\rho \sim \rho_0$. These high-density phases are *exotic* by the standards of terrestrial nuclear physics. Their possible existence above some threshold densities results from assumed specific features of strong (hadronic) interactions (pion condensation, kaon condensation) and the quark structure of baryons (quark deconfinement).

This chapter starts with an overview of exotic phases of dense matter, § 7.2. Pion condensation and kaon condensation, and their impact on the EOS, are studied in §§ 7.3 and 7.4. The quark matter EOS is described in § 7.5. Properties of possible mixed-phase states are studied in § 7.6. Models of solid neutron star cores are reviewed in § 7.7. The formation of exotic phases in neutron star cores is studied in § 7.8. Finally, in § 7.9 we describe the consequences of phase transitions in neutron star cores for stellar structure and dynamics.

## 7.2. Exotic phases – an overview

We will briefly describe the development of the theories of pion condensation, kaon condensation, and quark deconfinement in dense matter. At the first step we will consider one-phase (pure) states of dense matter, but we will also outline possible mixed phases advanced in the beginning of the 1990s. In addition, we will discuss a possible crystalline phase of dense matter. The most exotic self-bound states will be mentioned only briefly; their more detailed description will be given in Chapter 8.

### 7.2.1 Meson condensates

Pions are the lightest mesons. Therefore, negative pions which could replace electrons are natural candidates for a condensation. Bahcall & Wolf (1965a) treated the pions as free particles and showed that $\pi^-$ could be present at $\rho \gtrsim \rho_0$. However, pions strongly interact with baryons, and the inclusion of the s-wave $\pi N$ repulsion prevents the condensation of free pions. The idea of pion condensation as the condensation of pion-like medium excitations owing to the strong $\pi N$ attraction in the p-wave was advanced by Migdal (1971, 1972), Sawyer (1972b), and Scalapino (1972). These authors demonstrated that such excitations condense in a finite-momentum state, and the phase transition is accompanied by the loss of translational invariance. Further investigations have shown the possibility of various phases of pion condensate and the importance of nucleon correlations for the existence of the condensation itself. Pion condensation softens the equation of state (EOS); short-range correlations in dense matter make pion condensation in neutron star cores less likely than it was thought in the 1970s (§ 7.3).

Kaons are the lightest strange mesons. Because of their large mass, their possible condensation was not obvious and was pointed out rather late (Kaplan & Nelson, 1986). Using a simplified dense-matter model, Kaplan and Nelson showed that negative kaons (again, kaon-like excitations rather than real particles) could condense into a zero-momentum state at $\rho \gtrsim 3\rho_0$. $K^-$ condensation is energetically preferable provided the s-wave kaon-nucleon attraction in dense medium is sufficiently strong. Kaon condensation strongly softens the EOS. Calculations based on various many-body theories indicate the possibility of different condensates ($K^-$, $\overline{K}^0$, in a state of zero or finite momentum). However, these results do not seem very conclusive (see § 7.4).

### 7.2.2 Deconfinement of quarks

Quarks are the basic constituents of hadrons. Therefore, a fundamental description of dense matter should involve quark degrees of freedom. This picture implies that at sufficiently high densities baryons dissolve into quark matter. Ivanenko & Kurdgelaidze (1965, 1969) were the first who suggested that neutron stars could contain quark cores. Itoh (1970) constructed models of superdense low-mass stars built of a free degenerate gas of massive quarks. Perturbative calculations of the EOS of quark matter started in the 1970s after the formulation of quantum chromodynamics (QCD), but their region of validity was restricted to very high densities (Collins & Perry, 1975). The MIT Bag Model (Chodos *et al.*, 1974) represented non-perturbative effects of confinement via the bag constant. In the 1970–1980s the Bag Model was widely used for calculating the EOS of quark matter. Neutron-star densities are, unfortunately, too low for the validity of perturbative QCD calculations. Therefore, the question whether the quark matter exists in neutron stars cannot be answered on purely theoretical grounds (§ 7.5).

### 7.2.3 Mixed-phase state

Phase transitions can be of first or second order (with or without density jump). Before 1991 phase transitions of first order were considered in a simple way; only pure, homogeneous, electrically neutral states of phases were allowed. As the stellar equilibrium requires a non-vanishing pressure gradient, the two phases (lower-pressure and higher-pressure ones) were thought to occupy separate regions of the core, with a density jump at the phase interface. However, as was demonstrated by Glendenning (1991, 1992), this picture was a consequence of a tacit assumption that the matter is locally electrically neutral. Imposing the condition of zero average electric charge may open the possibility of the coexistence of two phases (of opposite electric charge) in some pressure interval. The structure of the mixed phase results from the interplay of the surface tension at the phase interface, the bulk nuclear energy, the kinetic energy of matter constituents, and the Coulomb energy (§ 7.6).

### 7.2.4 Crystallization

The s-wave (strong) NN interaction is the sum of the short-range repulsion and long-range attraction. It is therefore similar to the atom-atom interaction in noble gases which solidify at sufficiently high densities. Early studies indicated crystallization of high-density neutron matter (e.g., Cazzola *et al.*, 1966; Canuto & Chitre, 1974). However, more precise calculations demonstrated that the solidification due to the short-range neutron-neutron repulsion does not occur (see, e.g., Takemori & Guyer, 1975).

According to several authors, neutron matter at high densities could become a layered liquid crystal. This one-dimensional solidification could be favored by the in-medium enhancement of the tensor interaction via excitation of $\Delta$-resonances in intermediate nucleon states (Pandharipande & Smith, 1975). A one-dimensional localization combined with the spin-isospin ordering was also obtained in pion-condensate models (see Takatsuka & Tamagaki 1988a,b, and references therein). Some authors pointed out a possible localization of baryons in self-consistent potential wells created by other baryons. Localized baryons (protons or charged hyperons) could then form crystalline structures (Kutschera & Wójcik, 1995; Perez Garcia *et al.*, 2002). Finally, let us remind that each of two phases in a mixed-phase state is electrically charged (§ 7.6). The Coulomb interaction can then produce crystalline structures in the mixed-phase layer (Glendenning & Pei, 1995). Any crystalline structure in a neutron star core may induce shear strain (§ 7.7).

### 7.2.5 Exotic self-bound states of superdense matter

Large uncertainties in EOS calculations allowed Witten (1984) to propose an extreme hypothesis that the bulk quark matter in beta equilibrium might be the true ground state of the matter at zero pressure (§ 8.3). Earlier suggestions on the existence of superdense strange nuclei were advanced by Bodmer (1971). If this *strange matter* hypothesis is true, then *strange stars* could exists which are built exclusively of quark matter. The EOS of strange matter and the structure of strange stars are studied in detail in Chapter 8 together with other, "even stranger" hypothetical self-bound states of dense matter such as an abnormal state (Lee & Wick, 1974) and a more recent Q-matter (Bahcall *et al.*, 1990).

## 7.3. Pion condensation

Pions are ubiquitous in dense nucleon matter. An exchange of *virtual* pions generates the long-range component of the NN interaction (§ 5.6). The energy $\omega_\pi$ and momentum $k$ of virtual pions do not satisfy the standard formula $\omega_\pi^2 = m_\pi^2 c^4 + k^2 c^2$. The virtual nature of such pions is also reflected in their lifetime $\sim \hbar/m_\pi c^2 \approx 5 \times 10^{-24}$ s; they can travel over characteristic length-scales $\sim \hbar/m_\pi c \approx 1.4$ fm. Virtual pions generate the NN interaction but do not appear as constituents of standard nuclear structures (in particular, of nuclear matter).

Because of the strong pion-nucleon attraction in the p-wave, the ground state of dense nuclear matter could contain a Bose-Einstein "condensate of pions". This possibility was pointed out independently and using different arguments by Migdal (1971, 1972), Sawyer (1972b), and Scalapino (1972). The authors considered symmetric nuclear matter where the interaction of pions with nuclear medium is charge-independent (neglecting small effects associated with the difference of masses $m_{\pi^\pm}$ and $m_{\pi^0}$). However, in a strongly asymmetric nuclear matter of neutron star cores the pion interaction with nucleon medium essentially depends on the pion charge. Let us focus on a pion condensation in the $npe\mu$ matter; the effects of hyperons will be outlined in the end of this section.

There are two approaches to study the onset of pion condensation. The first one, developed by Migdal and his collaborators, is based on the Landau-Migdal theory of Fermi liquids. This theory and its applications for nuclear matter and neutron-star matter were reviewed by Migdal *et al.* (1990). The second approach is based on models of effective Lagrangians representing strong interactions. It was introduced by Sawyer (1972b) and Scalapino (1972). It has been significantly elaborated later, as reviewed by Brown & Weise (1976), Baym (1979), and Takatsuka *et al.* (1993). In what follows, we discuss the possibility of pion condensation in the framework of the Fermi-liquid theory (Migdal *et al.*, 1990).

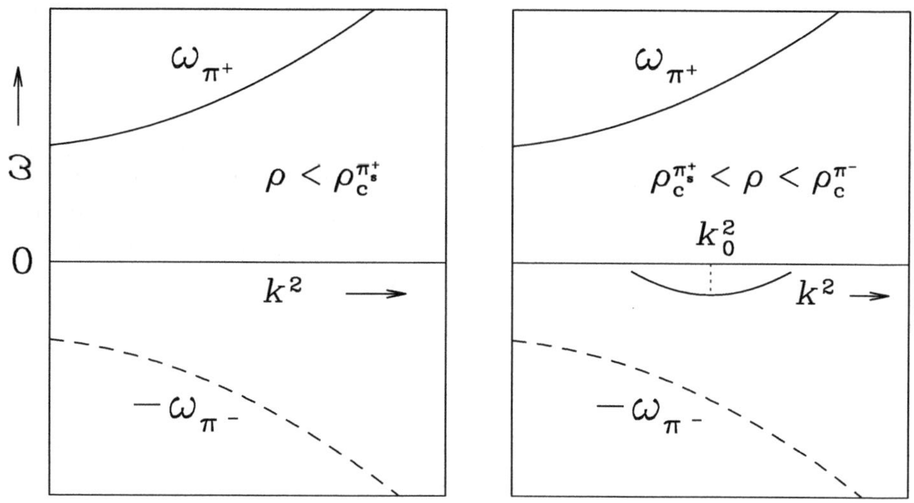


*Figure 7.1.* Schematic spectrum of lowest energy charged-pion excitations in the $npe\mu$ matter at densities lower than the $\pi^-$ condensation threshold. Elementary excitation energy $\omega$ is plotted versus $k^2$, square of excitation wavenumber. *Left*: Density is lower than the $\pi_s^+$ condensation threshold. *Right*: Density is higher than the $\pi_s^+$ condensation threshold, and at $k \sim k_0$ matter becomes unstable with respect to the formation of zero-sound-like $\pi_s^+$ excitations (see text for details). After Migdal *et al.* (1990).

Let us first reconsider the stability of a spatially uniform state of the $npe\mu$ matter in beta equilibrium. As shown in §3.5, such a state is *unstable* with respect to the formation of periodic "nuclear structures" (density waves) at $\rho \lesssim 0.5\rho_0$. This instability signals a phase transition to a spatially-ordered non-uniform state which is the neutron star crust. Now consider the stability of the uniform $npe\mu$ matter at $\rho \gtrsim \rho_0$. This can be done by studying *excitations* of the $npe\mu$ matter associated with perturbations of the ground state of the system. In the context of pion condensation, there are two types of relevant excitations. Both are boson-type excitations which differ in their low-momentum dispersion relations. Excitation quanta have well defined spin and charge quantum numbers $S$ and $q$. Pion condensation is related to excitations with the pion quantum numbers $S = 0$ and $q = \pm 1$ or $q = 0$.

Excitations of the first type correspond to collective modes (analogous to the well known zero-sound in liquid $^3$He). They have an acoustic large-wavelength dispersion relation $\omega_s \simeq c_s k$ at $k \to 0$, where $c_s$ is the zero-sound speed. These three zero-sound-like excitations will be denoted as $\pi_s^{\pm,0}$.

Excitations of the second type correspond to "real pions" and have a characteristic "resonance" ("optical") long-wavelength behavior. In the limit of a "dilute medium" ($\rho \longrightarrow 0$) their energy spectrum $\omega_\pi \simeq \sqrt{m_\pi^2 c^4 + k^2 c^2}$ is similar to the energy spectrum of real pions in vacuum. These three excitations will be denoted as $\pi^{\pm,0}$.

Using the Fermi-liquid theory one can calculate the dispersion relations $\omega_\pi(k)$ for all six excitation modes in the $npe\mu$ matter, provided the gas of excitations is diluted, so that the their number density $n_\pi \ll n_{\rm b}$. A schematic representation of these dispersion relations is given in Figs. 7.1 and 7.2.

Let us start with excitations of zero-sound type. The $npe\mu$ matter turns out to be most susceptible ("the softest") to $\pi_{\rm s}^+$ excitations. As seen from the right panel of Fig. 7.1, for this mode the function $\omega_{\pi_{\rm s}^+}(k)$ acquires a *minimum*, $\omega_{\pi_{\rm s}^+}^{\rm min} < 0$, at some $k = k_0(\rho)$. As soon as

$$\omega_{\pi_{\rm s}^+}^{\rm min} + \mu_n < \mu_p \, , \tag{7.1}$$

the $\pi_{\rm s}^+$ mode is spontaneously excited with some (or even all) protons converted into neutrons. This instability with respect to

$$p \longrightarrow \pi_{\rm s}^+ + n \, , \tag{7.2}$$

if it occurs at all, starts at a well defined critical density $\rho_{\rm c}^{\pi_{\rm s}^+}$. In the particle-hole language, the system becomes unstable with respect to the formation of pairs of protons and neutron-holes (which are just $\pi_{\rm s}^+$ excitations). Simultaneously, the $npe\mu$ matter looses translational invariance.

Let us turn to $\pi^\pm$ excitations of which $\pi^-$ is relevant for us. For sufficiently high densities the function $\omega = \omega_{\pi^-}(k)$ has a minimum at a finite $k = k_{\pi^-}$ (Fig. 7.2). If $\omega_{\pi^-}^{\rm min} < |\omega_{\pi_{\rm s}^+}^{\rm min}|$, then there exists such a value of $k$ at which pairs of excitations $\pi_{\rm s}^+\pi^-$ with opposite momenta of $|\boldsymbol{k}| = k$ can be spontaneously created,

$$\omega_{\pi_{\rm s}^+}(\boldsymbol{k}) + \omega_{\pi^-}(-\boldsymbol{k}) = 0 \, . \tag{7.3}$$

In this way the $\pi_{\rm s}^+$ instability initiates $\pi^-$ condensation.

As in a symmetric nuclear matter, the spontaneous creation of $\pi^0$ quasiparticles becomes possible when

$$\omega_{\pi^0}^{\rm min} \le 0 \, . \tag{7.4}$$

This instability, if it happens at all, starts at a well defined density $\rho_{\rm c}^{\pi^0}$.

In thermodynamic equilibrium, elementary excitations (boson quasiparticles) $\pi_{\rm s}^+$, $\pi^-$, and $\pi^0$ form Bose-Einstein condensates, and their chemical potentials are equal to the energy of the occupied momentum state with minimum quasiparticle energy ($\omega_{\pi_{\rm s}^+}$, $\omega_{\pi^-}$, and $\omega_{\pi^0} = 0$, respectively).

The mechanism of spontaneous transition to a pion-condensed state deserves a comment. As pointed out by Migdal many times, pion condensation in a pure neutron matter does not imply an instability of the form

$$n \longrightarrow p + \pi^- \, . \tag{7.5}$$

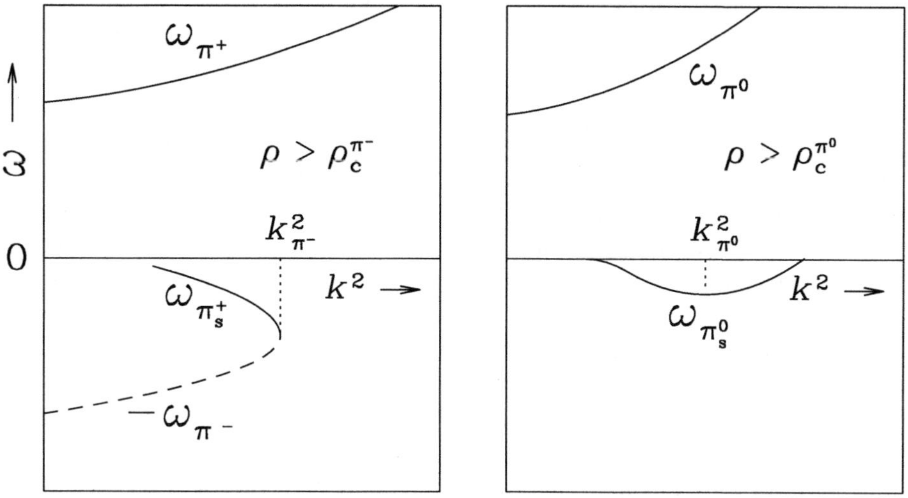


*Figure 7.2.* Schematic spectrum of lowest energy charged-pion excitations in the $npe\mu$ matter above the $\pi^-$ condensation threshold (*left*) and above the threshold for the formation of $\pi_s^0$ excitations (*right*); see text for details. After Migdal *et al.* (1990).

Actually, according to numerous calculations, one has $\omega_{\pi^-}^{\rm min} > \mu_n$ at any density and the reaction is prohibited in pure neutron matter (see, e.g., Migdal *et al.* 1990). Instead, the condensation consists (just as we have described) in the excitation of the zero-sound type $\pi_s^+$ mode followed by the formation of $\pi_s^+\pi^-$ pairs.

The Landau Fermi-liquid theory is valid only for a small-amplitude pion condensate with $n_\pi \ll n_{\rm b}$ ($\pi = \pi_s^+, \pi^-$, and $\pi^0$), particularly, in the vicinity of the threshold density $n_{\rm c}^\pi$ (from now on we omit the index of an unstable pion-like mode). On the normal-phase ($n_{\rm b} < n_{\rm c}^\pi$) and pion-condensed ($n_{\rm b} > n_{\rm c}^\pi$ and $n_{\rm b} - n_{\rm c}^\pi \ll n_{\rm c}^\pi$) sides one has

$$\text{normal phase} \quad : \quad \mathcal{E} = \mathcal{E}^{(\rm n)} \,, \tag{7.6}$$

$$\text{pion-condensed} \quad : \quad \mathcal{E} = \mathcal{E}^{(\rm n)} + (n_{\rm b} - n_{\rm c}^\pi)^2 \mathcal{A} \,, \tag{7.7}$$

where $\mathcal{A} \le 0$ and the label (n) indicates the normal (non-condensed) phase of the $npe\mu$ matter. The quantities $\mathcal{E}^{(n)}$ and $\mathcal{A}$ are functions of $n_{\rm b}$ continuous at $n_{\rm b} = n_{\rm c}^\pi$. Therefore, the pressure is continuous at $n_{\rm b} = n_{\rm c}^\pi$ but the sound speed $v_{\rm s} = ({\rm d}P/{\rm d}\rho)^{1/2}$ drops with increasing $n_{\rm b}$. These features are characteristic of second-order phase transition (see below).

As soon as $n_\pi/n_{\rm b}$ ceases to be small, one needs to go beyond the standard Fermi-liquid picture. In particular, one has to deal explicitly with condensate fields $\varphi_\pi(\boldsymbol{r}, t)$. In order to make the problem tractable, one introduces an effective Lagrangian for nucleons plus the condensate system, where $\varphi_\pi(\boldsymbol{r}, t)$ are replaced by their expectation values $\langle\varphi_\pi(\boldsymbol{r}, t)\rangle$. In this approach, the low-amplitude limit corresponds to $\mathcal{E} - \mathcal{E}^{(\mathrm{n})} \simeq a|\langle\varphi_\pi\rangle|^2$, where $a < 0$. The next term in the small pion-amplitude expansion is $b|\langle\varphi_\pi\rangle|^4$, where $b > 0$; it is needed to stabilize the condensed state at a finite (but still small) pion-condensate amplitude. The above picture is consistent with the Landau theory of phase transitions (Landau & Lifshitz, 1993), indicating that pion condensation is second-order phase transition.[1] The effective Lagrangian of the nucleon system containing pion condensate is a functional of nucleon number densities and $\langle\varphi_\pi\rangle$. Various models of $\mathcal{L}_{\mathrm{eff}}(n_n, n_p, \langle\varphi_\pi\rangle)$ have been considered in the literature. They involve different approximations of hadronic interactions and different treatments of spatial inhomogeneities of condensed states (see, e.g., Migdal *et al.*, 1990; Brown & Weise, 1976; Baym, 1979). Especially important ingredients are the s-state pion-nucleon interaction, the nucleon excited state in the form of $\Delta$ resonances, and the pion-pion interaction needed to stabilize the system at a finite pion-condensate amplitude. The pion-condensed ground state is usually determined in the variational approximation, using a plausible *ansatz* for a periodic spatial structure of the pion-condensate field.

The structure of a pion-condensed state is most important for calculating the neutrino emissivity (§ 1.3.5; Maxwell *et al.* 1977; for review see Migdal *et al.* 1990; Pethick 1992; Yakovlev *et al.* 2001). The pion-condensate field couples to nucleons and mixes neutron and proton states. Therefore, neutrino emission processes involve quasinucleons $\widetilde{n}$ and $\widetilde{p}$ which fill their Fermi seas and have comparable Fermi momenta $p_{\mathrm{F}\widetilde{n}} \sim p_{\mathrm{F}\widetilde{p}}$, in contrast to $p_{\mathrm{F}n} \gg p_{\mathrm{F}p}$ characteristic of normal nucleon matter. In view of this, the *direct Urca* process in pion-condensed matter,

$$\widetilde{n} \longrightarrow \widetilde{p} + e + \overline{\nu}_e\,, \quad \widetilde{p} + e \longrightarrow \widetilde{n} + \nu_e\,, \tag{7.8}$$

can be open even if it is blocked in normal phase of the $npe\mu$ matter by momentum conservation (see § 5.12). As the direct Urca process with quasinucleons, Eq. (7.8), involves only three degenerate fermions, the neutrino emissivity depends on the temperature as $Q_\nu^{(\pi)} \propto T^6$, to be contrasted with $Q_\nu \propto T^8$ characteristic of non-condensed matter (with the forbidden direct Urca process); see § 1.3.5. Calculations carried out in the 1980s give values of $Q_\nu^{(\pi)}$ which are a factor of $(10^3 - 10^4)\,T_9^{-2}$ larger (with $T_9 \equiv T/10^9$ K) than $Q_\nu$ in normal matter with forbidden direct Urca processes (see, e.g., Tatsumi, 1983, 1987; Muto & Tatsumi, 1988). Consequently, the presence of a pion-condensed matter leads

[1] The pion condensation model considered in the present section is obtained in the mean-field approximation. In particular, the pion field is treated neglecting quantum fluctuations around mean-field values. The inclusion of quantum fluctuations changes the character of the phase transition and makes it weakly first-order (Dyugaev 1975, also see Migdal *et al.* 1990 and references therein).

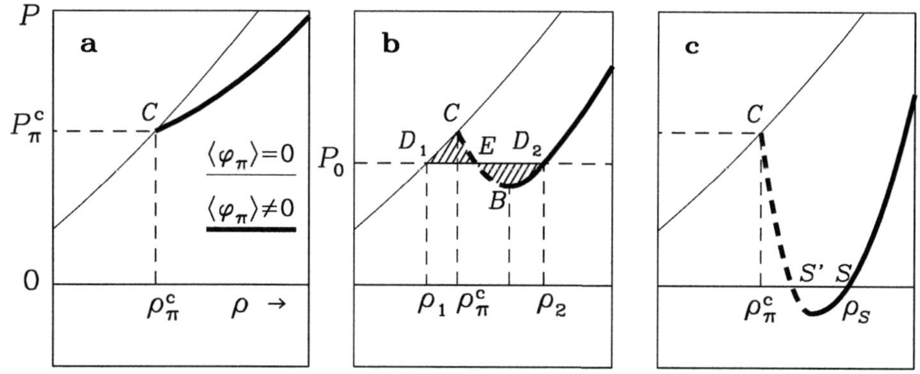


*Figure 7.3.* Three qualitatively different EOSs of pion-condensed matter. Thick solid lines show pion-condensed phase whereas thin solid lines refer to normal phase: (a) Second-order phase transition with moderate softening; (b) Strong softening implying first-order phase transition at $P = P_0$; (c) Very strong softening leading to a stable "abnormal" superdense self-bound state at $P = 0$ and $\rho = \rho_S$. For further explanations see the text.

to a *fast cooling* of neutron stars, to be contrasted with a *slow cooling* scenario without pion condensation (§ 1.3.7).

In the present section we focus on the effects of pion condensation on the EOS. After calculating the ground-state energy density of the hadronic component of the pion-condensed matter, we get a function of two independent variables which are the average baryon density $n_\mathrm{b}$ and the average net electric charge density $n_q^{(\mathrm{h})}$ (in the units of $e$). In order to impose the beta-equilibrium condition between hadrons and leptons, we should minimize the total energy density

$$\mathcal{E}(n_\mathrm{b}, n_q^{(\mathrm{h})}, n_e, n_\mu) = \mathcal{E}(n_\mathrm{b}, n_q^{(\mathrm{h})}) + \mathcal{E}_e(n_e) + \mathcal{E}_\mu(n_\mu)\,, \tag{7.9}$$

at a fixed average baryon density $n_\mathrm{b}$ and under electric charge neutrality,

$$n_n + n_p = n_\mathrm{b}\,, \qquad n_p + n_{\pi_\mathrm{s}^+} = n_{\pi^-} + n_e + n_\mu\,. \tag{7.10}$$

The procedure is identical to that used in § 5.11 for obtaining the EOS of the normal $npe\mu$ matter. In this way we get the EOS, $\mathcal{E} = \mathcal{E}(n_\mathrm{b})$ and $P = P(n_\mathrm{b})$, of the fully equilibrated pion-condensed $npe\mu$ matter. As already mentioned above, the sound speed drops at $n_\mathrm{b} = n_\mathrm{c}^\pi$ with increasing $n_\mathrm{b}$. The magnitude of this drop and the EOS at higher densities are very model-dependent. Three possibilities (with different consequences for the neutron-star structure) are shown schematically in Fig. 7.3.

Figure 7.3a refers to moderate softening with second-order phase transition at $\rho = \rho_\mathrm{c}^\pi$. Further stiffening at higher $\rho$ in the pion condensed phase is due to the $\pi\pi$ repulsion.

Figure 7.3b illustrates strong softening. Second-order phase transition to a pion-condensed state destabilizes the ground state of the matter, so that $v_s^2 < 0$ on the CB segment of the thick solid line. The EOS exhibits first-order phase transition at $P = P_0$, where the normal matter of the density $\rho = \rho_1$ can coexist with the condensed matter of the density $\rho = \rho_2 > \rho_1$. The densities $\rho_1$ and $\rho_2$ are determined from the equality of the baryon chemical potentials in the normal and pion-condensed phases:

$$\mu_b = (P + \mathcal{E})/n_b\,, \quad P(\rho_1) = P(\rho_2) = P_0\,, \quad \mu_b(\rho_1) = \mu_b(\rho_2)\,. \tag{7.11}$$

Notice a gradual stiffening of the EOS at highest densities resulting from the $\pi\pi$ repulsion. The Maxwell construction of a first order phase transition, depicted in Fig. 7.3b, assumes that both phases are separately electrically neutral. Relaxing this condition opens the possibility of mixed-phase structures, which will be studied in § 7.6.

Figure 7.3c shows the most dramatic and speculative case suggested by Migdal (1972, 1974) (also see Migdal *et al.*, 1990, for a review). We mention this case for the sake of completeness. Strong softening of the EOS at $\rho > \rho_c^\pi$ leads to the instability of the ground state on the CB segment. Moreover, the pressure can become negative (we have already encountered the negative pressure problem in § 5.4 in the context of the equilibrium of nuclear matter). Therefore, apart from the standard $^{56}$Fe ground state of "cold catalyzed matter" at $P = 0$ with the energy $E_0 = 930.4$ MeV per baryon (§ 3.2), two other zero-pressure equilibria appear at the points $S'$ and $S$. The point $S'$ describes an *unstable* self-bound equilibrium because $(\mathrm{d}P/\,\mathrm{d}\rho)_{S'} < 0$. On the contrary, the point $S$ corresponds to a stable equilibrium, with the energy per baryon $E_S$. If $E_S > E_0$, it is a metastable superdense state of matter, whose lifetime could be calculated if the energy barrier between this state and the standard ($^{56}$Fe) state were known. However, if $E_S < E_0$, the superdense state $S$ represents a true ground state of the matter! This would mean the existence of *abnormal nuclei*, much denser than the normal ones (Migdal, 1971, 1974). In the astrophysical context, one could then speculate on a possible existence of *abnormal stars* with a superdense surface of density $\rho_S$ and arbitrarily small radius $R \propto M^{1/3}$ at $M \ll M_\odot$. These are "golf-ball stars" considered by Hartle (1975). Other aspects of this shocking suggestion as well as other versions of "abnormal state" of dense matter and "abnormal stars" will be studied in §§ 8.9 and 8.20.

Up to now we have limited ourselves to the simplest model of pion condensation. Numerous studies of pion condensation have proved the complexity of the problem and large uncertainties in numerical results. However, they have clearly indicated specific physical effects which favor or disfavor the condensation. The inclusion of $\Delta$ resonances in intermediate nucleon states, which lowers the pion self-energy in nucleon matter, acts in favor of pion condensation.

On the contrary, many effects tend to prohibit it, particularly, the short-range repulsion between nucleons and the s-wave repulsion in the $\pi^- n$ interaction.

So far we have considered pion condensation at $T = 0$. At a finite $T$ one should calculate the free energy density $\mathcal{F} = \mathcal{E} - T\mathcal{S}$, where $\mathcal{S}$ is the entropy density. The mechanism of $\pi^-$ condensation is the same as at $T = 0$: a strong p-wave $\pi n$ attraction which counterbalances the kinetic energy increase due to the ordered structure of a condensed state. Notice that because $\mathcal{S}^{(\pi)} < \mathcal{S}^{(\mathrm{n})}$ (the normal state is *less ordered* than the condensed one) increasing $T$ makes the condensation less favorable and shifts it (if any) to higher densities $\rho_{\mathrm{c}}^{\pi}(T) > \rho_{\mathrm{c}}^{\pi}(0)$ (see Kunihiro *et al.*, 1993, and references therein). On the other hand, at $k_{\mathrm{B}}T \gtrsim 30$ MeV a gas of thermal pions also contributes to thermodynamic quantities (Kolehmainen & Baym, 1982).

Plotting the phase diagram in the $\rho - T$ plane one can get the phase-coexistence curve and find the value of the critical temperature, $T_{\mathrm{crit}}^{\pi}$, above which the phase transition does not occur at any density. The critical temperature seems to be very high by neutron star standards, $k_{\mathrm{B}}T_{\mathrm{crit}}^{\pi} \gtrsim 60$ MeV (Kolehmainen & Baym, 1982; Kunihiro *et al.*, 1993).

## 7.4. Kaon condensation

Till 1986 nobody expected kaon condensation in neutron star cores. The main reason was that kaons are too massive ($m_{K^\pm}c^2 \equiv m_K c^2 = 493.6$ MeV, $m_{K^0}c^2 = 497.7$ MeV in vacuum). As kaons are bosons, they can occupy one single quantum state to minimize the energy. If present, they could form, similarly to pions, a boson condensate. Consider first noninteracting $K^-$. The stability in dense matter implies

$$\mu_{K^-} = m_K c^2 = \mu_e \, , \tag{7.12}$$

assuming that kaons occupy their ground zero-momentum state. This condition can be satisfied only at densities much larger than those relevant for neutron stars,

$$n_{\mathrm{b}} > \frac{1}{3\pi^2}\left(\frac{m_K c}{\hbar}\right)^3 \approx 33\, n_0 \cdot \frac{x_e}{0.1} \, , \tag{7.13}$$

where $x_e = n_e/n_{\mathrm{b}}$.

However, kaons interact strongly with baryons. As shown by Kaplan & Nelson (1986), the in-medium effects may greatly reduce the effective mass of $K^-$. As a result, $K^-$ could appear at a few times normal nuclear density and form a *kaon condensate*. The condensation has been studied by many authors (see Ramos *et al.*, 2001, for a review).

Let us consider the $npe\mu$-matter. Negative kaons can be formed via the strangeness-changing processes

$$e + N \longrightarrow K^- + N + \nu_e \, , \qquad n + N \longrightarrow p + K^- + N \, , \tag{7.14}$$

where "spectator" nucleons $N$ are required to satisfy momentum conservation (respecting, of course, energy conservation as well). In view of the strong degeneracy of constituents of the $npe\mu$ matter, the energy $\omega_K$ of a created kaon should satisfy

$$\omega_K < \mu_e \ . \tag{7.15}$$

In analogy to pion condensation, we actually deal with $K^-$ excitations (which have $K^-$ quantum numbers) in the $npe\mu$ matter. This method of studying kaon condensation provides a correct description of strongly interacting many-body systems; it was developed by Kolomeitsev & Voskresensky (2003).

Terrestrial experiments indicate strong in-medium effects of kaons moving in the nuclear matter. Scattering of $K^-$ off atomic nuclei, observed in experiments, is described by an optical potential. The real part of this potential, $U_{K^-}$, turns out to be strongly attractive. It is a single-particle potential for a kaon in the nuclear matter at $n_\mathrm{b} \simeq n_0$ and $x_p \simeq 1/2$. The value of $U_{K^-}$, which is needed to reproduce the scattering experiments, ranges from $\approx -80$ MeV to $\approx -120$ MeV. Calculations show that this attraction occurs in the s-wave kaon-nucleon state and the potential well deepens significantly at $n_\mathrm{b} \approx (2-3)\, n_0$.

The processes described by Eq. (7.14) become allowed above some threshold density $n_\mathrm{c}^{K^-}$. It is the minimum density, at which the minimum value of $\omega_K$ (reached at zero kaon momentum for the s-wave kaon-nucleon attraction) satisfies

$$\omega_K^{\min} = \mu_e \ . \tag{7.16}$$

Notice, that in this case

$$\omega_K^{\min} = \mu_{K^-}^0 \ , \tag{7.17}$$

where $\mu_{K^-}^0$ is the limiting chemical potential of kaons defined by Eq. (5.116). As zero-momentum (non-moving) kaons replace ultrarelativistic electrons, the pressure becomes lower: the kaon condensate does not contribute to the pressure. However, kaon condensation is accompanied by the gain of binding energy. Because of the negative kaon charge, a growing kaon density increases the proton fraction. This, in turn, lowers the nucleon contribution to the energy density under the action of the symmetry term in the nucleon component $\mathcal{E}_\mathrm{N}$.

Kaon condensate strongly affects the nucleon component of dense matter. In particular, neutrino emission processes of the direct Urca type involve two types of quasiparticle states, which are superpositions of a nucleon state and a hyperon-like state. Such quasiparticles fill their own Fermi seas with Fermi momenta comparable to $p_{\mathrm{F}n}$ in normal matter. The dominating neutrino process is analogous to Eq. (7.8) and leads to the neutrino emissivity $Q_\nu^{(K)} \propto T^6$. Detailed calculations show that $Q_\nu^{(K)}$ is $\sim (10^2 - 10^4) T_9^{-2}$ times larger than the standard neutrino emissivity of normal $npe\mu$ matter (see Brown *et al.*, 1988; Muto & Tatsumi, 1990; Thorsson *et al.*, 1995). Therefore, the presence of kaon-condensation initiates *fast cooling* of neutron stars (§ 1.3.7).

The effect of kaon condensation on the EOS of dense matter is often discussed using the effective mean-field model. In this model, the energy density is

$$\mathcal{E} = \mathcal{E}_{K^-} + \mathcal{E}_{\rm N} + \mathcal{E}_e + \mathcal{E}_\mu \,, \tag{7.18}$$

where the condensate contribution is

$$\mathcal{E}_{K^-} = n_{K^-}\omega_K^{\rm min} \,. \tag{7.19}$$

A spatially uniform kaon condensate affects the condition of the electric-charge neutrality of dense matter, which implies

$$n_{K^-} = n_p - n_e - n_\mu \,. \tag{7.20}$$

The same mean-field model which describes kaons in dense matter can be used to calculate the real part of the kaon optical potential $U_{K^-}$ and properties of nuclear matter. By fitting the kaon-nucleus scattering data, one can fix the parameters of mean-field model. Other parameters, which describe the interaction of kaons with the nuclear matter, can be evaluated using the quark model of hadrons, and the remaining parameters are adjusted to the saturation properties of nuclear matter (see Ramos *et al.*, 2001, and references therein).

The attraction between kaons and nucleons in the p-wave may also lead to kaon condensation (Kolomeitsev *et al.*, 1996; Kolomeitsev & Voskresensky, 2003). In such a case, due to momentum dependence of the attraction, kaons condense in a state with a non-vanishing momentum. In analogy to pion condensation (§ 7.3), a kaon-condensed state loses then translational invariance.

Depending on the strength of the kaon-nucleon attraction, kaon condensation leads either to second-order or first-order phase transition. To be specific, consider a simple mean-field model of kaon condensation, where kaons condense owing to the s-wave attraction. If the real part of the kaon optical potential in nuclear matter is $U_{K^-} \gtrsim -80$ MeV then the transition to kaon-condensed phase is of second-order, without any density jump. However, for $U_{K^-} \lesssim -90$ MeV the kaon condensation implies first-order phase transition associated with a density jump which grows with increasing the depth of $U_{K^-}$ (Glendenning & Schaffner-Bielich, 1998).

Kaon condensation is sensitive to the presence of hyperons. In the presence of $\Sigma^-$ hyperons the electron fraction $x_e$ decreases with increasing density, because $\Sigma^-$ hyperons replace electrons (see § 5.14). In addition, $x_e$ decreases because new hyperon species appear with growing density. It is equivalent to the increase of the number of fermionic degrees of freedom which generally lowers the Fermi energies and momenta, as compared to the $npe\mu$ matter. Therefore, the hyperonization tends to hinder kaon condensation. Conversely, kaons replace negatively charged hyperons and decrease their abundance in dense matter.

Another effect which weakens or even prevents kaon condensation at neutron star densities is connected with kaon-nucleon and baryon-baryon correlations. Such correlations are neglected in the mean-field models of dense matter. The effect of kaon-nucleon and nucleon-nucleon correlations was studied by Pandharipande *et al.* (1995). In their model the correlations reduce the kaon-nucleon attraction and shift the condensation threshold to the densities which are much higher than those obtained in the mean-field approach, probably beyond the maximum densities in neutron star centers. Later the problem was reconsidered by Kolomeitsev & Voskresensky (2003) who included short-range baryon-baryon correlations using the Fermi-liquid theory. They obtained second-order phase transition at $n_{\rm b} \gtrsim 4\, n_0$ due to kaon condensation resulting from the s-wave attraction. They obtained also second-order phase transition to a kaon-condensed state due to the p-wave kaon-nucleon attraction. The transition occurs at $n_{\rm b} \sim (3-5)\, n_0$ depending on parameters of their model. In some models of Kolomeitsev and Voskresensky the preferable kaon-condensed state contains nearly symmetric nuclear matter without any hyperons; its appearance is accompanied by a strong first-order phase transition with a very large density jump (by a factor of two or larger). This dramatic softening of the EOS would lead to the existence of a separate family of superdense kaon-condensed stars (see § 7.9.5). As a large-amplitude kaon condensate implies large proton fraction to counterbalance negative charge of kaons, it has been suggested to call kaon-condensed stars "nuclear stars" in contrast to standard neutron stars (Lee *et al.*, 1994; Brown *et al.*, 1994). Some authors considered a simultaneous condensation of $K^-$ and $\bar{K}^0$, and obtained a very strong softening of the EOS (see, e.g., Banik & Bandyopadhyay 2001; Pal *et al.* 2000).

If kaon condensation is a first-order phase transition, then a mixed-phase state involving a mixture of the kaon-condensed and non-condensed matter could exist under favorable conditions (§ 7.6). Kaon condensation in a non-strange or a weakly-strange dense matter is necessarily connected with the production of strangeness. Therefore, it has to involve weak interactions, which can complicate the formation of kaon-condensed stellar cores (§ 7.8.2).

We restricted ourselves to the zero-temperature approximation. The effect of finite temperatures on the s-wave kaon condensation was studied for protoneutron stars (Pons *et al.*, 2000, 2001). For $k_{\rm B}T \lesssim 60$ MeV this effect is rather weak. It consists in some increase of of the condensation threshold $\rho_{\rm c}^K(T) > \rho_{\rm c}^K(0)$ and some effective "smearing" of the phase transition by thermally excited kaons. We are not aware of any calculation of the critical temperature for kaon condensation; in any case, it is expected to be very high.

## 7.5. Quark matter

The EOS of a degenerate hadronic matter at asymptotically high densities, where the hadronic energies $\gg 1$ GeV, is very simple. Under such conditions, quarks are no longer confined to hadrons. On the contrary, they constitute a weakly-interacting Fermi gas (Collins & Perry, 1975). Therefore, one can apply the QCD in the weak-coupling limit and calculate the reliable EOS in the one-gluon exchange approximation.

First let us consider this reliable asymptotic limit. Then we will discuss how to extrapolate this EOS to neutron star densities. This is the "top-down" approach (see, e.g., Rho, 2001), where one starts from the *top*, i.e., from a solid QCD result, and proceeds *down*wards in density, via an extrapolation combined with some phenomenology, to a phenomenological description of the deconfinement transition in neutron star cores.

Parameters needed for calculating the EOS in the weak-coupling limit are quark masses and the quark-gluon coupling constant. The quark masses cannot be measured directly but can be inferred indirectly from hadron properties. Therefore, they depend on a model used for inferring. We will restrict ourselves to three lightest quarks, because heavier quarks cannot appear in stable neutron star cores (§ 8.12). The masses of $u$, $d$, and $s$ quarks given by the Particle Data Group are the estimates of the so called *current quark masses* (see Yao *et al.*, 2006, and references therein) which will be marked by the upperscript "(c)". They are $m_u^{(c)}c^2 = (1.5 - 3.0)$ MeV, $m_d^{(c)}c^2 = (3 - 7)$ MeV, and $m_s^{(c)}c^2 = (70 - 120)$ MeV. As the chemical potentials of $u$ and $d$ quarks in the quark matter are much larger than $m_u^{(c)}c^2$ and $m_d^{(c)}c^2$, these quarks can be treated as ultra-relativistic and massless. However, one should account for the $s$ quark mass, reflecting in this way the SU(3) (flavor) symmetry breaking.

The flavor symmetry breaking implies the presence of electrons in an electrically neutral quark matter $T = 0$.[2] We are looking for a thermodynamic equilibrium of a four-component plasma, with four thermodynamic variables $n_i$, where $i = u,\ d,\ s$, and $e$. We assume the equilibrium with respect to the weak-interaction processes

$$d \longrightarrow u + e + \overline{\nu}_e\,, \qquad u + e \longrightarrow d + \nu_e\,, \tag{7.21a}$$

$$u + d \longrightarrow s + u\,, \qquad s + u \longrightarrow u + d\,. \tag{7.21b}$$

Because the matter is thought to be transparent to neutrinos ($\mu_{\nu_e} = \mu_{\overline{\nu}_e} = 0$), we come to the following relations between the chemical potentials:

$$\mu_d = \mu_u + \mu_e\,, \quad \mu_d = \mu_s\,. \tag{7.22}$$

It is advantageous to use the thermodynamic potential per unit volume

$$\Omega(\mu_u, \mu_d, \mu_s, \mu_e) = \mathcal{E} - \mu_u n_u - \mu_d n_d - \mu_s n_s - \mu_e n_e\,. \tag{7.23}$$

---

[2]This statement is valid in the weak-coupling regime. Color-flavor locked quark superconductivity with a large gap ($\Delta \gtrsim 100$ MeV at $m_s c^2 \lesssim 200$ MeV) will expel electrons (Rajagopal & Wilczek, 2000).

The electron gas can be treated as free and ultrarelativistic, so that

$$\Omega_e = -\mu_e^4/[12\pi(\hbar c)^3] \,. \tag{7.24}$$

For massless noninteracting quarks $u$ and $d$ we have an additional factor of three due to the color degree of freedom,

$$\Omega_i^{(0)} = -\mu_i^4/[4\pi(\hbar c)^3] \qquad \text{for } i = u, d. \tag{7.25}$$

In the weak-coupling limit, the quark contribution to $\Omega$, denoted as $\Omega_\mathrm{q}$, can be calculated using a perturbation expansion in the QCD (strong-interaction) coupling constant $\alpha_\mathrm{s} = g_\mathrm{c}^2/4\pi$, where $g_\mathrm{c}$ is the quark-gluon coupling constant.[3] The expansion has to be performed using a renormalization scheme. In particular, we should renormalize $\alpha_\mathrm{s}$ and the strange-quark mass $m_s$. The renormalized quantities depend on the selected value of the *renormalization point*, denoted by $\rho_\mathrm{R}$, which has the dimension of energy.[4] The renormalized constant $\alpha_\mathrm{s}$ decreases with the growth of the mean quark energy. Let us restrict ourselves to the first-order approximation valid for a sufficiently small $\alpha_\mathrm{s}$. In this case, the contributions of all three flavors to $\Omega_\mathrm{q}$ are additive,

$$\Omega_\mathrm{q} = \Omega_u(\mu_u) + \Omega_d(\mu_d) + \Omega_s(u_s). \tag{7.26}$$

The lowest-order formula for a plasma of ultrarelativistic $u$ and $d$ quarks is

$$\Omega_i/\Omega_i^{(0)} = 1 - 2\alpha_\mathrm{s}/\pi \qquad \text{for } i = u,\ d\,. \tag{7.27}$$

The contribution of the massive $s$ quark is more complicated (see, e.g., Farhi & Jaffe, 1984),

$$\begin{aligned}\frac{\Omega_s}{\Omega_s^{(0)}} &= \sqrt{1-y_s^2}\left(1-\frac{5}{2}y_s^2\right)+\frac{3}{2}\,y_s^4 L \\ &- \frac{2\alpha_\mathrm{s}}{\pi}\left[3\left(\sqrt{1-y_s^2}-y_s^2 L\right)^2-2(1-y_s^2)^2+3y_s^4\,(\ln y_s)^2\right] \\ &- \frac{\alpha_\mathrm{s}}{\pi}\ln\left(\frac{\rho_\mathrm{R}}{\mu_s}\right)\left[\sqrt{1-y_s^2}-y_s^4\,L\right]\,,\end{aligned} \tag{7.28}$$

where $y_s \equiv m_s c^2/\mu_s$ and $L = \ln\{(1+\sqrt{1-y_s^2}\,)/y_s\}$ . As we have already stressed, we use the renormalized values $\alpha_\mathrm{s}(\rho_\mathrm{R})$ and $m_s(\rho_\mathrm{R})$. To the lowest

[3] Two definitions of the QCD coupling constant are encountered in the literature. The first one, $\alpha_\mathrm{s} \equiv g_\mathrm{c}^2/4\pi$ (see, e.g., Baluni, 1978a,b) is adopted here. However, some authors define $\alpha_\mathrm{s} \equiv g_\mathrm{c}^2/16\pi$ (see, e.g., Freedman & McLerran, 1977).

[4] In principle, measurable quantities have to be independent of $\rho_\mathrm{R}$. In practice, especially if one works in the lowest-order approximation neglecting higher-order terms, the specific choice of $\rho_\mathrm{R}$ does matter. In applications to degenerate quark matter one usually puts $\rho_\mathrm{R} = \mu$, where $\mu$ is the quark chemical potential.

order, $\Omega_{\rm q}$ is independent of $\rho_{\rm R}$, provided $\alpha_{\rm s}(\rho_{\rm R})$ and $m_s(\rho_{\rm R})$ satisfy the lowest-order *renormalization group* equations

$$\frac{\partial m_s}{\partial \rho_{\rm R}} = -\frac{2\alpha_{\rm s}}{\pi}\frac{m_s c^2}{\rho_{\rm R}} + \mathcal{O}(\alpha_{\rm s}^2)\,, \qquad \frac{\partial \alpha_{\rm s}}{\partial \rho_{\rm R}} = \mathcal{O}(\alpha_{\rm s}^2)\,. \tag{7.29}$$

Using Eqs. (7.29), (7.28) and (7.27), one can check that

$$\frac{{\rm d}\Omega_{\rm q}}{{\rm d}\rho_{\rm R}} = \frac{\partial \Omega_{\rm q}}{\partial \rho_{\rm R}} + \frac{\partial m_s}{\partial \rho_{\rm R}}\frac{\partial \Omega_{\rm q}}{\partial m_s} + \frac{\partial \alpha_{\rm s}}{\partial \rho_{\rm R}}\frac{\partial \Omega_{\rm q}}{\partial \alpha_{\rm s}} = \mathcal{O}(\alpha_{\rm s}^2)\,. \tag{7.30}$$

Notice that putting $\rho_{\rm R} = \mu_s$ in Eq. (7.28) cancels the last term on the right-hand-side and removes explicit dependence of $\Omega_s$ on $\rho_{\rm R}$.

The renormalization of the strange-quark mass $m_s$ deserves a comment. Using the approximation linear in $\alpha_{\rm s}$ we deduce from Eq. (7.29) that $m_s > m_s^{\rm (c)}$. As $\alpha_{\rm s}$ is constant, the equation for $m_s(\rho_{\rm R})$ is easily integrated. The upper integration limit for the renormalization point, denoted by $\Lambda$, should satisfy $\Lambda \gg 1$ GeV; it belongs to quark asymptotic energies at which non-perturbative effects can be neglected. The lower integration limit should be set equal to the actual renormalization point $\rho_{\rm R} < \Lambda$. We get

$$m_s = m_s^{\rm (c)}\,(\Lambda/\rho_{\rm R})^{2\alpha_{\rm s}/\pi} > m_s^{\rm (c)}\,. \tag{7.31}$$

The number densities of all constituents can be calculated from the standard formula:

$$n_i = -\partial\Omega/\partial\mu_i \qquad \text{for } i = u, d, s, e\,. \tag{7.32}$$

This enables us to add two relations needed to close the system of equations for all four $\mu_i$ in an electrically neutral quark matter at a given value of $n_{\rm b}$:

$$\frac{1}{3}\,(n_u + n_d + n_s) = n_{\rm b}\,, \qquad n_e + \frac{1}{3}\,(n_d + n_s - 2n_u) = 0\,. \tag{7.33}$$

After solving this system [Eqs. (7.22) and (7.33)], one calculates the EOS of quark matter from the thermodynamic relations,

$$P(n_{\rm b}) = -\Omega = -\Omega_e - \Omega_{\rm q}\,, \tag{7.34a}$$

$$\mathcal{E}(n_{\rm b}) = \Omega + \sum_i n_i\mu_i = \Omega_e + \Omega_{\rm q} + \sum_i n_i\mu_i\,, \tag{7.34b}$$

where the sum is over $i = u, d, s$, and $e$.

The electron fraction in the quark matter, $n_e/n_{\rm b}$,vanishes for $m_s = 0$ but increases rapidly with increasing $m_s$. An approximate analytical formula for the electron fraction can be derived by expanding the difference $\mu_s - \mu_u$ in powers of $m_s$. Keeping the lowest power of $m_s$ for $\alpha_s \lesssim 0.2$ Haensel *et al.* (1986a) obtained

$$\frac{n_e}{n_b} \simeq \frac{0.002}{1-2\,\varepsilon}\left\{1-\frac{4}{3}\,\varepsilon-4\,\varepsilon\,\ln\left(\frac{2\mu_u}{\rho_R}\right)\right\}^3\left(\frac{n_0}{n_b}\right)^2\left(\frac{m_s c^2}{200\ \mathrm{MeV}}\right)^6, \quad (7.35)$$

where $\varepsilon = \alpha_s/\pi$. In particular, this estimate is valid for noninteracting quarks ($\alpha_s \to 0$). At $n_b \gg n_0$ and $m_s c^2 \sim 200$ MeV the electron fraction is too small ($n_e/n_b \ll 10^{-3}$) to affect the EOS.

Calculation of the EOS up to the second order in $\alpha_s$ (the fourth order in $g_c$) is more complicated. It involves the renormalization-group method for evaluating non-perturbative effects, responsible for the asymptotic freedom. Moreover, the plasmon and three-gluon coupling contributions should be added (Freedman & McLerran, 1977; Baluni, 1978b).[5] With the new terms included, the renormalized $\alpha_s(\rho_R)$ becomes density dependent. It exhibits the asymptotic freedom property $\alpha_s \to 0$ at $n_b \to \infty$ and grows with decreasing $n_b$. For simplicity, let us consider massless *uds* quarks. In this case, Baluni (1978b) and Freedman & McLerran (1977) calculated the thermodynamic potential $\Omega_q$ as a function of a unique quark chemical potential $\mu$ up to terms quadratic in $\alpha_s$. The second-order expression $\Omega_q^{(2)}$ depends on $\mu$ and $\rho_R$, $\Omega_q \simeq \Omega^{(2)}\left(\mu, \alpha_s(\rho_R), \rho_R\right)$, where the renormalized QCD coupling constant satisfies the renormalization-group equation, and the upperscript "(2)" reminds that terms $\sim \alpha_s^3(\rho_R)$ are neglected. Putting $\rho_R = \mu$ one has

$$\Omega_q^{(2)}(\mu) = -P(\mu) = -\frac{3\mu^4}{4\pi^2(\hbar c)^3}\left(1-2\varepsilon-3\varepsilon^2\ln\varepsilon-7.46\,\varepsilon^2\right), \quad (7.36)$$

where $\varepsilon = \alpha_s(\mu)/\pi$. A one-parameter EOS is obtained using

$$n_s = n_u = n_d = n_b(\mu) = -\frac{\partial\Omega_q(\mu)}{\partial\mu}, \quad \mathcal{E}(\mu) = \mu\, n_b(\mu) + \Omega_q(\mu). \quad (7.37)$$

Eliminating $\mu$ from Eqs. (7.36) and (7.37) one gets the standard form of the EOS, $P = P(\mathcal{E})$.

Extrapolating the expression for $\Omega_q$ to the densities in a neutron star core, where $\mu \lesssim$ GeV, one obtains the EOS of quark matter to be used for studying a phase transition between the baryon and quark phases. Let us parameterize the EOS of the baryon matter as $P = P^B(\mu_b)$, where $\mu_b$ is the baryon chemical potential. Let us define the baryon chemical potential in the quark matter

[5] One should keep in mind the definition of $\alpha_s$ to avoid confusion concerning the importance of the higher-order terms. The statement of Freedman & McLerran (1977) that "...the quantum-chromodynamic structure constant, $\alpha_s = g_c^2/16\pi$, is small, $\alpha_s < 1/4$...", does not sound convincing if one switches to the definition $\alpha_s \equiv g_c^2/4\pi$ used by Baluni (1978b).

(index Q) as $\mu_b^Q(P) = [\mathcal{E}^Q(P) + P]/n_b^Q(P)$. This is the energy associated with the change of the baryon number by one, at a constant volume and $T = 0$, assuming additionally electrical neutrality and weak-interaction equilibrium. The dependence of $\mu_b^Q$ on $P$ can be inverted to give the pressure in the quark matter as a function of $\mu_b$. The equilibrium phase transition takes place at such $P = P_{BQ}$, where the curves $P^B(\mu_b)$ and $P^Q(\mu_b)$ cross. The baryon-quark (B-Q) phase transition is of first order. It is accompanied by a density jump from $n_B$ at the baryon matter side to $n_Q$ at the quark matter side.

Early calculations indicated that the parameters of the B-Q phase transition are sensitive to the EOS of baryon matter. Actually, $P_{BQ}$ is uncomfortably model dependent, because the $P(\mu_b)$ curves intersect at a small angle. Typical transition values are $\mu_b \sim (1-2)$ GeV and $\rho \sim 10^{15}$ g cm$^{-3}$ (Baluni, 1978b; Morley & Kislinger, 1979). Freedman & McLerran (1977) obtained even lower transition densities using a simple "physical picture" where the confinement consists in a condensation of a gas of quarks into small baryon "droplets" of three quarks. As a rule, the B-Q phase transition is strongly first-order, with a large density jump so that stellar configurations with small quark cores could be unstable (see §7.9.5).

These results should be treated with caution. Typical quark energies in the B-Q transition region are *significantly less* than 1 GeV. Therefore, they are outside the weak-coupling regime, where the perturbative approach is valid. There is even a more fundamental doubt: It is well known that phase transitions in many-body systems cannot be described using perturbative approaches.

Confinement (or deconfinement) is a non-perturbative phenomenon. A simple phenomenological model of the confinement in the context of the hadronic structure was proposed as early as in 1974, when the QCD was still in its infancy. It was the famous MIT Bag Model suggested by Chodos *et al.* (1974). In this model non-perturbative effects responsible for the quark confinement are represented by the bag constant $\mathcal{B}$, the excess of the energy density of the QCD vacuum (where quarks can move freely) over the energy density of the ordinary vacuum outside hadrons. Quarks are assumed to be confined to a region of space called the "bag". They cannot escape from the bag because they become infinitely massive outside it. Using the MIT Bag Model and adjusting its parameters (the bag constant, quark masses, and $\alpha_s$) one can determine masses of mesons and baryons, as well as their magnetic moments.

In the MIT Bag Model a quark core in a neutron star is "a giant MIT Bag" (Baym & Chin, 1976). The energy density of the quark matter is therefore the sum of the bag constant, kinetic energy of quarks and their interaction energy,

$$\mathcal{E}_q = \mathcal{B} + \mathcal{E}_q^{(\rm kin)} + \mathcal{E}_q^{(\rm int)} , \tag{7.38}$$

where $\mathcal{E}_{\rm q}^{(\rm int)}$ should be calculated using the perturbative scheme of the QCD. The pressure is

$$P_{\rm q} = -\mathcal{B} + P_{\rm q}^{(\rm kin)} + P_{\rm q}^{(\rm int)} . \quad (7.39)$$

This expression contains the QCD coupling constant $\alpha_{\rm s}$ and the strange-quark mass $m_s$, both renormalized (while $u$ and $d$ quarks are treated as massless). The bag constant in Eqs. (7.38) and (7.39) can be treated as an effective constant pressure exerted by the normal vacuum on the QCD vacuum. The B-Q phase transition turns out to be strongly first-order, with a baryon density and mass density jump by a factor of ~two (!) (see Table 1 of Baym & Chin, 1976).

At the same $\alpha_{\rm s}$ and $m_s$ the bag constant $\mathcal{B}$ shifts $\mu^{\rm Q}(P)$ upwards, as compared to the $\mathcal{B} = 0$ case, shifting the B-Q transition to higher densities. As the $\mu^{\rm Q}(P)$ and $\mu^{\rm B}(P)$ curves intersect at a small angle, the increased phase-transition pressure $P_{\rm BQ}$ can easily exceed the maximum central pressure of stable neutron stars, $P_{\rm max}$. This circumstance was pointed out by Baym & Chin (1976). They assumed $\mathcal{B} = 56$ MeV fm$^{-3}$ and $\alpha_{\rm s} = 0.55$ (the "experimental values" of $\mathcal{B}$ and $\alpha_{\rm s}$ in the late 1970s) and concluded that it is unlikely to find the quark matter in stable neutron stars because typically $P_{\rm BQ} > P_{\rm max}$. The same conclusion has been obtained by other authors (see, e.g., Chapline & Nauenberg, 1976). Simultaneously, it has been realized that if $\mathcal{B}$ and especially $\alpha_{\rm s}$ were significantly smaller than the "experimental values" the B-Q phase transition would occur at a lower density (nuclear or even subnuclear). This would lead to the existence of "quark stars" consisting predominantly, or even exclusively, of quark matter (*obese "neutron" stars* of Brecher & Caporaso 1976, also see § 8.10). The bag-model calculations of the EOS of quark matter have been continued in the 1980s and 1990s. The condition $P_{\rm BQ} < P_{\rm max}$ and associated quark cores in neutron stars have been obtained only for sufficiently small $\mathcal{B}$ and $\alpha_{\rm s}$. Calculations have definitely excluded quark cores for the values of $\mathcal{B} \simeq 200$ MeV fm$^{-3}$ extracted from some QCD lattice calculations (see, e.g., Cleymans *et al.*, 1986; Karsch, 2002a,b).

The MIT Bag Model was constructed to describe the quark structure of hadrons. Therefore it has a built-in quark confinement. As $\mathcal{B}$ is assumed constant, the model does not exhibit the asymptotic freedom property of the QCD. In this model, quarks are, strictly speaking, confined within the bag even at $n_{\rm b} \to \infty$ and the pressure of the normal vacuum on the bag stays constant.

There have been numerous attempts to include the medium-dependence of the bag constant on a phenomenological level. One has to introduce the QCD vacuum dependent on $n_{\rm b}$ in order to determine a medium-dependent $\mathcal{B}$. One of the approaches utilizes the Quark-Meson Coupling Model, which describes nuclear matter as a system of non-overlapping MIT Bags bound by the exchange of scalar and vector mesons (see, e.g., Jin & Jennings, 1996; Müller & Jennings, 1997). Another approach is based on the Nambu–Jona-Lasinio model of the quark structure of hadrons (see Aguirre, 2003, and references therein).[6] In both cases one gets the bag constant which decreases with increasing density. Such approaches are certainly model-dependent.

Another group of models focused on the density dependence of quark masses (see, e.g., Chakrabarty *et al.*, 1989; Chakrabarty, 1991, and references therein).[7] In these models one mimics the strong non-perturbative interaction between quarks through the dependence of their *constituent* masses on the baryon density. In the limit $n_b \to \infty$ constituent masses tend to current quark masses.

Phenomenological models with density-dependent quark masses and the quark-quark interaction screened in dense medium can be constructed in such a way as to reproduce correct low-density and high-density limits (Dey *et al.*, 1998). The density dependence of quark masses can be introduced via a phenomenological scalar density-dependent potential, such that quark masses tend to current quark masses at $n_b \to \infty$. The quark-quark interaction is represented by a vector-interaction term which vanishes at $n_b \to \infty$ due to the screening in dense medium, reproducing the asymptotic freedom. Finally, there are other models of the quark matter based on different assumptions and techniques (e.g., Schmidt *et al.*, 1994; Blaschke *et al.*, 1998).

A brief and incomplete review of theories of the quark matter at $\rho \sim (10^{15} - 10^{16})$ g cm$^{-3}$, given above, demonstrates an overabundance of phenomenological models. This is a clear sign of fundamental difficulty in understanding the density regime which is far from the weak-coupling one. The quark energies at $\rho \sim (10^{15} - 10^{16})$ g cm$^{-3}$ are not high by the QCD standards, and we are far from the true asymptotic freedom regime (sometimes called the Asymptopia) which requires $\mu_f \gg 1$ GeV. In contrast to the Early Universe younger than a few microseconds, the interiors of compact stars do not belong to the Asymptopia.

A common feature of all models reviewed in the present section is the assumption of a gas-like quark matter. This is the natural state in the Asymptopia (except for the phenomena in the neighborhood of the Fermi surface, like *color superconductivity*, which have minor effect on the EOS, see below) but the situation at $\rho \sim 10^{15}$ g cm$^{-3}$ may be quite different. Another common feature is the presence of phenomenological ingredients which are not derived on a more fundamental ground. Additionally, as we stressed earlier, the value of $P_{BQ}$ is very sensitive to employed models of the confined (B) and deconfined

---

[6]Let us remind that the original Nambu–Jona-Lasinio model (Nambu & Jona-Lasinio, 1961a,b) was formulated for describing hadrons before the introduction of quarks into the particle physics.

[7]Let us recall that even in perturbative QCD calculations performed in the weak-coupling regime $\alpha_s$ and $m_s$ get some (weak) density dependence due to the renormalization. However, in standard calculations described earlier in this section one treats the renormalization point $\rho_R$ and the bag constant $\mathcal{B}$ as fixed parameters; this strongly simplifies the derivation of the EOS.

(Q) phases of matter. All in all, the degree of our ignorance on the existence of the B-Q phase transition at $\rho \sim (10^{15} - 10^{16})$ g cm$^{-3}$ is very high.

Notice that one should be careful in calculating the EOS of the quark matter using models with density-dependent $\mathcal{B}$, $\alpha_s$, $m_s$ (see, e.g., Peng *et al.*, 1999, 2000). The density dependence introduces additional terms in the pressure:

$$P_q = -\Omega_q + n_b \frac{\partial \Omega_q}{\partial \mathcal{B}} \left( \frac{\partial \mathcal{B}}{\partial n_b} \right)_{[\mu_j]} + n_b \sum_{f=u,d,s} \frac{\partial \Omega_q}{\partial m_f} \left( \frac{\partial m_f}{\partial n_b} \right)_{[\mu_j]} + n_b \frac{\partial \Omega_q}{\partial \alpha_s} \left( \frac{\partial \alpha_s}{\partial n_b} \right)_{[\mu_j]}, \tag{7.40}$$

where derivatives are calculated at fixed quark chemical potentials. The terms containing these derivatives have to be included in the EOS to make it consistent with the first law of thermodynamics.

Up to this point we have ignored possible superconductivity of quark matter. It can be shown that in the weak-coupling regime the quark matter is superconducting below some critical temperature (Barrois 1977; Alford *et al.* 1998; for review see Rajagopal & Wilczek 2000). This superconductivity arises because a weak attraction between two quarks (unavoidable for a quark pair) favors a rearrangement of quark states in the vicinity of the Fermi surface. A state composed of boson-like quark (Cooper) pairs is then energetically preferable over the Fermi-gas state. Quark pairs form a boson condensate. However, they cannot be color singlets, because one needs at least three quarks or a quark-antiquark pair to get a colorless hadron. Therefore, a condensate of quark pairs carries color, which leads to the names of *color conductivity* and *color currents*. The ground state of color superconductor is separated from the excited states by an energy gap, which can be as large as $\Delta \sim 100$ MeV (Alford *et al.*, 1998). However, the gap affects only quark states near the Fermi surface whereas bulk thermodynamic quantities such as the energy density or pressure result from the whole Fermi seas. Thus the effect of color superconductivity on the EOS is rather weak. For massless quarks in the weak-coupling regime one gets the following formula for the condensation energy density, released in a transition from a normal (n) state to a superconducting (s) state:

$$\Omega_s - \Omega_n = -\beta \, (\Delta/\mu)^2 \, |\Omega_n| \,, \tag{7.41}$$

where $\mu$ is the quark chemical potential, $\beta \sim 1$ is a numerical factor, and only the leading term in $\Delta/\mu$ is kept. The above estimate assumes $\mu \gtrsim 1$ GeV. Hence even at $\Delta \gtrsim 100$ MeV the relative modification of the EOS due to color superconductivity does not exceed a few percent (see, e.g., Alford *et al.* 1998; for a review see Rajagopal & Wilczek 2000; also see § 5.16). An extrapolation of Eq. (7.41) to $\rho \sim 10^{14} - 10^{15}$ g cm$^{-3}$, where $\mu \sim 300$–$600$ MeV and the implied effect could reach $\sim 6\%$, is risky because the weak-coupling description becomes invalid.

The models of the B-Q phase transition reviewed in this section suffer from many deficiencies. First, as a rule, the Q and B phases are described by different physical models. Therefore, they cannot give a consistent description of the phase transition. Some calculations based on a single physical model are not realistic in the low-density (nuclear) segment. Additionally, the phase of the quark matter is usually treated as a Fermi gas, whereas one should expect strong quark-quark correlations near the phase transition point. Ab initio QCD lattice calculations of the B-Q phase transition similar to those already performed for hot hadronic matter (see, e.g., Karsch 2002 and references therein) are still not available.

If the B-Q phase transition is first-order and the surface tension at the B-Q interface is not too high, both phases could coexist in a mixed-phase state in some range of pressures and densities (see § 7.6). The physical mechanism and astrophysical scenarios for the formation of the quark matter via the nucleation of the deconfined quark phase in a neutron star core are studied in § 7.8.3.

## 7.6. Mixed-phase state

In the previous sections we considered the thermodynamic equilibrium of matter assuming local electric neutrality. Let A and B be two phases of the matter. Fixing the baryon number density $n_{\rm b}$ we can calculate the minimum energy densities $\mathcal{E}^{\rm A}$ and $\mathcal{E}^{\rm B}$ of electrically neutral phases. Passing to the enthalpy per baryon $\mu_{\rm b} = (\mathcal{E} + P)/n_{\rm b}$, we can demonstrate that the thermodynamic equilibrium at any value of the pressure can be realized by a single (pure) phase of matter. Then the two phases can coexist only at one value of the pressure, where $\mu_{\rm b}^{\rm A}(P) = \mu_{\rm b}^{\rm B}(P)$.

Let $x$ be the fraction of baryons contained in the higher-density phase. For simplicity, the surface effects will be neglected. Therefore, the pressure $P$ is constant throughout the system. Moreover, we will assume the additivity of thermodynamic potentials of two phases. At a given pressure, the total enthalpy per baryon for a *mixture* of two phases is then

$$\mu_{\rm b} = x\mu_{\rm b}^{\rm B} + (1-x)\mu_{\rm b}^{\rm A} \,. \tag{7.42}$$

Both phases are assumed to be electrically neutral, reaching their own minimum values $\mu_{\rm b}^{\rm A}(P)$ and $\mu_{\rm b}^{\rm B}(P)$. However, we have $\mu_{\rm b}^{\rm B} > \mu_{\rm b}^{\rm A}$ for $P < P_0$ and $\mu_{\rm b}^{\rm B} < \mu_{\rm b}^{\rm A}$ for $P > P_0$. Therefore, $\mu_{\rm b}$ reaches minimum at $x = 0$ for $P < P_0$ and at $x = 1$ for $P > P_0$: thermodynamic equilibrium at $P \neq P_0$ corresponds to a single phase, and both phases coexist only at $P = P_0$. As the pressure is a monotonous function of the radial coordinate within the star, electrically neutral phases cannot mix in the stellar interior at thermodynamic equilibrium. The inclusion of the surface tension can only strengthen this conclusion.

Being obvious on macroscopic scales, electrical neutrality may be violated on microscopic scales. It can be too restrictive for a thermodynamic equilibrium of such a multi-component system like dense matter. Let us recall the case of the inner neutron star crust, particularly the Liquid Drop Model described in § 3.3.3. The nucleon component exists there in two phases. The denser phase consists of a nuclear matter and is positively charged. The less dense phase consists of a neutron gas. Neutrons in both phases are in equilibrium; their chemical potentials coincide (with an appropriate treatment of the interface between the two phases, § 3.3.3). The electron gas has nearly constant number density, so that the matter is electrically neutral only on average. In spite of a disadvantage resulting from the presence of Coulomb energy, a mixture of two nucleon phases is energetically preferable over a uniform electrically neutral phase, because of the nuclear energy gain connected with the nuclear matter binding.

Now consider a general first-order phase transition between the A and B phases of dense matter. As shown by Glendenning (1991, 1992), relaxing the microscopic charge neutrality condition makes a mixed-phase state energetically preferable provided the surface tension and Coulomb contributions are sufficiently small.

Let us focus on thermodynamic equilibrium of a multi-component and multi-phase dense matter, neglecting the Coulomb and surface contributions. The elementary constituents of the matter are hadrons ($h$) and leptons (electrons and muons). Hadrons are baryons, quarks, and strongly interacting meson condensates of § 7.3 and § 7.4. The energy densities in both phases (which are not necessarily in equilibrium) depend on number densities of the matter constituents in these phases,

$$\mathcal{E}^{\rm A} = \mathcal{E}^{\rm A}(\{n_h^{\rm A}\}, n_e^{\rm A}, n_\mu^{\rm A}) \ , \quad \mathcal{E}^{\rm B} = \mathcal{E}^{\rm B}(\{n_h^{\rm B}\}, n_e^{\rm B}, n_\mu^{\rm B}) \ . \tag{7.43}$$

As translational invariance may be broken within a phase, the number densities are actually the volume-averaged ones. We assume that the size of the region occupied by a non-uniform phase is larger than the characteristic length-scale of the non-uniformity. Then the volume averages within each phase are well defined.

The corresponding electric charge densities (in units of the elementary charge) and baryon number densities (in units of nucleon baryon charge) are

$$\rho_{\rm e}^{\rm A} = \sum_h n_h^{\rm A} q_h - n_e^{\rm A} - n_\mu^{\rm A} \ , \qquad n_{\rm b}^{\rm A} = \sum_h n_h^{\rm A} q_h \ , \tag{7.44a}$$

$$\rho_{\rm e}^{\rm B} = \sum_h n_h^{\rm B} q_h - n_e^{\rm B} - n_\mu^{\rm B} \ , \qquad n_{\rm b}^{\rm B} = \sum_h n_h^{\rm B} b_h \ , \tag{7.44b}$$

where $q_h$ and $b_h$ are, respectively, the electric and baryonic charges of a hadron $h$.

Let $\chi$ denote the fraction of volume occupied by phase B. The thermodynamic equilibrium of a mixture of phases A and B at a fixed average baryon density $n_\mathrm{b}$ can be calculated by minimizing the average energy density

$$\mathcal{E} = (1-\chi)\,\mathcal{E}^\mathrm{A} + \chi\,\mathcal{E}^\mathrm{B}\,, \tag{7.45}$$

under the condition

$$n_\mathrm{b} = (1-\chi)\,n_\mathrm{b}^\mathrm{A} + \chi\,n_\mathrm{b}^\mathrm{B}\,, \tag{7.46}$$

and under the constraint of average (macroscopic) electrical neutrality

$$\rho_\mathrm{e} = (1-\chi)\,\rho_\mathrm{e}^\mathrm{A} + \chi\,\rho_\mathrm{e}^\mathrm{B} = 0. \tag{7.47}$$

For example, let us consider a first-order phase transition associated with kaon condensation (§ 7.4). Let us assume that the characteristic length-scale of the hadron electric-charge inhomogeneities is much smaller than the electron and muon screening lengths (which is not always true – see, e.g., Norsen & Reddy 2001). Then the electron and muon number densities can be treated as uniform,

$$n_e^\mathrm{A} = n_e^\mathrm{B} \equiv n_e\,, \quad n_\mu^\mathrm{A} = n_\mu^\mathrm{B} \equiv n_\mu\,. \tag{7.48}$$

We have to determine the values of eight variables (four nucleon number densities, two lepton number densities, the kaon number density, and the volume fraction $\chi$) by minimizing $\mathcal{E}$, Eq. (7.45), under the conditions (7.46) and (7.47). This leads to a set of nonlinear equations relating thermodynamic variables; each of these equations has a clear physical meaning. Mechanical equilibrium between the two phases requires

$$P^\mathrm{A} = P^\mathrm{B}\,. \tag{7.49}$$

The strong interactions imply the equality of chemical potentials of nucleons in the two phases,

$$\mu_N^\mathrm{A} = \mu_N^\mathrm{B} = \mu_N \quad (N = n, p)\,. \tag{7.50}$$

Finally, the weak interactions involving hadrons and leptons lead to

$$\mu_n = \mu_p + \mu_e\,, \quad \mu_e = \mu_{K^-}\,, \quad \mu_\mu = \mu_e\,. \tag{7.51}$$

Together with Eqs. (7.46) and (7.47) we get eight equations for eight thermodynamic variables. The solution corresponds to thermodynamic equilibrium at a fixed $n_\mathrm{b}$ and under the constraint of macroscopic electrical neutrality. Let us stress that the equilibrium is calculated neglecting the Coulomb and surface (phase interface) effects and assuming a uniform lepton background. By analogy with the simplest model of the inner crust, described in § 3.4.1, we will use the terms the "bulk approximation" and "bulk equilibrium".

We will discuss the properties of a mixed phase in the bulk approximation using schematic (but generic) plots in Figs. 7.4 and 7.5. Let us first analyze the

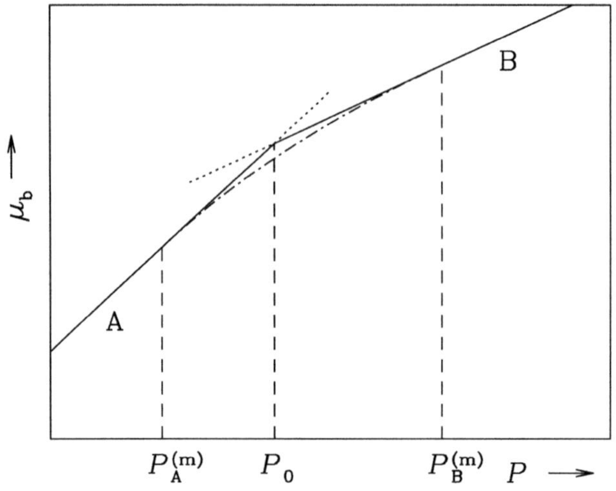


*Figure 7.4.* Baryon chemical potential $\mu_b$ versus pressure $P$ in the presence of an equilibrium first-order phase transition between phases A and B at $P = P_0$. Phase A, if stable with respect to the phase transition into phase B, is represented by the solid line; its dotted continuation corresponds to an *overcompressed* state, metastable with respect to the transformation into phase B. Analogous notation is used for phase B, where the dotted segment corresponds to a metastable *undercompressed* state. The mixed phase AB ($P_A^{(m)} < P < P_B^{(m)}$) is shown by the dot-and-dash line.

character of the bulk equilibrium as a function of the pressure (Fig. 7.4). For $P < P_A^{(m)}$, the equilibrium is realized by the pure phase A. For $P_A^{(m)} < P < P_B^{(m)}$, the equilibrium corresponds to a mixed state AB. The volume fraction occupied by phase B increases monotonously with $P$, from zero at $P = P_A^{(m)}$ to one at $P = P_B^{(m)}$. For $P > P_B^{(m)}$ we have the pure phase B.

A mixed-phase state, where kaon-condensed matter coexists with baryon matter, was studied by Glendenning & Schaffner-Bielich (1998, 1999); and Norsen & Reddy (2001). Models for a mixed phase of deconfined quark matter coexisting with baryon matter were constructed by Heiselberg *et al.* (1993) and Glendenning & Pei (1995). The importance of the phase interface (the surface tension and curvature energy) for creating a mixed phase was emphasized by Heiselberg *et al.* (1993); Christiansen & Glendenning (1997); Christiansen *et al.* (2000); and Norsen & Reddy (2001). The effect of electric charge screening and surface tension on the mixed phase with kaon condensate was studied by Maruyama *et al.* (2006).

A mixed-phase state affects the EOS as visualized in Fig. 7.5. For the sake of comparison, we also show a standard equilibrium first-order transition between pure phases A and B. For this transition, the densities $n_A < n_b < n_B$

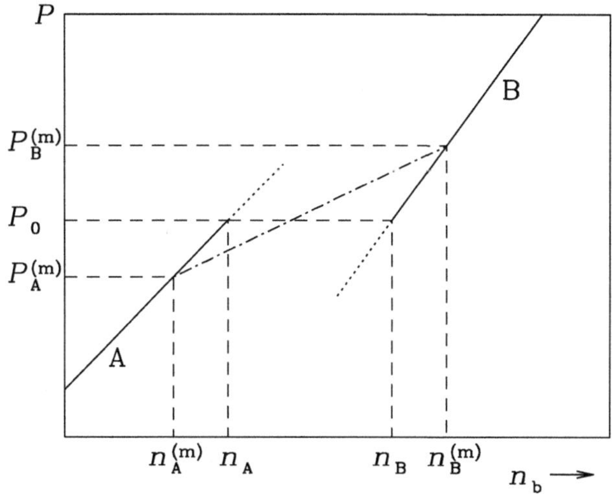


*Figure 7.5.* Same as in Fig. 7.4 but for $P$ versus $n_b$. The mixed phase AB ($P_A^{(m)} < P < P_B^{(m)}$) is represented by the dot-and-dash line. Equilibrium first-order phase transition between pure phases A and B takes place at $P = P_0$. It is accompanied by the density jump $n_A \to n_B$. For further explanations see the text.

could not exist in the stellar interior because $P$ should be monotonous there. On the contrary, the mixed-phase layer of density $n_A^{(m)} < n_b < n_B^{(m)}$ can exist in the star, with the pressure increasing from $P_A^{(m)}$ at the top of the layer to $P_B^{(m)}$ at the bottom. Mixing of the phases A and B softens the EOS as compared to the EOS of the pure phase A, but the softening is weaker than in the limiting case of the transition between pure phases A and B.

The surface and Coulomb effects increase $\mathcal{E}(n_b)$ and $\mu_b(P)$. They affect the size and shape of structures forming the mixed-phase layer. For a periodic structure, the virial theorem of §§ 3.3.3 and 3.4.2 tells us that the surface contribution is twice the Coulomb one. Both contributions push up the value of $\mu_b^{(m)}(P)$. They are especially important at the edges of the mixed-phase region, where the droplets of one phase are small within the dominating phase. It is clear that these effects increase $P_A^{(m)}$ and decrease $P_B^{(m)}$, narrowing the mixed-phase layer in the stellar interior. If the surface and Coulomb contributions are sufficiently large they can entirely remove the mixed phase (because the difference in $\mu_b(P)$ of the pure and mixed phases is usually small). In particular, this happens provided the surface tension $\sigma$ exceeds some critical value $\sigma_{\rm crit}$ so that $\mu_b^{(m)}(P) > \mu_b^A(P)$ for $P < P_0$ and $\mu_b^{(m)}(P) > \mu_b^B(P)$ for $P > P_0$. Heiselberg *et al.* (1993) obtained $\sigma_{\rm crit} \simeq 70$ MeV fm$^{-2}$ for a transition from nucleon to quark matter. The actual value of the surface tension for quark matter droplets in baryonic medium is very poorly known, $\sigma = (10\text{–}100)$ MeV fm$^{-2}$. In the case of kaon condensate, an inclusion of electric charge screening and of surface tension makes the EOS allowing for a mixed phase very similar to that with a first order phase transition between two pure phases (Maruyama *et al.*, 2006).

## 7.7. Solid cores of neutron stars

### 7.7.1 Physical origins and models

The history of theories of hypothetical crystalline neutron star cores, sketched in § 7.2.4, started with the idea of crystallization of neutron matter owing to the strong short-range repulsion of the neutron-neutron interaction. This repulsion was often represented by a hard-core one, so that $v_{ij} = \infty$ for $r_{ij} \leq r_{\rm core}$. The hard-core repulsion solidifies the neutron matter at $n_{\rm b} \sim (4\pi r_{\rm core}^3/3)^{-1}$; in this case neutrons become localized by the infinite potential walls. Other arguments in favor of crystallization were taken from the physics of noble gases. The short-range repulsion between two rare-gas atoms is well represented by the repulsive term of the Lennard-Jones potential $v_{ij} \propto (r_{ij})^{-12}$. It is well known that noble gases solidify at sufficiently high pressures. In the beginning of the 1970s some authors applied the so-called *law of corresponding states*, first suggested by Anderson & Palmer (1971), to deduce the solidification density of neutron matter by scaling the experimental results for $^3$He.

However, the short-range neutron-neutron repulsion occurs due to the exchange of vector mesons (see § 5.6). Therefore, it is of the Yukawa form $v_{ij} \propto \exp(-\mu r_{ij})/\ (\mu r_{ij})$, often called a "soft-core repulsion". In this case the hard-core and noble-gas arguments are invalid. Moreover, many-body calculation of the liquid-solid transition in dense neutron matter is a tremendous computational challenge. One has to calculate $\mathcal{E}^{(\rm liq)}$ and $\mathcal{E}^{(\rm sol)}$ with very high precision in order to find the density above which $\mathcal{E}^{(\rm liq)} > \mathcal{E}^{(\rm sol)}$. By 1974, the precision was still insufficient. Some calculations gave no solidification at all, while several authors obtained solidification at $(5 - 30) \times 10^{14}$ g cm$^{-3}$ (the review of the early, alas, unreliable calculations is given by Canuto 1975). The consensus was reached in the second-half of the 1970s, when the precision of many-body calculations became sufficiently high: for realistic $v_{ij}$ neutron matter does not solidify at densities expected in neutron star cores.

Another possibility of getting a solid structure is to increase the energy from the medium-range attractive tensor component (§ 5.5.1). This mechanism, related to the $\pi^0$-condensation in neutron matter (§ 7.3), was analyzed by Pandharipande & Smith (1975). The amplification of an already strong tensor interaction component was induced by the coupling of nucleons and pions to $\Delta$-isobars excited in intermediate nucleon states. Strong coupling at the $n\Delta\pi$

vertex amplified the tensor component of the many-body Hamiltonian, $\hat{V}_{\rm T}$. Its attractive contribution became sufficient to arrange neutrons in a cubic lattice and correlate their spins with crystal planes. The spin-quantization axis was chosen to be parallel to the cube side. All neutrons in a given plane perpendicular to the quantization axis had the same spin projection. Neutrons in adjacent planes had antiparallel spins. Such a spin and space structure gave a large attractive contribution to $\mathcal{E}$ from the tensor interaction component

$$\hat{V}_{\rm T} = \sum_{i<j} v_{\rm T}(r_{ij})\left(3\boldsymbol{\sigma}_j \cdot \boldsymbol{n}_{ij}\boldsymbol{\sigma}_i \cdot \boldsymbol{n}_{ij} - \boldsymbol{\sigma}_i \cdot \boldsymbol{\sigma}_j\right) , \tag{7.52}$$

where $\boldsymbol{n}_{ij} = \boldsymbol{r}_{ij}/r_{ij}$ (see § 5.5.1). The value of $\mathcal{E}^{(\rm sol)}$ was then obtained variationally by minimizing the expectation value of the Hamiltonian within a family of neutron wave functions, localized around the crystal lattice sites. Those wave functions, with alternating spin polarization in adjacent crystal planes, took full advantage of the tensor attraction (in contrast to the disordered liquid state, in which the first-order tensor contributions average to zero). In this way Pandharipande and Smith obtained $\mathcal{E}^{(\rm sol)} < \mathcal{E}^{(\rm liq)}$ for $\rho \gtrsim 3\rho_0$. The first-order liquid-solid phase transition was associated with a density jump, by $\sim$20%, which considerably softened the EOS. Because of its periodic spatial and spin structure, the ground state wave function led to a nonzero expectation value of the $\pi^0$-condensate. The neutral pion condensate had a standing-wave structure with a characteristic wave number $k_{\pi^0}$ (see § 7.3).

The first-order contribution $\hat{V}_{\rm T}^{\rm OPE}$ of the one-pion exchange potential (OPEP, see § 5.6) vanishes in an isotropic fluid. However, as shown by Takatsuka & Tamagaki (1976, 1977), it can dominate in a state of the so called alternating-spin layers (ALS, see below; also see Takatsuka *et al.*, 1978; Matsui *et al.*, 1979). In the ALS state, neutrons and protons are localized in parallel planes (one dimensional, 1D, localization). Let the planes be perpendicular to the spin quantization $z$-axis. Neutrons localized in one plane have spins aligned with the $z$-axis and protons in this plane have opposite spins, so that the isospin-spin states are $(n\uparrow)$ and $(p\downarrow)$. In adjacent planes one has $(n\downarrow)$ and $(p\uparrow)$. As we are dealing with the 1D localization (nucleons can freely move in the $xy$ plane), matter behaves as a *smectics A* phase in the nomenclature of liquid crystals (see § 3.7.2). If the binding gain due the tensor attraction overcomes the binding loss due to the kinetic energy increase (resulting from 1D localization), the dense matter undergoes the phase transition to the ALS-phase.[8] Calculations of the ALS structure of dense matter were further developed by the Kyoto

[8]The idea that a strong tensor component of the OPEP can lead to a solid-like structure of nucleon matter with a periodic spin-isospin ordering was first proposed by Calogero *et al.* (1973); also see Calogero *et al.* (1975).

group, with the emphasis on possible signatures of this phase in neutron stars (see Takatsuka *et al.* 1993 for review).

Moreover, Kutschera & Wójcik (1989, 1990) suggested that, for a sufficiently low proton fraction $x_p \lesssim 0.05$, nuclear matter could be unstable with respect to proton localization accompanied by a modulation of neutron density. At such low $x_p$, protons behave as impurities in neutron matter interacting mainly with the neutron background. While localization increases the proton kinetic energy, it increases also the proton binding in neutron matter by locking protons in potential wells corresponding to minima of the neutron density. Further work pointed out an analogy with *polarons*, well known in the physics of condensed matter.[9] Protons seem to behave as nuclear *polarons* and form a lattice at high densities (Kutschera & Wójcik, 1993, 1995). Of course, the proton localization occurs only for those model EOSs which predict the decrease of the proton fraction $x_p$ with growing $\rho$ at high densities. Moreover, calculations of $\mathcal{E}^{(\mathrm{sol})}$ involve many approximations. Proton localization takes place if the difference $\mathcal{E}^{(\mathrm{sol})} - \mathcal{E}^{(\mathrm{liq})}$ becomes negative for $\rho > \rho_{\mathrm{loc}}$. Calculated difference turns out to be so small that numerical results should be taken with a grain of salt. Recent calculations, performed for several models of the $npe\mu$ matter [with decreasing $x_\mathrm{p}$ at high $\rho$ and vanishing protons at still higher $\rho$!] give $\rho_{\mathrm{loc}} = (3 - 6)\ \rho_0$ (Kutschera *et al.*, 2002). The idea of localization of impurities was extended to the hyperonic matter by Perez Garcia *et al.* (2002).

In the mixed-phase state of dense matter ($P_\mathrm{A}^{(\mathrm{m})} < P < P_\mathrm{B}^{(\mathrm{m})}$, see § 7.6) each of the phases is electrically charged. To minimize the sum of the Coulomb and surface energies, the less dense and more dense phases (A and B) can be distributed into a periodic structure (see Glendenning 2001 and references therein). For a low volume fraction $\chi \ll 1$ of phase B, spherical droplets of this phase can form a cubic [most probably, body-centered cubic (bcc)] lattice immersed in the background of phase A. With increasing $\chi$, the three dimensional (3D) cubic lattice is thought to be replaced by the two dimensional columnar phase of B-rods immersed in phase A, analogous to the 2N phase in the bottom of the neutron star crust (§ 3.4.2). The difference from the crust is that the hadron component of phase A is electrically charged. For $\chi \sim 1/2$, 1D alternating A and B slabs appear; at higher $\chi$ they are replaced by the columnar phase of A-rods immersed in the liquid of phase B. At $\chi \to 1$, droplets of phase A form a bcc crystal in such a liquid.

[9] The formation of polarons in terrestrial crystals results from the electron-phonon interaction, which can lead to a strong increase of the electron effective mass. The strongest effect occurs in ionic crystals because of the strong Coulomb attraction between ions and electrons. Electrons can become self-trapped in local deformations of the ion lattice (see, e.g., Chapter 10 of Kittel 1986). The instability of crystals against density perturbations in terrestrial solids corresponds to the instability of the neutron component in the $npe\mu$ matter.

### 7.7.2 Elastic shear moduli

Consider shear moduli of a $\pi^0$-condensed 3D neutron crystal, proposed by Pandharipande & Smith (1975). A crude estimate of the $c_{44}$ shear modulus (see § 3.7.1), quoted by Pandharipande *et al.* (1976), is $c_{44} \sim 10^{35}$ erg cm$^{-3}$ at $\rho \sim 10^{15}$ g cm$^{-3}$. It is more than six orders of magnitude larger (!) than the shear modulus of the crust at $\rho \sim 10^{13}$ g cm$^{-3}$ (§ 3.7.1). Takatsuka & Tamagaki (1988b) evaluated the *average* shear modulus of the 1D ALS liquid crystal in the $\pi^0$-condensed matter at $\rho \sim 1.3 \times 10^{15}$ g cm$^{-3}$ and obtained $\sim 5 \times 10^{35}$ erg cm$^{-3}$. Let us stress that nucleon localization, crystal structure, and shear strain in such crystals are produced by *strong interactions*, in contrast to the neutron star crust, §§ 3.7.1 and 3.7.2, where they are due to *Coulomb interaction*.

In a mixed-phase state of dense matter ($P_{\rm A}^{(\rm m)} < P < P_{\rm B}^{(\rm m)}$, see § 7.6) each phase is electrically charged. Therefore, the analogy with crustal matter (§ 3.7.2) is not complete. Nevertheless, we expect that the elastic shear moduli are of the same order of magnitude as the Coulomb contribution $\mathcal{E}_{\rm Coul}$ to the energy density in the crustal matter. Let us remind that in an equilibrium state $\mathcal{E}_{\rm Coul} = \frac{1}{2}\,\mathcal{E}_{\rm surf}$ (see §§ 3.3.3 and 3.4.2). If the denser phase occupies a sizable fraction of the volume, then the surface contribution to the energy density can be estimated as the surface tension $\sigma$ divided by the surface thickness $\sim 1$ fm. This prescription gives $\mathcal{E}_{\rm surf} \simeq 30\,\sigma_{30}$ MeV fm$^{-3}$ and $\mathcal{E}_{\rm Coul} \simeq 2 \times 10^{34}\,\sigma_{30}$ erg cm$^{-3}$, where $\sigma_{30} \equiv \sigma/30$ MeV fm$^{-2}$. Therefore, we expect the elastic moduli $\sim 10^{34}$ erg cm$^{-3}$ for a mixed phase of quarks and nucleons and $\sim 10^{33}$ erg cm$^{-3}$ for a mixed phase of kaon condensate and nucleons. This is, respectively, one order and two orders of magnitude smaller than for the $\pi^0$-condensed solid.

The shear modulus of a solid formed by localized protons can be estimated from elementary consideration. The Coulomb energy per proton is roughly $e^2/r_{pp}$, where $r_{pp}$ is a mean interproton distance. Then the Coulomb energy density is $\mathcal{E}_{\rm Coul} \sim n_p e^2/r_{pp} \sim 10^{31}\ (n_{\rm b}/4n_0)^{4/3}(x_p/0.01)^{4/3}$ erg cm$^{-3}$. The elastic shear moduli for the matter with localized protons are, therefore, several orders of magnitude smaller than for the $\pi^0$-solid or the mixed-phase core.

A recent brief review of shear moduli of exotic neutron star cores and their impact on the elastic-strain support of deformations of pulsars can be found in Owen (2005).

## 7.8. Nucleation of exotic high-density phase

A new exotic phase may form during the neutron star formation and evolution. For example, an accretion of matter or pulsar spindown induce a gradual increase of the central density which may trigger the appearance of the exotic phase.

In §§7.3, 7.4, and 7.5 we have considered *equilibrium phase transitions* which take place in thermodynamic equilibrium. The equilibrium is realized only if the timescale $t_{\rm evol}$ of the local pressure variation is much longer than the timescale $t_{\rm react}$ of reactions which form and equilibrate the new phase. If the evolution timescale in the compressed matter ($\dot{P} > 0$) can be estimated as $t_{\rm evol} \sim P/\dot{P}$.

If the equilibrium phase transition is of second order and takes place at $P = P_0$, phase B appears, under equilibrium conditions, after the central pressure exceeds $P_0$. The actual scenario can be different. While the formation of phase B is driven by strong interactions, the slowness of weak interactions ($t_{\rm weak} > t_{\rm evol}$) may result in deviation from thermodynamic equilibrium. Then second-order phase transition can occur at some $P = P'_0$, slightly different from $P_0$. The formation of phase B is quasistatic, on a timescale $t_{\rm evol}$.

If the phase transition is of first order, the scenario is essentially different. The formation of phase B does not need to be quasistatic and can trigger a neutron star corequake. The kinetics of the first-order phase transition is complicated. Additional uncertainties arise from uncertainties in the physics of dense matter. In what follows we will discuss the models of the *nucleation* of the phase B in the metastable phase A. The nucleation is concerned with fluctuations of parameters (such as the local density, or the number of particles in a metastable drop of phase B which trigger the phase transition). Two simplest versions of nucleation theory refer to two extreme cases. In the classical regime, the temperature is assumed to be sufficiently high to trigger the phase transition by a thermal fluctuation. Thermodynamic aspects of such a model are described, for instance, in §162 of Landau & Lifshitz (1993), whereas its kinetic aspects are presented in §§99,100 of Lifshitz & Pitaevskiĭ (1981); also see Langer (1969). In the quantum regime, below a characteristic temperature $T_{\rm Q}$, thermal fluctuations are negligible compared to quantum ones. Quantum fluctuations initiate a phase transition via the quantum tunneling effect (Lifshitz & Kagan, 1972).

Let us consider fluctuations of density and composition at fixed $P$ and $T$. The relevant thermodynamic potential is the Gibbs free energy $G$. In the static approximation for $P > P_0$ the bulk phase B is preferable over phase A. On fluctuation timescales ($\sim 10^{-23}$ s, which is the characteristic timescale of strong interactions), weak interaction processes are frozen. This can shift the threshold condition, compared to the static one, leading to $P'_0 \neq P_0$. Moreover, this slowness can modify the properties of phases A and B, involved in the fluctuations. For the sake of simplicity, we will neglect these effects and put $P'_0 = P_0$. For a central stellar pressure higher than $P_0$, phase A is metastable within the neutron star core with $P > P_0$, whereas phase B is stable. However, the phase transition is hindered by the effects of the surface and Coulomb energies. Only at $P = P_{\rm crit}$ phase A becomes unstable against conversion into

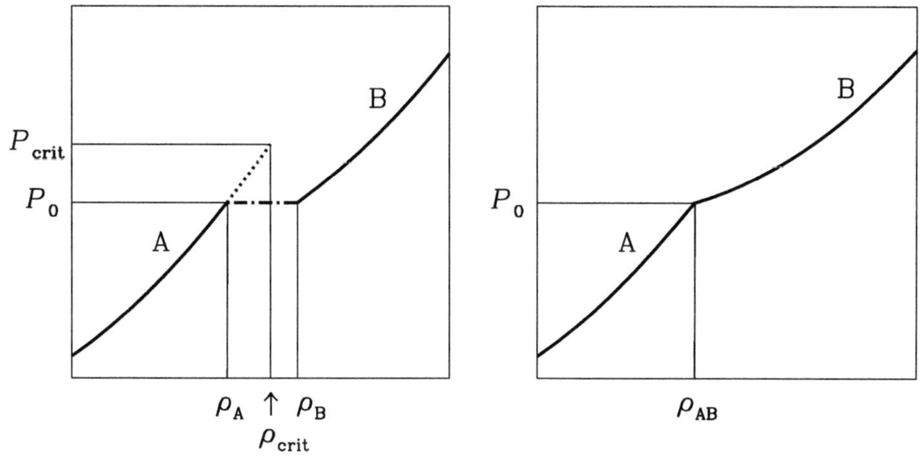


*Figure 7.6.* *Left*: First-order phase transition. The solid line denotes stable phases A and B. Equilibrium phase transition AB occurs at $P = P_0$, with the density jump from $\rho_A$ to $\rho_B$. The dotted line denotes a metastable phase A. At $P = P_{crit}$ this phase becomes unstable with respect to $A \to B$ transition. *Right*: Second-order phase transition at $\rho_{AB}$. No density jump, no metastability.

phase B (see Fig. 7.6). Of course, the formation of a droplet of phase B in the medium of phase A is possible either via thermal or via quantum fluctuations even at $P < P_0$. However, the Gibbs free energy associated with the droplet is $G^*(A_{drop}) > 0$, and such droplets always convert back to phase A. For $P > P_0$ the bulk phase B is stable. The probability of a local formation of a droplet of phase B via a fluctuation decreases very strongly with increasing the number of baryons $A_{drop}$ in the droplet. For small $A_{drop}$, however, the positive contribution of the surface energy to $G^*$ prevails over the gain in the bulk binding, $G^*(A_{drop}) > 0$. This makes the droplet unstable with respect to the reconversion to phase A. However, at some value $A_{drop} = A_{crit}$, the energy excess due to the drop appearance vanishes, $G^*(A_{crit}) = 0$. These "supercritical droplets" with $A_{drop} > A_{crit}$ grow spontaneously, destabilizing the metastable phase A and inducing the phase transition. The transition from a configuration of the metastable pure phase A to a configuration of the stable phase B in the stellar core goes via a combination of hydrodynamic and kinetic processes.

The basic parameters of the transition can be related to thermodynamic properties of both phases and their interface. For simplicity, we neglect Coulomb forces and treat the densities in each phase as constant. We restrict ourselves to fluctuations of $A_{drop} \equiv A$ and put $P_0' = P_0$. Let a droplet of phase B be spherical and have the baryon density $n_{\rm b}^{\rm B}$. Then $A = \frac{4}{3}\pi R^3 n_{\rm b}^{\rm B}$ implies

$$R = r_0 A^{\frac{1}{3}} f_{\rm B}^{-\frac{1}{3}} \; , \quad f_{\rm B} = \frac{n_{\rm b}^{\rm B}}{n_0} \; , \quad r_0 = \left(\frac{4\pi}{3n_0}\right)^{\frac{1}{3}} = 1.14 \text{ fm} \; . \tag{7.53}$$

At $P > P_0$ the bulk part of $G^*$ (proportional to $A_{\rm drop}$) is negative,

$$G^*_{\rm bulk} = -A\Delta\mu \; , \qquad \Delta\mu \equiv \mu^{\rm A}(P) - \mu_{\rm b}^{\rm B}(P) > 0 \; , \tag{7.54}$$

whereas the surface contribution is positive,

$$G^*_{\rm surf} = 4\pi r_0^2 A^{\frac{2}{3}} f_{\rm B}^{-\frac{2}{3}} \sigma \; , \tag{7.55}$$

where $\sigma$ is the surface tension. Therefore, $A_{\rm crit}$

$$A_{\rm crit} = \frac{1}{f_{\rm B}^2} \left(\frac{4\pi r_0^2 \sigma}{\Delta\mu}\right)^3 \; . \tag{7.56}$$

Another quantity of interest is the maximum value of $G^*(A)$. It is the height of the barrier to be crossed. This value is reached at $A = A_{\rm M}$:

$$A_{\rm M} = \frac{8}{27} A_{\rm crit} \; , \quad G^*_{\rm max} = G^*(A_{\rm M}) = 4\pi r_0^2 \, \sigma \, \frac{28}{81} \, A_{\rm crit}^{\frac{2}{3}} \, f_{\rm B}^{-\frac{2}{3}} \; . \tag{7.57}$$

The probability distribution $\mathcal{P}$ of fluctuations with a Gibbs free energy excess $G^*$ is

$$\mathcal{P}(G^*) = \frac{1}{k_{\rm B}T} \exp\left(-\frac{G^*}{k_{\rm B}T}\right) \; , \qquad \int_0^\infty \mathcal{P}(G^*) \; {\rm d}G^* = 1 \; . \tag{7.58}$$

The nucleation rate can be calculated in the quasiclassical approximation (Landau & Lifshitz, 1976). For fluctuations of $A_{\rm drop}$, the droplet radius $R$ can be treated as a generalized coordinate. The classical Lagrangian of the droplet is then

$$\mathcal{L}(R, \dot{R}) = \frac{1}{2} \mathcal{M}_{\rm eff}(R) \dot{R}^2 - G(R) \; , \tag{7.59}$$

where the potential energy is the Gibbs free energy excess expressed as a function of $R$, and $\mathcal{M}_{\rm eff}$ is the effective droplet mass. The effective mass can be calculated explicitly from the kinetic theory; the flow of matter is induced by the motion of the droplet surface owing to a fluctuation of $A_{\rm drop}$ (see, e.g., Lifshitz & Kagan 1972; for application to pion condensation see Haensel & Schaeffer 1982).

Let us assume that fluctuations are driven by strong interactions with characteristic timescale $\tau_{\rm s} \sim 10^{-23}$ s $\sim 1$ fm$/c$. Then the lifetime of a metastable state, equal to the time of the formation of a single supercritical drop, is

$$\tau = {\rm e}^W \tau_{\rm s} \; , \tag{7.60}$$

where $e^W$ is the *retardation factor* owing to the Gibbs free energy barrier. This factor can be calculated in the quasiclassical approximation,

$$e^W = \int_0^\infty dG^* \, \mathcal{P}(G^*) \exp\left[-\frac{2}{\hbar}\int_{R_1}^{R_2} \sqrt{2\mathcal{M}_{\rm eff}(R)\,[G(R) - G^*]}\, dR\right] , \tag{7.61}$$

where $R_1$ and $R_2$ are the classical turning points at which $G(R) = G^*$. The integral over $G^*$ can be evaluated using the saddle point approximation.

The expressions for $W$ are simplified in the thermal and quantum regimes. These regimes are separated by a characteristic temperature $T_{\rm Q}$, which depends on $P$. At $T \gg T_{\rm Q}$ the quantum tunneling is negligible and $W = G_{\rm max}/k_{\rm B}T$. On the contrary, at $T \ll T_{\rm Q}$ the quantum tunneling dominates and $W$ can be approximated by its $T = 0$ value which corresponds to $G^* = 0$:

$$W \simeq -\frac{2}{\hbar}\int_{R_1}^{R_2} \sqrt{2\mathcal{M}_{\rm eff}(R)G(R)}\, dR\,. \tag{7.62}$$

These ideas have been applied for describing the nucleation of exotic phases (pion condensate, kaon condensate, quark matter) in a supercompressed neutron star core. However, the neutron star matter is more complicated than liquid helium for which the original Lifshitz-Kagan theory was developed. Additional complications are introduced by the multicomponent structure of the matter, by the slowness of weak interaction processes, and by Coulomb interactions between charged particles.

### 7.8.1 Pion condensation

A model for the nucleation of pion condensate in a supercompressed neutron star core was proposed by Haensel & Schaeffer (1982), who discussed also possible astrophysical consequences of this phenomenon. A normal neutron star matter was approximated by a pure neutron matter. Different nucleation paths were considered; the optimal path was found to depend on the temperature of the metastable core. One possibility is to form a critical droplet of the pion-condensed phase via a density fluctuation. Another way is to grow a small pion-condensed drop to the critical size by a particle number fluctuation (as suggested in the original Lifshitz-Kagan model). The system evolves along the most probable path, which turns out to be a fluctuation of $A_{\rm drop}$.

Let us describe the main results in a way independent of the model for the pion-condensed phase. The value of $A_{\rm crit}$ strongly decreases with the growth of the overcompression $\Delta P_{\rm over} \equiv P - P_0$. Thus, the lifetime of an overcompressed state decreases rapidly with increasing $\Delta P_{\rm over}$. The nucleation occurs when $t_{\rm nucl}(P_{\rm c}) \sim t_{\rm evol}(P_{\rm c})$. Because the $t_{\rm nucl}(P)$ dependence is steep, the critical value of $P = P_{\rm nucl}$ for triggering the phase transition can be determined with a rather good precision. Haensel & Schaeffer (1982) and Muto & Tatsumi

(1990) applied this approach to the nucleation of pion-condensate in the supercompressed neutron star core made of a purely neutron matter. The authors used the expression for the surface tension $\sigma$ derived by Baym *et al.* (1971a). It gives $\sigma$ as a function of densities and energies of the drop and ambient matter. The original expression was derived to describe the surface tension of atomic nuclei in a neutron gas of the inner neutron star crust (see §§ 3.3.3). For the pion condensation problem, a nucleus is replaced by a pion-condensed droplet. This is reasonable because the $\pi^-$ condensation increases the proton fraction compared to the ordinary neutron star matter (§ 7.3). On the other hand, the neutron gas is replaced by the ordinary neutron star matter, which is also a good approximation as long as neutrons are most abundant particles. Crucial for the astrophysical applications is the overpressure $\Delta P_{\rm over}$ needed for the nucleation in the quantum regime. As mentioned before, the pressure $P_{\rm nucl}$ (or the corresponding density $\rho_{\rm nucl}$) is determined rather well. Then one can obtain the condition for the nucleation of pion condensate in the neutron star core during the Universe age, $t_{\rm Hubble} \simeq 1.5 \times 10^{10}$ years. Muto & Tatsumi (1990) obtain the required overcompression $\Delta P_{\rm over}/P_0 = 0.02$ or 0.05, depending on employed model. Notice that putting $t_{\rm nucl} = \tau$, where $\tau$ is given Eq. (7.60), we consider the nucleation via the formation of a single drop.

Under typical conditions in the neutron star core ($T \lesssim 10^9$ K) the nucleation proceeds via the quantum tunneling through the energy barrier. At much higher $T$ the thermal effects increase the nucleation rate through thermally excited droplet states. On the other hand, the growth of $T$ increases $P_{\rm crit}$. For a newly born neutron star with $T \gtrsim 10^{10}$ K the nucleation proceeds in the thermal (classical) regime (Haensel & Schaeffer, 1982).

Let us make one comment on the importance of weak interactions. Phase transitions are usually analyzed assuming the equilibrium with respect to weak interactions (§ 7.3). However, weak interaction processes are slow and lepton numbers can be frozen during the nucleation. The freezing, as well as the inclusion of the Coulomb energy in the droplet formation will decrease the nucleation rate compared to the rate obtained in beta-equilibrium. This will increase the overpressure $\Delta P_{\rm over}$ required for the nucleation.

### 7.8.2 Kaon condensation

The formation of a droplet of kaon condensate is connected with the production of strangeness. In this case the nucleation should involve weak interaction processes. For kaon condensation in the $npe\mu$ matter, whose equilibrium properties were described in § 7.4, the strange kaon-phase has to nucleate in the non-strange medium. In this respect kaon condensation is different from pion condensation or (B-Q) phase transition from baryon matter to quark matter, where the first step is a strong-interaction process. For example, the first step in the B-Q phase transition is the formation of a droplet of two flavor, non-strange $(ud)$ quark matter in the non-strange baryon matter.

The surface tension at the interface between the normal and condensed phases was calculated by Christiansen *et al.* (2000) using the non-uniform relativistic mean-field model. In the approximation in which only linear terms in the curvature of the droplet surface are kept, the surface contribution to the thermodynamic potential of a spherical droplet is $\sigma = \sigma_s + 2\sigma_c/R$ (see § 3.4.2), where $\sigma_s$ is the surface tension, $\sigma_c$ is the curvature coefficient, and $R$ is the droplet radius. Christiansen *et al.* (2000) obtain $\sigma_s = 30$ MeV for a small admixture of kaon-condensed droplets in a nucleon matter. Unfortunately, their curvature coefficient is negative, probably because of several oversimplified assumptions made by the authors. Let us remind that the calculation of $\sigma_c$ is ambiguous even for a basic problem of nuclear surface; $\sigma_c$ depends on the assumed position of the phase interface (see, e.g., Douchin *et al.*, 2000, and references therein).

The nucleation of kaon condensate in a non-strange $npe$ matter was studied by Norsen (2002). He considered only the thermal regime and used the nucleation theory of Langer (1969). He analyzed local density fluctuations at a fixed volume and a fixed nucleon number of a matter element. In addition, he assumed that the matter element is electrically neutral and contains one density and charge fluctuation (in the Wigner-Seitz cell approximation). The central quantities of this study are the free energy excess $F^*$ implied by the fluctuation, and the number of nucleons $A_{\rm crit}$ in the critical droplet. It was supposed that the critical drop has enough time to acquire strangeness corresponding to the kaon-condensed state. Following Christiansen *et al.* (2000), Norsen (2002) used $\sigma_s = (20\text{–}30)$ MeV fm$^{-2}$ and neglected the curvature term in the surface contribution to $F^*$.

It turns out that weak interaction processes producing strangeness via $e + N \to \nu_e + K^- + N$ (where an additional nucleon is needed for momentum conservation) and $n \to p + K^-$ are too slow to create a critical droplet of kaon condensate from a density fluctuation during the fluctuation lifetime. Norsen (2002) suggested that strangeness can be produced at a reasonable rate from thermal kaon-antikaon $(K^-K^+)$ pairs, but this mechanism can operate only at extremely high temperatures typical for protoneutron stars (§ 1.4.2). For $k_B T \lesssim 10$ MeV it could operate only at a very high overcompression, where $A_{\rm crit}$ and therefore $F^*$ are sufficiently small. However, as Norsen (2002) argues, a protoneutron star cools so rapidly that kaon condensate has actually no time to nucleate.

The conclusion of the above discussion is unfavorable for kaon condensation in neutron stars. Thus, a neutron star, whose central pressure at birth is too low for the nucleation, but exceeds $P_0$ later owing to accretion in a binary system, may remain in a metastable non-kaon condensed state forever. However, as soon as $\mu_e > \omega^0_{K^-}$ (where $\omega^0_{K^-}$ is the minimum energy of a single zero momentum kaon in dense matter), a spontaneous formation of kaons is possible. This occurs at $\rho_c = \rho_{\rm crit}$, Fig. 7.6, and triggers the kaon condensation. Kinetics of kaon condensation at $P > P_{\rm crit}$ was studied in detail by Muto *et al.* (1997), Muto *et al.* (2000a), and Muto *et al.* (2000b). The timescale for relaxation of dense matter to the equilibrium kaon-condensed state at $\rho \sim 4\rho_0$ and $T = 10^{11}$ K is only $10^{-4}$ s, but grows to $\sim 10$ s at $10^{10}$ K (Muto *et al.*, 2000b).

The "minimal model" of the $npe\mu$ matter may be an oversimplification. The matter at densities relevant for kaon condensation can contain some fraction of hyperons. As hyperons are strongly interacting carriers of strangeness, their presence seems favorable for kaon condensation. A hyperon density fluctuation with sufficiently large strangeness could be a seed to nucleate a droplet of kaon-condensed matter without any strangeness production via weak interactions, for instance, via $n + \Lambda \to p + n + K^-$. Such a strong interaction process is allowed provided the in-medium energy of $K^-$ satisfies $\omega_{K^-} \leq \mu_e$. However, the presence of hyperons in thermodynamic equilibrium reduces the electron number density and chemical potential. Consequently, it pushes up the equilibrium threshold pressure for kaon condensation or even blocks kaon condensation at any density. Thus, the presence of hyperons alleviates the problem of strangeness condensation but actually hinders kaon condensation in neutron star cores (see Ramos *et al.*, 2001, for a more detailed discussion).

### 7.8.3 Quark deconfinement

Quark deconfinement is a strong interaction process and therefore can be described by standard methods. After the $ud$ phase nucleates, weak interaction converts about half of $d$ quarks into $s$ quarks via $u + d \to s + u$ , producing a stable quark matter. Quark deconfinement is expected to occur at very high densities, so that the nucleation should proceed in the quantum tunneling regime.

Iida & Sato (1997) used the model of Lifshitz & Kagan (1972), extended to include the Coulomb contribution. The latter is available because the baryon component, which undergoes deconfinement, is charged. Moreover, Ida and Sato took into account the electron screening effect in the Thomas-Fermi approximation. The effect turned out to be important for the phase transition dynamics. The time $\tau$ needed to form a single droplet of $A = A_{\rm crit}$ was calculated as a function of the overpressure. The authors varied the surface tension $\sigma$ at the interface of the $ud$ quark matter and the nucleon matter in the range $\sigma = (5 - 20)$ MeV fm$^{-2}$ because $\sigma$ is very poorly known. The electron screening allows an $ud$ droplet to grow unrestrictedly. On a timescale $\sim 10^{-9}$ s the droplet converts, via weak-interaction processes, into the stable $uds$ matter, forming a central quark core of the neutron star. The surface tension $\sigma$ determines the overpressure needed to create a single supercritical droplet in a given point. In Fig. 7.7 we show the time of the droplet creation (Iida & Sato,

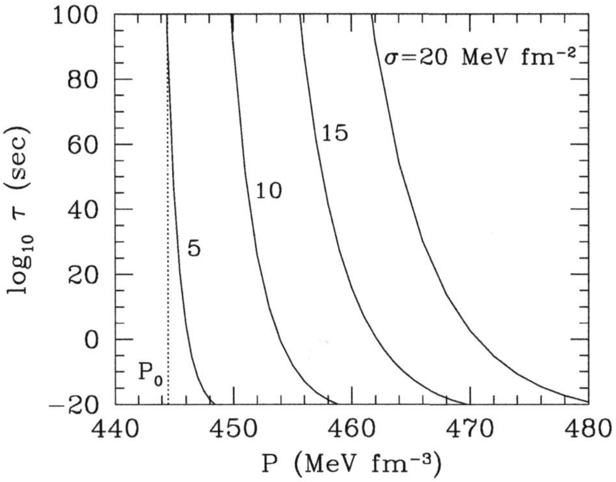


*Figure 7.7.* Time for the formation of a single droplet of the quark matter in the neutron star center versus the central pressure $P$ for four values of the surface tension $\sigma$. Equilibrium first order phase transition between the nucleon matter and the deconfined $ud$ phase takes place at $P = P_0$. From Iida & Sato (1997), with the kind permission of the authors.

1997) versus the pressure $P > P_0$ for several selected values of $\sigma$. Notice a paramount importance of $\sigma$. At a fixed value of $\Delta P_{\text{overv}} = 5$ MeV fm$^{-3}$, the time of the droplet formation changes from $\tau < 10^{-20}$ s for $\sigma = 5$ MeV fm$^{-2}$ to $\tau > 10^{100}$ s for $\sigma = 10$ MeV fm$^{-2}$! The formation of a single droplet over the Universe age $t_{\text{Hubble}} \sim 1.5 \times 10^{10}$ years requires an overpressure $\Delta P_{\text{overv}}/P_0$ from $\sim 0.01$ for $\sigma = 5$ MeV fm$^{-2}$ to $\sim 0.05$ for $\sigma = 20$ MeV fm$^{-2}$. Dissipative effects in the quantum tunneling decrease the nucleation probability at a given $\Delta P_{\text{over}}$. These effects result from excitations of nucleons (unimportant for superfluid nucleons) and from collisions of electrons with an expanding quark droplet, which increase $\Delta P_{\text{over}}$ compared to the non-dissipative value (Iida, 1997).

Because the nucleation of $ud$ droplets (if any) is expected at high densities $\rho \sim (5-10)\rho_0$, the baryon matter is likely to contain not only nucleons but also hyperons. As shown by Iida & Sato (1998), the presence of a substantial fraction of hyperons significantly lowers the density jump at the interface between the baryon matter and quark matter. This reduces an effective droplet mass, $\mathcal{M}_{\text{eff}}$, making the nucleation easier than in the absence of hyperons. The presence of hyperons also greatly reduces the lepton fraction as compared to a purely nucleonic case. Therefore, in the hyperonic matter the effect of electron screening, important for reducing the Coulomb term (that would otherwise limit the growth of an $ud$ droplet in the nucleon matter), is negligible. However, the charge densities of baryon components are also so small, that the Coulomb term is unimportant in the nucleation. All in all, $\Delta P_{\rm over}$ needed to get $t_{\rm nucl} \sim t_{\rm evol}$ in the hyperonic matter, turns out to be a few times smaller than in the absence of hyperons (Iida & Sato, 1998).

An important question is the outcome of the phase transition initiated by the nucleation of a supercritical droplet of $ud$ matter. The dynamics of the growth of such a droplet was studied by Iida & Sato (1998). They find it likely (due to a large speed of droplet expansion, combined with a relatively rapid conversion of $d$ quarks into $s$ quarks) that the inner stellar core composed of the quark matter originates from one supercritical quark droplet nucleated at the stellar center. Putting it differently, the formation of the mixed phase of quark matter and baryon matter in the neutron star core is unlikely.

It is clear from Fig. 7.7 that the surface tension in a quark droplet is crucial for calculating the nucleation rate. Alas, this quantity is poorly known. It is taken from theoretical studies of strangelets (droplets of self-bound strange quark matter) *in vacuum*, assuming additionally $\alpha_{\rm s} = 0$ (Madsen 1993a, 1994; for a review see §3.2 of Madsen 1999 and references therein). Additional complications and uncertainties stem from the curvature term in the surface energy (§ 3.3.3 of Madsen 1999), which seems to be very important for the stability of quark droplets with $A \lesssim 100$ (Madsen, 1993b) but is neglected in calculations of the nucleation rate.

### 7.8.4 The nucleation of a mixed normal-exotic phase

Let us assume that the phase transition at $P = P_0$ from a pure phase A to a pure phase B is first order. Let us further suppose, following Fig. 7.4, that in the pressure interval $P_{\rm A}^{\rm (m)} < P < P_{\rm B}^{\rm (m)}$ thermodynamic equilibrium is realized in the form of the mixed AB phase.

Consider a neutron star where the central pressure $P_{\rm c}$ increases at a rate $t_{\rm evol} \sim P_{\rm c}/\dot{P}_{\rm c}$, owing to accretion in a binary system or owing to pulsar spin-down. Let the stellar core initially consist of the pure phase A. If the internal stellar temperature is $\lesssim 10^9$ K (see, e.g., Miralda-Escudé *et al.*, 1990), thermal fluctuations are negligible and phase B can nucleate only via quantum fluctuations. However, as long as $P_{\rm c} < P_0$, the nucleation in the quantum regime takes too much time ($\tau \to \infty$). The actual nucleation will start at some $P_{\rm c} = P_{\rm nucl} > P_0$ by the formation of a single droplet. If the expansion rate of the first droplet is larger than the formation rate of other droplets in the core, then the pure phase B will fill the central stellar core with $P \leq P_0$. This core will then coexist with the outer layer of phase A, with the baryon density drop $n_{\rm B} - n_{\rm A}$ at the interface. Such a scenario seems likely in the nucleation of the quark matter (§ 7.8.3). Its consequences will be analyzed in §§ 7.9.4 and 7.10.

The central temperature of a protoneutron star stays high, $T_{\rm c} \gtrsim 10^{11}$ K, for some tens of seconds. This time may be sufficient to complete the nucleation of phase B at $P < P_0$ and mix phases A and B, achieving thermodynamic equilibrium. However the stellar core may have no possibility to remain in thermodynamic equilibrium because of too rapid cooling to $T \lesssim 10^{10}$ K. Thus, the final mixed state may be different from the strict ground state of the core at $T = 0$. This mixed phase may remain "frozen" in some metastable state.

Finally, let us mention two difficulties in forming a mixed phase of kaon-condensed matter. As discussed in § 7.8.2, it is difficult to nucleate kaon condensate because of slow weak interaction processes. One needs very high $T \gtrsim 10^{11}$ K and low kaon effective masses for the condensation via thermal fluctuations. This may happen only in massive newly born neutron stars with exclusively high central temperatures and densities, where a mixed-phase core could be formed. Medium-mass neutron stars have insufficiently high central densities to nucleate kaon condensate at their birth. On the other hand, high-mass neutron stars, which gain their mass by accretion, can remain in a metastable state forever, because their internal temperature is too low for nucleating kaon condensate in their cores.

## 7.9. Phase transitions and neutron star structure

In previous sections we analyzed the effects of first-order and second-order phase transitions on the EOS of dense matter. Now let us discuss possible impact of these effects on neutron star structure. Unfortunately, the existence of exotic phases cannot be strictly proven by the present theory. Therefore, the question *Is a specific exotic phase present in a neutron star core?* can be answered only by unambiguous identification of signatures of this phase in neutron star observations.

We will pay special attention to the behavior of global stellar parameters just after the formation of a new phase in the stellar core. The general relativistic theory was developed by Haensel *et al.* (1986a) and Zdunik *et al.* (1987); its simpler Newtonian version had been presented earlier by Schaeffer *et al.* (1983).

### 7.9.1 Linear response to a phase transition in the neutron star center

A stellar configuration with a small core of the higher-density phase B can be constructed by considering a perturbation of the "reference" configuration $\mathcal{C}_0$ with the central pressure $P_c = P_0$ (at the phase transition point) and the central density $\rho_A$ (of the lower-density phase A at this point). Global parameters of $\mathcal{C}_0$ will be marked by the subscript 0. Configurations built exclusively of phase A will be denoted by $\mathcal{C}$, whereas those containing the higher-density phase phase B will be denoted by $\mathcal{C}^*$. Let $r_B$ be the radius of the B-phase core (Fig. 7.8). Stellar parameters of configurations $\mathcal{C}^*$ will be marked by asterisk. The small parameter of the perturbative expansion is $\overline{r}_B \equiv r_B/R_0$. In view of its smallness, the only parameters characterizing the phase transition at $P = P_0$

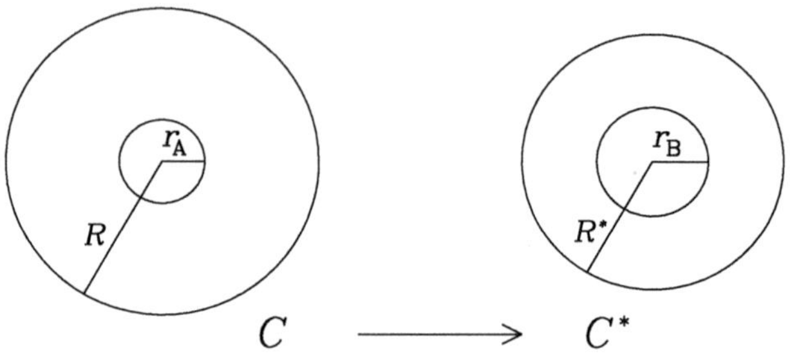


*Figure 7.8.* Metastable one-phase stellar configuration $\mathcal{C}$ with $P_c > P_0$ and stable configuration $\mathcal{C}^*$ with the same baryon number. Configuration $\mathcal{C}$ has an overcompressed stellar core of radius $r_A$, such that $\rho(r_c) = \rho_A$. Configuration $\mathcal{C}^*$ has a core of the phase B of radius $r_B$ with $P^*(r_B) = P_0$. For further explanations see the text.

are the densities $(\rho_A, \rho_B)$[10] and adiabatic indices $(\gamma_A, \gamma_B)$ at both sides of the phase transition. To the lowest order in $\overline{r}_B$, the central density of configuration $\mathcal{C}^*$ with a small core of the new phase B is

$$\rho_c^* \simeq \rho_B \left(1 + \frac{1}{6} \kappa_B \overline{r}_B^2\right) , \tag{7.63}$$

where $x_B = P_0/(\rho_B c^2)$ and the parameter $\kappa_B$ is defined by

$$\kappa_B^2 = (1 + x_B)(1 + 3x_B) \frac{4\pi G \rho_B R_0^2}{v_B^2} , \qquad v_B^2 = \left(\frac{dP}{d\rho}\right)_B , \tag{7.64}$$

where $v_B$ is the sound speed in phase B at $P = P_0$.[11] In the lowest-order approximation, the global parameters characterizing configuration $\mathcal{C}^*$ with a small core of phase B are:[12]

$$M^* = M_0 + (3 - 2\lambda + 3x_A)\, c_M \overline{r}_B^{\,2} , \tag{7.65a}$$
$$\mathcal{A}^* = \mathcal{A}_0 + (3 - 2\lambda + 3x_A)\, c_{\mathcal{A}} \overline{r}_B^{\,2} , \tag{7.65b}$$
$$R^* = R_0 + (3 - 2\lambda + 3x_A)\, c_R \overline{r}_B^{\,2} , \tag{7.65c}$$
$$I^* = I_0 + (3 - 2\lambda + 3x_A)\, c_I \overline{r}_B^{\,2} , \tag{7.65d}$$

where $x_A = P_0/(\rho_A c^2)$, and $\lambda = \rho_B/\rho_A$ specifies the density jump. Coefficients $c_M$, $c_{\mathcal{A}}$, $c_R$, and $c_I$ can be calculated numerically after constructing the reference configuration $\mathcal{C}_0$ (Haensel *et al.*, 1986b; Zdunik *et al.*, 1987).

[10] The density jump $(\rho_B > \rho_A)$ corresponds to first-order phase transition. There is no density jump $(\rho_A = \rho_B)$ for second-order phase transition or for transition to a mixed phase.
[11] Derivatives of any physical quantity $Q$ with respect to density (or any other variable) for configuration $\mathcal{C}^*$, taken at sides A and B of the phase interface, will be denoted by $(dQ/d\rho)_A$ and $(dQ/d\rho)_B$, respectively.
[12] To avoid confusion, in the rest of the present chapter the total baryon number of a stellar configuration will be denoted by $\mathcal{A}$.

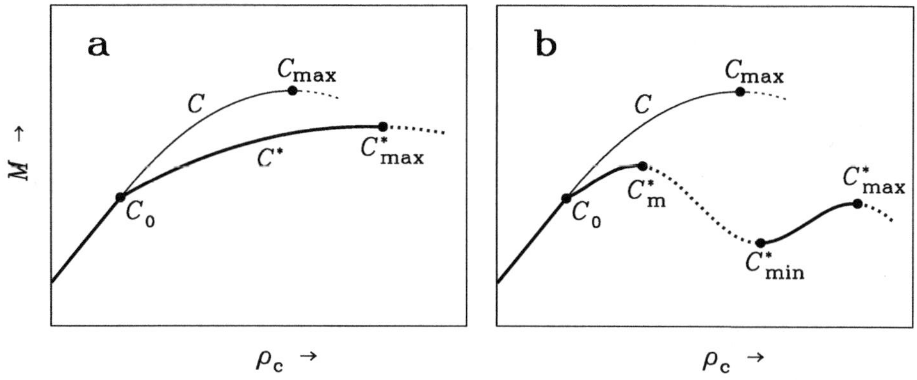


*Figure 7.9.* Stellar mass versus central density for EOSs with a phase transition at $P = P_0$ without any density jump (thick lines). $C_0$ is the last stellar configuration composed exclusively of phase A; its central pressure and density are $P_c = P_0$ and $\rho_c = \rho_A$. Thin lines represent configurations calculated for the EOS without any phase transition. Dotted segments show unstable configurations. *Panel* (a): Moderate softening of the EOS. *Panel* (b): Strong softening producing the unstable branch between $C^*_m$ and $C^*_{min}$ (the thick dotted line) and the *separate branch of superdense stars* between $C^*_{min}$ and $C^*_{max}$.

In the same way one can obtain the expressions for the global parameters of configuration $C$ of the pure phase A, with a small overcompressed core of radius $r_A$ and central density $\rho_c \simeq \rho_A(1 + \frac{1}{6}\,\kappa_A^2\,\overline{r}_A^2)$. It is sufficient to put $\lambda = 1$ and replace all indices B by A.

In what follows, we will distinguish the case of moderate and strong softening of the EOS by a phase transition. We will refer to the phase transition as *moderate*, if it does not produce an unstable segment in the $M - R$ and $M - \rho_c$ curves. In the opposite case of *strong softening*, the phase transition will produce this unstable segment, which separates a family of lower-density stars from a distinct family of superdense compact objects.

## 7.9.2 Moderate softening without any density jump

In this case $\rho(P)$ is continuous, $\lambda = 1$, but the adiabatic index $\gamma = (n_b/P)\,\mathrm{d}P/\,\mathrm{d}n_b$ drops when the new phase forms at $\rho = \rho_A$. The appropriate $M - \rho_c$ curve is displayed in Fig. 7.9a. This situation is typical for hyperonization and also for transitions to a mixed phase (baryons and pion or kaon condensate, baryons and quark matter). In the latter case, the mixed phase exists up to some limiting density, where phase A disappears and $\gamma$ undergoes the second discontinuous change (actually, an increase).

Derivatives of the global parameters ($M$, $I$, $R$, and $\mathcal{A}$) with respect to $\rho_c$ are easily calculated using Eqs. (7.65) in the vicinity of $C_0$. They are discontinuous at $\rho_c = \rho_A$. Let $(\mathrm{d}M/\mathrm{d}\rho_c)_A$ and $(\mathrm{d}M/\mathrm{d}\rho_c)_B$ denote $\mathrm{d}M/\mathrm{d}\rho_c$ taken at the lower-density and higher-density sides of $\rho_c = \rho_A$, respectively.

It is straightforward to calculate $(\mathrm{d}M/\mathrm{d}\rho_c)_B$ using the formulae of § 7.9.1. First we express both $\mathrm{d}M^*$ and $\mathrm{d}\rho_c^*$ in terms of $\overline{r}_B$,

$$\mathrm{d}\rho_c^* = \frac{1}{3}\,\rho_B\,\kappa_B\,\overline{r}_B\,\mathrm{d}\overline{r}_B\ , \qquad \mathrm{d}M^* = 2\,(1+3x_B)\,c_M\,\overline{r}_B\,\mathrm{d}\overline{r}_B\ , \tag{7.66}$$

which results in

$$\left(\frac{\mathrm{d}M^*}{\mathrm{d}\rho_c^*}\right)_B = 6\,(1+3x_B)\,\frac{c_M}{\rho_B\,\kappa_B}\ . \tag{7.67}$$

Similarly, we can calculate the derivative from side A taking into account that it is continuous along the one-phase ($\mathcal{C}$) family,

$$\left(\frac{\mathrm{d}M}{\mathrm{d}\rho_c}\right)_A = 6\,(1+3x_A)\,\frac{c_M}{\rho_A\,\kappa_A}\ . \tag{7.68}$$

Because $\rho_B = \rho_A$ and $x_B = x_A$ in a second-order phase transition, we finally get (omitting asterisks):

$$\left(\frac{\mathrm{d}M}{\mathrm{d}\rho_c}\right)_B = \frac{\gamma_B}{\gamma_A}\left(\frac{\mathrm{d}M}{\mathrm{d}\rho_c}\right)_A\ . \tag{7.69}$$

Similar relations hold for $\mathrm{d}I/\mathrm{d}\rho_c$ and $\mathrm{d}R/\mathrm{d}\rho_c$. Therefore, the derivatives $\mathrm{d}M/\mathrm{d}R$, $\mathrm{d}I/\mathrm{d}M$, and $\mathrm{d}I/\mathrm{d}M$ are continuous when the new phase forms. Using Eq. (7.69) one can also show that $\mathrm{d}M/\mathrm{d}P_c$ is also continuous at $P_c = P_0$.

### 7.9.3 Strong softening without any density jump: The third family of compact stars

This is the case of extremely strong decrease of $\gamma$ after the phase transition, followed by hardening of the EOS at still higher densities. In this case the $M - \rho_c$ curve has two stable branches of static equilibrium configurations (Fig. 7.9b) corresponding to two distinct families of neutron stars. The first (lower-density or normal) family is continuously connected with low-mass neutron stars. The second family contains "compact neutron stars". It will be called the "higher-density family" or "superdense family". The stable configurations belonging to the superdense family have central densities $\rho^*_{\rm min} < \rho^*_c < \rho^*_{\rm max}$ and masses $M^*_{\rm min} < M^* < M^*_{\rm max}$. They are more compact and tightly bound than configurations containing the same number of baryons and belonging to the normal branch.

The family of superdense stars deserves a general remark. These stars form actually the *third family of compact stars* built of degenerate matter, after the well known white dwarfs and "lower-density" neutron stars.

Two examples of phase transitions with strong softening of the EOS without any density jumps are given by Glendenning & Kettner (2000) and Schaffner-Bielich *et al.* (2002). In the model of Glendenning & Kettner (2000), the softening results from the appearance of a phase of quarks and baryons and takes place for a substantial quark fraction. Their EOS stiffens at higher densities when the mixed phase is replaced by the pure quark phase. In the model of Schaffner-Bielich *et al.* (2002) the softening is due to the copious appearance of hyperons. It is followed by a substantial stiffening of the EOS at higher densities. Similarly to the first example, it leads to the appearance of the "higher-density branch" of the $M - \rho_c$ curve; the models belonging to this branch are mostly composed of hyperons. In both examples, the maximum mass of the "higher-density branch" is lower than for the "lower-density" family, $M^*_{\rm max} < M_{\rm max}$.

At the same baryon number, the energy $M^* c^2$ of a high-density configuration is smaller than the energy of a lower-density one. Let $\mathcal{A}^*_{\rm min}$ and $\mathcal{A}^*_{\rm max}$ denote the limiting baryon numbers of stable configurations of the higher-density branch. For $\mathcal{A}^*_{\rm min} < \mathcal{A} < \mathcal{A}^*_{\rm max}$, true stable configurations are those belonging to the higher-density branch, whereas configurations consisting of the same number of baryons and belonging to the lower-density branch are *metastable*. Possible manifestations of these properties will be discussed in §§7.9.7 and 7.9.8.

## 7.9.4 Moderate softening with density jump

Now consider the softening of the EOS due to a phase transition accompanied by a density jump. The phase transition occurs at $P = P_0$, where the pure phase A of density $\rho_{\rm A}$ coexists with the pure phase B of higher density $\rho_{\rm B}$. Then the relative density jump is $\lambda = \rho_{\rm B}/\rho_{\rm A} > 1$. Because of the discontinuity of $\rho$ at $P = P_0$, equilibrium configurations have to be parameterized by the central pressure $P_c$. Let us remind that for any global parameter $Q$ of an equilibrium configuration

$$\frac{{\rm d}Q}{{\rm d}\rho_c} = \frac{c^2 P_c \gamma_c}{\mathcal{E}_c + P_c} \frac{{\rm d}Q}{{\rm d}P_c} . \tag{7.70}$$

Consider configurations in the vicinity of $\mathcal{C}_0$. As can be deduced from Eqs. (7.65), the appearance of a core of the new phase implies a discontinuous change of the derivatives ${\rm d}M/{\rm d}P_c$ and ${\rm d}R/{\rm d}P_c$. Using the same method as in §7.9.2, we get

$$\left(\frac{{\rm d}M}{{\rm d}P_c}\right)_{\rm B} = \frac{2(\lambda - \lambda_{\rm crit})}{1 + 3x_{\rm B}} \left(\frac{{\rm d}M}{{\rm d}P_c}\right)_{\rm A} , \quad \lambda_{\rm crit} \equiv \frac{3}{2}\left(1 + \frac{P_0}{\rho_{\rm A} c^2}\right) . \tag{7.71}$$

The equations for the derivatives of $I$ and $R$ are similar. Although these derivatives suffer jumps, the derivatives ${\rm d}M/{\rm d}R$, ${\rm d}I/{\rm d}R$, ${\rm d}I/{\rm d}M$ are continuous

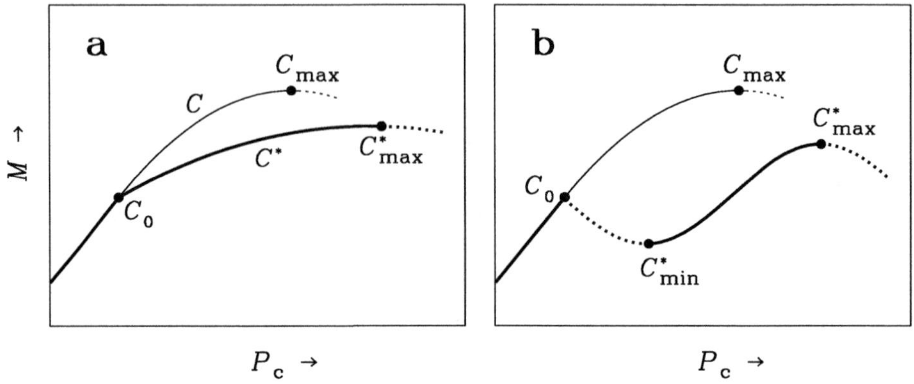


*Figure 7.10.* Mass versus central pressure for EOSs containing phase transition with a density jump (thick lines). $\mathcal{C}_0$ is the last configuration ($P_c = P_0$) composed exclusively of the lower-density phase. Thin lines show configurations calculated using EOSs without any phase transition. Dotted segments correspond to unstable configurations. *Panel* (a): Moderate softening of the EOS. *Panel* (b): Strong softening producing the unstable branch between $\mathcal{C}_0$ and $\mathcal{C}^*_{min}$ (the thick dotted line) and the separate branch of *superdense stars* between $\mathcal{C}^*_{min}$ and $\mathcal{C}^*_{max}$. For further explanation see the text.

at the phase transition.[13] Of course, Eq. (7.71) transforms into Eq. (7.69) in the limit of $\lambda \to 1$.

Let us stress the importance of the factor $\lambda - \lambda_{crit} = (3 - 2\lambda + 3x_A)/2$. In principle, it could be *negative*. Then the first-order phase transition could *destabilize* a stellar configuration (see the next subsection). However, for moderate softening with $\lambda \lesssim 1.5$, considered in the present subsection, this factor is positive.

### 7.9.5 Strong softening with density jump: The third family of compact stars

Such a phase transition corresponds to $\lambda = \rho_B/\rho_A > \lambda_{crit}$. In this case the appearance of a small core of phase B destabilizes the neutron star. Using Eqs. (7.65) one can see that in the presence of such a core $\mathrm{d}M/\mathrm{d}\rho_c < 0$. These configurations are therefore *unstable* and collapse into *stable* configurations with large cores of phase B. The *instability condition* $\lambda > \frac{3}{2}(1 + P_0/\rho_A c^2)$ had been first derived by Seidov (1971) using the static energy method. Ten

[13]This property stems from the the linear response theory formulated in 1986–1987 (Haensel *et al.*, 1986a; Zdunik *et al.*, 1987). Later it was rederived by Lindblom (1998) who used detailed and strict mathematical analysis of equilibrium configurations for an EOS exhibiting first-order phase transition.

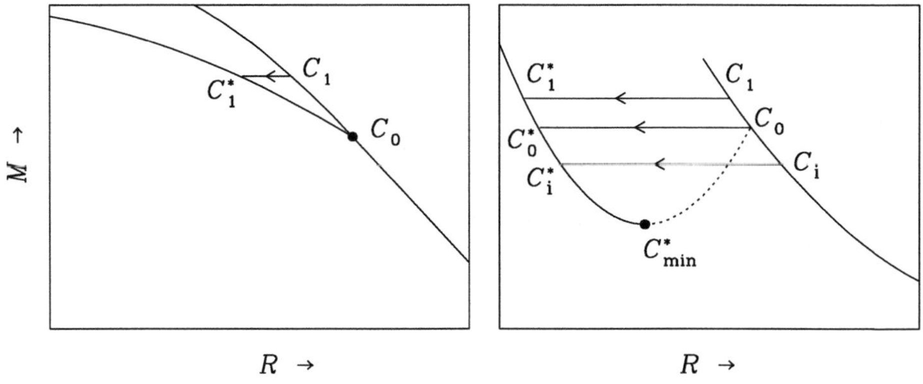


*Figure 7.11.* Vicinity of the "reference configuration" $\mathcal{C}_0$ with $P_c = P_0$ in the $M - R$ plane for phase transitions with $\lambda < \lambda_{\rm crit}$ (*left*) and $\lambda > \lambda_{\rm crit}$ (*right*). The dotted segment corresponds to unstable two-phase configurations. Arrows connect configurations with the same baryon numbers.

years later this condition was rediscovered by Kaempfer (1981) who studied the necessary condition for the onset of neutron star collapse initiated by a phase transition in its center. It is worth to mention that the Newtonian version of this criterion ($\lambda > \frac{3}{2}$) had been first obtained by Lighthill (1950) (see also Ramsey, 1950) in the context of stability of planets. Relativistic effects stabilize neutron stars with small cores of phase B by increasing $\lambda_{\rm crit}$. The increase can be as high as $\sim 0.2$. The dynamics of the collapse of configurations with a small core of phase B will be studied in § 7.9.8.

Stable equilibrium configurations of neutron stars split into two families visualized in Fig. 7.10b. The superdense branch $C^*_{\rm min}C^*_{\rm max}$ forms the *third family of compact stars*, apart from white dwarfs and lower-density neutron stars. Contrary to the instability triggered by second-order phase transition (associated with the hyperonization or the appearance of a mixed phase of quarks and baryons, § 7.9.3), for the instability at $\lambda > \lambda_{\rm crit}$ one typically has $M_{\rm max} < M^*_{\rm max}$ and $\mathcal{A}_{\rm max} < \mathcal{A}^*_{\rm max}$ (see, e.g., Brown & Weise, 1976; Haensel & Prószyński, 1982; Migdal *et al.*, 1990). Possible observational consequences of this behavior are described in § 7.9.8.

### 7.9.6 The proof of the inequality $M^*(\mathcal{A}) < M(\mathcal{A})$

Let us consider the right panel of Fig. 7.11 and introduce the dimensionless quantity $F = \sqrt{1 - 2GM/Rc^2}$. We will treat $F$ as a function of the baryon number of a neutron star configuration and distinguish three branches of the $M - R$ diagram. The branch of normal configurations has $\mathcal{A}(\mathcal{C}) < \mathcal{A}_{\rm max}$; the dependence of $F$ on $\mathcal{A}$ along this branch will be denoted as $F(\mathcal{A})$. The second branch between $\mathcal{C}_0$ and $\mathcal{C}^*_{\rm min}$ refers to unstable configurations with $\mathcal{A}^*_{\rm min} < \mathcal{A} < \mathcal{A}_0$, where $\mathcal{A}_0 \equiv \mathcal{A}(\mathcal{C}_0)$; the corresponding function will be denoted by $F_{\rm unst*}(\mathcal{A})$. Finally, the third branch describes stable compact configurations with $\mathcal{A}^*_{\rm min} < \mathcal{A} < \mathcal{A}^*_{\rm max}$; the associated function will be denoted as $F_*(\mathcal{A})$.

Consider configuration $\mathcal{C}_1$ with the same baryon number as $\mathcal{C}^*_1$ (Fig. 7.11). Using the small-increment theorem, Eq. (6.52), we can express the masses of both configurations as

$$M(\mathcal{C}_1) = M(\mathcal{C}_0) + \int_{\mathcal{A}_0}^{\mathcal{A}_1} F(\mathcal{A})\, \mathrm{d}\mathcal{A}\,, \tag{7.72}$$

$$M(\mathcal{C}^*_1) = M(\mathcal{C}^*_0) + \int_{\mathcal{A}_0}^{\mathcal{A}_1} F_*(\mathcal{A})\, \mathrm{d}\mathcal{A}\,. \tag{7.73}$$

On the other hand, the small-increment theorem enables us to relate $M(\mathcal{C}^*_0)$ and $M(\mathcal{C}_0)$,

$$M(\mathcal{C}^*_0) = M(\mathcal{C}_0) + \int_{\mathcal{A}^*_{\rm min}}^{\mathcal{A}_0} \left[F_*(\mathcal{A}) - F_{\rm unst.*}(\mathcal{A})\right]\, \mathrm{d}\mathcal{A}\,. \tag{7.74}$$

At a given $\mathcal{A}$, the differences in the values of $M(\mathcal{A})$ of the three branches are much smaller than the differences in $R$. Therefore, one has $F(\mathcal{A}) > F_*(\mathcal{A})$ and $F_{\rm unst*}(\mathcal{A}) > F_*(\mathcal{A})$ under the integrals, and thus $M^*(\mathcal{A}) < M(\mathcal{A})$. The proof is even simpler for a $\mathcal{C}_1 \to \mathcal{C}^*_1$ transition shown in the left panel of Fig. 7.11.

The inequality $M^*(\mathcal{A}) < M(\mathcal{A})$ is quite general. It means that the change in the EOS associated with the phase transition always lowers the total energy of the star. However, the reader should be warned that the energy excess $\Delta E = (M - M^*)c^2$ is usually very small. Therefore, while calculating $\Delta E$ one has to be sure that the EOS satisfies the condition of *thermodynamic consistency* discussed in § 6.4.1. Otherwise, there is a danger of violating the strict inequality $M^*(\mathcal{A}) < M(\mathcal{A})$, leading to an apparent paradox: a phase transition in a neutron star core is blocked by global energy conservation!

### 7.9.7 Nonequilibrium first-order phase transition and its consequences

Nucleation initiated by first-order phase transition was analyzed in § 7.8. Here we discuss the nucleation of a new phase in the stellar center associated with the compression of matter owing to accretion or pulsar spindown. The compression timescale in the stellar center is $t_{\rm comp} \sim P_{\rm c}/\dot{P}_{\rm c}$. As stressed in

§ 7.8, the problem of phase transitions is *kinetical*.[14] Let $P_0$ be the pressure of the equilibrium phase transition. In the quantum tunneling regime the nucleation time at $P_c = P_0$ is infinite. Therefore, in reality the phase transition will take place off equilibrium, in a metastable core at some $P_c > P_0$ at which $t_{\rm nucl} \sim t_{\rm comp}$. The nonequilibrium appearance of the new phase in a supercompressed stellar core will destroy hydrostatic equilibrium. It is expected to trigger a neutron star "corequake" (or "minicollapse") due to the rapid formation of a core of a denser phase. It produces a discontinuous drop of the moment of inertia $I$ and a jump of rotation frequency, i.e., a pulsar glitch (§ 1.4.4).

Even more serious complications arise in a transition to a mixed phase. An example is a mixed quark-baryon phase considered by Glendenning *et al.* (1997). In full equilibrium, this phase exists at $P > P_{\rm A}^{\rm (m)}$, lower than the pressure required for the equilibrium first-order phase transition from a pure baryon phase to a pure quark phase. Thus, the nucleation of the denser phase at $P = P_{\rm A}^{\rm (m)}$ is virtually impossible at temperatures of a few times $10^8$ K typical for interiors of middle-aged radio pulsars. Therefore, the phase transition will proceed in a nonequilibrium way at much higher pressure. It is expected to lead to the formation of the pure quark core (see § 7.8.4) and it will be accompanied by a strong stellar corequake. The drop in $I$ and the jump in spin frequency, occurring on a dynamical timescale, will be significant. Therefore, we expect that characteristic signatures of quark-deconfinement phase transition are different from those resulting from an equilibrium formation of the mixed-phase core considered for instance, by Glendenning *et al.* (1997) and Chubarian *et al.* (2000). Only a very high temperature in a newly born neutron star could allow for the formation of a mixed phase in quasi-equilibrium, because strong thermal fluctuations would produce then a very rapid nucleation of the quark phase.

### 7.9.8 Large-amplitude corequake and collapse

This phenomenon may occur if the softening of the EOS in a phase transition results in two families of stable configurations; see Fig. 7.10b. Consider configuration $\mathcal{C}_{\rm i}$ on the lower-density branch in Fig. 7.11b. Strictly speaking, $\mathcal{C}_{\rm i}$ is *metastable*, because configuration $\mathcal{C}_{\rm i}^*$ with the same baryon number has lower energy. However, the transition $\mathcal{C}_{\rm i} \to \mathcal{C}_{\rm i}^*$ would require overcoming a huge energy barrier. Therefore, it can occur only at the neutron star birth, when the required energy could be provided by large-amplitude motions. Otherwise, configuration $\mathcal{C}_{\rm i}$ can be treated as stable. If the neutron star accretes matter, its equilibrium configuration changes quasistatically, moving upward along the $\mathcal{C}$-

[14] Even for the simplest transition with the appearance of a new hyperon species one has to check if the system is in full equilibrium. Typically, it is not because at $T \lesssim 10^8$ K, characteristic of middle-aged pulsars, the timescale of establishing beta equilibrium in the $npe\mu$ matter is longer than the compression timescale.

branch. The central pressure will increase reaching at some moment the value $P_c = P_0$ corresponding to the *equilibrium* first-order phase transition. However, as the compression of the matter in the stellar center is quasistatic and takes place at relatively low temperature, the star will continue moving along the thin line $\mathcal{C}$ having a small growing core of metastable, supercompressed phase A. When the over-compression reaches the critical value $\Delta P_{\rm nucl}$ in the center of configuration $\mathcal{C}_1$, phase B will start to nucleate, initiating a transition to configuration $\mathcal{C}_1^*$ with a large core of phase B. The kinetics of the nucleation of pion condensate, kaon condensate, and quark matter has been analyzed in §7.8.

During the collapse $\mathcal{C}_1 \to \mathcal{C}_1^*$, the stellar radius shrinks and a large amount of energy is liberated. The moment of inertia decreases, accelerating stellar rotation. The dynamics of a large-amplitude collapse triggered by a first-order phase transition was studied by Migdal *et al.* (1979); Kaempfer (1982); Berezin *et al.* (1982, 1983); Berezin & Dmitrieva (1984); Haensel *et al.* (1990a). Neutron star corequakes (or mini-collapses) produced by first-order phase transitions in the neutron star cores were useful in modeling some gamma-ray bursts (Ramaty *et al.*, 1980, 1981; Ellison & Kazanas, 1983; Haensel & Prószyński, 1982; Muto & Tatsumi, 1990). However, detailed hydrodynamic simulations of such a mini-collapse show that while the collapse and subsequent bounce can easily impart the energy $\gtrsim 10^{50}$ erg in the ejected envelope, the rest mass of the ejecta is too high to consider this phenomenon as a viable gamma-ray burst mechanism (Fryer & Woosley, 1998). To get the mean Lorentz factor $[1 - (v/c)^2]^{-1/2} \gtrsim 100$ required for producing a fireball and subsequent observed gamma-ray burst at a cosmological distance, an energy $\sim 10^{51}$ erg must be imparted in less than $10^{-5}\ M_\odot$ of the ejected baryon mass; the calculated rest mass of the ejecta turns out to be orders of magnitude larger.

The case shown in Fig. 7.9b is different. Consider configuration $\mathcal{C}_1$ on the lower-density branch and assume the hyperonization of dense matter. Accretion will quasistatically change the configuration, increasing its mass, baryon number, and central density. The appearance of hyperons will soften the EOS and bend the $M - \rho_c$ line. After reaching $\mathcal{C}_{\rm max}$, the star is doomed to a collapse: further accretion triggers the dynamical instability. If $\mathcal{A}_{\rm max} > \mathcal{A}^*_{\rm max}$ the star collapses into a black hole. The families of neutron star configurations with $\mathcal{A}_{\rm max} > \mathcal{A}^*_{\rm max}$ were constructed by Glendenning & Kettner (2000) and Schaffner-Bielich *et al.* (2002). It is clear, that the high-density branch between $\mathcal{C}^*_{\rm min}$ and $\mathcal{C}^*_{\rm max}$ cannot by reached in the course of the quasistatic evolution of configurations belonging to the normal, lower-density branch. Consider configuration $\mathcal{C}_1$ of such a total baryon number $\mathcal{A}_1$ that $\mathcal{A}^*_{\rm min} < \mathcal{A}_1 < \mathcal{A}^*_{\rm max}$. It has lower binding energy than configuration $\mathcal{C}_1^*$ of the same total baryon number. While $\mathcal{C}_1$ is strictly speaking metastable, the transition $\mathcal{C}_1 \to \mathcal{C}_1^*$ requires overcoming a huge energy barrier. Therefore $\mathcal{C}_1$ is actually stable with respect to small-amplitude perturbations and a transition to the high-density branch requires large-amplitude motions with the kinetic energy exceeding the height of the relevant energy barrier. This situation may occur at the neutron star birth in a gravitational collapse of a massive stellar core with $\mathcal{A} < \mathcal{A}^*_{\rm max}$.

## 7.10. Changes in stellar parameters due to a corequake

Consider an EOS with a first-order phase transition at $P = P_0$, as shown in Fig. 7.6a. Let the pressure in the neutron star center be $P_{\rm c} < P_0$. Assume that $P_{\rm c}$ increases due to accretion or pulsar spindown. The compression timescale is $t_{\rm comp} \sim P_{\rm c}/\dot{P}_{\rm c}$. As long as $P_{\rm c} \leq P_0$, the stellar core is built of the pure, stable phase A; such a configuration will be denoted by $\mathcal{C}$. After $P_{\rm c}$ crosses $P_0$ two possibilities can occur. If the nucleation timescale $t_{\rm nucl}$ of phase B is much shorter than $t_{\rm comp}$ then the core of phase B starts to grow just after $P_{\rm c}$ exceeds $P_0$. However, because the star is relatively cold and the energy barrier is rather high, we typically have $t_{\rm nucl} \gg t_{\rm comp}$ and the compression to $P_{\rm c} > P_0$ will lead to the formation of a metastable core of supercompressed phase B. The radius of this core $r_{\rm A}$ is determined by the equation $P(r_{\rm A}) = P_0$. With the increase of $P_{\rm c}$ the value of $t_{\rm nucl}$ rapidly decreases. At some $P_{\rm c} = P_{\rm nucl}$ (corresponding to the central density $\rho_{\rm nucl}$) we have $t_{\rm nucl} \sim t_{\rm comp}$ and phase B will nucleate. This will lead to a local pressure deficit, violation of hydrostatic equilibrium, and a *corequake* resulting in the formation of a stable configuration $\mathcal{C}^*$ with a core of phase B. The core radius $r_{\rm B}$ is determined by $P(r_{\rm B}) = P_0$.

Let us assume that there is no mass ejection during the $\mathcal{C} \to \mathcal{C}^*$ transition, $\mathcal{A}(\mathcal{C})^* = \mathcal{A}(\mathcal{C})$. We will determine changes in stellar parameters, such as $R$, $I$, and $M$, for the case in which $r_{\rm A}$ and $r_{\rm B}$ are much smaller than the stellar radius and deviations from spherical symmetry are negligible. In this case we can apply the linear response method developed by Haensel *et al.* (1986b) and Zdunik *et al.* (1987). The main results are described below.

Let $\mathcal{C}_0$ be the last stable configuration of phase A ($P_{\rm c} = P_0$). It will be used as a reference configuration; its parameters will be labelled by "0". As already mentioned in this chapter, small dimensionless parameters of the problem are $\overline{r}_{\rm A} \equiv r_{\rm A}/R_0$ and $\overline{r}_{\rm B} \equiv r_{\rm B}/R_0$. The condition $\mathcal{A}(\mathcal{C}) = \mathcal{A}(\mathcal{C}^*)$ leads to the relation between $\overline{r}_{\rm A}$ and $\overline{r}_{\rm B}$. It can be obtained neglecting fourth and higher powers of $\overline{r}_{\rm A}$ and $\overline{r}_{\rm B}$:

$$(1 + 3x_0)\, \overline{r}^2_{\rm A} = (3 - 2\lambda + 3x_{\rm A})\, \overline{r}^2_{\rm B} - (\lambda - 1)\, a\, \overline{r}^3_{\rm B}\,. \qquad (7.75)$$

Here, $\lambda = \rho_{\rm B}/\rho_{\rm A} > 1$ and $x_A = P_0/\rho_{\rm A} c^2$; the dimensionless factor $a$ can be calculated after constructing the reference configuration $\mathcal{C}_0$.

For $\rho_{\rm c}(\mathcal{C}_0) \gtrsim 2\rho_0$ the factor $a$ increases with the growth of mass $M_0 = M(\mathcal{C}_0)$ (Haensel *et al.*, 1986b). In the linear-response approximation (which is valid at $M_0 \lesssim 0.9\, M_{\rm max}$) for a broad class of EOSs of phase A this dependence can be fitted by a simple formula (P. Haensel 2002, unpublished)

$$a(M_0) \simeq 15 \ (M_0/M_{\rm max})^2 \ . \tag{7.76}$$

In order to avoid the introduction of non-linear terms, we assume that configurations $\mathcal{C}^*$ with any arbitrarily small core of phase B are stable. This is valid for $3-2\lambda+3x_{\rm A} > 0$ (see § 7.9.4), when $r_{\rm A} \to 0$ implies $r_{\rm B} \to 0$. Then the linear response method enables us to derive simple lowest-order expressions for the changes in $R$ and $I$ due to the transition $\mathcal{C} \to \mathcal{C}^*$ associated with a corequake (see Fig. 7.8):

$$\Delta R = R(\mathcal{C}^*) - R(\mathcal{C}) = -(\lambda - 1)\, \alpha_R\, \overline{r}_{\rm B}^3\, R_0 + \mathcal{O}(\overline{r}_{\rm B}^5)\,, \tag{7.77}$$

$$\Delta I = I(\mathcal{C}^*) - I(\mathcal{C}) = -(\lambda - 1)\, \alpha_I\, \overline{r}_{\rm B}^3\, I_0 + \mathcal{O}(\overline{r}_{B}^5)\,, \tag{7.78}$$

where $\alpha_R$ and $\alpha_I$ are dimensionless parameters defined by the basic configuration $\mathcal{C}_0$. In the mass range $M_\odot \lesssim M_0 \lesssim 0.9\ M_{\rm max}$ they depend weakly on $P_0$. For a medium-stiff EOS of phase A Haensel *et al.* (1986b) obtained $\alpha_R \simeq 3$ and $\alpha_I \simeq 7$.

The appearance of the core of phase B is accompanied by the energy release

$$\Delta E = [M(\mathcal{C}) - M(\mathcal{C}^*)]\, c^2\ , \tag{7.79}$$

which results from the gain in the gravitational energy partly compensated by the increase of the proper (internal) energy (see § 6.3) due to the compression of the stellar matter. As shown by Zdunik *et al.* (1987), this compensation removes the terms proportional to $\overline{r}_{\rm B}^3$ and $\overline{r}_{\rm B}^4$, leaving $\Delta E \propto \overline{r}_{\rm B}^5$. The expression for $\Delta E$ is derived by combining the linear response approximation with the small-increment theorem, Eq. (6.52):

$$\Delta E = (\lambda - 1)\alpha_E\, \overline{r}_{\rm B}^5 \left[3 - 2\lambda + 3x_{\rm A} - \frac{5}{4}(\lambda - 1)a\overline{r}_{\rm B}\right] E_0 + \mathcal{O}(\overline{r}_{\rm B}^7)\ , \tag{7.80}$$

where $E_0 = M_0 c^2 = 1.788 \times 10^{54}\ M_0/M_\odot$ erg. For moderately stiff EOSs of phase A and $M_\odot \lesssim M_0 < 0.9\ M_{\rm max}$, one gets a remarkably constant $\alpha_E \simeq 0.2$ (Haensel *et al.*, 1986b).

### 7.10.1 Estimates of $\Delta R$, $\Delta I$, and $\Delta E$ in corequakes

The lowest-order approximation for $\Delta R$ can be rewritten as

$$\Delta R \simeq -3(\lambda - 1) \left(\frac{r_{\rm B}}{100\ {\rm m}}\right)^3 \left(\frac{10\ {\rm km}}{R_0}\right)^2 \ {\rm m}\ . \tag{7.81}$$

The actual shrinking of the stellar radius may be very small. Assuming $R_0 =$ 10 km, one gets the shrinking by $\Delta R \simeq 2$ m for a 100 m core and $\lambda - 1 =$

0.2. However, recall that the shrinking occurs in a huge gravitational field. Moreover, $\Delta R \propto r_{\rm B}^3$. For $r_{\rm B} = 500$ m and the same $\lambda$ we obtain $\Delta R \simeq 200$ m, an impressive neutron star quake which would be terrifying by our terrestrial standards.

The lowest-order approximation for $\Delta I$ is

$$\frac{\Delta I}{I} \simeq -7 \times 10^{-6}\,(\lambda - 1)\left(\frac{r_{\rm B}}{100\ {\rm m}}\right)^3 \left(\frac{10\ {\rm km}}{R_0}\right)^3 . \qquad (7.82)$$

Assuming again $R_0 = 10$ km , $r_{\rm B} = 100$ m, and $\lambda - 1 = 0.2$ the change in $I$ will induce a pulsar spin-up typical for pulsar glitches (§ 1.4.4). For $r_{\rm B} = 500$ m, one gets a spin-up with $\Delta\Omega/\Omega = |\Delta I|/I \simeq 10^{-4}$ which becomes a distinct manifestation of a phase transition in the stellar core.

The upper bound to the energy release $\Delta E$ due to a neutron star corequake in the lowest-order linear response approximation reads:

$$\begin{aligned}\Delta E \ &\simeq\ 5 \times 10^{43}\,(\lambda - 1)\,[3 - 2\lambda + 3x_{\rm A} + (\lambda - 1)\,a\,\overline{r}_{\rm B}] \\ &\quad\times \left(\frac{r_{\rm B}}{100\ {\rm m}}\right)^5 \left(\frac{R_0}{10\ {\rm km}}\right)^5 \frac{M_0}{M_\odot}\ {\rm erg}\ . \end{aligned} \qquad (7.83)$$

For $M_0 = 1.4\,M_\odot$, $a = 10$ and $\lambda - 1 = 0.3$, the formation of the core with radius $r_{\rm B} = 100$ m is accompanied by the energy release $\sim 10^{44}$ erg. The energy release grows rapidly with increasing $r_{\rm B}$. For $r_{\rm B} = 1$ km, we get $\Delta E \sim 10^{49}$ erg.

## 7.11. Mixed-phase core and neutron star corequakes

Consider a phase transition to a mixed phase AB (labeled by 'm'). Let the central pressure of a stellar configuration $\mathcal{C}$ be slightly lower than the threshold pressure $P_{\rm A}^{(\rm m)}$ for the appearance of the mixed phase (Fig. 7.11a). Let the central density and pressure increase owing to accretion or stellar spindown. After $P_{\rm c}$ crosses $P_{\rm A}^{(\rm m)}$, a metastable core of phase A with radius $r_{\rm A}$ starts to grow. The nucleation of phase B in the metastable phase A will trigger the transition $\mathcal{C} \to \mathcal{C}^*$, where the stable configuration $\mathcal{C}^*$ is assumed to have a mixed-phase core of radius $r_{\rm m}$. In contrast to the case described in § 7.10, we have no density jump at $P = P_{\rm A}^{(\rm m)}$, so that $\lambda = 1$ and it is only the adiabatic index $\gamma$ which suffers a drop at $P = P_{\rm A}^{(\rm m)}$. We have $\gamma_{\rm A} > \gamma_{\rm m}$, where $\gamma_{\rm A}$ and $\gamma_{\rm m}$ are the adiabatic indices of phases A and m at $P = P_{\rm A}^{(\rm m)}$. The leading-order expressions for the changes of $R$, $I$, and $E$ (the energy release) in a corequake initiated by the $\mathcal{C} \to \mathcal{C}^*$ transition can be obtained in the linear response approximation (as in § 7.10). The only parameter of the mixed-phase EOS, which enters leading-order expressions, is the adiabatic index calculated in the limit of vanishing fraction of phase B,

$$\gamma_{\rm m} = \left( \frac{n_{\rm b}}{P} \frac{{\rm d}P}{{\rm d}n_{\rm b}} \right)_{P=P_{\rm A}^{\rm (m)}+0} . \tag{7.84}$$

The drop of $\gamma$ at $P = P_{\rm A}^{\rm (m)}$ is as significant as the jump of $n_{\rm b}$ at $P = P_0$ for a transition between pure phases. Therefore, it is convenient to factor-out the dependence of leading quantities on $\gamma_{\rm A}$ and $\gamma_{\rm m}$ in such a way that the linear response coefficients vanish automatically at $\gamma_{\rm A} = \gamma_{\rm m}$. Following Bejger *et al.* (2005a), we can rewrite the leading-order formulae in the form:

$$\begin{aligned}
\Delta R &= R(\mathcal{C}^*) - R(\mathcal{C}) = -(\gamma_{\rm A}/\gamma_{\rm m} - 1)\,\beta_R\,\overline{r}_{\rm m}^5\,R_0\,, \\
\Delta I &= I(\mathcal{C}^*) - I(\mathcal{C}) = -(\gamma_{\rm A}/\gamma_{\rm m} - 1)\,\beta_I\,\overline{r}_{\rm m}^5\,I_0\,, \\
\Delta E &= [M(\mathcal{C}) - M(\mathcal{C}^*)]c^2 = (\gamma_{\rm A}/\gamma_{\rm m} - 1)\,\beta_E\,\overline{r}_{\rm m}^7\,M_0 c^2\,.
\end{aligned} \tag{7.85}$$

Lower-order terms available at $\lambda > 1$ in Eqs. (7.77), (7.78), and (7.80) are now absent because they contain the factor $(\lambda - 1)$ vanishing at $\lambda = 1$.

For a given EOS of non-exotic phase A, the coefficients $\beta_Q$ are well defined functions of $M_0$. Bejger *et al.* (2005a) calculated them versus $M_0$ for the SLy EOS. In the mass range $0.8\,M_\odot \lesssim M_0 \lesssim 1.5\,M_\odot$ they stay almost constant, $\beta_R \simeq 0.8$, $\beta_I \simeq 2.0$, and $\beta_E \simeq 0.5$. Notice that these constant values of $\beta_Q$ are close to one-third of the parameters $\alpha_Q$ (which determine linear response to a first-order phase transition considered in § 7.10 for a medium-stiff EOS).

For $1.5\,M_\odot \lesssim M_0 \lesssim 0.9\,M_{\rm max}$, the values of $\beta_Q$ smoothly increase with growing $M_0$. At $M_0 \sim 0.9\,M_{\rm max}$ they are larger than the constant values. All in all, $\beta_Q$ vary by less than a factor of two within a broad range of masses $0.5\,M_\odot \lesssim M_0 \lesssim 0.9\,M_{\rm max}$. Similar results were obtained also for the FPS EOS (Bejger *et al.*, 2005a).

Parameters $\beta_Q$ rapidly increase for $M_0 \to M_{\rm max}$ as well as in the limit of low $M_0$. A strong increase of $\beta_Q$ indicates a strong softening of the star with respect to radial perturbations when a core of the mixed phase appears, and implies the breakdown of the linear response approximation. In both cases of high and low $M_0$, one approaches an instability with respect to small radial perturbations. At $M_0 \to M_{\rm max}$ this is the relativistic maximum-mass instability leading to a collapse into a black hole (§ 6.5.1). For $M_0 \to M_{\rm min} \simeq 0.1\,M_\odot$, neutron stars become less bound, ready to be unstable with respect to explosion (see §§ 6.5.1 and 6.7.2).

The problem of the nucleation of phase B in the mixed-phase state has been discussed in § 7.8.4. For an accreting or spinning-down pulsar, thermal effects are expected to be small and phase B will nucleate at $P_{\rm c} = P_{\rm nucl} \gtrsim P_0 > P_{\rm A}^{\rm (m)}$. Then phase B will likely equilibrate with phase A by forming a mixed-phase state, which extends to pressures $P_{\rm A}^{\rm (m)} < P_0$. Therefore, at a fixed stellar baryon number $\mathcal{A}$, both $r_{\rm A}$ and the radius $r_{\rm m}$ of the mixed-phase core are expected to be larger than the radius $r_{\rm B}$ for the core of pure phase B.

# Chapter 8

# STRANGE MATTER AND STRANGE STARS

## 8.1. Introduction

This chapter is different from others. It deals with objects which, at the time of this writing, are hypothetical. We do not know whether strange matter and strange stars exist or not. We start with the discussion of the strange matter hypothesis (§ 8.3). Its history is described in § 8.4. Models of strange matter are presented in §§ 8.5–8.8. Finally, § 8.9 is devoted to other speculative self-bound phases of super-dense matter.

Assuming the strange matter hypothesis, we develop models of strange stars. The history of strange-star physics is presented in § 8.10. Models of non-rotating bare strange stars are studied in § 8.11. In § 8.12 we present the proof that strange stars do not contain heavy quarks ($c$, $b$, $t$). The surface of bare strange stars is studied in §§ 8.14 and 8.15. § 8.16 is devoted to strange stars with the crust of normal matter. Apparent radii of strange stars, important for their observational identifications, are studied in § 8.17. A conversion of neutron stars into strange stars is outlined in § 8.19. Basic properties of "even stranger" stars built of abnormal matter and Q-matter are reviewed in § 8.20. Rotating strange stars are studied in § 8.21.

## 8.2. Units

A comment on units may be useful. In the literature, two different conventions concerning the bag constant $\mathcal{B}$ of the MIT Bag Model are used. Particle physics theorists prefer to set $\hbar = c = 1$ and measure $\mathcal{B}$ in $\mathrm{MeV}^4$. We will denote this value by $\overline{\mathcal{B}}$. However, one often prefers to measure the bag constant in the "nuclear physics units," MeV $\mathrm{fm}^{-3}$, and we follow this convention. Our value of the bag constant, denoted by $\mathcal{B}$, is related to the particle-physics one via $\mathcal{B} = \overline{\mathcal{B}}/(\hbar c)^3$. For example, $\overline{\mathcal{B}}^{1/4} = 145$ MeV corresponds to $\mathcal{B} = 57.5$ MeV fm$^{-3}$.

## 8.3. The strange matter hypothesis

We are used to the fact that stable forms of matter do not contain strange quarks. In particular, under laboratory conditions hyperons and hypernuclei are unstable and decay into nucleons and ordinary nuclei, respectively. However, the situation changes if we consider the matter under extremely high pressure, at densities at which baryons loose their identity and transform into quasi-free $u$ and $d$ quarks (for simplicity, we neglect an admixture of $s$ quarks provided by hyperons which may appear at high densities). At such high densities, and for temperatures relevant for stellar objects, the matter is strongly degenerate and thermal effects are small. At a given pressure $P$, a stable state of the matter is realized at the smallest baryon chemical potential $\mu_{\rm b} = (\mathcal{E} + P)/n_{\rm b}$.

The situation presented schematically in the left panel of Fig. 8.1 corresponds to the deconfinement of quarks in a nucleon matter at $P = P_{\rm D}$. If $P > P_{\rm D}$, the baryon matter is unstable with respect to the transformation into the $ud$ plasma. However, the value of $\mu_{\rm b}$ in the deconfined phase can be further reduced by the conversion of about half of $d$ quarks into $s$ quarks. The conversion proceeds via weak-interaction process, in contrast to the conversion of baryons into quarks, which is a strong-interaction process. Actually, as shown in the left panel of Fig. 8.1, the $uds$ matter in weak-interaction equilibrium, with strangeness per unit baryon number $\approx -1$, becomes the *ground state of the matter* for $P > P_{\rm DS}$, where $P_{\rm DS}$ is smaller than $P_{\rm D}$. Let us assume that the models of $ud$ and $uds$ quark matter can be extrapolated to the zero pressure. Such an extrapolation is possible within the MIT Bag Model, described in § 8.5. The standard situation, presented schematically in the left panel of Fig. 8.1, is such that at small pressures, particularly at $P \to 0$, the stable matter consists of nucleons and does not contain strange quarks. The true ground state of the cold "ordinary" matter is the $^{56}$Fe crystal with $E_0 \equiv \mu_{\rm b}(P = 0) = 930.4$ MeV (§ 3.2).

The *strange matter* hypothesis is shown schematically in the right panel of Fig. 8.1. The $uds$ matter remains absolutely stable even at $P = 0$. This means that sufficiently heavy droplets of the $uds$ matter with baryon number $A > A_{\rm min} \gg 10$ (to reduce the surface effects) have lower energy per baryon at zero pressure than $^{56}$Fe. Then ordinary heavy nuclei are *metastable* with respect to the transition into the strange-matter. This does not contradict experimental data because the lifetime of the metastable nuclei could be extremely long. The transition to the true ground state requires, roughly speaking, an $A$th order weak-interaction process needed to *simultaneously* convert $A \sim 10^2$ of $d$ quarks into $s$ quarks. It is a simple exercise to show that the lifetime is longer than

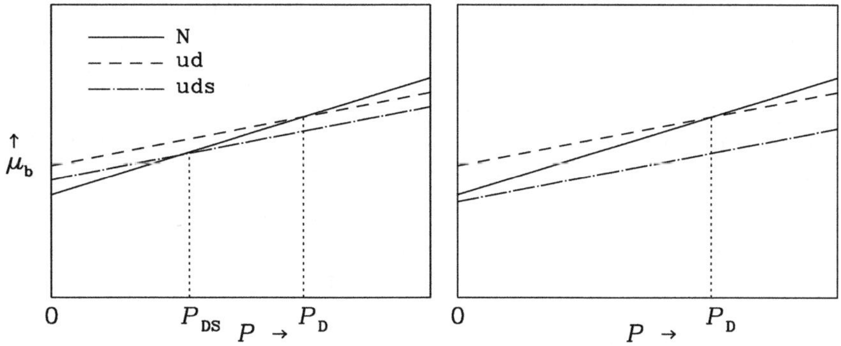


*Figure 8.1.* Baryon chemical potential versus pressure for baryonic matter (N), electrically neutral $ud$ matter, and $uds$ matter in weak-interaction equilibrium. $P_D$ is the threshold pressure for the deconfinement of the $ud$ matter, and $P_{DS}$ is the threshold pressure above which the $uds$ matter becomes more stable than the baryonic one. *Left:* The standard situation in which $^{56}$Fe is the ground state at zero pressure. *Right:* The strange-matter hypothesis; the true ground state at zero pressure is realized by a self-bound $uds$ matter.

$10^{100}$ yr. On the other hand, Fig. 8.1 shows, in agreement with experiments, that at low pressures the $ud$ matter is unstable with respect to the conversion into nucleons; this conversion takes place through strong interactions, on timescales $\sim 10^{-22}$ s.

While the strange-matter hypothesis may seem shocking at a first glance, its validity does not have any immediate dramatic consequence for the world around us. The strange matter will not be produced from the ordinary matter under terrestrial conditions, or under conditions prevailing in normal stars. The possibility of its production under extreme astrophysical conditions – in the cores of neutrons stars, supernovae, or in the early Universe – will be outlined in § 8.4. Now we would like to comfort the reader; even if nuggets of strange matter exist somewhere around us, they have positive electric charge (see §§ 8.14 and 8.15). Therefore, they will not represent a danger characteristic of horrifying ice-nine which absorbs all matter after touching it (Vonnegut, 1963). However, if the strange-matter hypothesis is true then some (or even all) neutron stars could actually be strange stars, built exclusively or predominantly of strange matter. Astrophysical arguments for and against the existence of strange matter and strange stars in the Universe are presented in § 9.8.

## 8.4. Strange matter – history

An intriguing possibility that atomic nuclei could be only an isomeric metastable state of a more general system of baryon number $A$ was first discussed by Bodmer (1971). He assumed that for a sufficiently large $A$ there exists a "collapsed nucleus" $C_A$ of much smaller radius and much higher density than the "normal nucleus" $N_A$. At a given $A$, the radius of $C_A$ was supposed to be $\sim 1/3$ of the normal nucleus, so that the density of the collapsed nuclei was about thirty times higher than of the normal ones. The hypothetical $C_A$ nucleus was much tighter bounded than $N_A$ and would correspond to the true ground state of the matter. Bodmer considered three models of $C_A$. Two of them were related to an abnormal state of nuclear and hyperonic matter with an enormous binding energy; such states are similar to the "abnormal matter" of Lee & Wick (1974), discussed in § 8.9. In the third model, $C_A$ was built of the $uds$ matter and had small positive electric charge. Using qualitative arguments, Bodmer showed that the existence of $C_A$ is not in conflict with experimental data, if *metastable* $N_A$ are sufficiently long-lived isomers. For example, in one specific case he obtained the lifetime of $N_A$ significantly longer than $10^{24}$ yr. The paper by Bodmer was of qualitative character; the MIT Bag Model of hadrons was formulated three years later.

Quantitative studies started only after the formulation of the MIT Bag Model (Chodos *et al.*, 1974; DeGrand *et al.*, 1975). Eight years after the Bodmer paper, the possibility of the existence of long-lived (as compared to the strong-interaction timescale $\sim 10^{-22}$ s) "quark nuclei" was studied within the Bag Model by Chin & Kerman (1979). They stressed the necessity of $s$ quarks for the stability of their "multiquark droplets". These $uds$ droplets of $A > 10$ were shown to be stable with respect to strong interactions and decayed via weak leptonic processes. With the lifetime longer than $10^{-4}$ s, the multiquark droplets of Chin & Kerman were actually metastable and converted eventually into nuclei, in contrast to the strict strange matter hypothesis which would made them absolutely stable. Similar ideas on the existence of droplets composed of $u$, $d$, and $s$ quarks were put forward independently and nearly simultaneously by Terazawa, who proposed to call them "super-hypernuclei" (Terazawa 1979, as quoted by Terazawa 1989a,b). However, his paper, published in symposium proceedings, was unnoticed at that time and had no impact on the later development of the idea of strange matter in the mid 1980s.

The strange-matter investigations got a real start after the classical paper by Witten (1984) who proposed a cosmological scenario, in which "quark nuggets" with equal numbers of $u$, $d$ and $s$ quarks could be formed. He suggested that these nuggets could appear during the hadronization epoch in the early Universe, $\sim 10^{-5}$ s after the Big Bang. During a first-order phase transition, associated with the hadronization, some fraction of matter could condense as quark nuggets which, if stable, could form a component of the dark matter in the today Universe. Using the simplest version of the MIT Bag Model with massless non-interacting quarks (§ 8.5), Witten showed that for reasonable values of the bag constant the $uds$ matter could be *absolutely stable* (the true ground state at zero pressure).

Shortly after the Witten's paper, Farhi & Jaffe (1984) published the paper entitled "Strange matter". They explored in detail properties of the $uds$ matter, including the effects of finite $s$ quark mass and lowest-order QCD interactions (see § 8.6). Using their version of the MIT Bag Model they demonstrated the existence of a wide region in the Bag-Model parameter space, in which $uds$ quarks could form the ground state of the matter.

Since then, the strange matter hypothesis has been addressed in numerous papers, from theoretical and experimental points of view. In particular, strange matter has been searched for in cosmic rays and in relativistic heavy-ion collisions. The physics of strange matter is reviewed by Glendenning (2000), Weber (1999), Madsen (1999), and Weber (2005). A detailed review of earlier work, done before 1991, is given by Madsen & Haensel (1991).

## 8.5. The simplest "toy" MIT Bag Model

The EOS of quark matter was described in § 7.5. Here, we consider the simplest MIT Bag Model which assumes massless and non-interacting quarks confined to a "bag" of a QCD vacuum. The confinement is described by the bag constant $\mathcal{B}$, which determines the difference between the energy densities of the standard and QCD vacua (§ 7.5). Let us consider the matter of $u$, $d$ and $s$ quarks. The equilibrium with respect to the weak interaction processes, Eq. (7.21), combined with the electric neutrality condition, Eq. (7.33b), implies

$$n_e = 0 \, , \qquad n_u = n_d = n_s \, , \tag{8.1}$$

so that the baryon number density is $n_{\rm b} = n_u$.

The energy density given by the MIT Bag Model is (§ 7.5)

$$\mathcal{E} = \mathcal{E}_{\rm kin} + \mathcal{B} \, , \tag{8.2}$$

where $\mathcal{E}_{\rm kin}$ is the contribution from the kinetic energy of quarks. The total number of internal quark degrees of freedom is $3_{\rm color} \times 3_{\rm flavour} \times 2_{\rm spin} = 18$. Using the formula for the energy density of an ultra-relativistic Fermi gas combined with the perfect flavor-color symmetry of our system we get

$$\mathcal{E}_{\rm kin} = b \, n_{\rm b}^{4/3} \, , \qquad b = (9/4)\, \pi^{2/3} \hbar c = 952.37 \ {\rm MeV \ fm} \, . \tag{8.3}$$

The pressure is readily obtained from

$$P = n_{\rm b}^2 \frac{{\rm d}}{{\rm d}n_{\rm b}} \left( \frac{\mathcal{E}}{n_{\rm b}} \right) = \frac{1}{3} \, b \, n_{\rm b}^{4/3} - \mathcal{B} \, . \tag{8.4}$$

It is the sum of the positive contribution from the quark kinetic energy and the negative contribution from the vacuum, which squeezes the quark plasma. Equation (8.4) implies the existence of a self-bound state at zero pressure with the baryon number density

$$n_{\rm b,s} = (3\mathcal{B}/b)^{3/4} = 0.28665\ (\mathcal{B}_{60})^{3/4}\ {\rm fm}^{-3}\ , \tag{8.5}$$

where $\mathcal{B}_{60} = \mathcal{B}/60$ MeV fm$^{-3}$, and the lower label "s" in $n_{\rm b,s}$ reminds us that it is the *surface* baryon density of an object built of the $uds$ matter. The energy per unit baryon number at $P = 0$ is

$$E_0 = 4\,(b/3)^{3/4}\,\mathcal{B}^{1/4} = 837.26\ (\mathcal{B}_{60})^{1/4}\ {\rm MeV}\ , \tag{8.6}$$

and the mass density

$$\rho_{\rm s} = 4\mathcal{B}/c^2 = 4.2785 \times 10^{14}\ \mathcal{B}_{60}\ {\rm g\ cm}^{-3}\ . \tag{8.7}$$

The strangeness per unit baryon number for this $uds$ matter is $S/A = n_s/n_{\rm b} = -1$. Its analog in the baryonic world is the matter composed of $\Lambda^0$ hyperons.

One can also contemplate a deconfined equivalent of a neutron matter, which is an electrically neutral $ud$ plasma. We will use the prime to distinguish the quantities referring to the $ud$ quark matter from those of the $uds$ matter. At a given baryon number density $n'_{\rm b}$ we have

$$n'_e = n'_s = 0\ , \qquad n'_d = 2n'_u = 2n'_{\rm b}\ . \tag{8.8}$$

It is easy to show, that thermodynamic formulae for the $ud$ matter remain the same as for the $uds$ matter, but with a different coefficient $b'$, related to $b$ by

$$b'/b = (1 + 2^{4/3})/3 = 1.1733\ . \tag{8.9}$$

It is also easy to see, that at any pressure (or baryon density) the $uds$ phase is energetically preferable over the $ud$ one. At a given $P$, a suitable thermodynamic potential is the enthalpy per unit baryon number, which is equal to the baryon chemical potential $\mu_{\rm b}(P)$,

$$\mu_{\rm b}(P) = \frac{\mathcal{E} + P}{n_{\rm b}} = \frac{3P + 4\mathcal{B}}{n_s}\left(\frac{P}{\mathcal{B}} + 1\right)^{-3/4}\ . \tag{8.10}$$

Therefore, at any $P$

$$\mu'_{\rm b}(P)/\mu_{\rm b}(P) = n_{\rm b}/n'_{\rm b} = (b'/b)^{3/4} = 1.1273\ . \tag{8.11}$$

In particular, at $P = 0$,

$$E'_0/E_0 = 1.1273\ , \quad n'_{\rm b,s}/n_{\rm b,s} = 0.88705\ , \quad \rho'_{\rm s} = \rho_{\rm s}\ . \tag{8.12}$$

Thus, the $ud$ phase is unstable with respect to weak interactions and transforms into the $uds$ phase via the strangeness-changing reaction $u{+}d \to s{+}u$. In other words, for the deconfined quark matter the "strange" state with $S/A = -1$ is preferable over the "non-strange" state with $S/A = 0$. This is in contrast with the observed "baryonic world", where stable forms of matter at zero pressure do not contain strange quarks and have $S/A = 0$; hyperons and/or hypernuclei decay into nucleons. The same is true for dense stellar matter at subnuclear densities.

The simplest Bag Model described above will be referred to as SQM0, with SQM being the acronym for "Strange Quark Matter" (or "Self-bound Quark Matter"). Of course, we have to specify the only parameter of the model, $\mathcal{B}$. For the hypothesis of strange matter to be valid, we should have $E_0 < 930.4$ MeV. Using Eq. (8.6), we get

$$E_0 < 930.4\,\text{MeV} \quad \Longrightarrow \quad \mathcal{B}_{60} < 1.53\,. \tag{8.13}$$

On the other hand, our SQM0 model should be consistent with reality. For instance, nuclei should be stable with respect to the formation of droplets of the $ud$ matter. The necessary condition is

$$E_0' > 930.4\,\text{MeV} \quad \Longrightarrow \quad \mathcal{B}_{60} > 0.944\,. \tag{8.14}$$

A very reasonable stronger condition is that neutrons should not coagulate into droplets of the $ud$ matter. It implies

$$E_0' > 939.6\,\text{MeV} \quad \Longrightarrow \quad \mathcal{B}_{60} > 0.982\,. \tag{8.15}$$

In what follows we will use this last condition to fix the lower bound on $\mathcal{B}$ in the SQM0 model.

Thus we have obtained the simplest SQM0 "toy model". Its EOS is given by Eqs. (8.2)–(8.4) with the bag constant in the range $0.982 < \mathcal{B}_{60} < 1.53$.

## 8.6. The Bag Model with massive strange quarks and QCD interactions

The model was described in § 7.5. The quark ($\Omega_\text{q}$) and vacuum energy ($\mathcal{B}$) contribution to the thermodynamic potential per unit volume is

$$\Omega_Q = \Omega_\text{q} + \mathcal{B}\,. \tag{8.16}$$

At sufficiently high densities, $\Omega_\text{q}$ can be calculated using a perturbation expansion in the QCD coupling constant $\alpha_\text{s}$. The expansion has to be treated within the renormalization scheme, § 7.5. In particular, we have to use the *renormalized* values of the QCD coupling constant and strange-quark mass, which will depend on the selected value of the *renormalization point* denoted by $\rho_\text{R}$. We will follow the prescription of Farhi & Jaffe (1984) who put the renormalization point at the energy $\rho_\text{R} = m_n c^2/3 = 313$ MeV typical for quark chemical potentials.

The renormalized coupling constant $\alpha_{\rm s}$ decreases with increasing mean quark energies. We restrict ourselves to the first-order approximation valid for a sufficiently small $\alpha_{\rm s}$. In this approximation, the contributions of all three flavors to $\Omega_{\rm q}$ are additive, Eq. (7.26). The $u$ and $d$ quarks can be treated as massless. The calculation of the renormalized values of $\alpha_{\rm s}$ and the strange quark mass $m_{\rm s}$ is somewhat ambiguous. As we have seen in § 7.5, $\alpha_{\rm s}$ does not need to be renormalized in the lowest-order approximation.

As for the renormalized $s$-quark mass, let us use Eq. (7.31). The numerical values of parameters are quite uncertain (§ 7.5). To be specific, let us put the current-quark mass $m_s^{\rm (c)}c^2 = 100$ MeV (Yao *et al.*, 2006), $\Lambda = 100$ GeV, $\alpha_{\rm s} = 0.2$, and $\rho_{\rm R} = m_n c^2/3$ (Farhi & Jaffe, 1984). In this way we get $m_s = 2.1 m_s^{\rm (c)} \simeq 200\ {\rm MeV}/c^2$, which will be our "standard renormalized mass" of the $s$ quark in this chapter.

The hypothetical self-bound state at $P = 0$ is determined from the condition

$$\Omega_{\rm q} + \Omega_e + \mathcal{B} = 0 \ . \tag{8.17}$$

It specifies the associated baryon number density $n_{\rm b,s}$ and the energy per baryon $E_0$.

In what follows, we will *assume* that $\alpha_{\rm c} \simeq 0.2$, so that the lowest-order approximation will be satisfactory. However, under the conditions relevant for compact stars typical quark energies $\epsilon \sim \mu_f$ are insufficiently large to reach the true Asymptotic Freedom regime (Asymptopia) which requires $\mu_f \gg 1$ GeV. Thus, the Bag Model is phenomenological, and its results should be taken with a grain of salt. Luckily, as we will see in §§ 8.7 and 8.9 and as can be verified for other even more exotic models of dense matter, the very hypothesis on a self-bound high-density ground state of the matter at zero pressure implies a generic shape of the EOS, similar to that given by the Bag Model. Therefore, qualitative results obtained within the Bag Model may have much broader meaning than the Bag Model itself.

In Fig. 8.2 we show lines of constant $E_0$ in the $\mathcal{B} - m_s$ plane. The region of the stable strange matter is bounded by the dash-dot line and the thin vertical line. The region to the right of the curve of $E_0 = 939$ MeV corresponds to the $uds$ matter which is *unstable* with respect to nucleon emission. At a fixed $m_s$, the strange matter exists for $\mathcal{B}_{\rm min} < \mathcal{B} < \mathcal{B}_{\rm max}$. The allowed range of $\mathcal{B}$ becomes narrower with the growth of $m_s$. The thin vertical line gives the value $\mathcal{B}_{\rm min}$ below which the (non-strange) $ud$ matter is preferable over an ordinary nucleon matter. This region is excluded, because it corresponds to a spontaneous coagulation of nucleons into droplets of the $ud$ matter, which would further convert into the SQM, destroying the nucleon matter around us.

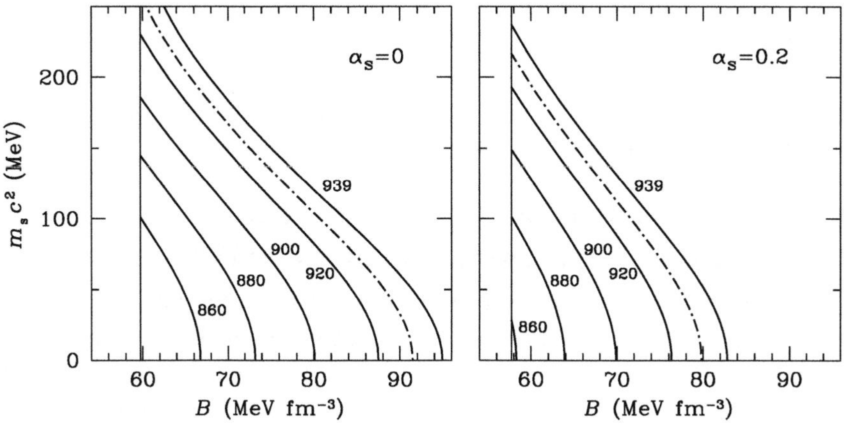


*Figure 8.2.* Lines (in the $m_s - \mathcal{B}$ plane) of constant energy per unit baryon number (in MeV, labeled near curves) for strange quark matter at zero pressure. The dot-and-dashed lines refer to $^{56}$Fe crystal (930.4 MeV). *Left*: Non-interacting quarks. *Right*: Lowest order calculation for $\alpha_s = 0.2$. For further explanations see the text.

The simultaneous effect of $\alpha_s$, $m_s$ and $\mathcal{B}$ on $E_0$ is illustrated in the right panel of Fig. 8.2. The main effect of $\alpha_c$ is to narrow the range of allowed values of $\mathcal{B}$, lowering both $\mathcal{B}_{\rm min}$ and $\mathcal{B}_{\rm max}$.

Thus, the Bag Model predicts a rather large window of reasonable values of $\mathcal{B}$ and $m_s$, where a self-bound SQM is the true ground state of cold dense matter at $P = 0$.

## 8.7. Other models of strange quark matter

There are many (possibly, less popular) models of self-bound stable quark matter different from the MIT Bag Model.

The Nambu–Jona-Lasinio model has already been mentioned in § 7.5. In contrast to the above models, it deals with massive "constituent quarks" and respects the chiral symmetry. In recent years, it has been frequently used for describing the phase transition to the quark matter at high densities. For some choice of parameters, the Nambu–Jona-Lasinio model predicts a stable self-bound state at $P = 0$, corresponding to strange matter (see, e.g., Hanauske *et al.*, 2000). A model with self-consistent density-dependent quark masses can also predict a stable strange matter in some region of the parameter space (Wang, 2000). Dey *et al.* (1998) proposed a model for quark matter using density-dependent quark masses and a color-dependent vector interquark potential. It predicts a strongly bound state of the stable quark matter at $P = 0$, with $\rho_s > 10^{15}$ g cm$^{-3}$. In the Cloudy Bag Model, a pion cloud is coupled to the quark-confining bag. The model parameters are determined by fitting the baryon masses. This model can also predict a stable state of the quark matter at $P = 0$ (Ng *et al.*, 2003).

This list is not complete. However, as we show in § 8.8, the variety of model EOSs of the strange quark matter in compact stars has a generic linear form described actually by only two parameters.

## 8.8. The equation of state of strange matter

### 8.8.1 The linear approximation of the EOS

An important feature of the EOS of the self-bound SQM is the vanishing pressure at $\rho < \rho_s$. For densities only slightly higher than $\rho_s$, the EOS has a linear form $P \propto (\rho - \rho_s)$. It is very fortunate that this simple form turns out to be a reasonable approximation at higher densities, up to the maximum density in stable strange stars. Linear approximation is excellent for the MIT Bag Model (Zdunik, 2000) and is quite reasonable for other models of strange matter(Gondek-Rosińska *et al.*, 2000).

Let us focus on the Bag Model. A linear approximation reads

$$P \approx ac^2(\rho - \rho_s) , \tag{8.18}$$

where *constants* $a$ and $\rho_s$ are determined by fitting the exact EOS (Zdunik, 2000).The value of the fit parameter $\rho_s$ is usually slightly different from the "surface density" $\rho(P = 0)$ for an *exact* EOS. Notice that the linear form is exact for massless quarks (either free or interacting); in which case $a = 1/3$. As shown by Zdunik (2000), for $m_s c^2 \leq 300$ MeV and $\alpha_c \leq 0.6$, at densities relevant for stable strange stars the linear fit reproduces the exact EOS within a few percent. The linear approximation leads to strange star models which at $M \gtrsim 0.3\ M_\odot$ coincide (in radius, mass, and other parameters) within better than 1% with models calculated using the exact EOS. Moreover, Zdunik (2000) obtained fitting formulae which express $a$, $\rho_s$, and $n_{b,s}$ in terms of $\mathcal{B}$, $m_s$, and $\alpha_c$ of the Bag Model.

Some examples of $a$ and $\rho_s$ obtained for the different models of SQM are given in Table 8.1. One notices significant difference between the Bag Model EOSs and those based on the model of Dey *et al.* (1998). For the Bag Model, $a$ is slightly lower that $1/3$. The effects of strange-quark mass and QCD interactions lower $a$ by less than ten percent. The effect of $m_s$ is dominant; we have $m_s c^2 = 200$ MeV and 100 MeV for the SQM1 and SQM2 EOSs, respectively. On the contrary, for the model of Dey et al., the interactions are important and increase $a$ by fifty percent compared to $a = 1/3$. The values of $\rho_s$ are two-three times larger than for the Bag Model. Therefore, the SS1 and SS2 EOSs give very compact and dense strange stars. Precise arguments based on the *scaling* properties of stellar configurations can be formulated using the results of Appendix E.

*Table 8.1.* Parameters $a$ and $\rho_s$ in the linear approximation (8.18) for four EOSs of SQM. The SQM1 and SQM2 EOSs are calculated by Zdunik *et al.* (2000) using the MIT Bag Model with $m_s c^2 = 200$ MeV, $\alpha_s = 0.2$, $\mathcal{B} = 56$ MeV fm$^{-3}$ (SQM1); and $m_s c^2 = 100$ MeV, $\alpha_s = 0.6$, $\mathcal{B} = 40$ MeV fm$^{-3}$ (SQM2). The SS1 and SS2 EOSs are obtained by Dey *et al.* (1998) within a vector-interaction model.

| Model EOS | $a$ | $\rho_s$ ($10^{14}$ g cm$^{-3}$) |
|---|---|---|
| SQM1 | 0.301 | 4.50 |
| SQM2 | 0.324 | 3.06 |
| SS1 | 0.463 | 11.54 |
| SS2 | 0.455 | 13.32 |

Two parameters, $\rho_s$ and $a$, completely determine the linear dependence of $P$ on $\rho$. The dependence of the baryon number density on the pressure is specified by an additional parameter, $n_{b,s}$. Analytical expressions are given in Appendix E.

### 8.8.2 The adiabatic index of strange matter

An important parameter, which characterizes the stiffness of the EOS, is the adiabatic index $\gamma = (n_b/P)\, dP/\, dn_b$. It determines the stability of stars with respect to small perturbations. In Fig. 8.3 we show an example of $\gamma$ calculated for the SQM0 EOS. The adiabatic index for strange stars is qualitatively different from that for ordinary neutron stars. The values of $\gamma$ in the outer layer of a strange star are very large. In contrast, we have $\gamma \simeq 4/3$ (see §3.6) in the outer crust of a neutron star. For typical densities in stable strange stars, the value of $\gamma$ is significantly higher than the value $4/3$, predicted by the Asymptotic Freedom of the QCD at $\rho \to \infty$.

### 8.8.3 The effect of quark superconductivity on the EOS

It is widely accepted that in an idealized case of massless quarks the ground state of high-density quark matter in the asymptotic region of weak QCD interactions corresponds to the color-flavor-locked (CFL) superconducting phase. Of course, one gets $n_e = 0$ for massless quarks in normal and superconducting phases of electrically neutral quark matter in beta equilibrium. However, as shown by Rajagopal & Wilczek (2001), the same is true in the CFL phase of quark matter with massive $s$ quarks, provided $m_s$ is not too large ($m_s \lesssim 200$ MeV$/c^2$) but the superfluid gap is sufficiently large ($\Delta \gtrsim 100$ MeV). Under such conditions, the binding energy gain due to CFL

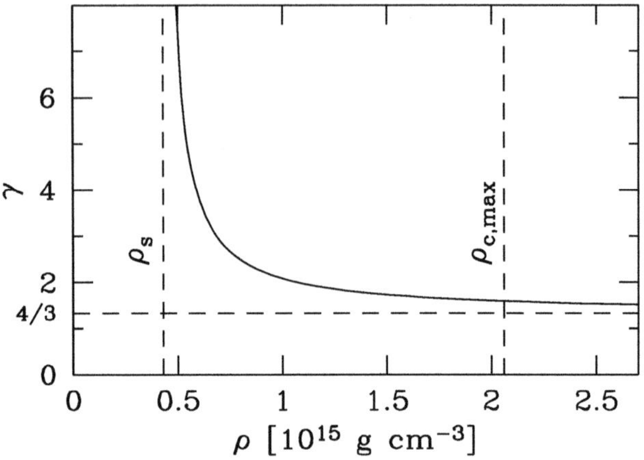


*Figure 8.3.* Adiabatic index $\gamma$ versus density for the SQM0 EOS. The left vertical dashed line shows the surface density of the bare strange star. The right dashed vertical line is the maximum central density of non-rotating strange stars. The horizontal line is $\gamma = 4/3$ characteristic of a free ultrarelativistic Fermi gas.

pairing of quarks prevails over the kinetic energy increase, and the ground state is realized for the flavor-symmetric quark matter with $n_u = n_d = n_s = n_b/3$, which implies an *enforced* absence of electrons.

The enforced condition $n_e = 0$ in the CFL superconducting state is easily derived in the case in which pairing is the only quark-quark interaction (Rajagopal & Wilczek, 2001). Then the normal state is a Fermi gas of $u$, $d$, and $s$ quarks. The CFL superconducting state is calculated using the Bardeen-Cooper-Schrieffer (BCS) theory assuming the weak-coupling regime, $\Delta/\overline{\mu} \ll 1$. Here, $\overline{\mu} = (\mu_d + \mu_u + \mu_s)/3$ is the mean quark chemical potential (one third of the baryon chemical potential of the quark matter). In order to study CFL states, one uses the thermodynamic potential $\Omega$ per unit volume. To be specific, let us consider the CFL condensate of green-color $d$ quarks and red-color $s$ quarks. Let $k_{Fd}$, $k_{Fs}$, $\mu_d$, and $\mu_s$ be the appropriate Fermi momenta and chemical potentials. The attractive pairing contribution to $\Omega$ is maximized for $k_{Fs} = k_{Fd}$. However, this introduces a nonzero $\delta\mu = (\mu_s - \mu_d)/2$, which acts in the inverse direction and increases $\Omega$. For $m_s = 0$ we have flavor symmetry and $\delta\mu \equiv 0$. With increasing $m_s$, the minimum of $\Omega$ is still realized by the $k_{Fs} = k_{Fd}$ state, but the increase of $\delta\mu$ shifts the minimum up. As shown by Rajagopal & Wilczek (2001), the paired CFL state remains preferable if $m_s$ is not too large as compared to $\Delta$,

$$|m_s^2 c^4/(4\overline{\mu}) - \delta\mu| < \Delta \, . \tag{8.19}$$

The above discussion can be generalized to the full $uds$ system. The main conclusion is that if $m_s$ obeys the inequality (8.19), the system is in the CFL superconducting state with $n_e = 0$. The CFL pairing enforces electrical neutrality of the quark component of dense matter.

Let us assume that all quarks are paired with the same gap $\Delta$. Let the gap be sufficiently large so that electrons are absent and $n_u = n_d = n_s$. The thermodynamic functions can be calculated including the terms linear and quadratic in $\Delta$ (Alford *et al.*, 2001). One can choose the mean quark chemical potential $\overline{\mu}$ as an independent thermodynamic variable. At a fixed $\overline{\mu}$, the change in the thermodynamic potential $\Omega(\overline{\mu})$ provided by CFL pairing is

$$\Omega'(\overline{\mu}) \equiv \Omega_{\rm s}(\overline{\mu}) - \Omega_{\rm n}(\overline{\mu}) = \Omega'_1(\overline{\mu}) + \Omega'_2(\overline{\mu}) , \tag{8.20}$$

where $\Omega'_1$ results from the enforced equality of all three Fermi momenta and the $\Omega'_2$ is due to a finite gap in the quark energy spectrum. The indices "n" and "s" refer to normal and superconducting states, respectively. The calculation of $\Omega'_1(\mu)$ is straightforward and will be omitted here. This term is usually insignificant, and we can set $\Omega'(\overline{\mu}) = \Omega'_2(\mu)$. Then

$$\Omega'(\overline{\mu}) \simeq -\frac{3}{\pi^2}\left(\frac{\Delta}{\overline{\mu}}\right)^2 \frac{\overline{\mu}^4}{(\hbar c)^3} . \tag{8.21}$$

The corresponding changes in pressure, energy density, and baryon density are

$$P'(\overline{\mu}) = \mathcal{E}'(n_{\rm b}) \simeq \frac{3}{\pi^2}\left(\frac{\Delta}{\overline{\mu}}\right)^2 \frac{\overline{\mu}^4}{(\hbar c)^3} , \quad n'_{\rm b}(\overline{\mu}) \simeq \frac{2}{\pi^2}\left(\frac{\Delta}{\overline{\mu}}\right)^2 \frac{\overline{\mu}^3}{(\hbar c)^3} . \tag{8.22}$$

Using these expressions, one can calculate the decrease of the energy per baryon in the strange quark matter at $P = 0$; it will be denoted by $E'_0$. Standard thermodynamical relations yield

$$E'_0 \simeq -\frac{3}{\pi^2}\left(\frac{\Delta}{E_0}\right)^2 \frac{E_0^4}{(\hbar c)^3 n_s} , \tag{8.23}$$

where $E_0$ is the energy per baryon in the normal quark matter at $P = 0$ and $n_s$ is the corresponding baryon number density; terms of higher order with respect to $\Delta/E_0$ are neglected. Thus, the additional binding of the quark matter at zero pressure grows quadratically with increasing $\Delta$ and widens the stability window in the $m_s - \mathcal{B}$ plane (as shown in Fig. 8.4). Although the effect of CFL pairing on the EOS of quark matter at high pressures is not large ($\propto (\Delta/\overline{\mu})^2$) it can be crucial for the very existence of a stable state at zero pressure.

## 8.9. Even stranger matter

Several other types of hypothetical self-bound state of dense matter, which could constitute the true ground state of the matter at $P = 0$, have been suggested in the past. In the 1970s, it was proposed that pion condensation could

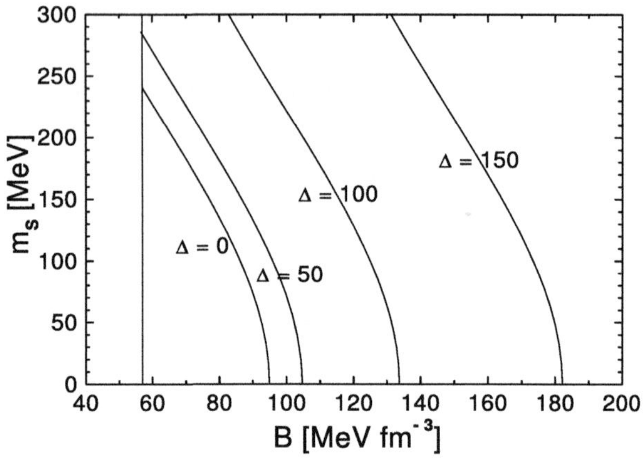


*Figure 8.4.* The increase of the stability region (between the vertical line and a curved line in the $m_s - \mathcal{B}$ plane) of strange quark matter at $P = 0$ due to CFL pairing with a gap $\Delta$ (in MeV). Non-strange quarks are massless and CFL pairing is the only quark-quark interaction. The left vertical boundary results from the condition that the baryonic world around us should be relatively stable (§§8.5, 8.6). From Lugones & Horvath (2003) with the kind permission of the authors.

lead to the appearance of a self-bound, superdense state of pion-condensed nucleon matter (Migdal, 1971, 1974; Hartle, 1975). This phase could exist in the form of "abnormal nuclei" with large atomic number $A$ and the density significantly higher than $\rho_0$ (Migdal, 1971, 1974).[1]

The idea of "abnormal state" of nuclear matter, advanced by Lee & Wick (1974; for a review see Lee 1975), was based on a schematic field-theoretical $\sigma$-model of strongly interacting nucleon matter. In the abnormal state, which appears at sufficiently high densities, nucleons could become nearly massless. This is because the $\sigma$-field term couples to nucleons as a negative addition to the nucleon rest (bare) mass, implying a nearly vanishing nucleon effective mass. This density-dependent effect could lead to the appearance of the second minimum in the dependence of the energy per baryon on $n_{\rm b}$. The minimum occurred at some $n_{\rm b}$ significantly higher than $n_0$, with the binding energy

[1]Let us mention the statement of Hartle (1975) on hypothetical "abnormal nuclei" :"... the problem of their nonobservance until now is apparently no worse than for the "ordinary" superheavy nuclei predicted from the shell model". This statement is no longer true. In the late 1990s, an experimental evidence was found for the formation of relatively stable nuclei with $Z = 112,\ 114,\ 116$, and $A \simeq 280-290$ (for a review, see Hoffmann & Münzenberg 2000, who stress that these results are preliminary). On the contrary, no experimental evidence on the existence of the "abnormal nuclei" has been found up to the time of this writing.

much larger than in the "normal state" at $n_b = n_0$. However, this original $\sigma$-model was schematic and could not pretend to describe quantitatively the normal nuclear matter at $n_b \approx n_0$. Were the $\sigma$-model more complicated to give a correct quantitative description of the nuclear matter at $n_b \approx n_0$, the abnormal state at supranuclear density would disappear (Pandharipande & Smith, 1975; Moszkowski & Källman, 1977).

Supersymmetric extensions of the Standard Model of elementary particles and their interactions predicts the existence of baryonic Q-balls. They are macroscopic self-bound superdense lumps of a scalar-field condensate with a well defined electric and baryonic charge. They were proposed as a hypothetical component of cosmological dark matter. It was also suggested that Q-balls of a macroscopic size could have energy per unit baryon number lower than $^{56}$Fe. Were the Q-matter the real ground state of matter at zero pressure, an absorption of a Q-ball could trigger the conversion of an ordinary neutron star into a Q-star on timescales ranging from $10^7$ yr to $10^{11}$ yr (Kusenko *et al.* 1998, also see Madsen 1998a). For longer conversion timescales, Q-stars would not exist in the today Universe, even if Q-balls existed as a relic from the supersymmetric epoch of the Big Bang.

EOSs of Q-matter were constructed by Bahcall *et al.* (1990). A common feature of these models, shared with the Lee-Wick model of abnormal matter, is that nucleons become nearly massless inside the condensed Q-phase. Two basic parameters of the model are the energy density $U_0$ of the scalar field in the Q-matter and the coupling strength $\alpha_v$ of the vector field to nucleons. As we will see in § 8.20, it is convenient to replace $\alpha_v$ by the dimensionless parameter $\zeta = \alpha_v U_0^{1/2} \pi / \sqrt{3}$. Let us consider the simplest case of $\zeta = 0$. Then the EOS for the Q-matter coincides with the EOS for the strange matter of massless, non-interacting quarks and $\mathcal{B} = U_0$. However, the standard value used by Bahcall *et al.* (1990) is $U_0 = 13.0$ MeV fm$^{-3}$, which corresponds to $\rho(P = 0) \equiv \rho_s = 10^{14}$ g cm$^{-3}$. In this model the density of Q-balls is subnuclear. With increasing $\zeta$, the EOS of the Q-matter becomes stiffer and the value of $\rho_s$ lower. In the limiting case of $\zeta = 16$ considered by Bahcall *et al.* (1990) (for the same value of $U_0 = 13.0$ MeV fm$^{-3}$) they get $\rho_s = 5.5 \times 10^{13}$ g cm$^{-3}$. The predicted density of a self-bound Q-matter at zero pressure is two to five times lower (!) than the normal nuclear density.

## 8.10. Strange stars – history

By strange stars we mean compact objects built entirely or predominantly of self-bound quark matter. These hypothetical objects will be our main topic in this chapter. In § 8.20, we will also outline models of stars built of a self-bound abnormal matter and Q-matter.

To our knowledge, compact stars built entirely of quark matter were first considered in a brief note by Itoh (1970). He used the EOS of free, degenerate

Fermi gas of $u$, $d$, and $s$ quarks of equal mass $m_q$. In the present terminology, his model (with para-Fermi statistics of order 3) corresponds to three colors of quarks. Assuming a huge quark mass $m_q = 10\ \mathrm{GeV}/c^2$, Itoh got the maximum mass of his quark stars $M_{\max} \sim 10^{-3}\, M_\odot$. This result can be easily understood, using a general property of the EOS of free, degenerate Fermi gas, first discussed in the classical paper by Oppenheimer & Volkoff (1939). Writing the Tolman-Oppenheimer-Volkoff equation of hydrostatic equilibrium in a dimensionless form, one sees that $M_{\max}$ scales with rest mass of fermions, $m_{\rm F}$, as $M_{\max} \propto m_{\rm F}^{-2}$. For neutrons, Oppenheimer & Volkoff (1939) got $M_{\max} = 0.7\, M_\odot$. Using the scaling and taking into account the increase of the number of internal degrees of freedom from two for neutrons to six for quarks, we recover the factor of $\sim 10^{-3}$. It should be stressed that the quark matter considered by Itoh (1970) is not self-bound at zero pressure, and his quark stars have $\rho(R) = 0$. They are, therefore, qualitatively different from contemporary strange stars.

First models of stars built exclusively of quark matter and possessing the surface density $\sim 10^{14}$ g cm$^{-3}$ were considered by Brecher & Caporaso (1976). Using the simplest SQM0 EOS of massless non-interacting quarks (§ 8.5) and assuming a rather low value of $\rho_{\rm s} = 2.28 \times 10^{14}$ g cm$^{-3}$, they obtained $M_{\max} = 2.8\, M_\odot$, much higher than for the EOSs of baryon matter available at that time; that is why their paper was entitled "Obese 'neutron' stars". Their bag constant was about half of the today standard value, which easily explains why they obtained such a large $M_{\max}$ for their quark stars (see the scaling property of strange star configurations with respect to $\rho_{\rm s}$, derived in Appendix E).

Eight years later the topic of quark stars was reconsidered in the seminal paper by Witten (1984), who also used the simplest model of the three-flavor quark matter with the EOS of the form $P = \frac{1}{3}(\rho c^2 - \mathcal{B})$. He showed that for the bag constant $\mathcal{B} \approx 60$ MeV fm$^{-3}$ (close to the value needed to reproduce experimental masses of baryons within the MIT Bag Model), the parameters of the maximum-mass configuration for strange stars are similar to those for realistic neutron stars built of baryonic matter ($M_{\max} \sim 2\, M_\odot$, $R_{M_{\max}} \sim 10$ km).

The first detailed models of strange stars, based on a more realistic EOS of strange quark matter, taking into account strange quark mass and the lowest-order QCD interactions, were constructed by Haensel *et al.* (1986a), who considered also specific features of accretion on strange stars. Similar results were presented somewhat later by Alcock *et al.* (1986), who additionally proposed scenarios of the formation of strange stars and analyzed strange stars with the normal crust. These two papers summarized the basic properties of strange stars and their astrophysical manifestations.

Further development of the strange star physics was focused on the refinement of the EOS of strange matter (particularly, beyond the MIT Bag Model) and on various properties of strange stars, such as neutrino emission, rotation,

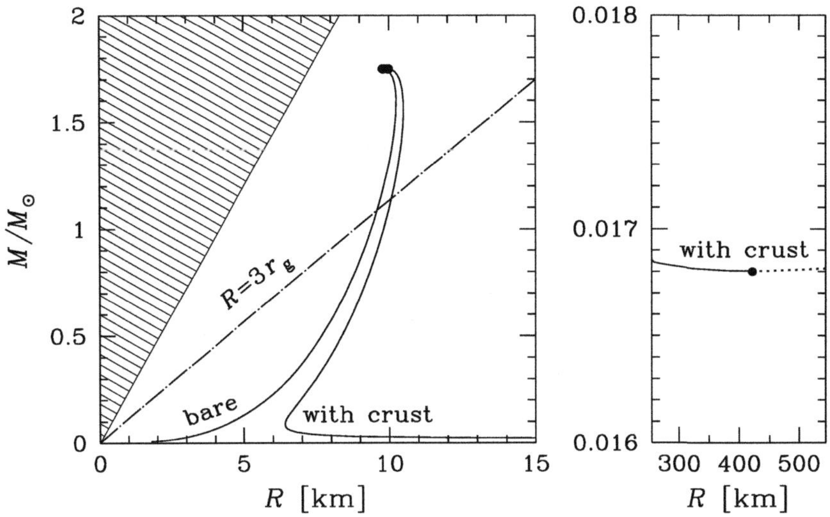


*Figure 8.5. Left:* Mass-radius relation for bare strange stars and for strange stars with the normal crust calculated for the MIT Bag Model EOS with $m_s c^2 = 200$ MeV, $\alpha_s = 0.2$, $\mathcal{B} = 60$ MeV fm$^{-3}$. *Right:* the vicinity of the minimum-mass configuration with the crust. Extremal (maximum and minimum mass) configurations are indicated by filled circles. Strange stars with the crust have the maximum crustal density of $4.3 \times 10^{11}$ g cm$^{-3}$. The dotted segment represents low-mass configurations unstable with respect to small perturbations. The hatched area is prohibited by General Relativity and by the condition $v_s \leq c$. The dash-and-dot line gives the radius of the (last) marginally stable circular orbit around a strange star.

superfluidity, pulsations, electromagnetic radiation, and cooling. The theory of strange stars is reviewed by Glendenning (2000), Weber (1999), Madsen (1999), Bombaci (2001), and Weber (2005). A detailed review of earlier work (before 1991) is given by Madsen & Haensel (1991).

## 8.11. Bare strange stars

Bare strange stars are built exclusively of self-bound strange quark matter (SQM) studied in §§ 8.5 and 8.6. Their surface density is equal to $\rho_s$, the SQM density at zero pressure. It is, therefore, fourteen orders of magnitude larger than the surface density of normal neutron stars.

In what follows, we will illustrate generic properties of strange stars using several MIT Bag Model EOSs.

The $M - R$ diagram for bare strange stars is shown in Fig. 8.5. For stars with $M \gtrsim M_\odot$, the radius changes very little with $M$, $R \simeq 9 - 11$ km, similar to neutron stars with a moderately stiff EOS. However, for lower $M$ the radius of bare strange stars behaves in a completely different way. It decreases

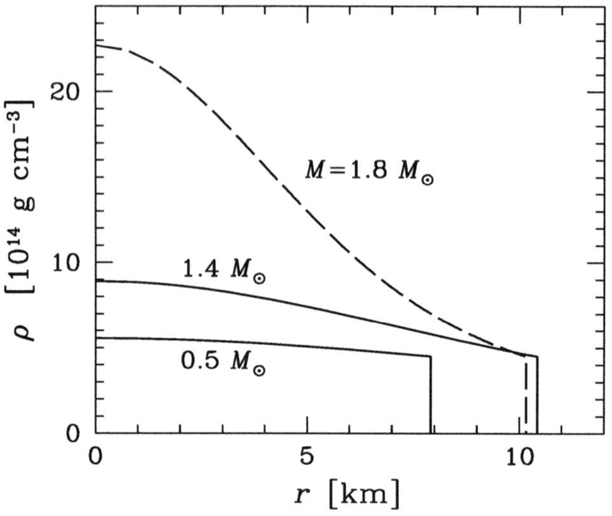


*Figure 8.6.* Mass density $\rho$ versus radial coordinate $r$ for three bare strange stars of different masses, calculated for the SQM1 EOS. The highest $M = 1.80\,M_\odot$ is the maximum mass for this EOS.

monotonically with decreasing $M$, with $R \propto M^{\frac{1}{3}}$ for $M \lesssim 0.3\,M_\odot$. This is easily explained using the Bag Model. Gravitational pull decreases rapidly with decreasing $M$ and can be neglected at $M \lesssim 0.3\,M_\odot$ compared to the pressure of the normal vacuum on the volume filled by the QCD vacuum. This pressure confines the SQM into a sphere of radius $R$. Due to a very high incompressibility of the strange matter, the density in a low-mass strange star is nearly constant $\approx \rho_\mathrm{s}$ (Fig. 8.6). On the other hand, low-mass strange stars can be described by the Newtonian theory, which gives $M \simeq \frac{4\pi}{3}\rho_\mathrm{s}R^3$ and $R \propto M^{\frac{1}{3}}$. The decrease of $R$ with decreasing $M$ is the unique feature of strange stars and Q-stars or stars built of abnormal matter (outlined in § 8.20).

Apart from the dramatic difference in radii between strange stars and neutron stars of $M \lesssim M_\odot$, one notices a systematic difference in the surface redshift $z_\mathrm{surf}$ as a function of $M$. Strange stars are more compact; for $1.4\,M_\odot$ their surface redshift is ~20% higher than for neutron stars (Fig. 8.7). This difference grows with decreasing $M$; it is ~30% for $M = M_\odot$ and reaches ~100% for $M = 0.5\,M_\odot$.

The internal structure of bare strange stars is very different from the structure of neutron stars. First of all, their surface density is huge, $\rho_\mathrm{s} \sim 10^{15}$ g cm$^{-3}$. The density profile in the interior of a bare strange star is very flat (Fig. 8.6). Even at the maximum mass (under the strongest gravitational compression), the central density is only five times higher than the surface one

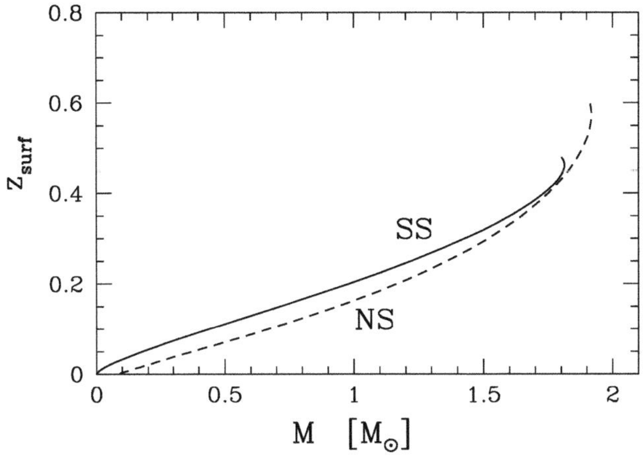


*Figure 8.7.* Surface redshift for bare strange stars (SS; the MIT Bag Model with $m_s c^2 =$ 150 MeV, $\alpha_s = 0.17$, $\mathcal{B} = 60$ MeV fm$^{-3}$) and for neutron stars (NS; the BBB2 EOS, Table 5.3) versus stellar mass.

(in contrast to fourteen orders of magnitude difference for neutron stars!). The density difference decreases rapidly with decreasing $M$. For a $1.4\ M_\odot$ bare strange star, the central density is only 40% higher than at the surface. In low-mass bare strange stars, $M \lesssim 0.3\ M_\odot$, the density is nearly constant.

As in the case of neutron stars, the ratio of the moment of inertia $I$ and $MR^2$ for slowly and rigidly spinning strange stars can be approximated by a simple function of the compactness parameter $r_g/R$, Eq. (1.4). However, this function is qualitatively different from that for neutron stars (Fig. 8.8). A good fit for strange stars is (Bejger & Haensel, 2002)

$$\frac{I}{MR^2} = \frac{2}{5}\left(1 + 0.34\,\frac{r_g}{R}\right) . \tag{8.24}$$

For low-mass strange stars, $I/MR^2 \to 2/5$, which is the value for constant-density Newtonian stars. Notice that except for $M \sim M_{\rm max}$, the ratio $I/MR^2$ for strange stars is larger than for neutron stars (at the same $r_g/R$), because of the flatness of the density profile of strange stars.

The interior of a strange star is composed of the SQM, studied in §§ 8.5, 8.6, and 8.8.3. In what follows we consider normal quarks, with a tiny admixture of electrons to outbalance the net positive charge of quarks (the case of CFL superconducting SQM with the enforced absence of electrons was discussed in § 8.8.3). A typical electron fraction is $n_e/n_b \lesssim 10^{-3}$. It decreases from the surface to the stellar center, as strange quarks become more relativistic, and it depends very strongly on $m_s$ (see § 7.5, in particular, Eq. (7.35)). The radial

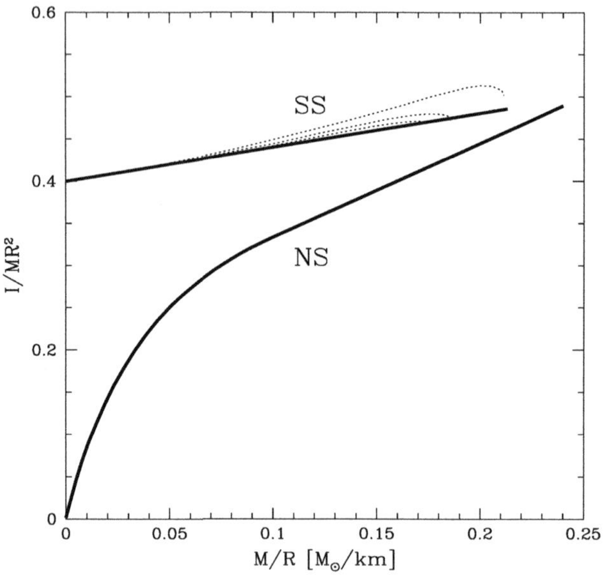


*Figure 8.8.* Dimensionless ratio $I/MR^2$ versus $M/R$ for several EOSs of SQM. The best-fit curve, Eq. (8.24), for strange stars (SS) is plotted by the upper thick line. For comparison, we show also the best-fit curve (NS) for neutron stars. From Bejger & Haensel (2002).

dependence of the fractions of all matter constituents in the interior of a $1.4\ M_\odot$ bare strange star is displayed in Fig. 8.9.

## 8.12. The nonexistence of quark stars with heavy quarks

Apart from the light $u$, $d$ quarks and the moderately massive $s$ quark, the complete set of known quarks includes also heavy $c$, $b$, and $t$ quarks. Their *running masses* are estimated as $m_c c^2 \simeq 1.2$ GeV, $m_b c^2 = 4$ GeV, and $m_t c^2 \simeq 170$ GeV (Yao *et al.*, 2006).

A $c$-quark could be produced in the $uds$ matter via the weak interaction process

$$d + u \longrightarrow c + d \,, \tag{8.25}$$

Physical constraints on the initial and final quark energies ($\epsilon$ and $\epsilon'$) result from the degeneracy of the SQM matter. The initial states of $d$ and $u$ quarks should be occupied, $\epsilon_d \leq \mu_d$ and $\epsilon_u \leq \mu_u$, while the final $d$ quark state should be empty, $\epsilon'_d \geq \mu_d$. The energy needed to create a single $c$ quark is $\epsilon'_c \geq m_c c^2$. The necessary condition resulting from the above constraints is

$$\mu_u \geq m_c c^2 \simeq 1.2 \text{ GeV} \,. \tag{8.26}$$

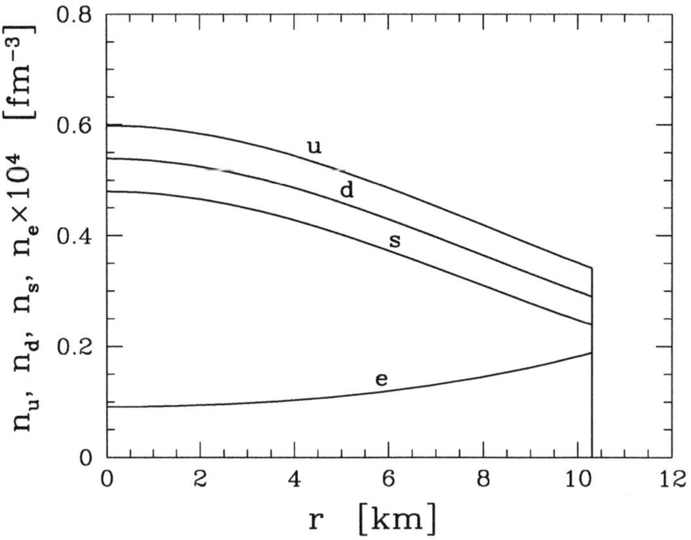


*Figure 8.9.* Number densities of $u$, $d$, $s$ quarks and electrons versus radial coordinate $r$ for a model of a bare strange star of $M = 1.4\ M_{\odot}$, calculated using the MIT Bag Model EOS with $m_s c^2 = 150$ MeV, $\alpha_s = 0.17$, $\mathcal{B} = 60$ MeV fm$^{-3}$.

It yields the lower limit on the number density of $u$ quarks,

$$n_u \geq \frac{1}{\pi^2}\left(\frac{m_c c^2}{\hbar c}\right)^3 = 22.8\ \mathrm{fm}^{-3}\ . \tag{8.27}$$

At such a high density even $s$ quarks are ultrarelativistic, so that

$$\rho \geq \rho_{\mathrm{crit},c} = \frac{9m_c}{4\pi^2}\left(\frac{m_c c^2}{\hbar c}\right)^3 = 1.1 \times 10^{16}\ \mathrm{g\ cm}^{-3}\ . \tag{8.28}$$

This critical density is much higher than the maximum central density of strange stars, $\rho_{\mathrm{c,max}}$ (§ 8.11). A detailed analysis of quark star models with central densities $\rho_{\mathrm{c}}$ higher than $\rho_{\mathrm{c,max}}$ was performed by Kettner *et al.* (1995). Let us consider the $M(\rho_{\mathrm{c}})$ curve in Fig. 8.10. For $\rho_{\mathrm{c}} > \rho_{\mathrm{c,max}}$, the stellar mass decreases and reaches a minimum at $\rho_{\mathrm{c}} \simeq 10^{17}$ g cm$^{-3}$. In the $\rho_{\mathrm{c}} > \rho_{\mathrm{c,crit}}$ part of the $S_1C_1$ segment of the curve $c$ quarks are present in stars, but stellar configurations are unstable with respect to the fundamental mode of radial pulsations because $\mathrm{d}M/\,\mathrm{d}\rho_{\mathrm{c}} < 0$. As we know from § 6.5.2 at the minimum at $C_1$ the stability of one of the radial modes has to change. The character of this change is determined by the static stability criterion in the $M - R$ plane (§ 6.5.1). The condition $\mathrm{d}M/\,\mathrm{d}\rho_{\mathrm{c}} > 0$ is only *necessary* for stability. As we

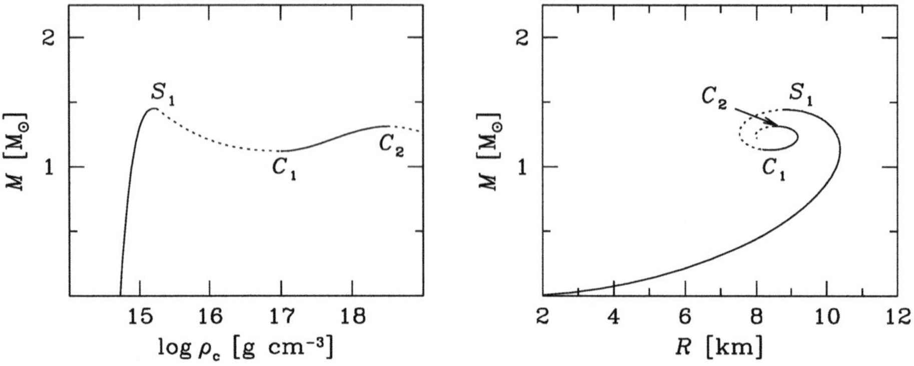


*Figure 8.10.* *Left*: Gravitational mass versus central density for stars built of SQM. The solid branch of the line ending at $S_1$ corresponds to equilibrium configurations of non-rotating strange stars, containing only $u$, $d$, and $s$ quarks. The dotted segment $S_1C_1$ refers to equilibrium configurations unstable with respect to the fundamental mode of radial pulsations. The solid segment $C_1C_2$ exhibits equilibrium configurations of *charmed* stars containing $u$, $d$, $s$, and $c$ quarks. They are unstable against first two modes of radial pulsations. The dotted segment to the right of $C_2$ describes equilibrium configurations of charmed stars which suffer additional instability against the next mode of radial pulsations. *Right*: The same configurations in the $M - R$ plane.

see in the right panel of Fig. 8.10, the radius increases with increasing $\rho_c$ on the $C_1C_2$ segment in the vicinity of $C_1$. Therefore, according to the static stability criterion, passing through $C_1$ leads to the instability of the first overtone of radial pulsations. Charmed stars are unstable with respect to the two lowest modes of radial pulsations and cannot exist in the Universe. This can be confirmed by the calculation of the spectrum of radial pulsations which shows that $\omega_0^2 < 0$ and $\omega_1^2 < 0$ in the $C_1C_2$-branch (Kettner *et al.*, 1995). In the dotted segment to the right of $C_2$ in the left panel of Fig. 8.10 the second overtone of radial pulsations becomes unstable. Superdense branches of quark stars containing $b$ and $t$ quarks are susceptible to more unstable radial modes.

## 8.13. Scaling properties

As we have shown in § 8.8, the approximate and reasonably precise linear representation of the EOS of the SQM is determined by two parameters, $a$ and $\rho_s$. Using this representation, one can rewrite the equations of hydrostatic equilibrium of strange stars in a dimensionless form, derived in Appendix E.

Consider first two EOSs with the same $a$ but with different values of the second parameter, $\rho_s$ and $\rho_s'$. The equilibrium configurations for these two EOSs form two different families, parameterized by the central density. Therefore, we can construct curves $M(R)$, $I(M)$, …, parameterized by $\rho_{\rm c}$. As we show in Appendix E, points of a curve obtained for a given $\rho_{\rm s}$ transform into points of a curve obtained for $\rho'_{\rm s}$ by a scaling transformation. For example, the transformation of $M(R)$ curves reads $R \to R' = (\rho_{\rm s}/\rho'_{\rm s})^{1/2}R$, $M \to M' = (\rho_{\rm s}/\rho'_{\rm s})^{1/2}M$. An extremum of the unprimed curve transforms into an extremum of the primed curve. Particularly interesting is the scaling of parameters of the maximum-mass configuration,

$$M'_{\rm max}/M_{\rm max} = R'_{M'_{\rm max}}/R_{M_{\rm max}} = (\rho_{\rm s}/\rho'_{\rm s})^{1/2} \,, \tag{8.29a}$$

$$\rho'_{\rm c,max}/\rho_{\rm c,max} = \rho'_{\rm s}/\rho_{\rm s} \,, \quad I'(M'_{\rm max})/I(M_{\rm max}) = (\rho_{\rm s}/\rho'_{\rm s})^{3/2} \,. \tag{8.29b}$$

According to the scaling of $M_{\rm max}$ and $R_{M_{\rm max}}$, the maximum surface redshift (reached at $M = M_{\rm max}$) does not depend on $\rho_{\rm s}$. The maximum value of the moment of inertia, $I_{\rm max}$, is reached at a mass slightly lower than $M_{\rm max}$, but of course it scales via the same factor $(\rho_{\rm s}/\rho'_{\rm s})^{3/2}$ as $I(M_{\rm max})$. At a fixed $a$, the ratio $I_{\rm max}/(M_{\rm max}R^2_{M_{\rm max}})$ does not depend on $\rho_{\rm s}$ and is just a number. This number depends weakly on $a$ (Appendix E) and is slightly lower than 1 for realistic EOSs of the SQM (Bejger & Haensel 2002 got the value 0.97).

There exists a maximum circumferential radius $R_{\rm max}$ of bare strange stars (Fig. 8.5). Its value scales according to the same relation as for $R_{M_{\rm max}}$.

Scaling properties are particularly simple (and actually exact) for the MIT Bag Model EOS with non-interacting massless quarks ($a = 1/3$, $\rho_{\rm s} = 4\mathcal{B}/c^2$, model SQM0 of § 8.5). The scaling formulae for $M_{\rm max}$, $R_{M_{\rm max}}$ and $\rho_{\rm c,max}$ were derived by Witten (1984), while those for $M_{\rm b,max}$ and $I_{\rm max}$ were obtained by Haensel *et al.* (1986a),

$$M_{\rm max} = \frac{1.96}{\sqrt{\mathcal{B}_{60}}}\, M_\odot \,, \quad \rho_{\rm c,max} = \frac{2.06 \times 10^{15}}{\mathcal{B}_{60}}\ {\rm g\ cm^{-3}} \,, \tag{8.30a}$$

$$M_{\rm b,max} = \frac{2.63}{(\mathcal{B}_{60})^{3/2}}\, M_\odot \,, \tag{8.30b}$$

$$R_{M_{\rm max}} = \frac{10.71}{\sqrt{\mathcal{B}_{60}}}\ {\rm km} \,, \quad I_{\rm max} = \frac{2.12 \times 10^{45}}{(\mathcal{B}_{60})^{3/2}}\ {\rm g\ cm^2} \,. \tag{8.30c}$$

Calculations performed for more realistic EOSs of the MIT Bag Model with massive $s$ quarks and QCD interactions show that the scaling with respect to $\mathcal{B} \to \mathcal{B}'$ is still very precise (Haensel *et al.*, 1986a). This results from a very weak nonlinearity of the dependence of $\rho_{\rm s}$ on $\mathcal{B}$. Actually, at a fixed interaction constant $\alpha_{\rm s}$ the value of $\rho_{\rm s}$ is *strictly proportional* to $\mathcal{B}$ provided we replace $\mathcal{B} \to \mathcal{B}'$ together with $m_s \to m'_s = m_{\rm s}(\mathcal{B}'/\mathcal{B})^{1/4}$ (Zdunik & Haensel, 1990; Zdunik, 2000).

The scaling formulae enable us to understand why the parameters of massive strange stars ($M \sim 1 - 2\ M_\odot$) are so similar to those for normal neutron stars. Consider first the simplest SQM0 model with massless, non-interacting quarks. For the hypothesis of strange matter to be correct, we should assume that $0.982 < \mathcal{B}_{60} < 1.52$ (§ 8.5). These constraints lead to $1.6\ M_\odot < M_{\rm max} < 2.0\ M_\odot$, 8.7 km $< R_{M_{\rm max}} <$ 10.8 km, which reminds us a typical $M - R$ range for massive neutron stars models built of baryon matter with soft or medium-stiff EOS. Taking a non-zero value of $\alpha_{\rm s}$ reduces the minimum value of the bag constant, $\mathcal{B}_{\rm min}$ (Fig. 8.2). This increases the upper bounds on $M_{\rm max}$ and $R_{M_{\rm max}}$, which to a good approximation are proportional to $(\mathcal{B}_{\rm min})^{-1/2}$.

## 8.14. The surface of a bare strange star with electrons

Let us consider a bare strange star built of normal quark matter with the admixture of electrons. It can be viewed as a huge MIT Bag of radius $R$. The characteristic energy scale in the quark matter is of the order of the quark chemical potential, $\mu_u \simeq \mu_d = \mu_s \simeq 300 - 400$ MeV. The thickness (diffuseness) of the quark surface is $h_{\rm q} \simeq \hbar c/\mu \sim 1$ fm, which is a typical scale of strong interactions. The quarks are bound to the quark surface by strong interactions. However, the SQM contains also a tiny fraction of electrons, $n_e/n_{\rm b} \sim 10^{-4}$. They do not participate in strong interactions, being attached to the quark matter only by Coulomb forces. These forces are provided by the net positive charge of quarks (as a consequence of mass differences, $m_s \gg m_u$, $m_d$; § 8.6). The maximum kinetic energy of degenerate electrons near the quark surface is $\epsilon_{e,{\rm kin}} = \mu_e - m_e c^2 \simeq$ 30–40 MeV. In order to be bound to quarks, the electrons should move in a positive electrostatic potential $V \gtrsim \epsilon_{e,{\rm kin}}/|e|$.

Locally, the quark surface can be treated as plane. Let it be the $xy$-plane, with $z < 0$ in the stellar interior. The electrostatic potential $V(z)$ can be calculated using the Thomas-Fermi approximation (Alcock *et al.*, 1986), with the boundary conditions $V(z \to -\infty) = V_{\rm q}$ and $V(z \to +\infty) = 0$. The calculation gives $V(0) = \frac{3}{4} V_{\rm q}$. The characteristic thickness of the electron surface (called the electrosphere) turns out to be $h_e \sim 100$ fm. The surface electric field is directed outward being of the order of $V_{\rm q}/h_e \sim 3 \times 10^{17}$ V/cm. This is the average electric field in the electrosphere; the local field can be higher. Notice, that the electron distribution extends above the quark surface and the surface electric field extends slightly above the electron distribution. In this sense, the surface of a strange star is not really "bare" (although it is perfectly sharp on macroscopic scales). Because of the positive electrostatic barrier, atomic nuclei (including protons) are repelled from the surface. The theory of the electrosphere should take into account effects of strange-quark mass and temperature (Usov *et al.*, 2005). The mass difference $m_s \gg m_u, m_d$ implies an additional positive charge excess of quarks near the surface. It increases the positive electrostatic potential at the surface.

The sharpness of the surface and a huge surface density of quarks strongly influence the spectrum and flux of emitted photons. Alcock *et al.* (1986) pointed out the importance of a very high plasma frequency in the SQM and predicted that the photon emissivity of the SQM surface should be extremely low. The plasma frequency near the surface is $\omega_{\rm pl} \simeq 20\,(n_{\rm b}/n_0)^{1/3}$ MeV/$\hbar$. No photons (transverse plasmons) can propagate in the SQM at energies $\hbar\omega_{\rm pl} \lesssim 20$ MeV. The temperature near the surface of a cooling strange star is orders of magnitude lower than $\hbar\omega_{\rm pl}/k_{\rm B} \simeq 2 \times 10^{11}$ K. Therefore, the flux $F_{\rm pl}$ of photons emergent from the surface should be low, with a very hard spectrum at typical photon energies $E_\gamma^{\rm pl} \sim 20$ MeV (Chmaj *et al.*, 1991). However, the total photon emissivity should include also a contribution $F_{\rm qq}$ from photon bremsstrahlung accompanying quark-quark collisions in a very thin photon-transparent surface layer. It turns out that the bremsstrahlung contribution has the temperature dependence $F_{\rm bb} \propto T^4$ similar to the black-body one, but is much more diluted, $F_{\rm qq}/F_{\rm bb} \simeq 10^{-4}$. It has a soft spectrum with a typical photon energy $E_\gamma^{\rm qq} \sim k_{\rm B}T$ (Chmaj *et al.*, 1991). It should be stressed, that both $F_{\rm pl}$ and $F_{\rm qq}$ were calculated neglecting the effects of the outer rarefied electron layer, which should be included in a more refined theory.

At sufficiently high temperatures, an additional important contribution to the photon flux comes from a thin outer layer ($\sim$ 100 fm) with huge electric field confining the electrons (Usov, 1998). As mentioned before, this electric field can be as high as $\sim 10^{17}$ V cm$^{-1}$. Let us recall, that vacuum is unstable and $e^+e^-$ pairs are created spontaneously in a constant electric field higher than the Schwinger limit, $E_{\rm crit} = m_e^2c^3/e\hbar \simeq 1.3 \times 10^{16}$ V cm$^{-1}$. Because of the degeneracy of electrons in the electrosphere, the $e^+e^-$ pair creation rate is strongly reduced as compared to the vacuum value (the energy threshold for a pair creation is now $\mu_e + m_ec^2$, much higher than the vacuum threshold $2m_ec^2$). However, the process can still go via quantum tunneling through the energy barrier. For a typical temperature $T \lesssim 10^9$ K, the contribution to the photon flux from $e^+e^-$ pair annihilation ($e^+e^- \to 2\gamma$) owing to the barrier penetration is (Usov, 1998) $F_{\rm bar} \propto F_{\rm bb}\exp(-\alpha/T_9)/T_9$, where $\alpha = 2m_ec^2/(k_{\rm B} \times 10^9\ {\rm K}) = 11.86$ and $T_9 = T/10^9$ K. Combining $F_{\rm bar}$ with $F_{\rm qq}$ and $F_{\rm pl}$, we find that the photon emissivity of cooling bare strange stars with $T < 10^{10}$ K is still many orders of magnitude lower than for neutron stars, and the photon spectrum at $T \gtrsim 10^9$ K can be very hard (Usov, 1998, 2001). Detailed studies of $e^+e^-$-pair and photon radiation from hot, bare strange stars were done by Aksenov *et al.* (2003, 2004).

The photon flux from stellar atmospheres is limited by the condition of hydrostatic equilibrium. The radiation pressure directed outward should be balanced by the gravitational pull acting on the plasma. For a spherically symmetric star, the radiation pressure is proportional to the photon luminosity $L_\gamma$. The condition of hydrostatic equilibrium can be satisfied only for $L_\gamma <$

$L_{\rm Edd}$, where $L_{\rm Edd}$ is the Eddington limit. For neutron stars, according to Eq. (1.3), $L_{\rm Edd}^{\rm NS} \sim 10^{38}$ erg s$^{-1}$.

However, the strange quark matter at the surface of the bare strange star is bound not by gravity but by strong interactions. The sharp quark surface separates the QCD vacuum, where quarks can move, from the outer ordinary vacuum, where quarks are confined to hadrons. Baryons can fall inside the SQM, but cannot escape from it.[2] Nevertheless, the surface cannot prevent the radiation of photons and $e^+e^-$-pairs. The confinement of quarks within the SQM makes the Eddington limit irrelevant for bare strange stars. The photon luminosity of bare strange stars can be as high as $10^{51}$ erg s$^{-1}$ if the surface temperature is $T_{\rm s} \sim 10^{11}$ K (Chmaj *et al.* 1991, Aksenov *et al.* 2004 and references therein). This makes them attractive for some models of inner engines to generate still mysterious gamma ray bursts at cosmological distances.

## 8.15. The surface of a bare strange star without electrons

This is the case of the CFL phase of quark matter with the enforced condition $n_e = 0$. The reason for the electric neutrality of the quark component is that the pairing is the most efficient (produces maximum binding) if all Fermi momenta of $u$, $d$, and $s$ quarks are equal. This in turn implies $n_u = n_d = n_s$.

In order to calculate the charge density near the strange star surface one should take into account, apart from the bulk contribution $\Omega_{\rm bulk}$, also the surface term $\Omega_{\rm surf} > 0$ in the thermodynamic potential of the system. To determine the equilibrium near the surface, one has to minimize the sum $\Omega_{\rm bulk} + \Omega_{\rm surf}$. For normal strange matter, the surface *increases* the positive charge of the quark component of the matter, because the additional reduction of the number of massive $s$-quarks near the surface is energetically preferable (Madsen, 2000b). For the CFL-strange matter, the reduction of the number of $s$-quarks near the surface makes the surface layer positively charged, creating a Coulomb barrier which prevents the absorption by the surface of atomic nuclei (Madsen, 2001).[3] This barrier allows CFL superconducting strange stars to have normal crusts.

More detailed studies of the electric field at the surface of the CFL-strange matter were performed by Usov (2004). The deficit of s-quark states near the surface implies the net positive charge of quarks in the surface layer (whose thickness is a few fm). This strongly localized positive charge produces a positive Coulomb potential $\sim 10^7$ V extending over $\sim 10^{-11}$ cm. It corresponds to

[2] In principle, colorless triplets of quarks could evaporate from SQM as baryons, but viewing a large binding energy of SQM, $\sim 50$ MeV, this evaporation is very inefficient at $T \lesssim 10^{11}$ K.

[3] The demonstration by Madsen (2001) that the reduction of the number of $s$-quarks near the surface of a strangelet (as near the surface of a star) implies a net *positive charge* of strangelets was crucial for ruling out a potential "disaster scenario" in which strangelets produced in heavy ion colliders could grow infinitely by absorbing nuclei, and swallow eventually the Earth. This demonstration was met with a relief and made some people state that "Madsen saved the world."

a huge electric field $\sim 10^{18}$ V cm$^{-1}$ directed outward. Thus, the electrosphere of a bare CFL superconducting strange star is similar to that of a normal strange star. As in the normal case, the superconducting star can radiate photon fluxes exceeding the Eddington limit by many orders of magnitude. The maximum mass and thickness of the normal crust around a superconducting core do not differ from those for normal strange stars (§ 8.14).

## 8.16. Strange stars with the crust

A strange star with the Coulomb barrier at its surface can have an envelope of normal matter, which could be formed via accretion (in a close binary or from the interstellar matter). The normal envelope could also be a leftover of a neutron star whose interior transformed into a strange star (§ 8.19). The envelope can coexist with the core of strange matter, provided it does not contain free neutrons (which would be absorbed by the SQM and dissolved into quarks). Therefore, an absolute upper bound on the bottom density of the crust, $\rho_{\rm b}$, is the neutron-drip density, $\rho_{\rm ND} \approx 4 \times 10^{11}$ g cm$^{-3}$. A superstrong electric field should separate normal nuclei from the quark matter, which is necessary for the envelope stability. However, even if the envelope contains only nuclei immersed in an electron gas, the force that supports the crust should be large enough to create a large gap between the nuclei and the SQM and prevent the absorption of the nuclei by the SQM via quantum Coulomb barrier penetration.

The calculation of the electrostatic potential $V(z)$ in the vicinity of the quark surface (at $z = 0$) differs from the calculation of $V(z)$ for a bare strange star by the boundary condition at $z \to +\infty$. Now the condition is $V(z \to +\infty) = V_{\rm cr}$, where $V_{\rm cr}$ is the potential at the bottom of the normal crust. The problem can be solved in the Thomas-Fermi approximation and has to involve the balance of forces (pressure, gravity, electrostatic) acting on the crust base. A simplified solution, in which both $V_{\rm q}$ and $V_{\rm cr}$ were treated as input parameters ($V_{\rm q} > V_{\rm cr}$), was presented by Alcock *et al.* (1986). They estimated the gap width $z_{\rm G}$ between the normal and SQM phases and calculated, within the quasiclassical WKB approximation, the probability for a crustal nucleus to tunnel through the gap. Their main conclusion was that for $\Delta V = V_{\rm q} - V_{\rm cr} \gtrsim 10\,{\rm MeV}/|e|$ ($z_{\rm G} \sim 10^2$ fm), the normal crust with the ground-state composition and the bottom density $\rho_{\rm b} \sim \rho_{\rm ND}$ is stable against the absorption of nuclei by the SQM. Therefore, the maximum crust density can be really set $\rho_{\rm b} = \rho_{\rm ND}$. For a strange star of $M \gtrsim M_{\odot}$, the maximum mass of the crust is $\sim 10^{-5}\ M_{\odot}$. However, the electron fraction in the SQM can be so small, that the crust stability conditions are violated at $\rho$ much lower than $\rho_{\rm ND}$. This would correspond to $\rho_{\rm b} \ll \rho_{\rm ND}$, with thinner and less massive crust (Huang & Lu, 1997).

For $M \gtrsim M_{\odot}$, the radius $R$ of a strange star with the crust is at most a few percent larger than for a bare strange star of the same mass (Fig. 8.5). For $M = 1.4\ M_{\odot}$ the maximum crust thickness is $\sim 300$ m (to be compared with

$R \sim 10$ km) and the crust mass is only $1.7 \times 10^{-5}\ M_\odot$. Both the mass and the thickness of the crust decrease with increasing $M$ and become very small at $M \simeq M_{\rm max}$. The presence of the crust implies a tiny increase of $M_{\rm max}$ (because the absence of the crust is equivalent to an infinitely soft EOS for $\rho < \rho_{\rm b}$, with $P(\rho < \rho_{\rm b}) = 0$). One can show that the increase of $R_{M_{\rm max}}$ is roughly linear in $P_{\rm b}/\rho_{\rm b}c^2 \sim 10^{-3}$, whereas the increase of $M_{\rm max}$ is quadratic in this small parameter.

The presence of the crust leads to a specific $R(M)$ dependence for low-mass strange stars shown in Fig. 8.5. The structure of the crust is determined by two factors: by the parameters of normal matter (particularly, by the bottom density $\rho_{\rm b}$ or, equivalently, by the bottom pressure $P_{\rm b}$) and by the gravitational pull of the SQM core specified by its mass, $M_{\rm core}$. Consider the right $R(M)$ curve in the left panel of Fig. 8.5. The central stellar density and pressure, as well as $M_{\rm core}$, decrease with decreasing $M$. On the contrary, the pressure at the crust bottom, $P_{\rm b}$, stays constant. For low-mass strange stars, Newtonian gravitation can be used. For all configurations in Fig. 8.5, $M_{\rm crust} \ll M$. The SQM core is self-bound by the confining QCD forces. On the contrary, the crust is not self-bound, but is bound to the SQM core by the gravitational pull. The hydrostatic equilibrium results from the balance of the gravitational pull and the pressure. This balance implies that $M_{\rm crust}$ has to increase sufficiently rapidly with growing $R$. Low-mass strange stars with the crust are large spheres of normal matter with $\rho \leq \rho_{\rm b}$, held together by the gravitational attraction of a small central SQM core. There is a minimum mass which can stabilize the crust. It corresponds to $M_{\rm min}$ (the filled dot) in the right panel of Fig. 8.5. For $\rho_{\rm b} = \rho_{\rm ND}$ we get $M_{\rm min} \simeq 0.02\ M_\odot$, about one fifth of the minimum mass of neutron stars. The value of $M_{\rm min}$ for $\rho_{\rm b} < \rho_{\rm ND}$ would be even smaller. Configurations to the right of the minimum in the $M - R$ diagram are unstable with respect to small perturbations (as follows from the static stability condition $\mathrm{d}M/\,\mathrm{d}\rho_{\rm c} > 0$, § 6.5, and confirmed by the linear dynamical analysis of Glendenning *et al.* 1995).

If the surface tension of droplets of SQM (strangelets) is sufficiently small, one could contemplate a possibility of a very exotic crusts on strange quark stars, built of a crystal lattice of positively charged strangelets immersed in an electron gas (Haensel *et al.* 1986b; Jaikumar *et al.* 2006). Such a crust would be rather thin (of the thickness $\sim 50$ m). We will not consider this model further.

### 8.16.1 The minimum radius of strange stars with the crust

Strange stars with the crust cannot have too small radii. They have the minimum radius $R_{\rm min}$ which depends on the pressure at the crust bottom and on the EOS of the SQM core. The problem of $R_{\rm min}$ was studied in detail by Zdunik (2002).

Strange stars with $R \sim R_{\rm min}$ can be analyzed in the Newtonian gravity. The equation of hydrostatic equilibrium in the crust is $\mathrm{d}P/\rho = -GM\ \mathrm{d}r/r^2$, where $M$ is the total stellar mass ($M \simeq M_{\rm core} \gg M_{\rm crust}$). Integrating from the surface ($r = R$, $P = 0$) to the crust bottom ($r = R_{\rm b}$, $P = P_{\rm b}$) we obtain

$$\chi \equiv \int_0^{P_{\rm b}} \frac{\mathrm{d}P}{\rho c^2} = \frac{GM}{c^2}\left(\frac{1}{R_{\rm b}} - \frac{1}{R}\right). \tag{8.31}$$

For the BPS EOS of the crust and $\rho_{\rm b} = \rho_{\rm ND} = 4.3 \times 10^{11}$ g cm$^{-3}$ one gets $\chi = 9.2 \times 10^{-3}$ (Zdunik 2002; notice that our *dimensionless* $\chi$ is $\chi = \chi_{\rm Zdunik}/c^2$).

Using the relation $M = \frac{4\pi}{3}\rho_{\rm s}R_{\rm b}^3$ valid for low-mass strange stars, we get

$$\frac{1}{R} = \frac{1}{R_{\rm b}} - \frac{3\chi c^2}{4\pi G \rho_{\rm s} R_{\rm b}^3}. \tag{8.32}$$

Hence the minimum of $R$ (the maximum of $1/R$) is

$$R_{\rm min} = \frac{3}{2}\,R_{\rm b} = \frac{9}{4}\sqrt{\frac{\chi c^2}{G\pi\rho_{\rm s}}}\,. \tag{8.33}$$

A typical error of this expression is $(1-3)\,\%$. For an MIT Bag Model EOS ($m_s c^2 = 200$ MeV, $\alpha_{\rm s} = 0.2$, $\mathcal{B} = 60$ MeV fm$^{-3}$) of the quark core, at $\rho_{\rm b} = \rho_{\rm ND} = 4.3 \times 10^{11}$ g cm$^{-3}$, we get $R_{\rm min} = 6.7$ km, in agreement with Fig. 8.5.

Note that $R_{\rm min} \propto \rho_{\rm s}^{-1/2}$. If $\rho_{\rm b} = \rho_{\rm ND}$, one typically has $R_{\rm min} \sim (5-7)$ km for the EOSs based on the MIT Bag Model and $R_{\rm min} \sim 4$ km for the EOSs of Dey *et al.* (1998) (Table 8.1). For $\rho_{\rm b} < \rho_{\rm ND}$, the minimum radius will be smaller, $R_{\rm min} \propto \sqrt{\chi(\rho_{\rm b})}$.

## 8.17. Apparent radii of strange stars

As in the case of neutron stars (§ 6.6.6), the difference between apparent radii $R_\infty(M)$ and circumferential radii $R(M)$ of strange stars increases with increasing $M$ and becomes large at the maximum mass. An example of the curves $R_\infty(M)$ for strange stars with and without the crust, calculated using an MIT Bag Model EOS ($m_s\ c^2 = 200$ MeV, $\alpha_{\rm s} = 0.2$, $\mathcal{B} = 60$ MeV fm$^{-3}$), are shown in Fig. 8.11. For comparison, we also show the $R_\infty(M)$ curves for neutron stars, calculated for a representative set of EOSs of baryonic matter. The curves for strange stars are qualitatively different from those for neutron stars. The apparent radius of neutron stars is always larger than 12 km, while there is no lower bound on $R_\infty$ for bare strange stars. This difference would be important to identify a strange star. It would be sufficient to show that $R_\infty$ is certainly below 12 km.

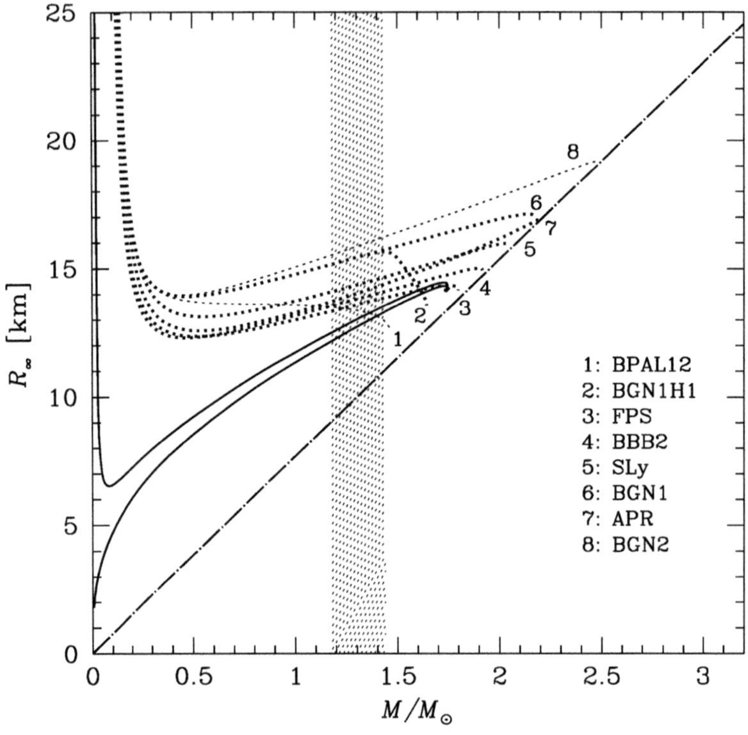


*Figure 8.11.* Apparent stellar radius $R_\infty$ versus gravitational mass $M$. The lower solid line is for bare strange stars calculated using an MIT Bag Model EOS ($m_s c^2 = 200$ MeV, $\alpha_s = 0.2$, $\mathcal{B} = 60$ MeV fm$^{-3}$). The upper solid line is for strange stars with the crust ($\rho_b = \rho_{ND}$). Dotted lines correspond to neutron star models calculated using EOSs from Table 5.3. The dot-and-dash line shows the minimum of $R_\infty$ at a given $M$ (§6.6.6). Shaded vertical band refers to the range of precisely measured masses of neutron stars (§9.1).

The shapes of the $R_\infty(M)$ and $R(M)$ curves for strange stars are different, especially for highest $M$. In contrast to the $R(M)$ curve (Fig. 8.5), the apparent radius $R_\infty$ of bare strange stars increases monotonically with growing $M$, except for a barely visible, tiny segment at $M \simeq M_{max}$. Within one percent, the maximum value of $R_\infty$ is reached at $M_{max}$. This maximum value is very close to the absolute lower bound on $R_\infty$ at $M = M_{max}$. The lower bound of $R_\infty(M)$ is reached at $r_g/R = 2/3$, at which the equator coincides with the closed circular photon orbit (§6.6.6). For the maximum-mass bare strange star with this EOS, we have $r_g/R = 0.54$, which is 19% lower than 2/3. Using the same arguments as in §6.6.6 (expanding $R_\infty/R$ around its minimum at $r_g/R = 2/3$) we estimate that $R_\infty(M_{max})$ exceeds the minimum value of

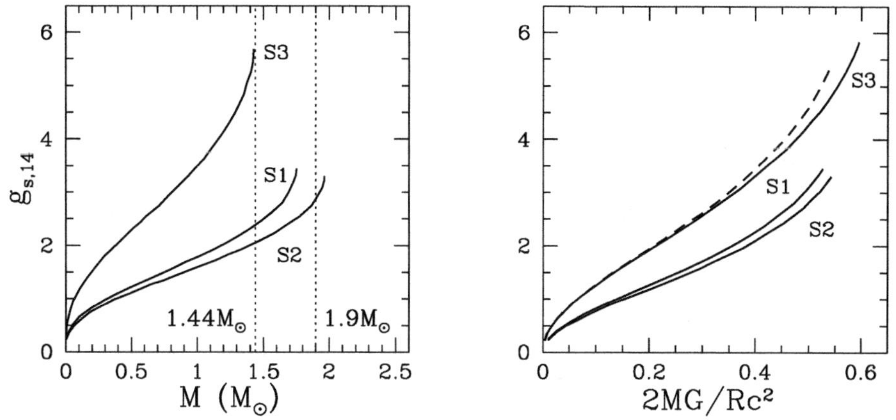


*Figure 8.12.* Surface gravity versus strange star mass (*left*) and compactness (*right*) for three EOSs of strange quark matter; S1 is the MIT Bag Model with $\alpha_s = 0.17$, $\mathcal{B} = 60$ MeV fm$^{-3}$, and $m_s c^2 = 200$ MeV; S2 is the SQM0 EOS of § 8.5 with $\mathcal{B} = 60$ MeV fm$^{-3}$; S3 is the SS1 EOS from Table 8.1. The dashed line in the right panel is obtained by transforming the S2 curve into the the S3 curve using Eq. (8.34).

$R_\infty$ at $M = M_{\rm max}$ by $\sim 2 \times (0.19)^2 \simeq 7\%$, in good agreement with precise numerical results shown in Fig. 8.11.

For strange stars with the crust extended to $\rho_{\rm b} = \rho_{\rm ND}$, the apparent radius reaches minimum of $\sim$ 6 km at $M \sim 0.02\ M_\odot$. At such a low mass the space-time curvature is small and $R_\infty \simeq R$. Therefore, all results concerning the minimum $R$ of strange stars with the crust (§ 8.16.1) apply – within a few percent – to the minimum value of $R_\infty$.

## 8.18. The surface gravity of strange stars

The dependence of the surface gravity $g_{\rm s}$ on $M$ for strange stars (Bejger & Haensel 2004, Fig. 8.12) is very different from that for baryonic stars (§ 6.6.5) because of the different $M - R$ diagrams. The surface gravity of strange stars is very model dependent. For $M = 1.44\ M_\odot$, the value of $g_{\rm s,14}$ ranges from 2.0 to 5.5. Such a scatter can be explained in terms of the scaling properties of strange star models.

As shown in § 8.8, various EOSs of strange matter can be quite well represented by the linear relation $P = ac^2(\rho - \rho_{\rm s})$, Eq. (8.18). The parameters of strange stars calculated for the different EOSs are connected via the scaling relations discussed in § 8.13 and Appendix E. The scaling implies that at a fixed $a$ the ratio $M/R$ does not depend on $\rho_{\rm s}$. Therefore, the maximum surface gravity of stable strange stars scales as

$$g'_{\rm s,max} = \left(\rho'_{\rm s}/\rho_{\rm s}\right)^{\frac{1}{2}} g_{\rm s,max} \,. \tag{8.34}$$

The values of $a$ for the EOSs S1, S2, and S3, used in Fig. 8.12, range from 0.30 to 0.46 (Table 8.1). The exact ratio $g_{\rm s,max}({\rm S3})/g_{\rm s,max}({\rm S2}) = 1.81$, while the scaling factor gives $[\rho_{\rm s}({\rm S3})/\rho_{\rm s}({\rm S2})]^{1/2} = 1.63$. The large difference in the maximum surface gravities is described by the scaling with respect to $\rho_{\rm s}$; the dependence on $a$ is much weaker and can be neglected in the first approximation. As we see in the right panel of Fig. 8.12, the precision of this scaling increases with decreasing the star compactness. The scaling becomes very precise for $r_{\rm g}/R < 0.2$ because the density within such strange stars is nearly constant and the dependence on $a$ is really negligible.

Although the dependence of $g_{\rm s}$ on $M$ and $r_{\rm g}/R$ for strange stars and baryonic stars is very different, the range of $g_{\rm s,max} = (3-6) \times 10^{14}$ cm s$^{-2}$ is similar.

## 8.19. The conversion of neutron stars into strange stars

Let us imagine that a small nugget of SQM appears in the neutron-star interior, for instance, as a high-energy strangelet from cosmic rays. It might also nucleate near the center of an accreting neutron star or a spinning down pulsar, due to the compression of the matter. Finally, it could appear owing to large thermal fluctuations in the hot and dense core of a newly born neutron star. In all these cases the neutron star will convert into a strange star via an exothermic burning of baryon matter into SQM.

The growth of the core of the SQM inside a neutron star was considered by Olinto (1987, 1991); Doroba (1989); Heiselberg & Pethick (1993) assuming that the conversion takes place in the diffusive (slow combustion) regime. The actual conversion timescale $\tau_{\rm conv}$, determined by the weak interaction rate (mainly by the strangeness changing process $u + d \to s + u$) and the diffusion rate, is uncertain and depends on an initial neutron star configuration. At $k_{\rm B}T \approx 10$ MeV, characteristic of a newly born neutron star, Heiselberg & Pethick (1993) got $\tau_{\rm conv} \sim 10$ min. On the other hand, Olesen & Madsen (1991) found that $\tau_{\rm conv}$ ranges from 0.1 s to a few minutes, depending on the neutron star temperature, the EOS of the baryon matter, and SQM parameters. Some authors (Horvath & Benvenuto, 1988; Benvenuto *et al.*, 1989) give arguments in favor of the transformation of slow (subsonic) combustion into a detonation shock in a newly born neutron star. They argue that the slow combustion front is hydrodynamically unstable due to the development of self-turbulence. This would strongly increase the front area and accelerate the conversion rate. Consequently, a slow conversion front would be transformed into a fast (supersonic) detonation front with $\tau_{\rm conv} \sim 0.1$ ms. In the detonation scenario, the outer layer of the neutron star would be ejected and a bare strange star would be formed.

Recently, however, Drago *et al.* (2005) have shown that while the conversion front is convectively unstable, it moves fast but subsonically in the baryon core: burning of hadronic matter into SQM is a strong deflagration process. For the Drago *et al.* (2005) models, a neutron star burns into a strange quark star in $\sim 10$ ms. Clearly, especially in view of a possible relation of the conversion process to the inner engines of long gamma ray bursts, the problem of the most likely scenario of the conversion of a neutron star into a hot, rapidly rotating strange quark star deserves further study.

## 8.20. Even stranger stars of abnormal matter and Q-stars

Several types of the hypothetical self-bound state of dense matter, different from the SQM, could constitute the true ground state of the matter at $P = 0$ (§ 8.9). Let us consider corresponding stellar models. We will mainly concentrate on the values of $M_{\rm max}$ and $R_{\rm max}$ for non-rotating stars using the scaling properties (§ 8.13 and Appendix E) of the maximum-mass configurations for the linear approximation of the EOSs (§ 8.8.1).

Employing an EOS of a self-bound pion-condensed matter, Hartle (1975) obtained $M_{\rm max} = 1.3\ M_\odot$, $R_{M_{\rm max}} = 5.4$ km, $\rho_{\rm c,max} = 8 \times 10^{15}$ g cm$^{-3}$, and $z_{\rm s}(M_{\rm max}) = 0.84$. Notice that the surface redshift $z_{\rm s}(M_{\rm max})$ is very close to the upper bound for the causality-limit EOS, $z_{\rm max}^{\rm CL} = 0.85$ (§ 6.6.4). Indeed, the parameters of the maximum-mass model are close to those for the causality-limit EOS with $\rho_{\rm s} = 2.7 \times 10^{14}$ g cm$^{-3}$ (see Appendix E for relevant formulae). This value of $\rho_{\rm s}$ is significantly lower than $\rho(P = 0) = 2 \times 10^{15}$ g cm$^{-3}$ for the EOS of Hartle (1975). The causality-limit EOS is, therefore, not a good fit to this EOS at $P \ll P_{\rm c,max}$, but this deficiency has no significant effect on the parameters of the maximum mass configuration. Notice that because $M_{\rm max} = 1.3\ M_\odot$ is lower than the mass of the Hulse-Taylor pulsar, the Hartle (1975) model is inconsistent with observations (§ 9.1). More exactly, the Hulse-Taylor pulsar cannot be an abnormal star of Hartle (1975).

Such a problem does not arise for self-bound models of Migdal (1974), provided one takes due account of large proton fraction in pion-condensed nucleon matter (Migdal *et al.*, 1990). A self-bound state was obtained for a large amplitude of the $\pi^-$ condensate. The negative charge density of pions is large and has to be balanced by a large positive charge of protons. Therefore, in contrast to a small proton fraction in the non-condensed matter, the pion-condensed matter is characterized by $n_p \sim n_n$. The EOS of symmetric nuclear matter, used by Migdal *et al.* (1990), at essentially supranuclear densities is much stiffer than the EOS of neutron matter, enlarging the maximum mass of self-bound pion condensed stars. However, this stiffening is model dependent. Also, the approximation of the nucleon contribution by the normal nuclear matter and the assumption of additivity of the pion-condensate contribution, $\mathcal{E} = \mathcal{E}_{\rm N} + \mathcal{E}_\pi$, is debatable (§ 7.3).

The models of hypothetical Q-stars were constructed by Bahcall *et al.* (1990). As we already mentioned in § 8.9, two basic parameters of the model are the energy density $U_0$ of the scalar field inside the Q-matter and the coupling strength $\alpha_v$ of the vector field to nucleons. Q-star parameters show a simple scaling with $U_0$ provided the dimensionless parameter $\zeta$ introduced in § 8.9 is kept constant. Then, the scaling of $M_{\rm max}$, $R_{M_{\rm max}}$, $I_{\rm max}$, and other quantities with $U_0$ is the same as with respect to the bag constant $\mathcal{B}$ for strange stars, Eq. (8.30) (although with different numerical coefficients). For $0 \leq \zeta \leq 16$ and $U_0 = 13.0$ MeV fm$^{-3}$, one gets an astonishingly high maximum stellar mass, $4.0 \leq M_{\rm max} \leq 8.3\ M_\odot$. It stems from a low value of the Q-matter density at zero pressure, $\rho(P = 0) \equiv \rho_{\rm s}$. Let us recall that nucleons are nearly massless in the Q-matter and consider the simplest case of $\zeta = 0$. Then the EOS for the Q-matter coincides with the EOS for the strange matter of massless, non-interacting quarks at $\mathcal{B} = U_0$. However, the standard value used in constructing Q-star models by Bahcall *et al.* (1990) is $U_0 = 13.0$ MeV fm$^{-3}$, which corresponds to $\rho_{\rm s} = 10^{14}$ g cm$^{-3}$. The maximum mass can then be calculated from Eq. (8.30); $M_{\rm max} \sim 4.0\ M_\odot$, in agreement with Figs. 4 and 8 of Bahcall *et al.* (1990). With increasing $\zeta$, the EOS of the Q-matter becomes stiffer. In the limiting case of $\zeta = 16$ considered by Bahcall *et al.* (1990) (at the same value of $U_0 = 13.0$ MeV fm$^{-3}$) they get $\rho(P = 0) = 5.5 \times 10^{13}$ g cm$^{-3}$ and $M_{\rm max} = 8.2\ M_\odot$. This is not surprising. For the causal-limit EOS, we can use Eq. (E.6) of Appendix E with $\rho_{\rm s} = 5.5 \times 10^{13}$ g cm$^{-3}$ and obtain $M_{\rm max}^{\rm CL} = 9.0\ M_\odot$. This estimate reproduces within 8% the value of $M_{\rm max}$, reported by Bahcall *et al.* (1990). Thus, while the arguments for the existence of the Q-matter stem from sophisticated supersymmetric extensions of field-theoretic models of dense nucleon matter, the practical reasons for a shockingly high $M_{\rm max}$ of Q-stars are very simple. Specifically, the predicted density of a self-bound Q-matter at zero pressure is two to five times lower (!) than the normal nuclear density.

## 8.21. Rotating strange quark stars

As we stressed many times in the present chapter, the matter distribution in strange stars is qualitatively different from that in neutron stars. Moreover, the mass-radius relation for strange stars with $M \lesssim M_\odot$ is completely different. These differences have strong impact on rotating strange stars.

It is worthwhile to recall that rapid rotation of strange stars became a topic of special interest in 1989, after the sensational detection of a pulsar with the spin period 0.5 ms in SN 1987A (withdrawn one year later, § 9.4). In several papers after the detection, a strange star was advanced as an appropriate and even unique model for this pulsar (Glendenning 1989a,b; Frieman & Olinto 1989; see, however, Zdunik & Haensel 1990).

Let us review properties of rotating strange stars. The formalism and notations will be the same as in § 6.12. However, numerical calculation of stationary configurations of rotating strange stars deserves an additional "technical" comment. As we will see, at $M \gtrsim M_{\odot}$ rotation affects strange stars stronger than neutron stars. A correct treatment of rapidly rotating strange stars requires exact two-dimensional calculations; the approximation of slow rotation is insufficient. On the other hand, bare strange stars have huge surface density $\rho_{\rm s} \sim (5 \times 10^{14} - 10^{15})$ g cm$^{-3}$. It implies a huge density discontinuity at the stellar surface, a difficult challenge for accurate two-dimensional solutions of the coupled partial differential equations for the stellar structure and metric functions. The early two-dimensional calculations, performed by Friedman in 1989 (quoted by Glendenning, 1989a,b) and Lattimer *et al.* (1991), were based on insufficiently precise methods. The most suitable method for exact two- (and three-) dimensional calculations of rapidly rotating strange stars was developed by Bonazzola *et al.* (1998b). It is a *multi-domain* spectral method with the *exact* treatment of the density discontinuity at the stellar surface (or in the interior, for example, at the core-crust interface in strange stars with the crust). A detailed description of this method is given by Gourgoulhon *et al.* (1999). Some examples are presented below.

### 8.21.1 Uniformly rotating strange stars

An example of a meridional cross-section of a rapidly rotating strange star is given in Fig. 8.13. The stellar baryon mass, the gravitational mass, and the rotation period are $M_{\rm b} = 2.00\, M_{\odot}$, $M = 1.60\, M_{\odot}$, and $P = 0.87$ ms, respectively. The oblateness is large, because the star rotates close to its Keplerian frequency. The density contrast in the star is small; the surface density is about 2/3 of the central one. The surface separates two domains of integration of the equations of hydrostatic equilibrium and space-time metric. The spectral method determines equilibrium configurations very precisely (Gourgoulhon *et al.*, 1999).

Rotating bare strange stars, stable with respect to axially symmetric deformations, occupy a region in the $M - R_{\rm eq}$ plane shown in Fig. 8.14; $R_{\rm eq}$ is the equatorial circumferential radius. The line $S$ corresponds to non-rotating stars (with the spin frequency $f = 0$), while the line $K$ is for the Keplerian rotation ($f = f_{\rm K}$). The central density increases along each line. $S_2$ and $K_3$ refer to the maximum mass non-rotating and rapidly rotating stars, respectively. The dashed line $S_2 - K_3$ shows the maximum-mass configurations with the spin frequency from $f = 0$ to $f_{\rm K}$. The dashed line $S_2 - K_2$ is the track of the star with the same fixed baryon number as the maximum-mass non-rotating star, which is spinning up from $f = 0$ to $f_{\rm K}$. The dashed line $S_{1.4} - K_{1.4}$ is a similar track of the star which has $M = 1.4\, M_{\odot}$ at $f = 0$.

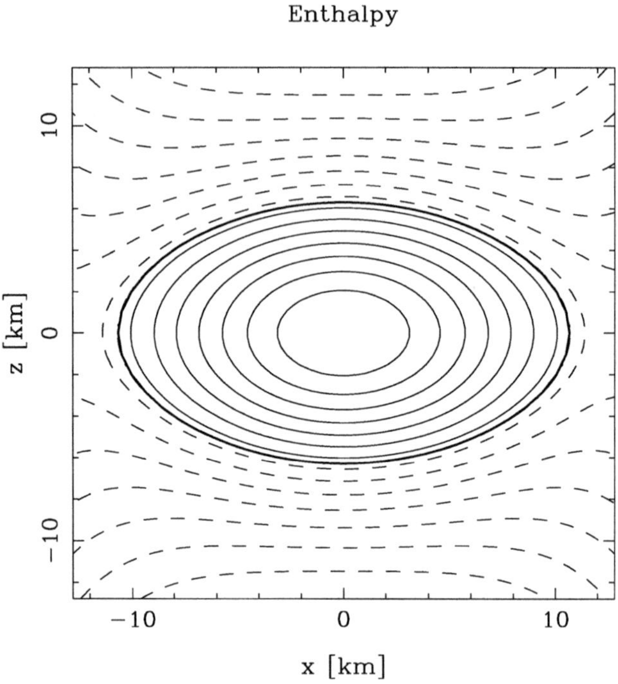


*Figure 8.13.* Cross section in the meridional plane $xz$ of a rapidly rotating strange star (for the SQM0 EOS with $\mathcal{B} = 60$ MeV fm$^{-3}$); $x = r \sin\theta \cos\phi$ and $z = r \cos\theta$, where $(r, \theta, \phi)$ are spherical coordinates in Eq. (6.95); $z$ is the rotational axis. Contours are isolines of the pseudo-enthalpy $H$. Solid lines refer to positive values of $H$, Eq. (6.99), in the star. Dashed lines refer to negative values outside the star, where $H$ is defined by Eq. (6.100). The thick solid line is the stellar surface. From Gourgoulhon *et al.* (1999).

One can see significant differences from the analogous plot for neutron stars (Fig. 6.18). The shape of the region for stably rotating strange stars differs from that for neutron stars, because of the different mass-radius relations. The hatched region $S_2K_2K_3$ contains supermassive strange stars ($M_b > M_{b,max}^{stat}$, where $M_{b,max}^{stat}$ is the maximum baryon mass for non-rotating stars). Generally, for $M \gtrsim M_\odot$ uniform rotation has stronger effect on strange stars than on neutron stars. The increase in $R_{eq}$ for strange stars is larger, because their flat density profile amplifies the effect of the centrifugal force in the equatorial ring. The area of supermassive region in the $M - R_{eq}$ plane is much larger; the centrifugal forces are more important than in neutron stars and can support larger mass. This will also be seen from our discussion of most massive spinning strange stars (§ 8.21.3). In contrast to neutron stars, rotation of bare strange stars does not become more important with the decrease of $M$. In particular, there

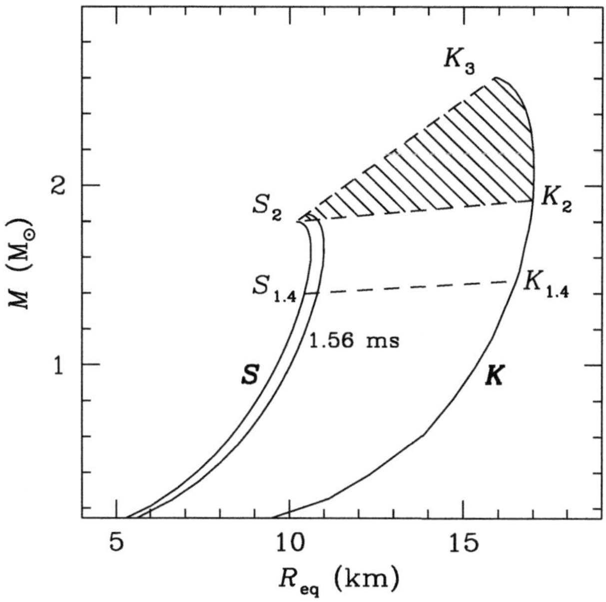


*Figure 8.14.* Families (in the $M - R_{eq}$ plane) of spinning bare strange stars (SQM1 EOS) stable with respect to axially symmetric perturbations. The central density increases upwards along each solid line. The line $S$ refers to non-rotating stars; the line $K$ refers to Keplerian rotation. The line "1.56 ms" corresponds to the period 1.56 ms of the second fastest rotator, PSR B1937+21 (§ 9.4). Other notations are explained in the text.

is no lower bound on the mass of spinning bare strange stars, because they are bound not by gravity but by the QCD confinement forces.

Now consider rotating strange stars with the crust extending to the neutron drip density ($\rho_b = \rho_{ND}$). The families of uniformly rotating configurations, stable with respect to axially symmetric perturbations, are shown in Fig. 8.15. For $M \gtrsim M_\odot$ the effect of the crust is small, and amounts to a few percent decrease of the Keplerian spin frequency $f_K$ at a given baryon mass. The mass of the crust is negligibly small ($\sim 10^{-5}\ M_\odot$) and its contribution to the stellar gravitational field can be neglected. The only effect of the crust is a shift of the equator outwards, by a few percent compared to the bare strange-star case; it results in the decrease of $f_K$. However, the crust becomes more important with decreasing stellar mass, especially at $M \lesssim 0.2\ M_\odot$. As in the case of neutron stars, the minimum mass increases rapidly with the growth of the spin frequency. For instance, for the spin frequency of PSR B1937+26, $f = 641$ Hz, and for our model of strange star with the crust, the rotation increases $M_{min}$ by a factor of five, from $0.017\ M_\odot$ to $0.09\ M_\odot$. For neutron star models based on

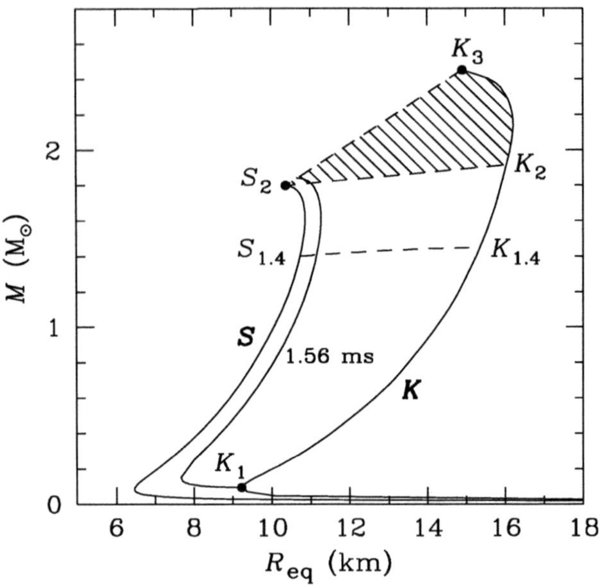


*Figure 8.15.* Families of rotating strange stars (the SQM1 EOS) with the crust ($\rho_b = \rho_{ND}$) in the $M - R_{eq}$ plane. Notations are the same as in Fig. 8.14. For other explanations see the text.

the SLy EOS, the effect was quite similar; $M_{min}$ increased by a factor of seven, from $0.09\ M_\odot$ to $0.61\ M_\odot$ (§ 6.12.2).

Now let us analyze the effect of rotation on strange stars (with the crust) at a fixed baryon mass $M_b$. These rotating configurations form a one-parameter family, along which a spinning down pulsar moves. For instance, we can start with a non-rotating star of $M(f = 0) = 1.4\ M_\odot$ or $1.75\ M_\odot$, and then construct a family of rotating configurations with $0 < f \leq f_K$. The results are displayed in Figs. 8.16 and 8.17. As seen from the upper panel of Fig. 8.16, the dependence of $r_{pol}/r_{eq}$ on $f$ is rather similar to that for neutron stars (Fig. 6.19); here $r_{eq}$ and $r_{pol}$ are equatorial and polar radial coordinates, respectively (not to be confused, for instance, with the circumferential radius $R_{eq}$ in Figs. 8.14 and 8.15). Notice, however, a significant difference in the rotational effect on $r_{eq}$ and $r_{pol}$ at $f \simeq f_K$. For strange stars, the decrease of $r_{pol}$ with respect to the non-rotating value $r_0$ is similar to the increase of $r_{eq}$. For neutron stars, the equatorial expansion was much larger than the polar flattening due to the importance of the crust.

At a fixed baryon mass, the central baryon number density $n_c$ decreases with increasing $f$. This is illustrated in the upper panel of Fig. 8.17. At the maximum frequency, $n_c$ lowers by 24% for the $M(f = 0) = 1.4\,M_\odot$ sequence, just as for neutron stars with the same $M(f = 0)$. However, for massive

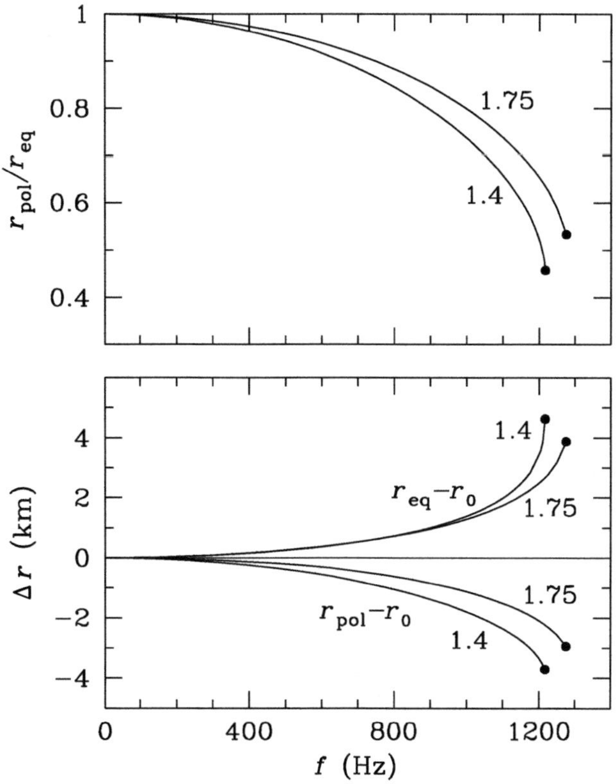


*Figure 8.16.* The effect of rotation on the shape of a strange star (the SQM1 EOS) with the crust ($\rho_b = \rho_{ND}$) versus spin frequency $f$. We show sequences of rotating strange stars with a fixed baryon number, starting with non-rotating stars of gravitational masses $1.4\ M_\odot$ and $1.75\ M_\odot$, and ending at the maximum rotation frequency (filled dots). *Top*: The ratio $r_{pol}/r_{eq}$ which determines the oblateness of a spinning star. *Bottom*: The equatorial expansion and polar flattening (with respect to the radius $s_0$ of the non-rotating star). All radii are defined as values of the radial coordinate in the space-time metric, Eq. (6.95).

strange stars with $M(f=0) = 1.75\ M_\odot$, one gets much larger decompression ($\approx 45\%$). This is because at $f = f_K$ a massive strange star has a different matter distribution than a non-rotating strange star of the same baryon mass. The centrifugal force leads to a significant central decompression, which is much larger than for neutron stars or medium-mass strange stars.

The most important difference between rotating strange stars and neutron stars concerns the parameter $\beta = E_{kin}/|E_{pot}|$, where the energies $E_{rot}$ and $E_{grav}$ are defined by Eqs. (6.105) and (6.106). For our rotating strange-star families $\beta$ is as large as $0.17 - 0.18$ at $f = f_K$, reaching the threshold value $\beta_{crit}$ for the secular viscosity-driven triaxial (bar) instability before the

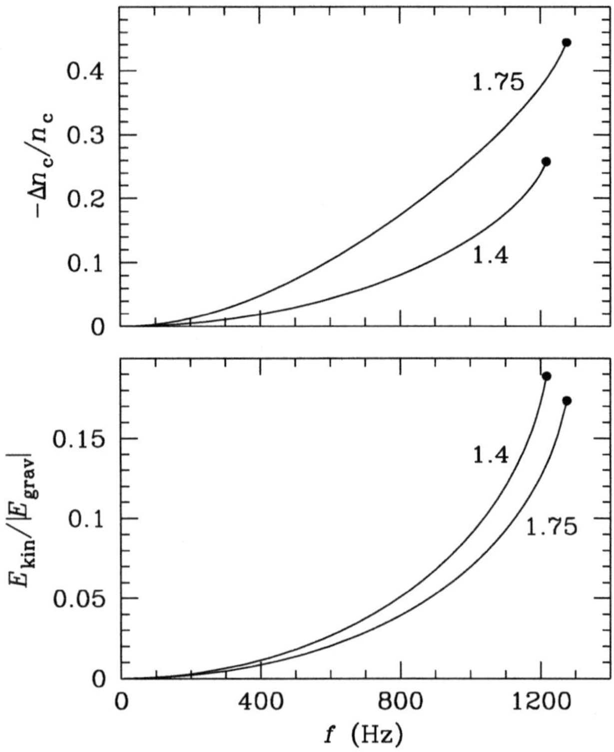


*Figure 8.17. Top*: Fractional decrease of the central density produced by rotation along the sequences of strange stars with fixed total baryon number (as in Fig. 8.16). *Bottom*: The ratio of the kinetic energy to the modulus of the gravitational energy along the same sequences.

mass-shedding limit. The situation with rotating neutron stars was different (Fig. 6.21). Rapidly rotating strange stars with $\beta > \beta_{crit}$ are unstable with respect to the triaxial deformation (§ 8.21.4) and might then emit gravitational waves. The secular viscosity-driven instability of rapidly rotating strange stars will be further discussed in § 8.21.4.

## 8.21.2 The crust of rotating strange stars

Preliminary studies of the crust on rotating strange stars were performed by Glendenning & Weber (1992). They used the approximation of slow rotation by Hartle (1967) supplemented by self-consistency conditions which allowed them to determine stellar parameters close to Keplerian frequencies.

It is obvious that rotation increases the baryon mass of the solid crust which can be supported by the strange star. It also increases the equatorial thickness of the crust and the stellar equatorial radius (Fig. 8.18). In what

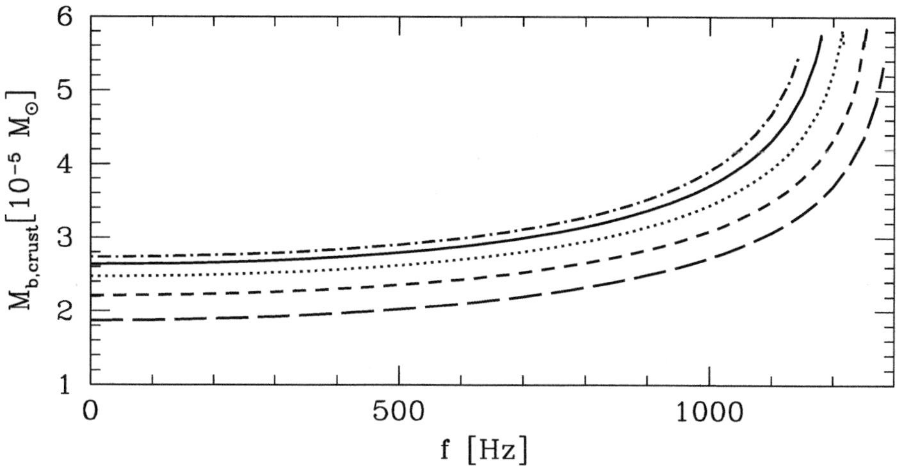


*Figure 8.18.* Baryon mass of the crust as a function of spin frequency (Zdunik *et al.*, 2001). The masses of non-rotating configurations are (from top to bottom) 1.75, 1.6, 1.4, 1.2, 1 $M_\odot$. The SQM1 EOS of the quark core is used.

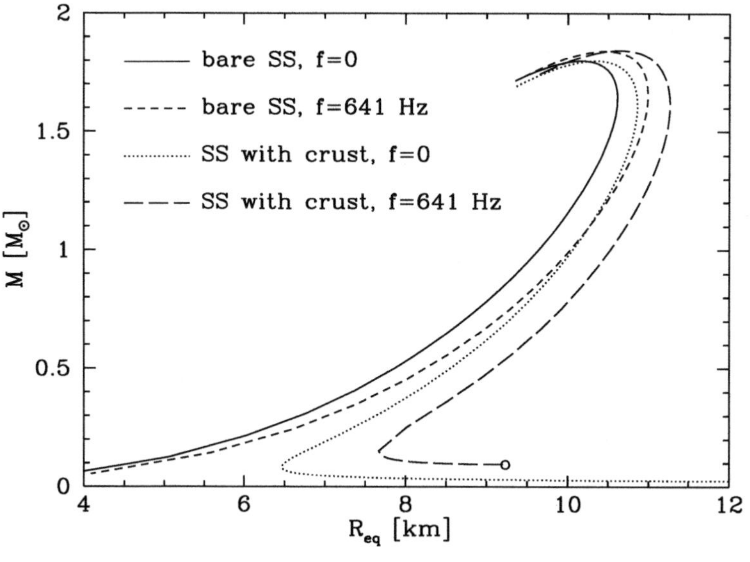


*Figure 8.19.* Gravitational mass versus circumferential equatorial radius. The SQM1 EOS for the quark core is used. The solid line is for non-rotating bare strange stars, the dotted line is for non-rotating strange stars with the crust ($\rho_b = \rho_{NS}$); the short- and long-dashed lines are, respectively, for bare stars and stars with the crust which rotate at the frequency of PSR B1937+26, 641 Hz; the open circle marks the point at which 641 Hz is the Keplerian frequency. After Zdunik *et al.* (2001).

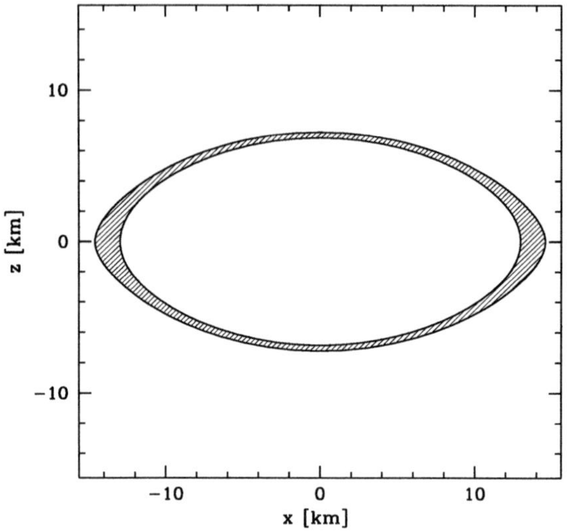


*Figure 8.20.* The cross section of the crust for a strange quark star with the baryon mass $M_B = 1.63\,M_\odot$ and the spin frequency $f = 1210$ Hz, close to the Keplerian frequency $f_K = 1217$ Hz. The SQM1 EOS (Table 8.1) is used. The coordinates are the same as in Fig. 8.13. From Zdunik (2004, unpublished), with the kind permission of the author.

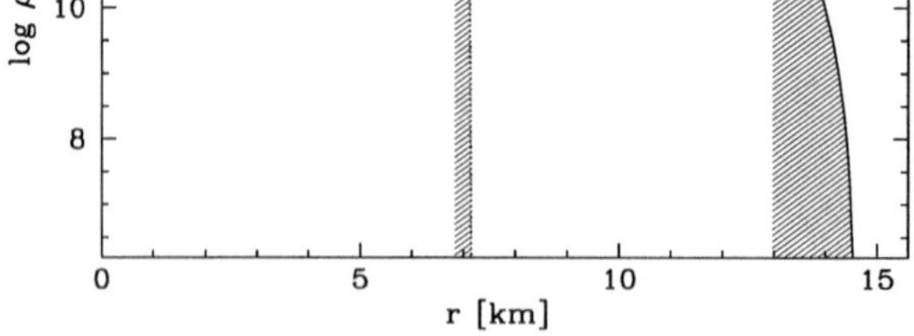


*Figure 8.21.* Logarithm of density versus the radial coordinate [defined in Eq. (6.95)] along the polar (dotted line) and equatorial (solid line) directions for the same rotating strange star model as in Fig. 8.20. Shaded areas correspond to the crust. From Zdunik *et al.* (2001).

follows, we will denote the baryon mass of the crust of a strange star rotating at a frequency $f$ by $M_{\rm b,crust}(f)$; for a non-rotating star this mass is $M_{\rm b,crust}(0)$. The circumferential equatorial radius of the star will be $R_{\rm eq}(f)$. Accurate two-dimensional calculations of the crust of rotating strange stars were performed by Zdunik *et al.* (2001). These authors showed that the quadratic approximation, typical for slow rotation, $R_{\rm eq}(f) - R(0) \propto f^2$ and $M_{\rm b,crust}(f) - M_{\rm b,crust}(0) \propto f^2$ is very precise for $f \lesssim 500$ Hz, but becomes less precise with the further increase of $f$. It badly underestimates $R_{\rm eq}$ and $M_{\rm b,crust}$ for $f \simeq f_{\rm K}$.

In Fig. 8.19 we present the mass-radius relation for non rotating strange stars and for stars rotating at the frequency $f = 641$ Hz of PSR B1937+26. The minimum-mass point for strange stars with the crust corresponds to the Keplerian frequency $f_{\rm K} = 641$ Hz. Strange stars with smaller mass have thicker crust and $f_{\rm K} < 641$ Hz.

For a given mass, a strange star with the crust is more oblate than a bare strange star rotating at the same $f$. This is because the crust is the most deformed part of the star. The effect is demonstrated in Figs. 8.20 and 8.21. As shown in Fig. 8.20, the crust thickness at the equator is about five times larger than at the pole. As a result, the density gradient in the equatorial ring of the crust is much less steeper than at the pole, Fig. 8.21.

Generally, the lower the baryon mass of the star, the steeper the increase of $r_{\rm eq}/r_{\rm pole}$ with the growth of $f$.

Rotation will increase both the crust mass and its equatorial thickness, $\Delta R_{\rm eq}$ (the difference of equatorial circumferential radii at the crust surface and bottom). For stars with $M \gtrsim M_\odot$, both quantities can be approximated by a quadratic function of $f$ up to $\sim$500 Hz. For higher $f$, this approximation is insufficient. The expressions fitting the accurate numerical results are

$$M_{\rm b,crust}(f)/M_{\rm b,crust}(0) = 1 + 0.24\, f_3^2 + 0.16\, f_3^8 \,, \qquad (8.35)$$

$$\Delta R_{\rm eq}(f)/\Delta R_{\rm eq}(0) = 1 + 0.4\, f_3^2 + 0.3\, f_3^6 \,, \qquad (8.36)$$

where $f_3 \equiv f/10^3$ Hz. Equation (8.35) is accurate nearly up to $f = f_{\rm K}$, but the equatorial radius increases very rapidly as $f \to f_{\rm K}$. In view of this one should be careful in using Eq. (8.36) at $f \lesssim 1$ kHz$\sim 0.8\, f_{\rm K}$, where the change of the equatorial thickness reaches $\sim 100\%$.

As pointed out by Glendenning & Weber (1992), the polar thickness also increases (albeit weakly) with the growth of $f$, due to the decrease of gravitational force at the pole (resulting from the oblateness of mass distribution). This effect is confirmed by calculations of Zdunik *et al.* (2001).

### 8.21.3 Maximum mass and maximum spin frequency of strange stars

Let us start with bare strange stars. The stably rotating configurations[4] with the maximum mass $M_{\rm max}^{\rm rot}$ and the minimum spin period $P_{\rm min}$ turn out to be very close (Gourgoulhon *et al.*, 1999). As already mentioned in the preceding section, the increase of the maximum mass and the corresponding equatorial radius by rotation is much larger than for neutron stars. It can be described by simple formulae

$$\text{bare strange stars:} \quad M_{\rm max}^{\rm rot} = 1.44\, M_{\rm max}^{\rm stat}\,, \quad R_{{\rm eq},M_{\rm max}}^{\rm rot} = 1.54\, R_{M_{\rm max}}^{\rm stat}. \quad (8.37)$$

These relations are *exact* for the SQM0 EOS with massless, non-interacting quarks (Gourgoulhon *et al.* 1999, see Appendix E for the derivation of the scaling relations). For this model, the EOS has the linear form (§ 8.8.1) $P = c^2 a(\rho - \rho_{\rm s})$, where $\rho_{\rm s} = 4\mathcal{B}/c^2$, $a = 1/3$, and $\mathcal{B}$ is the bag constant. For a finite strange-quark mass in the presence of QCD interactions, Eqs. (8.37) are approximate but still quite precise (see Appendix E). Let us notice a strong difference between spinning strange stars and neutron stars studied in § 6.12.4. The rotational increase of $M_{\rm max}$ by a factor of 1.44 for strange stars can be compared to the increase by a factor of 1.18 for neutron stars. The corresponding factors describing the rotational increase of $R_{\rm eq}$ are 1.54 for strange and 1.32 for neutron stars.

We can easily understand Eq. (8.37) using the *scaling* (Appendix E) of the general-relativistic equations for uniform stationary stellar rotation, § 6.12.1, applied to the EOS of the linear form. At a fixed $a$ in the linear approximation, equilibrium configurations can be obtained from a universal *dimensionless* solution of the equations for spinning stars. The scaling relations for the maximum mass configurations are

$$M_{\rm max}^{\rm rot\prime}/M_{\rm max}^{\rm rot} = R_{{\rm eq},M_{\rm max}}^{\rm rot\prime}/R_{M_{\rm max}}^{\rm rot} = \left(\rho_{\rm s}/\rho_{\rm s}^{\prime}\right)^{1/2}. \quad (8.38)$$

As the scaling for $M_{\rm max}$ and $R_{M_{\rm max}}$ for non-rotating stars is also given by Eq. (8.38), the quantities for rotating stars are proportional to those for non-rotating ones. The numerical coefficients depend rather weakly on $a$ (Appendix E) which implies good precision of Eq. (8.37) (within 2% for Bag Model EOSs).

The scaling relation for the minimum rotation period, which is also obtained from the dimensionless form of equations of spinning stars (Appendix E), reads

$$P_{\rm min}^{\prime}/P_{\rm min} = \left(\rho_{\rm s}/\rho_{\rm s}^{\prime}\right)^{1/2}. \quad (8.39)$$

[4]We restrict ourselves to configurations stable with respect to the mass shedding and axially symmetric perturbations. They may be secularly unstable, for instance, with respect to the viscosity driven triaxial instability or a gravitational radiation reaction driven instability (§ 8.21.4).

It gives an interesting formula which relates $P_{\rm min}$ with the mass and radius of the maximum-mass *non-rotating* configuration. It can be written in the form identical to the analogous formula (6.108) for neutron stars, but with a different numerical prefactor,

$$P_{\rm min} = C_{\rm SS}\, \left(M_\odot/M^{\rm stat}_{\rm max}\right)^{1/2} \left(R^{\rm stat}_{M_{\rm max}}/10\,{\rm km}\right)^{3/2}, \qquad C_{\rm SS} = 0.801 \;{\rm ms}. \tag{8.40}$$

This formula is *exact* for the SQM0 EOS (Gourgoulhon *et al.*, 1999), but remains precise (within 3%) for more realistic EOSs of SQM. Notice, that within 3% $C_{\rm SS} \simeq C_{\rm NS} = 0.82$ ms, where $C_{\rm NS}$ is the coefficient in the empirical formula $P_{\rm min}$ for neutron stars. Clearly, the formula for $P_{\rm min}$ is more universal (valid for neutron and strange stars) than the formulae for $M^{\rm rot}_{\rm max}$ and $R^{\rm rot}_{{\rm eq},M_{\rm max}}$.

## 8.21.4 Instabilities in rotating strange stars

Let us use the notations of §6.12.10 (devoted to instabilities in spinning neutron stars). As we stressed in §8.21.1, strange stars rotating at $f \sim f_{\rm K}$ have much (about twice) larger $\beta = E_{\rm kin}/|E_{\rm grav}|$ than neutron stars because of much flatter density profiles. Accordingly, strange stars may be susceptible to the secular viscosity driven triaxial (bar) instability, hereafter referred to as the triaxial instability.

The triaxial instability of rapidly rotating strange stars was studied in detail by Gondek-Rosińska *et al.* (2003) who constructed stationary axially symmetric stellar models. The stability with respect to a triaxial deformation was checked with a three-dimensional numerical code. The evolution of a deformed star was followed by constructing a series of triaxial quasi-equilibrium configurations (matter distribution and space-time metric) along which the perturbed configuration evolves. If the perturbed configuration relaxes to the equilibrium axially symmetric one, then the initial rotating equilibrium configuration is stable. In the opposite case, the perturbed configuration diverges from the axially symmetric one, which is a signal of the triaxial instability. This method can test the stability but does not give the timescale of the instability growth. The growth time is related to the timescale of angular momentum redistribution produced by the viscosity in a rotating star.

Consider a sequence of stationary configurations of rotating strange stars with a fixed baryon number and $0 < f < f_{\rm K}$ (similar to those shown in Figs. 8.17 and 8.16). If at some $f_{\rm crit} < f_{\rm K}$ a rotating axially symmetric configuration becomes unstable with respect to the triaxial deformations, then at $f > f_{\rm crit}$ the rotation is accompanied by the emission of gravitational waves. Therefore, an isolated strange star cannot rotate stably at $f > f_{\rm crit}$. However, a stationary rotation is still possible if the angular momentum loss due to the gravitational

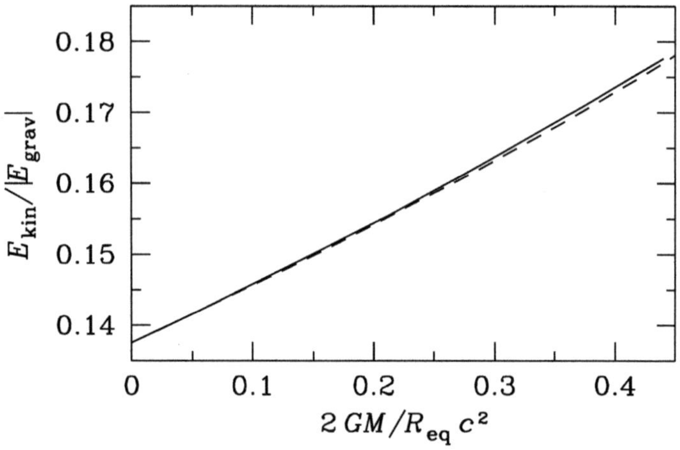


*Figure 8.22.* Critical value of $E_{\rm kin}/|E_{\rm grav}|$ for the triaxial instability versus the compactness parameter $2GM/R_{\rm eq}c^2$, for the SQM1 EOS (the solid line) and for the incompressible-fluid stars (the dashed line). For further explanations see the text.

waves emission is balanced by an external torque. This may happen for an accreting strange star in a compact binary. It could be a promising source of continuous gravitational radiation (Gondek-Rosińska *et al.*, 2003)

As noted by Gondek-Rosińska *et al.* (2003), the value of $\beta_{\rm crit}$ for strange stars is very close to that for idealized stars built of incompressible fluid and studied by Gondek-Rosińska & Gourgoulhon (2002). In both cases (incompressible-fluid and strange matter) the value of $\beta_{\rm crit}$ depends essentially on the only one stellar parameter, $x_{\rm c} \equiv 2GM/R_{\rm eq}c^2$, where $R_{\rm eq}$ is the circumferential equatorial radius of a spinning star. Figure 8.22 presents $\beta_{\rm crit}$ versus $x_{\rm c}$ for strange stars and incompressible stars. In the limit of $x_{\rm c} \to 0$ one recovers the classical Newtonian result for incompressible fluid stars, $\beta_{\rm crit}^{\rm Newt} = 0.1375$ (see, e.g., §7.3 of Shapiro & Teukolsky 1983). The two curves are almost identical and can be approximated by the expression

$$\beta_{\rm crit} = 0.1375 + 0.037 x_{\rm c}(x_{\rm c} + 2)\ . \tag{8.41}$$

General relativistic effects increase $\beta_{\rm crit}$ and stabilize rotation against the secular triaxial instability. For a given mass $M$, a strange star is unstable at $f > f_{\rm crit}$. We plot $f_{\rm crit}$ versus $M$ for three EOSs of the SQM in Fig. 8.23. We see that for $M \sim 1.5\ M_{\odot}$ the secular triaxial instability sets in at $f \simeq 900-1100$ Hz.

The function $f_{\rm crit}(M)$ shows a useful scaling property with respect to $\rho_{\rm s}$, the parameter of the linearized form of the EOS of strange matter, studied in § 8.8. Consider two EOSs of the SQM, a nonprimed and a primed one. Then,

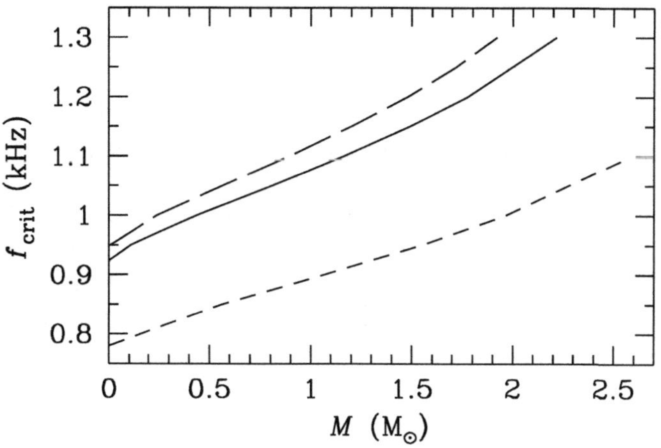


*Figure 8.23.* Critical frequency $f_{\rm crit}$ for the secular triaxial instability versus gravitational mass for three EOSs of the SQM based on the MIT Bag Model. They are SQM0 (the solid line); SQM1 (the long-dashed line); and the extreme model ($m_s = 100$ MeV $c^{-2}$, $\alpha_{\rm s} = 0.6$, $\mathcal{B} = 40$ MeV fm$^{-3}$; the short-dashed line), which gives very high $M_{\rm max}$.

to a very good approximation, one can express $f'_{\rm crit}(M')$ in terms of $f_{\rm crit}(M)$,

$$f'_{\rm crit}/f_{\rm crit} = M'/M = \left(\rho_{\rm s}/\rho'_{\rm s}\right)^{1/2} . \tag{8.42}$$

The dependence on the second parameter $a$ in the linear approximation is very weak and can be neglected (Gondek-Rosińska *et al.*, 2003).

As in the case of rapidly rotating neutron stars, one can study the stability of rotating strange stars with respect to non-axisymmetric, gravitational-radiation reaction driven instabilities of Chandrasekhar-Friedman-Schutz (CFS) type. First consider the case of normal (non-superfluid) SQM. Unfortunately, the existing investigations (Colpi & Miller, 1992) were based on the bulk viscosity underestimated by many orders of magnitude. It is expected that with much larger realistic bulk viscosity of normal SQM the CFS instabilities will be damped at any $f \leq f_{\rm K}$ (Madsen, 1992). In contrast to the normal SQM, a superconducting SQM has an exponentially suppressed bulk viscosity. Thus, superconducting rapidly rotating strange stars can be unstable to CFS modes.

The case of $r$-modes, which can be unstable at any $f$, is of particular interest. This problem was studied by Madsen (1998b, 2000a), who focused on the difference of rotating strange stars from neutron stars. Let us point out, that the bulk viscosity of the SQM, $\zeta_{\rm SQM}$, depends very strongly on the value of $m_s$ (Madsen, 1992). For instance, let us fix $m_s c^2 = 200$ MeV. Because of a very large bulk viscosity, $r$-modes in rotating non-superfluid strange stars are stable up to $f \sim f_{\rm K} > 1300$ Hz, provided the internal stellar temperature is

$5 \times 10^8$ K $\lesssim T \lesssim 2 \times 10^{10}$ K, the condition which is easily satisfied. However, because $\zeta_{\rm SQM} \propto m_s^4$ (Madsen, 1992), for $m_s c^2 = 100$ MeV the $r$-mode stability window of millisecond strange stars with $f \simeq f_{\rm K}$ is much narrower, $3 \times 10^8$ K $\lesssim T \lesssim 4 \times 10^9$ K. Bare strange stars with CFL superconductivity are unstable with respect to $r$-modes at any $f$. If a star possesses the crust (up to the neutron drip point), then $r$-modes become stable but only for $f \lesssim 100 - 200$ Hz, being damped by the electron shear viscosity in the crust and by crust-core rubbing (Madsen, 2000a). This excludes superconducting strange stars as models of millisecond pulsars. Further studies of the $r$-mode instability in rotating isolated and accreting strange stars confirmed distinct features of strange stars in this respect (Andersson *et al.*, 2002).

# Chapter 9

# THEORY VERSUS OBSERVATIONS

As discussed in Chapter 5, the fundamental problem of the EOS of neutron-star cores cannot be solved on purely theoretical basis: there is no strict theory but many theoretical models instead. Therefore, one can try to select (constrain) the true model using observations. In the present Chapter we will describe some results of this activity. The methods to constrain the EOS are numerous. The activity started just from the discovery of neutron stars in 1967 (Chapter 1). The results obtained by the time of this writing (2006) are a tremendous challenge of observational astrophysics, but look like a failure in their essence: the EOS has been only weakly constrained by observations.

Nevertheless, it is certainly not a failure on a longer time scale. First, the methods to study the neutron-star structure have been elaborated, and the most exciting observational discoveries have been made. Second, as we will discuss below, the situation drastically changed by the beginning of the new millennium: even current routine observations of already discovered neutron stars will certainly lead to a great progress in constraining the EOS, say, by 2010.

The majority of methods to constrain the EOS are based on the neutron star mass–radius relation (Fig. 6.5). A great difference of $M - R$ curves for soft, moderate and stiff EOSs will hopefully allow one to select the actual EOS. Basically, the problem looks fairly simple. If one could accurately measure the mass and radius of one neutron star, one would position it at the $M - R$ diagram and determine (constrain) the EOS. More "experimental" points would tighten this constraint.[1] As is clear from Fig. 6.5, one generally needs to measure

[1]In the idealized case of a complete "experimental" knowledge of the $M - R$ relation, we would be able to deduce the real EOS and solve thus the main mystery of neutron stars. It is an example of an "inverse problem", encountered in many domains of physics and astrophysics. Assume that the mass density is a non-

$M$ and $R$ with good accuracy (say, $\pm 5\%$) at a high significance level ($2\sigma$ and better) to distinguish between different theoretical predictions.

The attempts to solve the problem by confronting the $M - R$ diagram with a limited existing observational evidence are described in §§ 9.1–9.3. The methods of exploring the EOS using other ideas are outlined in §§ 9.4–9.7. Finally, in § 9.8 the arguments in favor and against the hypothesis of strange quark stars are critically reviewed.

## 9.1. Masses of neutron stars in binary systems

It is difficult to simultaneously measure the mass and radius of an isolated neutron star but it is easier to measure the mass of a neutron star in a binary system. Neutron stars enter binaries of several types (Chapter 1). One can distinguish X-ray and radio-pulsar binaries. X-ray binaries can be subdivided further into low-mass and high-mass ones; radio pulsars have been observed (Lorimer 2001) in binaries with neutron stars, white dwarfs, non-degenerate stars, and planets. PSR B1620–26, an 11-ms pulsar in the globular cluster M4, enters a triple system with a white dwarf and a high-mass planet (see, e.g., Thorsett *et al.* 1999, Richer *et al.* 2003, and references therein).

In this section we will describe numerous measurements of neutron-star masses $M$ in binary systems. By $M$ we will always mean the gravitational stellar mass. Any precisely measured mass is very important. However, it is clear from the consideration in Chapter 6 that one needs to search for most massive neutron stars to constrain the EOS.

Some selected observational results are plotted in Fig. 9.1 and discussed in subsequent sections. In binaries of different types the neutron star masses are measured using different techniques. X-ray binaries will be considered in § 9.1.1, double neutron star (radio pulsar – neutron star) binaries in § 9.1.2, radio pulsar – white dwarf binaries in § 9.1.3, and radio pulsars in binaries with non-degenerate stars in § 9.1.4. The results are summarized in § 9.1.5. The errorbars in Fig. 9.1 are mainly $2\sigma$ confidence limits. If observations provide $1\sigma$ limits only, we just widen them twice; this may be inaccurate but nevertheless is better than displaying unrealistically narrow $1\sigma$ limits.

### 9.1.1 X-ray binaries

We start with X-ray binaries, containing a neutron star (observed in X-rays but not in radio) and a companion star. Several such objects had been detected even before the discovery of neutron stars and were recognized as neutron

negative increasing function of the pressure, and equilibrium neutron star configurations are determined from the Tolman-Oppenheimer-Volkoff equation (§ 6.1). Then one can infer the EOS, $\rho = \rho(P)$, directly from the $M(R)$ curve (Gerlach 1968, Lindblom 1992). Unfortunately, with the observational basis, described in this Chapter, the project sounds purely academic.

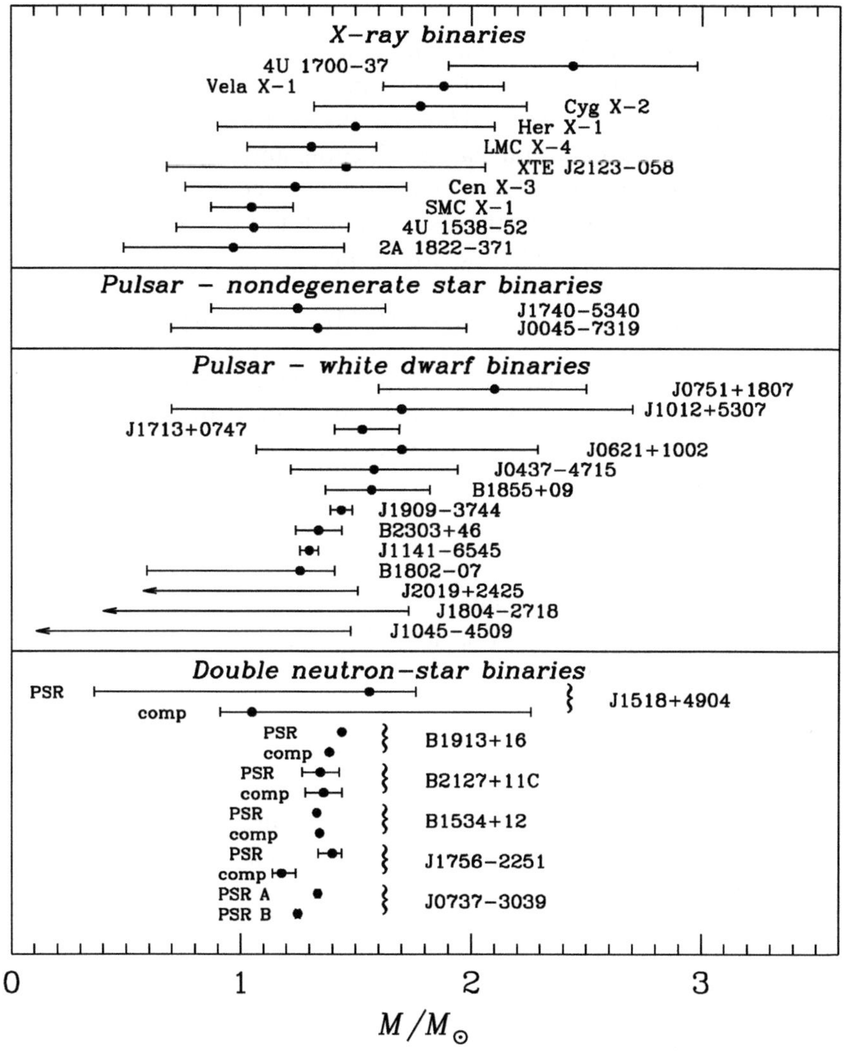


*Figure 9.1.* Neutron star masses inferred from observations of neutron stars in binary systems (by November 2005): X-ray binaries (§ 9.1.1), double neutron star binaries (§ 9.1.2), radio pulsar – white dwarf binaries (§ 9.1.3), and radio pulsar – non-degenerate star binaries (§ 9.1.4). The errorbars are $2\sigma$ confidence limits.

star systems later. They are mainly compact binaries; the X-ray emission is associated with an accretion from a companion to a neutron star. The orbits of the stars are close to Keplerian orbits. This enables one to employ the well elaborated astronomical methods to measure neutron-star masses.

### 9.1.1 a Keplerian orbits

Let us remind briefly the basic parameters of Keplerian orbits. Consider a binary system and approximate its components by point masses, $M_1$ and $M_2$ (with the total mass $M = M_1 + M_2$), moving in the $X - Y$ plane (Fig. 9.2). Star 1 will always be a neutron star. Each companion moves along an ellipse around the center of mass O. Let $e$ be the orbital eccentricity, while $a_1$ and $a_2$ be the major semi-axes of the ellipses for stars 1 and 2. From the definition of the center-of-mass one has $M_1 a_1 = M_2 a_2$, i.e.,

$$a_1 = aM_2/M, \qquad a_2 = aM_1/M, \qquad a = a_1 + a_2, \tag{9.1}$$

where $a$ is the major semi-axis of the ellipse of relative motion of companions.

The orbital motion is most easily analyzed in the center-of-mass reference frame, where conservation of the total energy $E$ and of the orbital angular momentum $J$ yields:

$$E = -GM_1M_2/(2a), \qquad J^2 = GM_1^2M_2^2 a(1 - e^2)/M, \tag{9.2}$$

$G$ being the gravitational constant.

The orbits of both companions are alike. It is sufficient to describe one of them, for instance, for star 1:

$$X_1 = r_1 \cos(\omega + \varphi), \quad Y_1 = r_1 \sin(\omega + \varphi), \quad r_1 = \frac{a_1\,(1 - e^2)}{1 + e\cos\varphi}, \tag{9.3}$$

where $\omega$ is the periastron longitude and $\varphi$ is the orbital phase.

The orbital period $P_b$ and the associated angular velocity $\Omega_b$ are given by the first Kepler law,

$$P_b = 2\pi/\Omega_b, \qquad \Omega_b^2 = GM/a^3. \tag{9.4}$$

The time evolution of the orbit, $\varphi(t)$, is given by the second Kepler law, which may be written as

$$\dot{\varphi} = \Omega_b\,(1 + e\,\cos\varphi)^2\,/(1 - e^2)^{3/2}. \tag{9.5}$$

This evolution is conveniently parameterized as

$$\begin{aligned} X_1 &= a_1 \left[(\cos\xi - e)\,\cos\omega - \sqrt{1 - e^2}\,\sin\xi\,\sin\omega\right], \\ Y_1 &= a_1 \left[(\cos\xi - e)\,\sin\omega + \sqrt{1 - e^2}\,\cos\xi\,\sin\omega\right], \end{aligned} \tag{9.6}$$

with $t - t_0 = (\xi - e\,\sin\xi)/\Omega_b$. Here, $t = t_0$ is some moment of periastron passage, and $\xi$ is a parameter which varies from $2\pi n$ to $2\pi(n+1)$ over an $n$-th orbital cycle. Clearly, one has $\dot{\varphi} = \Omega_b$ for a circular orbit ($e = 0$).

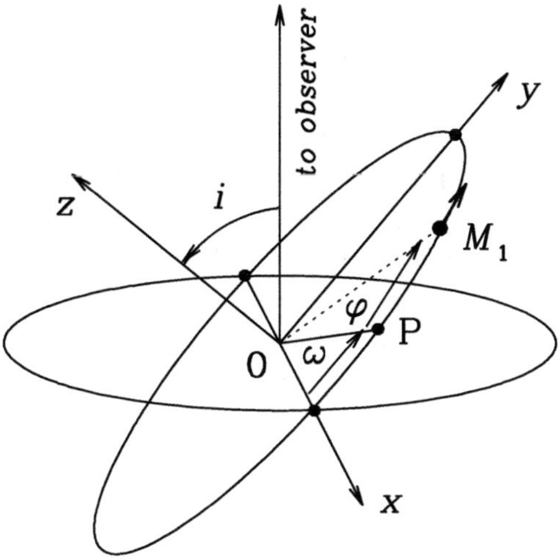


*Figure 9.2.* Keplerian orbit of the primary component $M = M_1$ of a binary; 0 is the center of mass; $X - Y$ is the orbital plane, $0X$ is the line of nodes, 0P is the periastron, and $i$ is the orbit inclination angle.

The orbital parameters, such as $P_b$, $a_1$, $a_2$, $e$, $\omega$, $t_0$, and the inclination $i$ of the orbital plane to the line of sight, fully specify Keplerian orbits of both companions and allow one to determine their masses $M_1$ and $M_2$. Thus, it is important to reconstruct these parameters from observations, which is usually a complicated problem.

To accomplish the task one has to observe the orbital variability of radiation from at least one of the companions. This gives the orbital period $P_b$. The next important step is to measure the orbital evolution of the radial velocity $v_{lj}$ (the projection of the orbital velocity on the line of sight) of the primary ($j = 1$) and/or secondary ($j = 2$). The phase dependence of the radial velocities is

$$v_{\mathrm{r}j} = K_j \left[\cos(\omega + \varphi_j) + e\cos\omega\right], \quad K_j = \frac{\Omega_b\, x_j}{\sqrt{1-e^2}}, \quad x_j \equiv a_j \sin i, \quad (9.7)$$

where $K_j$ is the velocity semi-amplitude, and $\varphi_j$ is the phase ($\varphi_1 \equiv \varphi$, $\varphi_2 = \varphi + \pi$, with $\omega$ referred to the primary as in Fig. 9.2).

Radial velocities of optical components are measured by observing linear Doppler shifts of spectral lines. Radial velocities of X-ray pulsars are measured by analyzing linear Doppler delay of X-ray pulse arrival times. Fitting an observed phase profile of the radial velocity $v_{\mathrm{r}j}(\varphi)$ by Eq. (9.7), one can determine $K_j$, $e$, $\omega$, $x_j$, and $T_0$, and calculate then the mass function $f_j$ of the primary or the secondary,

$$f_j \equiv (M_k \sin i)^3/M^2 = x_j^3\,\Omega_b^2/G, \tag{9.8}$$

where $M_k = M_2$ for $f_j = f_1$, and $M_k = M_1$ for $f_j = f_2$. The expression for $f_j$ is obtained using the Kepler law, Eq. (9.4), and is convenient for a mass determination, but basically it gives no new equation for $M_1$ and $M_2$ in addition to Eq. (9.4). The knowledge of the mass function gives constraints on the stellar masses and orbital parameters. If, for instance, $f_1$ is measured, then $M_2 \sin i = (f_1 M^2)^{1/3}$. From this relation one gets useful inequalities

$$M_2 \geq (f_1 M^2)^{1/3} \geq f_1, \quad M_1 \leq M - (f_1 M^2)^{1/3}, \quad \sin i \geq (f_1/M)^{1/3}. \tag{9.9}$$

By measuring radial velocities $v_{\mathrm{r}j}$ of one binary component $j$, one can determine five parameters: $P_b$, $x_j$, $e$, $\omega$, and $T_0$, which are often referred to as five *Keplerian parameters* of the orbit. They give us *two* independent equations [which come from the expressions for $P_b$ (or $f_j$) and $x_j$ (or $K_j$)] with *four* unknowns: $M_1$, $M_2$, $a$, and $\sin i$. Clearly, we need at least *two more* equations to measure the masses.

In this respect, any observation of the second binary component would be helpful. A measurement of its radial velocity is most desirable. It will give the second velocity semi-amplitude (the second value of $x_k$ and the second mass function $f_k$). The velocity semi-amplitudes $K_1$ and $K_2$ immediately yield the mass ratio, $q = M_1/M_2 = K_2/K_1$, the *third equation* of the problem. The last, *fourth* equation can be obtained, for instance, from observations of eclipses in eclipsing binaries or using some other tricks (e.g., performing multiband photometry of the optical companion in a compact binary §9.1.4).

A practical realization of this scheme is always complicated by numerous technical problems. One of the components may be faint, and difficult to observe. The measurement of radial velocities may be obscured, for instance, by a temporal variability of radiation from companions, accretion effects, tidal interaction, etc. If the radial velocity of the second component was not measured (and even if it was), any other information on the binary system would help. For instance, a determination of a spectral class and luminosity of an optical star might allow one to estimate its mass and radius. Observational indications that the optical companion fills its Roche lobe give additional constraints on the orbital parameters. Models for the reprocession of (pulsed) X-rays in an accretion disk around a neutron star and on the surface of an optical star can also be used to constrain the orbital parameters. A non-detection of eclipses in a binary with an unknown orbit inclination can serve for constraining the inclination angle. All in all, a study of every binary is a subject of art.

The orbital parameters of selected X-ray binaries are collected in Table 9.1. We list also the radii $R_2$ of companion stars which are useful for visualizing the compactness of the systems. We do not give statistical errors of the presented quantities (they may be found in cited references), just the central values, sufficient to understand the global parameters of the systems. The results of mass measurements are presented in Table 9.2 and Fig. 9.1.

### 9.1.1 b Vela X-1

Before discussing the overall results in § 9.1.1 c, let us describe an example: the mass measurements of Vela X-1 (=4U 0900−40), the famous X-ray pulsar in binary with GP Vel (=HD 77581, a B0.5 Ib supergiant). The pulsar spin period is 283 s, and the orbital period is nearly 9 days. The orbit is rather eccentric, $e \approx 0.09$, as compared to other X-ray binaries. The major semi-axis is $a \approx 50\,R_\odot$; the orbit inclination angle is $i \gtrsim 70°$; the radius of the optical star is $R_2 \approx 30\,R_\odot$. The star fills or nearly fills its Roche lobe.

Vela X-1 is a veteran source discovered by Chodil *et al.* (1967) in a rocket-borne X-ray observation. It attracted great attention as a bright variable X-ray source after the launch of the *Uhuru* orbital observatory (e.g., Giacconi *et al.* 1972). GP Vel was identified as the optical companion of the X-ray pulsar by Brucato & Kristian (1972) and Hiltner *et al.* (1972). It is widely accepted that the pulsar is a massive neutron star. The attempts to accurately measure its mass have been numerous (Fig. 9.3).

Initially, the accurate mass measurement was expected to be easy. The radial velocities of GP Vel in optics have been measured by many authors starting from Hiltner *et al.* (1972). The X-ray observations showed eclipsing nature of the system. Periodic 283-second X-ray pulsations were discovered by McClintock *et al.* (1976). Their timing allowed Rappaport *et al.* (1976) to measure radial velocities of Vela X-1; such measurements have been repeated afterward.

The results of some measurements of the Vela X-1 mass are presented in Fig. 9.3. Zuiderwijk *et al.* (1974) reported the lower limit of the Vela X-1 mass, $M_1 > 1.9\,M_\odot$. van Paradijs *et al.* (1976) obtained $M_1 = (1.6 \pm 0.3)\ M_\odot$ at the $1\sigma$ level. The next measurements gave: $M_1(2\sigma) = 1.85^{+0.35}_{-0.30}\ M_\odot$ (Rappaport & Joss 1983), $M_1(1.64\sigma) = 1.77^{+0.27}_{-0.21}\ M_\odot$ (Nagase 1989), $M_1(2\sigma) = 1.9^{+0.7}_{-0.5}\,M_\odot$ (van Kerkwijk *et al.* 1995a), and $M_1(2\sigma) = (1.86 \pm 0.32)\,M_\odot$ (Barziv *et al.* 2001). Let us mention that Stickland *et al.* (1997) reported noticeably lower values of $M_1$ but they turned out to be inaccurate as discussed by Barziv *et al.* (2001). Finally, Quaintrell *et al.* (2003) obtained $M_1(1\sigma) = (2.27 \pm 0.17)\ M_\odot$ assuming GP Vel fills its Roche lobe, i.e., the ratio of the GP Vel radius to the Roche lobe radius is $\beta = 1$; the inclination angle is then $i = (70.1 \pm 2.6)°$. Taking the maximum inclination, $i = 90°$, the same authors got $\beta = 0.89 \pm 0.03$ and $M_1(1\sigma) = (1.88 \pm 0.13)\ M_\odot$. By varying $i$ from 70° to 90°, they can get any value of $\beta$ between 1 and 0.89, and the values of $M_1$ intermediate between the two limiting cases. In reality, $\beta$

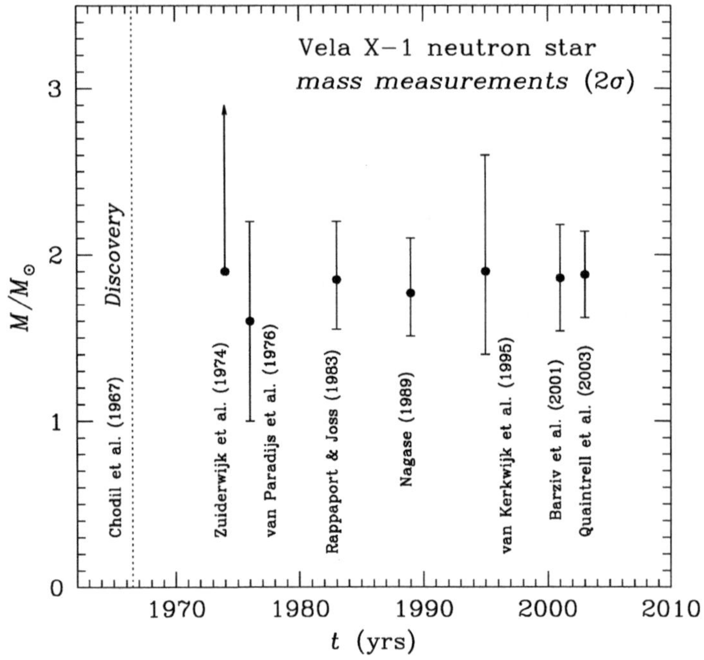


*Figure 9.3.* Some measurements of the Vela X-1 mass reported in the literature by 2006.

can vary over the eccentric orbit, and the situation is more complicated. Being conservative, in Fig. 9.3 we present the set of lower mass limits, not because it is more realistic but because an overestimation of $M_1$ can lead to wrong conclusions on the EOS in neutron-star cores. All mass limits of Vela X-1 in Fig. 9.3 are rescaled to $2\sigma$ levels. This simple rescaling may be inaccurate but the intrinsic errors of the measurements are evidently too large, i.e., the Vela X-1 mass is anyway not known precisely by the time of this writing.

The main obstacle of measuring $M_1$ consists in strong deviations from a pure Keplerian radial velocity curve of GP Vel (van Kerkwijk *et al.* 1995a, Barziv *et al.* 2001, Quaintrell *et al.* 2003). The deviations are probably caused by GP Vel oscillations excited by tidal forces in the eccentric orbit. The problem of extracting the Keplerian parameters is complicated and ambiguous.

There is no tendency of convergence of the reported mass limits in Fig. 9.3, although the mean mass values stay pretty high, $(1.8-2.0)\ M_\odot$, or higher. Wide errorbars cannot allow one to pinpoint the real mass of Vela X-1. For instance, using the results of Barziv *et al.* (2001) one can state that $M_1 > 1.54$ $M_\odot$ at the $2\sigma$ level but even this lower limit does not seem absolutely solid.

The absence of convergence of the mass limits in Fig. 9.3 does not mean the absence of any progress in understanding the binary. The observations are obviously progressing but reveal the complicated nature of the system.

### 9.1.1 c Neutron star masses in X-ray binaries

Let us outline briefly the results of the neutron-star mass measurements in the selected X-ray binaries (Fig. 9.1, Tables 9.1 and 9.2).

The binaries we have chosen are either galactic sources or the sources in the nearby galaxies, the Large Magellanic Cloud (LMC X-4) and the Small Magellanic Cloud (SMC X-1). Every binary consists of a neutron star, observed in X-rays, and an optical star, observed in optics and possibly in ultraviolet. All these binaries are compact. Optical companions fill or nearly fill their Roche lobes, so that the accretion on neutron stars sets in, either by outflow through the first Lagrange point or in the form of a strong wind from an optical companion. In all the cases, the radial velocities of optical companions have been measured. The accretion and tidal interaction made the orbits close to circular (the eccentricity $e \lesssim 0.1$, excluding 4U 1700–37).

The selected systems can be divided into three groups.

*(i) X-ray pulsars.* They are six classical X-ray pulsars observed since the beginning of X-ray astronomy: 4U 1538–52, SMC X-1, Cen X-3, LMC X-4, Her X-1, and Vela X-1. All of them are high-mass X-ray binaries (except for Her X-1 which is a medium-mass X-ray binary). The optical components are mainly O-B stars.

The periods of X-ray pulsations and the radial velocities of pulsars have been measured. All the binaries show X-ray eclipses. Therefore, the full set of observational data is available (radial velocities of both companions, eclipses, spectral class of optical companions, etc.). However, this does not allow one to accurately measure the masses (Table 9.2). The reason is the same as for Vela X-1: radial-velocity curves of optical companions often show pronounced deviations from the expected Keplerian shapes, presumably due to tidal effects and nonradial oscillations of the optical companions.

The most massive neutron star in these systems seems to be Vela X-1. By 2006 its mass remained rather uncertain, as it has been discussed in §9.1.1 b.

*(ii) The system 4U 1700–37.* It was discovered by the *Uhuru* satellite in December 1970 (Jones *et al.* 1973). This is also a high-mass X-ray binary, which demonstrates X-ray eclipses and highly variable X-ray emission but no strict X-ray periodicity. Accordingly, no radial velocity of the X-ray companion has been measured, only the radial velocity of the optical companion, the O6.5 Iaf$^+$ star HD 153919. A comparison of optical and ultraviolet observations of the optical star with stellar atmosphere models (taking into account intense mass outflow and deviations from local thermodynamic equilibrium) allowed Clark *et al.* (2002) to estimate the effective surface temperature, luminosity, and

*Table 9.1.* Parameters of selected X-ray binaries (by November 2005)

| System | $P_b$ (d) | $e$ | $x_1/c^{a)}$ (s) | $i$ (deg) | $R_2^{b)}$ ($R_\odot$) | Refs.[c)] |
|---|---|---|---|---|---|---|
| 2A 1822−371[d)] | 0.232 | <0.03 | 1.006 | 81–84 | | JK01, J03 |
| 4U 1538−52[e)] | 3.73 | 0.08 | 52.8 | $\sim 70$ | $\approx 15$ | K95 |
| SMC X-1[e)] | 3.89 | <0.00004 | 53.5 | $\approx 68$ | $\approx 15$ | K95, M05 |
| Cen X-3[e)] | 2.09 | <0.0008 | 39.6 | $\approx 73$ | $\approx 11$ | K95, A99, M05 |
| XTE J2123−058[d)] | 0.248 | | | $\approx 73$ | $\approx 0.6$ | T02 |
| LMC X-4[e)] | 1.41 | <0.01 | 26.3 | $\approx 65$ | $\approx 8$ | K95, M05 |
| Her X-1 | 1.70 | <0.0003 | 13.2 | 81 | $\approx 4.2$ | CVR95, R97 |
| Cyg X-2[d)] | 9.84 | 0.0 | | 49–73 | | OK99 |
| Vela X-1[e)] | 8.96 | 0.0898 | 114 | >70 | $\approx 30$ | Q03 |
| 4U 1700−37[e,f)] | 3.41 | 0.2 | | >55 | $\approx 22$ | C02, H03 |

[a)] $x_1/c = (a_1 \sin i)/c$ is the time of light propagation along the projected major semi-axis.
[b)] Radius of an optical component.
[c)] JK01: Jonker & van der Klis (2001); J03: Jonker *et al.* (2003); K95: van Kerkwijk *et al.* (1995b); M05: van der Meer *et al.* (2005); A99: Ash *et al.* (1999); T02: Tomsick *et al.* (2002); CVR95: Cheng *et al.* (1995); R97: Reynolds *et al.* (1997); OK99: Orosz & Kuulkers (1999); Q03: Quaintrell *et al.* (2003); C02: Clark *et al.* (2002); H03: Hammerschlag-Hensberge *et al.* (2003).
[d)] Low-mass X-ray binary.
[e)] High-mass X-ray binary.
[f)] The compact object may be a black hole.

surface gravity, but with rather large uncertainties. Without radial velocities of the X-ray companion, this information is insufficient to determine the masses.

In this situation, Clark *et al.* (2002) estimated the masses using Monte Carlo method. They randomly selected unknown or poorly known parameters (e.g., the radius of the optical star, the radius of the Roche lobe) within reasonable limits (in accordance with observations). They used the Kepler laws and familiar relations for the stars, which fill or nearly fill the Roche lobe, to determine the masses $M_1$ and $M_2$ for any set of selected parameters. In this way they obtained statistical distributions of masses over about one million of Monte Carlo selections, and estimated $M_1$ and $M_2$ (Table 9.2) from these statistical distributions. The mass of the X-ray companion is pretty high, $M_1 = (2.44 \pm 0.27)\ M_\odot$ (at $1\sigma$ level). This analysis was done for a circular orbit. Later Hammerschlag-Hensberge *et al.* (2003) reanalyzed high-resolution ultraviolet

*Table 9.2.* Masses of neutron stars and companions in X-ray binaries (by November 2005)

| System | $M_1/M_\odot$ | $M_2/M_\odot$ | Signif. | Reference |
|---|---|---|---|---|
| 2A 1822−371 | $0.97\pm0.24^{a)}$ | $0.33\pm0.05^{a)}$ | $1\sigma$ | Jonker *et al.* (2003) |
| 4U 1538−52 | $1.06^{+0.41}_{-0.34}$ | $16.4^{+5.2}_{-4.0}$ | $2\sigma$ | van Kerkwijk *et al.* (1995b) |
| SMC X-1 | $1.05\pm0.09$ | $15.5\pm1.5$ | $1\sigma$ | van der Meer *et al.* (2005) |
| Cen X-3 | $1.24\pm0.24$ | $19.7\pm4.3$ | $1\sigma$ | van der Meer *et al.* (2005) |
| XTE J2123−058 | $1.46^{+0.30}_{-0.39}$ | $0.53^{+0.28}_{-0.39}$ | $1\sigma$ | Tomsick *et al.* (2002) |
| LMC X-4 | $1.31\pm0.14$ | $15.6\pm1.8$ | $1\sigma$ | van der Meer *et al.* (2005) |
| Her X-1 | $1.5\pm0.3$ | $2.3\pm0.3$ | $1\sigma$ | Reynolds *et al.* (1997) |
| Cyg X-2 | $1.78\pm0.23$ | $0.60\pm0.13$ | $1\sigma$ | Orosz & Kuulkers (1999) |
| Vela X-1 | $2.27\pm0.17$ | $27.9\pm1.3$ | $1\sigma$ | Quaintrell *et al.* (2003)[b)] |
| Vela X-1 | $1.88\pm0.13$ | $23.1\pm0.2$ | $1\sigma$ | Quaintrell *et al.* (2003)[c)] |
| 4U 1700−37 | $2.44\pm0.27^{\,d)}$ | $58\pm11$ | $1\sigma$ | Clark *et al.* (2002) |

[a)] Is likely to be a lower mass limit.
[b)] Assuming the companion star fills its Roche lobe leading to inclination angle $i = 70.1^\circ \pm 2.6^\circ$.
[c)] Assuming $i = 90^\circ$ and the ratio of the radii of the companion star and the Roche-lobe is $\beta = 0.89 \pm 0.03$.
[d)] The compact object may be a black hole.

spectra of this binary and noticed that these spectra are in good agreement with optical spectra if the eccentricity is finite, $e \approx 0.22$. In that case $M_1$ and $M_2$ should be about 4% higher than reported by Clark *et al.* (2002). We have not introduced this correction in Table 9.2.

Unfortunately, one cannot exclude that 4U 1700–37 contains a black hole rather than a neutron star (see, e.g., Clark *et al.* 2002, for the arguments against and pro the black hole hypothesis). If future observations confirm the existence of a neutron star in 4U 1700–37, it will be a good candidate for a very massive neutron star.

(*iii*) *Low-mass X-ray binaries.* They are compact binaries containing a neutron star and a low-mass optical companion ($M_2 \lesssim M_\odot$). We present three such systems, 2A 1822–371, XTE J2123–058, and Cyg X-2.

2A 1822–371 was discovered by Jonker & van der Klis (2001) much later than the sources described above. It is an X-ray binary with the shortest orbital period, $P_b = 5.57$ h, among the selected sources. X-ray pulsations ($P = 0.59$ s) and X-ray eclipses have been detected; the radial velocities of the X-ray and optical companions have been measured. Unfortunately, the neutron star has a low mass, $M_1 \sim M_\odot$ (Jonker *et al.* 2003), insufficient to constrain the EOS of superdense matter.

XTE J2123–058 was also discovered later (Levine *et al.* 1998) than the majority of other sources. It shows type I X-ray bursts, quasiperiodic X-ray oscillations, partial X-ray eclipses, and behaves as an X-ray transient. The radial velocities of the optical companion have been measured but no strictly periodic X-ray pulsations have been detected. Nevertheless, the semi-amplitude $K_1$ of the radial velocity of the X-ray source has been determined (Casares *et al.* 2002) by measuring extreme wings of the rotationally broadened $H_\alpha$ line, assuming it originates in an accretion disk around the neutron star and shares its orbital motion. The neutron-star mass has been estimated by Tomsick *et al.* (2002) with high uncertainty, its central value $M \approx 1.46\, M_\odot$ is about the same as in double neutron star binaries (§ 9.1.2).

Finally, Cyg X-2 is one of the brightest and oldest known X-ray sources. The neutron star accretes at a nearly Eddington rate and shows type I X-ray bursts, but no X-ray eclipses. No strictly periodic X-ray pulsations (thus, no radial velocities of Cyg X-2) have been detected. The optical star, V1341 Cygni, is bright and good for a very high-resolution spectroscopy. Spectroscopic observations yielded its radial velocity and the mass function. In addition, they allowed Casares *et al.* (1998) to detect rotational broadening of lines in the spectrum of V1341. It is thought that the star fills its Roche lobe and its rotation is synchronized with the orbital motion. Under these conditions, the rotational broadening gives an independent relation for determining the masses (instead of the mass function of the primary). The orbit inclination angle can be constrained from general Keplerian relationships and from a non-detection of X-ray eclipses. Using this information, Orosz & Kuulkers (1999) determined the neutron star mass, $M_1(1\sigma) = 1.78 \pm 0.23\, M_\odot$. The central value is rather high and interesting for constraining the EOS in neutron star cores but large uncertainties hamper such an analysis.

Let us emphasize that real uncertainties of mass measurements in X-ray binaries can be larger than cited above, because of the complicated nature of the binaries. The best example is provided by two sets of possible masses of the neutron star in Vela X-1 (Quaintrell *et al.* 2003, Table 9.2). Another example – a reanalysis of light curves of optical components of X-ray binaries by Abubekerov *et al.* (2004) and Abubekerov (2004) using, particularly, the Roche model of optical stars. These studies lead to somewhat different values of $M_1$ and $M_2$. For instance, Abubekerov (2004) proposes, among other possibilities, an interpretation of the data which gives much lower mass of the compact object in 4U 1700–37 ($M_1 \sim 1.7\, M_\odot$ or even lower) and much lower (and more reasonable) mass of the companion ($M \sim 27\, M_\odot$).

Thus, the uncertainties of neutron-star mass measurements (by 2006) in X-ray binaries are too high for a definite conclusion on the EOS. Nevertheless, the observations give indirect evidence for the existence of massive neutron stars, with $M \sim 1.8\,M_\odot$ and higher, but the evidence is still not solid. The binaries most important for the EOS problem are: Cyg X-2, Vela X-1, and 4U 1700–37 (if its compact component *is* a neutron star).

## 9.1.2 Double neutron star binaries

High uncertainties of masses of neutron stars in X-ray binaries arise mostly due to deviations from Keplerian motion (accretion, tidal forces, oscillations of an optical companion) in compact binaries with an extended optical companion.

Luckily, Nature gave us a wonderful present in the form of binaries containing a radio pulsar and another compact star (a neutron star or a white dwarf). These systems have two great advantages over X-ray binaries. First, pulsar spin periods are measured with a very high precision (more than 10 correct digits) by methods of radio astronomy. Using these data, one can closely follow the evolution of the pulsar orbital velocity. Second, a close binary system composed of compact objects can accurately be treated as a system of two point masses. In the present section, we start with double neutron star binaries. The pulsar – white dwarf systems are analyzed in § 9.1.3. Precise analysis of orbital motion requires proper account of the effects of General Relativity, which is discussed in the next section.

### 9.1.2 a Relativistic orbits and pulsar timing

Let us outline relativistic orbital motion of two point masses. We adopt the standard theory of General Relativity which was proven to be a valid theory for this problem (§ 9.1.2 b). The relativistic evolution of a binary is accompanied by the losses of the total energy and the orbital momentum of the system due to the emission of gravitational waves. Gravitational radiation from two orbiting point masses was studied by Peters & Mathews (1963). The relativistic orbits were analyzed in a classical paper by Peters (1964) and refined later by many authors (see, e.g., Taylor 1987, Taylor & Weisberg 1989, and references therein). We will mainly follow Peters (1964).

Consider a binary with the velocity of both companions much lower than the speed of light. In the zero-order approximation, the orbital motion is Keplerian (§ 9.1.1 a), but the effects of General Relativity lead to long-term variations of the Keplerian parameters. Thus, the energy $E$, the momentum $J$, the orbital major semi-axis $a$, the eccentricity $e$, the orbital period $P_b$, and other Keplerian parameters become slowly varying functions of time. It should be stressed that, in principle, the relativistic effects renormalize the Keplerian relations. For instance, the first Kepler law, Eq. (9.4), transforms into (e.g., Taylor & Weisberg

1989)

$$\Omega_b = \left(\frac{GM}{a^3}\right)^{1/2} \left[1 + \left(\frac{M_1 M_2}{M^2} - 9\right)\frac{GM}{2ac^2}\right]. \tag{9.10}$$

The classical relation is reproduced by replacing $[\ldots] \to 1$. As long as the binary is not too compact, the deviation from the classical relation is small. For simplicity, we will ignore this deviation in subsequent analysis.

The loss rates of energy and of angular momentum in a binary due to the emission of gravitational waves are:

$$\begin{aligned}\dot{E} &= -\frac{32\, G^4 M_1^2 M_2^2 M}{5\, c^5 a^5 (1-e^2)^{7/2}}\left(1 + \frac{73}{24}e^2 + \frac{37}{96}e^4\right),\\ \dot{J} &= -\frac{32\, G^{7/2} M_1^2 M_2^2 M^{1/2}}{5\, c^5 a^{7/2}(1-e^2)^2}\left(1+\frac{7}{8}e^2\right).\end{aligned} \tag{9.11}$$

Variation rates of other quantities are easily determined by expressing them through the energy and angular momentum from Eqs. (9.2). In particular,

$$\begin{aligned}\dot{a} &= -\frac{64\, G^3 M_1 M_2 M}{5\, c^5 a^3 (1-e^2)^{7/2}}\left(1+\frac{73}{24}e^2+\frac{37}{96}e^4\right),\\ \dot{e} &= -\frac{304\, e\, G^3 M_1 M_2 M}{15\, c^5 a^4 (1-e^2)^{5/2}}\left(1+\frac{121}{304}e^2\right),\\ \dot{P}_b &= P_b\frac{3\dot{a}}{2a},\\ \dot{\omega} &= \frac{3\,\Omega_b GM}{a\,(1-e^2)\,c^2} = \frac{3\,\Omega_b^{5/3}(GM)^{2/3}}{(1-e^2)\,c^2},\end{aligned} \tag{9.12}$$

where $\dot{\omega}$ describes the periastron advance, a famous General Relativity effect. Strictly speaking, $E$, $J$ and other quantities experience low-amplitude short-term variations during every orbital period. These variations are neglected here, and the dot means the time derivative of a quantity averaged over an orbital period.

Dividing $\dot{a}$ by $\dot{e}$ one gets a differential equation which is analytically solved:

$$a(e) = \frac{c_0\, e^{12/19}}{1-e^2}\left(1+\frac{121}{304}e^2\right)^{870/2299}, \tag{9.13}$$

where $c_0$ is determined by the values of $a = a_0$ and $e = e_0$ in an initial moment of time $t = t_0$,

$$c_0 = a_0\,(1-e_0^2)\,e_0^{-12/19}\left(1+\frac{121}{304}e_0^2\right)^{-870/2299}. \tag{9.14}$$

From the expression for $\dot{a}$ one obtains

$$t - t_0 = \frac{15\, c^5 c_0^4}{304\, G^3 M_1 M_2 M}\, [F(e_0) - F(e)]\,, \tag{9.15}$$

where

$$F(e) = \int_0^e \frac{d\epsilon\; \epsilon^{29/19}}{(1-\epsilon^2)^{3/2}} \left(1 + \frac{121}{304}\,\epsilon^2\right)^{1181/2299} . \tag{9.16}$$

The integral cannot be calculated analytically. However, one can derive the asymptotes in the limits of small and large eccentricities:

$$\begin{aligned} F(e \to 0) &= \frac{19}{48}\, e^{48/19}, \\ F(e \to 1) &= \frac{1}{\sqrt{1-e^2}} \left(\frac{425}{304}\right)^{1181/2299} - 1.72892. \end{aligned} \tag{9.17}$$

We have computed $F(e)$ and fitted it by

$$\begin{aligned} F(e) &= \frac{e^{48/19}}{\sqrt{1-e^2}} \left(0.39583 + 0.79199\, e^{2.2876}\right) \\ &\quad - \frac{1.7289\, e^{5.1121}}{1 + 1.8062\, e^{0.30792}\,(1-e^2)^{0.51659}}. \end{aligned} \tag{9.18}$$

The fit error is $\lesssim 0.03\%$ for any $e$.

Now the calculation of a relativistic orbit is elementary. Taking $e = e_0$ and $a = a_0$ at $t = t_0$, we determine the evolution of the eccentricity, $e = e(t)$ from Eqs. (9.15) and (9.18), and obtain the major semi-axis, $a = a(t)$, from Eq. (9.13). The evolution of other parameters, for instance, the orbital period and periastron advance can be calculated then from Eq. (9.12). The evolution can be followed in the future ($t > t_0$) and in the past ($t < t_0$) down to the birth of our binary system.

For example, Fig. 9.4 shows the evolution of the orbit of the Hulse-Taylor pulsar, PSR B1913+16. The observations of this pulsar will be discussed in §9.1.2b. They have been performed over three decades (1974–2006) in the present epoch (the dotted vertical line) and enabled one to accurately determine $a_0$, $e_0$, $M_1$, and $M_2$. With these data, one can reconstruct the orbit. The major semi-axis $a(t)$ decreases with time: the binary emits gravitational waves and the companions fall onto each other. The ellipticity also decreases and the orbit becomes circular; the circularization goes generally quicker than the decrease of $a(t)$. In a finite period of time, $t = t_{\rm death}$, the semi-axis will become zero, $a(t_{\rm death}) = 0$, which means the merging of two neutron stars and the death of the system. The death time is given by

$$t_{\rm death} = t_0 + \frac{15\, c^5 c_0^4\, F(e_0)}{304\, G^3 M_1 M_2 M}. \tag{9.19}$$

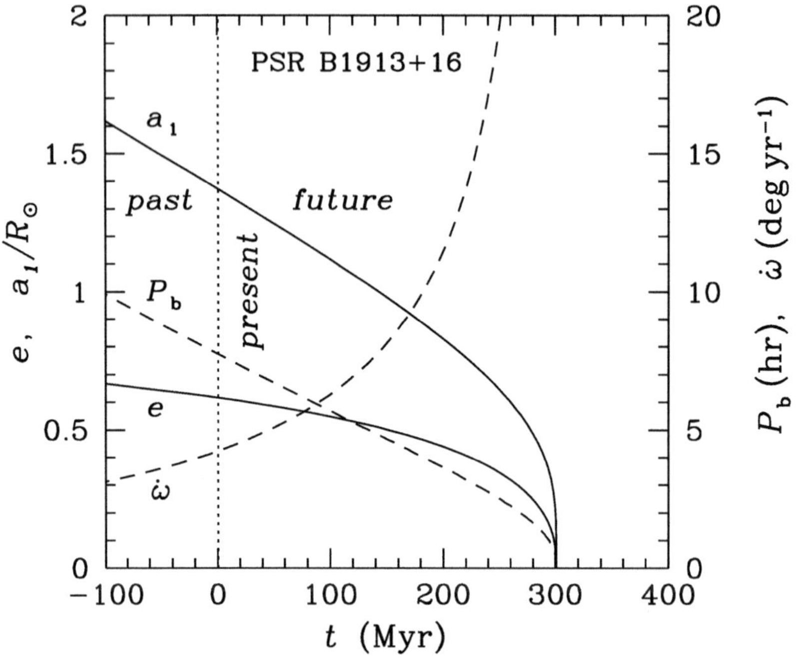


*Figure 9.4.* Orbit evolution of the Hulse-Taylor pulsar (with the full major semi-axis $a \approx 2.04\, a_1$). The parameters of the system are measured in the present epoch (vertical dotted line, $t_0 = 0$). The theory allows us to follow the evolution in the past and future. The pulsar was possibly born about 100 million years ago (§ 9.1.2 b).

The Hulse-Taylor pulsar will die in $t_{\rm death} - t_0 \approx 300$ Myr from now. Just before the death, the relativistic effects will be the strongest (§ 9.1.2 b). To be precise, our description of the relativistic orbit becomes invalid at some milliseconds before the official death. Specifically, it happens when $a(t)$ approaches the sum of neutron-star radii, $\sim (20-30)$ km. At that stage, the orbital velocities approach relativistic limit; the approximation of point masses fails; the orbital parameters greatly evolve over one orbital cycle. The next stage of *neutron-star merging* is the most interesting but a very complicated one; it is beyond the scope of our consideration. The merging and associated phenomena have been simulated in a number of papers (see Ruffert & Janka 2001, Rosswog & Davies 2002, Taniguchi & Gourgoulhon 2002, and references therein).

The case of circular (non-eccentric) orbits cannot be described by Eq. (9.15). It should be considered separately, and it is very simple. In this special case, $e(t) \equiv 0$, while $a(t)$ is determined directly from Eq. (9.12) for $\dot{a}$:

$$a(t) = \left[ \frac{256\, G^3 M_1 M_2 M}{5\, c^5} \, (t_{\rm death} - t) \right]^{1/4}, \qquad (9.20)$$

The death time is $t_{\rm death} = t_0 + 5\,c^5\,a_0^4/(256\,G^3 M_1 M_2 M)$; it can also be obtained from Eq. (9.19) in the limit of $e_0 \to 0$. The Keplerian angular velocity behaves as

$$\Omega_b(t) = \sqrt{GM/a^3} \propto (t_{\rm death} - t)^{-3/8}. \tag{9.21}$$

Actually, these solutions describe last evolutionary stages of any binary, which can be initially eccentric (because the circularization occurs usually quicker than the decrease of $a(t)$; e.g., Imshennik & Popov 1994). This will be discussed in § 9.1.2 b in more details.

One component of a double neutron star binary is usually observed as a radio pulsar, while its companion is not observed at all (although it can be a radio pulsar with an unfavorable orientation of pulsar beaming). Let us focus on this case. The important exclusion, the binary pulsar J0737–3039, will be discussed in § 9.1.2 d. The observed pulsar undergoes a traditional regular spindown in its comoving reference frame. Its orbital motion is studied by precise measuring of arrival times of pulses. The pulse periods, detected by a distant observer, are noticeably affected by the Doppler effect (linear and quadratic), and by general relativistic effects. The data have to be corrected for the motion of a binary (as a whole) and a detector (the Earth).

The interpretation of observations is done in several steps. First, even with a short set of observations, one can easily extract radial velocities and determine $P_b$, $e$, $x_1$, $\omega$, and hence the pulsar mass function $f_1$ (as for an ordinary binary system). These results are important: they give *two (out of four) relations* for measuring the neutron star masses $M_1$ and $M_2$.

*Two other relations* must be obtained at the next steps by accumulating data in new observation sets. In addition to the linear Doppler effect, one can measure the quadratic Doppler effect in the pulsar motion and general relativistic effects, such as the periastron advance, the gravitational redshift in the field of the secondary star, the general relativistic delay of signals in the gravitational field of the secondary (the so called Shapiro delay) and others. Of course, the gravitational redshift in the field of the primary is higher than in the field of the secondary, but it does not vary over an orbital cycle, being useless in timing analysis. The procedure of extracting relativistic orbital parameters from timing data is described by many authors (see Taylor & Weisberg 1989, Lorimer 2001, Stairs 2003, and references therein).

The periastron advance rate, $\dot{\omega}$, is measured as a secular variation of the periastron longitude. Naturally, a binary must be rather eccentric for its reliable determination. The expression for $\dot{\omega}$, Eq. (9.12), serves then as a new independent relation for the mass determination. It gives the total system mass, $M = M_1 + M_2$. Thus, if $\dot{\omega}$ is measured, then $M$ can be known with much better precision than the separate masses $M_1$ and $M_2$. The knowledge of $M$ sets useful constraints, Eq. (9.9), on stellar masses and the orbit inclination.

The relative shift of pulse arrival times, produced by the quadratic Doppler effect and the leading-order gravitational redshift in the field of the secondary, is given by $v^2/(2c^2) + GM_2/(r_{12}c^2)$, where $v(t)$ is the orbital velocity of the primary and $r_{12}(t)$ is the distance between the companions. Extracting such shifts from timing data, one can measure the parameter

$$\gamma = \frac{eGM_2(M_1 + 2M_2)}{\Omega_b c^2 aM}, \tag{9.22}$$

which can serve as an additional independent relation for mass measurements. Since $\gamma \propto e$, a binary has to be eccentric for determining $\gamma$.

Detecting Shapiro time delay, one can extract two other relativistic parameters,

$$s \equiv \sin i = \frac{\Omega_b^{2/3} M^{2/3} x_1}{G^{1/3} M_2}, \quad r = \frac{GM_2}{c^3}, \tag{9.23}$$

which characterize the "shape" and "range" of the Shapiro delay. They can serve as two other independent relations. The range parameter $r$ gives $M_2$, the companion mass, which causes the gravitational time dilatation. The shape of the Shapiro delay is determined by an impact parameter of light rays (the closest distance between a light trajectory and $M_2$); the shape parameter is just $s = \sin i$, expressed through binary parameters using the Kepler laws. Naturally, binaries observed "edge-on" (large $\sin i$) are most suitable for detecting the Shapiro effect. A non-detection of Shapiro delay in a binary with an unknown $i$ may be used to constrain $i$ (provided the binary is compact enough for the effect to be detectable at favorable $i$).

By following the evolution of Keplerian orbital elements in a set of observations one can obtain their time derivatives, such as $\dot{P}_b$, which are associated with the relativistic evolution of the orbit in the present epoch. For this purpose, one fits the observational data using Taylor expansions of orbital parameters. For instance, the expansion of the orbital phase can be written as $\varphi(t) = \Omega_b(t_0)(t - t_0) + \frac{1}{2}\dot{\Omega}_b(t - t_0)^2 + \ldots$, where $\Omega_b(t_0) = \Omega_b$ is a current value of the orbital angular velocity, and $\dot{\Omega}_b(t_0) = -\Omega_b \dot{P}_b/P_b$ is its time derivative. Such expansions are convergent for $|t - t_0| \ll \tau$, where $\tau$ is a characteristic time of the orbital evolution ($\tau \sim 100$ Myr for PSR B1913+16). By extracting $\dot{\Omega}_b$ from pulsar timing, one obtains $\dot{P}_b$. The derivatives inferred in this way ($\dot{\Omega}_b$ or $\dot{P}_b$ in our example) are also treated as relativistic parameters.

It is difficult to measure many relativistic parameters for the same binary, but even *two of them* are sufficient to infer the masses $M_1$ and $M_2$, while others can be used for checking the result.

There are other tricks (described, e.g., by Thorsett & Chakrabarty 1999) for measuring or constraining orbital parameters and neutron star masses from radio observations of pulsars in binary systems. Radio polarization measurements

*Table 9.3.* Parameters of double neutron star binaries (by November 2005)

| System | $P$ (ms) | $P_b$ (d) | $e$ | $x_1/c$ (s) | $i$ (deg) | $\dot{\omega}$ (deg/yr) | Other GR | Discovery[a)] |
|---|---|---|---|---|---|---|---|---|
| J1518+4904 | 40.9 | 8.63 | 0.249 | 20.0 | $> 20$ | 0.0111 | $\dot{P}_b$ | NST95 |
| J1811–1736 | 104.2 | 18.8 | 0.828 | 34.8 | | 0.009 | | L01 |
| J1829+2456 | 41.0 | 1.176 | 0.139 | 7.24 | | 0.28 | | C04 |
| B1534+12 | 37.9 | 0.421 | 0.274 | 3.73 | $\approx 77$ | 1.76 | $\gamma$, $\dot{P}_b$, $s$, $r$ | W91 |
| B1913+16 | 59.0 | 0.323 | 0.617 | 2.34 | $\approx 47$ | 4.22 | $\gamma$, $\dot{P}_b$ | HT75 |
| B2127+11C | 30.5 | 0.335 | 0.681 | 2.52 | $\approx 50$ | 4.46 | $\gamma$ | A90 |
| J0737–3039 | 22.7 | 0.102 | 0.088 | 1.42 | $\approx 88$ | 16.90 | $\gamma$, $s$, $r$ | B03 |
| J1756–2251 | 28.5 | 0.320 | 0.181 | 2.76 | $\approx 73$ | 2.58 | $\gamma$, $s$ | M01 |

[a)] NST95: Nice *et al.* (1995); L01: Lyne *et al.* (2001); C04: Champion *et al.* (2004); W91: Wolszczan (1991); HT75: Hulse & Taylor (1975); A90: Anderson *et al.* (1990); B03: Burgay *et al.* (2003); M01: Manchester *et al.* (2001)

can be helpful to determine the position of the pulsar magnetic axis which can be related to the spin axis and other parameters. Interstellar scintillations of pulsar radiation can also be used in the process of reconstructing the orbital elements. Proper motion of a binary across the sky induces a change of the inclination angle of the orbit, and hence a kinematic change in the projected semi-axis, $\dot{x}_1/x_1$. By measuring or constraining $\dot{x}_1/x_1$ from pulsar timing and by measuring proper motion, one can constrain the inclination angle.

It is easy to formulate the conditions under which a double neutron star binary is best suited for measuring the masses. *First*, the binary should be compact (with the orbital period $P_b \lesssim 1$ day) for the relativistic effects to be most pronounced. *Second*, it would excellent if our binary were eccentric; this would simplify the measurements of two parameters, the periastron advance $\dot{\omega}$ and the parameter $\gamma$, sufficient to determine $M_1$ and $M_2$. *Third*, even if the orbit is nearly circular (so that $\gamma$ and $\dot{\omega}$ are not detectable) but the binary is observed edge-on, one can expect to detect Shapiro delay (to measure $r$ and $s$) and infer the masses in this way. Other tricks can also work out.

By November 2005 eight double neutron star binaries have been discovered. Their orbital parameters are listed in Table 9.3; the neutron star masses have been inferred in six systems (Table 9.4 and Fig. 9.1). We will discuss these results in §§ 9.1.2 b, 9.1.2 c and 9.1.2 d.

*Table 9.4.* Masses of neutron stars (at $2\sigma$ levels) in double neutron star binaries (by November 2005)

| System | $M_1/M_\odot$ | $M_2/M_\odot$ | Reference |
|---|---|---|---|
| J1518+4904 | $1.56^{+0.20}_{-1.20}$ | $1.05^{+1.21}_{-0.14}$ | Thorsett & Chakrabarty (1999) |
| B1534+12 | 1.3332±0.0020 | 1.3452±0.0020 | Stairs *et al.* (2002) |
| B1913+16 | 1.4408±0.0006 | 1.3873±0.0006 | Weisberg & Taylor (2003) |
| B2127+11C | 1.349±0.080 | 1.363±0.080 | Thorsett & Chakrabarty (1999) |
| J0737–3039 | 1.337±0.010 | 1.250±0.010 | Lyne *et al.* (2004) |
| J1756–2251 | $1.40^{+0.04}_{-0.06}$ | $1.18^{+0.06}_{-0.04}$ | Faulkner *et al.* (2005) |

### 9.1.2 b The Hulse-Taylor pulsar

The famous *Hulse-Taylor pulsar*, PSR B1913+16, was the first radio pulsar discovered in a binary system (and the system turned out to be a double neutron-star binary!). The binary is so interesting and important that it requires a special attention.

The story started in 1973, six years after the discovery of pulsars. Joseph Taylor, at that time a professor of the Massachusetts University (at Amherst), suggested a new search for radio pulsars with the Arecibo radio telescope using a computer selection of sources and data analysis. He invited Russel Hulse, his graduate student, to assist him. The computer occupied two big "cupboards". It had a core memory of 16 K but no hard disk. Hulse made the drives that connected the computer with the outside world by himself. A teletype was used for interactive input and output. The data were stored on a tape drive.

PSR 1913+16 was discovered on July 2, 1974; its spin period was $P = 69$ ms. The next observation on August 25 was unbelievable: $P$ had changed by an "enormous" value of 27 microseconds. It was necessary to perform new observations to understand that the period variations were real and produced by Doppler effect due to the orbital motion of the pulsar in a binary system with the period of $P_b = 7$ h 45 min. This final conclusion was made on September 18, 1974. The discovery was announced in a paper (Hulse & Taylor 1975) published in the January 15, 1975, issue of the *Astrophysical Journal Letters*. In 1993, R. Hulse and J. Taylor were awarded the Nobel Prize for their discovery that has had a tremendous impact on physics.

The pulsar, a very weak radio source at a distance of 5–7 kpc from us, has been attracting permanent attention since the first publication and has been

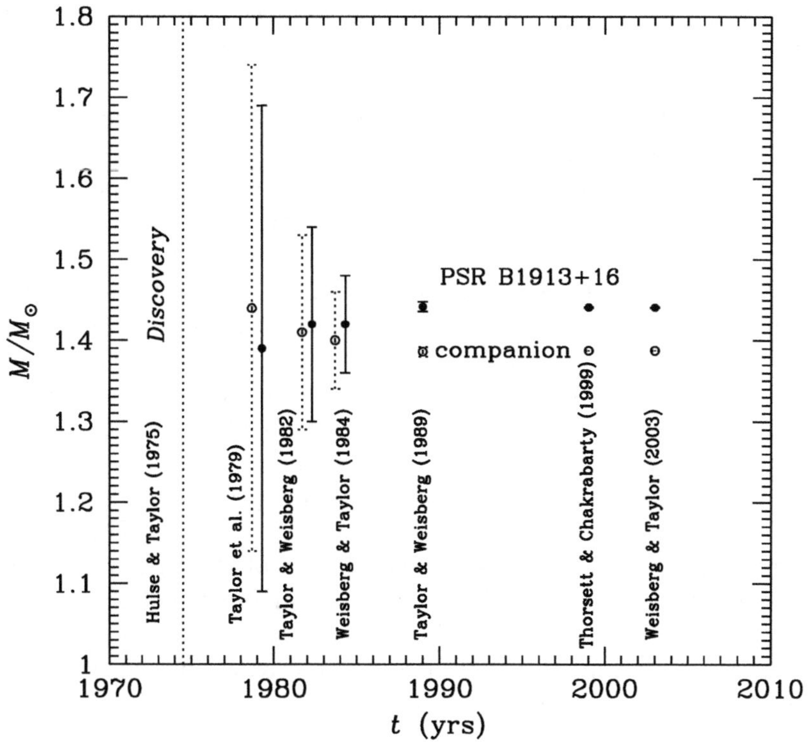


*Figure 9.5.* Mass measurements of the Hulse-Taylor pulsar and its companion (at the $2\sigma$ level).

observed numerous times (for instance, 5083 observations between 1981 and 2001, Weisberg & Taylor 2003).

Its orbit is highly eccentric ($e = 0.617$, see Table 9.3). Together with the short period, this makes the binary an ideal cosmic laboratory to study the effects of General Relativity. The major semi-axis is $a \approx 2 \times 10^6$ km. The smallest distance between two neutron stars (in periastron) is only $7.5 \times 10^5$ km (nearly the radius of the Sun), while the largest is $\approx 3.1 \times 10^6$ km. The maximum (periastron) orbital velocity reaches $\sim$400 km s$^{-1}$.

The system shows very pronounced effects of General Relativity (e.g., Weisberg & Taylor 2003), first of all, huge advance of periastron, $\dot{\omega} = 4.226607 \pm 0.000007$ deg yr$^{-1}$ (at $1\sigma$ level). In 30 years of the observations the periastron axis has turned by more than 125°. This can be compared with the same effect observed in the orbital motion of Mercury: 43″ in 100 yr! The relativistic parameter $\gamma$, Eq. (9.22), has also been measured, $\gamma(1\sigma) = 0.004294 \pm 0.000001$. As discussed in § 9.1.2 a, the measurement of two relativistic parameters ($\dot{\omega}$ and $\gamma$ in our example) is sufficient to find all the parameters of the binary system.

Moreover, it has been possible to accurately measure the third relativistic parameter, $\dot{P}_b(1\sigma) = -(2.4086 \pm 0.0052) \times 10^{-12}$ (corrected for the galactic acceleration). The detected decrease of the orbital period is attributed to the emission of gravitational waves, see Eq. (9.12). For the Hulse-Taylor system, the Einstein Theory of Relativity predicts $\dot{P}_b = -(2.40247 \pm 0.00002) \times 10^{-12}$, which is in excellent agreement with the observed value. The agreement leaves no doubt about the existence of gravitational waves and the validity of the Einstein Theory of Relativity. These results have ruled out several other theories of gravitation (and will likely rule out more) and constrain a possible cosmological variation of the gravitational constant with time (see, e.g., Taylor & Weisberg 1989, Damour & Taylor 1992, Stairs 2003, and references therein).

The current gravitational luminosity of the Hulse-Taylor system, given by Eq. (9.11), is $|\dot{E}| \approx 7.77 \times 10^{31}$ erg s$^{-1}$, about 2% of the (electromagnetic) luminosity of the Sun. Since gravitational radiation carries away the energy of orbital motion, the binary components are falling gradually one onto the other, as depicted in Fig. 9.4, and will merge in $\approx 3 \times 10^8$ yr. The event will be accompanied by a violent energy release, particularly, by a huge burst of gravitational radiation. It should be stressed the importance of eccentricity in the evolution of the system. Had the binary the same parameters ($M_1$, $M_2$, $a$, $P_b$) but a circular orbit in the present epoch, its gravitational radiation would be lower and the death time would be $\approx 1.64 \times 10^9$ years, more than 5 times higher.

The inclination angle of the orbit of the Hulse-Taylor pulsar, $i \approx 47°$, is unfavorable for measuring the Shapiro delay. By 2003, the Shapiro parameters, $r$ and $s$, Eq. (9.23), have been measured only marginally (e.g., Weisberg & Taylor 2003), but the results agree with the theoretical predictions (which can be made because all the orbital parameters are determined from the measurements of $\gamma$ and $\dot{\omega}$).

Needless to say, the masses of the pulsar and the companion have been determined with high accuracy. Some mass determinations are shown in Fig. 9.5. Taylor *et al.* (1979) reported $M_1 = (1.39 \pm 0.15)\, M_\odot$ and $M_2 = (1.44 \pm 0.15)\, M_\odot$. Further measurements gave $M_1 = (1.42 \pm 0.06)\, M_\odot$ and $M_2 = (1.41 \pm 0.06)\, M_\odot$ (Taylor & Weisberg 1982); $M_1 = (1.42 \pm 0.03)\, M_\odot$ and $M_2 = (1.40 \pm 0.03)\, M_\odot$ (Weisberg & Taylor 1984); $M_1 = (1.442 \pm 0.003)\, M_\odot$ and $M_2 = (1.386 \pm 0.003)\, M_\odot$ (Taylor & Weisberg 1989); $M_1 = (1.4411 \pm 0.00035)\, M_\odot$ and $M_2 = (1.3874 \pm 0.00035)\, M_\odot$ (Thorsett & Chakrabarty 1999). Finally, Weisberg & Taylor (2003) give

$$M_1 = (1.4408 \pm 0.0003)\, M_\odot, \qquad M_2 = (1.3873 \pm 0.0003)\, M_\odot. \tag{9.24}$$

All these results are presented at the $1\sigma$ level. In Fig. 9.5 we double the errorbars to show the mass limits at the $2\sigma$ level. One can see the rapid convergence of the results, from one observation to another (cf. Fig. 9.3). The Hulse-Taylor pulsar and its companion are the stars with the best determined masses (after the Sun). The success of their mass determination is mainly attributed to the radio astronomy (due to highly precise pulsar timing), to the pulsar nature (perfect rotators without phase shifts for many decades) and to General Relativity (which provides many relations to link orbital parameters). The Hulse-Taylor pulsar remains the most massive neutron star with an accurately measured mass till the time of this writing (since 1980 till 2006). Curiously, its mass nearly coincides with the Chandrasekhar mass limit of white dwarfs. The accuracy of its mass measurement is higher than the accuracy of measuring the gravitational constant (the latter is barely known with the three correct digits). Thus, its mass is better measured in solar masses than in grams (because one translates stellar masses into grams using the gravitational constant).

The observations of the Hulse-Taylor pulsar allowed one to measure another, extremely tiny effect of General Relativity – the geodetic precession of the pulsar spin axis about the vector of the total angular momentum of the binary. The total momentum can be accurately approximated by the orbital momentum. Thus, the pulsar spin axis should precess around the axis perpendicular to the orbital plane. The effect is analogous to the spin-orbital interaction in atomic physics with magnetic moments replaced by the curvature of space around a rotating body. The precession frequency is given by (e.g., Barker & O'Connell 1975)

$$\Omega_{\rm prec} = \Omega_b \, \frac{3GM_2}{2a\, c^2\, (1-e^2)} \left(1 + \frac{M_1}{3M}\right). \qquad (9.25)$$

It is of the same order of magnitude as the periastron advance rate $\dot{\omega}$, Eq. (9.12). The theory predicts $\Omega_{\rm prec} = 1.21$ deg yr$^{-1}$, for the Hulse-Taylor pulsar, which means the precession period of $\approx$300 years. This effect has been really observed (e.g., Konacki *et al.* 2003, and references therein). The angle between the pulsar spin and the precession axis has been determined to be $\approx 22^\circ$. Since the beaming of pulsar emission is related to the pulsar magnetic axis, the observations have allowed one to infer the angle between the pulsar spin axis and the magnetic axis ($\approx 27^\circ$). With these data, one can reconstruct the evolution of the pulsar magnetic axis and the associated detectability of the pulsar emission (e.g., Lorimer 2001, Stairs 2003). The pulsar beaming became favorable for detecting on the Earth in approximately 1940, 34 years before the pulsar discovery. It will become unfavorable by 2025, and the pulsar will disappear (what a loss for the astrophysical community!). It will reappear again approximately 240 years later.

Finally, using the theory of relativistic orbits (§ 9.1.2 a), we can extend our analysis of orbital parameters from the present epoch to the past and the future (Fig. 9.4). Since the pulsar spin frequency and its derivative have been precisely measured, $P = 0.05903$ s and $\dot{P} = 8.627 \times 10^{-18}$ (e.g., Taylor & Weisberg 1989), we can determine the characteristic pulsar spindown age

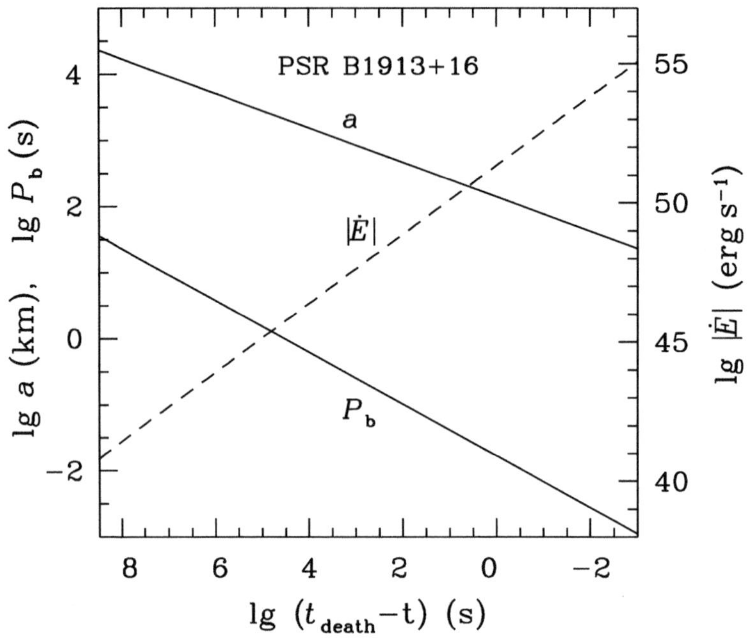


*Figure 9.6.* Orbital period, major semi-axis (left vertical axis), and gravitational luminosity (right vertical axis) of the Hulse-Taylor pulsar last ten years of its life before the final merging stage (a few milliseconds before the pulsar death).

$t = P/(2\dot{P}) \approx 10^8$ yr (§ 1.4). Thus, 100 Myr ago, when the pulsar was presumably born, the orbital parameters (Fig. 9.4) were equal to: $e = 0.666$, $a = 2.3 \times 10^{11}$ cm, $P_b = 9.93$ hr, and $\dot{\omega} = 3.12$ deg yr$^{-1}$. Looking into the future, for instance, in 200 Myr from now, we obtain: $e = 0.439$, $a = 1.2 \times 10^{11}$ cm, $P_b = 3.64$ hr, and $\dot{\omega} = 11.5$ deg yr$^{-1}$.

As discussed in § 9.1.2 b, the Hulse-Taylor system will die in about 300 Myr. Ten years before this event, when we should be fully prepared for its observation, the system parameters will be like this: $e = 0.00081$, $a = 17300$ km, $P_b = 23$ s, and $\dot{\omega} = 3.55 \times 10^5$ deg yr$^{-1}$= 0.011 deg s$^{-1}$. The orbit will be nearly circular; the separation between the neutron stars will still be much larger than their radii; the periastron advance will be overwhelming (although hardly detectable), and the gravitational luminosity of the system will be $\approx 1.2 \times 10^{41}$ erg s$^{-1}$, 30 million times larger than the electromagnetic luminosity of the Sun.

The last ten years of the Hulse-Taylor pulsar are shown in Fig. 9.6. The circular orbit can be described by Eqs. (9.20) and (9.21), which give $a(t) \propto (t_{\rm death} - t)^{1/4}$, $P_b(t) \propto (t_{\rm death} - t)^{3/8}$, and the gravitational luminosity behaves as $|\dot{E}| \propto (t_{\rm death} - t)^{-5/4}$. As discussed in § 9.1.2 a, this scaling fails at $(t_{\rm death} - t) \lesssim 1$ ms, at the onset of the final merging stage. By that time the distance between the companions becomes some tens kilometers, the orbital period decreases to $\sim$1 ms, and the gravitational luminosity is as high as $10^{55}$ erg s$^{-1}$. From that time on we have to observe the system most attentively.

We should not be surprised by the very huge gravitational luminosity of the system before the merging: the system turns into the most efficient gravitational-wave emitter. The total gravitational energy release can be as high as $\sim 10^{53}$ erg, the same as the total energy release in a supernova explosion that produces a neutron star. The huge energy release is natural: the energy released in both processes is of the order of the gravitational energy of a neutron star, Eq. (1.1).

#### 9.1.2 c Masses of neutron stars in double neutron star binaries

The parameters of eight double neutron star binaries, discovered by 2006, are listed in Table 9.3. Let us discuss seven of them, which are pulsar – neutron star binaries (the eight system, J0737–3039, will be described in the next section). All seven binaries are alike, particularly, rather eccentric. The pulsar spin periods are similar. The four of them – B1534+12, B1913+16, B2127+11C, and J1756–2251 – are very compact, with the orbital periods 8–10 hours, while the systems J1518+4904, J1811–1736, and J1829+2456 are wider. In all the systems but J1811–1736 and J1829+2456 at least two relativistic parameters have been determined (sufficient to measure the masses), one of them being the periastron advance. In the last column of Table 9.3 we list the publications which announced discoveries of the systems. A discovery date is particularly important because it takes some time (Fig. 9.5) to accurately measure the masses. The pulsars B1534+12 and B2127+11C were discovered (Wolszczan 1991, and Anderson *et al.* 1990) approximately fifteen years after the PSR B1913+16, while the pulsars J1518+4904 and J1811–1736 were discovered even later and in much wider binaries. That is why the masses of neutron stars in these two binaries are determined (constrained) less accurately. The pulsars J1756–2251 and J1829+2456 were recognized as members of double neutron star binaries just before the end of this writing.

PSR B1534+12 is an especially lucky finding: it is observed edge-on, so that the Shapiro effect has been measured. Accordingly, a collection of relativistic parameters has been determined for this pulsar ($\dot{\omega}$, $\gamma$, $s$, $r$, $\dot{P}_b$); see Stairs *et al.* (2002). Moreover, geodetic precession, discovered in observations of PSR B1913+16 (§ 9.1.2 b), has also been observed for PSR B1534+12, with $P_{\rm prec} \approx 700$ yr (e.g., Konacki *et al.* 2003).

The periastron advance of the pulsars J1518+4904 and J1811–1736 is much weaker than in other systems because the binaries are much wider. For the same reason $\dot{P}_b$ has been measured only marginally for J1518+4904. Similarly, no relativistic parameters (but $\dot{\omega}$) and no masses $M_1$ and $M_2$ have been determined for J1811–1736. However, the periastron advance in PSR J1811–1736 gives the total system mass $M(1\sigma) = (2.6 \pm 0.9)\ M_\odot$; its central value is the same as for other neutron-star binaries. One needs more time to observe PSR J1829+2456 and measure other relativistic parameters (but $\dot{\omega}$) in order to determine $M_1$ and $M_2$. The measured periastron advance gives the total mass $M(1\sigma) = (2.5 \pm 0.2)\ M_\odot$.

The inferred masses are presented in Table 9.4. The errorbars of the neutron star masses in J1518+4904 are still too large. Ignoring this binary and adding the system J0737–3039 (next section), by the end of 2005 we have ten neutron stars in the five binaries with wonderfully similar masses which range from $\approx$1.18 $M_\odot$ to 1.44 $M_\odot$. This might reflect a very specific evolutionary scenario in which such binaries were formed. Among these stars, the Hulse-Taylor pulsar is the most massive.

The neutron star masses in J1518+4904, J1811–1736, and J1829+2456 will accurately be measured in several years after 2006. Some of these neutron stars could appear more massive than the Hulse-Taylor pulsar. However, it would not be a surprise if the future observations gave nearly equal masses, $\sim 1.3\ M_\odot$, of all these stars.

### 9.1.2 d The binary pulsars J0737–3039A and B

It was a dream of many observers and theoreticians to discover a *compact binary of two (observed) radio pulsars*. And such a system, J0737–3039, has been really found at a distance of 500–600 pc from us.

The first pulsar, PSR J0737–3039A (the millisecond pulsar with the spin period of 22.6993 ms), was discovered by Burgay *et al.* (2003) in a high-latitude multibeam pulsar survey with the Parkes 64-m radio telescope in New South Wales, Australia. The first 4.5-min observation in August 2001 showed a large variation of apparent pulsar period, indicating that the pulsar is a member of a compact binary. The detection was followed by additional observations with the Parkes telescope and the Australia Telescope Compact Array. These observations started in May 2003. The first publication was based on the data collected during five months. The orbital period $P_b = 2.45$ hr turned out to be exceptionally short (Table 9.3) which means that the system is nearer to its death, than other systems (§ 9.1.2 b). Among the relativistic effects, the periastron advance was measured and appeared to be really huge, $\dot{\omega} \approx 17$ deg yr$^{-1}$. Thus, the orbital ellipse rotates by 180 degrees in 10.6 years. The measured $\dot{\omega}$ yielded the total system mass $M(1\sigma) \approx (2.58 \pm 0.02)\ M_\odot$, but the masses of neutron stars were not determined – one additional relation was needed.

The next results reported by Lyne *et al.* (2004) appeared in the Los Alamos electronic archive `astro-ph` only one month after the publication of the first paper (Burgay *et al.*, 2003). The new paper announced the discovery of the second pulsar, J0737–3039B, with the spin period 2.77346 s, in the same binary.

Timing of the second pulsar immediately gave its radial velocity semiamplitude, sufficient to measure masses and determine orbital parameters of the system. The masses are $M_1(1\sigma) = (1.337 \pm 0.005)\ M_\odot$ and $M_2(1\sigma) = (1.250 \pm 0.005)\ M_\odot$, lower than the mass of the Hulse-Taylor pulsar (but in line with accurately measured masses in other double neutron star binaries). In addition, Lyne *et al.* (2004) reported the detection of the relativistic parameter $\gamma$, and the Shapiro parameters *r* and *s*. The orbital inclination turns out to be $i \approx 87°$, very favorable for observing the Shapiro effect. Thus, the orbital parameters have been accurately determined in seven months since the beginning of systematic observations (since May 2003). The authors expected to measure $\dot{P}_b$, the last unknown basic post-Keplerian parameter, within one year.

The masses of both pulsars are too small to constrain the EOS directly. However, it seems possible to make use of tiny general relativistic spin-orbit contribution to the periastron advance in order to measure the moment of inertia of the more rapidly rotating pulsar A, in 5–10 years of radio observations. This result, combined with the mass measurements, will give stringent constraints on the EOS (Morrison *et al.*, 2004b; Lattimer & Schutz, 2005; Bejger *et al.*, 2005b). In any case the discovery of J0737–3039 is of utmost importance. Its lifetime, given by Eq. (9.19), is 86 Myr, the shortest among known double neutron star binaries. This is the fifth double neutron star binary (along with J1756–2251, B1534+12, B1913+16, and B2127+11C) whose lifetime is shorter than the Hubble time. The discovery increased the expected statistics of neutron star mergers (Burgay *et al.*, 2003; Kalogera *et al.*, 2004), important targets for gravitational detectors. Using Eq. (9.25) one can easily check that the periods of geodetic precession of the pulsars PSR J0737–3039 A and B are only 75 yr and 71 yr. Lyne *et al.* (2004) expected to detect this geodetic precession within a few years and to observe also some other effects of General Relativity. Moreover, Lyne *et al.* (2004) detected (radio) eclipses of J0737–3039A by J0737–3039B. This opens a fascinating possibility to use one pulsar for probing magnetospheric properties of the other.

## 9.1.3 Pulsar – white dwarf binaries

Observations of radio pulsar – white dwarf binaries are also very promising. Both companions of such a system, a neutron star and a white dwarf, are compact stars, i.e., the approximation of point masses is valid. A pulsar timing easily gives at least all Keplerian parameters of the orbit (particularly, $P_b$, $e$, and $x_1 = a_1 \sin i$). In order to measure the masses, one needs however two additional independent relations (§§ 9.1.1 a and 9.1.2 a). In a compact enough binary, the relativistic effects in pulsar motion can be detected as described in § 9.1.2 a. With some luck, two relativistic parameters can be determined and the stellar masses $M_1 = M_{\rm PSR}$ and $M_2 = M_{\rm WD}$ can be obtained. However, in wide binary systems, the relativistic effects are weak. One may reliably determine only one relativistic parameter or no parameters at all.

The great advantage of the pulsar – white dwarf binaries is that the additional relations for a mass measurement can be taken from optical observations of white dwarf companions, as described, for instance, by Thorsett & Chakrabarty (1999). For example, because a white dwarf mass and radius are theoretically related, an estimate of the radius (from measurements of the optical flux, effective surface temperature, and distance) can give the white dwarf mass. Another possibility is to measure the surface gravity by fitting an observed white dwarf spectrum with spectra given by theoretical atmosphere models.

However, the most powerful tool is provided by the so-called $P_b - M_2$ relation (e.g, Rappaport *et al.* 1995, Podsiadlowski *et al.* 2002, and references therein). It is expected that this relation holds for wide binaries containing millisecond pulsars in almost circular orbits. It assumes that in the past a binary contained a neutron star and a low-mass giant evolved later to a white dwarf. The evolution was accompanied by a mass transfer from the giant envelope to the pulsar. The mass exchange circularizes the orbit and recycles the pulsar to millisecond periods. The stellar evolution theory gives a strict relation between the mass of the giant core and the radius of its envelope. It is expected that the envelope fills its Roche lobe until the end of the mass transfer. The orbital separation at this phase is then a known function of the envelope radius and the giant-core mass (equal to the mass $M_2$ of the future white dwarf), which allows one to estimate $M_2$.

The parameters of some selected pulsar – white dwarf binaries are presented in Table 9.5; the inferred masses are given in Table 9.6. All the orbits, except for J1141–6545, B1802–07 and B2303+46, are nearly circular, which is probably a result of preceding evolution. PSR B1802–07 belongs to the globular cluster NGC 6539 – its orbit can be eccentric as a result of recent close encounter with one of the cluster stars. The two other eccentric binaries (J1141–6545 and B2303+46) are compact and contain rather young neutron stars born after white dwarfs – they had not enough time to circularize their orbits. Only two systems, B2320+46 and J1141–6545, contain slowly rotating pulsars; the spin periods of pulsars in other systems are short, from 3 ms to 30 ms. Only four systems, J0751+1807, J1012+5307, J1141–6545, and J1909–3744, are really compact ($P_b \lesssim 1$ day), i.e., the pulsars can show strong relativistic effects. The orbits of three of them (J0751+1807, J1012+5307, and J1909–3744) are nearly circular, which hampers the measurement of the periastron advance $\dot{\omega}$ and the parameter $\gamma$. In Table 9.5 we list the measured relativistic parameters. Let us mention that for some systems (particularly, for PSR J1713+0747 and PSR J2019+2425) kinematic variations of $\dot{x}_1/x_1$ induced by pulsar proper motion have been extracted from pulsar timing (Nice *et al.* 2003) to impose constraints

*Table 9.5.* Parameters of pulsar – white dwarf binaries (by November 2005)

| System | $P$ (ms) | $P_b$ (d) | $e$ | $x_1/c$ (s) | $i$ (deg) | $\dot{\omega}$ (deg/yr) | Other GR | Discovery[a)] |
|---|---|---|---|---|---|---|---|---|
| J0437−4715 | 5.76 | 5.74 | 0.000019 | 3.37 | $\approx 43$ | — | Sh[b)] | J93 |
| J0621+1002 | 28.9 | 8.32 | 0.002457 | 12.0 | $< 50$ | 0.0116 | — | C96 |
| J0751+1807 | 3.48 | 0.263 | 0.000003 | 0.397 | 65–85 | — | $\dot{P}_b$ | L95 |
| J1012+5307 | 5.26 | 0.605 | $< 10^{-6}$ | 0.582 | $\approx 52$ | — | Sh | N95 |
| J1045−4509 | 7.47 | 4.08 | $< 10^{-5}$ | 3.02 | | — | — | B94 |
| J1141−6545 | 394 | 0.198 | 0.172 | 1.86 | $\sim 76$ | 5.3 | $\gamma$, $\dot{P}_b$ | K00 |
| J1713+0747 | 4.75 | 67.83 | 0.000075 | 32.3 | $\approx 72$ | — | Sh | F93 |
| B1802−07 | 23.1 | 2.62 | 0.212 | 3.92 | $\gtrsim 10$ | 0.0578 | — | D93 |
| J1804−2718 | 9.34 | 11.1 | 0.00004 | 7.28 | | — | — | L96 |
| B1855+09 | 5.36 | 12.33 | 0.000022 | 9.23 | $\approx 87$ | — | Sh | S86 |
| J1909–3744 | 2.95 | 1.53 | $\sim 10^{-7}$ | 1.90 | 86.6 | — | Sh | J03 |
| J2019+2425 | 3.93 | 76.5 | 0.00011 | 38.8 | $\lesssim 70$ | — | — | N93 |
| B2303+46 | 1066 | 12.34 | 0.658 | 32.69 | | 0.0101 | — | D85 |

[a)] J93: Johnston *et al.* (1993); C96: Camilo *et al.* (1996); L95: Lundgren *et al.* (1995); N95: Nicastro *et al.* (1995); B94: Bailes *et al.* (1994); K00: Kaspi *et al.* (2000); F93: Foster *et al.* (1993); D93: D'Amico *et al.* (1993); L96: Lorimer *et al.* (1996); S86: Segelstein *et al.* (1986); J03: Jacoby *et al.* (2003); N93: Nice *et al.* (1993); D85: Dewey *et al.* (1985)
[b)] Shapiro delay (parameters $r$ and $s$)

on the orbit inclination angle (as mentioned in § 9.1.2 a). In Table 9.5 we also cite the publications that reported the discoveries of the sources. Discovery dates show how long the sources are observed (accurate measurements of relativistic effects take time, § 9.1.2). In Table 9.6 we mark [by ($P_b$)] those cases, where the $P_b - M_2$ relation has been used to estimate white dwarf masses.

The most accurate mass measurements have been done for J1141–6545. This compact ($P_b = 4.7$ hr) eccentric ($e = 0.17$) binary system was discovered by Kaspi *et al.* (2000). The system is excellent for observing the relativistic effects. Kaspi *et al.* (2000) measured the huge periastron advance $\dot{\omega} = 5.3$ deg yr$^{-1}$ and determined the total system mass $M \approx 2.3\, M_\odot$. Ord *et al.* (2002) managed to pinpoint the orbit inclination $i = 76 \pm 2.5°$ by observing interstellar scintillations of PSR J1141–6545 modulated by orbital motion. Thus, the measurements of $\dot{\omega}$ and $i$ gave two relations required to obtain the companion masses. In addition, Bailes *et al.* (2003) have measured two other relativistic

*Table 9.6.* Masses of stars in pulsar – white dwarf binaries (by November 2005)

| System | $M_1/M_\odot$ | | $M_2/M_\odot$ | | Ref.[a] |
|---|---|---|---|---|---|
| J0437−4715 | 1.58±0.18 | $(1\sigma)$ | 0.236±0.017 | $(1\sigma)$ | vS01 |
| J0621+1002 | $1.70^{+0.59}_{-0.63}$ | $(2\sigma)$ | $0.97^{+0.43}_{-0.24}$ | $(2\sigma)$ | S02 |
| J0751+1807 | $2.1^{+0.4}_{-0.5}$ | $(2\sigma)$ | $0.19 \pm 0.03$ | $(2\sigma)$ | NSS04, 05 |
| J1012+5307 | 1.7±1.0 | $(2\sigma)$ | $0.165 - 0.215$ | $(1\sigma)$ | TC99 |
| J1045−4509 | < 1.48 | | $\sim 0.13$ | $(P_b^{b)})$ | TC99 |
| J1141−6545 | 1.30±0.02 | $(1\sigma)$ | 0.99±0.02 | $(1\sigma)$ | B03 |
| J1713+0747 | $1.53^{+0.08}_{-0.06}$ | $(1\sigma)$ | $0.30-0.35$ | $(P_b)$ | S05 |
| B1802−07 | $1.26^{+0.15}_{-0.67}$ | $(2\sigma)$ | $0.36^{+0.67}_{-0.15}$ | $(2\sigma)$ | TC99 |
| J1804−2718 | < 1.73 | | $\sim 0.2$ | $(P_b)$ | TC99 |
| B1855+09 | $1.57^{+0.25}_{-0.20}$ | $(2\sigma)$ | $0.25 - 0.28$ | $(P_b)$ | NSS03 |
| J1909–3744 | 1.438±0.024 | $(1\sigma)$ | $0.2038 \pm 0.0022$ | $(1\sigma)$ | J05 |
| J2019+2425 | $< 1.51$ | | $0.32 - 0.35$ | $(P_b)$ | NSS01 |
| B2303+46 | $1.24 - 1.44$ | | $1.2 - 1.4$ | | KK99 |

[a] vS01: van Straten *et al.* (2001); S02: Splaver *et al.* (2002); NSS04: Nice *et al.* (2004); NSS05: Nice *et al.* (2005); NSS03: Nice *et al.* (2003); TC99: Thorsett & Chakrabarty (1999); B03: Bailes *et al.* (2003); S05: Splaver *et al.* (2005); J05: Jacoby *et al.* (2005); NSS01: Nice *et al.* (2001); KK99: van Kerkwijk & Kulkarni (1999)
[b] $M_2$ is estimated from the $P_b - M_2$ relation.

parameters, $\gamma$ and $\dot{P}_b$, which confirm the previous results and refine the mass determination. Moreover, geodetic precession of the pulsar axis has been observed (Hotan *et al.*, 2005). The system is so compact that its life time is shorter than the Universe age. Unfortunately, the pulsar mass $M_1 \approx 1.30\,M_\odot$ is noticeably lower than the mass of the Hulse-Taylor pulsar, and it does not impose new constraints on the EOS.

Another example of the accurate mass measurement is provided by PSR J1909–3754, a millisecond pulsar with a very short (3 ms) spin period. The pulsar was discovered by Jacoby *et al.* (2003) in a compact binary ($P_b \approx 1.5$ d). The companion was identified as a moderately hot white dwarf ($T_s \approx 8500$ K). The orbit is circular ($e \sim 10^{-7}$; no chances to measure $\dot{\omega}$ and $\gamma$) but, luckily, the system is observed edge-on. Only two years of timing allowed Jacoby *et al.* (2005) to measure Shapiro parameters and accurately determine the pulsar mass, $M_1(1\sigma) = (1.438 \pm 0.024)\,M_\odot$. It appears to be only slightly lower than the mass of the Hulse-Taylor pulsar. This mass value is interesting for the theory of pulsar spin-up to millisecond periods in binary systems.

The mass of PSR B2303+46 has also been determined with reasonable accuracy. The binary, discovered by Dewey *et al.* (1985), is the oldest known source listed in Tables 9.5 and 9.6. It was initially considered as a double neutron star binary. The eccentricity is exceptionally high, $e = 0.658$, but the orbital period is large and the relativistic effects are not very pronounced. Nevertheless, the periastron advance was measured, and it gave the total system mass $M \approx 2.64\, M_\odot$. The minimum mass limit of the companion star was determined then from the mass function, Eq. (9.9): $M_2 > 1.2\, M_\odot$. The last, fourth relation for fixing the masses was absent. Fortunately, van Kerkwijk & Kulkarni (1999) discovered a faint optical counterpart which implied that the second companion is most probably a white dwarf, but not a neutron star. Since the white dwarf mass cannot exceed the Chandrasekhar limit, one immediately concludes that $1.2\, M_\odot < M_2 < 1.4\, M_\odot$. Thus, the pulsar mass is nicely constrained, $1.24\, M_\odot < M_1 < 1.44\, M_\odot$, but it is again lower than the mass of the Hulse-Taylor pulsar.

Another example of a not too massive neutron star is provided by the PSR B1802–07. Its orbital period is not too large ($P_b = 2.62$ d) and the orbit is eccentric ($e = 0.21$). The periastron advance has been measured; it gives the total system mass $M = 1.62\, M_\odot$. The lower limit of the companion mass, as deduced from the mass function, yields $M_2 > 0.29\, M_\odot$, and the inclination is constrained as $\sin i > 0.18$. Assuming the uniform distribution over $\cos i$ in the allowed range $0.18 < \sin i < 1$, Thorsett & Chakrabarty (1999) obtained the mass limit, $M_1 < 1.41\, M_\odot$ (at the $2\sigma$ level). Other relativistic effects will hopefully be observed soon and improve the mass measurement (but not the constraint on the EOS).

The masses of neutron stars in other systems may be higher than the mass of the Hulse-Taylor pulsar, but they are measured (constrained) with less accuracy. In two rather wide binaries, J1713+0747 and B1855+09, Shapiro delay has been measured and the $P_b - M_2$ relation has been used. This gives two relations (the range parameter of Shapiro effect and the $P_b - M_2$ relation constrain actually the same quantity, $M_2$) to infer the masses. The central values $M_1(\text{J1713}) \approx 1.53\, M_\odot$ (Splaver *et al.*, 2005) and $M_1(\text{B1855}) = 1.57\, M_\odot$ (Nice *et al.*, 2003) are noticeably higher than the mass of the Hulse-Taylor pulsar, but the errors are still large (Table 9.6). In two other binaries, J0437–4715 and J1012+5307, Shapiro delay has also been measured, which is basically sufficient to determine $M_1$ and $M_2$. Again, the central values $M_1(\text{J0437}) = 1.58\, M_\odot$ and $M_1(\text{J1012}) = 1.7\, M_\odot$ are high and interesting for constraining the EOS, but the errors are too large. The last system is compact and therefore convenient for measuring $\dot{P}_b$, and that will considerably reduce the errors.

The relativistic effects have been observed also in the orbital motion of PSR J0621+1002. Six years of pulsar timing with three radio telescopes allowed Splaver *et al.* (2002) to detect the periastron advance and estimate the total system mass, $M = (2.8 \pm 0.3)\, M_\odot$. They combined their result with the mass limits provided by the mass function and by the Chandrasekhar limit of white-dwarf mass. In addition, they used the limits on the inclination angle $i$ given by the mass function and by a non-detection of Shapiro delay (Shapiro effect would be detectable for high $i$). In that way they constrained the neutron star mass (Table 9.6) with a high and interesting central value $M_1 = 1.7\, M_\odot$ but with large errorbars (to be reduced in several years).

It is very important to mention another pulsar, PSR J0751+1807. Its orbit is nearly circular (thus, $\dot{\omega}$ cannot be accurately measured). However, the orbital period is short ($P_b = 6.3$ hrs), so that $\dot{P}_b$ has been measured, and Shapiro delay has marginally been detected, constraining $i$ to intermediate values. Nice *et al.* (2004, 2005) have made a full relativistic analysis of the timing data obtained by 2004. This allowed them to conclude that the pulsar mass is very high, $M_1(2\sigma) = 2.1^{+0.4}_{-0.5}\, M_\odot$. The errorbar is still too wide. However, with high probability, $M_1 > 1.6\, M_\odot$, which is much higher than the mass of the Hulse-Taylor pulsar. Shapiro effect is expected to be accurately measured soon after 2006. This will strictly constrain the PSR J0751+1807 mass.

Finally, let us mention the systems J1045–4509, J1804–2718 and J2019+2425. They are rather wide binaries containing millisecond pulsars in almost circular orbits. The $P_b - M_2$ relation has been used to estimate $M_2$. No relativistic effects have been observed. The data are insufficient to measure the masses, but they give conservative upper limits: $M_1 < 1.48\, M_\odot$, $M_1 < 1.73\, M_\odot$, and $M_1 < 1.51\, M_\odot$, for PSR J1045–4509, PSR J1804–2718 (Thorsett & Chakrabarty 1999), and PSR J2019+2425 (Nice *et al.* 2001), respectively. PSR J1804–2718 is potentially interesting and may be a massive neutron star. There is a chance, that the pulsars J1045–4509 and J2019+2425 are slightly more massive than the Hulse-Taylor pulsar, but generally these two systems are less interesting for the EOS problem than other candidates.

In addition, let us mention two pulsars (Ter 5 I and J) discovered by Ransom *et al.* (2005) in the globular cluster Terzan 5 using the Green Bank Telescope. Both pulsars enter compact and eccentric binaries; their binary companions are possibly white dwarfs. The orbital periods of Ter 5 I and J are $P_b = 1.328$ and $1.102$ days, and the eccentricities are $e = 0.428$ and $0.350$, respectively. The pulsar spin periods are 9.57 and 80.34 ms. In both cases the periastron advance has been measured, $\dot{\omega} = 0.255$ deg yr$^{-1}$ and 0.327 deg yr$^{-1}$, and the total system mass has been determined, $M(1\sigma) = (2.17 \pm 0.02)\, M_\odot$(for Ter 5 I) and $(2.20 \pm 0.04)\, M_\odot$(for Ter 5 J). One equation is still missing to determine neutron star masses but the relativistic $\gamma$ parameter can be measured to solve this problem (in 1–3 years for Ter 5 I and in 5–10 years for Ter 5 J). Statistical estimates assuming constant probability distribution over cos $i$ indicate that both pulsars are massive, $M_1(2\sigma) \gtrsim 1.7\ M_\odot$, but further observations are required to confirm this result.

To summarize the above discussion, there are at least seven neutron star – white dwarf binaries (and several promising candidates, including Ter 5 I and J) potentially very interesting for the EOS problem. The pulsars J0437–4715, J0621+1002, J0751+1807, J1012+5307, J1713+0747, J1804–2718, and B1855+09 – may be more massive than the Hulse-Taylor pulsar. One pulsar, PSR J0751+1807, is almost certainly more massive than $1.6\ M_\odot$. The masses of all these pulsars will be refined in several years after 2006.

### 9.1.4 Pulsars in binaries with non-degenerate stars

Three radio pulsars have been found in binaries with non degenerate stars by 2005. In two systems the companions are massive main-sequence stars, while the companion of the third pulsar is less massive. Pulsar timing gives the pulsar spin period $P$, the orbital period $P_b$, and the orbit eccentricity $e$ (and naturally the pulsar mass function $f_1$).

The orbits in the systems with the main-sequence stars are extremely eccentric. Both main-sequence stars are relatively bright, so that their types have been determined and their masses have been estimated.

The first system (PSR B1259–63 and a Be companion) was discovered by Johnston *et al.* (1992). According to the pulsar timing, $P = 47.8$ ms, $P_b = 3.4$ yr, and $e = 0.87$. Unfortunately, the pulsar mass function is of little use to constrain the pulsar mass $M_1$ for systems, where $M_1 \ll M_2$. In other words, $M_1$ is highly uncertain.

The second system (PSR J0045–7319 and a B1 V star) was discovered by Kaspi *et al.* (1994) in the Small Magellanic Cloud. The pulsar timing gave: $P = 0.926$ s, $P_b = 51.169$ days, and $e = 0.808$ ($f_1 = 2.17\,M_\odot$). Bell *et al.* (1995) measured the radial velocity of the optical companion and obtained the mass ratio: $M_2/M_1 = 6.3 \pm 1.2$. Thorsett & Chakrabarty (1999) reported an estimate of the companion mass, $M_2 = (10 \pm 1)\,M_\odot$, based on the observed optical luminosity $1.2 \times 10^4\ L_\odot$, the effective surface temperature $2.4 \times 10^4$ K, and a grid of stellar atmosphere models. This gives $M_1 = (1.58 \pm 0.34)\,M_\odot$ at the $1\sigma$ level (Table 9.7). The central value is higher than the mass of the Hulse-Taylor pulsar and potentially interesting for the EOS problem, but the errorbar is too wide. New observations are required to pinpoint $M_1$.

The third system is quite different. The pulsar (PSR J1740–5340) was discovered in the globular cluster NGC 6397 by D'Amico *et al.* (2001a). The timing yielded (D'Amico *et al.* 2001a,b): $P = 3.6503$ ms, $P_b = 1.354$ d, $x_1/c = 1.653$ s, $e < 10^{-4}$ (and $f_1 = 0.002644\,M_\odot$). Thus, the orbit is nearly circular. The optical companion was identified by Ferraro *et al.* (2001) as a rather bright variable object that had been detected earlier by Taylor *et al.* (2001).

*Table 9.7.* Mass of neutron stars in binaries with non-degenerate stars (2005)

| System | $P_b$, d | $M_1/M_\odot$ | Signif. | Reference |
|---|---|---|---|---|
| J0045–7319 | 51.169 | 1.58±0.34 | $1\sigma$ | Thorsett & Chakrabarty (1999) |
| J1740–5340 | 1.354 | 1.53±0.19 | $1\sigma$ | Kaluzny *et al.* (2003) |

Photometric observations of both groups indicated an unusual position of the companion at the Hertzsprung-Russell diagram of the cluster: $\approx$0.2 mag to the red from the turnoff and $\approx$0.7 mag below the base of the red giant branch. This position possibly implies (e.g., Ergma & Sarna 2002) that the optical companion is an evolved star with a helium core surrounded by a hydrogen envelope. Detailed photometric and spectroscopic observations of this star were done by Kaluzny *et al.* (2003). Using the photometric data on the stellar variability in the $B$, $V$, and $I$ bands and assuming that the star fills its Roche lobe, the authors obtained the inclination angle $i = 43.9 \pm 2.1$ deg. Spectroscopic observations allowed them to measure the radial velocities. In this way Kaluzny *et al.* (2003) obtained two additional relations required for mass measurements. The result is: $M_1(1\sigma) = (1.53 \pm 0.19)\, M_\odot$ and $M_2(1\sigma) = (0.296 \pm 0.034)\, M_\odot$. The central value of $M_1$ is slightly higher than for the Hulse-Taylor pulsar.

### 9.1.5 Summary of neutron star mass measurements

Let us summarize the results of measurements of neutron star masses in binary systems made by the end of 2005.

(*1*) The most accurate measurements have been performed for eight neutron stars in four double neutron star binaries (§ 9.1.2). The most massive among them is the Hulse-Taylor pulsar, PSR B1913+16.

(*2*) It is likely that some X-ray binaries (§ 9.1.1), pulsar – white dwarf binaries (§ 9.1.3), and pulsar – non-degenerate star binaries (§ 9.1.4) contain more massive neutron stars. This can be a consequence of mass transfer on neutron stars during the evolution of these systems.

(*3*) Three X-ray binaries, Cyg X-2, Vela X-1, and 4U 1700–37, are good candidates for containing massive neutron stars. However, the progress in measuring the masses in X-ray binaries is slow (§ 9.1.1 b).

(*4*) Seven pulsar – white dwarf binaries (J0437–4715, J0621+1002, J0751+1807, J1012+5307, J1713+0747, J1804–2718, and B1855+09) contain possibly massive neutron stars. The stellar masses in these systems (or at least in some of them) are expected to be accurately measured in several years after 2006. The mass of PSR J0751+1807 is higher than $1.6\, M_\odot$ at the $2\sigma$ level.

Thus the Hulse-Taylor pulsar remains to be the neutron star with the highest accurately measured mass, but the mass of PSR J0751+1807 is noticeably higher. Accordingly, we come to the observational constraint

$$M_{\max}(\mathrm{EOS}) > M_{\mathrm{obs}}^{(\max)}, \qquad M_{\mathrm{obs}}^{(\max)} = 1.6\, M_{\odot} \quad (2006), \tag{9.26}$$

where $M_{\max}(\mathrm{EOS})$ is the maximum allowable mass for non-rotating neutron star models; for spin periods longer than, say, 5 ms the effects of rotation on $M_{\max}$ are negligibly small (see § 6.12.2).

Using Eq. (9.26), we can rule out all EOSs, which predict the maximum neutron star mass to be lower than $M_{\mathrm{obs}}^{(\max)}$. Unfortunately, this requirement does not solve the EOS problem: it excludes only the softest EOSs that appear in the literature (e.g., $npe\mu$ model BPAL12 of Bombaci 1995; some EOSs of dense matter with hyperons, e.g., Pandharipande 1971b; Balberg & Gal 1997; Vidaña *et al.* 2000b). All but one (BPAL12) in Table 5.3 in Chapter 5 and Table 6.1 in Chapter 6 are consistent with the current value of $M_{\mathrm{obs}}^{(\max)}$ (see Figs. 6.3 and 6.5), with BGN1H1 being only marginally consistent. It would be highly desirable to accurately measure the masses of more massive neutron stars, the more massive the better. A definite discovery of a 1.9 $M_{\odot}$ neutron star would rule out nearly all EOSs with hyperons, while a discovery of, say, a 2.1 $M_{\odot}$ neutron star would leave us with very stiff EOSs of dense matter containing nucleons only.

### 9.1.6 From November 2005 to August 2006 and beyond

The tremendous progress in neutron star mass measurements can be illustrated by a short description of new results obtained from November 2005 (when this book was mainly completed) to August 2006 (when last minute changes were introduced).

Jacoby *et al.* (2006) reported the direct measurement of the orbital decay $\dot{P}_b = (-3.95 \pm 0.13) \times 10^{-12}$ in the double neutron star binary PSR B2127+11C discussed in § 9.1.2c. New timing data allowed the authors to infer more accurate values of neutron star masses, $M_1(1\sigma) = (1.358 \pm 0.010)\, M_{\odot}$ and $M_2(1\sigma) = (1.354 \pm 0.010)\, M_{\odot}$ (compare with the values in Table 9.4).

Bassa *et al.* (2006) and Cocozza *et al.* (2006) presented new observations of a compact binary containing the millisecond pulsar J1911–5958A and a white dwarf. The pulsar was discovered in 1999 (D'Amico *et al.*, 2001b). The pulsar timing showed that the orbit is nearly circular and the orbital period is $P_b = 0.84$ days. The authors present new optical observations of the white dwarf which allowed them to measure its radial velocities. Moreover, comparing the data with white dwarf atmosphere models and theoretical mass-radius relations, Bassa *et al.* (2006) obtained the white dwarf mass $M_2(1\sigma) = (0.18 \pm 0.02)\, M_{\odot}$. The radial velocities and $M_2$ gave them two equations to determine the pulsar mass $M_1(1\sigma) = 1.40^{+0.16}_{-0.10}\,M_\odot$. No relativistic effects have been detected from this binary.

Lorimer *et al.* (2006) reported the discovery of PSR J1906+0746, a young pulsar in a highly relativistic orbit with a neutron star or a white dwarf ($P_b$ = 3.98 hr, $e = 0.085$). Huge periastron advance has been measured, $\dot{\omega} \approx 7.^\circ 57$ yr$^{-1}$, which gives the total binary mass $M(1\sigma) = (2.61 \pm 0.02)\,M_\odot$ and the binary merging time $t_{\rm death} \sim 300$ Myr. Further observations are required to resolve $M_1$ and $M_2$.

Finally, Freire *et al.* (2006) discovered a new millisecond pulsar J1741+1351 in a circular orbit with a white dwarf ($P_b = 16$ days). The pulsar timing shows the Shapiro effect which gives two parameters ($\sin i \approx 0.96$, $M_2(1\sigma) = (0.30 \pm 0.07)\,M_\odot$) and two respective equations to determine the pulsar mass, $M_1(1\sigma) = (1.8 \pm 0.3)\,M_\odot$, interesting for the EOS problem.

With this spectacular progress, new exciting and decisive neutron star mass measurements can be expected any day!

## 9.2. Gravitational surface redshift

Early expectations for a direct determination of gravitational redshift $z_{\rm surf}$ (§ 6.1) of spectral lines emitted from the surface of a neutron star arose from observations of spectral lines in gamma-ray bursts. One of the most popular hypotheses about the origin of the bursts assumed that they are associated with neutron stars in the Galaxy. The spectra of numerous gamma-ray bursts detected by several gamma-ray observatories at the end of 1970s and in the first half of 1980s revealed the presence of broad "emission line features" which were widely attributed to the redshifted gamma quanta from the $e^+e^- \to 2\gamma$ annihilation process near the neutron star surface (e.g., Golenetskii *et al.* 1986). The intrinsic energy of annihilation quanta, 0.511 MeV, was expected to be redshifted for a distant observer. An analysis of $\sim 40$ "emission line features" deduced from the apparent line center position, led to gravitational redshifts $z = 0.2$–$0.5$, with the centroid lying between 0.25 and 0.35 (Liang, 1986). However, in the 1990s, thousands of gamma ray bursts were detected by the BATSE detectors on board the *Compton* Gamma Ray Observatory, with no clear evidence of spectral features at 0.3–0.5 MeV.

Moreover, in the late 1990s it was proved that many observed gamma-ray bursts originate at cosmological distances, thus being unrelated to neutron stars in the Local Group of galaxies. A notable exception is provided by *soft gamma-ray repeaters* (SGRs), likely related to *magnetars* (ultra-magnetized neutron stars; § 1.4.5 The first gamma-ray burst useful for evaluating $z_{\rm surf}$ was the extraordinary gamma-ray burst (GRB 790305) on March 5, 1979 (Mazets *et al.*, 1979b) from a source in the Large Magellanic Cloud. The burst source is now called SGR 0526−66 (=RX J052600.3–660433). In quiescence, it behaves like an anomalous X-ray pulsar (Kulkarni *et al.*, 2003); see § 1.4.5. The spectrum of the GRB 790305 exhibited a prominent emission line at $430 \pm 30$ keV with the full width at half-maximum $\simeq 150$ keV (Mazets *et al.*, 1981, 1982). Assuming that the line originated from $e^+e^- \to 2\gamma$ and was broadened by thermal plasma motion, one gets $z_{\rm obs} = 0.23 \pm 0.07$ (e.g., Higdon & Lingenfelter 1990). This estimate is compared with theoretical neutron star models in Fig. 9.7. As one can see, for a broad set of EOSs, surface redshifts of neutron stars with masses within the range of precisely measured masses in double neutron star binaries are consistent with this estimate.

If, however, this line was formed far above the surface, then $z_{\rm obs}$ gives only the lower bound on the true surface redshift. For instance, the emission line at 6.4 keV, observed in a burst from another soft gamma repeater, 1900+14 (very similar to SGR 0526$-$66), was interpreted by Strohmayer & Ibrahim (2000) as the non-redshifted Fe K$\alpha$ line. In this case, the absence of a redshift implies that the line was formed at least 80 km away from the neutron star (possibly in an accretion disk).

Another method, successfully applied previously to white dwarfs, consists in measuring redshifts of identified spectral lines for a neutron star in a binary. First attempts to use absorption lines in X-ray spectra of X-ray binaries to deduce $z_{\rm surf}$ were plagued by difficulties of line identification and by large line widths. A successful measurement was probably made from the observations of absorption lines in the X-ray burst spectra of the low-mass X-ray binary EXO 0748$-$676 (Cottam *et al.*, 2002). The most significant absorption features were identified with transitions of H-like and He-like iron and H-like oxygen, all with the same redshift $z_{\rm obs} = 0.35$. A further theoretical analysis of Bildsten *et al.* (2003) lent support to this interpretation by showing how the strength of the observed H$\alpha$ Fe XXVI line can be reconciled with the non-identification of Ly$\alpha$ and photoabsorption features.

## 9.3. Neutron star radii

Models of dense matter can be tested by comparing neutron star radii $R$ determined from observations with theoretical predictions. Let us remind that $R$ denotes the circumferential stellar radius (§ 6.1). Unfortunately, for typical neutron star masses between 1 and 2 $M_\odot$, the radius is not very sensitive to the EOS (§ 6.6.1). As can be deduced from Fig.6.5, in order to really distinguish between the different EOSs, one has to measure the radius with a precision of $\sim 0.5$ km (within statistical significance of $2\sigma$ or better).

Let us outline several methods to evaluate the neutron star radius from observations. First of all, one can infer the radius by measuring $z_{\rm surf}$ and the stellar mass $M$:

$$R = \frac{(1 + z_{\rm surf})^2}{z_{\rm surf}(2 + z_{\rm surf})}\, r_{\rm g}, \quad r_{\rm g} = \frac{2GM}{c^2} \approx 2.95\, \frac{M}{M_\odot}\ \text{km}\,. \tag{9.27}$$

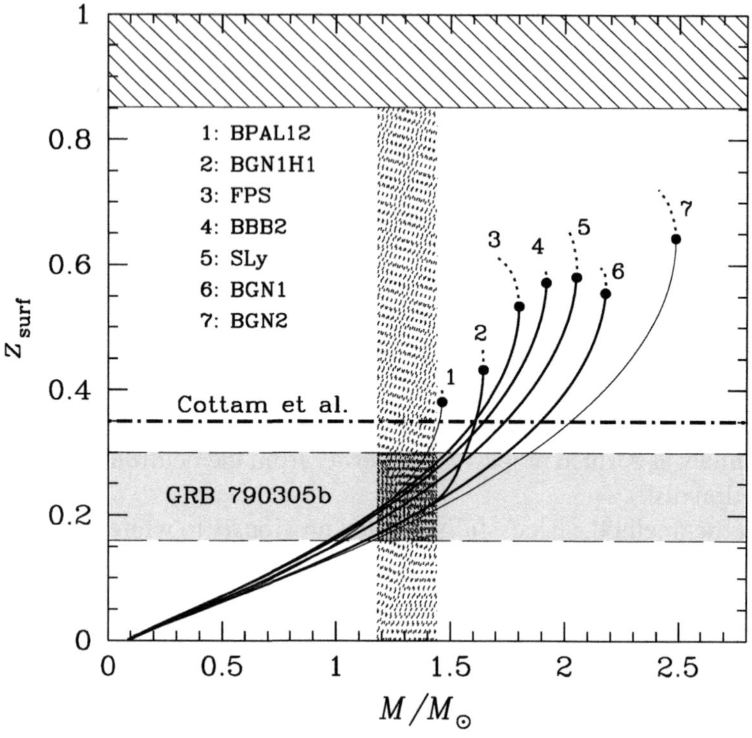


*Figure 9.7.* The gravitational surface redshift versus stellar mass for selected EOSs (see Table 5.3) compared with observations. The horizontal band limited by dashed lines corresponds to $1\sigma$ evaluation of the redshift for the $e^+e^-$ annihilation line in the spectrum of GRB 790305b. The thick dot-and-dashed line gives the value of $z_{surf}$ obtained in the analysis of spectra of X-ray bursts by Cottam *et al.* (2002). The hatched area is prohibited by General Relativity combined with the condition $v_{sound} \leq c$ (§6.6.3). Shadowed vertical band is the range of precisely measured masses of neutron stars in double neutron star binaries, § 9.1.2 c.

Thus, if we need to measure $R$ with an accuracy of 0.5 km, we must measure $M$ *and* $z_{surf}$ in such a way to determine $M/z_{surf}$ within $\sim 5\%$ at the $2\sigma$ level. Such measurements have not been performed by 2006.

## 9.3.1 Thermal emission of isolated neutron stars

Another way to measure neutron star radii consists in analyzing thermal emission from neutron star surfaces (§ 1.4.5). Choosing a theoretical model of the surface emission, one can try to adjust model parameters (in particular, the radius) to the observed spectrum. As we discussed in §§ 6.6.6 and 1.3.3, instead of the circumferential radius $R$ it is often convenient to introduce the *apparent* radius as would be measured by a distant observer.

Unfortunately, the non-thermal radiation from magnetospheres of neutron stars (§ 1.4.4) is usually so bright that the thermal radiation from their surfaces is hardly observable. In particular, this is so for young active pulsars like the Crab pulsar (age $\sim 10^3$ yr). The thermal radiation of middle-aged (age $\sim 10^4$–$10^6$ yr) and sufficiently warm neutron stars, with effective temperatures $(0.3 - 1) \times 10^6$ K, can dominate in soft X-ray and UV ranges. Surfaces of old neutron stars (age $\gtrsim 10^6$ yr) are too cold to emit observable thermal radiation, but polar caps of old active pulsars can be hot enough to be detected in X-rays.

That is why the thermal radiation from neutron star surface was detected only after the launch of X-ray satellites *Einstein* (1978–1981) and *EXOSAT* (1983–1986). First reliable spectra of the thermal radiation from several pulsars were obtained even later, with the sensitive X-ray observatory *ROSAT* (1990–1998) as reviewed, e.g., by Ögelman (1995) and Becker (1999). A new era in observing the thermal emission of neutron stars has started since 1999 with the launch of X-ray observatories of outstanding capabilities, *Chandra* and *XMM-Newton* (e.g., Pavlov *et al.* 2002).

Extracting neutron star parameters from these observations is a complicated task. It is not sufficient to obtain a good high-quality spectrum; its interpretation meets several serious difficulties. First, as a rule, a spectrum includes both thermal and non-thermal components, so that one has to carefully separate the thermal one. Second, theoretical interpretation of observations depends on the composition of a neutron star atmosphere, which is unknown *a priori*, so that one should try several possibilities. Third, even if the atmosphere composition is fixed, the model of the outgoing spectral flux depends on many parameters such as $M$, $R$, and the effective surface temperature $T_s$. For a strongly magnetized star, one should also take into account the effects of the magnetic field $B$ and its geometry and an associated non-uniform temperature distribution over the stellar surface. For instance, the polar cap of a pulsar can be heated by a stream of electrons or positrons moving along open field lines from the magnetosphere. This is what happens in the so called *inner gap model* of pulsar emission, which stems from the seminal paper by Sturrock (1971). The temperature of "hot spots" produced by this heating on the stellar surface may be much higher than outside them. Fourth, the detected spectrum depends on the distance $d$ to the star and the column density of interstellar hydrogen, which are usually poorly known. Therefore, it is not surprising that numerous attempts to determine neutron star radii from observations of thermal radiation have been largely inconclusive.

Generally, while fitting an observed spectrum, one infers possible values of $R_\infty$ within some confidence interval. An assumed EOS of dense matter can be ruled out, if the theoretical curve $R_\infty(M)$ does not intersect this interval. As a rule, these results should be taken with a grain of salt, because of difficulties in estimating the errorbars of $R_\infty$. The emission of active pulsars may be mainly produced by hot polar caps, rather than by entire colder surfaces. In these cases one will infer a polar cap radius (much smaller than $R_\infty$) from observations. For instance, Zavlin *et al.* (2002), using a hydrogen atmosphere model, estimated the radius of the polar cap of the millisecond pulsar J0437$-$4715 as $R_{\rm pc} = 2.0^{+0.3}_{-0.2}$ km. It can be taken as the observational *lower bound* to the stellar radius. Obviously, it is too low to be useful for constraining the EOS in neutron star cores.

However, there are several cooling (isolated) middle-aged neutron stars, which apparently show a thermal-like emission and whose distance has been determined from parallax measurements. The distance determination eliminates the largest uncertainty in constraining $R_\infty$. These neutron stars are Geminga (PSR B0633+17), PSR B0656+14, Vela (PSR B0833$-$45), and RX J1856$-$3754, which we consider in the rest of this section.

Golden & Shearer (1999) used *HST* and *BTA* optical photometry of Geminga and PSR B0656+14 to estimate their radii. Previous X-ray observations had indicated that a significant part of soft X-ray flux from these neutron stars is of thermal origin. Golden & Shearer (1999) compared the sum of the Rayleigh-Jeans tail of the thermal blackbody flux ($\propto \nu^2$, where $\nu$ is the photon frequency) and the synchrotron-emission flux ($\propto \nu^{-\alpha}$ with $\alpha = 1.9$ for Geminga and 1.4 for PSR B0656+14) with the pulsed optical flux detected from these pulsars and with the *upper limit* to the non-pulsed optical flux. Assuming that the thermal emission powered by neutron star cooling is non-pulsed, they obtained the upper bound on the *Rayleigh-Jeans parameter* $G \equiv T_6^\infty (R_{10}^\infty / d_{500})^2$, where $R_{10}^\infty \equiv R_\infty/10$ km, $T_6^\infty = T_{\rm s}^\infty/10^6$ K is the effective surface temperature (in MK) as detected by a distant observer, and $d_{500} \equiv d/500$ pc. For Geminga, the restriction reads $G \le 3.9$ (at the $3\sigma$ level), while $d = 159^{+59}_{-34}$ pc from the parallax measurement (Caraveo *et al.*, 1996). If one fixes $d \approx 160$ pc, then the blackbody fit of the X-ray spectrum gives $T_6^\infty \approx 0.4$–0.6 (Halpern *et al.*, 1996), so that $R_\infty \lesssim 10$ km, below the absolute lower limits $R_\infty \gtrsim 12$– 14 km for most realistic EOSs (§ 6.6.6). However, taking into account rather large errors in $d$, one obtains $R_\infty \lesssim 18$ km (Haensel, 2001), which is satisfied for any reasonable EOS and, therefore, is not useful for the EOS selection. This may indicate that $d$ is actually larger than the assumed 160 pc. On the other hand, one cannot exclude that other, non-blackbody models (e.g., magnetized atmosphere models) yield noticeably larger values of $R_\infty$.

The most comprehensive analysis of the data on PSR B0656+14 made in the pre-*Chandra* era was done by Edelstein *et al.* (2000). *Chandra* X-ray observations of PSR B0656+14 were analyzed by Marshall & Schulz (2002). Combined optical, X-ray, and UV data gave the best estimate $T_6^\infty = 0.47$–0.74 (associated with the soft spectrum component) and $R_{10}^\infty/d_{500} = 1.4$–4.0 at the 99% confidence level, with the most probable distance $d \approx 200$ pc (to be compared with the estimate $d = 760$ pc based on the dispersion measure,

Manchester *et al.* 2005). Later, Brisken *et al.* (2003) measured the parallax of this pulsar and obtained $d = 288^{+33}_{-27}$ pc. Analyzing previous $R/d$ estimates, the latter authors concluded that, within the model uncertainties, any radius $R_\infty$ is between $\sim 13$ and $\sim 20$ km, which is not useful for the EOS problem.

The parallax measurements of the Vela pulsar gave the distance $d = 293^{+19}_{-17}$ pc (Caraveo *et al.*, 2001; Dodson *et al.*, 2003). The X-ray spectrum of this pulsar observed with *Chandra* (Pavlov *et al.*, 2001) and *XMM-Newton* (Mori *et al.*, 2004) can be fitted by a sum of non-thermal (power-law) and thermal (blackbody or hydrogen-atmosphere) components. The blackbody fit yields $T_s^\infty \approx 1.5 \times 10^6$ K and $R_\infty = 2.5 \pm 0.2$ km. It may describe the thermal radiation of a hot polar cap on the pulsar surface. On the other hand, a magnetized ($B = 10^{12}$ G) hydrogen atmosphere model allows the neutron star to be "canonical" (see Eq. (1.6)), yielding the temperature $T_s^\infty = 0.674^{+0.034}_{-0.033} \times 10^6$ K and the distance $d_{\rm fit} = 256^{+44}_{-43}$ pc (Mori *et al.*, 2004). Using the approximate scaling $R_\infty \propto d$ and the parallax distance $d$, the canonical $R_\infty = 13$ km can be translated into $R_\infty = 14.9^{+4.1}_{-2.9}$ km.

Another nearby neutron star with the thermal-like spectrum is RX J1856–3754, discovered in the *ROSAT* observations by Walter *et al.* (1996) and extensively studied in subsequent years. A faint optical counterpart of this star was found by Walter & Matthews (1997). The distance determined from the parallax measurements is $d = 117 \pm 12$ pc (Walter & Lattimer, 2002). One-component blackbody and atmosphere models failed to explain the spectrum in both optical and X-ray ranges. The X-ray spectrum is well fitted by the blackbody spectrum with $T_6^\infty = 0.73$ and $R_{10}^\infty = 0.44\,(d/120\ \mathrm{pc})$, whereas the optical data obey the Rayleigh-Jeans law with the flux $\sim 7$ times higher than that obtained by extrapolating the X-ray spectrum into the optical range (e.g., Burwitz *et al.* 2003 and references therein). This may indicate the presence of a softer thermal component, produced by the emission from the entire surface, while the "hard" component may be emitted from a hot polar cap. The constraints obtained by Burwitz *et al.* (2003) for the "soft" component read $4\ \mathrm{eV} \lesssim k_\mathrm{B} T_s^\infty < 33.6\ \mathrm{eV}$ (i.e., $0.046 \lesssim T_6^\infty < 0.39$) and $1.63 < R_{10}^\infty \lesssim 4.6$ (at $d = 120$ pc). However, the hot polar cap model implies pulsations of radiation from a rotating star. Many attempt to find these pulsations failed. Burwitz *et al.* (2003) proposed an alternative model, according to which the radiation comes from a solid or liquid surface (rather than a gaseous atmosphere), which emits the blackbody radiation but with reduced emissivity in the X-rays. This yields the lower bound $R_\infty > 12.3$ km at the $4\sigma$ level consistent with the majority of the EOSs of dense matter (the upper bound is uncertain because of an unknown reflectivity in the optical).

The latter result stimulated extensive studies of the emissivity and spectrum of magnetized neutron stars with a condensed surface (van Adelsberg *et al.* 2005, and references therein). Simultaneously, the first self-consistent models of partially ionized hydrogen neutron star atmospheres with strong magnetic fields have been constructed (Ho *et al.*, 2003; Potekhin *et al.*, 2004).

### 9.3.2 X-ray emission from accreting neutron stars in binaries

**X-ray bursts.** Several attempts have been made to constrain the EOS in neutron star cores by analyzing observations of type I X-ray bursts accompanied by neutron star atmosphere expansion (see, e.g., Shaposhnikov & Titarchuk 2002, 2004, and references therein). The analysis is based on the well formulated theory of X-ray spectrum formation in atmospheres of bursting neutron stars, where the radiation transfer is mainly determined by electron scattering. A standard assumption is that X-ray bursts (§ 1.4.6) originate from thermonuclear flashes in the surface layers of accreting neutron stars with weak magnetic fields ($B \lesssim 10^8$–$10^9$ G). The model parameters in the spectral fits are neutron star mass $M$, radius $R$, distance $d$, atmosphere composition, and a possible anisotropy of the atmospheric emission. For example, Titarchuk & Shaposhnikov (2002) inferred the values of $M$ and $R$ from an analysis of three X-ray bursts from Cygnus X-2. Assuming the helium mass fraction $Y = 0.3$ and the distance $d = 11$ kpc, they obtained $R = 9.0 \pm 0.5$ km and $M = 1.44 \pm 0.06\, M_\odot$ (at the $1\sigma$ level). With the same $Y$ and $d = 9$ kpc they got $R = 7.7 \pm 0.4$ km and $M = 0.97 \pm 0.04\, M_\odot$. If true, this would favor very soft EOSs of dense matter. Shaposhnikov *et al.* (2003) analyzed a set of type I X-ray bursts from the low-mass X-ray binary (LMXB) 4U 1728–34. The data are well explained by their theory if the burster atmosphere is dominated by helium, the distance ranges from 4.5 to 5.0 kpc, $R = 8.7 - 9.7$ km and $M = 1.2 - 1.6\ M_\odot$. These results also favor soft EOSs. In the same framework, Shaposhnikov & Titarchuk (2004) derived $R = 11$–$13$ km and $M \approx 1.3$–$2.0\, M_\odot$ by analyzing an expansion stage of a type I X-ray burst of the brightest LMXB 4U 1820–30 (employing the distance $d = 6.4 \pm 0.6$ kpc reported by Vacca *et al.* 1986). This analysis favors moderately soft EOSs. Unfortunately, the accuracy of $R$ and $M$ determination by this method is insufficiently high to impose stringent constraint on the EOS but the method could be improved with new observations of X-ray bursts and with better determination of $d$.

Nath *et al.* (2002) simulated X-ray burst oscillations observed from two LMXBs using a model that includes emission from either a single hot spot or a pair of antipodal hot spots on the neutron star surface. They modeled the spreading of thermonuclear burning over the stellar surface by assuming that the hot-spot size grows linearly with time, and calculated the radiation flux as a function of rotational phase of hot spots. For LMXB 4U 1636−53, they constrained the neutron star compactness as $GM/(c^2R) < 0.163$ at 90% confidence which requires a relatively stiff EOS of the stellar interior. For example, for a neutron star with the canonical mass $1.4\,M_{\odot}$ this gives $R > 12.8$ km.

There are promising perspectives of constraining $R$ or $M/R$ from bursting LMXBs. For example, a discovery of quiescent thermal radiation from X-ray bursters in globular clusters (where the distance and interstellar column density are known rather well) opens new opportunities for precise measurements of neutron star radii (Rutledge *et al.* 2002a, see also Heinke *et al.* 2005 and references therein).

**Quasiperiodic oscillations (QPOs).** Kilohertz quasiperiodic oscillations (QPOs) manifest themselves in power energy spectrum of many LMXBs as twin peaks at frequencies $\sim 0.5$–1 kHz. The characteristics of kilohertz QPOs are similar for many sources. Several models have been put forward to explain these features. The model most fashionable in the 1990s assumed that QPOs originate at the inner edge of accretion disks when infalling matter elements rotate with the Keplerian frequency at their last stable orbit. The existence of such an orbit is predicted by General Relativity; its circumferential radius is $3r_{\rm g} = 8.85\,M/M_{\odot}$ km. The corresponding frequency of the orbital motion is $\nu_0 \approx 2.21\,M_{\odot}/M$ kHz. The theory predicted two QPO peaks at nearly the same frequency $\nu_0$, separated by the neutron star spin frequency (*the beat-frequency model*, or BFM, proposed by Alpar & Shaham, 1985). If neutron stars, which produce QPOs in LMXBs, are localized inside their marginally stable orbits ($R < 3\,r_{\rm g}$), they have masses $M \sim 2\,M_{\odot}$, which favors stiff EOSs (Zhang *et al.*, 1997).

The BFM has been elaborated in different ways. One of the modifications was the sonic-point model (Miller *et al.* 1998 and references therein). The sonic point is located in the disk layer, where the radial inflow regime changes from subsonic to supersonic. Its position is determined by the luminosity of the source. In this model, kilohertz QPOs are generated in the vicinity of the sonic point, rather than near the last stable orbit.

However, all these models meet serious difficulties when confronted with observations (e.g., van der Klis, 2000). In particular, the peak frequency separation turns out to be variable. Lamb & Miller (2001) modified the model to allow for such variability. However, Jonker *et al.* (2002) found that in the LMXB 4U 1636−53 the difference of QPO peak frequencies varies from less than half the spin frequency (inferred from burst oscillations) to *more* than the spin frequency, which strongly challenges the BFM. In addition, the neutron star mass in Cyg X-2, evaluated from the BFM, clearly contradicts the mass inferred from the burst model (Titarchuk & Shaposhnikov, 2002, see above). Moreover, taking into account the correlation of QPO frequencies of white dwarf, neutron star, and black hole binaries, Mauche (2002) argued that the data exclude a very wide class of QPO models, particularly, those which involve the QPO production at the stellar surface or in the presence of a magnetic field. An alternative model of QPOs suggested by Titarchuk *et al.* (1998) (see also Titarchuk 2002) interprets the observed QPO frequencies in terms of relaxation oscillations of a hot inner boundary region of an accretion disk.

Wijnands *et al.* (2003) detected kHz QPOs from a pulsar, whose spin frequency is known from *independent* observations. Their analysis shows that the observations contradict the existing BFM versions and pose "a severe challenge to all other QPO models."

Thus, at the time of this writing (2006), the model uncertainties obstruct obtaining firm bounds on neutron star radii from observations of X-ray bursts and QPOs.

**Quiescent thermal emission.** Brown *et al.* (1998) proposed a model for quiescent emission of soft X-ray transients – LMXBs containing transiently accreting neutron stars. According to their model, the thermal emission is powered by the heat released in the accreted matter (*deep crustal heating*, see § 1.4.6) which sinks deeply in the crust under the weight of newly accreted material. Along with other mechanisms (such as a residual accretion or a pulsar shock), the thermal surface emission due to the deep crustal heating can be an important source of the quiescent emission of soft X-ray transients. Using this model, Rutledge *et al.* (2001a,b) analyzed quiescent spectra of Aql X-1 ($4 \lesssim d \lesssim 6.5$ kpc) after an accretion outburst. Assuming a pure hydrogen atmosphere and a hard power-law spectrum of nonthermal radiation component, they derived the neutron star radius $R_\infty = 15.9^{+0.8}_{-2.9}\,(d/5\text{ kpc})$ km. A similar analysis of observations of KS 1731−260, another soft X-ray transient located near the Galactic center, gave the estimate $R_\infty = 23^{+30}_{-15}\,(d/8\text{ kpc})$ km (Rutledge *et al.*, 2002b). Several other results of this type have also been rather inconclusive.

**Accretion disk – magnetosphere restrictions.** A discovery of the first accreting millisecond X-ray pulsar, SAX 1808.4–3658, raised hope to constraint its radius. Psaltis & Chakrabarty (1999) considered four disk-magnetosphere interaction models of this source and associated constraints on neutron star parameters. In particular, they obtained the upper limits on $R$ as a function of $M$ (but neglecting space-curvature effects, crucial for the difference between $R$ and $R_\infty$). Their results strongly depend on an employed model and do not give a firm unambiguous limit on $R$.

### 9.3.3 Final remarks on radius measurements

As the reader could easily conclude, the constraints on neutron star radii made by the time of this writing (2006) are insufficiently precise to restrict the EOS in neutron star cores. However, the progress is spectacular in recent years

*Table 9.8.* Millisecond pulsars with spin periods shorter than 2 ms

| No. | PSR | $P$ (ms) | No. | PSR | $P$ (ms) |
|---|---|---|---|---|---|
| 1 | J1748–2446ad | 1.396 | 6 | J1748–2446P | 1.729 |
| 2 | B1937+21 | 1.558 | 7 | J1843–1113 | 1.846 |
| 3 | B1957+20 | 1.607 | 8 | J0034–0534 | 1.877 |
| 4 | IGR J00291+5934 | 1.670 | 9 | A1744–361 | 1.887 |
| 5 | J1748–2446O | 1.677 | | | |

and we expect that observations and improved theoretical models will lead to useful constraints on neutron star radii in near future.

## 9.4. Millisecond and submillisecond pulsars

To be consistent with observations of millisecond pulsars, an EOS of dense matter should give stable neutron star models at the shortest observed pulsar period, $P_{\rm obs}^{\rm min}$. We described theoretical models of spinning neutron stars in § 6.12. The rotation cannot be too fast and exceed the mass-shedding limit; the hydrostatic equilibrium of the star has to be stable with respect to small axially symmetric perturbations (§ 6.12.2). The "theoretical" minimum period of uniform rotation, $P_{\rm min}({\rm EOS})$, consistent with these stability conditions, depends on the EOS of dense matter. Fortunately, the values of $P_{\rm min}({\rm EOS})$ obtained in extensive accurate calculations for *causal* EOSs are reproduced with a surprisingly good precision (typically better than 5%) by the simple "empirical formula" (6.108),

$$P_{\rm min}({\rm EOS}) \approx 0.82 \left(\frac{M_\odot}{M_{\rm max}^{\rm stat}({\rm EOS})}\right)^{\frac{1}{2}} \left(\frac{R_{M_{\rm max}}^{\rm stat}({\rm EOS})}{10~{\rm km}}\right)^{\frac{3}{2}} {\rm ms}~, \tag{9.28}$$

where $M_{\rm max}^{\rm stat}$ and $R_{M_{\rm max}}^{\rm stat}$ are, respectively, the mass and radius of the most massive stable *non-rotating* star. Equation (9.28) gives the necessary condition for a rotating neutron star with a given EOS to be consistent with the shortest observed pulsar period:

$$\left(\frac{10~{\rm km}}{R_{M_{\rm max}}^{\rm stat}({\rm EOS})}\right)^3 \frac{M_{\rm max}^{\rm stat}({\rm EOS})}{M_\odot} > 0.67 \left(\frac{1~{\rm ms}}{P_{\rm obs}^{\rm min}}\right)^2 . \tag{9.29}$$

In order to rotate stably with the period $P_{\rm obs}^{\rm min}$, a star must be sufficiently massive and compact.

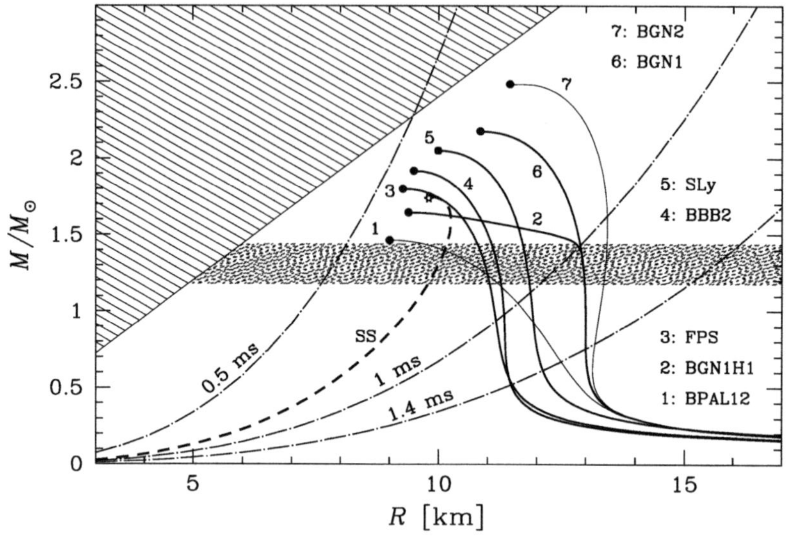


*Figure 9.8.* EOS constraints from observation of rapidly rotating pulsars. The dashed line, labeled by SS, corresponds to *bare strange stars*, studied in § 8.11 (with the EOS of strange quark matter described in § 8.11). Precise condition discussed in the text implies that at a given spin period (1.4 ms, 1 ms, 0.5 ms) a maximum-mass non-rotating star has to lie above a dot-and-dashed line. For $M < M_{\rm max}$ this condition becomes approximate.

The theoretical predictions can be confronted with observations of millisecond pulsars (which, by definition, are the pulsars with spin periods $P \lesssim 30$ ms – see Lorimer 2001). By 2006 approximately 150 such pulsars have been discovered. A list of nine fastest rotators, with $P < 2$ ms, is given in Table 9.8. All of them but IGR J00291+5934 and A1744–361 are observed as radio pulsars, while IGR J00291+5934 and A1744–361 are accreting millisecond X-ray pulsars in binary systems (Galloway *et al.*, 2005; Bhattacharyya *et al.*, 2006). The first millisecond pulsar discovered, PSR B1937+214 (Backer *et al.*, 1982), with $P = 1.5578$ ms, remained the most rapidly rotating neutron star till 2005. (Until its discovery, the fastest rotation, $P = 33.1$ ms, was demonstrated by the Crab pulsar.) In 2005 a new spinning champion was discovered, PSR J1748–2446ad, with $P = 1.396$ ms (Hessels *et al.*, 2006). As seen from Fig. 9.8, the rotational stability curve, above which a pulsar with $P = 1.4$ ms is stable, lies too low to produce any noticeable constraint on the EOS of dense matter.

A very strong constraint would be given by a much faster submillisecond pulsar, with $P = 0.5$ ms. Such a pulsar was discovered on January 18, 1998, in a 7 hour optical observation of the supernova remnant 1987A in the Large Magellanic Cloud with the Cerro Tololo 4-meter telescope in Chile. The discovery ruled *almost all* theoretical EOSs, which were believed to be realistic – they became forbidden by that observation! As the reader can see, the pulsar rules out all EOSs used to calculate the $M - R$ curves in Fig. 9.8. That nightmare for theorists lasted for one year. In that year, all theoretical concepts of dense matter physics were revisited in attempts to construct a realistic EOS which would be able to explain a 0.5 ms pulsar (see, e.g., § 9.8.4), but no convincing resolution of the apparent conflict between theory and observation was found. The story finished in February, 1990, with the discovery of exactly four times more rapid optical oscillations from the Crab pulsar with one of the telescopes of the Las Campanas observatory in Chile, near Cerro Tololo. Both telescopes had similar TV cameras for transmitting images of what the telescopes were seeing to observatory control rooms. In January and February, the hottest summer period in Chile, the TV transmitters could produce falsely modulated signals, the effect that had not been known before. The discovery of 0.5 ms pulsar turned out to be an artifact, and realistic theoretical EOSs were rehabilitated. Nevertheless, the story was a good push to the neutron star theory.

## 9.5. The Crab Nebula and the moment of inertia of the Crab pulsar

The AD 1054 supernova remnant, the Crab Nebula, was discovered by amateur astronomer John Bevis in 1731 and rediscovered by Charles Messier in 1758. John Duncan found in 1921 that the nebula is expanding (see Duncan 1939). Now it is probably the most often observed nebula in the sky. Optical observations of its filaments indicate that filaments are accelerated. This acceleration as well as the nebula emission are powered by the Crab pulsar which was discovered in the center of the nebula in 1968 (§ 1.4.2). The energy reservoir is constituted by the pulsar rotational energy, which the pulsar loses at a rate $\dot{E}_{\rm rot} = I\Omega\dot{\Omega}$. Here, $I$ is the pulsar moment of inertia, while $\Omega$ and $\dot{\Omega}$ are, respectively, the angular frequency and its time derivative, obtained from the pulsar timing. Assuming the balance between $\dot{E}_{\rm rot}$ and the power of the nebula radiation and accelerated expansion, one may constrain $I$, which may be useful to constrain the EOS of dense matter.

The idea was proposed and realized by Manchester & Taylor (1977). Later the problem was reconsidered using more recent data on the Crab Nebula (Haensel, 1990). In both cases it has been assumed that the acceleration rate $\dot{v}$ is constant during the nebula expansion. The constraints derived by Manchester & Taylor (1977) were weak and did not eliminate any of EOSs. The later analysis by Haensel (1990) revealed the crucial dependence of the result on the mass $M_{\rm neb}$ of the expanding nebula. The maximum limits on $M_{\rm neb}$ obtained in the 1980s ruled out the softest EOSs.

Recent estimates of the mass contained in optical filaments are significantly higher than older ones. As shown by Bejger & Haensel (2002, 2003), this opens the possibility of setting stronger constraints on the EOS.

### 9.5.1 The energy balance of the Crab pulsar and its nebula

The measured spin period of the Crab pulsar and its derivative are $P_{\rm p} = 33.41$ ms and $\dot{P}_{\rm p} = 4.228 \times 10^{-13}$, which correspond to the current spin frequency $\Omega_{\rm p} = 2\pi/P_{\rm p} = 188\ {\rm s}^{-1}$ and $\dot{\Omega}_{\rm p} = -2.38 \times 10^{-9}\ {\rm s}^{-2}$. The pulsar spin energy powers the acceleration of the nebula, the emission of electromagnetic waves, and the sweeping away the interstellar matter. The value of $\dot{\Omega}$ can be related to $\Omega$ by

$$\dot{\Omega}(t) = -K\Omega^n(t), \tag{9.30}$$

where $K$ and $n$ are constants determined from the pulsar timing (see, e.g., Shapiro & Teukolsky 1983; also see § 1.4.4). For the Crab pulsar, we get $K = 4.68 \times 10^{-15}$ (in CGS units). The braking index $n$ is obtained through the measurable timing parameters $\Omega$, $\dot{\Omega}$, and $\ddot{\Omega}$ as $n = \Omega\ddot{\Omega}/\dot{\Omega}^2$ . From the timing of the Crab pulsar in the period from 1982 to 1987 one has $n = 2.509 \pm 0.001$ (Lyne *et al.*, 1988). We adopt the standard assumption that $n$ and $K$ are fixed after the pulsar formation (e.g., in less than a few months after the neutron star birth). In what follows we will count the pulsar age from that moment. The integration of Eq. (9.30) from $t = 0$ gives the well known pulsar spindown law

$$\Omega(t) = \frac{\Omega_0}{\{1 + Kt(n-1)\Omega_0^{n-1}\}^{1/(n-1)}}, \tag{9.31}$$

where $\Omega_0$ is the initial spin frequency. Putting the current pulsar age $t_{\rm p} = 950$ years we get $\Omega_0 = 326\ {\rm s}^{-1}$ and the initial period $P_0 = 2\pi/\Omega_0 = 19.3$ ms. The spin-down energy loss rate is

$$\dot{E}_{\rm rot} = \frac{\rm d}{{\rm d}t}\left(\frac{1}{2} I\Omega^2\right) = -I\Omega|\dot{\Omega}|, \tag{9.32}$$

where $I$ is independent of $\Omega$ because $\Omega$ is much smaller than the mass-shedding limit (§ 6.10). The pulsar spin energy transforms into electromagnetic radiation of the nebula (with the luminosity $\dot{E}_{\rm rad}$) and into the accelerated expansion of the nebula. When calculating the power of this expansion, $\dot{E}_{\rm exp}$, one should take into account that the nebula sweeps the interstellar medium:

$$\dot{E}_{\rm exp} = \frac{\rm d}{{\rm d}t}\int \frac{1}{2}\rho v^2\ {\rm d}V\ ; \tag{9.33}$$

the integration goes over the nebula volume $V$. The nebula mass $M_{\rm neb}$ is thought to be mostly contained in luminous filaments. The velocity vectors

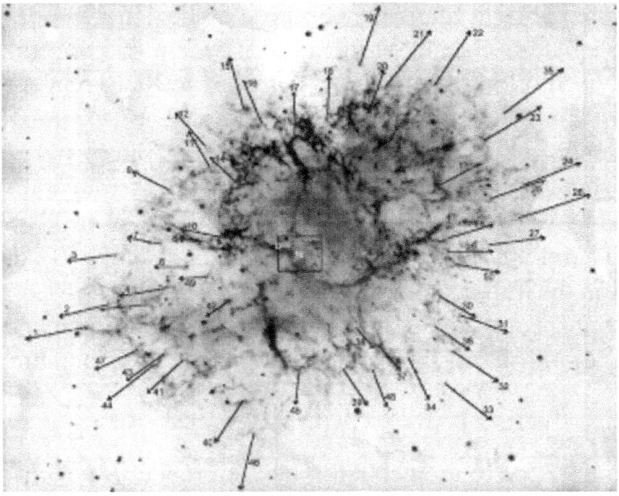

*Figure 9.9.* Expansion of the Crab Nebula. Arrows show motion of 50 optical filaments in the next 250 years at the current expansion speed. From Nugent (1998), with the kind permission of the author.

of the filaments are nearly radial and have approximately the same magnitude (Fig. 9.9). Treating $v$ as constant in space, we get

$$\begin{aligned} \dot{E}_{\rm exp} &= \frac{1}{2}\frac{\mathrm{d}}{\mathrm{d}t}\left(M_{\rm neb}v^2\right) = \dot{E}_{\rm acc} + \dot{E}_{\rm sweep}; \\ \dot{E}_{\rm acc} &\equiv M_{\rm neb}v\dot{v}, \quad \dot{E}_{\rm sweep} \equiv \frac{1}{2}\dot{M}_{\rm neb}v^2 , \end{aligned} \tag{9.34}$$

where $\dot{M}_{\rm neb}$ is the mass of the interstellar medium accumulated by the expanding nebula per unit time.

## 9.5.2 Observational evaluations of the Crab Nebula parameters

The current mass of the Crab Nebula is very important for our analysis (Bejger & Haensel, 2002, 2003) but its observational estimation is difficult. In the last two decades the reported values of the mass (contained in optically shining filaments) varied from $(2-3)\,M_\odot$ (Davidson & Fesen, 1985), through $(1-2)\,M_\odot$ (MacAlpine & Uomoto, 1991) to $(4.6 \pm 1.8)\,M_\odot$ (Fesen *et al.*, 1997).

Optical filaments form actually an expanding shell. Its motion can be studied by comparing positions of filaments on high-resolution photographs taken now and some time ago (Duncan, 1939; Trimble, 1968; Wyckoff & Murray, 1977; Nugent, 1998). We will use the results obtained by Nugent (1998) from the data on 50 bright filaments observed earlier (in 1939, 1960, 1976, and 1992). His results are visualized in Fig. 9.9. By projecting the straight-line constant velocity motion of filaments backward in time, Nugent obtained the convergence of filament trajectories at AD $1130 \pm 16$ yr, in agreement with the previous analysis by Trimble (1968). Had the Crab Nebula expanded at a constant $v$, this would be the moment of the supernova explosion. However, the date known from historical records is 1054 (see § 1.4.2, page 32), which is $\sim 75$ years earlier. Therefore, the expansion was accelerated ($\dot{v} > 0$), owing to which $v$ increased from the initial value $v_0$ to the current value $v_{\rm p} \sim 1.5 \times 10^8$ km s$^{-1}$, also known from spectroscopic observations (e.g., Sollerman *et al.* 2000).

The total luminosity of the nebula is estimated as (Peterson, 1998)

$$\dot{E}_{\rm rad}(d) \simeq 1.25 \ (d/d_{\rm DF})^2 \times 10^{38} \ \text{erg s}^{-1}, \tag{9.35}$$

where $d$ is the distance to the nebula. The "reference value" $d_{\rm DF} = 1.83$ kpc comes from Davidson & Fesen (1985).

### 9.5.3 Bound on the moment of inertia

The constraint on the moment of inertia of the Crab pulsar results from the condition that, at any moment, the loss of the pulsar spin energy is sufficient to power the nebula,

$$\Omega|\dot{\Omega}|I_{\rm Crab} \geq \dot{E}_{\rm rad} + \dot{E}_{\rm acc} + \dot{E}_{\rm sweep}. \tag{9.36}$$

Using the observational data and simple physical arguments one can easily estimate (e.g., Bejger & Haensel 2002, 2003) that $\dot{E}_{\rm acc}$ is at least several times higher than $\dot{E}_{\rm rad}$ and $\dot{E}_{\rm sweep}$. Hence, the pulsar mainly powers the nebula acceleration at a nearly constant $M_{\rm neb}$; $\dot{E}_{\rm sweep}$ will dominate later.

One usually constrained the moment of inertia assuming that the nebula acceleration $\dot{v}$ is independent of time (Manchester & Taylor 1977; Haensel 1990; Bejger & Haensel 2002). Here we go beyond this model using the approach by Bejger & Haensel (2003) and slightly improving it. We assume that $\dot{E}_{\rm acc}$ dominated the pulsar energy loss all the time till now, after some initial short-term period ($\lesssim$3 yr) in which the Crab Nebula was powered by supernova itself (i.e., by $^{56}$Ni radioactive decay heating), but we do not assume a constant acceleration. In our approximation

$$I\Omega(t)|\dot{\Omega}(t)| \approx M_{\rm neb}\rho v(t)\dot{v}(t). \tag{9.37}$$

This immediately gives

$$M_{\rm neb} \left(v^2(t) - v_0^2\right) = I \left(\Omega_0^2 - \Omega^2(t)\right), \tag{9.38}$$

which is just energy conservation of the spinning pulsar and expanding nebula. Because the velocity variation during the acceleration is relatively small, we approximately have

$$v(t) = v_0 + \frac{I}{2M_{\rm neb}v_0}(\Omega_0^2 - \Omega(t)^2). \tag{9.39}$$

In order to complete our analysis let us approximate the expanding nebula by a sphere and calculate its current ($t = t_{\rm p}$) radius:

$$R_{\rm neb} = \int_0^{t_{\rm p}} v(t)\ {\rm d}t. \tag{9.40}$$

It is sufficient to substitute $v(t)$ from Eq. (9.39), set $v_0 \approx v_{\rm p}$ in the denominator, and take $\Omega(t)$ from the spindown equation (9.31). The integral is then taken analytically. Putting, in addition, $t = t_{\rm p}$ in Eq. (9.39) in the same approximation, we come to a set of two equations

$$\begin{aligned} v_{\rm p} &= v_0 + \frac{I\left(\Omega_0^2 - \Omega_{\rm p}^2\right)}{2M_{\rm neb}v_{\rm p}}, \\ R_{\rm neb} &= v_0\, t_{\rm p} - \frac{I\Omega_0^{3-n}}{2(3-n)(1-n)KM_{\rm neb}v_{\rm p}} \\ &\times \left\{2 + (\Omega_{\rm p}/\Omega_0)^{3-n}\left[1 - n - (3-n)\left(\Omega_0/\Omega_{\rm p}\right)^2\right]\right\}. \end{aligned} \tag{9.41}$$

The Crab pulsar timing constants $n$ and $K$ are given in § 9.5.1. The present expansion velocity $v_{\rm p} \approx 1.5 \times 10^8$ cm s$^{-1}$ has been discussed in § 9.5.2.

Thus, the two equations allow us to find $v_0$ and $I = I_{\rm Crab} \equiv 10^{45}\, I_{\rm Crab,45}$ g cm$^2$ for any values of $M_{\rm neb}$ and $R_{\rm neb}$ determined from observations. Let us take the values $M_{\rm neb} = (4.6 \pm 1.8)\, M_\odot$ reported by Fesen *et al.* (1997). The shell radius $R_{\rm neb}$ is also very uncertain. The spherical-shell model is the simplest representation of the expanding filament fireball of the Crab Nebula, which is famous for its rather complicated crab-like shape. The value of $R_{\rm neb}$ can be estimated as a mean for an ellipsoid which is a more precise model of the Crab Nebula shape. Assuming $d = 1.83$ kpc, one gets $R_{\rm neb} = 1.25$ pc (see, e.g., Douvion *et al.* 2001).

Actually, the moment of inertia of the Crab pulsar $I_{\rm Crab}$ has to be constrained taking into account observational uncertainties of all parameters (especially, of $M_{\rm neb}$, $R_{\rm neb}$, and $v_{\rm p}$) as well uncertainties of the underlying model (e.g., the spherical shell of the envelope). Such an analysis would give too large scatter of $I_{\rm Crab}$. For illustration, we adopt the values $v_{\rm p} \approx 1.5 \times 10^8$ cm s$^{-1}$ and $R_{\rm neb} = 1.25$ pc, and infer $I_{\rm Crab}$ assuming the central and maximum values of $M_{\rm neb}$ given above. We obtain

$$\begin{aligned} M_{\rm neb} &= 4.6\, M_\odot \Longrightarrow I_{\rm Crab,45} = 2.2; \\ M_{\rm neb} &= 6.4\, M_\odot \Longrightarrow I_{\rm Crab,45} = 3.1\ . \end{aligned} \tag{9.42}$$

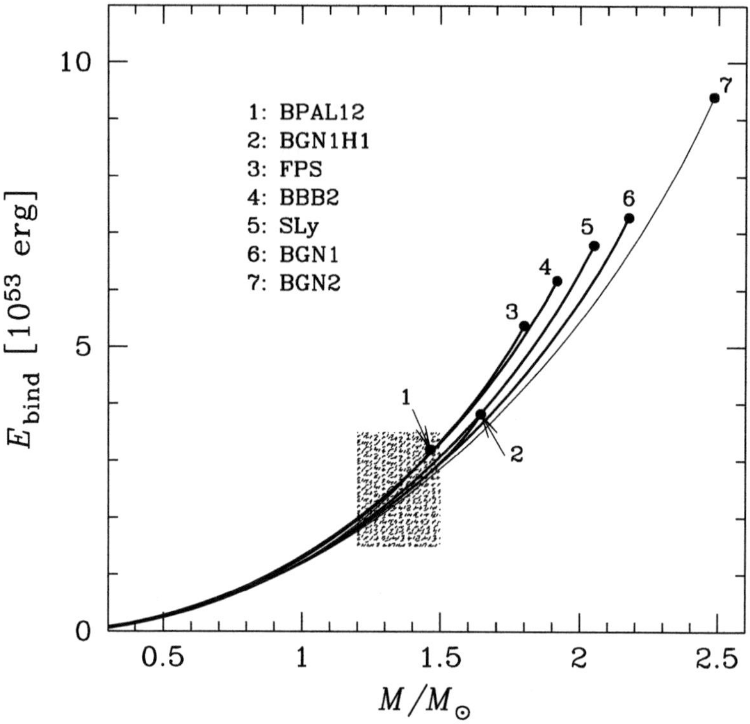


*Figure 9.10.* Binding energy with respect to the presupernova core (§§ 6.2,6.7) versus neutron star mass for selected EOSs of dense matter (Table 5.3). Shaded box is acceptable for the SN 1987A event neglecting neutrino oscillations. For further explanations see the text.

In order to get $I_{\rm Crab,45} = 3.1$, which corresponds to $M_{\rm neb} = 6.4\ M_\odot$, a very stiff EOS is required (§ 6.10). However, the uncertainties stemming from large errors in other input parameters weaken the real implications of this finding. It is important to notice that the present results differ significantly from those obtained assuming the constant nebula acceleration $\dot{v}$.

## 9.6. Neutrinos from Supernova 1987A and binding energy of neutron stars

The detection of the neutrino burst from SN 1987A had enormous impact on the theory of core-collapsed (type II) supernovae (§ 1.4.2). Basically, it confirmed the main idea that such supernovae result from the gravitational collapse of a degenerate (mass $\sim 1.4\ M_\odot$) core of a massive ($\sim 15\ M_\odot$) star at the end of its thermonuclear evolution.

The theory of type II supernovae predicts the gravitational collapse of a degenerate core on timescales $\sim 0.1$ s and the formation of a hot ($k_{\rm B}T \sim 50$ MeV) protoneutron star of radius $\sim 50$ km, composed of $A_b \sim 10^{57}$ baryons, with the electron lepton number $L_e \simeq 0.35 A_b$ (§ 1.4.2). The initial gravitational mass of the collapsing core is $M_{in} \simeq A_b m_0$, where $m_0 = 1.6586 \times 10^{-24}$ g is the mass of the $^{56}$Fe nucleus divided by 56 (§6.2). The gravitational mass of the protoneutron star is only slightly lower than $M_{in}$, because the gain in the gravitational energy is compensated by the increase of the internal energy contained mostly in strongly degenerate neutrinos $\nu_e$ trapped in the stellar interior. On timescales $\sim 10$ s electron neutrinos diffuse out of the protoneutron star core and the star shrinks into an ordinary neutron star of gravitational mass $M$. The shrinking is accompanied by the energy release $\sim (Am_0 - M)c^2$. Approximately 99% of the energy is carried away by neutrinos and antineutrinos of all flavors ($\nu_e$, $\nu_\mu$, $\nu_\tau$), $E_\nu \simeq E_{bind}$.

Let us analyze briefly the neutrino outburst of SN 1987A neglecting neutrino oscillations (for a review see, e.g., Imshennik & Nadyozhin 1988 and Burrows 1990). The total of 25 detected captures of $\bar{\nu}_e$ by protons within $\sim 10$ s enabled one to reconstruct the parameters of the original neutrino burst. Spherical symmetry was assumed to make such an analysis feasible. The total energy of electron antineutrinos radiated during the explosion was estimated as $E_{\bar{\nu}_e} \simeq (5 \pm 2) \times 10^{52}$ erg. Numerical simulations of the neutron star birth showed equipartition of energies emitted in all neutrino flavors. Accordingly, $E_{\bar{\nu}_e} \simeq E_\nu/6$, and the total radiated neutrino energy can be estimated as $E_\nu \simeq (3 \pm 1.5) \times 10^{53}$ erg (the value we quote is subject to additional uncertainties, see Lattimer & Yahil 1989; Burrows 1990 and references therein). This could be considered as the "measurement" of the binding energy of the newly born neutron star, $E_{bind} \simeq E_\nu$. The "measured" value of $E_{bind}$ could be compared with theoretical predictions for various EOSs of dense matter. From the evolutionary scenarios one expects that the neutron star born in SN 1987A had $M = (1.2 - 1.5)M_\odot$. In this way we obtain a shaded acceptable rectangle in the $E_{bind} - M$ plane shown in Fig. 9.10. All EOSs in Table 5.3 (and nearly all which can be found in the literature) satisfy this "observational" constraint.

This analysis has to be revised taking into account neutrino oscillations discovered and confirmed several years after the SN 1987A event. Oscillations $\overline{\nu}_e \longleftrightarrow \overline{\nu}_\mu, \overline{\nu}_\tau$ during the neutrino propagation to the detector distort (harden) the $\overline{\nu}_e$ energy spectrum (see Takahashi & Sato 2003, and references therein). However, the total neutrino energy $E_\nu$ released from SN 1987A is not strongly affected by oscillations, taking into account large uncertainties in the measured value of $E_\nu$ (Kachelriess *et al.*, 2002).

New giant neutrino detectors built in the last decade, such as the Super-Kamiokande Observatory, the Sudbury Neutrino Observatory and the Kamioka Liquid-scintillator Anti-Neutrino Detector, offer much better possibilities for neutrino detection than in 1987. To be specific, let us imagine that a galactic type II supernova explodes at a distance of 10 kpc from us. The calculations performed by Takahashi & Sato (2003) who included neutrino oscillations and used a standard model for a neutrino outburst predict $\sim 10^4$ of $\overline{\nu}_e + p \to n + e^+$ absorptions registered by the SuperKamiokande detector. Neutrino oscillations harden the energy spectra. Because the cross-section of $\overline{\nu}_e + p \to n + e^+$ increases with neutrino energy, the inclusion of neutrino oscillations increases the predicted number of registered events by $(20-30)$ % (Takahashi & Sato, 2003). Combined with the (hopefully) refined theory of type II supernovae and a more precise model for neutrino mixing, such data could yield precious information on the binding energy of neutron stars, and provide us with more restrictive constraint on the EOS of dense matter.

## 9.7. Pulsar glitches and crustal moment of inertia

It is widely believed that sudden spin-ups of radio pulsars, called *glitches* (§ 1.4.4), are produced by angular momentum transfer from the superfluid component of the stellar interior to the (normal) crust. Sudden frequency jumps of magnitude $\Delta\Omega/\Omega \sim 10^{-9} - 10^{-5}$ are accompanied by jumps in the spin-down rate, of magnitude $\Delta\dot{\Omega}/\dot{\Omega} \sim 10^{-3} - 10^{-2}$. In the standard scenario (Alpar *et al.*, 1984), a neutron star crust (a lattice of nuclei coupled by Coulomb forces to electron fluid) spins down under the action of magnetic torque. A superconducting proton fluid in the core is coupled to electrons via electron-fluxoid scattering (assuming that protons form a type II superconductor). On the other hand, neutron superfluid in the core is coupled to proton superfluid via superfluid drag effects, resulting from the Fermi-liquid interactions. As a consequence, the crust is thought to be rather strongly coupled to the core, with the coupling time of the core to the magnetic field and the crust of a few seconds. Therefore, it could be a good approximation to describe the core as rigidly coupled to both the magnetic field and the crust. The total moment of inertia of this rigidly rotating component will be denoted by $I_{\rm c}$. On the contrary, neutron superfluid in the inner crust is only weakly coupled to crustal lattice (with typical coupling times of weeks to months). Consequently, crustal neutron superfluid has an excess of angular velocity, as compared to the rigidly rotating component (the latter component constitutes typically more than 95% of the stellar mass, § 6.8).

The magnetic spindown of the crust increases the difference of angular frequencies and induces a mechanical strain between the neutron superfluid vortices pinned to nuclei in the crust. At a certain threshold value of the angular velocity difference, some of neutron vortices ($\sim 10^{11}$ for a macroglitch) undergo a violent unpinning and move outward. The excess of angular momentum is transferred to the crust leading to a pulsar glitch. The glitch is followed by a relaxation towards a steady rotational state (on timescales of weeks to months). Then, a new excess of the angular velocity in crustal superfluid is built up and continues until the threshold value is reached.

*Table 9.9.* Glitches of the Vela pulsar in the period from 1969 to 2000. Glitch dates are approximate (within one month); $\Delta\dot{\Omega}$ is a change in the spin down rate that survives long after a glitch

| $i$ | Date | $\Delta\Omega/\Omega$ $(10^{-6})$ | $\Delta\dot{\Omega}/\dot{\Omega}$ $(10^{-2})$ | Relaxation (d) |
|---|---|---|---|---|
| 1 | 1969 Feb | 2.3 | 0.71 | 35 |
| 2 | 1971 Aug | 2.0 | 0.72 | 33 |
| 3 | 1974 Sep | 2.0 | 0.72 | 30 |
| 4 | 1978 July | 3.1 | 0.66 | 39 |
| 5 | 1981 Oct | 1.1 | 0.63 | 32 |
| 6 | 1982 Aug | 2.0 | 0.60 | 31 |
| 7 | 1985 July | 1.3 | 0.65 | 31 |
| 8 | 1988 Dec | 1.8 | 0.47 | 29 |
| 9 | 1991 July | 2.7 | 0.74 | 29 |
| 10 | 1994 July | 0.86 | – | – |
| 11 | 1994 Aug | 0.20 | – | – |
| 12 | 2000 Jan | 3.1 | 0.67 | 19 |

Since the discovery of pulsar glitches in 1969 till 2005, about 100 of them have been observed from nearly 30 pulsars (§ 1.4.4). We will restrict ourselves to "giant glitches" (macroglitches) with $\Delta\Omega/\Omega \gtrsim 10^{-6}$, which demonstrate largest acceleration of pulsar rotation and are, therefore, useful for studying the neutron star structure. The largest number of macroglitches has been detected from the Vela pulsar (twelve in the period from 1969 to 2000). One Vela's macroglitch implies approximately one extra pulse (pulsar rotation) per day.

Observations of macroglitches have been used by numerous authors to test models of neutron star interior and to derive constraints on stellar parameters (see, e.g., Alpar *et al.* 1984; Datta & Alpar 1993; Alpar *et al.* 1993; Abney *et al.* 1996; Link *et al.* 1999, and references therein). In what follows, we will use the approach of Link *et al.* (1999), who proposed a general scheme for constraining the parameters of the Vela pulsar from a minimal set of assumptions. Let $\Delta\Omega_i$ be the angular frequency jump in an $i$th glitch, and $\overline{\Omega}$ be the average value of $\Omega$ over the entire period of $t = 31$ years. The cumulated angular momentum imparted to the rigidly rotating component during a sequence of glitches is

$$J_{\rm glitch} = I_{\rm c} \sum_i \Delta\Omega_i \; . \tag{9.43}$$

As one sees from Fig. 9.11, to a good approximation, this quantity increases linearly with time. Let us start counting $J_{\rm glitch}$ at some time moment

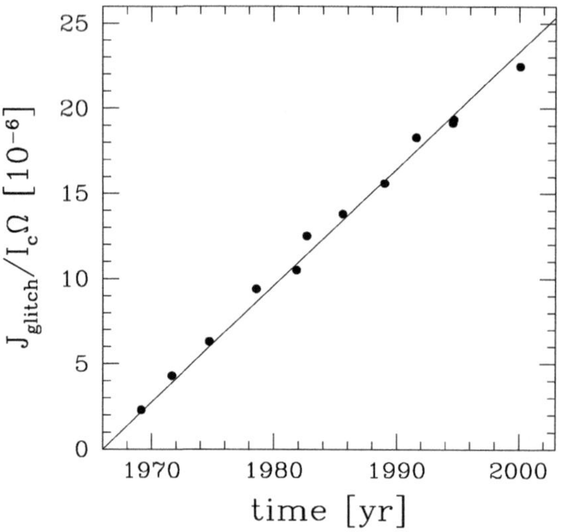


*Figure 9.11.* Time evolution of the cumulated angular momentum transferred during glitches to the strongly coupled crust-core component of the Vela pulsar. Filled circles correspond to observed individual glitches (Table 9.9). The straight line is the linear fit, Eq. (9.44).

$t_0$, preceding the first observed glitch. Then, the angular momentum transferred to the rigidly rotating component up to a moment $t$ is

$$J_{\rm glitch}^{\rm obs}(t, t_0) \simeq (t - t_0)\, I_{\rm c}\overline{\Omega} A \; . \tag{9.44}$$

An average time separation between successive glitches is a few years. Putting $t_0$ on January 1, 1966, three years before the first observed glitch, we obtain a good linear fit, with $A = 0.685 \times 10^{-6}\ {\rm yr}^{-1}$ (Fig. 9.11).

The crustal neutron superfluid component, weakly coupled to the crust, represents a reservoir of the angular momentum, $J_{\rm res}$, accumulated during interglitch periods. Let $I_{\rm res} = I - I_{\rm c}$ be the moment of inertia of this component. The angular momentum excess accumulated in this component during a time interval $t - t_0$ is bounded from above,

$$J_{\rm res}(t, t_0) \leq (t - t_0)\, I_{\rm res} |\dot{\Omega}| \; , \tag{9.45}$$

because crustal neutron superfluid spins down between glitches. Here, $\dot{\Omega}$ is the average (smooth) Vela's spin-down rate, after eliminating short term effects produced by glitches (see Fig. 6.6 of Lyne & Graham-Smith 1998). The actual average rate of the angular momentum transfer can be calculated using Eq. (9.44),

$$\dot{J}_{\rm res} = I_{\rm c}\overline{\Omega} A \; . \tag{9.46}$$

Therefore, Eq. (9.45) gives the inequality

$$(I_{\rm res}/I_{\rm c}) \geq (\overline{\Omega}/|\dot{\Omega}|)\, A\,. \tag{9.47}$$

For the Vela pulsar we have $\overline{\Omega}/|\dot{\Omega}| = 2.26 \times 10^4$ yr (see, e.g., Lyne & Graham-Smith 1998) so that $I_{\rm res}/I_{\rm c} \geq 1.5\%$. The bound on $I_{\rm res}/I_{\rm c}$ can be translated into a lower bound on $I_{\rm crust}/I$. Under our assumptions, $I_{\rm res} < I_{\rm crust}$ and $I_{\rm c} > I - I_{\rm crust}$. This gives the constraint for the Vela pulsar in the two-component glitch model, $I_{\rm crust}/(I - I_{\rm crust}) \simeq I_{\rm crust}/I > 1.5\%$. Although this constraint is based on general assumptions, it is not strong. For a $1.4\ M_\odot$ neutron star, it is satisfied even for the softest EOS in Table 5.3.

Let us mention that much stronger constraints on $I_{\rm crust}/I$ for the Vela pulsar were derived in the past. Analyzing first eight Vela glitches, Alpar *et al.* (1993) obtained $I_{\rm crust}/I > 2.4\%$. Even higher lower bound was given by Datta & Alpar (1993), $I_{\rm crust}/I > 3.4\%$. Such constraints would rule out the $1.4\ M_\odot$ neutron star models based on soft and even medium stiff EOSs. However, they implied a specific model of *individual* Vela macroglitches (even in the "minimal" phenomenological model of Alpar *et al.* 1993). Thus, these stronger lower bounds are valid only if underlying models are *quantitatively* correct, a very strong requirement for so complex phenomenon as a neutron star glitch.

The bounds on $I_{\rm crust}/I$ for several other glitching pulsars (PSR B1737–30, PSR B1758–23, PSR B1338–62) are less stringent and precise, but generally similar to the Vela's one. The bound for the Crab pulsar is too low to be interesting. According to Link *et al.* (1999) this may indicate that the Crab pulsar loses a significant amount of the angular momentum contained in crustal neutron superfluid between glitches, for instance, via thermal creep of superfluid vortices (Alpar *et al.*, 1984; Link *et al.*, 1999; Chau & Cheng, 1993). The Crab pulsar is much younger (and can be significantly hotter) than other pulsars considered in this section. Higher internal temperature could greatly accelerate thermal creep of neutron vortices in the crust.

## 9.8. Pros and cons of the existence of strange stars

If the strange matter hypothesis is true and the strange matter exists in the Universe, then some (or maybe all) neutron stars could actually be strange stars. Ever since the idea of strange quark stars has been advanced, a lively debate concerning its validity has taken place. Below we discuss some pros and cons of the existence of strange stars.

### 9.8.1 Con: Macroglitches of radio pulsars

It is widely believed that pulsar glitches require the presence of the solid crust and a superfluid component weakly bound to the crust. The analysis of macroglitches indicates (§ 9.7) that the fractional crustal moment of inertia exceeds $10^{-2}$. However, as we have seen in § 8.16, the maximum fractional crustal moment of inertia of strange stars with $M > 0.5\ M_{\odot}$ is lower than $10^{-4}$.[2] Therefore, it was stated long ago, that strange stars with the crust cannot explain pulsar macroglitches; glitching pulsars are not strange stars (Alcock *et al.*, 1986; Alpar, 1987). Of course, it was obvious from the very beginning, that the standard two-component model of pulsar glitches (§ 9.7), in which superfluid neutrons in the inner crust transfer an accumulated excess of angular momentum to the rest of the pulsar body, does not apply to strange stars. As noted by Alpar (1987), a large value of $\Delta\dot{\Omega}/|\dot{\Omega}| \sim 10^{-3} - 10^{-2}$ in glitches requires the existence of a strange star component distinct from the quark matter. The effective moment of inertia $I_{\rm d}$ of this component should satisfy $I_{\rm d}/I \approx \Delta\dot{\Omega}/|\dot{\Omega}|$. However, the moment of inertia of the strange star crust is two orders of magnitude smaller. This statement of Alpar (1987) is based on observational estimates for macroglitches, $\Delta\Omega/\Omega \sim 10^{-6} - 10^{-5} \ll \Delta\dot{\Omega}/|\dot{\Omega}|$. According to Alpar, such estimates indicate that glitches are produced by the angular momentum transfer from the non-quark matter to the rest of the star coupled to the magnetosphere. In his model macroglitches *are not* associated with sudden changes of the moment of inertia.

The lower bound on $I_{\rm crust}/I$ might be weakened, if one uses the arguments based exclusively on conservation of the total angular momentum $J$ during a glitch (Glendenning & Weber, 1992). A tacit assumption is that the angular momentum transferred during a glitch comes from a sudden decrease of the moment of inertia of the crust in a so called crust-quake, $\Delta I_{\rm crust} = -f I_{\rm crust}$, where $f$ is the fraction of the decreasing moment of inertia. Notice that this assumption is just the opposite to that made by Alpar (1987). Retaining only terms linear in $\Delta\Omega$, we see that conservation of the total angular momentum implies $\Delta J = \Delta\Omega I_0 - \Omega f I_{\rm crust} = 0$, where $I_0$ is the moment of inertia of the strange star component accelerated during the quake. This results in $f I_{\rm crust}/I_0 \sim \Delta\Omega/\Omega \sim 10^{-6}$ for macroglitches, while from the strange star models $I_{\rm crust} \sim 10^{-5}\ I$. All matter constituents (quarks, electrons, as well as nuclei in the crust) are charged and rather strongly coupled by Coulomb forces, so that $I_0 \simeq I$. Therefore, one should require $f \sim 0.01 - 0.1$ for macroglitches. In view of the low shear modulus (§ 3.7) and the associated low critical elastic strain of the outer crust, such an amplitude of crust quakes seems unrealistically high.

[2]Strictly speaking, this is true for strange stars with $M > 0.5\ M_{\odot}$ and spin periods longer than 1 ms. We do not consider too low-mass strange stars rotating close to the Keplerian limit, where $I_{\rm crust}/I$ can be as high as $10^{-3}$ (Glendenning & Weber, 1992; Zdunik *et al.*, 2001).

### 9.8.2 Con: Strangelets in galaxies

General astronomical arguments against strange matter were presented by Madsen (1988) and Caldwell & Friedman (1991). They argued that if some objects built of strange matter are present in a galaxy (e.g., in our Galaxy) in the form of strange stars or as nuggets of strange matter (*strangelets*) in cosmic rays, then essentially all neutron stars must be strange stars. The argument is based on the dramatic effects of a tiny abundance of strangelets. It was suggested, that even if strangelets had not survived as a leftover of hadronization process in the Big Bang, they could be produced by tidal disruption of strange stars in collisions/coalescence with another strange star, neutron star or a black hole. If strangelets are created in this way, the expected abundance of strangelets in the interstellar medium would be much higher than the minimum value required to convert all neutron stars into strange stars. Hence, all neutron stars would actually be strange stars. But, as we argued in § 9.8.1, pulsars exhibiting macroglitches *are not* strange stars, and therefore the hypothesis that all neutron stars are strange stars seems to be in conflict with reality.

However, it is far from obvious, that the fragmentation of a strange star will necessarily produce a swarm of strangelets. Let us remind that, in contrast to a neutron star, fragments of a strange star will not explode after the decompression. A fission of strange stars into a few fragments of comparable mass may be more natural. A coalescence of a strange star with a black hole in a close binary has been numerically simulated by Kluźniak & Lee (2003). They find the striking difference as compared to the merging of a neutron star and a black hole. In the latter case, the neutron star is disrupted by the tidal forces created by the black hole. The fragments form a disk around the black hole. A fraction of matter is engulfed by the black hole, while the other fraction is ejected from the binary. The fragments of the strange star do not explode. According to Kluźniak & Lee, they are totally engulfed by the black hole; no strangelets are injected into the Galaxy during the merging of a strange star and a black hole. This would indicate that the pollution of the Galaxy by strangelets is not an unavoidable consequence of the existence of strange stars, contrary to the assumption of Madsen (1988) and Caldwell & Friedman (1991).

Let us summarize. If strangelets, produced during the hadronization phase in the expanding Universe after the Big Bang, do not evaporate completely into nucleons, then the Galactic flux of strangelets would be too large to allow for ordinary glitching neutron stars to exist. If strangelets could survive (e.g., due to an additional binding resulting form the CFL superconductivity, § 8.8.3), then observations of macroglitches in some pulsars provide an argument against the strange matter hypothesis. If, on the contrary, "primordial strangelets" evaporated before the Universe cooled down, then the existence of strange stars did not necessarily led to the pollution of the Galaxy by strangelets. In this case strange stars might coexist with ordinary neutron stars which can exhibit macroglitches.

### 9.8.3 Pro: Too small radii for some neutron stars

Numerous authors used the mass-radius relation to suggest that at least some compact X-ray sources, observed as X-ray bursters or sources of X-ray millisecond quasiperiodic oscillations (QPOs), are strange stars rather than neutron stars (Li *et al.* 1999a,b; see also Bombaci 1997 and review articles of Bombaci 2001 and Weber 2005). Using a phenomenological model, which reproduces the main observational properties of the transient X-ray source SAX J1808.4–3658, Li *et al.* (1999a) argued that the source cannot be a neutron star built of baryonic matter. Specifically, the radius was too small for a standard neutron star. For instance, the authors got the radius $R < 8$ km for $M = 1.3\ M_\odot$. Such a small radius could be obtained for some models of strange stars, and the authors concluded that "SAX J1808.4–3658 is a likely strange star candidate".

Li *et al.* (1999b) analyzed observations of the compact object in the low-mass X-ray binary 4U 1728–34 which is a source of kilohertz QPOs. The X-ray data were obtained with the Rossi X-ray Timing Explorer and were analyzed using a model for kilohertz QPOs proposed by Osherovich & Titarchuk (1999). The deduced constraints on mass and radius hinted towards strange stars: $M < 1.1\ M_\odot$ and $R < 9$ km. However, this conclusion was far from being a clear-cut one. First, the obtained "observational" mass-radius range was obviously model-dependent, with a complex underlying model of the X-ray source. In such a case, one should carefully evaluate the uncertainties of extracted stellar parameters. Second, the constraint on the mass-radius relation was severe even for strange stars. The obtained value $M_{\rm max} \simeq M_\odot$ for strange stars was uncomfortably low, inconsistent with the $1.44\ M_\odot$ measured mass of the Hulse-Taylor pulsar, Eq. (9.24).[3]

The present observations imply that we are far from a clear-cut situation of, e.g., $R_\infty < 9$ km (Fig. 8.11) which would *unambiguously* indicate a strange star. Such a situation seemed to become true on April 10, 2002, when a NASA News Release appeared together with an e-print in `arXiv/astro-ph` (Drake *et al.*, 2002). It was stated that the isolated nearby neutron star RX J1856–3754 radiates like a black body of $T_{\rm s}^\infty = 7 \times 10^5$ K, shows no sign of pulsations (which would indicate a surface temperature anisotropy and stellar rotation), and

[3]The conflict might be resolved by postulating the existence of different families of compact stars with the different formation scenarios and different values of $M_{\rm max}$.

has an apparent radius of $(3.8–8.2)$ km.[4] However, as was shown in subsequent papers by other authors, Drake *et al.* (2002) and their sensational result referred only to the "hard" X-ray component of the spectrum. Actually, the spectrum contains also the "soft" component in the visual and UV bands (Walter & Lattimer 2002; Braje & Romani 2002; Burwitz *et al.* 2003; these results are discussed in § 9.3). The soft component can be modeled by an isotropic blackbody radiation with $T^{\infty}_{\rm s,soft} \simeq 3.5 \times 10^5$ K and $R^{\rm soft}_{\infty} \simeq 16$ km; the latter value is typical for an ordinary neutron star with a stiff EOS.

The spectrum of RX J1856–3754 is still a puzzle, especially because of the absence of any pulsations. The lack pulsations in the given case might indicate a very special orientation of the symmetry axis, strictly coincident with the line of sight.

### 9.8.4 Pro: Submillisecond pulsars

Let us consider the simplest MIT bag model of strange quark matter, described in § 8.5. As we have seen in § 8.21.3, the minimum stable rotation period of strange stars is then given by a remarkably precise formula $P_{\rm min} \simeq 0.63\ (\mathcal{B}_{60})^{-1/2}$ ms , where $\mathcal{B}_{60}$ is the bag constant $\mathcal{B}$ measured in $60$ MeV fm$^{-3}$. Thus, for sufficiently large values of $\mathcal{B}$, massive strange stars (contrary to baryonic neutron stars, § 6.12.6) could be compact enough to sustain rotation at submillisecond periods. After the announcement of the discovery of the 0.5 ms pulsar in the SN 1987A in January 1989 (§ 9.4), it was suggested to treat that pulsar as a strange star (Glendenning, 1989a). However, this would require a "tuning" of the MIT bag model parameters. For example, it is easy to see, that the simplest MIT bag model with massless, non-interacting quarks at the required value of $\mathcal{B}$ would not give the three-flavor quark matter favored energetically over $^{56}$Fe. A similar difficulty arises for more realistic models of strange quark matter (Zdunik & Haensel, 1990; Prakash *et al.*, 1990). These problems might be alleviated if one does not wish to treat the stability conditions imposed on the parameters of the MIT bag model as strict and precise (Glendenning, 1989a, 2000).

The scaling formula for $P_{\rm min}$ can also be obtained for other models of strange matter; the scaling with respect to $\mathcal{B}$ can be replaced by the scaling with respect to the parameter $\rho_{\rm s}$ in the linear approximation of the strange matter EOS, $P/c^2 = a(\rho - \rho_{\rm s})$ (§ 8.8). We get then $P_{\rm min} \simeq 0.41\ (10^{15}\ {\rm g\ cm^{-3}}/\rho_{\rm s})^{1/2}$ ms. Therefore, models of a self-bound quark matter with $\rho_{\rm s} \simeq 10^{15}$ g cm$^{-3}$ can sustain half-millisecond spin periods (Dey *et al.*, 1998). However, because

[4]The *NASA News Release 02-082* at `http://www.msfc.nasa.gov/news/` was entitled "Cosmic X-rays reveal evidence for new form of matter" and resulted in headlines in many newspapers and TV news broadcasts.

of the scaling $M_{\rm max} \propto \rho_{\rm s}^{-1/2}$ these models lead to $M_{\rm max} \lesssim 1.4\,M_\odot$, in conflict with the measurements of neutron star masses (§ 9.1).

The discovery of a half-millisecond pulsar was withdrawn after one year of heated debates. As for now, the shortest observed pulsar period of 1.4 ms (Table 9.8) does not require an introduction of strange stars. Were a half-millisecond pulsar be discovered in the future, it would undoubtedly be used as an argument in favor of strange stars.

# Appendix A
# TABLES OF EOSs IN NEUTRON STAR CRUST

In this Appendix we present tables of the EOS for the ground-state matter in the neutron star crust for $\rho > 10^7$ g cm$^{-3}$. At lower densities, the EOS can be influenced by the presence of strong magnetic fields (Chapter 4) and by the thermal effects (Chapter 2). The analytical description of properties of atomic nuclei in neutron star crusts is given in the Appendix B, and the analytic parameterization of the EOS is presented in the Appendix C.

**The outer crust.** Up to $\rho \simeq 10^{11}$ g cm$^{-3}$, the EOS in the outer crust is determined by experimental masses of neutron rich nuclei. This fact was exploited by Haensel & Pichon (1994, referred to as HP), who made maximal use of the experimental data. At higher densities, they used the semiempirical massformula of Möller (1992).[1] The HP EOS is given in Table A.1 for the same pressure grid as in the tabulated EOS of Baym *et al.* (1971b, referred to as BPS), except for the last line, which corresponds to the neutron drip point. The EOS in Table A.1 is very similar to the BPS one; typical differences in density at the same pressure do not exceed a few percent. One can refine the EOS by introducing density discontinuities which accompany changes of nuclides. It can be done by inserting additional pairs of lines $(n_i, \rho_i, P_i)$, $(n_{i+1}, \rho_{i+1}, P_i)$, corresponding to the density jumps given in Table 3.1. This would complicate the integration of the equations of hydrostatic equilibrium while constructing neutron star models. On the other hand, these weak first-order phase transitions soften the EOS. The softening is well reproduced by interpolation between tabulated points in Table A.1, which leads to a smoothed EOS, easy to use in calculations. The analytical representation of such a smoothed EOS is given in the Appendix C.

**The inner crust.** Out of several existing EOSs of the inner crust, we selected a recent SLy model of Douchin & Haensel (2001). This EOS is given in Table A.2. For a better presentation of this EOS in the vicinity of neutron drip density $\rho_{\rm ND}$ and the crust-core interface, $\rho = \rho_{\rm cc}$, Table A.2 is somewhat extended, using the SLy model to $\rho < \rho_{\rm ND}$ and $\rho > \rho_{\rm cc}$.

The EOSs in Tables A.1 and A.2 are based on different dense-matter models and give different neutron drip points. Because $\rho_{\rm ND}({\rm SLy}) < \rho_{\rm ND}({\rm HP})$, we recommend to use $\rho_{\rm ND} = \rho_{\rm ND}({\rm SLy})$. At the first glance, the SLy EOS for $\rho < \rho_{\rm ND}({\rm SLy})$ nearly coincides with the HP one. Let us remind, however, that the SLy EOS (as well as the FPS one) has been calculated within the Compressible Liquid Drop Model, with no shell or pairing effects. Therefore, the density there is a smooth function of the pressure, except for a narrow vicinity of the neutron drip point and the crust-core transition. Except for these two narrow regions, it can be nicely fitted by analytical functions, as described in the Appendix C.

[1] After this book was almost completed, new EOSs of the outer crust appeared (Rüster *et al.*, 2006), based on different nuclear models and up-to-date experimental data for very neutron rich nuclei. There are some model-dependent differences in $(Z, A)$ at higher densities, but the smoothed EOSs of Rüster *et al.* (2006) and HP are very similar.

*Table A.1.* The EOS of the outer crust derived by Haensel & Pichon (1994). The last line with a nucleus observed in laboratory and present in the ground state of dense matter, as well as the line corresponding to the neutron drip point, are printed in boldface.

| $\rho$ (g cm$^{-3}$) | $P$ (dyn cm$^{-2}$) | $n_b$ (cm$^{-3}$) | $\rho$ (g cm$^{-3}$) | $P$ (dyn cm$^{-2}$) | $n_b$ (cm$^{-3}$) |
|---|---|---|---|---|---|
| 3.303E7 | 3.833E24 | 1.991E31 | 2.091E10 | 1.938E28 | 1.257E34 |
| 6.592E7 | 1.006E25 | 3.973E31 | 2.533E10 | 2.503E28 | 1.522E34 |
| 1.315E8 | 2.604E25 | 7.926E31 | 3.315E10 | 3.404E28 | 1.991E34 |
| 2.625E8 | 6.676E25 | 1.581E32 | 4.174E10 | 4.628E28 | 2.507E34 |
| 3.305E8 | 8.738E25 | 1.991E32 | 5.039E10 | 5.949E28 | 3.025E34 |
| 5.239E8 | 1.629E26 | 3.156E32 | 6.619E10 | 8.089E28 | 3.973E34 |
| 8.303E8 | 3.029E26 | 5.001E32 | 8.337E10 | 1.100E29 | 5.002E34 |
| 1.0455E9 | 4.129E26 | 6.296E32 | **9.631E10** | **1.450E29** | **5.777E34** |
| 1.212E9 | 5.036E26 | 7.299E32 | 1.091E11 | 1.495E29 | 6.545E34 |
| 1.606E9 | 6.860E26 | 9.667E32 | 1.4155E11 | 2.033E29 | 8.485E34 |
| 2.545E9 | 1.272E27 | 1.532E33 | 1.701E11 | 2.597E29 | 1.0195E35 |
| 4.166E9 | 2.356E27 | 2.507E33 | 2.096E11 | 3.290E29 | 1.256E35 |
| 6.606E9 | 4.362E27 | 3.974E33 | 2.730E11 | 4.473E29 | 1.635E35 |
| 8.031E9 | 5.662E27 | 4.830E33 | 3.325E11 | 5.816E29 | 1.990E35 |
| 1.011E10 | 7.702E27 | 6.081E33 | 4.188E11 | 7.538E29 | 2.506E35 |
| 1.319E10 | 1.048E28 | 7.930E33 | 4.299E11 | 7.805E29 | 2.572E35 |
| 1.661E10 | 1.425E28 | 9.982E33 | **4.321E11** | **7.857E29** | **2.585E35** |

*Table A.2.* The SLy EOS of the ground state of the inner crust, together with the adjacent segments of the SLy EOS of the outer crust and the core (calculated by Douchin & Haensel, 2001). The first and last lines corresponding to the inner crust are printed in boldface.

| $n_b$ ($cm^{-3}$) | $\rho$ ($g\ cm^{-3}$) | $P$ ($dyn\ cm^{-2}$) | $n_b$ ($cm^{-3}$) | $\rho$ ($g\ cm^{-3}$) | $P$ ($dyn\ cm^{-2}$) |
|---|---|---|---|---|---|
| 1.7590E35 | 2.9398E11 | 5.0926E29 | 7.6609E35 | 1.2831E12 | 1.3370E30 |
| 1.8297E35 | 3.0582E11 | 5.3344E29 | 1.2616E36 | 2.1141E12 | 2.1547E30 |
| 1.9024E35 | 3.1800E11 | 5.5843E29 | 1.8947E36 | 3.1766E12 | 3.4272E30 |
| 1.9772E35 | 3.3052E11 | 5.8426E29 | 2.6726E36 | 4.4827E12 | 5.2679E30 |
| 2.0540E35 | 3.4338E11 | 6.1094E29 | 3.6062E36 | 6.0511E12 | 7.7976E30 |
| 2.0791E35 | 3.4759E11 | 6.1968E29 | 4.7097E36 | 7.9058E12 | 1.1147E31 |
| 2.0823E35 | 3.4810E11 | 6.2078E29 | 7.4963E36 | 1.2593E13 | 2.0894E31 |
| **2.0905E35** | **3.4951E11** | **6.2150E29** | 1.1197E37 | 1.8824E13 | 3.5841E31 |
| 2.1604E35 | 3.6121E11 | 6.3573E29 | 1.5999E37 | 2.6920E13 | 5.7611E31 |
| 2.2306E35 | 3.7296E11 | 6.4675E29 | 2.2073E37 | 3.7170E13 | 8.8117E31 |
| 2.3114E35 | 3.8650E11 | 6.5813E29 | 2.9477E37 | 4.9677E13 | 1.2947E32 |
| 2.4014E35 | 4.0158E11 | 6.6998E29 | 4.2684E37 | 7.2017E13 | 2.1620E32 |
| 2.4997E35 | 4.1805E11 | 6.8228E29 | 6.2200E37 | 1.0509E14 | 3.8475E32 |
| 2.6426E35 | 4.4199E11 | 6.9945E29 | 7.3174E37 | 1.2372E14 | 5.0462E32 |
| 3.0533E35 | 5.1080E11 | 7.4685E29 | **7.5959E37** | **1.2845E14** | **5.3711E32** |
| 3.5331E35 | 5.9119E11 | 8.0149E29 | 7.7100E37 | 1.3038E14 | 5.3739E32 |
| 4.0764E35 | 6.8224E11 | 8.6444E29 | 9.7100E37 | 1.6441E14 | 9.2059E32 |
| 4.6800E35 | 7.8339E11 | 9.3667E29 | 1.1710E38 | 1.9854E14 | 1.5028E33 |
| 5.3414E35 | 8.9426E11 | 1.0191E30 | 1.3710E38 | 2.3281E14 | 2.3136E33 |
| 6.0594E35 | 1.0146E12 | 1.1128E30 | 1.5710E38 | 2.6722E14 | 3.4072E33 |

# Appendix B
# ANALYTICAL MODELS OF NUCLEAR DENSITY PROFILES

Here we present analytical formulae which fit the results of calculations of microscopic density profiles of neutrons and protons, associated with nuclear structures in the ground state of a neutron star crust. These formulae were derived and used by Kaminker *et al.* (1999) (although not published there) and elaborated further. They are useful, for instance, for calculating neutrino emission or electron transport properties of crustal matter which depend on proton charge distribution within atomic nuclei (e.g., Kaminker *et al.* 1999; Gnedin *et al.* 2001). We can warn the reader that the formulae are not meant to be used for constructing any EOS of dense matter.

The higher the density in the neutron star crust, the more important finite nuclear sizes. Near the crust bottom, the shape of the nuclei may change from roughly spherical to cylindrical or plane-parallel (Chapter 3). Let us consider local density profiles of neutrons and protons within the nuclei, $n_n(r)$ and $n_p(r)$, where $r$ is the distance either from the center of a spherical nucleus, or from the symmetry axis of a cylindrical nuclei, or from the symmetry plane of a slablike nucleus. Let $n_j^{\rm out}$ be the number densities of neutrons or protons (for $j = n$ or $p$) outside the nucleus. The mean-square radii of the neutron and proton distributions are

$$\overline{r_j^2} = \frac{\int_0^{r_{\rm c}} (n_j(r) - n_j^{\rm out}) r^{d+1}\, {\rm d}r}{\int_0^{r_{\rm c}} (n_j(r) - n_j^{\rm out}) r^{d-1}\, {\rm d}r}, \tag{B.1}$$

where $r_{\rm c}$ is the Wigner-Seitz cell radius (Chapter 3), and $d$ is the space dimension of the nuclear phase ($d = 3, 2$, and 1 for spherical, cylindrical, and plane nuclei, respectively). Let us remind that for simple three-dimensional crystals and liquids, $r_{\rm c}$ equals the ion-sphere radius $a_{\rm i}$.

## B.1. Steplike profile model

Far from the bottom of the inner crust, at densities much below the nuclear saturation density, a steplike approximation of the nucleon density profile may be good,

$$n_j(r) = \begin{cases} n_j^{\rm in} & \text{at} \quad r < r_j, \\ n_j^{\rm out} & \text{at} \quad r \geq r_j, \end{cases} \tag{B.2}$$

where $n_j^{\rm in}$ are the number densities of neutrons or protons (for $j = n$ or $p$) inside a nucleus, and $r_n$ and $r_p$ are the radii of neutron and proton distributions, respectively. In the case of spheres, the form factor for the steplike profile $n_p(r)$ is

$$F_q = \frac{3}{(qr_p)^3} \left[\sin(qr_p) - qr_p \cos(qr_p)\right]. \tag{B.3}$$

The mean-square radius (for any $d$) equals

$$\overline{r_j^2} = \frac{d}{d+2}\, r_j^2. \tag{B.4}$$

In the outer crust (at $\rho \lesssim \rho_{\rm ND}$), the radii of the nuclei are unaffected by the pressure of ambient medium and one can use the standard formula $r_p = 1.15\, A^{1/3}$ fm (e.g., Pethick & Ravenhall 1995). At higher densities, the values of $r_p$ obtained numerically (Negele & Vautherin, 1973) can be approximated as $r_p = 1.83\, Z^{1/3}$ fm (Itoh & Kohyama, 1983).

The importance of the nuclear size effects depends on the fraction of volume the nuclei occupy. A relevant parameter is the ratio of the proton core radius to the Wigner-Seitz cell radius,

$$x_{\rm nuc} = r_p / r_{\rm c}. \tag{B.5}$$

For spherical nuclei, we have

$$x_{\rm nuc} = \begin{cases} 0.00155\,(A/Z)^{1/3}\,x_{\rm r} & \text{at} \quad \rho < \rho_{\rm ND}, \\ 0.00247\,x_{\rm r} & \text{at} \quad \rho > \rho_{\rm ND}, \end{cases} \tag{B.6}$$

where $x_{\rm r}$ is the relativity parameter defined by Eq. (2.2).

## B.2. Smooth Composition Model

At $\rho \gtrsim 10^{13}$ g cm$^{-3}$ the nucleon density profiles deviate significantly from the steplike distribution (Chapter 3). Oyamatsu (1993) calculated the local neutron and proton number density distributions within a Wigner-Seitz cell and fitted them in the form

$$n_j(r) = \begin{cases} (n_j^{\rm cen} - n_j^{\rm out})\left[1 - (r/R_j)^{t_j}\right]^3 + n_j^{\rm out} & \text{at} \quad r < R_j, \\ n_j^{\rm out} & \text{at} \quad r \geq R_j, \end{cases} \tag{B.7}$$

where $n_j^{\rm cen}$, $n_j^{\rm out}$, $t_j$, and $R_j$ are the fit parameters. The parameter $t_j$ controls the sharpness of the local density profile, while $R_j$ determines the size of a nucleus. These parameters, as well as the sizes of Wigner-Seitz cells, are presented in Table 6 of Oyamatsu (1993) for several values of the mean baryon number density $n_{\rm b}$ (for spherical and nonspherical nuclei). With increasing $n_{\rm b}$ the profiles become smoother, approaching the limit of uniform matter; therefore, the parameters $t_j$ decrease.

Real local density distributions of neutrons and protons are not cut off at a certain distance from the center. Therefore, $R_n$ and $R_p$ can be treated only as convenient fit parameters. Near the bottom of the crust, the local density distribution is rather smooth, and the boundary between the free neutrons and the neutrons bound within the nuclei becomes rather uncertain. On the other hand, while describing properties of neutron star crust, one often uses such quantities as the radii of neutron and proton distributions ($r_n$ and $r_p$) and the number of nucleons within a nucleus ($A$). To determine them, let us consider a nucleus as a combination of imaginary neutron and proton spheres of *equivalent radii* $r_n$ and $r_p$ and *equivalent neutron and proton densities* $n_n^{\rm in}$ and $n_p^{\rm in}$. We define the equivalent radius $r_j$ as the radius of the imaginary steplike density distribution that reproduces the mean-square radii (B.4) of the real distribution. In this case $r_j = [(1+2/d)\overline{r_j^2}]^{1/2}$. For the distribution (B.7), the equivalent radii become

$$\begin{aligned} r_j \;=\; & \left[1 - 3\,\frac{d+2}{d+2+t_j}\,\gamma_j^{t_j} + 3\,\frac{d+2}{d+2+2t_j}\,\gamma_j^{2t_j} - \frac{d+2}{d+2+3t_j}\,\gamma_j^{3t_j}\right]^{1/2} \\ & \times \left[1 - \frac{3d}{d+t_j}\,\gamma_j^{t_j} + \frac{3d}{d+2t_j}\,\gamma_j^{2t_j} - \frac{d}{d+3t_j}\,\gamma_j^{3t_j}\right]^{-1/2} \gamma_j R_j, \end{aligned} \tag{B.8}$$

where $\gamma_j = r_j^{\rm max}/R_j$ and $r_j^{\rm max} = \min(R_j, r_{\rm c})$. The parameter $\gamma_j$ equals 1 (and can be dropped out) in all the cases except for the case where the fit parameter $R_j$ exceeds the Wigner-Seitz cell radius $r_c$, that may happen near the crust-core boundary. In the latter case, the local density profile (B.7) must be truncated at $r = r_c$, and we get $\gamma_j < 1$. Our model satisfies the natural requirement that the equivalent radii $r_j$ are smaller than $r_c$.

We define the equivalent neutron and proton densities within the nucleus in such a way to reproduce the total number of nucleons (protons or neutrons) implied by Eq. (B.7):

$$n_j^{\rm in} = n_j^{\rm out} + \frac{d}{r_j^d} \int_0^{r_j^{\rm max}} (n(r) - n_j^{\rm out})\, r^{d-1}\, {\rm d}r$$

$$= n_j^{\rm out} + \left[1 - \frac{3d}{d+t_j}\gamma_j^{t_j} + \frac{3d}{d+2t_j}\gamma_j^{2t_j} - \frac{d}{d+3t_j}\gamma_j^{3t_j}\right] \frac{R_j^d}{r_j^d}\,(n_j^{\rm cen} - n_j^{\rm out}). \quad \text{(B.9)}$$

The steplike profile of the previous section is recovered from Eqs. (B.7)–(B.9) in the limit of $t_j \to \infty$.

## B.2.1 Spherical nuclei in the inner crust

Oyamatsu (1993) presented the fit parameters for spherical nuclei in the ground-state matter at three values of the mean baryon number density $n_{\rm b}$= 0.01, 0.03, and 0.055 fm$^{-3}$ (i.e., $\rho = 1.66 \times 10^{13}$, $4.98 \times 10^{13}$, and $9.13 \times 10^{13}$ g cm$^{-3}$) in the inner crust of a neutron star. These parameters are quite consistent with those presented by Negele & Vautherin (1973) for nearly the same $n_{\rm b}$. Some of these parameters can also be deduced from Figs. 3 and 4 and from Table 3 of Negele & Vautherin (1973) for several other values of $n_{\rm b}$ in the inner crust. The parameters appear to be smooth functions of $n_{\rm b}$, and we interpolated them between the given points at $\rho_{\rm ND} \leq \rho \leq 1.4 \times 10^{14}$ g cm$^{-3}$. We present $r_c$ and the parameters of Eq. (B.7) as functions of the dimensionless argument

$$\nu = \ln(n_{\rm b} \times 100 \text{ fm}^3).$$

The interpolation reads

$$r_c = (31.68 - 8.4\,\nu - 0.238\,\nu^2 + 0.1152\,\nu^3)\ \text{fm}, \quad \text{(B.10a)}$$

$$t_n = (0.2027 + 0.004\,506\,{\rm e}^{\nu})^{-1}, \quad \text{(B.10b)}$$

$$R_n = (9.406 + 1.481\,\nu + 0.4625\,\nu^2 + 0.057\,38\,\nu^3)\ \text{fm}, \quad \text{(B.10c)}$$

$$n_n^{\rm cen} - n_n^{\rm out} = (0.097\,61 - 0.013\,22\,\nu - 0.005\,544\,\nu^2 - 7.624 \times 10^{-4}\,\nu^3)\ \text{fm}^{-3}, \quad \text{(B.10d)}$$

$$t_p = (0.1558 + 0.002\,225\,\nu + 9.452 \times 10^{-4}\,\nu^2)^{-1}, \quad \text{(B.10e)}$$

$$R_p = (8.345 + 0.7767\,\nu + 0.1333\,\nu^2 + 0.008\,707\,\nu^3)\ \text{fm}, \quad \text{(B.10f)}$$

$$n_p^{\rm cen} = (0.0404 - 0.010\,97\,\nu - 7.23 \times 10^{-4}\,\nu^2 + 2.25 \times 10^{-4}\,\nu^3)\ \text{fm}^{-3}. \quad \text{(B.10g)}$$

This interpolation smears out the jumps in the nuclear composition with increasing $\rho$, but it allows one to calculate the parameters of spherical nuclei at any density in the ground-state matter of the inner crust.

The number of protons ($Z$) and the total number of nucleons ($A$) within a nucleus are most easily found using the equivalent radii and densities defined by Eqs. (B.8) and (B.9) with $d = 3$:

$$Z = \frac{4\pi}{3}\, r_p^3\, n_p^{\rm in}, \quad A = Z + \frac{4\pi}{3}\, r_n^3\, n_n^{\rm in}. \quad \text{(B.11)}$$

Now the parameter $n_n^{\rm out}$ is determined from the relation

$$A' = A + \frac{4\pi}{3}\,(r_c^3 - r_n^3)\, n_n^{\rm out} = \frac{4\pi}{3}\, r_c^3 n_{\rm b}, \quad \text{(B.12)}$$

where $A'$ is the number of nucleons within a Wigner-Seitz cell.

The nucleus mass is assumed to be $m_{\rm i} = Am_n + Zm_p$.

### B.2.2 Spherical nuclei in the outer crust

Making use of the results of Haensel & Pichon (1994), we have also obtained an analytic description of atomic nuclei in the ground-state matter for lower densities,

$$10^8 \text{ g cm}^{-3} \le \rho \le \rho_{\rm ND}.$$

We have adopted the same parameterization (B.7) and constructed analytic expressions for the nuclear parameters versus

$$\nu = \ln[1 + 2\,n_{\rm b}/(10^{-8} \text{ fm}^{-3})].$$

These expressions read:

$$R_n = (5.788 + 0.020\,77\,\nu + 0.014\,89\,\nu^2) \text{ fm}, \tag{B.13a}$$

$$n_n^{\rm cen} = (0.0808 + 1.688 \times 10^{-4}\,\nu + 9.439 \times 10^{-5}\,\nu^2) \text{ fm}^{-3}, \tag{B.13b}$$

$$R_p = 5.688 + 0.026\,28\,\nu + 0.009\,468\,\nu^2 \text{ fm}, \tag{B.13c}$$

$$n_p^{\rm cen} = (0.0738 + 1.22 \times 10^{-4}\,\nu - 1.641 \times 10^{-4}\,\nu^2) \text{ fm}^{-3}, \tag{B.13d}$$

$$n_n^{\rm out} = n_p^{\rm out} = 0, \tag{B.13e}$$

$$t_n = t_p = 6. \tag{B.13f}$$

Equation (B.13f) just formally sets $t_j$ values at $\rho < \rho_{\rm ND}$ close to those at $\rho = \rho_{\rm ND}$. In many applications it is sufficient to use steplike profiles (i.e., $t_j \to \infty$) at $\rho \le \rho_{\rm ND}$.

The free nucleons outside nuclei are absent in this regime, $n_n^{\rm out} = n_p^{\rm out} = 0$, and the Wigner-Seitz radius is

$$r_{\rm c} = \left(\frac{4\pi}{3}\frac{n_{\rm b}}{A}\right)^{-1/3}, \tag{B.14}$$

$A$ and $Z$ being determined by Eq. (B.11).

At low densities in the outer crust, Eqs. (B.13) reproduce the parameters of $^{56}$Fe-nuclei.

### B.2.3 Exotic nuclei

According to model I of Oyamatsu (1993), the phase with spherical nuclei in the inner crust is realized up to a density $n_{\rm b} = 0.0586$ fm$^{-3}$ ($\rho = 0.973 \times 10^{14}$ g cm$^{-3}$). It is followed by the phase with rodlike nuclei up to $n_{\rm b} = 0.0749$ fm$^{-3}$ ($\rho = 1.24 \times 10^{14}$ g cm$^{-3}$) and the phase with slablike nuclei (up to $n_{\rm b} = 0.0827$ fm$^{-3}$, $\rho = 1.37 \times 10^{14}$ g cm$^{-3}$). Subsequently there are two phases with the roles of nuclear matter and neutron matter reversed, the rodlike one (up to $n_{\rm b} = 0.0854$ fm$^{-3}$, $\rho = 1.42 \times 10^{14}$ g cm$^{-3}$), and the "Swiss cheese" (inverted-spheres) one, which is the analog of the phase with spherical nuclei and is the last phase in the neutron star crust (up to $n_{\rm b} = 0.0861$ fm$^{-3}$, $\rho = 1.43 \times 10^{14}$ g cm$^{-3}$). At higher densities the nuclei dissolve into the uniform matter of the neutron star core.

In each crystalline phase of matter the Wigner-Seitz cell has its own geometry, but we assume that in the phases of (body centered cubic) crystals of ordinary or inverted spherical nuclei it may be approximated by a sphere, and in rodlike phases by a right circular cylinder. Let the nucleon density distributions be described by Eq. (B.7). In the phases with rods and slabs, $n_p^{\rm out} = 0$, $n_n^{\rm out}$ describes the number density of free neutrons, and the region $r < R_n$ is occupied by the nucleus itself (with $n_n^{\rm cen} > n_n^{\rm out}$). In the two "bubble" phases with the roles of nuclear matter and neutron matter reversed, $n_p^{\rm out} \neq 0$, and $n_j^{\rm out} > n_j^{\rm cen}$, i.e., the local number density of neutrons and protons increases with increasing distance $r$ from the center of the Wigner-Seitz cell. We interpolate the parameters of Eq. (B.7) as functions of $n_{\rm b}$ within each phase separately. In the rest of this Appendix, we introduce

$$\nu \equiv n_{\rm b} \times \text{fm}^3. \tag{B.15}$$

**Rodlike nuclei.** For the cylindrical nuclei, we have

$$r_{\rm c} = (8.3014 + 764.026\,\nu - 16\,827.2\,\nu^2 + 100\,759\,\nu^3)\ \text{fm}, \tag{B.16a}$$
$$t_n = -0.122\,016 + 163.6626\,\nu - 2\,751.439\,\nu^2 + 15\,238.15\,\nu^3, \tag{B.16b}$$
$$R_n = (-40.383\,47 + 2\,328.248\,\nu - 37\,345.32\,\nu^2 + 207\,924.7\,\nu^3)\ \text{fm}, \tag{B.16c}$$
$$n_n^{\rm cen} = (0.11371 - 0.611\,5979\,\nu + 9.431\,739\,\nu^2 - 63.742\,37\,\nu^3)\ \text{fm}^{-3}, \tag{B.16d}$$
$$n_n^{\rm out} = (-0.019\,130\,93 + 1.706\,435\,\nu - 12.511\,92\,\nu^2 + 59.524\,78\,\nu^3)\ \text{fm}^{-3}, \tag{B.16e}$$
$$t_p = -2.521\,390 + 304.3897\,\nu - 4966.492\,\nu^2 + 25\,571.19\,\nu^3, \tag{B.16f}$$
$$R_p = (-41.773\,61 + 2331.504\,\nu - 37674.33\,\nu^2 + 212\,689.7\,\nu^3)\ \text{fm}, \tag{B.16g}$$
$$n_p^{\rm cen} = (11.371 - 0.611\,5979\,\nu + 9.431\,739\,\nu^2 - 63.742\,37\,\nu^3)\ \text{fm}^{-3}, \tag{B.16h}$$
$$n_p^{\rm out} = 0. \tag{B.16i}$$

The numbers of protons and nucleons ($Z$ and $A$) inside a nucleus, and the total number of nucleons $A'$ *per unit length* of a cylindrical Wigner-Seitz cell equal

$$Z = \pi r_p^2 n_p^{\rm in}, \quad A = Z + \pi r_n^2 n_n^{\rm in}, \quad A' = A + \pi (r_{\rm c}^2 - r_n^2) n_n^{\rm out}, \tag{B.17}$$

where $r_p$, $n_p^{\rm in}$, $r_n$, and $n_n^{\rm in}$ are defined by Eqs. (B.8) and (B.9) with $d = 2$.

**Slablike nuclei.** For the plane-parallel nuclei,

$$r_{\rm c} = (-245.4595 + 11\,168.62\,\nu - 157\,290.7\,\nu^2 + 722\,159\,\nu^3)\ \text{fm}, \tag{B.18a}$$
$$t_n = -267.2904 + 10\,459\,\nu - 135\,445.7\,\nu^2 + 585\,206\,\nu^3, \tag{B.18b}$$
$$R_n = (9\,831.081 - 371\,401.1\,\nu + 4\,675\,343\,\nu^2 - 19\,591\,770\,\nu^3)\ \text{fm}, \tag{B.18c}$$
$$n_n^{\rm cen} = (0.247\,6255 - 6.583\,347\,\nu + 91.630\,22\,\nu^2 - 425.2562\,\nu^3)\ \text{fm}^{-3}, \tag{B.18d}$$
$$n_n^{\rm out} = (-0.805\,9552 + 31.998\,28\,\nu - 401.3776\,\nu^2 + 1\,723.221\,\nu^3)\ \text{fm}^{-3}, \tag{B.18e}$$
$$t_p = 0.002\,725\,985 + 253.6894\,\nu - 5\,499.141\,\nu^2 + 33\,259.03\,\nu^3, \tag{B.18f}$$
$$R_p = (-714.6039 + 28\,584.02\,\nu - 380\,503.3\,\nu^2 + 1\,703\,796\,\nu^3)\ \text{fm}, \tag{B.18g}$$
$$n_p^{\rm cen} = (0.082\,536\,46 - 2.548\,742\,\nu + 31.836\,15\,\nu^2 - 147.7704\,\nu^3)\ \text{fm}^{-3}, \tag{B.18h}$$
$$n_p^{\rm out} = 0. \tag{B.18i}$$

The numbers of protons and nucleons ($Z$ and $A$) inside a nucleus, and the total number of nucleons $A'$ *per unit surface area* of a slablike Wigner-Seitz cell equal

$$Z = 2\,r_p n_p^{\rm in}, \quad A = Z + 2\,r_n n_n^{\rm in}, \quad A' = A + 2\,(r_{\rm c} - r_n)\,n_n^{\rm out}, \tag{B.19}$$

where $r_p$, $n_p^{\rm in}$, $r_n$, and $n_n^{\rm in}$ are defined by Eqs. (B.8) and (B.9) with $d = 1$.

**Cylindrical "bubbles".** For the phase of "inverse cylindrical" nuclei,

$$r_c = (430.966\,65 - 9\,710.2218\,\nu + 56\,422.005\,\nu^2)\ \text{fm}, \tag{B.20a}$$
$$t_n = 64.947\,03 - 1\,034.690\,\nu + 3\,501.129\,\nu^2, \tag{B.20b}$$
$$R_n = (271.654 - 6\,015.092\,\nu + 35\,000.53\,\nu^2)\ \text{fm}, \tag{B.20c}$$
$$n_n^{\text{cen}} = (0.443\,7022 - 9.905\,772\,\nu + 65.004\,63\,\nu^2)\ \text{fm}^{-3}, \tag{B.20d}$$
$$n_n^{\text{out}} = (-0.323\,9546 + 9.926\,548\,\nu - 59.979\,51\,\nu^2)\ \text{fm}^{-3}, \tag{B.20e}$$
$$t_p = 94.08485 - 1\,457.401\,\nu + 4\,499.405\,\nu^2, \tag{B.20f}$$
$$R_p = (441.3152 - 10\,140.12\,\nu + 60\,000.7\,\nu^2)\ \text{fm}, \tag{B.20g}$$
$$n_p^{\text{cen}} = 0, \tag{B.20h}$$
$$n_p^{\text{out}} = (-0.077\,336\,78 + 2.235\,163\,\nu - 15.000\,97\,\nu^2)\ \text{fm}^{-3}. \tag{B.20i}$$

The number of protons outside the "bubbles" and the total number of nucleons ($Z$ and $A'$) *per unit length* of a cylindrical Wigner-Seitz cell equal

$$Z = \pi(r_c^2 n_p^{\text{out}} + r_p^2 n_p^{\text{in}}), \quad A' = Z + \pi r_c^2 n_n^{\text{out}} - \pi r_n^2\,(n_n^{\text{out}} - n_n^{\text{in}}), \tag{B.21}$$

where $r_p$, $n_p^{\text{in}}$, $r_n$, and $n_n^{\text{in}}$ are defined by Eqs. (B.8) and (B.9) with $d = 2$. In this case, $n_p^{\text{in}}$ is negative, which corresponds to a deficit of protons inside the bubble relative to the surrounding medium. Contrary to the case of the ordinary nuclei, the proton core radius $r_p$ is now greater than $r_n$, since the skin of the neutron bubble remains composed of neutrons, as in ordinary nuclei.

**Spherical "bubbles".** For the "inverse spherical" nuclei,

$$r_c = (36.6584 - 248.1623\,\nu)\ \text{fm}, \tag{B.22a}$$
$$t_n = 73.420\,26 - 830.003\,\nu, \tag{B.22b}$$
$$R_n = (10.989\,02 + 50.06982\,\nu)\ \text{fm}, \tag{B.22c}$$
$$n_n^{\text{cen}} = (-0.039\,683\,58 + 1.299\,808\,\nu)\ \text{fm}^{-3}, \tag{B.22d}$$
$$n_n^{\text{out}} = (0.235\,124 - 1.749\,754\,\nu)\ \text{fm}^{-3}, \tag{B.22e}$$
$$t_p = 105.9954 - 1\,204.998\,\nu, \tag{B.22f}$$
$$R_p = (2.004\,501 + 150.005\,8\,\nu)\ \text{fm}, \tag{B.22g}$$
$$n_p^{\text{cen}} = 0, \tag{B.22h}$$
$$n_p^{\text{out}} = (0.033\,66631 - 0.350\,0152\,\nu)\ \text{fm}^{-3}. \tag{B.22i}$$

The number of protons $Z$ outside the "bubbles" and the total number of nucleons $A'$ in a Wigner-Seitz cell equal

$$Z = (4\pi/3)(r_c^3 n_p^{\text{out}} + r_p^3 n_p^{\text{in}}), \quad A' = Z + \frac{4\pi}{3} r_c^3 n_n^{\text{out}} - \frac{4\pi}{3} r_n^3 (n_n^{\text{out}} - n_n^{\text{in}}), \tag{B.23}$$

where $r_p$, $n_p^{\text{in}}$, $r_n$, and $n_n^{\text{in}}$ are defined by Eqs. (B.8) and (B.9) with $d = 3$. As for cylindrical "bubbles", $n_p^{\text{in}}$ is negative and $r_p > r_n$.

Thus, we have a simple analytic description of the neutron and proton local density profiles for the ground-state matter throughout the outer and inner neutron star crusts including nonspherical phases of atomic nuclei. This description is referred to as the *smooth composition model* (SCM) of ground-state matter.

# Appendix C
# ANALYTICAL REPRESENTATIONS OF UNIFIED EOSs

EOSs are usually tabulated, and subsequently interpolated between mesh points in computer codes. Interpolation introduces ambiguities in calculated parameters of neutron star models. Moreover, interpolation should respect exact thermodynamic relations which turned out to be especially serious in high-precision two-dimensional modeling of rapidly spinning neutron stars (Nozawa *et al.*, 1998). In three-dimensional calculations of stationary configurations in a close neutron star binary one needs derivatives of the pressure with respect to the enthalpy; tabulated EOSs become even less useful (see, e.g., Gourgoulhon *et al.* 2001). The problems of using tabulated EOSs are particularly serious for the EOSs constructed by matching different EOS segments (e.g., the crust and the core).

In view of all these problems, it is of great interest to derive analytical representations of EOSs. They introduce no ambiguity of interpolation; the derivatives can be precisely calculated; they can be constructed fulfilling exactly the thermodynamic relations. Here we present, following Haensel & Potekhin (2004), analytical representations of two *unified* EOSs, FPS and SLy (see Chapters 3 and 5).

The outer and inner crusts as well as the inner crust and the core of a neutron star are separated by phase transitions. There may also be phase transitions in the core (Chapter 7) and weak density jumps between layers containing different nuclei in the crust (Chapter 3). These weak jumps in the crust will be neglected; the EOSs we consider do not contain any phase transitions in the core. We will approximate the EOSs by fully analytical functions. However, the different character of the EOS in the outer crust, inner crust and the core is reflected by the complexity of the fit, which consists of several fractional-polynomial parts, matched together by virtue of the function

$$f_0(x) = \frac{1}{\mathrm{e}^x + 1} \,. \tag{C.1}$$

We employ two tabulated unified EOSs, FPS[1] or SLy, at $\rho > 5 \times 10^{10}$ g cm$^{-3}$. At lower densities, $10^8$ g cm$^{-3}$ $\lesssim \rho < 5 \times 10^{10}$ g cm$^{-3}$, the crustal matter is described by the EOS of Haensel & Pichon (1994) (HP94), based on experimental nuclear data. This EOS is supplemented by the BPS EOS for cold catalyzed matter at $\rho \lesssim 10^8$ g cm$^{-3}$(Chapter 3). The lowest-density parts of the tables at $\rho < 10^5$ g cm$^{-3}$ have not been used in the fitting. At such low $\rho$ the EOS is no longer one-parametric, but depends also on temperature (see Fig. 1.3).

[1]The FPS table has been kindly provided by N. Stergioulas.

*Table C.1.* Parameters of the fit (C.2)

| i | $a_i$(FPS) | $a_i$(SLy) | i | $a_i$(FPS) | $a_i$(SLy) |
|---|---|---|---|---|---|
| 1 | 6.22 | 6.22 | 10 | 11.8421 | 11.4950 |
| 2 | 6.121 | 6.121 | 11 | −22.003 | −22.775 |
| 3 | 0.006004 | 0.005925 | 12 | 1.5552 | 1.5707 |
| 4 | 0.16345 | 0.16326 | 13 | 9.3 | 4.3 |
| 5 | 6.50 | 6.48 | 14 | 14.19 | 14.08 |
| 6 | 11.8440 | 11.4971 | 15 | 23.73 | 27.80 |
| 7 | 17.24 | 19.105 | 16 | −1.508 | −1.653 |
| 8 | 1.065 | 0.8938 | 17 | 1.79 | 1.50 |
| 9 | 6.54 | 6.54 | 18 | 15.13 | 14.67 |

## C.1. Representation convenient for non-rotating stars

For non-rotating star, it is instructive to parameterize the pressure as function of density. Let us introduce $\xi = \lg(\rho/\text{g cm}^{-3})$ and $\zeta = \lg(P/\text{dyn cm}^{-2})$. Then the parameterization reads

$$\begin{aligned}\zeta &= \frac{a_1 + a_2\xi + a_3\xi^3}{1 + a_4\,\xi}\, f_0(a_5(\xi - a_6)) + (a_7 + a_8\xi)\, f_0(a_9(a_{10} - \xi)) \\ &\quad + (a_{11} + a_{12}\xi)\, f_0(a_{13}(a_{14} - \xi)) + (a_{15} + a_{16}\xi)\, f_0(a_{17}(a_{18} - \xi))\,. \qquad \text{(C.2)}\end{aligned}$$

The parameters $a_i$ for the FPS and SLy EOSs are given in Table C.1. The typical fit error of $P$ is (1–2)% (for $\xi \gtrsim 5$). The maximum error is associated with the phase transitions available in the original tabulated EOSs but smoothed by the fit (C.2). For the FPS EOS, the maximum error is 3.6% at $\xi = 14.22$ (the crust-core interface). For the SLy EOS, the maximum error is 2.9% at $\xi = 8.42$ (the $^{62}$Ni-$^{64}$Ni phase transition in the HP94 table).

The overall EOS throughout the neutron star is presented in Fig. 1.3 in Chapter 1. The figure shows $\log P$ against $\log\rho$ ($\log \equiv \lg \equiv \log_{10}$) for the tabulated EOSs (symbols) and the corresponding fit (the solid line). Triangles correspond to the BPS data, stars to HP94, and dots to the SLy data. By construction, the fit is accurate at $\rho \gtrsim 10^5$ g cm$^{-3}$. As stated above, at lower $\rho$ the EOS becomes temperature-dependent. This is illustrated by the dashed lines, that show the OPAL EOS of iron plasma (Rogers *et al.* 1996; see §2.4) for $T = 10^6$, $10^7$, and $10^8$ K. However, a reasonable continuation of the fit to lower densities can be constructed by a simple interpolation. For instance, the dotted line in Fig. 1.3 corresponds to $P = 10^{\zeta} + P_0$, where $\zeta$ is given by Eq. (C.2) (and $\xi$ should be positive), and $P_0 = 3.5 \times 10^{14}\,\rho$ approximates the OPAL EOS for $\rho \sim \rho_s$ at $T = 10^7$ K ($P$ is in dyn cm$^{-2}$, $\rho$ in g cm$^{-3}$, and $\rho_s = 7.86$ g cm$^{-3}$ is the lowest density in the BPS table).

In Fig. C.1 we compare the FPS and SLy EOSs. Symbols on the top panel show the data (triangles, stars, dots, and open circles for BPS, HP94, SLy, and FPS, respectively) and lines show the fits (the solid line is for SLy and the dot-dashed line for FPS). In order to make the differences between the data and fits and between SLy and FPS EOSs visible, we plot the *difference* $\log P - 1.4\log\rho$, where $P$ is in dyn cm$^{-2}$ and $\rho$ in g cm$^{-3}$. The bottom panel shows the relative difference between the tabulated and fitted EOSs (solid and dot-dashed lines for SLy and FPS, respectively). It illustrates the accuracy of the fit (C.2).

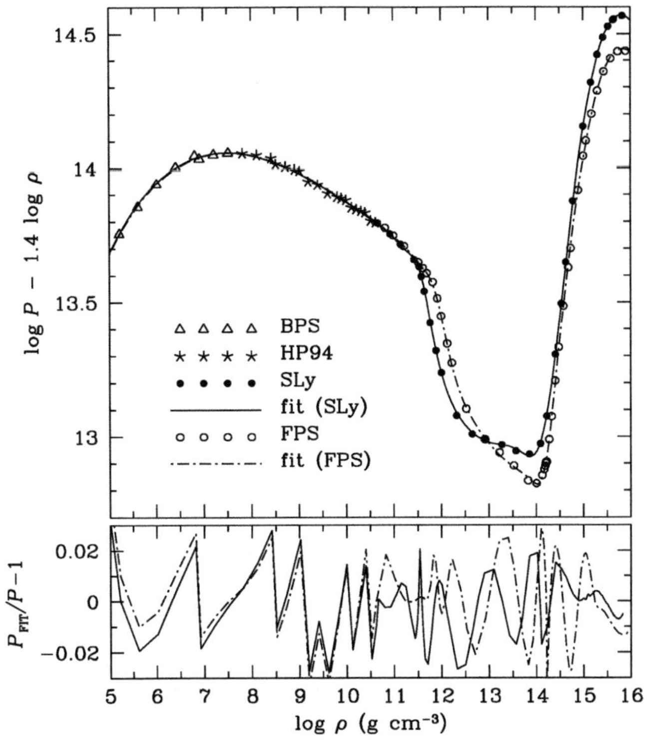


*Figure C.1.* Comparison of the data and fits for the SLy and FPS EOSs (Haensel & Potekhin, 2004). *Top*: Rarefied tabular data (symbols) and the fit (C.2) (lines). *Bottom*: Relative difference between the data and fit. Filled dots and the solid line are for the SLy EOS; open circles and the dot-dashed line are for the FPS EOS (triangles and stars on the top panel are the BPS and HP94 data at $\rho < 5 \times 10^{10}$ g cm$^{-3}$).

Now the baryon number density $n_\mathrm{b}(\rho)$ can be easily obtained from the integral form of Eq. (5.97):

$$\ln\left(\frac{n_\mathrm{b}}{n_\mathrm{bs}}\right) = c^2 \int_{\rho_\mathrm{s}}^{\rho} \frac{\mathrm{d}\rho'}{P(\rho') + \rho' c^2} , \tag{C.3}$$

where $\rho_\mathrm{s}$ and $n_\mathrm{bs}$ are the values of $\rho$ and $n_\mathrm{b}$ at some low-density ("surface") point. Substituting $P(\rho')$ from Eq. (C.2), we recover the original tabular values with maximum errors $< 0.4\%$ and $< 0.12\%$ for the FPS and SLy EOSs, respectively.

In some applications, it may be convenient to use $n_\mathrm{b}$ as an independent variable, and treat $\rho$ and $P$ as functions of $n_\mathrm{b}$. For this purpose one can use the fit:

$$\begin{aligned} \frac{\rho}{n_\mathrm{b} m_0} &= 1 + \frac{p_1 n_\mathrm{b}^{p_2} + p_3 n_\mathrm{b}^{p_4}}{(1 + p_5 n_\mathrm{b})^2}\, f_0(-p_6(\log n_\mathrm{b} + p_7)) \\ &\quad + \frac{n_\mathrm{b}}{8 \times 10^{-6} + 2.1\, n_\mathrm{b}^{0.585}}\, f_0(p_6(\log n_\mathrm{b} + p_7)) , \end{aligned} \tag{C.4}$$

*Table C.2.* Parameters of the fits (C.4) and (C.5)

| $i$ | $p_i$(FPS) | $p_i$(SLy) | $q_i$(FPS) | $q_i$(SLy) |
|---|---|---|---|---|
| 1 | 0.320 | 0.423 | 0.608 | 0.183 |
| 2 | 2.17 | 2.42 | 2.41 | 1.26 |
| 3 | 0.173 | 0.031 | 2.39 | 6.88 |
| 4 | 3.01 | 0.78 | 3.581 | 3.612 |
| 5 | 0.540 | 0.238 | 1.681 | 2.248 |
| 6 | 0.847 | 0.912 | 0.850 | 0.911 |
| 7 | 3.581 | 3.674 | 11.64 | 11.56 |

where $n_\mathrm{b}$ is in fm$^{-3}$ and $m_0 = 1.66 \times 10^{-24}$ g. The inverse fit $n_\mathrm{b}(\rho)$ is given by

$$\begin{aligned} \frac{x}{n_\mathrm{b}} &= 1 + \frac{q_1 x^{q_2} + q_3 x^{q_4}}{(1+q_5 x)^3}\, f_0(q_6(q_7 - \log\rho)) \\ &\quad + \frac{x}{8\times 10^{-6} + 2.1\, x^{0.585}}\, f_0(q_6(\log\rho - q_7)), \end{aligned} \qquad \text{(C.5)}$$

where $x = \rho/m_0$ and $\rho$ is in g cm$^{-3}$. Coefficients $p_i$ and $q_i$ of the fits (C.4) and (C.5) are given in Table C.2. The difference $(\rho - nm_0)$ is approximated by these equations with the error of a few percent.

It should be stressed that thermodynamics requires Eq. (5.97) to be satisfied exactly. To achieve this, one should not totally rely on the fits (C.4) and (C.5); otherwise thermodynamic consistency will be violated on the scale of fit errors (a fraction of percent). Thus, if $\rho$ is used as an input, then $n_\mathrm{b}(\rho)$ should be calculated from Eq. (C.3). Alternatively, if the input is $n_\mathrm{b}$, then, after calculating $\rho_\mathrm{fit}(n_\mathrm{b})$ from Eq. (C.4) and $P(n_\mathrm{b}) = P(\rho_\mathrm{fit}(n_\mathrm{b}))$ from Eq. (C.2), one should refine $\rho(n_\mathrm{b})$ using the relation

$$\frac{\rho(n_\mathrm{b})}{n_\mathrm{b}} = \frac{\rho_\mathrm{s}}{n_\mathrm{bs}} + \int_{n_\mathrm{bs}}^{n_\mathrm{b}} \frac{P(n'_\mathrm{b})}{n'^2_\mathrm{b} c^2}\, \mathrm{d}n'_\mathrm{b}\,, \qquad \text{(C.6)}$$

which also follows from Eq. (5.97).

## C.2. Representation convenient for rotating stars

For rotating stars, it is most useful to parameterize the density and pressure as functions of the pseudo-enthalpy $H$, Eq. (6.99). The latter can be written in terms of the enthalpy per baryon $h$ according to Eq. (6.102). Let us define $\eta \equiv h/m_0c^2 - 1$. In view of the relation (6.101), the function $\xi(\eta)$ (to be parameterized) is not independent of the function $\zeta(\xi)$ parameterized by Eq. (C.2). In order to fulfill Eq. (6.101) as accurately as possible, we first calculate $\eta(\xi)$ using Eqs. (C.2) and (6.101) and then find the inverse fit $\xi(\eta)$. The best fit reads:

$$\begin{aligned} \xi &= \left(b_1 + b_2 \lg\eta + \frac{b_3\eta^{b_4}}{1+b_5\eta}\right) f_0(b_6(\lg\eta - b_7)) \\ &\quad + \frac{b_8 + b_9\lg\eta + (b_{10} + b_{11}\lg\eta)(b_{12}\eta)^7}{1 + b_{13}\eta + (b_{12}\eta)^7}\, f_0(b_6(b_7 - \lg\eta)) \\ &\quad + b_{14}\, f_0(b_{15}(b_{16} - \lg\eta))\,, \end{aligned} \qquad \text{(C.7)}$$

*Table C.3.* Parameters of the fit (C.7)

| $i$ | $b_i$(FPS) | $b_i$(SLy) | $i$ | $b_i$(FPS) | $b_i$(SLy) |
|---|---|---|---|---|---|
| 1 | 5.926 | 5.926 | 9 | 11.97 | 34.96 |
| 2 | 0.4704 | 0.4704 | 10 | 15.432 | 15.328 |
| 3 | 19.92 | 20.13 | 11 | 0.6731 | 0.621 |
| 4 | 0.2333 | 0.2347 | 12 | 49.4 | 63.1 |
| 5 | 2.63 | 3.07 | 13 | 11.47 | 68.5 |
| 6 | 54.7 | 97.8 | 14 | 1.425 | 2.518 |
| 7 | −1.926 | −2.012 | 15 | 3.0 | 2.6 |
| 8 | 36.89 | 89.85 | 16 | 0.913 | 1.363 |

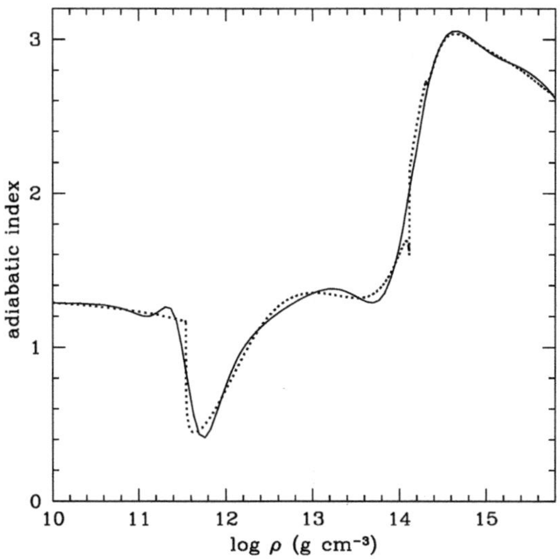


*Figure C.2.* Adiabatic index for the SLy EOS. The solid line is the fit, the dotted line shows precise values.

where the parameters $b_i$ are given in Table C.3. The typical fit error of $\rho$, provided by Eq. (C.7), is about 1% at $\eta \gtrsim 10^{-7}$ (i.e., at $\xi \gtrsim 3$); the maximum fit error $< 4\%$ occurs near the neutron drip and near the crust-core interface.

Combining the fits (C.2) and (C.7) with Eq. (C.3) or Eq. (C.5) we get the parameterizations of $\rho(H)$, $P(H)$, and $n_b(H)$ needed for calculating stationary rotating neutron star models. In this case, the function $P(H) = P(\rho(H))$ obtained from Eqs. (C.2) and (C.7) reproduces the tabular values with a typical error ~(1–2)% and with the maximum error within 10% near the crust-core interface.

The remark on the thermodynamic consistency, made at the end of §C.C.1, applies here as well. One should refine fitted values of either $n_b$ or $\rho$, using the exact relations (C.3) or (C.6).

## C.3. Adiabatic index

An important dimensionless parameter characterizing the stiffness of the EOS at a given density is the adiabatic index, defined by Eq. (5.109). Using our fit (C.2), we obtain the analytical expression

$$(1+P/\rho c^2)^{-1}\gamma = \mathrm{d}\zeta/\,\mathrm{d}\xi$$
$$= \left[\frac{a_2 - a_1 a_4 + 3a_3\xi^2 + 2a_3 a_4 \xi^3}{(1+a_4\xi)^2} - a_5 \frac{1_1 + a_2\xi + a_3\xi^3}{1+a_4\xi} f_0(a_5(a_6-\xi))\right] f_0(a_5(\xi - a_6))$$
$$+ \sum_{i=2}^{4} f_0(a_{4i+1}(a_{4i+2}-\xi))\left[a_{4i} + a_{4i+1}(a_{4i-1} + a_{4i}\xi) f_0(a_{4i+1}(\xi - a_{4i+2}))\right] . \quad \text{(C.8)}$$

The behavior of $\gamma$ in different neutron star layers is displayed in Fig. C.2. Precise values of $\gamma$ calculated by Douchin & Haensel (2001) are shown by the dotted line, and the fit, given by Eqs. (C.2) and (C.8), is shown by the solid line.

# Appendix D
# SEMI-ANALYTICAL EOSs IN NEUTRON STAR CORES

In this Appendix we describe a class of EOSs for uniform matter in neutron star cores composed of nucleons, electrons and muons. These EOSs are based on analytic expressions for the energy per nucleon (excluding the rest-mass energy) quadratic in neutron excess,

$$E_{\rm N} = W(u) + S(u)\,(1 - 2x_p)^2\,, \tag{D.1}$$

where $u \equiv n_{\rm b}/n_0$ is the dimensionless baryon number density, $x_p = n_p/n_{\rm b}$ is the proton fraction; $W(u)$ and $S(u)$ are, respectively, the energy per nucleon in symmetric nuclear matter and the symmetry energy (assumed to be given by analytic functions). The total energy per nucleon is then $E = E_{\rm N} + E_{\rm N0} + E_e + E_\mu$, where $E_{\rm N0}$ is the nucleon rest-mass contribution, while $E_e$ and $E_\mu$ are the electron and muon contributions (also given analytically because electrons and muons constitute almost free Fermi gases). In this case, the total energy $E$ is presented in an analytic form which allows one to avoid ambiguities of interpolation (of otherwise tabulated values of $E_{\rm N}$) and to strictly satisfy thermodynamic relations and conservation laws.

The beta equilibrium conditions are given by relations between the chemical potential of nucleons and leptons,

$$\mu_n = \mu_p + \mu_e\,, \qquad \mu_e = \mu_\mu\,. \tag{D.2}$$

The local electric neutrality requires $x_p = x_e + x_\mu$, where $x_e = n_e/n_{\rm b}$ and $x_\mu = n_\mu/n_{\rm b}$. The electron and muon chemical potentials are equal to the appropriate Fermi energies,

$$\mu_e \approx c p_{{\rm F}e}\,, \qquad \mu_\mu = \sqrt{m_\mu^2\, c^4 + p_{{\rm F}\mu}^2 c^2}\,, \tag{D.3}$$

where $p_{{\rm F}j} = \hbar\,(3\pi^2 n_{\rm b} x_j)^{1/3}$ with $j = e$ or $\mu$. At a fixed $n_{\rm b}$ under the natural simplified assumption that $m_p = m_n$ the beta equilibrium conditions reduce to a set of two equations

$$x_\mu + x_e - \frac{1}{2} + \mathcal{A}\, x_e^{1/3} = 0\,, \tag{D.4a}$$

$$x_e^{2/3} - x_\mu^{2/3} - \mathcal{B} = 0\,, \tag{D.4b}$$

where $\mathcal{A}$ and $\mathcal{B}$ are dimensionless functions of $n_{\rm b}$,

$$\mathcal{A} = \hbar c\,(3\pi^2 n_{\rm b})^{1/3}/(8S(n_{\rm b}))\,, \quad \mathcal{B} = (m_\mu c/\hbar)^2/(3\pi^2 n_{\rm b})^{2/3}\,. \tag{D.5}$$

Beta equilibrium depends on $S(u)$ but not on $W(u)$. For a given $n_{\rm b}$, one can easily solve Eqs. (D.4a) and (D.4b) and determine all particle fractions. After that one can use standard thermodynamic relations, derive the analytic expressions for the energy density ($\rho c^2$) and the pressure, and calculate $\rho$ and $P$ at given $n_{\rm b}$ and the particle fractions. In this way one constructs a semi-analytical EOS; the only simple numerical procedure consists in solving Eqs. (D.4a) and (D.4b). The numerical accuracy of this EOS for an employed nuclear interaction model (D.1) can be formally very high.

*Table D.1.* Three sets of parameters for $W(u)$ models of Prakash *et al.* (1988)

| $K_0$ MeV | $A$ MeV | $B$ MeV | $B'$ | $\sigma$ | $C_1$ MeV | $C_2$ MeV |
|---|---|---|---|---|---|---|
| 120 | 75.94 | –30.88 | 0 | 0.498 | –83.84 | 23.0 |
| 180 | 440.94 | –213.41 | 0 | 0.927 | –83.84 | 23.0 |
| 240 | –46.65 | 39.54 | 0.3 | 1.663 | –83.84 | 23.0 |

The muons appear only in sufficiently dense matter in which $x_e > \mathcal{B}^{3/2}$. At lower densities the muons are absent ($x_\mu = 0$) and Eq. (D.4b) can be disregarded. Then Eq. (D.4a) reads $2\,x_e - 1 + 2\,\mathcal{A}\,x_e^{1/3} = 0$ and can be solved analytically. In this case the procedure of constructing the EOS becomes purely analytical.

**Model PAL.** Prakash *et al.* (1988) proposed a model (PAL) of $W(u)$ which fits experimental values of the energy per nucleon and the density of symmetric nuclear matter at the saturation point $u = 1$: $(\mathrm{d}W/\mathrm{d}u)_1 = 0$, $W(1) = -16$ MeV, and $n_0 = 0.16\ \mathrm{fm}^{-3}$. They suggested three versions corresponding to three values of the compression modulus at saturation, $K_0$=120, 180, and 240 MeV. The functional form of $W(u)$ is

$$W(u) = E_0^{\mathrm{FFG}} u^{2/3} + \frac{Au}{2} + \frac{Bu^\sigma}{1 + B'u^{\sigma-1}} + 3 \sum_{i=1,2} C_i \alpha_i^3 \left[ \frac{u^{1/3}}{\alpha_i} - \mathrm{Artan}\left( \frac{u^{1/3}}{\alpha_i} \right) \right], \quad \text{(D.6)}$$

where $\alpha_1 = 1.5$ , $\alpha_2 = 3$, and the energy of free Fermi gas (FFG) is $E_0^{\mathrm{FFG}} = \frac{3}{5}\epsilon_{\mathrm{F}}(n_0) = 0.3\,p_{\mathrm{N0}}^2/m_n = 22.1$ MeV, where $p_{\mathrm{N0}} = \hbar\,(1.5\pi^2 n_0)^{1/3}$. The three sets of parameters for three models of $W(u)$ are given in Table D.1.

Prakash *et al.* (1988) proposed $S(u)$ of the form

$$S(u) = (2^{2/3} - 1)\, E_0^{\mathrm{FFG}} \left[ u^{2/3} - F(u) \right] + S_0 F(u)\ . \quad \text{(D.7)}$$

Putting $F \equiv 0$ we recover the value of $S$ for a free Fermi gas model, Eq. (5.103) (see a discussion following Eq. (5.103)). Actually, the function $F(u)$ is defined in such a way to reproduce the experimental value $S_0$, so that $S(1) = S_0$ and therefore $F(1) = 1$. Prakash *et al.* (1988) assumed $S_0 = 30$ MeV, and proposed three models (I,II, and III) of $F(u)$,

$$F_{\mathrm{I}}(u) = u\ , \quad F_{\mathrm{II}}(u) = 2u^2/(u+1)\ , \quad F_{\mathrm{III}}(u) = \sqrt{u}\ . \quad \text{(D.8)}$$

Thus, they get nine PAL EOSs, which differ in the stiffness and in the density dependence of the symmetry energy.

**Model PAPAL.** Page & Applegate (1992) proposed one very simple power-law density dependence of the symmetry energy

$$S(u) = 30 u^{0.7}\ \mathrm{MeV}\ . \quad \text{(D.9)}$$

They combined the above model for $S(u)$ with $K_0 = 180$ MeV model for $W(u)$ of Prakash *et al.* (1988). Accordingly they obtained what we call the PAPAL EOS of the $npe\mu$ matter, which yields $M_{\rm max} = 1.7\ M_\odot$ and a direct Urca core for $M > 1.35\ M_\odot$. One can also implant this form of $S(u)$ into other PAL models, with $K_0 = 120$ and 240 MeV, and obtain thus softer and stiffer EOSs (see, e.g., Yakovlev *et al.* 2001).

**Model HHJ.** Heiselberg & Hjorth-Jensen (2000) constructed a two-parameter fit to the EOS of nuclear matter proposed by Akmal *et al.* (1998) (hereafter APR, with boost corrections and three-body forces; $V_{18} + \delta v + \text{UIX}^*$),

$$W(u) = E_0\, u\,(2 + \delta - u)/(1 + \delta u)\ , \tag{D.10}$$

where $\delta$ is the "softness" parameter important for $u \gg 1$. By construction, $W(1) = E_0 = -15.8$ MeV (the value adopted by Akmal *et al.* 1998). The free parameter $\delta$ is related to the incompressibility of the symmetric nuclear matter at saturation point,

$$K_0 = 9\left(\frac{\mathrm{d}^2 W}{\mathrm{d}u^2}\right)_{u=1} = \frac{18\,|E_0|}{1+\delta}\ . \tag{D.11}$$

As far as the symmetry energy is concerned, Heiselberg & Hjorth-Jensen (2000) fitted the APR results with a simple formula of Page & Applegate (1992) type,

$$S(u) = 32\ u^\zeta\ \text{MeV}\ . \tag{D.12}$$

Heiselberg & Hjorth-Jensen (2000) suggested the basic values $\delta = 0.2$ and $\zeta = 0.6$ which make their EOS similar to the APR EOS. In contrast to the PAL and PAPAL EOSs, which are largely phenomenological and relatively old, the HHJ EOS is based on the recent realistic APR EOS.

The HHJ EOS has a very simple analytic form and can be made slightly softer (or stiffer) by increasing (decreasing) the value of $\delta$ with respect to 0.2 (at a fixed $\zeta$). Fixing $\delta$ one can regulate the symmetry energy (D.12) by increasing (decreasing) $\zeta$ with respect to $\zeta = 0.6$. This would slightly decrease (increase) the threshold density for opening the direct Urca process (Gusakov *et al.*, 2005).

# Appendix E
# SCALING OF STELLAR MODELS FOR LINEAR EOSs

Let us outline scaling relations of stellar models built of the matter with the linear EOS of the form

$$P = ac^2(\rho - \rho_s). \tag{E.1}$$

## E.1. The causal limit EOS with $a = 1$

It is convenient to introduce the dimensionless variables,

$$\widetilde{\rho} = \frac{\rho}{\rho_s}, \quad \widetilde{P} = \frac{P}{\rho_s c^2} = \widetilde{\rho} - 1, \quad \widetilde{r} = \frac{r}{r_0}, \quad \widetilde{m} = \frac{m}{M_0}, \quad \widetilde{n}_b = \frac{n_b}{n_s}, \tag{E.2}$$

where $r_0 = c/\sqrt{G\rho_s}$, $M_0 = \rho_s r_0^3$, and $n_s$ is the value of the baryon number density at the stellar surface $\rho = \rho_s$. These variables allow one to rewrite the relativistic equations of hydrostatic equilibrium, Eqs. (6.7)-(6.8), in a dimensionless form. Using the thermodynamic relation

$$\mathrm{d}\widetilde{\rho}/\,\mathrm{d}\widetilde{n}_b = (\widetilde{P} + \widetilde{\rho})/\widetilde{n}_b, \tag{E.3}$$

one gets

$$\widetilde{n}_b = \left[2\widetilde{P} + 1\right]^{1/2} = (2\widetilde{\rho} - 1)^{1/2}. \tag{E.4}$$

**Non-rotating stars.** The dimensionless Tolman-Oppenheimer-Volkoff and mass-balance equations read

$$\begin{aligned} \frac{\mathrm{d}\widetilde{\rho}}{\mathrm{d}\widetilde{r}} &= -\frac{\widetilde{m}}{\widetilde{r}^2}\frac{(2\widetilde{\rho} - 1)}{(1 - 2\widetilde{m}/\widetilde{r})}\left(1 + 4\pi\widetilde{r}^3\frac{\widetilde{\rho} - 1}{\widetilde{m}}\right), \\ \frac{\mathrm{d}\widetilde{m}}{\mathrm{d}\widetilde{r}} &= 4\pi\widetilde{r}^2\widetilde{\rho}. \end{aligned} \tag{E.5}$$

The boundary conditions are $\widetilde{\rho}(0) = \widetilde{\rho}_c$ and $\widetilde{m}(0) = 0$. The radius $\widetilde{R}$ is determined by $\widetilde{\rho}(\widetilde{R}) = 1$ and the dimensionless total gravitational mass $\widetilde{M} = \widetilde{m}(\widetilde{R})$. The solutions of Eqs. (E.5) form a one-parameter family of configurations labeled by $\widetilde{P}_c = \widetilde{\rho}_c + 1$. The dependence $\widetilde{M}(\widetilde{P}_c)$ is shown in Fig. E.1. The dimensionless maximum mass is $\widetilde{M}^{\rm CL}_{\rm max} = 0.0851$; it is reached for $\widetilde{P}_{\rm c,max} = 2.03$ ($\widetilde{\rho}_{\rm c,max} = 3.03$). The corresponding mass-radius relation is shown in Fig. E.2. Numerical values of the surface redshift (Fig. E.3) are independent of the choice of units.

A mass-radius diagram at any given $\rho_s$ can be obtained from Fig. E.2 by coming back to ordinary units. Consequently, the points of a curve obtained for a given $\rho_s$ transform into points of a curve calculated for another $\rho_s'$. For example, the $M'(R')$ curve is obtained via scaling of the $M(R)$ one, namely, $R \to R' = (\rho_s/\rho_s')^{-1/2}R$, $M \to M' = (\rho_s/\rho_s')^{-1/2}M$. In geometrical terms, $M - R$ curves are self-similar. Any extremum of an unprimed curve transforms into an

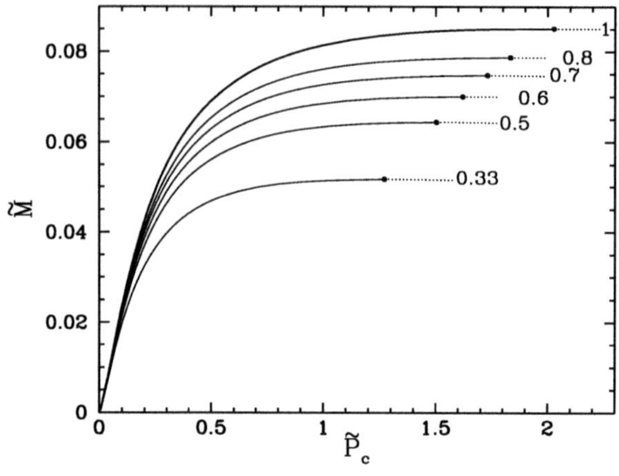


*Figure E.1.* The $\widetilde{M}(\widetilde{P}_\mathrm{c})$ curves labeled by the values of $a$. Filled circles mark maximum-mass configurations. Solid and dotted segments correspond to stable and unstable configurations, respectively. Prepared by J.L. Zdunik (2006, unpublished); with the kind permission of the author.

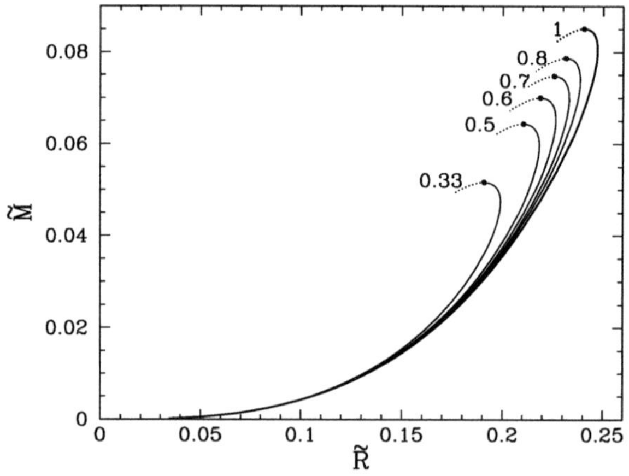


*Figure E.2.* The $\widetilde{M}(\widetilde{R})$ curves labeled by the values of $a$. Notations are the same as in Fig. E.1. Prepared by J.L. Zdunik (2006, unpublished); with the kind permission of the author.

extremum of a primed curve. In particular, the maximum mass configurations scale as

$$M_\mathrm{max}^\mathrm{CL} = M_0\,\widetilde{M}_\mathrm{max} = 2.116\,(\rho_\mathrm{s,15})^{-1/2}\,M_\odot\,, \tag{E.6a}$$

$$R_{M_\mathrm{max}}^\mathrm{CL} = r_0\widetilde{R}_{M_\mathrm{max}} = 8.825\,(\rho_\mathrm{s,15})^{-1/2}\ \mathrm{km}\,, \tag{E.6b}$$

$$\rho_\mathrm{c,max}^\mathrm{CL}/\rho_\mathrm{s} = 3.03\,, \tag{E.6c}$$

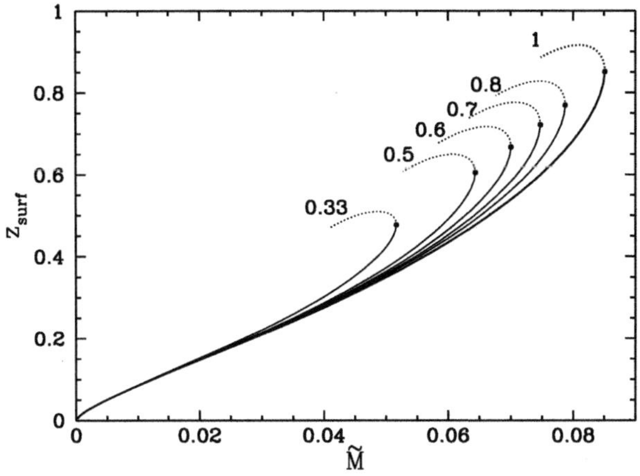


*Figure E.3.* The $z_{\rm surf}(\widetilde{M})$ curves labeled by the values of $a$. Notations are the same as in Fig. E.1. Prepared by J.L. Zdunik (2006, unpublished); with the kind permission of the author.

where $\rho_{\rm s,15} \equiv \rho_{\rm s}/10^{15}$ g cm$^{-3}$. The density contrast within stellar models based on the causal limit (CL) EOS is very low. Even at the maximum mass, the central density is only three times higher than the surface one.

The maximum surface redshift $z_{\rm surf}$ for stable configuration is reached at $M = M_{\rm max}$. It is independent of $\rho_{\rm s}$ and can be readily obtained from Eqs. (E.6a) and (E.6b),

$$z_{\rm max}^{\rm CL} = 0.8509 . \qquad \text{(E.7)}$$

The equation which determines the moment of inertia for a slow rigid rotation, Eq. (6.65), can be written in a dimensionless form provided one expresses angular frequencies in $\sqrt{G\rho_{\rm s}}$. This form, together with the matching conditions explained in § 6.10.1, yields then the dimensionless moment of inertia $\widetilde{I}$, so that $I = M_0 r_0^2 \widetilde{I}$. The maximum moment of inertia for the CL EOS is given by

$$I_{\rm max}^{\rm CL} = M_0 r_0^2 \, \widetilde{I}_{\rm max}^{\rm CL} = 1.979 \, (\rho_{\rm s,15})^{-3/2} \; 10^{45} \, {\rm g \, cm}^2 . \qquad \text{(E.8)}$$

The maximum of $I$ is reached for $M$ slightly (by $\sim$ 0.5%) lower than $M_{\rm max}$. The value of $I_{M_{\rm max}}^{\rm CL}$, which is lower than $I_{\rm max}^{\rm CL}$ by $\approx 1.6\%$, scales with $\rho_{\rm s}$ in the same way as $I_{\rm max}^{\rm CL}$.

**Rotating stars.** In order to transform Eq. (6.98), which describes the structure of a rotating star, into a dimensionless form one should supplement the dimensionless quantities (E.2) with the dimensionless frequencies

$$\widetilde{\Omega} = \Omega/\sqrt{G\rho_{\rm s}} \, , \qquad \widetilde{\omega} = \omega/\sqrt{G\rho_{\rm s}} \, . \qquad \text{(E.9)}$$

In this way, one gets a dimensionless solution of Eq. (6.98). Solutions corresponding to any $\rho_{\rm s}$ can then be obtained by returning to physical units. Any two solutions for $\rho_{\rm s}'$ and $\rho_{\rm s}$ are self-similar. In particular, this is true for extremal configurations. For a rotating maximum-mass star one gets the same scaling as in the non-rotating case,

$$M_{\rm max}^{\rm CL,rot'} = M_{\rm max}^{\rm CL,rot} \, (\rho_{\rm s}/\rho_{\rm s}')^{1/2} . \qquad \text{(E.10)}$$

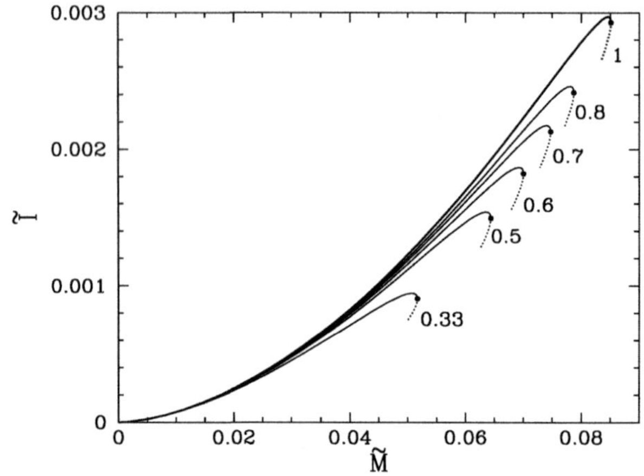


*Figure E.4.* The $\widetilde{I}(\widetilde{M})$ curves, labeled by the values of $a$. Notations are the same as in Fig. E.1. Prepared by J.L. Zdunik (2006, unpublished); with the kind permission of the author.

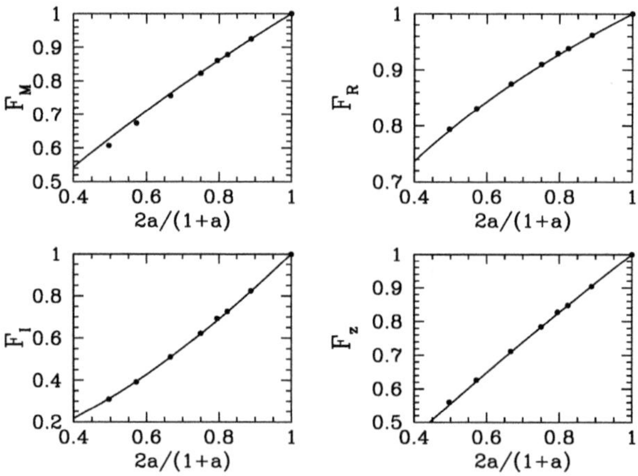


*Figure E.5.* Functions $F_Q$ ($Q = M, R, z, I$) which determine the scaling for maximum-mass rotating stars with respect to variations of $a$. Filled circles show exact results, solid lines are the fits (E.17). Prepared by J.L. Zdunik (2006, unpublished); with the kind permission of the author.

A maximally rotating configuration has the shortest spin period $P = P_{\rm min}$ (the highest frequency $\Omega = \Omega_{\rm max}$) of all stably rotating configurations. This configuration is only slightly different from a maximum mass configuration. The shortest spin period $P_{\rm min}$ scales as

$$P_{\rm min}^{\rm CL\,\prime} = P_{\rm min}^{\rm CL}\,(\rho_{\rm s}/\rho_{\rm s}')^{1/2}\,. \tag{E.11}$$

After integrating numerically the relevant dimensionless equations and returning to the ordinary physical units, we get the formulae for the CL EOS in an explicit form

$$M_{\max}^{\rm CL,rot} = 2.75\ (\rho_{\rm s,15})^{-1/2}\ M_\odot\ , \qquad \text{(E.12a)}$$

$$P_{\min}^{\rm CL} = 0.415\ (\rho_{\rm s,15})^{-1/2}\ \text{ms}\ . \qquad \text{(E.12b)}$$

## E.2. The case of $a < 1$

The dimensionless form of the EOS is

$$\widetilde{P} = a(\widetilde{\rho} - 1)\ . \qquad \text{(E.13)}$$

The formula for $\widetilde{n}_{\rm b}(P)$ can be derived using Eq. (E.3). One gets then a generalization of Eq. (E.4),

$$\widetilde{n}_{\rm b} = [(a+1)\widetilde{\rho} - a]^{1/(1+a)} = \left[(a+1)\widetilde{P}\,a^{-1} + 1\right]^{1/(1+a)}\ . \qquad \text{(E.14)}$$

**Non-rotating stars.** The dimensionless equations read

$$a\frac{\mathrm{d}\widetilde{\rho}}{\mathrm{d}\widetilde{r}} = -\frac{\widetilde{m}}{\widetilde{r}^2}\frac{[(1+a)\widetilde{\rho} - a]}{(1 - 2\widetilde{m}/\widetilde{r})}\left(1 + 4\pi\widetilde{r}^3 a\,\frac{\widetilde{\rho} - 1}{\widetilde{m}}\right)\ ,$$

$$\frac{\mathrm{d}\widetilde{m}}{\mathrm{d}\widetilde{r}} = 4\pi\widetilde{r}^2\widetilde{\rho}\ . \qquad \text{(E.15)}$$

Now we get a dimensionless solution at any $a$. At a fixed $a$, the curves calculated using normal units and representing solutions with different $\rho_{\rm s}$ scale with the same power of $\rho_{\rm s}$ as in the case of $a = 1$. However, the numerical coefficients in these scaling relations depend on $a$. As shown by J.L. Zdunik (2006, unpublished), this dependence can be described by functions $F_Q(y)$, where $y = 2a/(a+1)$ is more convenient than just $a$, and $Q = M, R, z, I, \ldots$ In particular, the parameters of maximum-mass configurations are related to those obtained at $a = 1$ by

$$M_{\max} = F_M(y)\ M_{\max}^{\rm CL}\ , \quad R_{M_{\max}} = F_R(y)\ R_{M_{\max}}^{\rm CL}\ ,$$

$$z_{M_{\max}} = F_z(y)\ z_{M_{\max}}^{\rm CL}\ , \quad I_{M_{\max}} = F_I(y)\ I_{M_{\max}}^{\rm CL}\ . \qquad \text{(E.16)}$$

To a very good approximation (Fig. E.5), $F_Q(y)$ are simple power-laws (J.L. Zdunik, 2006, unpublished),

$$F_M(y) = y^{2/3}, \quad F_R(y) = y^{1/3}, \quad F_z(y) = y^{0.85}, \quad F_I(y) = y^{5/3}\ . \qquad \text{(E.17)}$$

The scaling with respect to $a \to a'$ takes simple form in the Newtonian limit. To make the equations of hydrostatic equilibrium independent of $a$, it is sufficient to "include" $a$ into the gravitational constant $G$. Equilibrium configurations for any pair of values of $a$ and $\rho_{\rm s}$ can then be obtained from the dimensionless solution (independent of $a$) by multiplying radius and mass by $a^{1/2}r_0$ and $a^{3/2}M_0$, respectively. This gives new scaling relations.

**Rotating stars.** Solutions of dimensionless equations of stationary motion depend parametrically on $a$. The power in the scaling under $\rho_{\rm s} \to \rho_{\rm s}'$ is the same as for $a = 1$, but the numerical prefactor is modified and depends on $a$.

# BIBLIOGRAPHY


Abbott B., Abbott R., Adhikari R., *et al.* (LIGO Scientific Collaboration), Kramer M., Lyne A., 2005, "Limits on gravitational wave emission from selected pulsars using LIGO data," *Phys. Rev. Lett.* **94**, 181103 (6 pages).

Abe R., 1959, "Giant cluster expansion theory and its application to high-temperature plasma," *Progr. Theor. Phys.* **22**, 213–226.

Abney M., Epstein R.I, Olinto A.V., 1996, "Observational constraints on the internal structure and dynamics of the Vela Pulsar," *Astrophys. J.* **466**, L91–L94.

Abrahams A.M., Shapiro S.L., 1991a, "Equation of state in a strong magnetic field: finite temperature and gradient corrections," *Astrophys. J.* **374**, 652–667.

Abrahams A.M., Shapiro S.L., 1991b, "Molecules and chains in a strong magnetic field: statistical treatment," *Astrophys. J.* **382**, 233–241.

Abrikosov A.A., 1960, "Some properties of strongly compressed matter," *Zh. Eksp. Teor. Fiz.* **39**, 1797–1805 [in Russian].

Abubekerov M.K., 2004, "The mass of the compact object in the X-ray binary 4U 1700–37," *Astron. Rep.* **48**, 649–658.

Abubekerov M.K., Antokhina E.A., Cherepashchuk A.M., 2004, "Masses of X-ray pulsars in binary systems with OB supergiants," *Astron. Rep.* **48**, 89–102.

Adams F.C., Laughlin G., 1997, "A dying universe: the long-term fate and evolution of astrophysical objects," *Rev. Mod. Phys.* **69**, 337–372.

Adams F.C., Laughlin G., Mbonye M., Perry M.J., 1998, "Gravitational demise of cold degenerate stars," *Phys. Rev. D* **58**, 083003 (7 pages).

Aguirre R., 2003, "Chiral symmetry and strangeness content in nuclear physics parameterized by a medium dependent bag constant," *Phys. Lett. B* **559**, 207–213.

Akaishi Y., Harada T., Shinmura S., Myint K.S., 2000, "Coherent $\Lambda$–$\Sigma$ coupling in $s$-shell hypernuclei," *Phys. Rev. Lett.* **84**, 3539–3541.

Akiyama S., Wheeler J.C., Meier D.L., Lichtenstadt, 2003, "The magnetorotational instability in core-collapse supernova explosions," *Astrophys. J.* **584**, 954–970.

Akhiezer A.I., Berestetskiĭ V.B., 1965, *Quantum Electrodynamics* (New York: Interscience).

Akmal A., Pandharipande V.R., 1997, "Spin-isospin structure and pion condensation in nucleon matter," *Phys. Rev. C* **56**, 2261–2279.

Akmal A., Pandharipande V.R., Ravenhall D.G., 1998, "Equation of state of nucleon matter and neutron star structure," *Phys. Rev. C* **58**, 1804–1828.

Aksenov A.G., Milgrom M., Usov V.V., 2003, "Radiation from hot, bare, strange stars," *Mon. Not. R. Astron. Soc.* **343**, L69–L72.

Aksenov A.G., Milgrom M., Usov V.V., 2004, "Structure of pair winds from compact objects with application to emission from hot bare strange stars," *Astrophys. J.* **609**, 363–377.

Alastuey A., Jancovici B., 1980, "Magnetic properties of a nearly classical one-component plasma in three and two dimensions. II. Strong field," *Physica A* **102**, 327–343.

Albers R.C., Gubernatis J.E., 1986, "Low-order anharmonic contributions to the internal energy of the one-component plasma," *Phys. Rev. B* **33**, 5180–5185; erratum: *ibid.*, **42**, 11373.

Alcock C., Farhi E., Olinto A.V., 1986, "Strange stars," *Astrophys. J.* **310**, 261–272.

Alford M., Rajagopal K., Wilczek F., 1998, "QCD at finite baryon density: nucleon droplets and color superconductivity," *Phys. Lett. B* **422**, 247–256.

Alford M., Rajagopal K., Reddy S., Wilczek F., 2001, "Minimal color-flavor-locked-nuclear interface," *Phys. Rev. D* **64**, 074017 (14 pages).

Allen B., Woan G., 2004, "Upper limits on the strength of periodic gravitational waves from PSR J1939+2134," *Class. Quant. Grav.* **21**, S671–S676.

Alpar M.A., 1977, "Pinning and threading of quantized vortices in the pulsar crust superfluid," *Astrophys. J.* **213**, 527–530.

Alpar M.A., 1987, "Comment on strange stars," *Phys. Rev. Lett.* **58**, 2152.

Alpar M.A., Shaham J., 1985, "Is GX5-1 a millisecond pulsar?" *Nature* **316**, 239–241.

Alpar M.A., Langer S.A., Sauls J.A., 1984, "Rapid postglitch spin-up of the superfluid core in pulsars," *Astrophys. J.* **282**, 533–541.

Alpar M.A., Chau H.F., Cheng K.S., Pines D., 1993, "Postglitch relaxation of the Vela pulsar after its first eight large glitches: a reevaluation with the vortex creep," *Astrophys. J.* **409**, 345–359.

Ambartsumyan V.A., Saakyan G.S., 1960, "The degenerate superdense gas of elementary particles," *Astron. Zh.*, **37**, 193–209 [Engl. transl.: *Sov. Astron. – AJ*, **4**, 187].

Anderson P.W., Itoh N., 1975, "Pulsar glitches and restlessness as a hard superfluidity phenomenon," *Nature* **256**, 25–27.

Anderson P.W., Palmer R.G., 1971, "Solidification pressure of nuclear and neutron star matter," *Nature (Phys. Sci.)* **231**, 145–146.

Anderson S.B., Gorham P.W., Kulkarni S.R., Prince T.A., & Wolszczan A., 1990, "Discovery of two radio pulsars in the globular cluster M15," *Nature* **436**, 42–44.

Andersson N., Kokkotas K.D., 2001, "The R-mode instability in rotating neutron stars," *Int. J. Mod. Phys. D* **10**, 381–442.

Andersson N., Jones D.I., Kokkotas K.D., 2002, "Strange stars as persistent sources of gravitational waves," *Mon. Not. R. Astron. Soc.* **337**, 1224–1232.

Antia H.M., 1993, "Rational function approximations for Fermi-Dirac integrals," *Astrophys. J. Suppl. Ser.* **84**, 101–108.

Arnett D., 1996, *Supernovae and Nucleosynthesis* (Princeton: Princeton University Press).

Arnett W.D., Bowers R.L., 1977, "A microscopic interpretation of neutron star structure," *Astrophys. J. Suppl. Ser.* **33**, 415–436.

Arponen J., 1972, "Internal structure of neutron stars," *Nucl. Phys.* **A 191**, 257–282.

Arzoumanian Z., Chernoff D.F., Cordes J.M., 2002, "The velocity distribution of isolated radio pulsars," *Astrophys. J.* **568**, 289–301.

Ash T.D.C., Reynolds A.P., Roche P., Norton A.J., Still M.D., & Morales-Rueda L., 1999, "The mass of the neutron star in Centaurus X-3," *Mon. Not. R. Astron. Soc.* **307**, 357–364.

Audi G., Bersillon O., Blachot J., Wapstra A.H., 1997, "The NUBASE evaluation of nuclear and decay properties," *Nucl. Phys.* **A 624**, 1–124.

Audi G., Wapstra A.H., Thibault C., 2003, "The AME atomic mass evaluation (II). Tables, graphs and references," *Nucl. Phys.* **A 729**, 337–676.

Avron J.E., Herbst I.W., Simon B., 1978, "Separation of center of mass in homogeneous magnetic fields," *Ann. Phys. (N.Y.)* **114**, 431–451.

Baade W., Zwicky F., 1934a, "Supernovae and cosmic rays," *Phys. Rev.* **45**, 138.

Baade W., Zwicky F., 1934b, "On super-novae," *Proc. National Acad. Sci.* **20**, 254–259.

Baade W., Zwicky F., 1934c, "Remarks on super-novae and cosmic rays," *Phys. Rev.* **46**, 76–77.

Backer D.C., Kulkarni S.R., Heiles C., Davis M.M., Goss W.M., 1982, "A millisecond pulsar," *Nature* **300**, 615–618.

Bahcall J.N., Wolf R.A.,1965a, "Neutron stars. II. Neutrino-cooling and observability," *Phys. Rev.* **140**, B1452–B1466.

Bahcall J.N., Wolf R.A., 1965b, "An observational test of theories of neutron-star cooling," *Astrophys. J.* **142**, 1254–1256.

Bahcall S., Lynn B.W., Selipsky S.B., 1990, "New models for neutron stars," *Astrophys. J.* **362**, 251–255.

Baiko D.A., 2000, *Kinetic phenomena in cooling neutron stars*, PhD thesis (St. Petersburg: Ioffe Phys.-Tech. Inst.) [in Russian], unpublished.

Baiko D.A., 2002, "Effect of the electron gas polarizability on the specific heat of phonons in Coulomb crystals," *Phys. Rev. E* **66**, 056405 (10 pages).

Baiko D.A., Kaminker A.D., Potekhin A.Y., Yakovlev D.G., 1998, "Ion structure factors and electron transport in dense Coulomb plasmas," *Phys. Rev. Lett.* **81**, 5556–5559.

Baiko D.A., Yakovlev D.G., DeWitt H.E., Slattery W.L., 2000, "Coulomb crystals in the harmonic lattice approximation," *Phys. Rev. E* **61**, 1912–1919.

Baiko D.A., Haensel P., Yakovlev D.G., 2001a, "Thermal conductivity of neutrons in neutron star cores," *Astron. Astrophys.* **374**, 151–163.

Baiko D.A., Potekhin A.Y., Yakovlev D.G., 2001b, "Thermodynamic functions of harmonic Coulomb crystals," *Phys. Rev. E* **64**, 057402 (4 pages).

Bailes M., Harrison P.A., Lorimer D.R., Johnston S., Lyne A.G., Manchester R.N., D'Amico N., Nicastro L., Tauris T.M., & Robinson C., 1994, "Discovery of three binary millisecond pulsars," *Astrophys. J.* **425**, L41–L44.

Bailes M., Ord S.M., Knight H.S., Hotan A.W., 2003, "Self-consistency of relativistic observables with general relativity in the white dwarf-neutron star binary PSR J1141–6545," *Astrophys. J.* **595**, L49–L52.

Bailin D., Love A., 1984, "Superfluidity and superconductivity in relativistic fermion systems," *Phys. Rep.* **107**, 325–385.

Balberg S., Barnea N., 1998, "S-wave pairing of Lambda hyperons in dense matter," *Phys. Rev. C* **57**, 409–416.

Balberg S., Gal A., 1997, "An effective equation of state for dense matter with strangeness," *Nucl. Phys.* **A 625**, 435–472.

Baldo M., 1999, "The many-body theory of the nuclear equation of state," in *Nuclear Methods and Nuclear Equation of State*, edited by M. Baldo (Singapore: World Scientific),1–120.

Baldo M., Burgio F., 2001, "Microscopic theory of the nuclear equation of state and neutron star structure," in *Physics of Neutron Star Interiors*, edited by D. Blaschke, N.K. Glendenning, & A. Sedrakian, *Lecture Notes in Physics* **578**, 1–29.

Baldo M., Bombaci I., Burgio G.F., 1997, "Microscopic nuclear equation of state with three-body forces and neutron-star structure," *Astron. Astrophys.* **328**, 274–282.

Baldo M., Giansiracusa G., Lombardo U., Song H.Q., 2000, "Bethe-Brueckner-Goldstone expansion in neutron matter," *Phys. Lett. B* **473**, 1–5.

Baldo M., Fiasconaro A., Song H.Q., Giansiracusa G., Lombardo U., 2001, "High density symmetric nuclear matter in Bethe-Brueckner-Goldstone approach," *Phys. Rev. C* **65**, 017303 (4 pages).

Baldo M., Lombardo U., Saperstein E.E., Tolokonnikov S.V., 2005, "The role of superfluidity in the structure of the neutron star inner crust," *Nucl. Phys.* **A 750**, 409–424.

Baluni V., 1978a, "(Non)-Abelian gauge field theories of the Fermi gas. Neutron-quark stars," *Phys. Lett. B* **72**, 381–384.

Baluni V., 1978b, "Non-Abelian gauge theories of Fermi systems: Quantum-chromodynamic theory of highly condensed matter," *Phys. Rev. D* **17**, 2092–2121.

Banik S., Bandyopadhyay D., 2001, "A third family of super dense stars in the presence of antikaon condensates," *Phys. Rev. C* **64**, 055805 (11 pages).

Bardeen J., Cooper L.N., Schrieffer J.R., 1957, "Theory of superconductivity," *Phys. Rev.* **108**, 1175–1204.

Bardeen J.M., Thorne K.S., Meltzer D.W., 1966, "A catalogue of methods for studying the normal modes of general-relativistic stellar models," *Astrophys. J.* **145**, 505–513.

Barkat Z., Buchler J.-R., Ingber L., 1972, "Equation of state of neutron-star matter at subnuclear densities," *Astrophys. J.* **176**, 723–738.

Barker B.M., O'Connell R.F., 1975, "Relativistic effects in the binary pulsar PSR 1913+16," *Astrophys. J.* **199**, L25–L26.

Barrois B., 1977, "Superconducting quark matter," *Nucl. Phys.* **B 129**, 390–396.

Barziv O., Kaper L., van Kerkwijk M.H., Telting J.H., van Paradjis J., 2001, "The mass of the neutron star in Vela X-1," *Astron. Astrophys.* **377**, 925–944.

Basko M.M., 1985, "Equation of state of metals in the mean-ion approximation," *High Temp. (USSR)* **23**, 388–396.

Bassa C.G., van Kerkwijk M.H., Koester D., Verbunt F., 2006, "The masses of PSR J1911–5958A and its white dwarf companion," *Astron. Astrophys.* **456**, 295–304.

Batty C.J., Friedman E., Gal A., 1994, "Density dependence of the $\Sigma$ nucleus optical potential derived from $\Sigma^-$ atom data," *Phys. Lett. B* **335**, 273–278.

Baumgarte T.W., Shapiro S.L., Teukolsky S.A., 1996a, "Computing the delayed collapse of hot neutron stars to black holes," *Astrophys. J.* **458**, 680–691.

Baumgarte T.W., Janka H.-T., Keil W., Shapiro S.L., Teukolsky S.A., 1996b, "Delayed collapse of hot neutron stars to black holes via hadronic phase transition," *Astrophys. J.* **468**, 823–833.

Bauer M., 1976, "An *ab initio* shell-model formula," *At. Data Nucl. Data Tables* **17**, 442–449.

Baus M., Hansen J.P., 1980, "Statistical mechanics of simple Coulomb systems," *Phys. Rep.* **59**, 1–94.

Baym G., 1979, "Neutron stars and the physics of matter at high density," in *Nuclear Physics with Heavy Ions and Mesons*, edited by R. Balian, M. Rho, & G. Ripka (Amsterdam: North-Holland), 745–837.

Baym G., 1982, "Neutron stars: the first fifty years," in *The Neutron and its Application*, edited by P. Schofield, *IOP Conf. Ser.* **64**, 45–50.

Baym G., Chin S.A., 1976, "Can a neutron star be a giant MIT bag?" *Phys. Lett. B* **62**, 241–244.

Baym G., Pines D., 1971, "Neutron starquakes and pulsar speedup," *Ann. Phys. (N.Y.)* **66**, 816–835.

Baym G., Pethick C., Pines D., 1969, "Electrical conductivity in neutron star matter," *Nature*, **224**, 674–675.

Baym G., Bethe H.A., Pethick C.J., 1971a, "Neutron star matter," *Nucl. Phys.* **A 175**, 221–271.

Baym G., Pethick C., Sutherland P., 1971b, "The ground state of matter at high densities: Equation of state and stellar models," *Astrophys. J.* **170**, 299–317.

Becker W., 1999, "An X-ray study of isolated neutron stars and their wind nebulae," in *Highlights in X-ray Astronomy*, edited by B. Aschenbach & M.J. Freyberg, MPE Report 272 (Garching: MPE), 49–56.

Bejger M., Haensel P., 2002, "Moments of inertia for neutron and strange stars: limits derived for the Crab pulsar," *Astron. Astrophys.* **396**, 917–921.

Bejger M., Haensel P., 2003, "Accelerated expansion of the Crab Nebula and evaluation of its neutron-star parameters," *Astron. Astrophys.* **405**, 747–751.

Bejger M., Haensel P., 2004, "Surface gravity of neutron stars and strange stars," *Astron. Astrophys.* **420**, 987–991.

Bejger M., Haensel P., Zdunik J.L., 2005a, "Mixed-phase induced core-quakes and the changes in neutron star parameters," *Mon. Not. R. Astron. Soc.* **359**, 699–706.

Bejger M., Bulik T., Haensel P., 2005b, "Constraints on the dense matter equation of state from the measurements of PSR J0737–3039A moment of inertia and PSR J0751+1807 mass," *Mon. Not. R. Astron. Soc.* **364**, 635–639.

Bell J.F., Bessell M.S., Stappers B.W., Bailes M., Kaspi V.M., 1995, "PSR J0045–7319: A dual-line binary radio pulsar," *Astrophys. J.* **447**, L117–L119.

Bellwied R., 2005, "Evidence for a new phase of matter measured with the STAR experiment at RHIC," *Nucl. Phys.* **A 752**, 398c–406c.

Benvenuto O.G., Horvath J.E., Vucetich H., 1989, "Strange matter, detonations and supernovae," *Int. J. Modern Phys.* **4**, 257–265.

Berestetskiĭ V.B., Lifshitz E.M., Pitaevskii L.P., 1982, *Quantum Electrodynamics* (Oxford: Butterworth-Heinemann).

Berezin Yu.A., Dmitrieva O.E., 1984, "Dynamics of a developing neutron star in general relativity," *Pisma v Astron. Zh.* **10**, 417–421 [Engl. transl.: *Sov. Astron. Lett.* **10**, 175–177].

Berezin Yu.A., Dmitrieva O.E., Yanenko N.N., 1982, "Calculation of model neutron stars with pion condensation," *Pisma v Astron. Zh.* **8**, 86–89 [Engl. transl.: *Sov. Astron. Lett.* **8**, 43–45].

Berezin Yu.A., Mukanova B.G., Fedoruk M.N., 1983, "The effect of nonzero temperature upon the dynamics of neutron stars with pion condensation, " *Pisma v Astron. Zh.* **9**, 116–119 [Engl. transl.: *Sov. Astron. Lett.* **9**, 63–65].

Beskin V.S., 1999, "Radio pulsars," *Physics – Uspekhi* **42**, 1071–1098.

Beskin V.S., Gurevich A.V., Istomin Ya.N., 1993, *Physics of the Pulsar Magnetosphere* (Cambridge: Cambridge University Press).

Bethe H.A., 1953, "What holds the nucleus together," *Scientific American* **189**, 58–63.

Bethe H.A., 1965, "Three-body correlations in nuclear matter," *Phys. Rev.* **138**, B804–B822.

Bethe H.A., Critchfield C.L., 1938, "The formation of deuterons by proton combination," *Phys. Rev.* **54**, 248–254.

Bethe H.A., Goldstone J., 1957, "Effect of a repulsive core in the theory of complex nuclei," *Proc. Roy. Soc. (London) A* **239**, 551-567.

Bethe H.A., Johnson M.B., 1974, "Dense baryon matter calculations with realistic potentials," *Nucl. Phys.* **A 230**, 1–58.

Bethe H.A., Börner G., Sato K., 1970, "Nuclei in neutron matter," *Astron. Astrophys.* **7**, 279–288.

Bezchastnov V.G., Potekhin A.Y., 1994, "Transitions between shifted Landau states and photoionization of the hydrogen atom moving in a strong magnetic field," *J. Phys. B* **27**, 3349–3361.

Bezchastnov V.G., Pavlov G.G., Ventura J., 1998, "Discrete eigenstates of the $He^+$ ion moving in a strong magnetic field," *Phys. Rev. A* **58**, 180–185.

Bezkrovniy V., Filinov V.S., Kremp D., Bonitz M., Schlanges M., Kraeft W.D., Levashov P.R., Fortov V.E., 2004, "Monte Carlo results for the hydrogen Hugoniot," *Phys. Rev. E* **70**, 057401 (4 pages).

Bhattacharyya S., Strohmayer T.E., Miller M.C., Markwardt C.B., 2005, "Constraints on compact star parameters from burst oscillation light curves of the accreting millisecond pulsar XTE J1814–338," *Astrophys. J.* **619**, 483–491.

Bhattacharyya S, Strohmayer T.E., Markwardt C.B., Swank J.H., 2006, "The discovery of a neutron star with a spin frequency of 530 Hz in A1744–361," *Astrophys. J.* **639**, L31–L34.

Bildsten L., Ushomirsky G., 2000, "Viscous boundary layer damping of r-modes in neutron stars," *Astrophys. J.* **529**, L33–L36.

Bildsten L., Chang P., Paerels F., 2003, "Atomic spectral features during thermonuclear flashes of neutron stars," *Astrophys. J.* **591**, L29–L32.

Biot E.C., 1846, "Catalogue des étoiles extraordinaires observées en Chine depuis les temps anciens jusqu'à l'an 1203 de notre ère," *Connaissance des temps, ou des mouvements célestes à l'usage des astronomes et des navigateurs pour l'an*, Additions.

Bisnovatyi-Kogan G.S., 2001, *Stellar physics. Vol. 1: Fundamental concepts and stellar equilibrium* (Berlin: Springer).

Bisnovatyi-Kogan G.S., 2002, *Stellar physics. Vol. 2: Stellar evolution and stability* (Berlin: Springer).

Bisnovatyi-Kogan G.S., Chechetkin V.M., 1979, "Nonequilibrium shells of neutron stars and their role in sustaining X-ray emission and nucleosynthesis," *Uspekhi Fiz. Nauk* **127**, 263–296 [Engl. transl.: *Sov. Phys. Uspekhi* **22**, 89–108].

Blackett P.M.S., 1947, "The magnetic field of massive rotating bodies," *Nature* **159**, 658–666.

Blaizot J.P., 1980, "Nuclear compressibilities," *Phys. Rep.* **64**, 171–248.

Blandford R.D., Hernquist L., 1982, "Magnetic susceptibility of a neutron star crust," *J. Phys. C: Solid State Phys.* **15**, 6233–6243.

Blandford R.D., Applegate J.H., Hernquist L., 1983, "Thermal origin of neutron star magnetic fields," *Mon. Not. R. Astron. Soc.* **204**, 1025–1048.

Blaschke D., Grigorian H., Poghosyan G., Roberts C.D., Schmidt S., 1998, "A dynamical, confining model and hot quark stars," *Phys. Lett. B* **450**, 207–214.

Blinnikov S.I., Novikov I.D., Perevodchikova T.V., Polnarev A.G., 1984, "Exploding neutron stars in close binaries," *Pis'ma v Astron. Zh.* **10**, 422–428 [Engl. transl.: *Sov. Astron. Lett.* **10**, 177–179].

Blinnikov S.I., Dunina-Barkovskaya N.V., Nadyozhin D.K., 1996, "Equation of state of a Fermi gas: Approximations for various degrees of relativism and degeneracy," *Astrophys. J. Suppl. Ser.* **106**, 171–203.

Bludman S.A., Ruderman M.A., 1968, "Possibility of the speed of sound exceeding the speed of light in ultradense matter," *Phys. Rev.* **170**, 1176–1184.

Bludman S.A., Ruderman M.A., 1970,"Noncausality and instability in ultradense matter," *Phys. Rev. D* **1**, 3243–3246.

Bocquet M., Bonazzola S., Gourgoulhon E., Novak J., 1995, "Rotating neutron star models with a magnetic field," *Astron. Astrophys.* **301**, 757–775.

Bodmer A.R., 1971, "Collapsed nuclei," *Phys. Rev. D* **4**, 1601–1606.

Boguta, J., 1981, 'Remarks on the beta stability in neutron stars," *Phys. Lett. B* **106**, 255–258.

Boguta J., Bodmer A.R., 1977, "Relativistic calculations of nuclear matter and the nuclear surface," *Nucl. Phys.* **A 292**, 413–428.

Bohr A., Mottelson B.R., Pines D., 1958, "Possible analog between the excitation spectra of nuclei and those of superconducting metal state," *Phys. Rev.* **110**, 936–938.

Bombaci I., 1995, "An equation of state for asymmetric nuclear matter and the structure of neutron stars," in *Perspectives on Theoretical Nuclear Physics*, edited by I. Bombaci, A. Bonaccorso, A. Fabrocini, *et al.*(Pisa: Edizioni ETS), 223–237.

Bombaci I., 1997, "Observational evidence for strange matter in compact objects from the X-ray burster 4U 1820–30," *Phys. Rev. C* **55**, 1587–1590.

Bombaci I., 2001, "Strange quark stars: structural properties and possible signatures for their existence," *Lecture Notes in Physics* **578**, 253–284.

Bonazzola S., Gourgoulhon E., Salgado M., 1993, "Axisymmetric rotating relativistic bodies: new numerical approach for 'exact' solutions," *Astron. Astrophys.* **278**, 421–443.

Bonazzola S., Frieben J., Gourgoulhon E., 1996, "Spontaneous symmetry breaking of rapidly rotating stars in general relativity," *Astrophys. J.* **460**, 379–389.

Bonazzola S., Frieben J., Gourgoulhon E., 1998a, "Spontaneous symmetry breaking of rapidly rotating stars in general relativity: influence of the 3D-shift vector," *Astron. Astrophys.* **331**, 280–290.

Bonazzola S., Gourgoulhon E., Marck J.A., 1998b, "Numerical approach for high precision 3D relativistic star models," *Phys. Rev. D* **58**, 104020 (14 pages).

Bonche P., Vautherin D., 1981, "A mean field calculation of the equation of state of supernova matter," *Nucl. Phys.* **A 372**, 496–526.

Bonche P., Vautherin D., 1982, "Mean-field calculations of the equation of state of supernova matter II," *Astron. Astrophys.* **112**, 268–272.

Born M., 1940, "On the stability of crystal lattices. I." *Proc. Cambridge Phil. Soc.* **36**, 160–172.

Boynton P.E., Groth E.J., Partridge R.B., Wilkinson D.T., 1969, *IAU Circular* No. 2179.

Bowyer S., Byram E.T., Chubb T.A., Friedman H., 1964, "Lunar occulation of X-ray emission from the Crab nebula," *Science* **146**, 912–917.

Boynton P.E., Groth E.J., Hutchinson D.P., Nanos G.P., Partridge R.B., Wilkinson D.T., 1972, "Optical timing of the Crab Pulsar, NP 0532," *Astrophys. J.* **175**, 217–421.

Brack M., Guet C.,Håkansson H.-B., 1985, "Selfconsistent semiclassical description of average nuclear properties – a link between microscopic and macroscopic models," *Phys. Rep.* **123**, 275–364.

Braje T.M., Romani R.W., 2002, "RX J1856-3754: evidence for a stiff equation of state," *Astrophys. J.* **580**, 1043–1047.

Brecher K., Caporaso G., 1976, "Obese 'neutron' stars," *Nature* **259**, 377–378.

Brinkmann W. 1980, "Thermal radiation from highly magnetized neutron stars," *Astron. Astrophys.* **82**, 352–361.

Brisken W.F., Thorsett S.E., Golden A., Goss W.M., 2003, "The distance and radius of the neutron star PSR B0656+14," *Astrophys. J.* **593**, L89–L92.

Broderick A., Prakash M., Lattimer J.M., 2000, "The equation of state of neutron star matter in strong magnetic fields," *Astrophys. J.* **537**, 351–367.

Brown E.F. & Bildsten L., 1998, "The ocean and crust of a rapidly accreting neutron star: Implications for magnetic field evolution and thermonuclear flashes," *Astrophys. J.* **496**, 915–933.

Brown G.E., Weise W., 1976, "Pion condensates," *Phys. Rep.* **27**, 1–34.

Brown G.E., Kubodera K., Page D., Pizzochero P., 1988, "Strangeness condensation and cooling of neutron stars," *Phys. Rev. D* **7**, 2042–2046.

Brown G.E., Chang H.L., Rho M., Thorsson V., 1994, "From kaon-nuclear interactions to kaon condensation," *Nucl. Phys.* **A 567**, 937–956.

Brown E.F., Bildsten L., Rutledge R.E., 1998, "Crustal heating and quiescent emission from transiently accreting neutron stars," *Astrophys. J.* **504**, L95–L98.

Brucato R.J., Kristian J., 1972, "Optical candidates for two X-ray sources," *Astrophys. J.* **173**, L105–L107.

Brueckner K.A., 1954, "Nuclear saturation and two-body forces. II. Tensor forces," *Phys. Rev.* **96**, 508–516.

Brueckner K.A., 1955, "Nuclear saturation and two-body forces. III. Details of the structure of the nucleus," *Phys. Rev.* **97**, 1353–1366.

Brueckner K.A., Levinson C.A., 1955, "Approximate reduction of the many-body problem for strongly interacting particles to a problem of self-consistent fields," *Phys. Rev.* **97**, 1344–1352.

Brueckner K.A., Levinson C.A., Mahmoud H.M., 1954, "Two-body forces and nuclear saturation. I. Central forces," *Phys. Rev.* **95**, 217–228.

Brueckner K.A., Coon S., Dabrowski J., 1968, "Nuclear symmetry energy," *Phys. Rev.* **168**, 1184–1188.

Brueckner K.A., Buchler J.R., Clark R., Lombard R.J., 1969, "Statistical theory of nuclei. II. Medium and heavy nuclei," *Phys. Rev.* **181**, 1543–1551.

Brush S.G., Sahlin H.L., Teller E., 1966, "Monte Carlo study of a one-component plasma. I," *J. Chem. Phys.* **45**, 2102–2118.

Buchler J.-R., Barkat Z., 1971a, "Properties of low-density neutron-star matter," *Phys. Rev. Lett.* **27**, 48–51.

Buchler J.-R., Barkat Z., 1971b, "Clustering of nucleons in low density neutron star matter," *Astrophys. Letters* **7**, 167–170.

Buchler J.-R., Ingber L., 1971, "Properties of the neutron gas and application to neutron stars," *Nucl. Phys.* **A 170**, 1–11.

Burgay M., D'Amico N., Possenti A., Manchester R.N., Lyne A.G., Joshi B.C., McLaughlin M.A., Kramer M., Sarkisian J.M., Camilo F., Kalogera V., Kim C., Lorimer D.R., 2003, "An increased estimate of the merger rate of double neutron stars from observations of a highly relativistic system," *Nature* **426**, 531–533.

Burkova L.A., Dzyaloshinskii I.E., Drukarev S.F., Monozon B.S., 1976, "Hydrogen-like system in crossed electric and magnetic fields," *Zh. Eksp. Teor. Fiz.* **71**, 526–530 [Engl. transl.: *Sov. Phys. JETP* **44**, 276–278].

Burrows A., 1990, "Neutrinos from supernova explosions," *Annu. Rev. Nucl. Part. Sci.* **40**, 181–212.

Burrows A., Lattimer J.M., 1984, "On the accuracy of the single-nucleus approximation in the equation of state of hot, dense matter," *Astrophys. J.* **285**, 294–303.

Burwitz V., Haberl F., Neuhäuser R., Predehl P., Trümper J., Zavlin V. E., 2003, "The thermal radiation of the isolated neutron star RX J1856.5−3754 observed with *Chandra* and *XMM-Newton*," *Astron. Astrophys.* **399**, 1109–1114.

Caldwell R.R., Friedman J.L., 1991, "Evidence against a strange ground state for baryons," *Phys. Lett. B* **264**, 143–148.

Calogero F., Palumbo F., Ragnisco O., 1973, "Density fluctuations in nuclear matter," in *Proceedings of the International Conference on Nuclear Physics - vol. 1*, edited by J. de Boer & H.-J. Mang (Amsterdam: North-Holland), 717.

Calogero F., Palumbo F., Ragnisco O., 1975, "Density fluctuations and nuclear structure," *Nuovo Cimento A* **29 A** ser. 2 (4), 509–547.

Cameron A.G.W., 1959, "Pycnonuclear reactions and nova explosions," *Astrophys. J.* **130**, 916–940.

Cameron A.G.W., 1999, "Adventures in cosmogony," *Annu. Rev. Astron. Astrophys.* **37**, 1–36.

Camilo F., Nice D.J., Shrauner J.A., Taylor J.H., 1996, "Princeton-Arecibo declination-strip survey for millisecond pulsars," *Astrophys. J.* **469**, 819–827.

Canal R., 1994, "Type I supernovae, white dwarfs and neutron stars," in *Supernovae*, edited by S. Bludman, R. Mochkovich, & J. Zinn-Justin (Amsterdam: North-Holland), 157–198.

Canuto V., 1975, "Equation of state at ultrahigh densities. Part 2," *Annu. Rev. Astron. Astrophys.* **13**, 335–380.

Canuto V., Chitre S.M., 1974, "Crystallization of dense neutron matter," *Phys. Rev. D* **9**, 1587–1613.

Canuto V., Chiu H.Y., 1971, "Intense magnetic fields in astrophysics," *Space Sci. Rev.* **12**, 3–74.

Canuto V., Ventura J., 1977, "Quantizing Magnetic Fields in Astrophysics," *Fundam. Cosm. Phys.* **2**, 203–353.

Caporaso G., Brecher K., 1979, "Must ultrabaric matter be superluminal?" *Phys. Rev. D* **20**, 1823–1831.

Caraveo P.A., Bignami G.F., Mignani R., Taff L.G., 1996, "Parallax observations with the Hubble Space Telescope yield the distance to Geminga," *Astrophys. J.* **461**, L91–L94.

Caraveo P.A., De Luca A., Mignani R., Bignami G.F., 2001, "The distance to the Vela pulsar gauged with Hubble Space Telescope parallax observations," *Astrophys. J.* **561**, 930–937.

Cardall C., Prakash M., Lattimer J.M., 2001, "Effects of strong magnetic fields on neutron star structure," *Astrophys. J.* **554**, 322–339.

Carlini A., Treves A., 1989, "A precessing neutron star model for E 2259+586," *Astron. Astrophys.* **215**, 283–286.

Carnahan N.F., Starling K.E., 1969, "Equation of state for nonattracting rigid spheres," *J. Chem. Phys.* **51**, 635–636.

Carr W.J., Jr., 1961, "Energy, specific heat, and magnetic properties of the low-density electron gas," *Phys. Rev.* **122**, 1437–1446.

Carr W.J., Jr., Coldwell-Horsfall R.A., Fein A.E., 1961, "Anharmonic contribution to the energy of a dilute electron gas – Interpolation for the correlation energy," *Phys. Rev.* **124**, 747–752.

Carter B., Quintana H., 1972, "Foundations of general relativistic high-pressure elasticity theory," *Proc. Roy. Soc. London Ser. A* **331**, 57–83.

Carter B., Quintana H., 1975a, "Relativistic formulation of the neutron starquake theory of pulsar glitches," *Ann. Phys. (N.Y.)* **95**, 74–89.

Carter B., Quintana H., 1975b, "Stationary elastic rotational deformation of a relativistic neutron star model," *Astrophys. J.* **202**, 511–522.

Carter B., Chamel N., Haensel P., 2005, "Entrainment coefficient and effective mass for conduction neutrons in neutron star crust: simple microscopic models," *Nucl. Phys.* **A 748**, 675–697.

Casares J., Charles P., Kuulkers E., 1998, "The mass of the neutron star in Cygnus X-2 (V1341 Cygni)," *Astrophys. J.* **493**, L39–L42.

Casares J., Dubus G., Shahbaz T., Zurita C., Charles, P.A., 2002, "VLTE spectroscopy of XTE J2123–058 during quiescence: the masses of the two components," *Mon. Not. R. Astron. Soc.* **329**, 29–36.

Cavedon J.M., Frois B., Goutte D., *et al.*, 1987 "Measurement of charge-density differences in the interior of Pb isotopes," *Phys. Rev. Lett.* **58**, 195–198.

Cazzola P., Lucaroni L., Scaringi C., 1966, "On the properties of nuclear matter at high densities," *Nuovo Cimento B* **43**, 250–258.

Centelles M., Viñas X., Barranco M., Marcos S., Lombard R.J., 1992, "Semiclassical approximations in non-linear $\sigma\omega$ models," *Nucl. Phys.* **A 537**, 486–500.

Centelles M., Viñas X., Barranco M., Schuck P., 1993, "A semiclassical approach to relativistic mean field theory," *Ann. Phys. (N.Y.)* **221**, 165–204.

Ceperley D.M., Alder B.J., "Ground state of the electron gas by a stochastic method," *Phys. Rev. Lett.* **45**, 566–569.

Chabanat E., Bonche P., Haensel P., Meyer J., Schaeffer R., 1997, "A Skyrme parameterization from subnuclear to neutron star densities," *Nucl. Phys.* **A 627**, 710–746.

Chabanat E., Bonche P., Haensel P., Meyer J., Schaeffer R., 1998, "A Skyrme parameterization from subnuclear to neutron star densities. Part II. Nuclei far from stabilities," *Nucl. Phys.* **A 635**, 231–256.

Chabrier G., 1990, "An equation of state for fully ionized hydrogen," *J. Phys. (Paris)* **51**, 1607–1632.

Chabrier G., 1993, "Quantum effects in dense Coulombic matter – Application to the cooling of white dwarfs," *Astrophys. J.* **414**, 695–700.

Chabrier G., 1999, "Review on white dwarf cooling theory," in *11th European Workshop on White Dwarfs*, edited by S.-E. Solheim & E.G. Meistas, *ASP Conf. Ser.* **169**, 369–377.

Chabrier G., Ashcroft N.W., 1990, "Linear mixing rule in screened binary ionic mixtures," *Phys. Rev. A* **42**, 2284–2291.

Chabrier G. & Baraffe I., 2000, "Theory of low-mass stars and substellar objects," *Annu. Rev. Astron. Astrophys.* **38**, 337–377.

Chabrier G., Potekhin A.Y., 1998, "Equation of state of fully ionized electron-ion plasmas," *Phys. Rev. E* **58**, 4941–4949.

Chabrier G., Ashcroft N.W., DeWitt H.E., 1992, "White dwarfs as quantum crystals," Nature **360**, 48–50.

Chadwick J., 1932, "Possible existence of a neutron," *Nature* **129**, 312.

Chakrabarty S., 1991, "Equation of state of strange quark matter and strange star," *Phys. Rev. D* **43**, 627–630.

Chakrabarty S., Raha S., Sinha B., 1989, "Strange quark matter and the mechanism of confinement," *Phys. Lett. B* **229**, 112–116.

Champion D.J., Lorimer D.R., McLaughlin M.A., Cordes J.M., Arzoumanian Z., Weisberg J.M., Taylor J.H., 2004, "PSR J1829+2456: a relativistic binary pulsar," *Mon. Not. R. Astron. Soc.* **350**, L61–L65.

Chandrasekhar S., 1931, "The maximum mass of ideal white dwarfs," *Astrophys. J.* **74**, 81–82.

Chandrasekhar S., 1935, "Stellar configurations with degenerate cores. (Second paper)," *Mon. Not. R. Astron. Soc.* **95**, 676–693.

Chandrasekhar S., 1939, *An Introduction to the Study of Stellar Structure* (Chicago: University of Chicago Press).

Chandrasekhar S., 1964, "The dynamical instability of gaseous masses approaching the Schwarzschild limit in General Relativity," *Astrophys. J.* **140**, 417–433.

Chandrasekhar S., 1970, "Solutions of two problems in the theory of gravitational radiation," *Phys. Rev. Lett.* **24**, 611–615.

Chandrasekhar S., Fermi E., 1953, "Problems of gravitational stability in the presence of a magnetic field," *Astrophys. J.* **118**, 116–141.

Chanmugam G., Gabriel M., 1971, "Neutron star models and pulsars," *Astron. Astrophys.* **13**, 374–379.

Chapline G., Nauenberg M., 1976, "Phase transition from baryon to quark matter," *Nature* **264**, 235–236.

Chatterjee S., Cordes J.M., 2004, "Smashing the Guitar: An evolving neutron star bow shock," *Astrophys. J.* **600**, L51–L54.

Chau H.F., Cheng K.S., 1993, "Correlated creep rate of a vortex line under the effect of vortex tension and its relation to the glitches of pulsars," *Phys. Rev. B* **47**, 2707–2714.

Cheng F.H., Vrtileck S.D. & Raymond J.C., 1995, "An archival study of Hubble Space Telescope observations of Hercules X-1/HZ Herculis," *Astrophys. J.* **452**, 825–832.

Cheng K.S., Yao C.C., Dai Z.G., 1997, "Properties of nuclei in the inner crusts of neutron stars in the relativistic mean-field theory," *Phys. Rev. C* **55**, 2092–2100.

Chen Z., Goldman S.P., 1992, "Relativistic and nonrelativistic finite-basis-set calculations of low-lying levels of hydrogenic atoms in intense magnetic fields," *Phys. Rev. A* **45**, 1722–1731.

Chin S.A., Kerman A.K, 1979, "Possible long-lived hyperstrange multiquark droplets," *Phys. Rev. Lett.* **43**, 1292–1295.

Chiu H.-Y., 1964, "Supernovae, neutrinos and neutron stars," *Ann. Phys.* **26**, 364–410.

Chiu H.-Y., Salpeter E.E., 1964, "Surface X-ray emission from neutron stars," *Phys. Rev. Lett.* **12**, 413–415.

Chmaj T., Haensel P., S lomiński W., 1991, "Photon emissivity of strange matter," *Nucl. Phys.* **B 24**, 40–44.

Chodil G., Mark H., Rodrigues R., Seward F.D., Swift C.D., 1967, "X-Ray intensities and spectra from several cosmic sources," *Astrophys. J.* **150**, 57–65.

Chodos A., Jaffe R.L., Johnson K., Thorn C.B., Weisskopf V.F., 1974, "New extended model of hadrons," *Phys. Rev. D* **9**, 3471–3495.

Christiansen M.B., Glendenning N.K., 1997, "Finite size effects and the mixed quark-hadron phase in neutron stars," *Phys. Rev. C* **56**, 2858–2864.

Christiansen M.B., Glendenning N.K., Schaffner-Bielich J., 2000, "Surface tension between a kaon condensate and the normal nuclear matter phase," *Phys. Rev. C* **62**, 025804 (6 pages).

Chubarian E., Grigorian H., Poghosyan G., Blaschke D., 2000, "Deconfinement transition in rotating compact stars," *Astron. Astrophys.* **357**, 968–976.

Clark J.P., Goodwin S.P., Crowther P.A., Kaper L., Fairbairn M., Langer N., Brocksopp C., 2002, "Physical parameters of the high-mass X-ray binary 4U 1700−37," *Astron. Astrophys.* **392**, 909–920.

Cleymans J., Gavai R.V., Suhonen E., 1986, "Quarks and gluons at high temperatures and densities," *Phys. Rep.* **130**, 217–292.

Cocozza G., Ferraro F.R., Possenti A., D'Amico N., 2006, "The puzzling properties of the helium white dwarf orbiting the millisecond pulsar PSR J1911-5958A in NGC 6752," *Astrophys. J.* **641**, L129–L132.

Coelho H.T., Das T.K., Robilotta M.S., 1983, " Two-pion-exchange three-nucleon force and the $^3$H and $^3$He nuclei," *Phys. Rev. C* **29**, 1812–1828.

Cohen E.G.D., Murphy T.J., 1969, "New results in the theory of the classical electron gas," *Phys. Fluids* **12**, 1403–1411.

Cohen J.M., Cameron A.G.W., 1971, "Neutron star models including the effects of hyperon formation," *Astrophys. Space Sci.* **10**, 227–245.

Cohen R., Lodenquai J., Ruderman, M., 1970, "Atoms in Superstrong Magnetic Fields," *Phys. Rev. Lett.* **25**, 467–469.

Coldwell-Horsfall R.A., Maradudin A.A., 1960, "Zero-point energy of an electron lattice," *J. Math. Phys.* **1**, 395–404.

Collins J.C., Perry M.J., 1975, "Superdense matter: neutrons or asymptotically free quarks?" *Phys. Rev. Lett.* **34**, 1353–1356.

Collins G.W. II, Claspy W.P., Martin J.C., 1999, "A reinterpretation of historical references to the supernova of A.D. 1054," *Publ. Astron. Soc. of Pacific* **111**, 871–880.

Colpi M., Miller J.C., 1992, "Rotational properties of strange stars," *Astrophys. J.* **388**, 513–520.

Colpi M., Shapiro S.L., Teukolsky S.A., 1989, "Exploding neutron stars near the minimum mass," *Astrophys. J.* **339**, 318–338.

Colpi M., Shapiro S.L., Teukolsky S.A., 1989, "Explosion of neutron stars near the minimum mass," *Astrophys. J.* **369**, 422–439.

Comella J.M., Craft H.D., Lovelace R.V.E., Sutton J.M., Tyler G.L., 1969, "Crab Nebula pulsar NP 0532," *Nature* **221**, 453–454.

Cook G.B., Shapiro S.L., Teukolsky S.A., 1994, "Rapidly rotating neutron stars in general relativity: Realistic equations of state," *Astrophys. J.* **424**, 823–845.

Coon S.A., Barrett R.C., 1987, "$\rho$–$\omega$ mixing in nuclear charge asymmetry," *Phys. Rev. C* **36**, 2189–2194.

Coon S.A., Peña M.T., 1993, "Momentum and coordinate space three-nucleon potentials," *Phys. Rev. C* **48**, 2559–2575.

Coon S.A., Scadron M.D., 1998, "Vector meson dominance and $\rho - \omega$ mixing," *Phys. Rev. C* **58**, 2958–2962.

Coon S.A., Peña M.T., Ellis R.G., 1984, "Three-nucleon force and the triton binding energy," *Phys. Rev. C* **30**, 1366–1369.

Cordes J.M., 1993, "The detectability of planetary components to radio pulsars," in *Planets Around Pulsars, ASP Conf. Ser.* **36**, 43–60.

Cordes J.M., Romani R.W., Lundgren S.C., 1993, "The Guitar nebula – A bow shock from a slow-spin, high-velocity neutron star," *Nature* **362**, 133–135.

Cordes J.M., Greenstein G., 1981, "Pulsar timing. IV – Physical models for timing noise processes," *Astrophys. J.* **245**, 1060–1079.

Cornelisse R., Heise J., Kuulkers E., Verbunt F., in't Zand J.J.M., 2000, "The longest thermonuclear X-ray burst ever observed? A BeppoSAX Wide Field Camera observation of 4U 1735–44," *Astron. Astrophys.* **357**, L21–L24.

Cornu F., 1998, "Quantum plasmas with or without a uniform magnetic field. I – III," *Phys. Rev. E* **58**, 5268–5346.

Cottam J., Paerels F., Mendez M., 2002, "Gravitationally redshifted absorption lines in the X-ray burst spectra of a neutron star," *Nature* **420**, 51–54.

Cowan R.D., Ashkin J., 1957, "Extension of the Thomas–Fermi–Dirac statistical theory of the atom to finite temperatures," *Phys. Rev.* **105**, 144–157.

Cumming A., Zweibel E., Bildsten L., 2003, "Magnetic screening in accreting neutron stars," *Astrophys. J.* **557**, 958–966.

Cumming A., Arras P., Zweibel E.G., 2004, "Magnetic field evolution in neutron star crusts due to the Hall effect and Ohmic decay," *Astrophys. J.* **609**, 999–1017.

Cutler C., Ushomirsky, G., Link B., 2003, "The crustal rigidity of a neutron star and implications for PSR B1828−11 and other precession candidates," *Astrophys. J.* **588**, 975–991.

Dall'Osso S., Israel G.L., Stella L., Possenti A., Perozzi E., 2003, "The glitches of the anomalous X-ray pulsar 1RXS J170849.0–400910," *Astrophys. J.* **599**, 485–497.

D'Amico N., Bailes M., Lyne A.G., Manchester R.N., Johnston S., Fruchter A.S., Goss W.M., 1993, "PSR B1802–07 – A globular cluster pulsar in an eccentric binary orbit," *Mon. Not. R. Astron. Soc.* **260**, L7–L10.

D'Amico N., Lyne A.G., Manchester R.N., Possenti A., Camilo F., 2001a, "Discovery of short-period binary millisecond pulsars in four globular clusters," *Astrophys. J.* **548**, L171–L174.

D'Amico N., Possenti A., Manchester R.N., Sarkissian J., Lyne A.G., Camilo F., 2001b, "An eclipsing millisecond pulsar with a possible main-sequence companion in NGC 6397," *Astrophys. J.* **561**, L89–L92.

Damour T., Taylor J.H., 1992, "Strong-field tests of relativistic gravity and binary pulsars," *Phys. Rev. D* **45**, 1840–1868.

Dandrea R.D., Ashcroft N.W., Carlsson A.E., 1986, "Electron liquid at any degeneracy," *Phys. Rev. B* **34**, 2097–2111.

Danz R.W., Glasser M.L., 1971, "Exchange energy of an electron gas in a magnetic field," *Phys. Rev. B* **4**, 94–99.

Däppen W., 1992, "The equation of state for stellar envelopes: Comparison of theoretical results," *Rev. Mex. Astron. Astrofis.* **23**, 141–149.

Däppen W., Anderson L., Mihalas D., 1987, "Statistical mechanics of partially ionized stellar plasma – The Planck-Larkin partition function, polarization shifts, and simulations of optical spectra," *Astrophys. J.* **319**, 195–206.

Datta B., Alpar M.A., 1993, "Implications of the crustal moment of inertia for neutron-star equations of state," *Astron. Astrophys.* **275**, 210–212.

Datta B., Thampan A.V., Bombaci I., 1998, "Equilibrium sequences of rotating neutron stars for new realistic equations of state," *Astron. Astrophys.* **334**, 943–952.

Davidson K., Fesen R. A., 1985, "Recent developments concerning the Crab Nebula," *Annu. Rev. Astron. Astrophys.* **23**, 119–146.

De Blasio F.V., 1995, "Nucleation of a dense plasma and crystallization of neutron star crusts," *Astrophys. J.* **452**, 359–363.

De Blasio F.V., 1998, "Crustal impurities and the internal temperature of a neutron star crust," *Mon. Not. R. Astron. Soc.* **299**, 118–122.

De Blasio F.V., 2000, "A dense two-component plasma in a strong gravity field and thermal conductivity of neutron stars," *Astron. Astrophys.* **353**, 1129–1133.

Debye P., Hückel E., 1923, "Zur Theorie der Elektrolythe," *Physikalische Z.* **24**, 185–206.

de Gennes P.G., Prost J., 1993, *The Physics of Liquid Crystals*, 2nd ed. (Oxford: Clarendon Press).

DeGrand T., Jaffe R.L., Johnson K., Kiskis J., 1975, "Masses and other parameters of light baryons," *Phys. Rev. D* **12**, 2060–2076.

Demeur M., Heenen P.H., Godefroid M., 1994, "Hartree-Fock study of molecules in very intense magnetic fields," *Phys. Rev. A* **49**, 176–183.

Detmer T., Schmelcher P., Cederbaum L.S., 1998, "Hydrogen molecule in a magnetic field: The lowest states of the Π manifold and the global ground state of the parallel configuration," *Phys. Rev. A* **57**, 1767–1777.

Detweiler S.L., Ipser J.R., 1973, "A variational principle an a stability criterion for the nonradial modes of pulsation of stellar models in general relativity," *Astrophys. J.* **185**, 685–707.

Dewey R.J., Taylor J.H., Weisberg J.M., Stokes G.H., 1985, "A search for low-luminosity pulsars," *Astrophys. J.* **294**, L25–L29.

DeWitt H.E., Slattery W.L., Yang J., 1993, "Monte Carlo simulations of the OCP freezing transition," in *Strongly Coupled Plasma Physics*, edited by H.M. Van Horn & S. Ichimaru (Rochester: University of Rochester Press), 425–434.

DeWitt H.E., Schlanges M., Sakakura A.Y., Kraeft W.D., 1995, *Phys. Lett. A* **197**, 326–329.

DeWitt H., Slattery W., Chabrier G., 1996, "Numerical simulation of strongly coupled binary ionic plasmas," *Physica B* **228**, 21–26.

DeWitt H., Slattery W., 1999, "Screening enhancement of thermonuclear reactions in high density stars," *Contrib. Plasma Phys.* **39**, 97–100.

Dey M., Bombaci I., Dey J., Ray S., Samanta B.C., 1998, "Strange stars with realistic quark vector interaction and phenomenological density-dependent scalar potential," *Phys. Lett. B* **438**, 123–128.

Dodson R., Legge D., Reynolds J.E., McCulloch P.M., 2003, "The Vela Pulsar's proper motion and parallax derived from VLBI observations," *Astrophys. J.* **596**, 1137–1141.

Doroba W., 1989, "Strange stars: are they bare?" *Acta Physica Polonica B* **20**, 967–979.

Douchin F., 1999, PhD thesis [in French] (École Normale Supérieure de Lyon: Lyon), unpublished.

Douchin F., Haensel P., 2000, "Inner edge of neutron-star crust with SLy effective nucleon-nucleon interactions," *Phys. Lett. B* **485**, 107–114.

Douchin F., Haensel P., 2001, "A unified equation of state of dense matter and neutron star structure," *Astron. Astrophys.* **380**, 151–167.

Douchin F., Haensel P., Meyer J., 2000, "Nuclear surface and curvature properties for SLy Skyrme forces and nuclei in the inner neutron-star crust," *Nucl. Phys.* **A 665**, 419–446.

Douvion, T., Lagage, P. O., Cesarski, C. J., Dwek, E., 2001, "Dust in the Tycho, Kepler and Crab supernova remnants," *Astron. Astrophys.* **373**, 281–291.

Drago A., Lavagno A., Parenti I., 2005, "Burning of an hadronic star into a quark star or a hybrid star" [astro-ph/0512652].

Drake J.J., Marshall H.L., Dreizler S., *et al.*, 2002, "Is RX J1856.5–3754 a quark star?" *Astrophys. J.* **572**, 996–1001.

Dreizler R.M., Gross E.K.U., 1990, *Density Functional Theory: An Approach to the Quantum Many-Body Problem* (Berlin: Springer).

Duerr H.-P., 1956, "Relativistic effects in nuclear forces," *Phys. Rev.* **103**, 469–480.

Duez M.D., Liu Y.T., Shapiro S.L., Shibata M., Stephens B.C., 2006, "Collapse of magnetized hypermassive neutron stars in general relativity," *Phys. Rev. Lett.* **96**, 031101 (4 pages).

Duncan J., 1939, "Second report on the expansion of the Crab Nebula," *Astrophys. J.* **89**, 482–485.

Duncan R.C., Thompson C., 1992, "Formation of very strongly magnetized neutron stars — Implications for gamma-ray bursts," *Astrophys. J.* **392**, L9–L13.

Dutta A.K., Onsi M., Pearson J.M., 2004 "Proton-shell effects in neutron-star matter," *Phys. Rev. C* **69**, 052801 (4 pages).

Dyugaev A.M., 1975, "Nature of phase transition during $\pi$-condensation," *Pis'ma v Zh. Eksp. Teor. Fiz.* **22**, 181–185 [Engl. transl.: 1976, *Sov. Phys.–JETP Lett.* **22**, 83–85].

Ebeling W., Kraeft W.D., Kremp D., 1977, *Theory of Bound States and Ionization Equilibrium of Plasmas and Solids* (Berlin: Akademie).

Ebeling W., Richert W., 1985, "Plasma phase transition in hydrogen," *Phys. Lett. A* **108**, 80–82.

Edelstein J., Seon K.I., Golden A., Min K.-W., 2000, "Extreme Ultraviolet Explorer observations of PSR B0656+14," *Astrophys. J.* **539**, 902–907.

Ellison D.C., Kazanas D., 1983, "Corequake and shock heating model of the 5 March 1979 gamma ray burst," *Astron. Astrophys.* **28**, 102–109.

Engelmann Ch., *et al.*, 1995, "Production and identification of heavy Ni isotopes: evidence for the doubly magic nucleus $^{78}_{28}$Ni," *Z. f. Physik* **A 352**, 351–352.

Ergma E., Sarna M.J., 2002, "Eclipsing binary millisecond pulsar PSR J1740–5340 — evolutionary considerations and observational test," *Astron. Astrophys.* **399**, 237–241.

Evans W.D., Klebesadel R.W., Laros J.G., Cline T.L., Desai U.D., Teegarden B.J., Pizzichini G., Hurley K., Niel M., Vedrenne G., 1980, "Location of the gamma-ray transient event of 1979 March 5," *Astrophys. J.* **237**, L7–L9.

Ewald P.P., 1921, "Die Berechnung optischer und elektrostatischer Gitterpotentiale," *Ann. Phys. (Leipzig)* **64**, 253–287.

Faddeev L.D., 1960, "Scattering theory for a three-particle system," *Zh. Eksp. Teor. Fiz.* **39**, 1459–1467 [Engl. transl.: *Sov. Phys. JETP*, **12**, 1014].

Faddeev, L.D., 1965, *Mathematical Aspects of the Three-Body Problem in Quantum Scattering Theory* (New York: Davey).

Faulkner A.J., Kramer M., Lyne A.G., Manchester R.N., McLaughlin M.A., Stairs I.H., Hobbs G., Possenti A., Lorimer D.R., D'Amico N., Camilo F., Burgay M., 2005, "PSR J1756–2251: a new relativistic double neutron star system," *Astrophys. J.* **618**, L119–L122.

Farhi E., Jaffe R.L., 1984, "Strange matter," *Phys. Rev. D* **30**, 2379–2390.

Farouki R.T., Hamaguchi S., 1993, "Thermal energy of the crystalline one-component plasma from dynamical simulations," *Phys. Rev. E* **47**, 4330–4336.

Fermi E., 1924, "Über die Wahrscheinlichkeit der Quantenzustände," *Z. f. Physik* **26**, 54–56.

Ferraro F.R., Possenti A., D'Amico N., Sabbi, E., 2001, "The bright optical companion to the eclipsing millisecond pulsar in NGC 6397," *Astrophys. J.* **561**, L93–L96.

Fesen R.A., Shull J.M., Hurdford A.P., 1997, "An optical study of the circumstellar environment around the Crab Nebula," *Astron. J.* **113**, 354–363.

Fetter A.L., Walecka J.D., 1971, *Quantum Theory of Many-Particle Systems* (New York: McGraw-Hill).

Feynman R.P., Metropolis N., Teller E., 1949, "Equations of state of elements based on the generalized Fermi–Thomas theory," *Phys. Rev.* **75**, 1561–1573.

Fishman G.J., Kouveliotou C., van Paradijs J., Harmon B.A., Paciesas W.S., Briggs M.S., Kommers J., Lewin W.H.G., 1995, "Galactic center," *IAU Circular* No. 6272.

Flowers E., Itoh N., 1976, "Transport properties of dense matter," *Astrophys. J.* **206**, 218–242.

Flowers E., Itoh N., 1981, "Transport properties of dense matter. III. Analytic formulae for thermal conductivity," *Astrophys. J.* **250**, 750–752.

Flowers E.G., Ruderman M., Sutherland P.G., 1976, "Neutrino-pair emission from finite-temperature neutron superfluid and the cooling of young neutron stars," *Astrophys. J.* **205**, 541–544.

Fontaine G., Graboske H.C., Jr., Van Horn H.M., 1977, "Equation of state for stellar partial ionization zones," *Astrophys. J. Suppl. Ser.* **35**, 293–358.

Forster H., Strupat W., Rösner W., Wunner G., Ruder H., Herold H., 1984, "Hydrogen atoms in arbitrary magnetic fields. II. Bound-bound transitions," *J. Phys. B: At. Mol. Opt. Phys.* **17**, 1301–1319.

Foster R.S., Wolszczan A., Camilo F., 1993, "A new binary millisecond pulsar," *Astrophys. J.* **410**, L91–L94.

Freedman B.A., McLerran L.D., 1977, "Fermions and gauge vector mesons at finite temperature and density. III. The ground-state energy of a relativistic quark gas," *Phys. Rev. D* **16**, 1169–1185.

Freire P., Jacoby B., Bailes M., Stairs I., Mott A., Ferdman R., Nice D., Backer D.C., 2006, "Discovery and timing of the PSR J1741+1351 binary pulsar," *AAS Meeting* **208**, No. 72.06.

Frenkel J., 1928, "Andwendung der Pauli-Fermischen Elektronengastheorie auf das Problem der Kohäsionkräfte," *Z. f. Physik* **50**, 234–248.

Friar J.L., 1975, "Relativistic effects on the wave function of a moving system," *Phys. Rev. C* **12**, 695–698.

Friedman B., Pandharipande V.R., 1981, "Hot and cold, nuclear and neutron matter," *Nucl. Phys.* **A 361**, 502–520.

Friedman J.L., Schutz B.F., 1978, "Secular instability of rotating Newtonian stars," *Astrophys. J.* **222**, 281–296.

Friedman J.L., Ipser J.R., Sorkin R.D., 1988, "Turning-point method for axisymmetric stability of rotating relativistic stars," *Astrophys. J.* **325**, 722-724.

Friedman J.L., Ipser J.R., Parker L., 1989, "Implications of a half-millisecond pulsar," *Phys. Rev. Lett.* **62**, 3015–3019.

Frieman J.A., Olinto A.V., 1989, "Is the sub-millisecond pulsar strange?" *Nature* **341**, 633–635.

Fryer C.L., Woosley S.E., 1998, "Gamma-ray bursts from neutron star phase transitions," *Astrophys. J.* **501**, 780–786.

Fryer C., Benz W., Herant M., Colgate S.A., 1999, "What can the accretion-induced collapse of white dwarfs really explain?" *Astrophys. J.* **516**, 892–899.

Fuchs K., 1935, "A quantum mechanical investigation of the cohesive forces of metallic copper," *Proc. Roy. Soc. London A* **151**, 585–602.

Fuchs K., 1936, "Quantum-mechanical calculation of the elastic constants of monovalent metals," *Proc. Roy. Soc. London A* **153**, 622–639.

Fuchs C., 2004, "The relativistic Dirac-Brueckner approach to nuclear matter," *Lecture Notes in Physics* **641**, 119–146.

Fujimoto M.Y., Hanawa T., Iben I., Jr., Richardson M.B., 1984, "Thermal evolution of accreting neutron stars," *Astrophys. J.* **278**, 813–824.

Fushiki I., Gudmundsson E.H., Pethick C.J., 1989, "Surface structure of neutron stars with high magnetic fields," *Astrophys. J.* **342**, 958–975.

Galam S., Hansen J.P., 1976, "Statistical mechanics of dense ionized matter. VI. Electron screening corrections to the thermodynamic properties of the one-component plasma," *Phys. Rev. A* **14**, 816–832.

Galloway D.K., Morgan E.H., Levine A.M., 2004, "A frequency glitch in an accreting pulsar," *Astrophys. J.* **613**, 1164–1172.

Galloway D.K., Markwardt C.B., Morgan E.H., Chakrabarty D., Strohmayer T.E., 2005, "Discovery of the accretion-powered millisecond X-ray pulsar IGR J00291+5934," *Astrophys. J.* **622**, L45–L48.

Gamow G., 1937, *Structure of Atomic Nuclei and Nuclear Transformations* (Oxford: Oxford University Press).

Gavriil F.P., Kaspi V.M., Woods P.M., 2002, "Magnetar-like X-ray bursts from an anomalous X-ray pulsar," *Nature* **419**, 142–144.

Gerlach U.H., 1968, "Equation of state at supranuclear densities and the existence of a third family of superdense stars," *Phys. Rev.* **172**, 1325–1330.

Giacconi R., Gursky H., Paolini F.R., Rossi B.B., 1962, "Evidence for X-rays from sources outside the solar system," *Phys. Rev. Lett.* **9**, 439–443.

Giacconi R., Murray S., Gursky H., Kellogg E., Schreier E., Tananbaum H., 1972, "The UHURU catalog of X-ray sources," *Astrophys. J.* **178**, 281–308.

Giles A.B., Swank J.H., Jahoda K., Zhang W., Strohmayer T., Stark M.J., Morgan E.H., 1996, "The main characteristics of GRO J1744-28 observed by the proportional counter array experiment on the *Rossi X-Ray Timing Explorer*," *Astrophys. J.* **469**, L25–L28.

Ginzburg V.L., 1964, "Magnetic fields of collapsing masses and the nature of superstars," *Dokl. Akad. Nauk SSSR* **156**, 43–46 [Engl. transl.: *Sov. Phys. Doklady* **9**, 329–332].

Ginzburg V.L., 1969, "Superfluidity and superconductivity in the Universe," *Uspekhi Fiz. Nauk* **97**, 601–619.

Ginzburg V.L., Kirzhnits D.A., 1964, "On the superconductivity of neutron stars," *Zh. Teor. Eksper. Fiz.* **47**, 2006–2007.

Glendenning N.K., 1985, "Neutron stars are giant hypernuclei?" *Astrophys. J.* **293**, 470–493.

Glendenning N.K., 1989a, "Fast pulsar in SN 1987A: Candidate for strange-quark matter," *Phys. Rev. Lett.* **63**, 2629–2632.

Glendenning N.K., 1989b, "PSR 1987A: the case for strange-quark stars," *J. Phys. G: Nucl. Part. Phys.* **15**, L255–L260.

Glendenning N.K., 1991, "Fast pulsars, variational bounds, other facets of compact stars," *Nucl. Phys. B – Proc. Suppl.* **24 B**, No. 2, 110–118.

Glendenning N.K., 1992, "First-order phase transitions with more than one conserved charge: Consequences for neutron stars," *Phys. Rev. D* **46**, 1274–1287.

Glendenning N.K., 2000, *Compact Stars: Nuclear Physics, Particle Physics, and General Relativity*, 2nd ed. (New York: Springer).

Glendenning N.K., Kettner C., 2000, "Possible third family of compact stars more dense than neutron stars," *Astron. Astrophys.* **353**, L9–L12.

Glendenning N.K., Moszkowski S.A., 1991, "Reconciliation of neutron-star masses and binding of the Λ in hypernuclei," *Phys. Rev. Lett.* **67**, 2414–2417.

Glendenning N.K., Pei S., 1995, "Crystalline structure of the mixed confined-deconfined phase in neutron stars," *Phys. Rev. C* **52**, 2250–2253.

Glendenning N.K., Schaffner-Bielich J., 1998, "Kaon condensation and dynamical nucleons in neutron stars," *Phys. Rev. Lett.* **81**, 4564–4567.

Glendenning N.K., Schaffner-Bielich J., 1999, "First order kaon condensate," *Phys. Rev. C* **60**, 025803 (13 pages).

Glendenning N.K., Weber F., 1992, "Nuclear solid crust on rotating strange stars," *Astrophys. J.* **400**, 647–658.

Glendenning N.K., Banerjee B., Gyulassy M., 1983a, "Normal and pion-condensed states in neutron star matter in a relativistic theory constrained by bulk nuclear properties. I," *Ann. Phys. (N.Y.)* **149**, 1–21.

Glendenning N.K., Hecking P., Ruck V., 1983b, "Normal and pion-condensed states in neutron star matter in a relativistic theory constrained by bulk nuclear properties. II," *Ann. Phys. (N.Y.)* **149**, 22–43.

Glendenning N.K., Weber F., Moszkowski S.A., 1992, "Neutron stars in the derivative coupling model," *Phys. Rev. C* **45**, 844–855.

Glendenning N.K., Kettner Ch., Weber F., 1995, "Possible class of dense white dwarfs," *Phys. Rev. Lett.* **74**, 3519–3521.

Glendenning N.K., Pei S., Weber F., 1997, "Signal of quark deconfinement in the timing structure of pulsar spin-down," *Phys. Rev. Lett.* **9**, 1603–1606.

Gnedin Yu.N., Sunyaev R.A., 1974, "Polarization of optical and X-radiation from compact thermal sources with magnetic field," *Astron. Astrophys.* **36**, 379–394.

Gnedin Yu.N., Pavlov G.G., Tsygan A.I., 1974, "Photoeffect in strong magnetic fields and the X-rays from neutron stars," *Zh. Eksp. Teor. Fiz.* **66**, 421–432 [Engl. transl.: *Sov. Phys. JETP* **39**, 301].

Gnedin O.Y., Yakovlev D.G., Potekhin A.Y., 2001, "Thermal relaxation in young neutron stars," *Mon. Not. R. Astron. Soc.* **324**, 725–736.

Gold T., 1968, "Rotating neutron stars as the origin of the pulsating radio sources," *Nature* **218**, 731–732.

Golden A., Shearer A., 1999, "Radius and distance estimates of the isolated neutron stars Geminga and PSR B0656+14 using optical photometry," *Astron. Astrophys.* **342**, L5–L8.

Goldman I., 1989, "New general-relativistic expression for the baryon number of a cold star," *Phys. Rev. D* **40**, 327–328.

Goldreich P., Julian W.H., 1969, "Pulsar electrodynamics," *Astrophys. J.* **157**, 869–880.

Goldstone J., 1957, "Derivation of the Brueckner many-body theory," *Proc. Roy. Soc. (London) A* **239**, 267–279.

Golenetskii S.V., Mazets E.P., Aptekar R.L., Gurian Yu.A., Ilinskii V.N., 1986, "Annihilation radiation in cosmic gamma-ray bursts," *Astrophys. Space Sci.* **124**, 243–278.

Gondek D., Zdunik J.L., 1995, "The effects of the phase transition on the maximum redshift of neutron stars," *Acta Astron.* **45**, 319–325.

Gondek-Rosińska D., Gourgoulhon E., 2002, "Jacobi-like bar mode instability of relativistic rotating bodies," *Phys. Rev. D* **66**, 044021 (11 pages).

Gondek-Rosińska D., Bulik T., Zdunik J.L., Gourgoulhon E., Ray E., Dey J., Dey M., 2000, "Rapidly rotating compact strange stars," *Astron. Astrophys.* **363**, 1005–1012.

Gondek-Rosińska D., Gourgoulhon E., Haensel P., 2003, "Are rotating strange stars good sources of gravitational waves?" *Astron. Astrophys.* **412**, 777–790.

Gourgoulhon E., Bonazzola S., 1994, "A formulation of the virial theorem in general relativity," *Class. Quantum Grav.* **11**, 443–452.

Gourgoulhon E., Haensel P., 1993, "Upper bounds on the neutrino burst from collapse of a neutron star into a black hole," *Astron. Astrophys.* **271**, 187–208.

Gourgoulhon E., Haensel P., Gondek D., 1995, "Maximum mass instability of neutron stars and weak interaction processes in dense matter," *Astron. Astrophys.* **294**, 747–756.

Gourgoulhon E., Haensel P., Livine R., Paluch E., Bonazzola S., Marck J.-A., 1999, "Fast rotation of strange stars, " *Astron. Astrophys.* **349**, 851–862.

Gourgoulhon E., Grandclément P., Taniguchi K., Marck J.-A., Bonazzola S., 2001, "Quasiequilibrium sequences of synchronized and irrotational binary neutron stars in general relativity: Method and tests," *Phys. Rev. D* **63**, 064029 (27 pages).

Goussard J.-O., Haensel P., Zdunik J.L., 1998, "Rapid differential rotation of protoneutron stars and constraints on radio pulsars periods," *Astron. Astrophys.* **330**, 1005–1016.

Green D.A., Stephenson F.R., 2002, *Historical supernovae and their remnants* (Oxford: Clarendon Press).

Green D.A., Stephenson F.R., 2003, "The historical supernovae," in *Supernovae and Gamma Ray Bursters*, edited by K.W. Weiler, *Lecture Notes in Physics* **598**, 7–19.

Grimm H.-J., Gilfanov M., Sunyaev R., 2003, "High-mass X-ray binaries as a star formation rate indicator in distant galaxies," *Mon. Not. R. Astron. Soc.* **339**, 793–809.

Grindlay J., Gursky H., Schnopper H., Parsignault D.R., Heise J., Brinkman A.C., Schrijver J., 1976, "Discovery of intense X-ray bursts from the globular cluster NGC 6624," *Astrophys. J.* **205**, L127–L130.

Groote V., Hilf E.R., Takahashi K., 1976, "A new semiempirical shell correction to the droplet model," *At. Data Nucl. Data Tables* **17**, 418–427.

Gudmundsson E.H., Pethick C.J., Epstein R.I., 1983, "Structure of neutron star envelopes," *Astrophys. J.* **272**, 286–300.

Gusakov M.E., Kaminker A.D., Yakovlev D.G., Gnedin O.Y., 2005, "The cooling of Akmal-Pandharipande-Ravenhall neutron star models," *Mon. Not. R. Astron. Soc.* **363**, 563–580.

Haberl F., Zavlin V.E., Trümper J., Burwitz V., 2004, "A phase-dependent absorption line in the spectrum of the X-ray pulsar RX J0720.4-3125," *Astron. Astrophys.* **419**, 1077–1085.

Haberl F., Turolla R., de Vries C.P., Zane S., Vink J., Méndez M., Verbunt F., 2006, "Evidence for precession of the isolated neutron star RX J0720.4−3125," *Astron. Astrophys.* **451**, L17–L21.

Haensel P., 1977, "Charge symmetry breaking nuclear forces and the properties of nuclear matter," *J. of Phys. G: Nuclear Physics* **3**, 373–380.

Haensel P., 1990, "The maximum moment of inertia of neutron stars and pulsar observations," in *The Magnetospheric Structure and Emission Mechanisms of Radio Pulsars*, edited by T.H. Hankins, J.M. Rankin, & J. Gil (Zielona Góra: Pedagogical University Press), 217–219.

Haensel P., 2001, "Apparent radii of neutron stars and equation of state of dense matter," *Astron. Astrophys.* **380**, 186–189.

Haensel P., Pichon B., 1994, "Experimental nuclear masses and the ground state of cold dense matter," *Astron. Astrophys.* **283**, 313–318.

Haensel P., Potekhin A.Y., 2004, "Analytical representations of unified equations of state of neutron-star matter," *Astron. Astrophys.* **428**, 191–197.

Haensel P., Prószyński M., 1982, "Pion condensation in cold dense matter and neutron stars," *Astrophys. J.* **258**, 306–320.

Haensel P., Schaeffer R., 1982, "Metastability of dense neutron matter," *Nucl. Phys.* **A 381**, 519–543.

Haensel P., Zdunik J.L., 1989, "A submillisecond pulsar and the equation of state of dense matter," *Nature* **340**, 617–619.

Haensel P., Zdunik J.L., 1990a, "Non-equilibrium processes in the crust of an accreting neutron star," *Astron. Astrophys.* **227**, 431–436.

Haensel P., Zdunik J.L., 1990b, "Equation of state and structure of an accreting neutron star," *Astron. Astrophys.* **229**, 117–122.

Haensel P., Zdunik J.L., 2003, "Nuclear composition and heating in accreting neutron-star crusts," *Astron. Astrophys.* **404**, L33–L36.

Haensel P., Zdunik J.L., Schaeffer R., 1986a, "Strange quark stars," *Astron. Astrophys.* **160**, 121–128.

Haensel P., Zdunik J.L., Schaeffer R., 1986b, "Changes in neutron star parameters implied by a neutron star corequake," *Astron. Astrophys.* **160**, 251–258.

Haensel P., Zdunik J.L., Dobaczewski J., 1989, "Composition and equation of state of cold catalyzed matter below neutron drip," *Astron. Astrophys.* **222**, 353–357.

Haensel P., Denissov A., Popov S., 1990a, "Neutron star corequake implied by pion condensation. Dynamic, neutrino and thermal effects," *Astron. Astrophys.* **240**, 78–84.

Haensel P., Urpin V.A., Yakovlev D.G., 1990b, "Ohmic decay of internal magnetic fields in neutron stars," *Astron. Astrophys.* **229**, 133–137.

Haensel P., Salgado M., Bonazzola S., 1995, "Equation of state of dense matter and maximum rotation frequency of neutron stars," *Astron. Astrophys.* **296**, 745–751.

Haensel, P., Kaminker, A.D., Yakovlev, D.G., 1996, "Electron $\nu\bar{\nu}$ bremsstrahlung in a liquid phase of neutron star crusts," *Astron. Astrophys.* **314**, 328–340.

Haensel P., Lasota J.P., Zdunik J.L., 1999, "On the minimum period of uniformly rotating neutron stars," *Astron. Astrophys.* **344**, 151–153.

Haensel P., Levenfish K.P., Yakovlev D.G., 2002a, "Bulk viscosity in superfluid neutron star cores. III. Effects of $\Sigma^-$ hyperons," *Astron. Astrophys.* **381**, 1080–1089.

Haensel P., Levenfish K.P., Yakovlev D.G., 2002b, "Adiabatic index of dense matter and damping of neutron star pulsations," *Astron. Astrophys.* **394**, 213–217.

Haensel P., Zdunik J.L., Douchin F., 2002b, "Equation of state of dense matter and the minimum mass of cold neutron stars," *Astron. Astrophys.* **385**, 301–307.

Halpern J.P., Martin C, Marshall H.L., 1996, "The Geminga Pulsar: Soft X-Ray Variability and an EUVE Observation," *Astrophys. J.* **473**, L37–L40.

Hamaguchi S., Farouki R.T., Dubin D.H.E., "Triple point of Yukawa systems," *Phys. Rev. E* **56**, 4671–4682.

Hammerschlag-Hensberge G., van Kerkwijk M.H., Kaper L., 2003, "The radial velocity curve of HD 153919 (4U 1700–37) revisited," *Astron. Astrophys.* **407**, 685–690.

Hanauske M., Zschiesche D., Pal S., Schramm S., Stöcker H., Greiner W., 2000, "Neutron star properties in a chiral SU(3) model," *Astrophys. J.* **537**, 958–963.

Hankins T.H., Kern J.S., Weatherall J.C., Eilek J.A., 2003, "Nanosecond radio bursts from strong plasma turbulence in the Crab pulsar," *Nature* **422**, 141–143.

Hansen J.P., 1973, "Statistical mechanics of dense ionized matter. I. Equilibrium properties of the classical one-component plasma," *Phys. Rev. A* **8**, 3096–3109.

Hansen B.M.S., 2004, "The astrophysics of cool white dwarfs," *Phys. Rep.* **399**, 1–70.

Hansen J.P., McDonald I.R., 1976, *Theory of Simple Liquids* (New York: Academic Press).

Hansen C.J., Van Horn H.M., 1975, "Steady-state nuclear fusion in accreting neutron-star envelopes," *Astrophys. J.* **195**, 735–741.

Hansen J.P., Torrie G.M., Vieillefosse P., 1977, "Statistical mechanics of dense ionized matter. VII. Equation of state and phase separation of ionic mixtures in a uniform background," *Phys. Rev. A* **16**, 2153–2168.

Harding A.K., Lai D., 2006, "Physics of strongly magnetized neutron stars," *Reports on Progress in Physics*, accepted [astro-ph/0606674].

Harries J.R., McCracken K.G., Francey R.J., Fenton A.J., 1967, "A strong X-ray source in the vicinity of the constellation Crux," *Nature* **215**, 38–40.

Harrison B.K., Wakano M., Wheeler J.A., 1958, "Matter-energy at high density: end point of thermonuclear evolution," in *La structure et évolution de l'univers* (Brussells: R. Stoops), 124–140.

Harrison B.K., Thorne K.S., Wakano M., Wheeler J.A., 1965, *Gravitation Theory and Gravitational Collapse* (Chicago: University of Chicago Press).

Hartle J.B., 1967, "Slowly rotating relativistic stars. I. Equations of structure," *Astrophys. J.* **150**, 1005–1030.

Hartle J.B., Sawyer R.F., Scalapino D.J., 1975, "Pion condensed matter at high densities: equation of state and stellar models," *Astrophys. J.* **199**, 471–481.

Hartle J.B., 1978, "Bounds on the mass and moment of inertia of non-rotating neutron stars," *Phys. Rep.* **46**, 201–247.

Hashimoto M., Seki H., Yamada M., 1984, "Shape of nuclei in the crust of neutron stars," *Prog. Theor. Phys.* **71**, 320–326.

Heinke C.O., Grindlay J.E., Edmonds P.D., 2005, "Three additional quiescent low-mass X-ray binary candidates in 47 Tucanae," *Astrophys. J.* **622**, 556–564.

Heiselberg H., Hjorth-Jensen M., 2000, "Phases of dense matter in neutron stars, " *Phys. Rep.* **328**, 237–327.

Heiselberg H., Pethick C.J., 1993, "Transport and relaxation in degenerate quark plasmas," *Phys. Rev. D* **48**, 2916–2928.

Heiselberg H., Pethick C.J., Staubo E.F., 1993, "Quark matter droplets in neutron stars," *Phys. Rev. Lett.* **70**, 1355–1359.

Hessels J.W.T., Ransom S.M., Stairs I.H.,Freire P.C.C., Kaspi V.M., Camilo, F., 2006, "A radio pulsar spinning at 716 Hz," *Science* **311**, 1901–1904.

Hewish A., 1975, "Pulsars and high density physics," *Rev. Mod. Phys.* **47**, 567–572.

Hewish A., Okoye S.E., 1965, "Evidence for an unusual source of high radio brightness temperature in the Crab nebula," *Nature* **207**, 59–60.

Hewish A., Bell S.J., Pilkington J.D.H., Scott P.F., Collins R.A., 1968, "Observation of a rapidly rotating radio source," *Nature* **217**, 709–713.

Higdon J.M., Lingenfelter R.E., 1990, "Gamma-ray bursts," *Annu. Rev. Astron. Astrophys.* **28**, 401–436.

Hiltner W.A., Werner J., Osmer P., 1972, "Binary nature of the B supergiant in the error box of the Vela X-ray source," *Astrophys. J.* **175**, L19–L22.

Ho W.C.G., Lai D., Chabrier G., Potekhin A.Y., 2003, "Atmospheres and spectra of strongly magnetized neutron stars – III. Partially ionized hydrogen models," *Astrophys. J.* **599**, 404–418.

Ho W.C.G., Kaplan D.L., Chang P., van Adelsberg M., Potekhin A.Y., 2006, "Magnetic hydrogen atmosphere models and the neutron star RX J1856.5−3754," *Mon. Not. R. Astron. Soc.*, submitted.

Hobbs G., Lyne A.G., Joshi B.C., Kramer M., Stairs I.H., Camilo F., Manchester R.N., D'Amico N., Possenti A., Kaspi V.M., 2002, "A very large glitch in PSR J1806–2125," *Mon. Not. R. Astron. Soc.* **333**, L7–L10.

Hoffberg M., Glassgold A.E., Richardson R.W., Ruderman M., 1970, "Anisotropic superfluidity in neutron star matter," *Phys. Rev. Lett.* **24**, 775–777.

Hoffmann S., Münzenberg G., 2000, "The discovery of the heaviest elements," *Rev. Mod. Phys.* **72**, 733–767.

Hohenberg P., Kohn W., 1964, "Inhomogeneous electron gas," *Phys. Rev.* **136**, B864–B871.

Hones E.W., Bergeson J.E. 1965, "Electric field generated by a rotation of magnetized sphere," *J. Geophys. Res.* **70**, 4951–4958.

Horvath J.E., Benvenuto O.G., 1988, "On the stability of slow neutron combustion in astrophysical objects," *Phys. Lett. B* **213**, 516–520.

Hotan A.W., Bailes M., Ord S.M., 2005, "Geodetic precession in PSR J1141-6545," *Astrophys. J.* **624**, 906–913.

Huang Y.F., Lu T., 1997, "Strange stars: how dense can their crust be?" *Astron. Astrophys.* **325**, 189–194.

Hubbard W.B., Macfarlane J.J., 1985, "Statistical mechanics of light elements at high pressure. VIII – Thomas-Fermi-Dirac theory for binary mixtures of H with He, C, and O," *Astrophys. J.* **297**, 133–144.

Hubble E., 1928, "Novae or temporary stars," *Astron. Soc. of Pacific Leaflet* **14**, 55–58.

Hulse R.A., Taylor J.H., 1975, "Discovery of a pulsar in a binary system," *Astrophys. J.* **195**, L51–L53.

Hummer D.G., Mihalas D., 1988, "The equation of state for stellar envelopes. I. An occupation probability formalism for the truncation of internal partition functions," *Astrophys. J.* **331**, 794–814.

Hund F., 1936, "Materie unter sehr hohen Drucken und Temperaturen," *Erg. exact. Naturwiss.* **15**, 189 [English translation in *Matter at High Densities in Astrophysics*, 1996, edited by Riffert H., Müther H., Herold H., & Ruder H. (Berlin: Springer), 217–257].

Hurley K., Cline T., Mazets E., Barthelmy S., Butterworth P., Marshall F., Palmer D., Aptekar R., Golenetskii S., Il'inskii V., Frederiks D., Mc Tiernan J., Gold R., Trombka J., 1999, "A giant periodic flare from the soft gamma-ray repeater SGR 1900+14," *Nature* **397**, 41–43.

Hurley K., Boggs S.E., Smith D.M., Duncan R.C., Lin R., Zoglauer A., Krucker S., Hurford G., Hudson H., Wigger C., Hajdas W., Thompson C., Mitrofanov I., Sanin A., Boynton W., Fellows C., von Kienlin A., Lichti G., Rau A., Cline T., 2005, "An exceptionally bright flare from SGR 1806–20 and the origins of short-duration $\gamma$-ray bursts," *Nature* **434**, 1098–1103.

Ichimaru S., Iyetomi H., Tanaka S., 1987, "Statistical physics of dense plasmas: thermodynamics, transport coefficients and dynamic correlations," *Phys. Rep.* **149**, 91–205.

Iida K., 1997, "Quantum kinetics of deconfinement transitions in dense nuclear matter – dissipation effects at low temperatures," *Prog. Theor. Phys.* **98**, 739–744.

Iida K., Sato K., 1997, "Quantum nucleation of two-flavor quark matter in neutron stars," *Prog. Theor. Phys.* **98**, 277–282.

Iida K., Sato K., 1998, "Effect of hyperons on the dynamical deconfinement transition in cold neutron star matter," *Phys. Rev. C* **58**, 2538–2559.

Imshennik V.S., Nadyozhin D.K., 1988, "Supernova 1987A in the Large Magellanic Cloud: Observations and theory," *Uspekhi Fiz. Nauk* **156**, 561–651 [Engl. transl.: 1989 (Amsterdam: Harwood)].

Imshennik V.S., Popov D.V., 1994, "Evolution of eccentric orbits of neutron star binaries emitting gravitational waves," *Astron. Lett.* **20**, 529–537.

Ingber L., 1968 "Nuclear forces," *Phys. Rev.* **174**, 1250–1263.

Inglis D.R., Teller E., 1939, "Ionic depression of series limits in one-electron spectra," *Astrophys. J.* **90**, 439–448.

Itzykson C., Zuber J.-B., 1980, *Quantum Field Theory* (New York: McGraw-Hill).

Itoh N., 1970, "Hydrostatic equilibrium of hypothetical quark stars," *Prog. Theor. Phys.* **44**, 291–292.

Itoh N., Kohyama Y., 1983, "Neutrino-pair bremsstrahlung in dense stars. I. Liquid metal case," *Astrophys. J.* **275**, 858–866.

Itoh N., Mitake S., Iyetomi H., Ichimaru S., 1983, "Electrical and thermal conductivities of dense matter in the liquid metal phase. I – High-temperature results," *Astrophys. J.* **273**, 774–782.

Ivanenko D., Kurdgelaidze D.F., 1965, "Hypothesis on quark stars," *Astrofizika* **1**, 479–482 [in Russian].

Ivanenko D., Kurdgelaidze D.F., 1969, "Remarks on quark stars," *Lett. Nuovo Cimento* **2**, 13–16.

Ivanov M.V., 1994, "Hartree-Fock mesh calculations of the energy levels of the helium atom in magnetic fields," *J. Phys. B: At. Mol. Opt. Phys.* **27**, 2663–2667.

Ivanov M.V., Schmelcher P., 2000, "Ground states of the atoms H, He,. . . , Ne and their singly positive ions in strong magnetic fields: The high field regime," *Phys. Rev. A* **61**, 022505 (13 pages).

Ivanov M.V., Schmelcher P., 2001, "The boron atom and boron positive ion in strong magnetic fields," *J. Phys. B: At. Mol. Phys.* **34**, 2031–2044.

Jacoby B.A., Bailes M., van Kerkwijk M.H., Ord S., Hotan A., Kulkarni S.R., Anderson S.B., 2003, "PSR J1909-3744: A binary millisecond pulsar with a very small duty cycle," *Astrophys. J.* **599**, L99–L102.

Jacoby B.A., Hotan A., Bailes M., Ord S., Kulkarni S.R., 2005, "The mass of a millisecond pulsar," *Astrophys. J.* **629**, L113–L116.

Jacoby B.A., Cameron P.B., Jenet F.A., Anderson S.B., Murty R.N., Kulkarni S.R., 2006, "Measurement of orbital decay in the double neutron star binary PSR B2127+11C," *Astrophys. J.* **644**, L113–L116.

Jaikumar P., Reddy S., Steiner A.W., 2006, "Strange star surface: A crust with nuggets," *Phys. Rev. Lett.* **96**, 041101 (4 pages).

Jancovici B., 1962, "On the relativistic degenerate electron gas," *Nuovo Cimento* **25**, 428–455.

Jänecke J., Eynon B.P., 1976, "Updated mass predictions from the Garvey-Kelson mass relations," *At. Data Nucl. Data Tables* **17**, 467–462.

Janka H.-T., 2004, "Neutron star formation and birth properties," in *Young Neutron Stars and Their Environments*, edited by F. Camilo & B.M. Gaensler (San Francisco: Astron. Soc. Pacific), 3–12.

Jin X., Jennings B.K., 1996, "Modified quark-meson coupling model for nuclear matter," *Phys. Rev. C* **54**, 1427–1436.

Jog C.J., Smith R.A., 1982, "Mixed lattice phases in cold dense matter," *Astrophys. J.* **253**, 839–841.

Johnson M.H., Teller E., 1955, "Classical field theory of nuclear forces," *Phys. Rev.* **98**, 783–787.

Johnson B.R., Hirschfelder J.O., Yang K.H., 1983, "Interaction of atoms, molecules and ions with constant electric and magnetic fields," *Rev. Mod. Phys.* **55**, 109–153.

Johnston S., Manchester R.N., Lyne A.G., Bailes M., Kaspi V.M., Qiao G., D'Amico N., 1992, "PSR 1259–63 – A binary radio pulsar with a Be star companion," *Astrophys. J.* **387**, L37–L41.

Johnston S., Lorimer D.R., Harrison P.A., Bailes M., Lyne A.G., Bell J.F., Kaspi V.M., Manchester R.N., D'Amico N., Nicastro L., 1993, "Discovery of a very bright nearby millisecond pulsar," *Nature* **361**, 613–615.

Jones P.B., 1985, "Density functional calculations of the ground-state energies and infinite linear molecules in very strong magnetic fields," *Mon. Not. R. Astron. Soc.* **216**, 503–510.

Jones P.B., 1986, "Properties of condensed matter in very strong magnetic fields," *Mon. Not. R. Astron. Soc.* **218**, 477–485.

Jones P.B., 1999, "Amorphous and heterogeneous phase of neutron star matter," *Phys. Rev. Lett.* **83**, 3589–3592.

Jones P.B., 2001a, "Comment on 'Gravitational radiation instability in hot young neutron stars," *Phys. Rev. Lett.* **86**, 1384.

Jones, P.B. 2001b, "Bulk viscosity of neutron-star matter," *Phys. Rev. D* **64**, 084003 (7 pages).

Jones P.B., 2001c, "First-principles point-defect calculations for solid neutron star matter," *Mon. Not. R. Astron. Soc.* **321**, 167–175.

Jones P.B., 2005, "Endpoint of the rp process and periodic gravitational wave emission," *Phys. Rev. D* **72**, 083006 (9 pages).

Jones C., Forman W., Tananbaum H., Schreier E., Gursky H., Kellogg E., Giacconi R., 1973, "Evidence for the binary nature of 2U 1700–37," *Astrophys. J.* **181**, L43–L48.

Jones M.D., Ceperley D.M., 1996, "Crystallization of the one-component plasma at finite temperature," *Phys. Rev. Lett.* **76**, 4572–4575.

Jones M.D., Ortiz G., Ceperley D.M., 1996, "Hartree-Fock studies of atoms in strong magnetic fields," *Phys. Rev. A* **54**, 219–231.

Jones M.D., Ortiz G., Ceperley D.M., 1998, "Spectrum of neutral helium in strong magnetic fields," *Phys. Rev. A* **59**, 2875–2885.

Jonker P.G., van der Klis M., 2001, "Discovery of an X-ray pulsar in the low-mass X-ray binary 2A 1822–371," *Astrophys. J.* **553**, L43–L46.

Jonker P.G., Méndez M., van der Klis M., 2002, "Kilohertz quasi-periodic oscillations difference frequency exceeds inferred spin frequency in 4U 1636-53," *Mon. Not. R. Astron. Soc.* **336**,

L1–L5.

Jonker P.G., van der Klis M., Groot P.G., 2003, "The mass of the neutron star in the low-mass X-ray binary 2A 1822–371," *Mon. Not. R. Astron. Soc.* **339**, 663–668.

Kachelriess M., Strumia A., Tomàs R., Valle J.W.F., 2002, "SN1987A and the status of oscillation solutions to the solar neutrino problem," *Phys. Rev. D* **65**, 073016 (15 pages).

Kaempfer B., 1981, "On stabilizing effect of relativity in cold spherical stars with a phase transition in the interior," *Phys. Lett. B* **101**, 366–368.

Kaempfer B., 1982, "On the collapse dynamics of cold stars with a phase transition in the interior," *Astron. Nachr.* **303**, 231–236.

Kalogera V., Baym G., 1996, "The maximum mass of a neutron star," *Astrophys. J.* **470**, L61–L64.

Kalogera V., Kim C., Lorimer D.R., Burgay M., D'Amico N., Possenti A., Manchester R.N., Lyne A.G., Joshi B.C., McLaughlin M.A., Kramer M., Sarkissian J.M., Camilo F., 2004, "The cosmic coalescence rates for double neutron star binaries," *Astrophys. J.* **601**, L179–L182; erratum: *ibid.*, **614**, L137–138.

Kaluzny J., Rucinski S.M., Thompson I.B., 2003, "Photometry and spectroscopy of the optical companion to the pulsar PSR J1740–5340 in the globular cluster NGC 6397," *Astron. J.* **125**, 1546–1553.

Kaminker A.D., Yakovlev D.G., 1981, "Description of a relativistic electron in a quantizing magnetic field. Transverse transport coefficients of an electron gas," *Theor. Math. Phys.* **49**, 1012–1020.

Kaminker A.D., Pethick C.J., Potekhin A.Y., Thorsson V., Yakovlev D.G., 1999, "Neutrino-pair bremsstrahlung by electrons in neutron star crusts," *Astron. Astrophys.* **343**, 1009–1024.

Kaplan J.L., Glasser M.L., 1972, "Electron gas in superstrong magnetic fields: Wigner transition," *Phys. Rev. Lett.* **28**, 1077–1079.

Kaplan D.B., Nelson A.E., 1986, "Strange goings in dense nucleonic matter," *Phys. Lett. B* **175**, 57–63.

Kaplan D.L., van Kerkwijk M.H., Anderson J., 2002, "The parallax and proper motion of RX J1856.5−3754 revisited," *Astrophys. J.* **571**, 447–457.

Kappes U., Schmelcher P., Pacher T., 1994, "Influence of a strong magnetic field on the chemical bond of the excited $H_2{}^+$ ion," *Phys. Rev. A* **50**, 3775–3781.

Kappes U., Schmelcher P., 1996, "Adiabatic potential-energy surfaces of the $H_2{}^+$ ion in a strong magnetic field," *Phys. Rev. A* **53**, 3869–3883.

Kardashev N.S., 1964, "Magnetic collapse and the nature of powerful sources of cosmic radio emission," *Astron. Zh.* **41**, 807–813.

Karsch F., 2002a, "Lattice QCD at high temperatures and density," *Lecture Notes Phys.* **583**, 209–249.

Karsch F., 2002b, "Lattice results on QCD thermodynamics," *Nucl. Phys.* **A 698**, 199–208.

Kaspi V.M., 2004, "Soft gamma repeaters and anomalous X-ray pulsars: Together forever," in *Young Neutron Stars and Their Environments*, edited by F. Camilo & B.M. Gaensler (San Francisco: Astron. Soc. Pacific), 231–238.

Kaspi V.M., Johnston S., Bell J.F., Manchester R.N., Bailes M., Bessell M., Lyne A.G., D'Amico N, 1994, "A massive radio pulsar binary in the Small Magellanic Cloud," *Astrophys. J.* **423**, L43–L45.

Kaspi V.M., Lyne A.G., Manchester R.N., Crawford F., Camilo F., Bell J.F., D'Amico N., Stairs I.H., McKay N.P.F., Morris D.J., Possenti A., 2000, "Discovery of a young radio pulsar in a relativistic binary orbit," *Astrophys. J.* **543**, 321–327.

Kaspi V.M., Gavriil F.P., Woods P.M., Jensen J.B., Roberts M.S.E., Chakrabarty D., 2003, "A major soft gamma repeater-like outburst and rotation glitch in the no-longer-so-anomalous X-ray pulsar 1E 2259+586," *Astrophys. J.* **588**, L93–L96.

Kaspi V.M., Roberts M.S.E., Harding A.K., 2004, "Isolated neutron stars," in *Compact Stellar X-ray Sources*, edited by W.H.G. Lewin & M. van der Klis, in press [astro-ph/0402136].

Kellogg J.M.B., Rabi I.I., Ramsey N.F., Zacharias J.R., 1939, "An electrical quadrupole moment of the deuteron," *Phys. Rev. 55* , 318–319.

Kettner Ch., Weber F., Weigel M.K., Glendenning N.K., 1995, "Structure and stability of strange and charm stars at finite temperatures," *Phys. Rev. D* **51**, 1440–1457.

Khersonskii V.K., 1987a, "On the ionization equilibrium of the atomic hydrogen in strong magnetic field," *Astron. Zh.* **64**, 433–436 [Engl. transl.: *Sov. Astron.* **31**, 225].

Khersonskii V.K., 1987b, "Dissociative equilibrium of the $H_2{}^+$ molecular ion in the magnetic field of a neutron star," *Astron. Zh.* **64**, 1233–1242 [Engl. transl.: *Sov. Astron.* **31**, 646].

Kirshner R.P., 1977, "Continuum light from supernovae," *Ann. New York Acad. Sci.* **302**, 81–89.

Kirzhnits D.A., 1958, "Correlation energy of a non-ideal Fermi gas," *Zh. Eksp. Teor. Fiz.* **35**, 1198–1208 [Engl. transl.: 1960, *Soviet Phys.–JETP* **8**, 1081].

Kirzhnits D.A., 1967, *Field-Theoretical Methods in Many-Body Systems* (Oxford: Pergamon).

Kirzhnits D.A., Shpatakovskaya G.V., 1995, "Statistical model of matter, corrected in the vicinity of nuclei," *Zh. Eksp. Teor. Fiz.* **108**, 1238–1252.

Kittel C., 1963, *Quantum Theory of Solids* (New York: Wiley).

Kittel C., 1986, *Introduction to Solid State Physics* (New York: Wiley).

Kluźniak W., Lee W.H., 2003, "The swallowing of a quark star by a black hole," *Mon. Not. R. Astron. Soc.* **335**, L29–L32.

Knorren R., Prakash M., Ellis P.J., 1995, "Strangeness in hadronic stellar matter," *Phys. Rev. C* **52**, 3470–3482.

Kohanoff J., Hansen J.P., 1996, "Statistical properties of the dense hydrogen plasma: An *ab initio* molecular dynamics investigation," *Phys. Rev. E* **54**, 768–781.

Kohn W., Sham L.J., 1965, "Self-consistent equations including exchange and correlation effects," *Phys. Rev.* **140**, A1133–A1138.

Kokkotas K.D., Apostolatos T.A., Andersson N., 2001, "The inverse problem for pulsating neutron stars: a 'fingerprint analysis' for the supranuclear equation of state," *Mon. Not. R. Astron. Soc.* **320**, 307–315.

Kolehmainen K., Baym G., 1982, "Pion condensation at finite temperature (II). Simple models including thermal excitations of the pion field," *Nucl. Phys.* **A 382**, 528–541.

Kolehmainen K., Prakash M., Lattimer J.M., Treiner J.R., 1985, "Surface and curvature properties of neutron-rich nuclei," *Nucl. Phys.* **A 439**, 535–572.

Kolomeitsev E.E., Voskresensky D.N., 2003, "Negative kaons in dense baryonic matter," *Phys. Rev. C* **68**, 015803 (31 pages).

Kolomeitsev E.E., Voskresensky D.N., Kaempfer B., 1996, "The impact of kaon polarization in nuclear matter on the $K^-$ production in heavy-ion collisions," *Int. J. Mod. Phys. E* **5**, 313–328.

Konacki M., Wolszczan A., Stairs I.H., 2003, "Geodetic precession and timing of the relativistic binary pulsars PSR B1534+12 and PSR B1913+16," *Astrophys. J.* **589**, 495–502.

Kondratyev V. N., 2002, "Statistics of Magnetic Noise in Neutron Star Crusts," *Phys. Rev. Lett.* **88**, 221101 (4 pages).

Kopidakis N., Ventura J., Herold H., 1996, "Atomic ionization in magnetic neutron star atmospheres: transverse motion effects," *Astron. Astrophys.* **308**, 747–762.

Koranda S., Stergioulas N., Friedman J.L., 1997, "Upper limits set by causality on the rotation and mass of uniformly rotating relativistic stars," *Astrophys. J.* **488**, 799–806.

Kouveliotou C., van Paradijs J., Fishman G.J., Briggs M.S., Kommers J., Harmon B.A., Meegan C.A., Lewin W.H.G., 1996, "A new type of transient high-energy source in the direction of the Galactic Centre," *Nature* **379**, 799–801.

Kouveliotou C., Dieters S., Strohmayer T., van Paradijs J., Fishman G.J., Meegan C.A., Hurley K., Kommers J., Smith I., Frail D., Murakami T., 1998a, "An X-ray pulsar with a superstrong magnetic field in the soft $\gamma$-ray repeater SGR 1806−20," *Nature* **393**, 235–237.

Kouveliotou C., Strohmayer T., Hurley K., van Paradijs J., Finger M.H., Dieters S., Woods P., Thompson C., Duncan R.C., 1999, "Discovery of a magnetar associated with the soft gamma repeater SGR 1900+14," *Astrophys. J.* **510**, L115–L118.

Kravchenko Yu.P., Liberman M.A., Johansson B. 1997, "Exact solution for a hydrogen atom in a magnetic field of arbitrary strength," *Phys. Rev. A* **54**, 287–305.

Kravchenko Yu.P., Liberman M.A., 1997, "Hydrogen molecular ion in a strong parallel magnetic field," *Phys. Rev. A* **55**, 2701–2710.

Kubis S., 2001, PhD thesis [in Polish] (Crakow: Institute for Nuclear Physics), unpublished.

Kulkarni S.R., Kaplan D.L., Marshall H.L., Frail D.A., Murakami T., Yonetoku D., 2003, "The quiescent counterpart of the soft gamma-ray repeater SGR 0526-66," *Astrophys. J.* **585**, 948–954.

Kunihiro T., Takatsuka T., Tamagaki R., 1993, "Neutral pion condensation in hot and dense nuclear matter," *Prog. Theor. Phys. Suppl.* **112**, 197–219.

Kusenko A., Shaposhnikov M., Tinyakov P.G., Tkachev I.I., 1998, "Star wreck," *Phys. Lett. B* **423**, 104–108.

Kutschera M., Wójcik W., 1989, "Magnetic properties of strongly asymmetric nuclear matter," *Phys. Lett. B* **223**, 11–15.

Kutschera M., Wójcik W., 1990, "A Thomas-Fermi model of localization of proton impurities in neutron matter," *Acta Physica Polonica B* **21**, 823–839.

Kutschera M., Wójcik W., 1993, "Proton impurity in the neutron matter: A nuclear polaron problem," *Phys. Rev. C* **47**, 1077–1085.

Kutschera M., Wójcik W., 1995, "Self-consistent proton crystallization in dense neutron-star matter," *Nucl. Phys.* **A 581**, 706–724.

Kutschera M., Stachniewicz S., Szmagliński A., Wójcik W., 2002, "Structure of proton component of neutron star matter for realistic nuclear models," *Acta Phys. Polonica B* **33**, 743–759.

Lagaris I.E., Pandharipande V.R., 1980, "Variational calculation of $v_8$ models of nuclear matter," *Nucl. Phys.* **A 334**, 217–228.

Lagaris I.E., Pandharipande V.R., 1981a, "Phenomenological two-nucleon interaction operator," *Nucl. Phys.* **A 359**, 331–348.

Lagaris I.E., Pandharipande V.R., 1981b, "Variational calculations of realistic models of nuclear matter," *Nucl. Phys.* **A 359**, 349–364.

Lagaris I.E., Pandharipande V.R., 1981c, "Variational calculations of asymmetric nuclear matter," *Nucl. Phys.* **A 369**, 470–482.

Lai D., 2001, "Matter in strong magnetic fields," *Rev. Mod. Phys.* **73**, 629–661.

Lai D., Salpeter E.E., 1995, "Motion and ionization equilibrium of hydrogen atoms in a superstrong magnetic field," *Phys. Rev. A* **52**, 2611–2623.

Lai D., Salpeter E.E., 1996, "Hydrogen molecules in a superstrong magnetic field: Excitation levels," *Phys. Rev. A* **53**, 152–167.

Lai D., Salpeter E.E., 1997, "Hydrogen phases on the surface of a strongly magnetized neutron star," *Astrophys. J.* **491**, 270–285.

Lai D., Shapiro E.E., 1991, "Cold equation of state in a strong magnetic field – Effects of inverse beta-decay," *Astrophys. J.* **383**, 745–751.

Lai D., Abrahams A.M., Shapiro S.L., 1991, "Equation of state in metals and cold stars: evaluation of statistical models," *Astrophys. J.* **377**, 612–628.

Lai D., Salpeter E.E., Shapiro S.L., 1992, "Hydrogen molecules and chains in a superstrong magnetic field," *Phys. Rev. A* **45**, 4832–4847.

Lai D., Chernoff D.F., Cordes J.M., 2001, "Pulsar jets: Implications for neutron star kicks and initial spins," *Astrophys. J.* **549**, 1111–1118.

Lamb D.Q., Lamb F.K., 1977, "Neutron star and degenerate dwarf models of X-ray bursts," *Ann. New York Acad. Sci.* **302**, 261.

Lamb F.K., Miller M.C., 2001, "Changing frequency separation of kilohertz quasi-periodic oscillations in the sonic-point beat-frequency model," *Astrophys. J.* **554**, 1210–1215.

Landau L.D., 1932, "On the theory of stars," *Phys. Z. Sowjetunion* **1**, 285–288.

Landau L.D., 1937, "Origin of stellar energy," *Doklady Akad. Nauk SSSR* **17**, 301–302 [in Russian]; *Nature*, 1938, **141**, 333–334.

Landau L.D., Lifshitz E.M., 1976, *Quantum Mechanics* (Oxford: Pergamon).

Landau L.D., Lifshitz E.M., 1984, *Theory of Elasticity* (Oxford: Pergamon).

Landau L.D., Lifshitz E.M., 1993, *Statistical Physics, Part 1* (Oxford: Pergamon).

Landau L.D., Lifshitz E.M., 1999, *The Classical Theory of Fields* (Oxford: Butterworth-Heinemann).

Landau L.D., Zeldovitch Ya.B., 1943, *Acta Phys. Chem. (USSR)* **18**, 194.

Langacker P., Sparrow D.A., 1982, "Implications of anomalous isospin violation for the low-energy nucleon-nucleon interaction," *Phys. Rev. C* **25**, 1194–1214.

Langer J.S., 1969, "Statistical theory of the decay of metastable states," *Ann. Phys. (N.Y.)* **54**, 258–275.

Langer W.D., Rosen L.C., 1970, "Hyperonic equation of state," *Astrophys. Space Sci.* **6**, 217–227.

Langer W.D., Rosen L.C., Cohen J.M., Cameron A.G.W., 1969, "An equation of state at subnuclear densities," *Astrophys. Space Sci.* **5**, 259–271.

Larson M.B., Link B., 2002, "Simulation of glitches in isolated pulsars," *Mon. Not. R. Astron. Soc.* **333**, 613–622.

Lasota J.-P., Haensel P., Abramowicz M.A., 1996, "Fast rotation of neutron stars" *Astrophys. J.* **456**, 300–304.

Lattimer J.M., Prakash M., 2001, "Neutron star structure and the equation of state," *Astrophys. J.* **550**, 426–442.

Lattimer J.M., Schutz B.F., 2005, "Constraining the equation of state with moment of inertia measurements," *Astrophys. J.* **629**, 979–984.

Lattimer J.M., Swesty F.D., 1991, "A generalized equation of state for hot, dense matter," *Nucl. Phys.* **A 535**, 331–376.

Lattimer J.M., Yahil A., 1989, "Analysis of the neutrino events from supernova 1987A," *Astrophys. J.* **340**, 426–434.

Lattimer J.M., Pethick C.J., Ravenhall D.G., Lamb D.Q., 1985, "Physical properties of hot, dense matter: The general case," *Nucl. Phys.* **A 432**, 646–742.

Lattimer J.M., Pethick C.J., Prakash M., Haensel P., 1991, "Direct URCA process in neutron stars," *Phys. Rev. Lett.* **66**, 2701–2704.

Lee T.D., 1975, "Abnormal nuclear states and vacuum excitation," *Rev. Mod. Phys.* **47**, 267–275.

Lee T.D., Wick G.C., 1974, "Vacuum stability and vacuum excitation in a spin-0 field theory," *Phys. Rev. D* **9**, 2291–2316.

Lee U., Yoshida S., 2003, "r-modes of neutron stars with superfluid cores," *Astrophys. J.* **586**, 403–418.

Lee C.-H., Brown G.E., Rho M., 1994, "Kaon condensation in 'nuclear star' matter," *Phys. Lett. B* **335**, 266–272.

Lejeune A., Grangé P., Martzloff M., Cugnon J., 1986, "Hot nuclear matter in an extended Brueckner approach," *Nucl. Phys.* **A 453**, 189–219.

Leonard P.J.T., Tremaine S., 1990, "The local Galactic escape speed," *Astrophys. J.* **353**, 486–493.

Levine A., Swank J., Smith E., 1998, "XTE J2123–058," *IAU Circular* No. 6955.

Levinger J.S., Simmons L.M., 1961, "Neutron gas," *Phys. Rev.* **124**, 916–922.

Lewin W.H.G., Doty J., Clark G.W., Rappaport S.A., Bradt H.V.D., Doxsey R., Hearn D.R., Hoffman J.A., Jernigan J.G., Li F.K., Mayer W., McClintock J., Primini F., Richardson J., 1976, "The discovery of rapidly repetitive X-ray bursts from a new source in Scorpius," *Astrophys. J.* **207**, L95–L99.

Lewin W.H.G., Rutledge R.E., Kommers J.M., van Paradijs J., Kouveliotou C., 1996, "A comparison between the Rapid Burster and GRO J1744-28," *Astrophys. J.* **462**, L39–L42.

Lewin W.H.G., van Paradijs J., van den Heuvel E.P.J., 1997, *X-ray Binaries* (Cambridge: Cambridge University Press).

Li X.-D., Bombaci I., Dey M., Dey J., van den Heuvel E.P.J., 1999a, "Is SAX J1808.4–3658 a strange star?" *Phys. Rev. Lett.* **83**, 3776–3779.

Li X.-D., Ray S., Dey J., Dey M., Bombaci I., 1999b, "On the nature of the compact star in 4U 1728–34," *Astrophys. J.* **527**, L51–L54.

Liang E.P., 1986, "Gamma-ray burst annihilation lines and neutron star structure," *Astrophys. J.* **304**, 682–687.

Lifshitz I.M., Kagan Yu., 1972, "Quantum kinetics of phase transitions at temperatures close to absolute zero," *Zh. Eksp. Teor. Fiz.* **62**, 385–402 [Engl. transl.: *Sov. Phys.–JETP* **35**, 206–214].

Lifshitz E.M., Pitaevskiĭ L.P., 1980, *Statistical Physics, Part 2* (Oxford: Pergamon).

Lifshitz E.M., Pitaevskiĭ L.P., 1981, *Physical Kinetics* (Oxford: Butterworth-Heinemann) (*PK*).

Lighthill M.J., 1950, "On the instability of small planetary cores (II)," *Mon. Not. R. Astron. Soc.* **110**, 339–342.

Lindblom L., 1984, "Limits on the gravitational redshift of neutron stars," *Astrophys. J.* **278**, 364–368.

Lindblom L., 1992, "Determining the nuclear equation of state from neutron-star masses and radii," *Astrophys. J.* **398**, 569–573.

Lindblom L., 1998, "Phase transitions and the mas-radius curves of relativistic stars," *Phys. Rev. D* **58**, 024008 (8 pages).

Lindblom L., Detweiler S.L., 1983, "The quadrupole oscillations of neutron star," *Astrophys. J. Suppl. Ser.* **53**, 73–92.

Lindblom L., Owen B.J., 2002, "Effect of hyperon viscosity on neutron-star r-modes," *Phys. Rev. D* **65**, 063006 (15 pages).

Lindgren K.A.U., Virtamo J.T., 1979, "Relativistic hydrogen atom in a strong magnetic field," *J. Phys. B: At. Mol. Phys.* **12**, 3465–3472.

Lindhard J., 1954, "On the properties of a gas of charged particles," *K. Danske Vidensk. Selsk., Mat.-Fys. Medd.* **28**, no. 8 (57 pages).

Link B., 2003, "Precession of isolated neutron stars," in *Radio Pulsars*, edited by M. Bailes, D.J. Nice, & S.E. Thorsett, *ASP Conf. Ser.* **302**, 241–247.

Link B., Epstein R.I., Lattimer J.M., 1999, "Pulsar constraints on neutron star structure and equation of state," *Phys. Rev. Lett.* **83**, 3362–3365.

Lipunov V.M., 1992, *Astrophysics of Neutron Stars* (Berlin: Springer).

Lipunov V.M., Postnov K.A., Prokhorov M.E., 1996, "The Scenario Machine: restrictions on key parameters of binary evolution," *Astron. Astrophys.* **310**, 489–507.

Liu J., di Stefano R., McClintock J., Kong A., Bregman J., Kuntz K., 2006, "Discovery of an eclipsing X-ray binary with a 32.69 hour period in M101: An analog of Her X-1 or LMC X-4?" *Astrophys. J.*, in press [astro-ph/0608354].

Livingstone M.A., Kaspi V.M., Gotthelf E.V., Kuiper L., 2006, "A braking index for the young, high magnetic field, rotation-powered pulsar in Kesteven 75," *Astrophys. J.* **647**, 1286–1292.

Lombardo U., Schulze H.-J., 2001, "Superfluidity in neutron star matter," in *Physics of Neutron Star Interiors*, edited by D. Blaschke, N. Glendenning, & A. Sedrakian (Berlin: Springer), 30–53.

López Vieyra J.C., Turbiner A.V., 2002, "$H_3^{2+}$ in a magnetic field: Triangular configuration," *Phys. Rev. A* **66**, 023409 (6 pages).

Lorenz C.P., 1991, *Dense Matter and the Compressible Liquid Drop Model*, PhD thesis (University of Illinois at Urbana-Champaign), unpublished.

Lorenz C.P., Ravenhall D.G., Pethick C.J., 1993, "Neutron star crusts," *Phys. Rev. Lett.* **70**, 379–382.

Lorimer D.R., 2001, "Binary and millisecond pulsars at the New Millennium," *Living Reviews in Relativity* **4**, 5 [http://www.livingreviews.org/lrr-2001-5].

Lorimer D.R., Lyne A.G., Bailes M., Manchester R.N., D'Amico N., Stappers B.W., Johnston S., Camilo F., 1996, "Discovery of four binary millisecond pulsars," *Mon. Not. R. Astron. Soc.* **283**, 1383–1387.

Lorimer D.R., Stairs I.H., Freire P.C., Cordes J.M., Camilo F., Faulkner A.J., Lyne A.G., Nice D.J., Ransom S.M., Arzoumanian Z., *et al.*, 2006, "Arecibo Pulsar Survey using ALFA. II. The young, highly relativistic binary pulsar J1906+0746," *Astrophys. J.* **640**, 428–434.

Lozovik Yu.E., Volkov S.Yu., 2004, "Hydrogen atom moving across a magnetic field," *Phys. Rev. A* **70**, 023410 (8 pages).

Lugones G., Horvath J.E., 2003, "High-density QCD pairing in compact star structure," *Astron. Astrophys.* **403**, 173–178.

Lundgren S.C., Zepka A.F., Cordes J.M., 1995, "A millisecond pulsar in a 6 hour orbit: PSR J0751+1807," *Astrophys. J.* **453**, 419–423.

Lundmark K., 1921, "Suspected new stars reported in old chronicles and among recent meridian observations," *Publ. Astron. Soc. of Pacific* **33**, 225–238.

Lunney D., Pearson J.M., Thibault C., 2003, "Recent trends in the determination of nuclear masses," *Rev. Mod. Phys.* **75**, 1021–1082.

Lyne A.G., Graham-Smith F., 1998, *Pulsar Astronomy*, 2nd ed. (Cambridge: Cambridge University Press).

Lyne A.G., Pritchard R.S., Smith F.G., 1988, "Crab pulsar timing 1982–87," *Mon. Not. R. Astron. Soc.* **233**, 667–676.

Lyne A.G., Camilo F., Manchester R.N., Bell J.F., Kaspi V.M., D'Amico N., McKay N.P.F., Crawford F., Morris D.J., Sheppard D.C., Stairs I.H., 2001, "The Parkes Multibeam Pulsar Survey: PSR J1811–1736, a pulsar in a highly eccentric binary system," *Mon. Not. R. Astron. Soc.* **312**, 698–702.

Lyne A.G., Burgay M., Kramer M., Possenti A., Manchester R.N., Camilo F., McLaughlin M.A., Lorimer D.R., D'Amico N., Joshi B.C., Reynolds J., Freire P.C.C., 2004, "A double-pulsar system: a rare laboratory for relativistic gravity and plasma physics," *Science*, **303**, 1153–1157.

MacAlpine G. M., Uomoto A., 1991, "Photometry of the Crab Nebula: Variability and the mass of the emitting gas," *Astron. J.* **102**, 218–223.

Machleidt R., 1989, "The meson theory of nuclear forces and nuclear structure," *Adv. Nucl. Phys.* **19**, 189–377.

Machleidt R., Sammarucca F., Song Y., 1996, "Nonlocal nature of the nuclear force and its impact on nuclear structure," *Phys. Rev. C* **53**, R1483–R1487.

Madsen J., 1988, "Astrophysical limits on the flux of of quark nuggets," *Phys. Rev. Lett.* **61**, 2909–2912.

Madsen J., 1992, "Bulk viscosity of strange quark matter, damping of quark star vibration, and the maximum rotation rate of pulsars," *Phys. Rev. D* **46**, 3290–3295.

Madsen J., 1993a, "Mass formula for strange and nonstrange quark matter," *Phys. Rev. D* **47**, 5156–5160.

Madsen J., 1993b, "Curvature contributions to the mass of strangelets," *Phys. Rev. Lett.* **70**, 391–393.

Madsen J., 1994, "Shell model versus liquid drop model for strangelets," *Phys. Rev. D* **50**, 3328–3331.

Madsen J., 1998a, "Detecting supersymmetric Q-balls with neutron stars," *Phys. Lett. B* **435**, 125–130.

Madsen J., 1998b, "How to identify a strange star," *Phys. Rev. Lett.* **81**, 3311–3314.

Madsen J., 1999, "Physics and astrophysics of strange quark matter," in *Hadrons in Dense Matter and Hadrosynthesis*, edited by J. Cleymans (Berlin: Springer), 162–203.

Madsen J., 2000a, "Probing strange stars and color superconductivity by $r$-mode instabilities in millisecond pulsars," *Phys. Rev. Lett.* **85**, 10–13.

Madsen J., 2000b, "Intermediate mass strangelets are positively charged," *Phys. Rev. Lett.* **85**, 4687–4690.

Madsen J., 2001, "Color-flavor locked strangelets," *Phys. Rev. Lett.* **87**, 172003.

Madsen J. & Haensel P. (editors), 1991, "Strange quark matter in physics and astrophysics," *Nucl. Phys. (Proc. Suppl.)* **B 24**.

Maekawa M., Tamagaki R., 1968, "An attempt to explain the origin of magnetic field in superdense stars," in *Proceedings of the Symposium on Cosmology* (Kyoto: Research Institute for Fundamental Physics), 80 [in Japanese].

Maessen P.M.M., Rijken Th.A., de Swart J.J., 1989, "Soft-core baryon-baryon one-boson-exchange potentials. II. Hyperon-nucleon potential," *Phys. Rev. C* **40**, 2226–2245.

Magierski P., Heenen P.-H., 2002, "Structure of the inner crust neutron stars: crystal lattice or disordered phase," *Phys. Rev. C* **65**, 045804 (13 pages).

Magierski P., Bulgac A., Heenen P.-H., 2001, "Neutron stars and the fermionic Casimir effect," *Int. J. Mod. Phys. A* **17**, 1059–1064.

Magierski P., Bulgac A., Heenen P.-H., 2003, "Exotic nuclear phases in the inner crust of neutron stars in the light of the Skyrme-Hartree-Fock theory," *Nucl. Phys.* **A 719**, C217–C220.

Malone R.C., Johnson M.B., Bethe H.A., 1975, "Neutron star models with realistic high-density equations of state," *Astrophys. J.* **199**, 741–748.

Manchester R. N., Taylor J. H., 1977, *Pulsars* (San Francisco: Freeman & Co.).

Manchester R.N., Lyne A.G., Camilo F., Bell J.F., Kaspi V.M., D'Amico N., McKay N.P.F., Crawford F., Stairs I.H., Possenti A., Kramer M., Sheppard D.C., 2001, "The Parkes multi-beam pulsar survey – I. Observing and data analysis systems, discovery and timing of 100 pulsars," *Mon. Not. R. Astron. Soc.* **328**, 17–35.

Manchester R.N., Hobbs G.B., Teoh A., Hobbs M., 2005, "The Australia Telescope National Facility Pulsar Catalogue," *Astron. J.* **129**, 1993–2006 [http://www.atnf.csiro.au/research/pulsar/psrcat/].

Mareš J., Friedman E., Gal A., Jenning B.K, 1995, "Constraints on the $\Sigma$-nucleus dynamics from Dirac phenomenology of the $\Sigma^-$ atoms," *Nucl. Phys.* **A 594**, 311–324.

Marshall H.L., Schulz N.S., 2002, "Using the high-resolution X-ray spectrum of PSR B0656+14 to constrain the chemical composition of the neutron star atmosphere," *Astrophys. J.* **574**, 377–381.

Marshall F.E., Gotthelf E.V., Middleditch J., Wang Q.D., Zhang W., 2004, "The big glitcher: The rotation history of PSR J0537–6910," *Astrophys. J.* **603**, 682–689.

Mansoori G.A., Carnahan N.F., Starling K.E., Leland T.W., 1971, "Equilibrium thermodynamic properties of the mixture of hard spheres," *J. Chem. Phys.* **54**, 1523–1525.

Marcos S., Barranco M., Buchler J.-R., 1982, "Low entropy adiabats for stellar collapse," *Nucl. Phys.* **A381**, 507–518.

Martin P.C., Schwinger J., 1958, "Theory of many-particle systems. I," *Phys. Rev.* **115**, 1342–1373.

Matsui T., Sakai K., Yasuno M., 1979, "Solid-like structure of pion condensed nuclear matter," *Prog. Theor. Phys.* **61**, 1093–1106.

Maruyama T., Tatsumi T., Voskresensky D.N., Tanigawa T., Endo T., Chiba S., 2006, "Finite size effects on kaonic 'pasta'," *Phys. Rev. C* **73**, 035802 (10 pages).

Mauche C.W., 2002, "Correlation of the quasi-periodic oscillation frequencies of white dwarf, neutron star, and black hole binaries," *Astrophys. J.* **580**, 423–428.

Maxwell O., Brown G.E., Campbell D.K., Dashen R.F., Manassah J.T., 1977, "Beta decay of pion condensates as a cooling mechanism for neutron stars," *Astrophys. J.* **216**, 77–85.

Mayall N.U., 1939, "The Crab Nebula, a probable supernova," *Astron. Soc. of Pacific Leaflet* **119**, 145–154.

Mazets E.P., Golenetskii S.V., Il'inskii V.N., Aptekar R.L., Guryan Y.A., 1979a, "Observations of a flaring X-ray pulsar in Dorado," *Nature* **282**, 587–589.

Mazets E.P., Golenetskii S.V., Il'inskii V.N., Panov V.N., Aptekar' R.L., Gur'yan Yu.A., Sokolov I.A., Sokolova Z.Ya., Kharitonova T.V., 1979b, "A flaring X-ray pulsar in Dorado," *Pis'ma v Astron. Zh.* **5**, 307 [Engl. transl.: *Sov. Astron. Lett.* **5**, 163–165].

Mazets E.P., Golenetskii S.V., Aptekar R.L., Guryan Yu.A., Il'inskii V.N., 1981, "Cyclotron and annihilation lines in gamma-ray bursts," *Nature* **290**, 378–382.

Mazets E.P., Golenetskii S.V., Guryan Yu.A., Il'inskii V.N., 1982, "The 5 March 1979 event and the distinct class of short gamma bursts: are they of the same origin?" *Astrophys. Space Sci.* **84**, 173–189.

Mazets E.P., Cline T.L., Aptekar R.L., Frederiks D.D., Golenetskii S.V., Il'inskii V.N., Pal'shin V.D., 2005, "The Konus-Wind and Helicon-Coronas-F detection of the giant $\gamma$-ray flare from the soft $\gamma$-ray repeater SGR 1806–20," *Astron. Rep.*, in press [astro-ph/0502541].

McClintock J.E., Rappaport S., Joss P.C., Bradt H., Buff J., Clark G.W., Hearn D., Lewin W.H.G., Matilsky T., Mayer W., Primini F., 1976, "Discovery of a 283-second periodic variation in the X-ray source 3U 0900–40," *Astrophys. J.* **206**, L99–L102.

McDermott P.N., Van Horn H.M., Hansen C.J., 1988a, "Nonradial oscillations of neutron stars," *Astrophys. J.* **325**, 725–748.

McDermott P.N., Van Horn H.M., Hansen C.J., 1988b, "The shear modulus of the neutron star crust and nonradial oscillations of neutron stars," *Astrophys. J.* **375**, 679–686.

Medin Z., Lai D., 2006a, "Density-functional-theory calculations of matter in strong magnetic fields: I. Atoms and molecules," *Phys. Rev.*, submitted [astro-ph/0607166].

Medin Z., Lai D., 2006b, "Density-functional-theory calculations of matter in strong magnetic fields: II. Infinite chains and condensed matter," *Phys. Rev.*, submitted [astro-ph/0607277].

Melrose D.B., 2000, "The status of pulsar emission theory," in *Pulsar Astronomy – 2000 and Beyond*, edited by M. Kramer, N. Wex, & N. Wielebinski, *ASP Conf. Ser.* **202**, 721.

Meltzer D.W., Thorne K.S., 1966, "Normal modes of radial pulsations of stars at the end point of thermonuclear evolution," *Astrophys. J.* **145**, 514–543.

Messiah A., 1961, *Quantum Mechanics*, vol. I and II (Amsterdam: North-Holland).

Mestel L., Ruderman M.A., 1967, "The energy content of a white dwarf and its rate of cooling," *Mon. Not. R. Astron. Soc.* **136**, 27–38.

Michel F.C., 2004, "The state of pulsar theory," *Adv. Sp. Res.* **33**, 542–551.

Migdal A.B., 1959, "Superfluidity and the moments of inertia of nuclei," *Nucl. Phys.* **13**, 655–674.

McLaughlin M.A., Arzoumanian Z., Cordes J.M., Backer D.C., Lommen A.N., Lorimer D.R., Zepka A.F., 2002, "PSR J1740+1000: A young pulsar well out of the Galactic plane," *Astrophys. J.* **564**, 333–342.

Migdal A.B., 1971, "Stability of vacuum and limiting fields," *Zh. Eksp. Teor. Fiz.* **61**, 2209–2224.

Migdal A.B., 1972, "Phase transitions in nuclear matter and non-pair nuclear forces," *Zh. Eksp. Teor. Fiz.* **63**, 1993–1999 [Engl. transl.: *Sov. Phys.–JETP* **36**, 1052–1055].

Migdal A.B., 1974, "Meson condensation and anomalous nuclei," *Phys. Lett.* **52 B**, 172–174.

Migdal A.B., Chernoutsan, A.I., Mishustin I.N., 1979, "Pion condensation and dynamics of neutron stars," *Phys. Lett. B* **83**, 158–160.

Migdal A.B., Saperstein E.E., Troitsky M.A., Voskresensky D.N., 1990, "Pion degrees of freedom in nuclear matter," *Phys. Rep.* **192**, 179–437.

Mihalas D., Däppen W., Hummer D.G., 1988, "The equation of state for stellar envelopes. II. Algorithm and selected results," *Astrophys. J.* **331**, 815–825.

Militzer B., Pollock E.L., 2005, "Equilibrium contact probabilities in dense plasmas," *Phys. Rev. B* **71**, 134303 (10 pages).

Miller M.C., Lamb F.K., Psaltis D., 1998, "Sonic-point model of kilohertz quasi-periodic brightness oscillations in low-mass X-ray binaries," *Astrophys. J.* **508**, 791–830.

Miller M.C., Neuhauser D., 1991, "Atoms in very strong magnetic fields," *Mon. Not. R. Astron. Soc.* **253**, 107–122.

Miralda-Escudé J., Haensel P., Paczyński B., 1990, "Thermal structure of accreting neutron stars and strange stars," *Astrophys. J.* **362**, 572–583.

Miralles J.A., Urpin V.A., Van Riper K.A., 1997, "Convection in the surface layers of neutron stars," *Astrophys. J.* **480**, 358–363.

Misner C.W., Thorne K.S., Wheeler J.A., 1973, *Gravitation* (San Francisco: Freeman & Co.).

Mochkovich R., Hansen J.P., 1979, "Fluid-solid coexistence curve of dense Coulombic matter," *Phys. Lett. A* **73**, 35–38.

Moiseenko S.G., Ardeljan N.V., Bisnovatyi-Kogan G.S., 2003, "Supernova type II: Magnetorotational explosion," *Rev. Mex. Astron. Astrofis.* **15**, 231–233.

Möller P., Nix J.R., 1988, "Nuclear masses from a unified macroscopic-microscopic model," *Atom. Data Nucl. Data Tables* **39**, 213–223.

Morales J., Jr., Pandharipande V.R., Ravenhall D.G., 2002, "Improved variational calculations of nucleon matter," *Phys. Rev. C* **66**, 054308 (13 pages).

Mori K., Hailey C.J., Paerels F., Zane S., 2004, "*XMM-Newton* observations of the Vela pulsar," *Adv. Sp. Res.*, **33**, 503–506.

Morley P.D., Kislinger M.B., 1979, "Relativistic many-body theory, quantum chromodynamics, and neutron stars/supernova," *Phys. Rep.* **51**, 63–110.

Morrison I.A., Baumgarte T.W., Shapiro S.L., 2004a, "Effect of differential rotation on the maximum mass of neutron stars: realistic nuclear equations of state," *Astrophys. J.* **610**, 941–947.

Morrison I.A., Baumgarte T.W., Shapiro S.L., Pandharipande V.R., 2004b, "The moment of inertia of the binary pulsar J0737–3039A: Constraining the nuclear equation of state," *Astrophys. J.* **617**, L135–L138.

Morton D.C., 1964, "Neutron stars and X-ray sources," *Nature* **201**, 1308–1309.

Moszkowski S.A., 1974, "Energy of neutron-star matter," *Phys. Rev. D* **9**, 1613–1625.

Moszkowski S.A., Källman C.G., 1977, "Abnormal neutron star matter at ultrahigh densities," *Nucl. Phys.* **A 287**, 495–500.

Müller E., 1984, "Variational calculation of iron and helium atoms and molecular chains in superstrong magnetic fields," *Astron. Astrophys.* **130**, 415–418.

Müller H., Jennings B.K., 1997, "Nuclear matter properties of the modified quark-meson coupling model," *Nucl. Phys.* **A 626**, 966–986.

Muslimov A.G. & Tsygan A.I., 1990, "Influence of General Relativity effects on electrodynamics in the vicinity of a magnetic pole of a neutron star," *Astron. Zh.* **67**, 263–173 [Engl. transl.: *Sov. Astron.* **34**, 133].

Muslimov A.G., Tsygan A.I., 1992, "General relativistic electric potential drops above pulsar polar caps," *Mon. Not. R. Astron. Soc.* **255**, 61–70.

Muto T., Tatsumi T., 1988, "Extra neutrino emissivity of neutron stars under the combined condensation of neutral and charged pions," *Prog. Theor. Phys.* **79**, 461–479.

Muto T., Tatsumi T., 1990, "Static and dynamical properties of neutron stars with combined $\pi^0 - \pi^c$ condensation," *Prog. Theor. Phys.* **83**, 499–521.

Muto T., Tatsumi T., Iwamoto N., 1997, "Nonequilibrium weak processes in kaon condensation," *Australian J. Phys.* **50**, 13–22.

Muto T., Tatsumi T., Iwamoto N., 2000, "Nonequilibrium weak processes in kaon condensation. I. Reaction rate for the thermal kaon process," *Phys. Rev. D* **61**, 063001 (14 pages).

Muto T., Tatsumi T., Iwamoto N., 2000, "Nonequilibrium weak processes in kaon condensation. II. Kinetics of condensation," *Phys. Rev. D* **61**, 083002 (15 pages).

Myers W.D., 1976, "Development of the semiempirical droplet model," *At. Data Nucl. Data Tables* **17**, 411–417.

Myers W.D., Swiatecki W.D., 1998, "Nuclear equation of state," *Phys. Rev. C* **57**, 3020–3025.

Nagai T., Fukuyama H., 1982, "Ground state of a Wigner crystal in a magnetic field. I. Cubic structure," *J. Phys. Soc. Japan* **51**, 3431–3442.

Nagai T., Fukuyama H., 1983, "Ground state of a Wigner crystal in a magnetic field. II. Hexagonal close-packed structure," *J. Phys. Soc. Japan* **52**, 44–53.

Nagara H., Nagata Y., Nakamura T., 1987, "Melting of the Wigner crystal at finite temperature," *Phys. Rev. A* **36**, 1859–1873.

Nagase F., 1989, "Accretion-powered X-ray pulsars," *Publ. Astron. Soc. Japan* **41**, 1–79.

Nagels M.M., Rijken Th.A., de Swart J.J., 1977, "Baryon-baryon scattering in a one-boson-exchange potential approach. II. Hyperon-nucleon scattering," *Phys. Rev. D* **15**, 2547–2564.

Nagels M.M., Rijken Th.A., de Swart J.J., 1979, "Baryon-baryon scattering in a one-boson-exchange potential approach. III. A nucleon-nucleon and hyperon-nucleon analysis including contributions of a nonet of scalar mesons," *Phys. Rev. D* **20**, 1633–1645.

Nambu Y., Jona-Lasinio G., 1961a, "Dynamical model of elementary particles based on an analogy with superconductivity. I," *Phys. Rev.* **122**, 345–358.

Nambu Y., Jona-Lasinio G., 1961b, "Dynamical model of elementary particles based on an analogy with superconductivity. II," *Phys. Rev.* **124**, 246–254.

Nath N.R., Strohmayer T.E., Swank J.H., 2002, "Bounds on compactness for low-mass X-ray binary neutron stars from X-ray burst oscillations," *Astrophys. J.* **564**, 353–360.

Negele J.W., Vautherin D., 1973, "Neutron star matter at subnuclear densities," *Nucl. Phys.* **A 207**, 298–320.

Neuhauser D., Koonin S.E., Langanke K., 1987, "Structure of matter in strong magnetic fields," *Phys. Rev. A* **36**, 4163–4175.

New K.C., Chanmugam G., Johnson W.W., Tohline J.E., 1995, "Millisecond pulsars: detectable sources of continuous gravitational waves?" *Astrophys. J.* **450**, 757–762.

Ng C.Y., Cheng K.S., Chu M.C., 2003, "Cooling properties of Cloudy Bag strange stars," *Astropart. Phys.* **19**, 171–192.

Nicastro L., Lyne A.G., Lorimer D.R., Harrison P.A., Bailes M., Skidmore B.D., 1995, "PSR J1012+5307 – a 5.26 millisecond pulsar in a 14.5-hour binary system," *Mon. Not. R. Astron. Soc.* **273**, L68–L70.

Nice D.J., Taylor J.H., Fruchter A.S., 1993, "Two newly discovered millisecond pulsars," *Astrophys. J.* **402**, L49–L52.

Nice D.J., Sayer R.W., Taylor J.H., 1995, "Binary pulsar PSR J1518+4904: Orbital precession and mass estimates," *BAAS* **27**, 879.

Nice D.J., Splaver E.M., Stairs I.H., 2001, "On the mass and inclination of the PSR J2019+2425 binary system," *Astrophys. J.* **549**, 516–521.

Nice D.J., Splaver E.M., Stairs I.H., 2003, "Neutron star masses from Arecibo timing observations of five pulsar – white dwarf binary systems," in *Radio Pulars*, edited by M. Bailes, D.J. Nice, & S.E. Thorsett, *ASP Conf. Ser.* **302** 75–80.

Nice D.J., Splaver E.M., Stairs I.H., 2004, "Heavy neutron stars? A status report on Arecibo timing of four pulsar – white dwarf systems," in *Young Neutron Stars and Their Environments*, edited by F. Camilo & B.M. Gaensler (San Francisco: Astron. Soc. Pacific), 49.

Nice D.J., Splaver E.M., Stairs I.H., Löhmer O., Jessner A., Kramer M., Cordes J.M., 2005, "A 2.1 $M_\odot$ pulsar measured by relativistic orbital decay," *Astrophys. J.* **634**, 1242–1249.

Nishizaki S., Yamamoto Y., Takatsuka T., 2002, "Hyperon-mixed neutron star matter and neutron stars," *Prog. Theor. Phys.* **108**, 703–718.

Nogga A., Kamada H., Glöckle W., 2000, "Modern nuclear force predictions for the $\alpha$ particle," *Phys. Rev. Lett.* **85**, 944–947.

Nogga A., Kamada H., Glöckle W., 2002, "The hypernuclei ${}^4_\Lambda$He and ${}^4_\Lambda$H: challenges for modern hyperon-nucleon forces," *Phys. Rev. Lett.* **88**, 172501 (4 pages).

Nomoto K., 1987, "Accretion induced collapse of white-dwarfs," in *13th Texas Symposium on Relativistic Astrophysics* (Singapore: World Scientific), 519–528.

Nomoto K., Kondo Y., 1991, "Conditions for accretion-induced collapse of white dwarfs," *Astrophys. J.* **367**, L19–L22.

Nomoto K., Yamaoka H., Shigeyama T., Kumagai S., Tsujimoto T., 1994, "Type I supernovae and evolution of interacting binary systems," in *Supernovae*, edited by S.A. Bludman, R. Mochkovitch, & J. Zinn-Justin (Amsterdam: North-Holland), 201–249.

Norman G.E., Starostin A.N., 1968, "Insufficiency of the classical description of a nondegenerate dense plasma," *Teplofiz. Vysokikh Temp. (USSR)* **6**, 410–415 [Engl. transl.: *High Temp. (USSR)* **6**, 394–399].

Norsen T., 2002, "Strangeness nucleation in neutron star matter," *Phys. Rev. C* **65**, 045805 (13 pages).

Norsen T., Reddy S., 2001, "First order kaon condensation in neutron stars: Finite size effects in the mixed phase," *Phys. Rev. C* **63**, 065804 (10 pages).

Nozawa T., Stergioulas N., Gourgoulhon E., Eriguchi Y., 1998, "Construction of highly accurate models of rotating neutron stars – comparison of three different numerical schemes," A&AS, 132, 431–454.

Nugent R.L., 1998, "New measurements of the expansion of the Crab Nebula," *Publ. Astron. Soc. Pacific*, **110**, 831–836.

Ogasawara R., Sato K., 1983, "Nuclei in the neutrino-degenerate dense matter. II," *Prog. Theor. Phys.* **70**, 1569–1582.

Ogata S., Ichimaru S., 1990, "First-principles calculations of shear moduli for Monte Carlo-simulated Coulomb solids," *Phys. Rev. A* **42**, 4867–4870.

Ögelman H., 1995, "X-ray observations of cooling neutron stars," in *The Lives of the Neutron Stars*, edited by M.A. Alpar, Ü. Kiziloğlu, & J. van Paradijs (Dordrecht: Kluwer), 101–120.

Olesen M.L., Madsen J., 1991, "Burning of neutron star into a strange star," *Nucl. Phys.* **B 24**, 170–174.

Olinto A.V., 1987, "On the conversion of neutron stars into strange stars," *Phys. Lett. B* **192**, 71–75.

Olinto A., 1991, "Converting neutron stars into strange stars," *Nucl. Phys.* **B 24**, 103–109.

Olson T.S., 2000, "Maximally incompressible neutron star matter," *Phys. Rev. C* **63**, 015802 (7 pages)

Onsi M., Pearson J.M., 2002, "Equation of state of stellar nuclear matter and the effective nucleon mass," *Phys. Rev. C* **65**, 047302 (3 pages).

Oppenheimer J.R., Serber R., 1938, "On the stability of stellar neutron cores," *Phys. Rev.* **54**, 540.

Oppenheimer J.R., Volkoff G.M., 1939, "On massive neutron cores," *Phys. Rev.* **55**, 374–381.

Ord S.M., Bailes M., van Straten W., 2002, "The scintillation velocity of the relativistic binary pulsar PSR J1141−6545," *Astrophys. J.* **574**, L75–L78.

Orosz J.A., Kuulkers E., 1999, "The optical light curves of Cygnus X-2 (V1341 Cyg) and the mass of its neutron star," *Mon. Not. R. Astron. Soc.* **305**, 132–142.

Osherovich V., Titarchuk L., 1999, "Kilohertz quasi-periodic oscillations in neutron star binaries modeled as Keplerian oscillations in a rotating frame of reference," *Astrophys. J.* **522**, L113–L116.

Owen B.J., 2005, "Maximum elastic deformations of compact stars with exotic equations of state," *Phys. Rev. Lett.* **95**, 211101 (4 pages).

Oyamatsu K., 1993, "Nuclear shapes in the inner crust of a neutron star," *Nucl. Phys.* **A 561**, 431–452.

Oyamatsu K., Yamada M., 1994, "Shell energies of non-spherical nuclei in the inner crust of a neutron star," *Nucl. Phys.* **A 578**, 181–203.

Oyamatsu K., Hashimoto M., Yamada M., 1984, "Further study of the nuclear shape in high-density matter," *Prog. Theor. Phys.* **72**, 373–375.

Pacini F., 1967, "Energy emission from a neutron star," *Nature* **216**, 567–568.

Paczyński B., 1992, "GB 790305 as a very strongly magnetized neutron star," *Acta Astron.* **42**, 145–153.

Page D., Applegate J.H., 1992, "The cooling of neutron stars by the direct URCA process," *Astrophys. J.* **394**, L17–L21.

Page D., Cumming A., 2005, "Superbursts from strange stars," *Astrophys. J.* **635**, L157–L160.

Page D., Lattimer J.M., Prakash M., Steiner A.W., 2004, "Minimal cooling of neutron stars: A new paradigm," *Astrophys. J. Suppl. Ser.* **155**, 623–650.

Page D., Geppert U., Weber F., 2005, "The cooling of compact stars," *Nucl. Phys.* **A**, in press [astro-ph/0508056].

Pal S., Bandyopadhyay D., Greiner W., 2000, "Antikaon condensation in neutron stars," *Nucl. Phys.* **A 674**, 553–577.

Pandharipande V.R., 1971a, "Dense neutron matter with realistic interactions," *Nucl. Phys.* **A 174**, 641–656.

Pandharipande V.R., 1971b, "Hyperonic matter," *Nucl. Phys.* **A 178**, 123–144.

Pandharipande V.R., Bethe H.A.,1973, "Variational method for dense systems," *Phys. Rev. C* **7**, 1312–1328.

Pandharipande V.R., Garde V.K., 1972, "Effect of tensor force on composition of dense matter," *Phys. Lett. B.* **39**, 608–610.

Pandharipande V.R., Ravenhall D.G., 1989, "Hot nuclear matter," in *Nuclear Matter and Heavy Ion Collisions*, edited by M. Soyeur, H. Flocard, B. Tamain, & M. Porneuf (Dordrecht: Reidel), 103–132.

Pandharipande V.R., Smith R.A., 1975, "Nuclear matter calculations with mean scalar fields," *Phys. Lett. B* **59**, 15–18.

Pandharipande V.R., Pines D., Smith R.A., 1976, "Neutron star structure: theory, observation, and speculation," *Astrophys. J.* **208**, 550–566.

Pandharipande V.R., Pethick C.J., Thorsson V., 1995, "Kaon energies in dense matter," *Phys. Rev. Lett.* **75**, 4567–4570.

Pasteur L., 1854, Inaugural Address at Lille, quoted in *La vie de Pasteur*, R. Vallery-Radot (Paris, 1903), 88.

Pavlov G.G., Mészáros P., 1993, "Finite-velocity effects on atoms in strong magnetic fields and implications for neutron star atmospheres," *Astrophys. J.* **416**, 752–761.

Pavlov G.G., Potekhin A.Y., 1995, "Bound-bound transitions in strongly magnetized hydrogen plasma," *Astrophys. J.* **450**, 883–895.

Pavlov G.G., Zavlin V.E., 2003, "Thermal radiation from cooling neutron stars," in *Texas in Tuscany. XXI Texas Symposium on Relativistic Astrophysics*, edited by R. Bandiera, R. Maiolino, & F. Mannucci (Singapore: World Scientific), 319–328.

Pavlov G.G., Shibanov Yu.A., Zavlin V.E., Meyer R.D., 1995, "Neutron Star Atmospheres," in *The Lives of the Neutron Stars*, edited by M.A. Alpar, Ü. Kiziloğlu, & J. van Paradijs (Dordrecht: Kluwer), 71–90.

Pavlov G.G., Zavlin V.E., Sanwal D., Burwitz V., Garmire G.P., 2001, "The X-Ray spectrum of the Vela pulsar resolved with the *Chandra X-Ray Observatory*," *Astrophys. J.* **552**, L129–L133.

Pavlov G.G., Zavlin V.E., Sanwal D., 2002, "Thermal radiation from neutron stars: Chandra results," in Proc. of *270. WE-Heraeus Seminar on Neutron Stars, Pulsars and Supernova remnants*, edited by W. Becker, H. Lesch, & J. Trümper, *MPE-Report* **278** (Garching: MPE), 273–286.

Pearson J.M., Aboussir Y., Dutta A.K., Nayak R.C., Farine M., 1991, "Thomas-Fermi approach to nuclear mass formula. III. Force fitting and construction of mass table," *Nucl. Phys.* **A 528**, 1–47.

Peng G.X., Chiang H.C., Yang J.J., Li L., Liu B., 1999, "Mass formulas and thermodynamic treatment in the mass-density-dependent model of strange quark matter," *Phys. Rev. C* **61**, 015201 (6 pages).

Peng G.X., Chiang H.C., Zou B.S., Ning P.Z., Luo S.J., 2000, "Thermodynamics, strange quark matter, and strange stars," *Phys. Rev. C* **62**, 025801 (7 pages).

Perez Garcia M.A., Corte Rodriguez N., Mornas L., Suarez Curieses J.P., Diaz Alonso J., 2002, "Formation of an ordered phase in neutron matter," *Nucl. Phys.* **A 699**, 939–962.

Pérez-Azorín J.F., Miralles J.A., Pons J.A., 2005, "Thermal radiation from magnetic neutron star surfaces," *Astron. Astrophys.* **433**, 275–283.

Perkins D.H., 2000, *Introduction to High Energy Physics* (Cambridge: Cambridge University Press).

Perrot F., Dharma-wardana C., 1984, "Exchange and correlation potentials for electron-ion systems at finite temperatures," *Phys. Rev. A* **30**, 2619–2626.

Perrot F., Dharma-wardana C., 1994, "Equation of state of dense hydrogen and the plasma phase transition; A microscopic calculational model for complex fluids," in *The Equation of State in Astrophysics*, edited by G. Chabrier & E. Schatzman (Cambridge: Cambridge University Press), 272–286.

Peters P.C., 1964, "Gravitational radiation and the motion of two point masses," *Phys. Rev.* **136**, B1124–B1232.

Peters P.C., Mathews J., 1963, "Gravitational radiation from point masses in a Keplerian orbit," *Phys. Rev.* **131**, 435–440.

Peterson, L.E., 1998, "The Crab Nebula at high energies – a historical perspective," *Rev. Mex. Astron. Astrofis. (Ser. Conf.)* **7**, 81–89.

Pethick C.J., 1992, "Cooling of neutron stars," *Rev. Mod. Phys.* **64**, 1133–1140.

Pethick C.J., Potekhin A.Y., 1998, "Liquid crystals in the mantles of neutron stars," *Phys. Lett. B* **427**, 7–12.

Pethick C.J., Ravenhall D.G., 1995, "Matter at large neutron excess and the physics of neutron-star crusts," *Annu. Rev. Nucl. Sci.* **45**, 429–484.

Pethick C.J., Ravenhall D.G., Lorenz C.P., 1995, "The inner boundary of a neutron-star crust," *Nucl. Phys.* **A 584**, 675–703.

Petkov I.Zh., Stoitsov M.V., 1991, *Nuclear Density Functional Theory* (Oxford: Clarendon Press).

Pierleoni C., Ceperley D.M., Bernu B., Magro W.R., 1994, "Equation of state of the hydrogen plasma by path integral Monte Carlo simulation," *Phys. Rev. Lett.* **73**, 2145–2149.

Pietsch W. & Haberl F., 2005, "XMM-Newton detection of type I X-ray bursts in M 31," *Astron. Astrophys.* **430**, L45–L48.

Pizzella G., Di Virgilio A., Bender P., Fucito F., 2001, "Gravitational-wave detectors," in *Gravitational Waves*, edited by I. Ciufolini, V. Gorini, U. Moschella, & P. Fre (Bristol: IOP Publishing), 89–176.

Pokrant M.A., 1977, "Thermodynamic properties of the nonzero-temperature, quantum-mechanical, one-component plasma," *Phys. Rev. A* **16**, 413–423.

Polcaro V.F., Martocchia A., 2006, "Supernovae astrophysics from Middle Age documents," in *Populations of High-Energy Sources in Galaxies*, edited by E.J.A. Meurs & G. Fabbiano (Cambridge: Cambridge University Press), 264–268.

Pollock L.E., Hansen J.P., 1973, "Statistical mechanics of dense ionized matter. II. Equilibrium properties and melting transition of the crystallized one-component plasma," *Phys. Rev. A* **8**, 3110–3122.

Pons J.A., Reddy S., Ellis P.J., Prakash M., Lattimer J.M., 2000, "Kaon condensation in proto-neutron star matter," *Phys. Rev. C* **62**, 035803 (20 pages).

Pons J.A., Miralles J.A., Prakash M., Lattimer J.M., 2001, "Evolution of proto-neutron stars with kaon condensates," *Astrophys. J.* **553**, 382–393.

Podsiadlowski Ph., Rappaport S., Pfahl E.D., 2002, "Evolutionary sequences for low- and intermediate-mass X-ray binaries," *Astrophys. J.* **565**, 1107–1133.

Potekhin A.Y., 1994, "Structure and radiative transitions of the hydrogen atom moving in a strong magnetic field," *J. Phys. B: At. Mol. Opt. Phys.* **27**, 1073–1090.

Potekhin A.Y., 1996a, "Electron conduction along quantizing magnetic fields in neutron star crusts. I. Theory," *Astron. Astrophys.* **306**, 999–1010; erratum: 1997, *ibid.*, **327**, 441.

Potekhin A.Y., 1996b, "Ionization equilibrium of hot hydrogen plasma," Phys. Plasmas **3**, 4156–4165.

Potekhin A.Y., 1998, "Hydrogen atom moving across a strong magnetic field: analytical approximations," *J. Phys. B: At. Mol. Opt. Phys.* **31**, 49–63.

Potekhin A.Y., Chabrier G., 2000, "Equation of state of fully ionized electron-ion plasma. II. Extension to high densities," *Phys. Rev. E* **62**, 8554–8563.

Potekhin A.Y., Chabrier G., 2003, "Equation of state and opacities for hydrogen atmospheres of neutron stars with strong magnetic fields," *Astrophys. J.* **585**, 955–974.

Potekhin A.Y., Chabrier G., 2004, "Equation of state and opacities for hydrogen atmospheres of magnetars," *Astrophys. J.* **600**, 317–323.

Potekhin A.Y., Pavlov G.G., 1997, "Photoionization of hydrogen in atmospheres of magnetic neutron stars," *Astrophys. J.* **483**, 414–425.

Potekhin A.Y., Chabrier G., Yakovlev D.G., 1997a, "Internal temperatures and cooling of neutron stars with accreted envelopes," *Astron. Astrophys.* **323**, 415–428.

Potekhin A.Y., Pavlov G.G., Ventura J., 1997b, "Ionization of the hydrogen atom in strong magnetic fields," *Astron. Astrophys.* **317**, 618–629.

Potekhin A.Y., Chabrier G., Shibanov Yu.A., Ventura J., 1999a, "Ionization equilibrium and equation of state of hydrogen plasmas in strong magnetic fields," *Contrib. Plasma Phys.* **39**, 101–104.

Potekhin A.Y., Chabrier G., Shibanov Yu.A., 1999b, "Partially ionized hydrogen plasma in strong magnetic fields," *Phys. Rev. E* **60**, 2193–2208.

Potekhin A.Y., Yakovlev D.G., Chabrier G., Gnedin O.Y., 2003, "Thermal structure and cooling of superfluid neutron stars with accreted magnetized envelopes," *Astrophys. J.* **594**, 404–418.

Potekhin A.Y., Lai D., Chabrier G., Ho W.C.G., 2004, "Electromagnetic polarization in partially ionized plasmas with strong magnetic fields and neutron star atmosphere models," *Astrophys. J.* **612**, 1034–1043.

Prakash M., Ainsworth T.L., Lattimer J.M., 1988, "Equation of state and the maximum mass of neutron stars," *Phys. Rev. Lett.* **61**, 2518–2521.

Prakash M., Baron E., Prakash M., 1990, "Rotation of stars containing strange quark matter," *Phys. Lett. B* **243**, 175–180.

Prakash M., Prakash M., Lattimer J.M., Pethick C.J., 1992, "Rapid cooling of neutron stars by hyperon and $\Delta$-isobars," *Astrophys. J.* **390**, L77–L80.

Preston M.A., Bhaduri R., 1975, *Structure of the Nucleus* (Reading, MA: Addison-Wesley).

Psaltis D., Chakrabarty D., 1999, "The disk-magnetosphere interaction in the accretion-powered millisecond pulsar SAX J1808.4-3658," *Astrophys. J.* **521**, 332–340.

Pudliner B.S., Pandharipande V.R., Carlson J., Wiringa R.B., 1995, "Quantum Monte Carlo calculations of $A \leq 6$ nuclei," *Phys. Rev. Lett.* **74**, 4396–4399.

Quaintrell H., Norton A.J., Ash T.D.C., Roche P., Willems B., Bedding T.R., Baldry I.K., Fender R.P., 2003, "The mass of the neutron star in Vela X-1 and tidally induced non-radial oscillations in GP Vel," *Astron. Astrophys.* **401**, 313–323.

Radhakrishnan V., Manchester R.N., 1969, "Detection of a change of state in the pulsar PSR 0833–45," *Nature* **222**, 228–229.

Raikh M.E., Yakovlev D.G., 1982, "Thermal and electrical conductivities of crystals in neutron stars and degenerate dwarfs," *Astrophys. Space Sci.* **87**, 193–203.

Rajagopal M., Romani R., 1996, "Model atmospheres for low-field neutron stars," *Astrophys. J.* **461**, 327–333.

Rajagopal K., Wilczek F., 2000, "The condensed matter physics of QCD," in *At the frontier of particle physics/Handbook of QCD*, edited by M. Shifman, vol. III (Singapore: World Scientific), 2061–2151

Rajagopal K., Wilczek F., 2001, "Enforced electrical neutrality of the color-flavor locked phase," *Phys. Rev. Lett.* **86**, 3492–3495.

Rajagopal M., Romani R., Miller M.C., 1997, "Magnetized iron atmospheres for neutron stars," *Astrophys. J.* **479**, 347–356.

Ramaty R., Bonazzola S., Cline T.L., Kazanas D., Mészáros P., Lingenfelter R.E., 1980, "Origin of the 5 March 1979 gamma-ray transient – A vibrating neutron star," *Nature* **287**, 122–124.

Ramaty R., Lingenfelter R.E., Bussard R.V., 1981, "Synchrotron cooling and annihilation of an $e^+e^-$ plasma: the radiation mechanism for the 5th March, 1979 transient," *Astrophys. Space Sci.* **75**, 193–203.

Ramos A., Schaffner-Bielich J., Wambach J., 2001, "Kaon condensation in neutron stars," in *Physics of Neutron Star Interiors*, edited by D. Blaschke, N.K. Glendenning, & A. Sedrakian, *Lecture Notes in Phys.* **578**, 175–202 [nucl-th/0011003].

Ramsey W.H., 1950, "On the instability of small planetary cores (I)," *Mon. Not. R. Astron. Soc.* **110**, 325–338.

Ransom S.M., Hessels J.W.T., Stairs J.H., Freire P.C.C., Camilo F., Kaspi V.M., Kaplan D.L., 2005, "Twenty-one millisecond pulsars in Terzan 5 using the Green Bank Telescope," *Science* **307**, 892–896.

Rappaport S., Joss P.C., McClintock J.E., 1976, "The 3U 0900–40 binary system — Orbital elements and masses," *Astrophys. J.* **206**, L103–L106.

Rappaport S.A., Joss P.C., 1983, "X-ray pulsars in massive binary systems," in *Accretion-Driven Stellar X-ray Sources*, edited by W.H.G. Lewin & E.P.J. van den Heuvel (Cambridge: Cambridge University Press), 1–39.

Rappaport S., Podsiadlowski Ph., Joss P.C.,Di Stefano R., Han Z., 1995, "The relation between white dwarf mass and orbital period in wide binary radio pulsars," *Mon. Not. R. Astron. Soc.* **273**, 731–741.

Ravenhall D.G., Pethick C.J., 1994, "Neutron star moments of inertia," *Astrophys. J.* **424**, 846–851.

Ravenhall D.G., Bennett C.D., Pethick C.J., 1972, "Nuclear surface energy and neutron-star matter," *Phys. Rev. Lett.* **28**, 978–981.

Ravenhall D.G., Pethick C.J., Wilson J.R., 1983, "Structure of matter below nuclear saturation density," *Phys. Rev. Lett.* **50**, 2066–2069.

Reichley P.E.& Downs G.S., 1969, "Observed decrease in the periods of pulsar PSR 0833–45," *Nature* **222**, 229–230.

Reid R.V., 1968, "Local phenomenological nucleon-nucleon potential," *Ann. Phys. (N.Y.)* **50**, 411–448.

Reisenegger A., 1995, "Deviations from chemical equilibrium due to spin-down as an internal heat source in neutron stars," *Astrophys. J.* **442**, 749–757.

Reisenegger A., 2003, "Origin and evolution of neutron star magnetic fields," in Proceedings of the *International Workshop on Strong Magnetic Fields and Neutron Stars*, edited by H.J. Mosquera Cuesta, H. Peréz Rojas, & C.A. Zen Vasconcellos (La Habana, Cuba: ICIMAF), 33–49.

Reisenegger A., Prieto J.P., Benguria R., Lai D., Araya P.A., 2005, "Magnetic fields in neutron stars: A theoretical perspective," in *Magnetic Fields in the Universe: From Laboratory and Stars to Primordial Structures*, edited by E.M. de Gouveia dal Pino, G. Lugones, & A. Lazarian, *AIP Conf. Proc.* **784**, 263–273.

Relovsky B.M., Ruder H., 1996, "Multichannel density-functional calculations for atoms and atomic chains in magnetic fields of compact stars," *Phys. Rev. A* **53**, 4068–4074.

Rembges F., Freiburghans C., Rauscher T., Thielemann F.-K., Schatz H., Wiescher M., 1997, "An approximation for the $rp$-process," *Astrophys. J.* **484**, 412–423.

Reuber A., Holinde K., Speth J., 1994, "Meson-exchange hyperon-hyperon interactions in free scattering and nuclear matter," *Nucl. Phys.* **A 570**, 543–579.

Reynolds A.P., Quaintrell H., Still M.D., Roche P., Chakrabarty D., Levine S.E., 1997, "A new mass estimate of Her X-1," *Mon. Not. R. Astron. Soc.* **288**, 43–52.

Rho M., 2001, "Physics of dense and superdense matter," in *Explosive Phenomena in Astrophysical Compact Objects*, edited by H.-Y. Chang, I. Yi, M. Rho, & C.-H. Lee, *AIP Conf. Proc.* **556**, 160–171.

Rhoades C.E., Jr., Ruffini R., 1974, "Maximum mass of a neutron star," *Phys. Rev. Lett.* **32**, 324–327.

Richer H.B., Ibata R., Fahlman G.G., Huber M., 2003, "The pulsar/white dwarf/planet system in M4: improved astrometry," *Astrophys. J.* **597**, L45–L47.

Rikovska-Stone J., Miller J.C., Koncewicz R., Stevenson P.D., Strayer M.R., 2003, *Phys. Rev. C* **68**, 034324 (16 pages).

Ring, P., Schuck P., 1980, *The Nuclear Many-Body Problem* (New York: Springer).

Rijken Th.A., Stoks V.G.J., Yamamoto Y., 1999, "Soft-core hyperon-nucleon potentials," *Phys. Rev. C* **59**, 21–40.

Roberts P.H., Stewartson K., 1963, "On the stability of a MacLaurin spheroid of small viscosity," *Astrophys. J.* **137**, 777–790.

Rogers F.J., 1986, "Occupation numbers for reacting plasmas – The role of the Planck-Larkin partition function," *Astrophys. J.* **310**, 723–728.

Rogers F.J., DeWitt H.E., "Statistical mechanics of reacting Coulomb gases," *Phys. Rev. A* **8**, 1061–1076.

Rogers F.J., Swenson F.J., Iglesias C.A., 1996, "OPAL equation-of-state tables for astrophysical applications," *Astrophys. J.* **456**, 902–908.

Rögnvaldsson, Ö.E., Fushiki I., Gudmundsson E.H., Pethick C.J., Yngvason J., 1993, "Thomas-Fermi calculations of atoms and matter in magnetic neutron stars: Effects of higher Landau bands," *Astrophys. J.* **416**, 276–290.

Romani R., 1987, "Model atmospheres for cooling neutron stars," *Astrophys. J.* **313**, 718–726.

Rosenfeld L., 1974, "Discussion of the report of D. Pines," in *Astrophysics & Gravitation*, Proc. 16th Solvay Conf. on Physics (Brussels: Universite de Bruxelles), 174.

Rösner W., Wunner G., Herold H., Ruder H., 1984, "Hydrogen atoms in arbitrary magnetic fields. I. Energy levels and wave functions," *J. Phys. B: At. Mol. Phys.* **17**, 29–52.

Rosswog S., Davies M.B., 2002, "High-resolution calculations of merging neutron stars – I. Model description and hydrodynamic evolution," *Mon. Not. R. Astron. Soc.* **334**, 481–497.

Ruder H., Wunner G., Herold H., Geyer F., 1994, *Atoms in Strong Magnetic Fields* (Berlin: Springer).

Ruderman M.A., 1967, "States of matter in stars," in *Proceedings of the Fifth Annual Eastern Theoretical Physics Conference*, edited by D. Feldman (New York: Benjamin), 25–41.

Ruderman M., 1968, "Causes of sound faster than light in classical models of ultradense matter," *Phys. Rev.* **172**, 1286–1290.

Ruderman M.A., 1971, "Matter in superstrong magnetic fields: the surface of a neutron star," *Phys. Rev. Lett.* **27**, 1306–1308.

Ruderman M.A., 1992, "Neutron star crust breaking and magnetic field evolution," in *The Structure and Evolution of Neutron Stars*, edited by D. Pines, R. Tamagaki, & S. Tsuruta (Redwood: Addison-Wesley:), 353–370.

Ruffert M., Janka H.-Th., 2001, "Coalescing neutron stars – A step towards physical models. III. Improved numerics and different neutron star masses and spins," *Astron. Astrophys.* **380**, 544–577.

Rüster S.B., Hempel M., Schaffner-Bielich J., 2006, "Outer crust of nonaccreting cold neutron stars," *Phys. Rev. C* **73**, 035804 (16 pages).

Rutledge R.E., Bilsten L., Brown E.F., Pavlov G.G., Zavlin V.E., 2001a, "Quiescent thermal emission from the neutron star in Aquila X-1," *Astrophys. J.* **559**, 1054–1059.

Rutledge R.E., Bilsten L., Brown E.F., Pavlov G.G., Zavlin V.E., 2001b, "Variable thermal emission from Aquila X-1 in quiescence," *Astrophys. J.* **577**, 346–358.

Rutledge R.E., Bilsten L., Brown E.F., Pavlov G.G., Zavlin V.E., 2002a, "A possible transient neutron star in quiescence in the globular cluster NGC 5139," *Astrophys. J.* **578**, 405–412.

Rutledge R.E., Bilsten L., Brown E.F., Pavlov G.G., Zavlin V.E., Ushomirsky G., 2002b, "Crustal emission and the quiescent spectrum of the neutron star in KS 1731-260," *Astrophys. J.* **580**, 413–422.

Sahakian G.S., Vartanyan Yu.L., 1963, "On the possible phase states of matter at extremely high densities," *Nuovo Cimento* **30**, 82–104.

Salgado M., Bonazzola S., Gourgoulhon E. & Haensel P., 1994, "High precision rotating neutron star models. II. Large sample of neutron star properties," *Astron. Astrophys.* **108**, 455–459.

Salpeter E.E., 1960, "Matter at high densities," *Ann. Phys. (N.Y.)* **11**, 393–413.

Salpeter E.E., 1961, "Energy and pressure of a zero-temperature plasma," *Astrophys. J.* **134**, 669–682.

Sandage A., Osmer P., Giacconi R., Gorenstein P., Gursky H., Waters J., Bradt H., Garmire G., Sreekantan B., Oda M., Osawa K., Jugaku J., 1996, "On the optical identification of Sco X-1," *Astrophys. J.* **146**, 316–322.

Sandulescu N., Nguyen Van Giai, Liotta R.J., 2004, "Superfluid properties of the inner crust of neutron stars," *Phys. Rev. C* **69**, 045802 (6 pages).

Santangelo, A., Segreto, A., Giarrusso, S., dal Fiume, D., Orlandini, M., Parmar, A. N., Oosterbroek, T., Bulik, T., Mihara, T., Campana, S., Israel, G. L., Stella, L., 1999, "A *Beppo SAX* study of the pulsating transient X0115+63: The first X-ray spectrum with four cyclotron harmonic features," *Astrophys. J.* **523**, L85–L88.

Sanwal D., Pavlov G.G., Zavlin V.E., Teter M.A., 2002, "Discovery of absorption features in the X-ray spectrum of an isolated neutron star," *Astrophys. J.* **574**, L61–L64.

Sarsa A., Fantoni S., Schmidt K.E., Pederiva F., 2003, "Neutron matter at zero temperature with auxiliary field diffusion Monte Carlo," *Phys. Rev. C* **68**, 024308 (15 pages).

Sato K., 1979, "Nuclear compositions in the inner crust of neutron stars," *Prog. Theor. Phys.* **62**, 957–968.

Saumon D., Chabrier G., Van Horn H.M., 1995, "An equation of state for low-mass stars and giant planets," *Astrophys. J. Suppl. Ser.* **99**, 713–741.

Sawyer R.F., 1972a, "Energy shifts of excited nucleons in neutron-star matter," *Astrophys. J.* **176**, 205–211; erratum: *ibid.*, **178**, 279.

Sawyer R.F., 1972b, "Condensed $\pi^-$ phase in neutron-star matter," *Phys. Rev. Lett.* **29**, 382–385; erratum: *ibid.*, **29**, 823.

Scalapino D.J., 1972, "$\pi^-$ condensate in dense nuclear matter," *Phys. Rev. Lett.* **29**, 386–388.

Schaeffer R., Haensel P., Zdunik J.L., 1983, "Phase transitions in neutron star cores I. Equilibrium configurations in general relativity," *Astron. Astrophys.* **126**, 121–145.

Schaffner J., Mishustin I.N., 1996, "Hyperon-rich matter in neutron stars," *Phys. Rev. C* **53**, 1416–1429.

Schaffner-Bielich J., Hanauske M., Stöcker H., Greiner W., 2002, "Phase transition to hyperon matter in neutron stars," *Phys. Rev. Lett.* **89**, 171101 (4 pages).

Schatz H., Bildsten L., Cumming A., Wiescher M., 1999, "The rapid proton process ashes from stable nuclear burning on an accreting neutron star," *Astrophys. J.* **524**, 1014–1029.

Schatz H., Aprahamian A., Barnard V., Bildsten L., Cumming A., Quellette M., Rauscher T., Thielemann F.-K., Wiescher M., 2001, "End point of the $rp$ process on accreting neutron stars," *Phys. Rev. Lett.* **86**, 3471–3474.

Schiavilla R., Pandharipande V.R., Wiringa R.B., 1986, "Momentum distributions in $A = 3$ and 4 nuclei," *Nucl. Phys.* **A 449**, 219–242.

Schiff L.I., 1968, *Quantum Mechanics* (New York: McGraw-Hill).

Schmid-Burgk J., 1973, "Temperature aspects of Landau orbital ferromagnetism in white dwarfs and neutron stars," *Astron. Astrophys.* **26**, 335–341.

Schmidt S., Blaschke D., Kalinovsky Y.L., 1994, "Scalar-pseudoscalar meson masses in nonlocal effective QCD at finite temperature," *Phys. Rev. C* **50**, 435–446.

Schreier E., Levinson R., Gursky H., Kellogg E., Tananbaum H., Giacconi R., 1972, "Evidence for the binary nature of Centaurus X-3 from UHURU X-ray observations," *Astrophys. J.* **172**, L79–L89; erratum: *ibid.*, **173**, L151.

Schulze H.-J., Baldo M., Lombardo U., Cugnon J., Lejeune A., 1998, "Hyperonic nuclear matter in Brueckner theory," *Phys. Rev. C* **57**, 704–712.

Schwarzschild M., 1958, *Structure and Evolution of the Stars* (Princeton: Princeton University Press).

Shabanova T.V., Lyne A.G., Urama J.O., 2001, "Evidence for free precession in the pulsar B1642–03," *Astrophys. J.* **552**, 321–325.

Shaposhnikov N., Titarchuk L., 2002, "Spectra of the expansion stage of X-ray bursts," *Astrophys. J.* **567**, 1077–1090.

Shaposhnikov N., Titarchuk L., 2004, "On the nature of the flux variability during an expansion stage of a type I X-ray burst: Constraints on neutron star parameters for 4U 1820–30," *Astrophys. J.* **606**, L57–L60.

Shaposhnikov N., Titarchuk L., Haberl F., 2003, "The bursting behavior of 4U 1728–34: The parameters of a neutron star and the geometry of a neutron star-disk system," *Astrophys. J.* **593**, L35–L38.

Shen H., 2002, "Complete relativistic equation of state for neutron stars," *Phys. Rev. C* **65**, 035802 (7 pages).

Seeger P.A., Havard W.M., 1976, "A semiempirical liquid-drop plus shell-correction formula," *At. Data Nucl. Data Tables* **17**, 428–430.

Segelstein D.J., Rawley L.A., Stinebring D.R., Fruchter A.S., Taylor J.H., 1986, "New millisecond pulsar in a binary system," *Nature*, **322**, 714–717.

Seidov Z.F., 1971, "The stability of a star with a phase change in general relativity theory," *Soviet Astronomy* **15**, 347–348.

Serot B.D., 1977, "A relativistic nuclear field theory with $\pi$ and $\rho$ mesons," *Phys. Lett. B* **86**, 146–150.

Shapiro S.L., Teukolsky S.A., 1983, *Black Holes, White Dwarfs, and Neutron Stars: The Physics of Compact Objects* (New York: Wiley).

Shapiro S.L., Teukolsky S.A., Wasserman I., 1989, "Testing nuclear theory using the 0.5 MS pulsar," *Nature* **340**, 451–452.

Shaviv G., Kovetz A., 1972, "The thermodynamics of white dwarf matter. II," *Astron. Astrophys.* **16**, 72–76.

Shklovsky I.S., 1967, "On the nature of the source of X-ray emission of Sco XR-1," *Astrophys. J.* **148**, L1–L4.

Shternin P.S., Yakovlev D.G., 2006, "Electron thermal conductivity owing to collisions between degenerate electrons," *Phys. Rev. D* **74**, 043004 (8 pages).

Siemens P.J., 1970, "Nuclear-matter reaction matrix," *Nucl. Phys.* **A 141**, 225–248.

Siemens P.J., Pandharipande V.R., 1971, "Neutron matter computations in Brueckner and variational theories," *Nucl. Phys.* **A 173**, 561–570.

Singwi W.L., Tosi M.P., Land R.H., Sjölander A., 1968, "Electron correlations at metallic densities," *Phys. Rev.* **176**, 589–599.

Sitenko A., Tartakovskii V., 1997, *Theory of Nucleus. Nuclear Structure and Nuclear Interactions* (Dordrecht: Kluwer).

Skyrme T.H.R., 1956, "The nuclear surface," *Phil. Mag. (London)*, **1**, 1043–1054.

Skyrme T.H.R., 1959, "The effective nuclear potential," *Nucl. Phys.* **9**, 615–634.

Soglasnov V.A., Popov M.V., Bartel N., Cannon W., Novikov A.Yu., Kondratiev V.I., Altunin V.I., 2004, "Giant pulses from PSR B1937+21 with widths $\leq$ 15 nanoseconds and $T_b \geq 5 \times 10^{39}$ K, the highest brightness temperature observed in the Universe," *Astrophys. J.* **616**, 439–451.

Sokolov A.A., Ternov I.M., 1974, *Relativistskiĭ Elektron* (Moscow: Nauka) [English translation: 1986, *Radiation from Relativistic Electrons*, 2nd rev. ed. (New York: AIP)].

Sollerman J., Lundqvist P., Lindler D., *et al.*, 2000, "Observations of the Crab Nebula and its pulsar in the far-ultraviolet and in the optical," *Astrophys. J.* **537**, 861–874.

Splaver E.M., Nice D.J., Arzoumanian Z., Camilo F., Lyne A.G., Stairs I.H., 2002, "Probing the masses of the PSR J0621+1002 binary system through relativistic apsidal motion," *Astrophys. J.* **581**, 509–518.

Splaver E.M., Nice D.J., Stairs I.H., Lommen A.N., Backer D.C., 2005, "Masses, parallax, and relativistic timing of the PSR J1713+0747," *Astrophys. J.* **620**, 405–415.

Stabler R., 1960, *Ph.D. thesis* (Cornell University), unpublished.

Stairs I.H., 2003, "Testing General Relativity with pulsar timing," *Living Reviews in Relativity* **6**, 5 [http://www.livingreviews.org/lrr-2003-5/].

Stairs I.H., Lyne A.G., Shemar S.L., 2000, "Evidence for free precession in a pulsar," *Nature* **406**, 484–486.

Stairs I.H., Thorsett S.E., Taylor J.H., Wolszczan A., 2002, "Studies of the relativistic binary pulsar PSR B1534+12. I. Timing analysis," *Astrophys. J.* **581**, 501–508.

Steinberg M., Ortner J., Ebeling W., 1998, "Equation of state of a strongly magnetized hydrogen plasma," *Phys. Rev. A* **58**, 3806–3816.

Stergioulas N., 2003, "Rotating stars in relativity," *Living Reviews in Relativity* **6**, 3 [http:/www.livingreviews.org/lrr-2003-3/].

Sterne T.E., 1933, "The equilibrium theory of the abundance of the elements: a statistical investigation of assemblies in equilibrium in which transmutations occur," *Mon. Not. R. Astron. Soc.* **93**, 736–777.

Stickland D., Lloyd C., Radziun-Woodham A., 1997, "The orbit of the supergiant component of VELA X-1 derived from IUE radial velocities," *Mon. Not. R. Astron. Soc.* **286**, L21–L24.

Stoks V.G.J., Klomp R.A.M., Terheggen C.P.F, de Swart J.J., 1994, "Construction of high-quality *NN* potential models," *Phys. Rev. C* **49**, 2950–2962.

Stolzmann W., Blöcker T., 1996, "Thermodynamic properties of stellar matter. I. Equation of state for stellar interiors," *Astron. Astrophys.* **314**, 1024–1040.

Stoner E.C., 1930, "The equilibrium of dense stars," *Phil. Mag.* **9**, 944–963.

Stoner E.C., 1932, "The minimum pressure of a degenerate electron gas," *Mon. Not. R. Astron. Soc.* **92**, 651–661.

Strohmayer T.E., 2004, "Future probes of the neutron star equation of state using X-ray bursts," in *X-Ray Timing 2003: Rossi and Beyond*, edited by P. Kaaret, F.K. Lamb, & J.H. Swank (Melville: AIP), 245–252.

Strohmayer T.E., Bildsten L., 2004, "New views of thermonuclear bursts," in *Compact Stellar X-Ray Sources*, edited by W.H.G. Lewin & M. van der Klis (Cambridge: Cambridge University Press), in press [astro-ph/0301544].

Strohmayer T.E. & Ibrahim A.I., 2000, "Discovery of a 6.4 keV emission line in a burst from SGR 1900+14," *Astrophys. J.* **537**, L111–L114.

Strohmayer T.E., Markwardt C.B., 2002, "Evidence for a millisecond pulsar in 4U 1636–53 during a superburst," *Astrophys. J.* **577**, 337–345.

Strohmayer T., Ogata S., Iyetomi H., Ichimaru S., Van Horn H.M., 1991, "The shear modulus of the neutron star crust and nonradial oscillations of neutron stars," *Astrophys. J.* **375**, 679–686.

Strohmayer T.E., Markwardt C.B., Swank J.H., in't Zand J., 2003, "X-Ray bursts from the accreting millisecond pulsar XTE J1814–338," *Astrophys. J.* **596**, L67–L70.

Strobel K., Schaab Ch., Weigel M.K., 1999, "Properties of non-rotating and rapidly rotating protoneutron stars," *Astron. Astrophys.* **350**, 497–512.

Sturrock P.A., 1971, "A model of pulsars," *Astrophys. J.* **164**, 529–556.

Sugahara Y., Toki H., 1994, "Relativistic mean-field theory for unstable nuclei with non-linear $\sigma$ and $\omega$ terms," *Nucl. Phys.* **A 579**, 557–572.

Suh I.-S., Mathews G.J., 2001, "Cold ideal equation of state for strongly magnetized neutron star matter: Effects on muon production and pion condensation," *Astrophys. J.* **546**, 1126–1136.

Sumiyoshi K., Oyamatsu K., Toki H., 1995, "Neutron star profiles in the relativistic Brueckner-Hartree-Fock theory," *Nucl. Phys.* **A 595**, 327–345.

Sumiyoshi K., Yamada S., Suzuki H., Hillebrandt W., 1998, "The fate of a neutron star just below the minimum mass: does it explode?" *Astron. Astrophys.* **334**, 159–168.

Swesty F.D., 1996, "Thermodynamically consistent interpolation for equation of state tables," *J. Comp. Physics* **127**, 118–127.

Taam R.E. & Picklum R.E., 1978, "Nuclear fusion and carbon flashes on neutron stars," *Astrophys. J.* **224**, 210–216.

Takahashi K., Sato, K., 2003, "Effects of neutrino oscillation on supernova neutrino. Inverted mass hierarchy," *Prog. Theor. Phys.* **109**, 919–931.

Takatsuka T., Tamagaki R., 1976, "A note on crystallization of dense neutron matter," *Prog. Theor. Phys.* **55**, 624–626.

Takatsuka T., Tamagaki R., 1977, "One-dimensional localized [ALS] structure in nuclear system and $\pi^0$-condensation," *Prog. Theor. Phys.* **58**, 694–696.

Takatsuka T., Tamagaki R., 1988a, "Pion-condensed neutron star and pulsar glitch ," *Nucl. Phys.* **A 478**, 785–791.

Takatsuka T., Tamagaki R., 1988b, "Corequake model of pulsar glitches for neutron stars with pion-condensed core," *Prog. Theor. Phys.* **79**, 274–279.

Takatsuka T., Tamagaki R., 1995, "Nucleon superfluidity in kaon-condensed neutron stars," *Progr. Theor. Phys.* **94**, 457–461.

Takatsuka T., Tamagaki R., 1997a, "Effects of charged-pion condensation on neutron $^3P_2$ superfluidity," *Progr. Theor. Phys.* **97**, 263–281.

Takatsuka T., Tamagaki R., 1997b, "Nucleon superfluidity in neutron star core with direct URCA cooling," *Progr. Theor. Phys.* **97**, 345–350.

Takatsuka T., Tamiya K., Tatsumi T., Tamagaki R., 1978, "Solidification and pion condensation in nuclear medium," *Prog. Theor. Phys.* **59**, 1933–1955.

Takatsuka T., Tamagaki R., Tatsumi T., 1993, "Characteristic aspects of pion-condensed phases," *Prog. Theor. Phys. Suppl.* **112**, 67–106.

Takemori M.T., Guyer R.A., 1975, "Solid neutron matter," *Phys. Rev. D* **11**, 2696–2723.

Tamagaki R., 1969, "Pairing effects due to LS force in high density neutron matter," *Soryuusiron Kenkyu*, **40**, D26–D32 [in Japanese].

Tamagaki R., 1970, "Superfluid state in neutron star matter. I. – Generalized Bogoliubov transformation and existence of $^3P_2$ gap at high density," *Prog. Theor. Phys.* **44**, 905–928.

Tanaka S., Mitake S., Ichimaru S., 1985, "Parameterized equation of state for electron liquids in the Singwi-Tosi-Land-Sjölander approximation," *Phys. Rev. A* **32**, 1896–1899.

Taniguchi K., Gourgoulhon E., 2002, "Quasiequilibrium sequences of synchronized and irrotational binary neutron stars in general relativity. III. Identical and different mass stars with $\gamma$=2," *Phys. Rev. D* **66**, 104019 (14 pages).

Tatsumi T., 1983, "A realistic calculation of the cooling rate of neutron stars with charged pion condensate," *Prog. Theor. Phys.* **69**, 1137–1145.

Tatsumi T., 1987, "Pion condensations and neutron star phenomena," *Prog. Theor. Phys. Suppl.* **91**, 299–313.

Taylor J.H., 1987, "Astronomical and space experiments to test relativity," in *General Relativity and Gravitation*, edited by M.A.H. MacCallum (Cambridge: Cambridge University Press), 209–222.

Taylor J.H., Weisberg J.M., 1982, "A new test of general relativity – Gravitational radiation and the binary pulsar PSR 1913+16," *Astrophys. J.* **253**, 908–920.

Taylor J.H., Weisberg J.M., 1989, "Further experimental tests of relativistic gravity using the binary pulsar PSR 1913+16," *Astrophys. J.* **345**, 434–450.

Taylor J.H., Fowler L.A., McCulloch P.M., 1979, "Measurements of general relativistic effects in the binary pulsar PSR 1913+16," *Nature* **277**, 437–440.

Taylor J.M., Grindlay J.E., Edmonds P.D., Cool A.M., 2001, "Helium white dwarfs and BY Draconis binaries in the globular cluster NGC 6397," *Astrophys. J.* **553**, L169–L172.

Terazawa H., 1989a, "Super-hypernuclei in the quark-shell model," *J. Phys. Soc. Japan* **58**, 3555–3563.

Terazawa H., 1989b, "Super-hypernuclei in the quark-shell model. II," *J. Phys. Soc. Japan* **58**, 4388–4393.

ter Haar B., Malfiet R., 1987, "Nucleons, mesons and deltas in nuclear matter – a relativistic Dirac-Brueckner approach," *Phys. Rep.* **149**, 207–286.

Thompson C., 2002, "Nature of the soft gamma repeaters and anomalous X-ray pulsars," *Mem. Soc. Astron. Ital.* **73**, 477–484.

Thorne K.S., 1967, "Relativistic stellar structure and dynamics," in *High Energy Astrophysics*, vol. III, edited by C. DeWitt, E. Schatzman, & P. Véron (New York: Gordon & Breach), 261–441.

Thorne K.S., 1977, "The relativistic equations of stellar structure and evolution," *Astrophys. J.* **212**, 825–831.

Thorne K.S., Campolattaro A., 1967, "Non-radial pulsation of general-relativistic stellar models. I. Analytic analysis for $l \geq 2$," *Astrophys. J.* **149**, 591–611; erratum: 1968, *ibid.*, **152**, 673.

Thorolfsson A., Rögnvaldsson Ö.E., Yngvason J., Gudmundsson E.H., 1998, "Thomas-Fermi calculations of atoms in magnetic neutron stars. II. Finite temperature effects," *Astrophys. J.* **502**, 847–857.

Thorsett S.E., Chakrabarty D., 1999, "Neutron star mass measurements. I. Radio pulsars," *Astrophys. J.* **512**, 288–299.

Thorsett S.E., Arzoumanian Z., Camilo F., Lyne A.G., 1999, "The triple pulsar system PSR B1620–26 in M4," *Astrophys. J.* **523**, 763–770.

Thorsson V., Prakash M., Tatsumi T., Pethick C.J., 1995, "Neutrino emission from dense matter containing meson condensates," *Phys. Rev. D* **52**, 3739–3741.

Thurner G., Körbel H., Braun M., Herold H., Ruder H., Wunner G., 1993, "Hartree–Fock calculations for excited states of two-electron systems in strong magnetic fields," *J. Phys. B: At. Mol. Opt. Phys.* **26**, 4719–4750.

Titarchuk L., 2002, "Effects of resonance in quasi-periodic oscillators of neutron star binaries," *Astrophys. J.* **578**, L71–L74.

Titarchuk L., Shaposhnikov N., 2002, "Three Type I X-ray bursts from Cygnus X-2: Application of analytical models for neutron star mass and radius determination," *Astrophys. J.* **570**, L25–L28.

Titarchuk L., Lapidus I., Muslimov A., 1998, "Mechanisms for high-frequency quasi-periodic oscillations in neutron star and black hole binaries," *Astrophys. J.* **499**, 315–328.

Tolman R.C., 1939, "Static solutions of Einstein's field equations for spheres of fluid," *Phys. Rev.* **55**, 364–373.

Tomsick J.A., Heindl W.A., Chakrabarty D., Kaaret P., 2002, "Rotational broadening measurement for the neutron star X-ray transient XTE J2123–058," *Astrophys. J.* **581**, 570–576.

Trimble V.L., 1968, "Motions and structure of the filamentary envelope of the Crab Nebula," *Astron. J.* **73**, 535–547.

Trümper J., Pietsch W., Reppin C., Voges W., Staubert R., Kendziorra E., 1978, "Evidence for strong cyclotron line emission in the hard X-ray spectrum of Hercules X-1," *Astrophys. J.* **219**, L105–L110.

Tsuruta S., Cameron A.G.W., 1965, "Composition of matter in nuclear statistical equilibrium at high densities," *Canadian J. Phys.* **43**, 2056–2077.

Tsuruta S., Cameron A.G.W., 1966a, "Cooling and detectability of neutron stars," *Canadian J. Phys.* **44**, 1863–1894.

Tsuruta S., Cameron A.G.W., 1966b, "Some effects of nuclear forces on neutron-star models," *Canadian J. Phys.* **44**, 1895–1922.

Turbiner A.V., López Vieyra J.C., 2003, "$H_2{}^+$ molecular ion in a strong magnetic field: Ground state," *Phys. Rev. A* **68**, 012504 (19 pages).

Turbiner A.V., López Vieyra J.C., 2004, "$H_2{}^+$ molecular ion in a strong magnetic field: Lowest excited states," *Phys. Rev. A* **69**, 053413 (10 pages).

Turbiner A.V., López Vieyra J.C., Guevara N.L., 2005, "Exotic $H_2{}^+$ ion in a strong magnetic field: Linear configuration," *Phys. Rev. A* **72**, 023403 (13 pages).

Turolla R., Zane S., Drake J.J., 2004, "Bare quark stars or naked neutron stars? The case of RX J1856.5–3754," *Astrophys. J.* **603**, 265–282.

Urpin V., 2004, "Neutron star oceans: Instability, mixing, and heat transport," *Astron. Astrophys.* **421**, L5–L8.

Urpin V., 2005, "Instabilities, turbulence, and mixing in the ocean of accreting neutron stars," *Astron. Astrophys.* **438**, 643–651.

Urpin V.A., Yakovlev D.G., 1979, "On temperature growth inside neutron stars," *Astrofysika*, **15**, 647–655 [Engl. Transl.: 1980, *Astrophysics* **15**, 429].

Usov V.V., 1998, "Bare quark matter surfaces of strange stars and $e^+e^-$ emission," *Phys. Rev. Lett.* **80**, 230–233.

Usov V.V., 2001, "Thermal emission from bare quark matter surfaces of hot strange stars," *Astrophys. J.* **550**, L179–L182.

Usov V.V., 2004, "Electric fields at the quark surface of strange stars in the color-flavor locked phase," *Phys. Rev. D* **70**, 067301 (3 pages).

Usov N.A., Grebenshchikov Yu.B., Ulinich F.R., 1980, "Three-dimensional Wigner crystal in a magnetic field," *Zh. Eksp. Teor. Fiz.* **78**, 296–306 [Engl. transl.: *Sov. Phys.–JETP* **51**, 148].

Usov V.V., Harko T., Cheng K.S., 2005, "Structure of the electrospheres of bare strange stars," *Astrophys. J.* **620**, 915–921.

Vacca W.D., Lewin W.H.G., van Paradijs J., 1986, "Edddington luminosities and photospheric radius expansion during X-ray bursts from 4U/MXB 1820–30," *Mon. Not. R. Astron. Soc.* **220**, 339–349.

van Adelsberg M., Lai D., Potekhin A.Y., Arras P., 2005, "Radiation from condensed surface of magnetic neutron stars," *Astrophys. J.* **628**, 902–913.

van der Klis M., 2000, "Millisecond oscillations in X-ray binaries," *Annu. Rev. Astron. Astrophys.* **38**, 717–760.

van der Klis M., Jansen F., van Paradijs J., Lewin W.H.G., van den Heuvel E.P.J., Trümper J.E., Sztajno M., 1985, "Intensity-dependent quasi-periodic oscillations in the X-ray flux of GX5-1," *Nature* **316**, 225–230.

van der Meer A., Kaper L., van Kerkwijk M.H., van den Heuvel E.P.J., 2005, "On the mass distribution of neutron stars in HMXBs," in *Massive Stars in Interacting Binaries*, edited by L. Burderi, L.A. Antonelli, F. D'Antona, T. Di Salvo, G.L. Israel, L. Piersanti, A. Tornambè, & O. Straniero (Melville: AIP), 623–626.

Van Horn H.M., 1968, "Crystallization of white dwarfs," *Astrophys. J.* **151**, 227–238.

Van Horn H.M., 1969, "Crystallization of a classical one-component Coulomb plasma," *Phys. Lett. A* **28**, 706–707.

van Kerkwijk M.H. & Kulkarni S.R., 1999, "A massive white dwarf companion to the eccentric binary pulsar system B2303+46," *Astrophys. J.* **516**, L25–L28.

van Kerkwijk M.H., van Paradijs J., Zuiderwijk E.J., Hammerschlag-Hensberge G., Kaper L., Sterken C., 1995a, "Spectroscopy of HD77581 and the mass of Vela X-1," *Astron. Astrophys.* **303** 483–496.

van Kerkwijk M.H., van Paradijs J., Zuiderwijk E.J., 1995b, "On the masses of neutron stars," *Astron. Astrophys.* **303**, 497–501.

van Leeuwen H.-J., 1921, "Problèmes de la theorie électronique du magnétisme," *J. de Physique et le Radium*, Ser. VI, **2**, 361–377.

van Paradijs J.A., Hammerschlag-Hensberge G., van den Heuvel E.P.J., Takens R.J., Zuiderwijk E.J., de Loore C., 1976, "Mass determination for the X-ray binary system Vela X-1," *Nature* **259**, 547–549.

van Paradijs J., van den Heuvel E.P.J., Kouveliotou C., Fishman G.J., Finger M.H., Lewin W.H.G., 1997, "Evidence for neutron star formation from accretion induced collapse of a white dwarf," *Astron. Astrophys.* **317**, L9–L12.

van Straten W., Bailes M., Britton M.C., Kulkarni S.R., Anderson S.B., Manchester R.N., Sarkissian J., 2001, "A test of general relativity from the three-dimensional orbital geometry of a binary pulsar," *Nature* **412**, 158–160.

Vartanyan Yu.L., Ovakimova N.K., 1976, "Cold evaporation of neutrons from nuclei in superdense matter," *Soobshcheniya Byurakanskoi Observatorii* **49**, 87–95 [in Russian].

Ventura J., Potekhin A.Y., 2001, "Neutron star envelopes and thermal radiation from the magnetic surface," in *The Neutron Star – Black Hole Connection*, NATO Science Ser. C, **567**, edited by C. Kouveliotou, E.P.J. van den Heuvel, & J. Ventura (Dordrecht: Kluwer), 393–414.

Ventura J., Herold H., Ruder H., Geyer F., 1992, "Photoabsorption in magnetic neutron star atmospheres," *Astron. Astrophys.* **261**, 235–244.

Vidaña I., Polls A., Ramos A., Hjorth-Jensen M., Stoks V.G.J., 2000a, "Strange nuclear matter within Brueckner-Hartree-Fock theory," *Phys. Rev. C* **61**, 025802 (12 pages).

Vidaña I., Polls A., Ramos A., Engvik L., Hjorth-Jensen M., 2000b, "Hyperon-hyperon interactions and properties of neutron star matter," *Phys. Rev. C* **62**, 035801 (8 pages).

Vieillefosse P., Hansen J.P., 1975, "Statistical mechanisms of dense ionized matter. V. Hydrodynamic limit and transport coefficients of the classical one-component plasma," *Phys. Rev. A* **12**, 1106–1116.

Vincke M., Baye D., 1988, "Centre-of-mass effects on the hydrogen atom in a magnetic field," *J. Phys. B: At. Mol. Opt. Phys.* **21**, 2407–2424.

Vincke M., Le Dourneuf M., Baye D., 1992, "Hydrogen atom in crossed electric and magnetic fields: transition from weak to strong electron-proton decentring," *J. Phys. B: At. Mol. Opt. Phys.* **25**, 2787–2807.

Vonnegut K., 1963, *Cat's Cradle* (New York: Holt, Rinehart, & Winston).

Walecka J.D., 1974, "A theory of highly condensed matter," *Ann. Phys. (N.Y.)* **83**, 491–529.

Walter F.M., Lattimer, J.M., 2002, "A revised parallax and its implications for RX J185635–3754," *Astrophys. J.* **576**, L145–L148.

Walter F.M., Matthews, L.D., 1997, "The optical counterpart of the isolated neutron star RX J185635–3754," *Nature* **389**, 358–360.

Walter F.M., Wolk S.J., Neuhäuser R., 1996, "Discovery of a nearby isolated neutron star," *Nature* **379**, 233–235.

Wang P., 2000, "Strange matter in a self-consistent quark mass-density-dependent model," *Phys. Rev. C* **62**, 015204 (5 pages).

Watanabe G., Iida K., Sato K., 2001, "Effects of neutrino trapping on thermodynamic properties of nuclear "pasta"," *Nucl. Phys.* **A 687**, 512–531.

Watanabe G., Sato K., Yasuoka K., Ebisuzaki T., 2003, "Structure of cold nuclear matter at subnuclear densities by quantum molecular dynamics," *Phys. Rev. C* **68**, 035806 (20 pages).

Watanabe G., Sato K., Yasuoka K., Ebisuzaki T., 2004, "Phases of hot nuclear matter at subnuclear densities," *Phys. Rev. C* **69**, 055805 (14 pages).

Weber F., 1999, *Pulsars as Astrophysical Laboratories for Nuclear and Particle Physics* (Bristol: IOP Publishing).

Weber F., 2005, "Strange quark matter and compact stars," *Progress in Particle and Nuclear Physics* **54**, 193–288.

Weber F., Weigel M.K., 1985, "Ground-state properties of nuclear matter using the $\Lambda$ approximations of the Green's function theory," *Phys. Rev. C* **32**, 2141–2147.

Weber F., Weigel M.K., 1989a, "Neutron star properties and the relativistic nuclear equation of state of many baryon matter," *Nucl. Phys.* **A 493**, 549–582.

Weber F., Weigel M.K., 1989b, "Baryon composition and macroscopic properties of neutron stars," *Nucl. Phys.* **A 505**, 779–822.

Weinberg S., 1972, *Gravitation and Cosmology: Principles and Applications of the General Theory of Relativity* (New York: Wiley).

Weisberg J.M., & Taylor J.H., 1984, "Observations of post-Newtonian timing effects in the binary pulsar PSR 1913+16," *Phys. Rev. Lett.* **52**, 1348–1350.

Weisberg J.M., Taylor J.H., 2003, "The relativistic binary pulsar B1913+16," in *Radio Pulars*, edited by M. Bailes, D.J. Nice, & S.E. Thorsett, *ASP Conf. Ser.* **302**, 93–98.

v. Weizsäcker C.F., 1935, "Theory of nuclear masses," *Z. f. Physik* **96**, 431–458.

Wigner E.P., Eisenbud L., 1947, "Higher angular momenta and long range interaction in resonance reactions," *Phys. Rev.* **72**, 29–41.

Wigner E., Huntington H.B., 1935, "Possibility of a metallic modification of hydrogen," *J. Chem. Phys.* **3**, 764–770.

Wijnands R., van der Klis M., 1998, "A millisecond pulsar in an X-ray binary system," *Nature* **394**, 344–346.

Wijnands R., van der Klis M., Homan J., Chakrabarty D., Markwardt C. B., Morgan E. H., 2003, *Nature* **424**, 44–47.

Wilets L., 1979, "Green's functions method for the relativistic field theory many-body problem," in *Mesons in Nuclei*, vol. III, edited by M. Rho & D. Wilkinson (Amsterdam: North-Holland), 790–837.

Will C.M., 1994, "The binary pulsar, gravitational waves, and the Nobel Prize," *Physics – Uspekhi* **37**, 697–704.

Wille U., 1986, "Vibrational and rotational properties of the $H_2{}^+$ molecular ion in a strong magnetic field," *J. Phys. B: At. Mol. Phys.* **20**, L417–L422.

Wille U., 1987, "Magnetically dressed one-electron molecular orbitals," *Phys. Rev. A* **38**, 3210–3235.

Wiringa R.B., Fiks V., Fabrocini A., 1988, "Equation of state for dense nucleon matter," *Phys. Rev. C* **38**, 1010–1037.

Wiringa R.B., Stoks V.G.J., Schiavilla R., 1995, "Accurate nucleon-nucleon potential with charge-independence breaking," *Phys. Rev. C* **51**, 38–51.

Witten E., 1984, "Cosmic separation of phases," *Phys. Rev. D* **30**, 272–285.

Wolf R.A., 1966, "Some effects of strong interactions on the properties of neutron star matter," *Astrophys. J.* **145**, 834–841.

Wolff R.G., 1983, PhD thesis [in German] (Munich: Technische Universität München), unpublished.

Wolszczan A., 1991, "A nearby 37.9-ms radio pulsar in a relativistic binary system," *Nature* **350**, 688–690.

Woosley S.E., Taam R.E., 1976, "Gamma-ray bursts from thermonuclear explosions on neutron stars," *Nature* **263**, 101–103.

Woosley S.E., Heger A., Cumming A., Hoffman R.D., Pruet J., Rauscher T., Fisker J.L., Schatz H., Brown B.A., Wiescher M., 2004, "Models for type I X-ray bursts with improved nuclear physics," *Astrophys. J. Suppl. Ser.* 151, 75–102.

Wunner G., Ruder H., Herold H., 1981, "Energy levels and oscillator strengths for the two-body problem in magnetic fields," *Astrophys. J.* **247**, 374–381.

Wyckoff S., Murray C. A., 1977, "Proper motion of the Crab pulsar," *Mon. Not. R. Astron. Soc.* **180**, 717–729.

Yakovlev D.G., 1991, "Electrical conductivity of neutron star cores and evolution of internal magnetic fields," in *Neutron Stars: Theory and Observation*, edited by J. Ventura & D. Pines (Dordrecht: Kluwer), 235–244.

Yakovlev D.G., 1993, "Kinetic properties of neutron star cores," in *Strongly Coupled Plasma Physics*, edited by H.M. Van Horn & S. Ichimaru (Rochester: University of Rochester Press), 157–166.

Yakovlev D.G., Kaminker A.D., 1994, "Neutron star crusts with magnetic fields," in *The Equation of State in Astrophysics*, edited by G. Chabrier & E. Schatzman (Cambridge: Cambridge University Press), 214–238.

Yakovlev D.G., Pethick C.J., 2004, "Neutron star cooling," *Annu. Rev. Astron. Astrophys.* **42**, 169–210.

Yakovlev D.G., Shalybkov D.A., 1989, "Degenerate cores of white dwarfs and envelopes of neutron stars: thermodynamics and plasma screening in thermonuclear reactions," *Sov. Sci. Rev., Ser. E: Astrophys. Space Phys.* **7**, 311–386.

Yakovlev D.G., Levenfish K.P., Shibanov Yu.A., 1999, "Cooling of neutron stars and superfluidity in their cores," *Uspekhi Fiz. Nauk* **169**, 825–868 [Engl. transl.: *Physics – Uspekhi* **42**, 737–778].

Yakovlev D.G., Kaminker A.D., Gnedin O.Y., Haensel P., 2001, "Neutrino emission from neutron stars," *Phys. Rep.* **354**, 1–155.

Yakovlev D.G., Gnedin O.Y., Kaminker A.D., Potekhin A.Y., 2002, "Cooling of superfluid neutron stars," in *270. WE-Heraeus Seminar on Neutron Stars, Pulsars and Supernova Remnants*, edited by W. Becker, H. Lesh, & J. Trümper (Garching: MPE), 287–299.

Yakovlev D.G., Gnedin O.Y., Kaminker A.D., Levenfish K.P., Potekhin A.Y., 2004, "Neutron star cooling: theoretical aspects and observational constraints," *Adv. Space Res.* **33**, 523–530.

Yao W.-M. *et al.*., 2006, "Review of particle physics," *J. of Phys. G: Nucl. Part. Phys.* **33**, 1–1232.

Young D.A., Corey E.M., DeWitt H.E., 1991, "Analytic fit to the one-component-plasma structure factor," *Phys. Rev. A* **44**, 6508–6512.

Youngblood D.H., Clark H.L., Y.-W. Lui, 1999, "Incompressibility of nuclear matter from the giant monopole resonance," *Phys. Rev. Lett.* **82**, 691–694.

Yukawa H., 1935, "Interaction of elementary particles," *Proc. Phys. Math. Soc. Japan* **17**, 48–57.

Zavlin V.E., Pavlov G.G., 2002, "Modeling neutron star atmospheres," in *270. WE-Heraeus Seminar on Neutron Stars, Pulsars and Supernova Remnants*, edited by W. Becker, H. Lesh, & J. Trümper (Garching: MPE), 263–272.

Zavlin V.E., Pavlov G.G., Sanwal D., Manchester R.N., Trümper J., Halpern J., Becker W., 2002, "X-radiation from the millisecond pulsar J0437−4715," *Astrophys. J.* **569**, 894–902.

Zdunik J.L., 2000, "Strange stars – linear approximation of the EOS and minimum QPO frequency," *Astron. Astrophys.* **359**, 311–315.

Zdunik J.L., 2002, "On the minimum radius of strange stars with crust," *Astron. Astrophys.* **394**, 641–645.

Zdunik J.L., Haensel P., 1990, "Maximum rotation frequency of strange stars," *Phys. Rev. D* **42**, 710–711.

Zdunik J.L., Haensel P., Schaeffer R., 1987, "Phase transitions in neutron star cores II. Equilibrium configurations in general relativity," *Astron. Astrophys.* **172**, 95–110.

Zdunik J.L., Haensel P., Gondek-Rosinska D., Gourgoulhon E., 2000, "Innermost stable circular orbits around strange stars and kHz QPOs in low-mass X-ray binaries," *Astron. Astrophys.* **356**, 612–618.

Zdunik J.L., Haensel P., Gourgoulhon E., 2001, "The crust of rotating strange quark stars," *Astron. Astrophys.* **372**, 535–543.

Zeldovich Ya.B., 1961, "The equation of state at ultrahigh densities and its relativistic limitations," *Zh. Eksp. Teor. Fiz.* **41**, 1609–1615 [Engl. transl.: 1962, *Sov. Phys.–JETP* **14**, 1143–1147].

Zeldovich Ya.B., 1962, "Static solutions with energy excess in the general theory of relativity," *Zh. Eksp. Teor. Fiz.* **42**, 1667–1671 [Engl. transl.: *Sov. Phys.–JETP*, **15**, 1158–1160].

Zeldovich Ya.B., Guseynov O.H., 1966, "Collapsed stars in binaries," *Astrophys. J.* **144**, 840–841.

Zeldovich Ya.B., Novikov I.D., 1971, *Relativistic Astrophysics. Vol.1: Stars and relativity* (Chicago: University of Chicago Press).

Zhang W., Strohmayer T.E., Swank J.H., 1997, "Neutron star masses and radii as inferred from kilohertz quasi-periodic oscillations," *Astrophys. J.* **482**, L167–171.

Zink J.W., 1968, "Shell structure and the Thomas-Fermi equation of state," *Phys. Rev.* **176**, 279–284.

Zuiderwijk E.J., van den Heuvel E.P.J., Hensberge G., 1974, "Orbit, spectrum and H$\alpha$ variations of HD 77581 (3U 0900-40)," *Astron. Astrophys.* **35**, 353–360.

Zwicky F., 1938, "On collapsed neutron stars," *Astrophys. J.* **88**, 522–525.

Zwicky F., 1939, "On the theory and observation of highly collapsed stars," *Phys. Rev.* **55**, 726–743.

# LIST OF SYMBOLS

*This list is not comprehensive*: for instance, we omit notations that are used only a few times in a particular section, and we do not list all notations which differ only by self-explanatory subscripts or superscripts. Standard mathematical notations (e, $\pi$, etc.) are also not listed.

$A$, $A_{\rm i}$ – ion mass number
$\boldsymbol{A}$ – vector potential
$A'$ – effective mass number including bound and free neutrons in the inner crust (Chapters 2, 4)
$A''$ – number of free neutrons per nucleus in the inner crust (Chapter 2)
$A_{\rm b}$ – total number of baryons in the star
$A_j$ – mass number of ions of species $j$
$a_0 = 0.5291772108 \times 10^{-8}$ cm – Bohr radius
$a_{\rm i}$ – ion-sphere radius
$a_{\rm m}$ – magnetic length
$\boldsymbol{B}$, $B$ – magnetic field
$\mathcal{B}$ – bag constant
$b$ – magnetic field in relativistic units (Chapter 4); binding energy per nucleon (Chapter 3)
$C$, $C_V$ – heat capacity at constant volume
$C_{\rm M}$ – Madelung constant (Chapters 2, 3)
$C_P$ – heat capacity at constant pressure
$c = 2.99\,792\,458 \times 10^{10}$ cm s$^{-1}$ – speed of light in vacuum
$d$, $D$ – distance
$d$ – dimensionality of a nuclear phase (Chapter 3)
$E$ – energy; energy per nucleon (Chapter 5)
$E_\kappa$ – binding energy of a quantum state $\kappa$ (Chapters 2, 4)
$E_{\rm grav}$ – gravitational energy
$E_{\rm int}$ – total internal energy of a star
$E_{\rm rot}$ – rotational energy
$\mathcal{E}$ – energy density
$e = 4.8032044 \times 10^{-10}$ esu – elementary charge
$F$ – (Helmholtz) free energy (Chapters 2, 4)
$F_{\rm id}$ – ideal-gas free energy
$F_{\rm ex}$ – excess free energy

$F_{\rm q}$ – quantum (Wigner) correction to the free energy
$F_{\rm xc}$ – exchange-correlation contribution to the free energy
$f^{(0)}(\epsilon - \mu, T)$ – Fermi-Dirac distribution function
$G = 6.674 \times 10^{-8}$ cm$^3$ g$^{-1}$ s$^{-2}$ – gravitational constant
$G(k)$ – local field correction
$\hat{G}, G_{NN'}, G_{BB'}$ – G-matrix of the Brueckner-Bethe-Goldstone theory (Chapter 5)
$g$ – gravitational acceleration
$g(r)$ – radial pair-correlation function
$g_e = 1.001\,159\,6522$ – electron gyromagnetic factor
$g_p = 5.585\,6947$ – proton gyromagnetic factor
$g_{\rm n}$ $({\rm n} = 1, 2, 3, \dots)$ – Green's functions (Chapter 5)
$g_{\rm s}, g_{\rm v}, g_\phi, \dots$ – coupling constants in a Lagrangian (Chapter 5)
$g_\kappa$ – statistical weight of a quantum state $\kappa$
$g_{ik}$ – spacetime metric tensor
$H$ – Hamiltonian function; pseudo-enthalpy (§ 6.12.1, Appendix C)
$\hat{H}$ – Hamiltonian operator
$H_n(\xi)$ – Hermite polynomial (Chapter 4)
$\mathcal{H}_n(\xi)$ – harmonic-oscillator function (Chapter 4)
$h$ – enthalpy per nucleon (Chapters 3, 6; Appendix C); gravitational wave strain (§ 6.11.2)
$h_e$ – thickness of the electron surface of the strange star (Chapter 8)
$h_{ij}$ – metric perturbation (Chapter 6)
$\hbar = 1.05457168 \times 10^{-27}$ erg s - Planck constant over $2\pi$
$I$ – moment of inertia
$I_\nu$ – Fermi-Dirac integral
$J$, $\boldsymbol{J}$ – angular momentum
$\hat{K}$ – kinetic energy operator
$\boldsymbol{K}$ – atomic pseudomomentum (Chapter 4)
$K_c$ – critical value of the atomic pseudomomentum (Chapter 4)
$K_0$ – incompressibility of nuclear matter
$\boldsymbol{k}$ – wave vector
$k_{\rm B} = 1.380\,6505 \times 10^{-16}$ erg K$^{-1}$ – Boltzmann constant
$k_{\rm F}$ – Fermi wavenumber
$k_{\rm TF}$ – Thomas-Fermi wave number
$\ell$ – leptons: electrons, muons
$L$ – stellar luminosity
$\hat{\boldsymbol{L}}$ – orbital angular momentum operator (Chapter 5)
$L_{\rm Edd}$ – Eddington luminosity limit
$L_\odot = 3.846 \times 10^{33}$ erg s$^{-1}$ – solar luminosity
$\mathcal{L}$ – Lagrangian density
$l_\kappa$ – rms size of an atom or ion in a quantum state $\kappa$

$M$ – stellar mass; gravitational stellar mass
$M_\odot = 1.9889 \times 10^{33}$ g – solar mass
$m$ – mass of a particle; magnetic quantum number (Chapter 4); azimuthal mode number (Chapter 6)
$m = m(r)$ – gravitational mass inside a sphere with radial coordinate $r$
$m_e = 9.109\,3826 \times 10^{-28}$ g – electron mass
$m^*$ – nucleon effective mass (Chapter 5)
$m_e^*$ – effective dynamic mass of an electron
$m_\mathrm{i}$ – (mean) ion mass
$m_j$ – mass of particle (species) $j$
$m_\mathrm{u} = 1.660\,5388 \times 10^{-24}$ g – unified atomic mass unit
$m_0 = 1.658\,610 \times 10^{-24}$ g – mass of the $^{56}$Fe atom divided by 56
$N_j$ – number of particles of type $j$
$n$ – number density; pulsar braking index (Chapters 1, 9); polytropic index (Chapters 2, 6); Landau quantum number (Chapter 4)
$n_\mathrm{b}$ – number density of baryons
$n_B$ – critical value of $n_e$ in a strong magnetic field (Chapter 2); number density of baryon species $B$ (Chapter 5)
$n_\mathrm{b,s}$ – baryon density at surface of bare strange star (Chapter 8)
$n_\mathrm{c}$ – baryon density at the stellar center
$n_\mathrm{cc}$ – baryon number density at the crust-core interface
$n_e$ – the electron number density
$n_j$ – number density of ions of species $j$
$n_\mathcal{N}$ – total number density of ions (atomic nuclei)
$n_n$ – number density of free neutrons
$n_\mathrm{s}$ – number density of neutrons in the neutron skin (Chapter 3)
$n_s$ – number density of strange quarks (Chapter 8)
$n_0$ – normal nucleon (baryon) density $= 0.16\ \mathrm{fm}^{-3}$
$\mathcal{N}_B(\epsilon)$ – $n_e$ in the approximation of strong degeneracy (Chapter 4)
$p$, $\boldsymbol{p}$ – particle momentum
$\hat{\boldsymbol{p}}$ – momentum operator
$\tilde{p}_i$ – electron momentum components in relativistic units (Chapter 4)
$P$ – pressure; star rotation period (Chapters 1, 9)
$P_\mathrm{c}$ – pressure at the stellar center
$P_\mathrm{id}^{(e)}$ – pressure of the ideal electron gas
$P_\mathrm{r} = 1.421\,775 \times 10^{25}\ \mathrm{dyn\ cm}^{-2}$ – relativistic unit of pressure
$Q$ – neutrino emissivity (Chapter 1); critical wave number of density perturbations (Chapter 3)
$q$ – particle charge
$q_\mathrm{D}$ – Debye wave number (Chapter 2)
$q'_\mathrm{D}$ – plasma screening wave number (Chapter 2)
$\boldsymbol{r}_c$ – guiding-center coordinate vector (Chapter 4)

$R$ – stellar radius
$R$ – circumferential radius of spherical star
$R_{\rm eq}$ – circumferential equatorial radius of rotating star
$R_S$ – ion density parameter
$R_\odot = 6.960 \times 10^{10}$ cm – solar equatorial radius
$R_\infty$ – apparent (radiation) stellar radius
$r_{\rm c}$ – equivalent cell radius (Chapter 3)
$r_{\rm D}$ – Debye length (Chapter 2)
$r_{\rm g}$ – gravitational (Schwarzschild) radius
$r_e$ – electron screening length
$r_{\rm eq}$ – equatorial radial coordinate of rotating star (Chapters 6,8)
$r_{\rm pol}$ – radial coordinate of the pole of rotating star (Chapters 6,8)
$\mathcal{R}_{ik}$ – Ricci tensor
$r_s$ – (plasma) density parameter
Ry $= 2.179\,872 \times 10^{-11}$ erg – Rydberg energy unit ($= 0.5$ Hartree)
$S$ – entropy
$s$ – spin quantum number (Chapter 4)
$S(q)$ – static structure factor (Chapter 2)
$S(\boldsymbol{q}, \omega)$ – dynamic structure factor (Chapter 2)
$\hat{\boldsymbol{S}}$ – total spin operator (Chapter 5)
$\hat{S}_{ij}$ – tensor coupling operator of $ij$ nucleon pair (Chapter 5)
$S_0$ – symmetry energy at saturation density
$T$ – temperature
$\hat{T}$, $T_{NN'}$ – in-medium T-matrix (Chapter 5)
$\hat{\boldsymbol{T}}$ – total isospin operator (Chapter 5)
$T_B$, $T_{\rm cycl}$ – critical values of $T$ in a magnetic field (Chapter 2)
$T_{\rm c}$, $T_{\rm crit}$ – critical temperature of a phase transition
$T_{\rm F}$ – Fermi temperature
$T_{\rm l}$ – temperature of gas-liquid transition
$T_{\rm m}$ – melting temperature
$T_{{\rm p}e}$ – electron plasma temperature
$T_{\rm pi}$ – ion plasma temperature
$T_{\rm r} = 5.929\,889 \times 10^9$ K – relativistic temperature unit
$T_{\rm s}$ – effective surface temperature
$T_{\rm s}^\infty$ – $T_{\rm s}$ as detected by a distant observer
$T_{ik}$ – stress-energy tensor
$t$ – time variable
$t_{\rm p}$ – quantum plasma parameter (Chapter 2); pulsar age (Chapter 9)
$t_{\rm r}$ – temperature in relativistic units
$U$ – internal energy; fluid velocity in the azimuthal direction (Chapter 6)
$\hat{U}$ – single-particle potential operator (Chapter 5)

$U(\sigma)$ – self-interaction contribution of $\sigma$ field to Hamiltonian density (Chapter 5)
$\mathcal{U}$ – potential energy of an ensemble of particles
$\boldsymbol{u}$ – displacement vector (Chapter 3)
$u_{ik}$ – components of the strain tensor (Chapter 3)
$V$ – volume; potential function
$\hat{V}$ – potential energy operator
$V^{\text{eff}}(k)$ – Fourier transform of the Coulomb potential
$\hat{V}_{ijk}$ – three-nucleon interaction potential (Chapter 5)
$v_\text{F}$ – Fermi velocity
$\hat{v}_{ij}$ – potential acting between a nucleon pair $ij$ (Chapter 5)
$v_\text{s}$ – speed of sound
$W_\mathcal{N}$ – energy of the nucleus
$w$ – fraction of volume occupied by atomic nuclei
$w_\kappa$ – occupation probability of a quantum state $\kappa$
$X_\nu$ – inverse function to the Fermi integral
$x_B$ – relativity parameter in a quantizing magnetic field
$x_\text{GR}$ – compactness parameter
$x_j$ – fraction of ion species $j$
$x_\text{r}$ – relativity parameter
$Z$, $Z_\text{i}$ – ion charge number
$\mathcal{Z}$ – partition function
$Z_\text{eff}$ – effective charge number
$Z_j$ – charge number of ions of species $j$
$z$ – gravitational redshift; proper depth (Chapter 6); starting energy parameter (Chapter 5)
$z_\text{surf}$ – surface gravitational redshift (Chapter 6)
$\alpha_\text{f} = 0.007\,297\,352\,57$ – fine-structure constant
$\alpha_\text{s}$ – strong interaction (QCD) coupling constant
$\alpha_v$ – coupling strength of the vector field to nucleons
$\beta_\text{r}$ – relativistic electron velocity parameter
$\Gamma$ – ion Coulomb coupling parameter (Chapters 2, 4); Lorenz factor (Chapters 6,7,8)
$\Gamma_e$ – (nondegenerate) electron Coulomb coupling parameter (Chapters 2, 4)
$\Gamma_j$ – Coulomb coupling parameter for species $j$ (Chapters 2, 4)
$\Gamma_\text{m}$ – value of $\Gamma$ at melting (Chapters 2, 4)
$\gamma$, $\gamma_\text{ad}$ – adiabatic index (polytrope exponent)
$\gamma$ – magnetic field in atomic units (Chapter 4); relativistic parameter of a binary system (Chapter 9)
$\gamma_B$ – electron Lorentz factor in a quantizing magnetic field
$\gamma_\text{r}$ – relativistic electron energy parameter (fiducial electron Lorentz factor)
$\Delta$ – superfluid energy gap (Chapters 1, 5, 7, 8); resonance (Chapters 5, 7)

$\delta$ – Dirac's delta function; neutron excess (Chapters 3, 5); quantum defect (Chapter 4)
$\epsilon$ – electron energy (Chapters 2, 4); oblateness parameter (Chapter 6); quark energy (Chapter 6)
$\epsilon_{\rm F}$ – Fermi energy
$\epsilon_k$ – kinetic energy of nucleon of momentum $k$ (Chapter 5)
$\varepsilon$ – dielectric (screening) function (Chapter 2)
$\varepsilon_k$ – quasiparticle energy (Chapter 5)
$\zeta$ – bulk viscosity (Chapter 1); dimensionless coupling parameter (Chapter 8)
$\eta$ – shear viscosity (Chapter 1); inverse quantum plasma parameter (Chapter 2)
$\theta$ – polar angle; electron degeneracy parameter (Chapter 2)
$\kappa$ – thermal conductivity (Chapter 1); set of quantum numbers (Chapters 2, 4); surface curvature (Chapter 3)
$\lambda\!\!\!^{-}_{\rm C} = 3.86\,159\,268 \times 10^{-11}$ cm - Compton wavelength over $2\pi$
$\lambda = \lambda(r)$ – metric function (Chapter 6)
$\lambda$ – squared oscillation frequency (Chapter 6); relative density jump (Chapter 7)
$\lambda_e$ – electron thermal wavelength (Chapters 2, 4)
$\lambda_{\rm H}$ – thermal wavelength of the H atom (Chapters 2, 4)
$\lambda_j$ – thermal wavelength of particle species $j$ (Chapters 2, 4); Lagrange multiplier (Chapter 5); $j$th eigenvalue of $\lambda$ (Chapter 6)
$\lambda_Q$ – critical wavelength of density perturbations (Chapter 3)
$\mu$ – chemical potential; shear modulus (Chapter 3)
$\mu_{\rm b}$ – baryon chemical potential
$\mu_e$ – electron chemical potential
$\nu$ – "longitudinal" quantum number (Chapter 4)
$\xi^i$ – Lagrangian displacements in a perturbed star (Chapter 6)
$\rho$ – mass density
$\rho_{\rm c}$ – mass density at the stellar center
$\rho_{\rm cc}$ – mass density at the crust-core interface
$\rho_{\rm m}$ – density at quantum melting
$\rho_{\rm s}$ – mass density at the surface of bare strange star (Chapter 8)
$\rho_B$ – critical value of $\rho$ in a strong magnetic field (Chapter 2)
$\rho_{\rm ND}$ – neutron-drip density
$\rho_0 = 2.8 \times 10^{14}$ g cm$^{-3}$ – normal nuclear density
$\sigma$ – electrical conductivity (Chapter 1); standard rms deviation (Chapters 6, 9)
$\sigma$, $\sigma_{\rm s}$ – surface tension (Chapter 7)
$\boldsymbol{\sigma}$ – spin
$\sigma^k$ – Pauli matrix
$\sigma_{ij}$ – stress tensor (Chapters 3, 6)
$\sigma_{\rm SB} = 5.67\,040 \times 10^{-5}$ erg cm$^{-2}$ s$^{-1}$ K$^{-4}$ – Stefan-Boltzmann constant
$\sigma_{\rm T} = 6.652\,4587 \times 10^{-25}$ cm$^2$ – Thomson scattering cross section

$\boldsymbol{\tau}$ – isospin (Chapter 5)

$\tau$ – mean lifetime of a baryon (Chapter 5) or of a nuclear state (Chapter 7); local proper time (Chapter 6)

$\Phi$, $\Phi(r)$ – ground-state wave-function (Chapter 5); metric function (Chapter 6)

$\Phi_{n,s}(r)$ – Landau function

$\phi$ – azimuthal angle

$\phi(r)$ – electrostatic potential (Chapter 2)

$\chi$ – normalized chemical potential of electrons; volume fraction of denser phase (Chapter 7)

$\chi_T$, $\chi_\rho$ – temperature and density logarithmic derivatives of pressure

$\psi_{ns}$ – basic bispinors (Chapter 4)

$\boldsymbol{\Omega}$, $\Omega$ – stellar spin frequency

$\Omega$ – thermodynamic potential (Chapters 4, 7)

$\omega$ – angular frequency of a stellar oscillation; single-particle energy (Chapter 5); spin frequency of the local inertial reference frame (§ 6.10); metric function (§ 6.12); photon frequency

$\overline{\omega}$ – local spin frequency of a star, as measured in a local inertial reference frame

$\omega_B$ – electron gyrofrequency

$\omega_{\rm c}$ – electron cyclotron frequency

$\omega_{\rm ci}$ – ion cyclotron frequency

$\omega_{cp}$ – proton cyclotron frequency

$\omega_{\rm g}$ – electron gyrofrequency

$\omega_{\rm pe}$ – electron plasma frequency

$\omega_{\rm pl}$ – plasma frequency in quark matter (Chapter 8)

$\omega_\infty$ – proton frequency as measured by a distant observer

$\nabla_{\rm ad}$ – adiabatic gradient

# LIST OF ABBREVIATIONS

APR – Akmal-Pandharipande-Ravenhall (EOS)
ALS – alternating-spin layers
AXP – anomalous X-ray pulsar
BB – baryon-baryon (interaction)
BZ – Brillouin zone
BBG – Brueckner-Bethe-Goldstone (theory)
BBP – Baym-Bethe-Pethick EOS; model; paper
bcc – body-centered cubic (crystal)
BCS – Bardeen-Cooper-Schrieffer (model, theory)
BPS – Baym-Pethick-Sutherland model; paper
CFL – color-flavor-locked (phase)
CFS – Chandrasekhar-Friedman-Schutz (instability)
CL – causality limit
CLDM – compressible liquid drop model
DFT – density functional theory
$ee$ – electron-electron (interaction)
eip – electron-ion plasma
EOS – equation of state
ETF – extended Thomas-Fermi (approximation)
fcc – face-centered cubic (crystal)
FFG – free Fermi gas
FPS – Friedman-Pandharipande-Skyrme (model; EOS)
GFT – Green function theory
hcp – hexagonal close-packed (crystal)
HFB – Hartree-Fock-Bogoliubov (approximation)
HH – hyperon-hyperon (interaction)
HMXB – high mass X-ray binary
HNC – hypernetted chain (appoximation)
HP – Haensel-Pichon model; EOS; paper
$ie$ – ion-electron (interaction)
$ii$ – ion-ion (interaction)
IS – intermediate- and short-range (component of a potential)
LMXB – low mass X-ray binary
LOFER – Landau orbital ferromagnetism

MEM – Meson Exchange Model
NH – nucleon-hyperon (interaction)
NN – nucleon-nucleon (interaction)
NNN – three-nucleon (interaction)
$npe$-matter – uniform matter of neutrons, protons, and electrons
$npe\mu$-matter – uniform matter of neutrons, protons, electrons, and muons
OBE – one-boson-exchange
OCP – one-component plasma
OPAL – Opacity Library (project)
OPEP – one-pion exchange potential
PSN – pre-supernova
PWN – pulsar-wind nebula
QCD – quantum chromodynamics
QPO – quasiperiodic oscillation
RBHF – relativistic Brueckner-Hartree-Fock (approximation) $\equiv$ DBHF – Dirac-Brueckner-Hartree-Fock (approximation)
RETF – relativistic extended Thomas-Fermi (approximation)
RMF – relativistic mean-field (model)
RPA – random-phase approximation SGR – soft gamma repeater
SLy – Skyrme Lyon effective interaction; EOS model
SN – supernova
SNR – supernova remnant
SQM – strange quark matter (self-bound quark matter)
SXT – soft X-ray transient

# SUBJECT INDEX